Review
of
Human
Physiology

H. FRANK WINTER, Ph.D.

Associate Professor, Department of Physiology,
School of Dental Medicine, Washington University,
St. Louis

MELVIN L. SHOURD, Ph.D.

Assistant Professor, Department of Physiology,
School of Dental Medicine, Washington University,
St. Louis

1978

W. B. SAUNDERS COMPANY · Philadelphia · London · Toronto

W.B. Saunders Company: West Washington Square
Philadelphia, PA 19105

1 St. Anne's Road
Eastbourne, East Sussex BN21 3UN, England

1 Goldthorne Avenue
Toronto, Ontario M8Z 5T9, Canada

Review of Human Physiology   ISBN 0-7216-9467-5

Last digit is the print number: 9   8   7   6   5   4   3   2   1

PREFACE

The primary purpose of this review text is to provide a systematic means of review and self-evaluation of the student's comprehension of major concepts of Arthur C. Guyton's *Textbook of Medical Physiology*. The format is that of a series of cognitive objectives, followed by content review statements for each objective in the form of multiple choice questions and answers. Specific page references, included with the question answers, provide an efficient reference source for reinforcing or correcting concepts as difficulties are encountered.

We wish to extend our deep appreciation to Mr. Jeffrey H. Schneider for his painstaking efforts in producing many of the illustrations used in the text, Mary V. Heil for her assistance in the preparation of a typed manuscript, and the editorial staff at W. B. Saunders Co. for their patient, co-operative, and productive attitudes.

We acknowledge, with special thanks and tribute, Arthur C. Guyton's superb style of writing and keen sense of organization, which have greatly facilitated the production of this review text and without which this effort would have not been feasible.

H. FRANK WINTER
MELVIN L. SHOURD

NOTE
TO THE STUDENT

This review text has been specifically designed to provide a combined means of self-evaluation and content review of major concepts of physiology as contained in Arthur C. Guyton's *Textbook of Medical Physiology*. Cognitive objectives, arranged sequentially as they are encountered in the text, serve to identify major learning concepts. The series of questions following each objective is intended to test your level of comprehension, identify potential areas of weakness, and when answered correctly to provide a series of review statements for the corresponding objective.

It is suggested that you undertake to answer the test questions in the review text only after reading the corresponding chapters in Guyton's text. For those questions where difficulty is encountered, utilize the reference pages to Guyton's text to reread the information upon which the question was based in order to better understand the underlying concepts.

Students using Guyton's *Basic Human Physiology* will not find specific page references; however, the following list provides a chapter-by-chapter correlation between this review text and *Basic Human Physiology*.

W & S (Winter and Shourd — Review of Human Physiology)
A.C.G. (A.C. Guyton — Basic Human Physiology)

W & S	A.C.G.	W & S	A.C.G.	W & S	A.C.G.
1	1	18	14	31, 32	23
2	2	19	15	33	25
3	3	20	16	34, 35	24
4	4	21	17	36	25
5, 6	5	22	18	37, 38	26
7, 8	6	23	19	39	27
9	7	24	15	40, 41	28
10	8	25	20	42, 43	29
11	9	26	20	44, 45	30
12	10	27	20	46, 47	31
13	11	28	19	48, 49	32
14	12	29	21	50	33
15,16,17	13	30	22	51	34

W & S	A.C.G.	W & S	A.C.G.
52, 53	35	68, 69	46
54	37	70	43
55	36	71, 72	47
56	37	73, 74	48
57	38	75	49
58, 59	39	76	50
60	40	77	51
61, 62	41	78	52
63	42	79	53
64	43	80	54
65, 66	44	81	55
67	45	82, 83	56

CONTENTS

1

Functional Organization of the Human Body and Control of the "Internal Environment"

OBJECTIVE 1–1.
Recognize the scope of physiology.

1. Physiology, or the study of _____ (S, structure; F, function) in living matter, deals with the organization of life processes at _____ (C, only cellular; T, only tissue; M, many) levels of organization.

   a. S, C   c. S, M   e. F, T
   b. S, T   d. F, C   f. F, M

2. Physiology in its broadest scope deals with _____ (N, normal; P, pathological) aspects of tissues of _____ (M, only mammals; V, only vertebrates; A, all living organisms).

   a. N, M   c. N, A   e. P, V
   b. N, V   d. P, M   f. P, A

OBJECTIVE 1–2.
Recognize cells as fundamental basic living units of the body, their numerical magnitude, and their general comparative characteristics.

3. The entire body is composed of about 75 _____ (M, million; B, billion; T, trillion) cells, of which perhaps the most abundant cell type is the _____ (N, neuron; RBC, erythrocyte).

   a. M, N   c. T, N   e. B, RBC
   b. B, N   d. M, RBC   f. T, RBC

4. _____ (O, oxidation; R, reduction) of carbohydrate, fat, or protein by _____% of mammalian cell types provides for metabolic energy required for cell function.

   a. O, 35   c. O, 100   e. R, 65
   b. O, 65   d. R, 35   f. R, 100

OBJECTIVE 1–3.
Identify and contrast the general composition and functional significance of intracellular and extracellular fluids.

5. The most abundant compound in the body, and the approximate percentage of the body weight that it represents, is:

   a. Protein, 56%   d. Protein, 26%
   b. Water, 56%   e. Water, 26%
   c. Inorganic salts, 56%   f. Inorganic salts, 26%

6. The "milieu interieur," or "internal environment," of the body pertains to the _____ (E, extracellular; I, intracellular) fluid compartment located _____ (W, within; B, between) cells.

   a. E, W   c. I, W
   b. E, B   d. I, B

7. The highest intracellular to extracellular concentration ratio for generalized mammalian cells occurs for:

   a. Glucose   d. Bicarbonate ions
   b. Sodium ions   e. Calcium ions
   c. Magnesium ions   f. Carbon dioxide

8. Intracellular, in contrast to extracellular, fluid contains higher concentrations of:

   a. Na, Mg, & phosphate ions   d. K, Mg, & phosphate ions
   b. Na, Mg, Ca, & Cl ions   e. K, Ca, Mg, & Cl ions
   c. Na, Ca, K, & Cl ions   f. K, Na, & phosphate ions

OBJECTIVE 1–4.

Identify homeostasis, the major functional systems of the body, and their functional homeostatic mechanisms.

9. The term _____ (E, hemostasis; O, homeostasis) refers to the maintenance of static or constant conditions of the "internal environment" or _____ (X, extracellular; I, intracellular) fluid.

    a. E, X   c. O, X
    b. E, I   d. O, I

10. Fluid exchange between the cardiovascular system and the interstitial fluid of the _____ (I, intracellular; E, extracellular) fluid compartment occurs primarily at the level of _____ (A, arterioles; C, capillaries; V, venules).

    a. I, A   c. I, V   e. E, C
    b. I, C   d. E, A   f. E, V

11. Cells, generally located no more than 25 to 50 _____ (A, angstroms; M, microns; MM, millimeters) from a capillary, receive a rapid equilibration of fluid from capillaries by the process of _____ (AT, active transport; PD, passive diffusion).

    a. A, AT   c. MM, AT   e. M, PD
    b. M, AT   d. A, PD   f. MM, PD

12. All blood in the circulation traverses the entire circuit of the circulation an average of once every _____ when a person is _____ (R, at rest; A, extremely active).

    a. Minute, R   c. 3 minutes, R
    b. Minute, A   d. 3 minutes, A

13. The most abundant end-product of body metabolism, _____ (L, lactic acid; CO_2, carbon dioxide; E, urea), is eliminated from the body through the _____ (U, urinary; R, respiratory; G, gastrointestinal) system.

    a. L, U   c. E, U   e. CO_2, U
    b. CO_2, R   d. L, R   f. E, G

14. The _____ (A, autonomic; S, somatic) nervous system operates largely at a subconscious level, and controls many functions of the internal organs including the gastrointestinal system _____ the heart.

    a. A, and   c. S, and
    b. A, but not   d. S, but not

15. The hormonal system is responsible for regulation of _____ (R, rapidly; S, slowly) reacting metabolic functions and mediates its effects predominately through the _____ (CV, cardiovascular; N, nervous) system.

    a. R, CV   c. S, CV
    b. R, N   d. S, N

OBJECTIVE 1–5.

Recognize the extensive use of homeostatic control systems by the body as exemplified by the control of oxygen and carbon dioxide concentrations in extracellular fluids and regulation of arterial pressure.

16. Control systems regulating interstitial fluid concentrations of constituents more directly involve the _____ for glucose and the _____ for electrolytes. (K, kidneys; L, liver and pancreas; G, gastrointestinal tract)

    a. K, K   c. G, K   e. G, G
    b. L, K   d. G, L   f. L, L

17. The regulatory mechanism of the oxygen concentration of extracellular fluid is dependent upon the chemical characteristics of the _____ (A, albumin; H, hemoglobin) content of _____ (P, plasma; R, red blood cells; W, white blood cells)

    a. A, P   c. A, W   e. H, R
    b. A, R   d. H, P   f. H, W

18. Elevated concentrations of the metabolic _____ (S, substrate; E, end-product) carbon dioxide are _____(A, augmented; O, opposed) by the action of carbon dioxide of _____ (I, increasing; D, decreasing) respiration.

    a. S, A, D   c. S, O, I   e. E, A, I
    b. S, O, D   d. E, O, I   f. E, A, D

19. Reflex effects, resulting from a rise in arterial pressure, _____(R, relaxation; S, stretch) of arterial walls, and_____ (A, activation; I, inactivation) of arterial baroreceptors, serve to_____(G, augment; O, oppose) the rise in arterial pressure.

    a. R, A, G   c. S, A, G   e. S, I, G
    b. R, I, O   d. S, A, O   f. S, I, O

OBJECTIVE 1–6.

Identify the general characteristics of homeostatic control mechanisms and their underlying basic physical principles.

20. Regulatory processes of the body functioning in homeostasis usually may be described as processes of:

    a. Adaptation   c. Positive feedback
    b. Accommodation   d. Negative feedback

21. A homeotherm with a normal temperature of $100°$ F moves from an environment of $90°$ F to $110°$ F with only a $1°$ F rise in body temperature. The amplification of the control system is:

    a. −1   c. −20
    b. −19   d. −21

22. _____ feedback is better known as a "vicious cycle" because it leads to _____ cycles of instability that may lead to the death of the organism.

    a. Positive, diminished   d. Negative,
    b. Positive, progressive   progressive
    c. Negative, diminished

23. A natural consequence of the organization of many regulatory systems of the body is the innate capability under appropriate conditions to develop:

    a. Fasciculations   d. Convulsions
    b. Fibrillations   e. Rebound
    c. Oscillations   f. None of the above

24. A high degree of damping of a control system causes _____ oscillation and greater _____ within the system.

    a. Diminished, instability   d. Greater,
    b. Diminished, stability   stability
    c. Greater, instability

2

The Cell and Its Function

OBJECTIVE 2–1.
Using the following diagram, identify the various structural components of a typical cell.

Directions: Match the lettered headings with the diagram and numbered list of descriptive words and phrases.

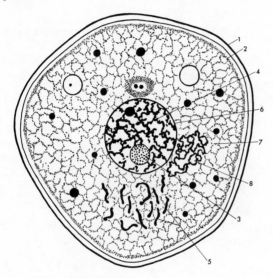

1. _____ Boundary separating intracellular and extracellular fluids.
2. _____ Outer layer of ectoplasm immediately beneath the cell membrane.
3. _____ A vacuole-like body within the nucleus, rich in RNA.
4. _____ Network of nuclear fibrils, rich in DNA.
5. _____ Infolded organelles found in cytoplasm; principal sites of the generation of energy (ATP).
6. _____ Condensed double layer of lipids and proteins that enclose the nucleoplasm.
7. _____ The protoplasm of a cell exclusive of that of the nucleus.
8. _____ Continuous system of membrane-bound cavities that ramify throughout the cytoplasm.

a. Cytoplasm
b. Chromatin material
c. Cell membrane
d. Mitochondria

e. Endoplasmic reticulum
f. Nuclear membrane
g. Nucleolus
h. Cortex

OBJECTIVE 2–2.
Identify cellular protoplasm, its general chemical constituents, and their general functional characteristics.

9. Next to water, present in cells in a concentration of_____ %, the second most abundant compound in most cells is_____(C, carbohydrate; L, lipid; P, protein).

   a. 40–55, C   c. 40–55, P   e. 70–85, L
   b. 40–55, L   d. 70–85, C   f. 70–85, P

10. Enzymes, controlling metabolic functions of the cell, are composed primarily of _____ (P, protein; S, steroids) of the _____ (F, fibrillar; G, globular) type.

   a. P, F   c. S, F
   b. P, G   d. S, G

11. Lipids typically comprise _____% of the cell weight, of which the most abundant lipid of animal tissues is_____(P, phospholipid; C, cholesterol; T, triglyceride).

   a. 2–3, P   c. 2–3, T   e. 10–15, C
   b. 2–3. C   d. 10–15, P   f. 10–15, T

12. The larger percentage of carbohydrates of cells, present in the form of _____ (C, cellulose; G, glycogen; L, glucose), functions in a _____(S, structural; M, metabolic) functional role.

   a. C, S   c. L, S   e. G, M
   b. G, S   d. C, M   f. L, M

OBJECTIVE 2–3.

Identify the physical and chemical properties of the cell membrane and their functional significance.

13. Cell membranes, approximately_____ Angstroms thick, generally contain a higher percentage composition of_____ (L, lipid; P, protein; O, polysaccharides).

a. 10, L   c. 10, O   e. 100, P
b. 10, P   d. 100, L   f. 100, O

14. A feature generally shared by surfaces of cell membranes, cell organelles, and small dispersed cytoplasmic granules relates to:

a. Single, "unit" membranes
b. Hydrophilia
c. Hydrophobia
d. 100-angstrom pore diameters
e. Glycoprotein coatings
f. Pinocytotic activity

15. The presence of_____(P, protein; L, lipid) on the outer layer of the unit membrane of cell surfaces renders it_____ .

a. P, hydrophobic   c. L, hydrophobic
b. P, hydrophilic   d. L, hydrophilic

16. _____(C, cholesterol; P, phospholipid) molecules, comprising a higher percentage of cell membrane lipids, contain charged polar groups that make them soluble in_____ (A, aqueous; L, lipid) solvents, and a hydrocarbon portion that renders them soluble in_____ (A, aqueous; L, lipid) solvents.

a. C, A, L   c. P, A, L
b. C, L, A   d. P, L, A

17. Cell membranes behave as though they are permeated by_____ (P, protein; L, lipid) -lined pores, approximately _____ angstroms in diameter.

a. P, 8   c. P, 800   e. L, 80
b. P, 80   d. L, 8   f. L, 800

18. Large holes or "pores" of the_____ (S, single; D, double) "unit membrane" structure of the nuclear membrane connect more directly with the_____ (M, mitochondria; N, nucleolus; E, endoplasmic reticulum).

a. S, M   c. S, E   e. D, N
b. S, N   d. D, M   f. D, E

19. _____(G, granular; A, agranular) endoplasmic reticulum, more notably functioning in the synthesis of lipid substances, differs from its counterpart with regard to the presence or absence of_____(L, lysosomes; R, ribosomes; M, microtubules).

a. G, L   c. G, M   e. A, R
b. G, R   d. A, L   f. A, M

20. The Golgi complex, prominent among _____(S, secretory; P, phagocytic) cells, is a specialized portion of the_____(N, nucleus; M, mitochondria; E, endoplasmic reticulum).

a. S, N   c. S, E   e. P, M
b. S, M   d. P, N   f. P, E

OBJECTIVE 2–4.

Recognize the colloidal nature of cytoplasm. Identify the types of cell organelles and inclusion bodies of cytoplasm and their general functional roles.

21. The dispersal of largely_____(L, hydrophilic; B, hydrophobic) particles and organelles within the cytoplasm is dependent upon their_____(N, noncharged; S, similarity of charged; D, dissimilarity of charged) surfaces.

a. L, N   c. L, D   e. B, S
b. L, S   d. B, N   f. B, D

22. "Oxidative phosphorylation" of_____(G, glucose; A, adenosine compounds) within the _____(H, hyaloplasm; N, nucleoplasm, M, mitochondria) of cells generates _____% of high energy substances utilized by cells for cellular functions.

a. G, H, 95   c. A, N, 5   e. A, H, 95
b. G, M, 5   d. A, M, 95   f. A, H, 5

23. The "digestive organelle" of cells, or the _____ (M, mitochondrion; G, Golgi apparatus; L, lysosome), contains enzymatic constituents that are more frequently characterized as _____ (K, alkaline hydrolases; C, acid hydrolases; E, esterases).

    a. M, C   c. L, K   e. G, K
    b. G, E   d. M, E   f. L, C

24. The lysosome, bordered by a _____ (S, single; D, double) unit membrane, _____ (I, is; N, is not) commonly found within cells that are actively phagocytic.

    a. S, I   c. D, I
    b. S, N   d. D, N

25. Microtubules, approximately _____ Angstroms in diameter, probably function as an intracellular _____ (C, cytoskeleton; R, contractile mechanism).

    a. 2.5, R   c. 250, R   e. 25, C
    b. 25, R   d. 2.5, C   f. 250, C

OBJECTIVE 2–5.
Identify the structural components of the nucleus and their general functional roles.

26. A cell component composed largely of _____ , and that does not appear to have a limiting membrane, is the _____ .

    a. DNA, nucleus   c. RNA, nucleus
    b. DNA, nucleolus   d. RNA, nucleolus

27. Chromosomes, containing genetic information in the form of _____ , are more readily identifiable with light microscopy during periods of _____ (P, increased protein synthesis; M, mitosis).

    a. RNA, P   c. ATP, P   e. DNA, M
    b. DNA, P   d. RNA, M   f. ATP, M

28. Ribosomes contain larger amounts of_____, which is synthesised in the _____.

    a. RNA, nucleolus   d. DNA, nucleolus
    b. RNA, nucleus   e. DNA, nucleus
    c. RNA, mitochondria   f. DNA, mitochondria

OBJECTIVE 2–6.
Contrast the animal cell with precellular forms of life.

29. The generalized animal cell differs from precellular forms of life in that animal cells possess:

    a. Reproductive   d. RNA
       capability   e. Organelles
    b. Cell membranes   f. Nuclei
    c. DNA

30. Typical animal cells are about_____ times the diameter of a small virus and about _____ times the diameter of a typical bacterium.

    a. 10^{12}, 10^4 d. 10^{12}, 10^6
    b. 10^6, 10^2 e. 10^6, 10^4
    c. 10^3, 10   f. 10^3, 10^2

OBJECTIVE 2–7.

Identify the functional roles of the endoplasmic reticulum.

31. Cellular processes leading to the secretion of glycoprotein by appropriate cells probably involve the _____ for protein synthesis, the _____ for conjugation with carbohydrates to form glycoprotein, and the _____ for membranous enclosure of the secretory granule. (G, Golgi complex; N, nucleus; M, mitochondria; ER, endoplasmic reticulum; R, Ribosomes)

   a. ER, R, G   c. R, G, G   e. N, ER, M
   b. ER, G, ER   d. R, ER, M   f. N, ER, G

32. Synthesis of _____ (G, glycogen granules; L, lipid secretory granules), which more probably requires the presence of endoplasmic reticulum, occurs in its _____ (A, agranular; R, granular) portions.

   a. G, A   c. L, A
   b. G, R   d. L, R

33. The Golgi complex, a specialized portion of the _____ (N, nucleus; ER, endoplasmic reticulum) is known to function in _____ (E, only external; I, only internal; EI, both external and internal) secretory processes of cells.

   a. N, E   c. N, EI   e. ER, I
   b. N, I   d. ER, E   f. ER, EI

OBJECTIVE 2–8.

Identify the processes of pinocytosis and phagocytosis, and their functional significance. Recognize the roles of lysosomes in phagocytosis and pinocytosis, regression of tissue, and autolysis of cells.

34. Protein transport to the interior of a cell from the extracellular solution _____ generally occur by passive diffusion; _____ generally occur by carrier-mediated active transport, and _____ generally occur by pinocytosis. (D, does; N, does not)

   a. D, D, D   c. D, N, D   e. N, N, D
   b. D, D, N   d. N, D, D   f. N, N, N

35. Pinocytotic vesicles which generally _____ (A, remain attached to; D, detach from) the cell membrane are _____ (S, smaller; L, larger) than the lower limit of resolution of the light microscope.

   a. A, S   c. D, S
   b. A, L   d. D, L

36. Recently formed _____ (G, phagocytic; N, pinocytotic) vesicles more generally contain particulate matter and fluid originating in the _____ (E, extracellular; I, intracellular) fluid.

   a. G, E   c. N, E
   b. G, I   d. N, I

37. Combinations of phagocytic vesicles with _____ (M, mitochondria; G, the Golgi apparatus; L, lysosomes) result in _____ (D, digestive; P, pinocytotic) vesicles.

   a. M, D   c. L, D   e. G, P
   b. G, D   d. M, P   f. L, P

38. Agents which promote tissue injury _____ (I, increase; D, decrease) the susceptibility of lysosomal membranes to rupture, and _____ (I, increase; D, decrease) the probability of autolysis.

   a. I, I   c. D, I
   b. I, D   d. D, D

OBJECTIVE 2–9.

Identify the general mechanisms by which cells extract energy from nutrients.

39. "Energy currency" in the form of _____ "high energy" bonds of _____ kilocalories per mole is most immediately available for cellular energy purposes.

    a. Glucose, 36   d. Glucose, 8
    b. Adenosine, 36   e. Adenosine, 8
    c. ATP, 36   f. ATP, 8

40. Approximately 90 per cent of all ATP is formed _____ (O, outside; I, inside) the mitochondria by means of _____ (A, aerobic; N, anaerobic) metabolism.

    a. O, A   c. I, A
    b. O, N   d. I, N

41. Cells _____ utilize ATP energy for membrane transport, _____ utilize ATP energy for synthesis of chemical compounds, and _____ utilize ATP energy for mechanical work. (D, do; N, do not)

    a. D, D, D   c. D, D, N   e. N, N, D
    b. D, N, D   d. D, N, N   f. N, N, N

42. Oxidation of metabolite hydrogen, derived from _____ (W, water; F, foodstuffs), is accomplished by the enzymes of the electron transport system located in the _____ (L, lysosomes; N, nuclei; M, mitochondria).

    a. W, L   c. W, M   e. F, N
    b. W, N   d. F, L   f. F, M

43. Red blood cells, lacking the ongoing functions of generalized cellular organelles while retaining glycolytic enzymes, would _____ (P, possess; L, lack) the ability to produce ATP, would _____ (R, require; N, not require) oxygen; and would _____ (P, possess; L, lack) the enzymes of the tricarboxylic acid or Krebs cycle.

    a. P, R, P   c. P, N, L   e. L, R, L
    b. P, N, P   d. L, N, L   f. L, R, P

OBJECTIVE 2–10.

Recognize the significance of ameboid, ciliary, and specialized muscle types of cell movement. Identify the processes and mechanisms involved in ameboid and ciliary forms of cell movement.

44. Myxomyosin, located in the gel of the _____ (N, endoplasm; C, ectoplasm) of cells exhibiting ameboid motion, contracts in the presence of ATP and _____ (Mg, magnesium; Ca, calcium; Fe, ferrous) ions.

    a. N, Mg   c. N, Fe   e. C, Ca
    b. N, Ca   d. C, Mg   f. C, Fe

45. The _____ (I, increased; D, decreased) thickness of ectoplasm in a developing cellular pseudopodium closest to a chemical source is indicative of _____ (P, positive; N, negative) chemotaxis.

    a. I, P   c. D, P
    b. I, N   d. D, N

46. A _____ (U, unidirectional; B, bidirectional; M, multidirectional) fluid movement over stationary cellular surfaces is more readily accomplished through _____ (A, ameboid; C, ciliary) forms of movement.

    a. U, A   c. M, A   e. B, C
    b. B, A   d. U, C   f. M, C

47. The _____ (R, reproductive; C, contractile) _____ (A, axoneme; P, basal body) portion of the cilium contains nine double tubular filaments and two singular tubular filaments.

    a. R, A   c. C, A
    b. R, P   d. C, P

3

Genetic Control of Cell Function — Protein Synthesis and Cell Reproduction

OBJECTIVE 3–1.
Identify the general mechanisms by which genes control cellular functions. Identify the chemical constituents, chemical pattern of organization, and means of "genetic coding" of DNA.

1. Genetically coded information necessary for cellular replication is vested in the sequence of _____(P, peptides; N, nucleotides) of _____.

   a. P, DNA   c. N, DNA
   b. P, RNA   d. N, RNA

2. Basic building blocks of nucleotides in the formation of DNA include two purines, _____ pyrimidines, _____ (R, ribose; D, deoxyribose), and _____(P, phosphoric acid; U, uracil).

   a. 2, R, P   c. 2, D, U   e. 4, R, U
   b. 2, D, P   d. 4, D, U   f. 4, R, P

3. Two strands of the DNA helix, attaching to each other by_____(P, peptide; H, hydrogen) bonding, contain the complementary base pairs adenine and _____(C, cytosine; T, thymine), and guanine and_____(C, cytosine; T, thymine; U, uracil).

   a. P, C, T   c. H, C, T   e. H, T, C
   b. P, T, U   d. H, C, U   f. H, T, U

4. _____ successive bases of DNA, constituting three "code words," are more probably responsible for the placement of_____amino acids during protein synthesis.

   a. 3, 3   c. 20, 3   e. 9, 3
   b. 9, 1   d. 3, 9   f. 20, 9

OBJECTIVE 3–2.
Identify three separate types of RNA and their chemical composition, synthesis, manner of coding, and functional significance.

5. The chemical composition of RNA differs from DNA in that RNA contains:

   a. Thymine & glucose   d. Uracil & glucose
   b. Thymine & deoxyribose   e. Uracil & deoxyribose
   c. Thymine & ribose   f. Uracil & ribose

6. The process of transferring the genetic code from DNA to _____(M, messenger; T, transfer) RNA is called _____ .

   a. M, translation   c. T, translation
   b. M, transcription   d. T, transcription

7. "Chain-initiating" and "chain-terminating" _____(A, anticodons; C, codons) are more likely to be located within _____(M, messenger; T, transfer) RNA.

   a. A, M   c. C, M
   b. A, T   d. C, T

8. RNA polymerase results in bond formation between _____(BR, RNA base and ribose; RP, ribose and phosphoric acid radicals; BP, RNA base and phosphoric acid radicals) during _____ synthesis.

   a. BR, protein   d. BR, RNA
   b. RP, protein   e. RP, RNA
   c. BP, protein   f. BP, RNA

9. The anticodon is that portion of a _____ (M, messenger; R, ribosomal; T, transfer) RNA molecule that attaches to a specific _____ (A, amino acid; C, RNA codon).

   a. M, A   c. T, A   e. R, C
   b. R, A   d. M, C   f. T, C

10. Each _____ (L, lysosome; H, mitochondrion; R, ribosome) is a specific locus for the synthesis of _____ (S, a single type; M, multiple types) of protein.

    a. L, S   c. R, S   e. H, M
    b. H, S   d. L, M   f. R, M

OBJECTIVE 3–3.

Describe the process of translation during protein synthesis.

11. Polyribosomes are frequently attached together by a single _____ (T, transfer; M, messenger) RNA molecule during the process of _____ (L, translation; C, transcription).

    a. T, L   c. M, L
    b. T, C   d. M, C

12. The proper sequence of amino acids within a forming protein molecule is achieved by the combination of activated amino acid- _____ (T, transfer; M, messenger) RNA complexes with appropriate _____ (A, anticodons; C, codons).

    a. T, A   c. M, A
    b. T, C   d. M, C

13. _____ (H, hydrostatic; P, peptide) bonds are formed between adjacent amino acids during protein synthesis through the _____ (E, enzymatic; N, nonenzymatic) removal of _____ (W, water; A, ammonia).

    a. H, E, W   c. P, E, W   e. P, N, W
    b. H, N, A   d. P, E, A   f. P, N, A

OBJECTIVE 3–4.

Identify the general cellular mechanisms of control of genetic expression and biochemical activities.

14. Cells are known to control proper proportions and quantities of different cellular constituents via _____ (A, activation only; R, repression only; AR, both activation and repression) of _____ (G, genes only; GE, both genes and enzymes).

    a. A, G   c. AR, G   e. R, GE
    b. R, G   d. A, GE   f. AR, GE

15. Regulatory genes more probably regulate genetic activity by controlling the formation of _____ (A, activator; R, repressor) substances which act upon structural genes via _____ (D, a direct; I, an indirect) mechanism.

    a. A, D   c. R, D
    b. A, I   d. R, I

16. The majority of controllable groups of structural genes or _____ (C, clones; O, operons) within a cell are more probably _____ (A, active; I, inactive) at any given time.

    a. C, A   c. O, A
    b. C, I   d. O, I

17. Approximately half of the chromosome mass is _____ , a suggested _____ (A, activator; R, repressor) substance for DNA.

    a. RNA, A   d. RNA, R
    b. Cyclic AMP, A   e. Cyclic AMP, R
    c. Histone, A   f. Histone, R

18. A _____ (P, positive; N, negative) feedback mechanism for the regulation of enzymatic activity more generally operates to regulate the _____ (F, first; L, last) enzyme within an enzymatic sequence.

    a. P, F   c. N, F
    b. P, L   d. N, L

19. Reduced ATP reserves of cells, associated with _____ (L, lower; H, higher) than normal cyclic AMP levels, regulate glycogen breakdown through the action of cyclic AMP as an enzyme _____ (A, activator; I, inhibitor).

    a. L, A   c. H, A
    b. L, I   d. H, I

OBJECTIVE 3–5.

Describe the events involved in replication and separation of chromosomal DNA for cell division.

20. DNA replication of the _____ pairs of chromosomes of human cells occurs _____ (P, prior to; D, during; F, following) mitosis.

    a. 23, P   c. 23, F   e. 46, D
    b. 23, D   d. 46, P   f. 46, F

21. Cells which are rapidly dividing in an uninhibited manner exhibit periods of about _____ for interphase and _____ for mitosis.

    a. 1–3 hr, ½ hr.   d. 1–3 hr, 3 hr.
    b. 10–30 hr, ½ hr.   e. 10–30 hr, 3 hr.
    c. 1–3 days, ½ hr.   f. 1–3 days, 3 hr.

22. Microtubules of the forming mitotic apparatus attach to the _____ (C, centrioles; E, centromeres) of the chromosomes during _____ (PM, prometaphase; A, anaphase; M, metaphase).

    a. C, PM   c. C, M   e. E, A
    b. C, A   d. E, PM   f. E, M

23. Replication of the original _____ of centrioles of the parent cell is more likely to occur during _____ (P, prophase; PM, prometaphase; T, telophase).

    a. One pair, P   d. Two pairs, P
    b. One pair, PM   e. Two pairs, PM
    c. One pair, T   f. Two pairs, T

OBJECTIVE 3–6.

Recognize the existence of mechanisms which regulate cell reproduction, growth, size, and differentiation.

24. The process of cellular differentiation, probably _____ (I, initiated; N, not initiated) by unequal division of genetic material during mitosis, is thought to be associated with a selective _____ (L, loss; R, repression) of genetic operons.

    a. I, L   c. N, L
    b. I, R   d. N, R

25. Colchicine, when applied to a proliferative pool of cells, results in a _____ (G, greater; L, lesser) average amount of DNA per cell and _____ (G, a greater; N, no significant change in; L, a lesser) average cell size.

    a. G, G   c. G, L   e. L, N
    b. G, N   d. L, G   f. L, L

26. Most cells of the body when grown in tissue culture media proliferate _____ (I, indefinitely; F, for a predetermined number of mitoses), provided that secretory products of the cells _____ (A, are; N, are not) allowed to accumulate.

    a. I, A   c. F, A
    b. I, N   d. F, N

4

Transport Through the Cell Membrane

OBJECTIVE 4-1.

Distinguish extracellular and intracellular fluids on the basis of the relative concentrations of their principle chemical constituents.

1. In contrast to intracellular fluid, extracellur fluid contains _____ quantities of sodium, _____ quantities of potassium, and _____ quantities of chloride. (S, smaller; L, larger)

   a. S, S, S   c. S, L, L   e. L, S, L
   b. S, L, S   d. L, S, S   f. L, L, L

2. In contrast to intracellular fluid, extracellular fluid contains _____ concentrations of glucose, _____ concentrations of amino acids, and _____ hydrogen ion concentrations. (G, greater; S, smaller)

   a. S, S, S   c. S, L, L   e. L, S, L
   b. S, L, S   d. L, S, S   f. L, L, L

3. Extracellular constituents with concentrations less than 10 mEq./liter include _____ , whereas extracellular constituents with concentrations greater than 100 mEq./liter include

   _____ .

   a. Na^+ & HCO_3^-, K & Cl^-
   b. Mg^{++} & Cl^-, Na^+
   c. K^+ & Ca^{++}, Na^+ & HCO_3^-
   d. Na^+ & HCO_3^-, K^+ & Ca^{++}
   e. Mg^{++} & Cl^-, K^+
   f. K^+ & Ca^{++}, Na^+ & Cl^-

4. The extracellular to intracellular fluid concentration ratio of bicarbonate ions, approximating _____ , is _____ (G, greater; S, smaller) than the comparable concentration ratio for chloride ions.

   a. 3, G   c. ½, G   e. 30, S
   b. 30, G   d. 3, S   f. ½, S

5. The concentration ratio of _____ divided by _____ most closely approximates unity. (E, extracellular; I, intracellular)

   a. $E Mg^{++}$, $I Mg^{++}$ d. $I K^+$, $I Cl^-$
   b. $E Na^+$, $E Cl^-$ e. $I Na^+$, $E Cl^-$
   c. $E Na^+$, $E HCO_3^-$ f. $E Ca^{++}$, $I Ca^{++}$

OBJECTIVE 4-2.

Identify the terms concentration difference or diffusion gradient and the factors that affect the rate of diffusion of a substance.

6. The total concentration change along the axis of a change or the _____ , divided by the distance, is equal to the _____ (CD, concentration difference; DG, diffusion gradient; ZP, zeta potential).

   a. CD, DG   c. DG, CD
   b. CD, ZP   d. DG, ZP

7. The diffusion rate of a dissolved substance is _____ proportional to the concentration gradient and _____ proportional to the molecular radius. (D, directly; I, inversely)

   a. D, D   c. I, D
   b. D, I   d. I, I

8. Diffusion rates of a dissolved substance _____ with increasing cross-sectional area and _____ with increasing absolute termperature. (I, increase; C, remain constant; D, decrease)

   a. I, I   c. I, D   e. D, C
   b. I, C   d. D, I   f. D, D

OBJECTIVE 4–3.

Identify the differing routes by which substances can diffuse through the cell membrane.

9. The largest percentage of the surface area of cell membranes is _____ (E, easily; N, not easily) penetrated by intracellular and extracellular fluids as a consequence of its _____ (C, carbohydrate; L, lipid; P, protein) constituent.

   a. E, C   c. E, P   e. N, L
   b. E, L   d. N, C   f. N, P

10. Oxygen, which is more soluble in _____ (L, lipids; W, water), passes through cell membranes _____ (S, less; M, more) readily than water, and has a higher rate of diffusion through membrane _____ (L, lipids; P, "pores").

   a. L, S, L   c. L, M, P   e. W, S, P
   b. L, M, L   d. W, M, P   f. W, S, L

11. Cellular membrane "pores" more probably owe their properties to _____ molecules and function more effectively to permit passive diffusion of _____ soluble substances. (L, lipid; W, water; P, protein)

   a. L, L   c. P, L
   b. L, W   d. P, W

OBJECTIVE 4–4.

Identify facilitated diffusion and contrast its properties with those of simple passive diffusion.

12. The process of facilitated diffusion, occuring through the _____ (A, aqueous pores; L, lipid portions) of cellular membranes, _____ (D, does; N, does not) involve carrier-mediated diffusion.

   a. A, D   c. L, D
   b. A, N   d. L, N

13. Rates of facilitated diffusion through cellular membranes _____ (I, increase; R, remain constant) with increasing concentration gradients within a low range, _____ (D, do; N, do not) exhibit "saturation," and _____ (D, do; N, do not) move substances "uphill" against their concentration gradients.

   a. I, D, D   c. I, N, D   e. R, N, D
   b. I, D, N   d. R, N, N   f. R, D, N

14. _____ (G, glucagon; I, insulin) more effectively increases the rate of glucose transport into the majority of cells, primarily through its effects upon _____ (EC, extracellular glucose concentrations; IC, intracellular glucose concentrations; R, rates of facilitated diffusion).

   a. G, EC   c. G, R   e. I, IC
   b. G, IC   d. I, EC   f. I, R

OBJECTIVE 4–5.

Identify membrane permeability. Characterize membrane pores on the basis of their permeability characteristics and identify the effects of different factors upon pore permeability.

15. Membrane permeability, or the rate of transport through the membrane surface per unit of _____ (CD, concentration difference; L, length; T, absolute temperature) is generally greater for _____ (Cl^-, chloride ions; U, urea; W, water)

    a. CD, Cl^- c. T, U   e. L, U
    b. L, W   d. CK, W   f. T, U

16. Cell membrane pores behave as though they were about _____ in diameter and occupy _____ (M, more; L, less) than 1% of the total cell membrane surface.

    a. 0.8 angstroms, M   d. 0.8 angstroms, L
    b. 8 angstroms, M   e. 8 angstroms, L
    c. 8 microns, M   f. 8 microns, L

17. Cell membranes behave as though they were permeated by minute pores each approximating the diameter of a _____ , with a pore lining of a surplus of _____ charges.

    a. Hydrated K ion, +   d. Hydrated K ion, –
    b. Na ion, +   e. Na ion, –
    c. Ribose molecule, +   f. Ribose molecule, –

18. Potassium ions as compared to sodium ions have a _____ ionic diameter, _____ solvated or hydrated diameter, and a _____ membrane permeability. (G, greater; L, lesser)

    a. L, G, G   c. L, L, G   e. G, L, G
    b. L, G, L   d. G, L, L   f. G, G, L

19. Other factors remaining constant, membrane permeabilities tend to be _____ for cations as opposed to anions, _____ for divalent as opposed to monovalent ions, and _____ for decreasing molecular sizes for a series of homologous substances. (H, higher; L, lower)

    a. L, L, L   c. L, H, H   e. H, H, L
    b. L, L, H   d. H, H, H   f. H, L, L

20. Diminished calcium ion concentrations within the extracellular fluid results in _____ permeabilities of cell membranes and _____ activity of neural tissues. (I, increased; D, decreased)

    a. I, I   c. D, I
    b. I, D   d. D, D

21. Antidiuretic hormone, secreted by the _____ (A, adrenal glands; H, hypothalamus), _____ (I, increases; D, decreases) the membrane permeability of cells lining the collecting ducts of the kidney and consequently promotes water _____ (E, excretion; R, reabsorption).

    a. A, I, E   c. A, D, R   e. H, I, R
    b. A, D, E   d. H, D, R   f. H, I, E

OBJECTIVE 4–6.

Recognize the significance of net diffusion, and identify the consequences of transmembrane concentration, electrical potential, and pressure differences upon net diffusion through cellular membranes.

22. The rate of net diffusion into a cell is _____ (K, directly; I, inversely) proportional to the concentration on the outside, _____ (P, plus; M, minus; D, divided by) the concentration on the inside of the cell.

    a. K, P   c. K, D   e. I, M
    b. K, M   d. I, P   f. I, D

23. An electrical potential difference imposed across a cell membrane, in which the intracellular fluid is made increasingly negative with respect to the extracellular fluid, will tend to result in _____ net movement of anions and _____ net movement of cations. (I, an inward; O, an outward; N, no)

    a. I, I   c. O, I   e. N, O
    b. N, I   d. O, O   f. I, O

24. An electrochemical equilibrium for an ion species is established across cellular membranes when the transmembrane potential is proportional to the _____ (A, antilogarithm; L, logarithm) of the transmembrane ion concentration _____ (D, difference; R, ratio).

    a. A, D
    b. A, R
    c. L, D
    d. L, R

25. Pressure is expressed in units of _____ (KE, kinetic energy; M, momentum; F, force) of molecular impact per unit of _____ (T, time; SA, surface area).

    a. KE, T
    b. M, T
    c. F, T
    d. KE, SA
    e. M, SA
    f. F, SA

OBJECTIVE 4–7.

Identify: osmosis; conditions which cause and oppose osmosis; osmotic pressure and its means of determination; and osmolarity and osmolality. Identify the nature of "bulk flow" of water through membrane pores.

26. Osmosis is a net movement of _____ (I, ions; W, water) in the _____ (S, same; O, opposing) direction as the transmembrane concentration gradient for water.

    a. I, S
    b. I, O
    c. W, S
    d. W, O

27. A hydrostatic pressure gradient, which exactly balances the tendency for osmosis across a semipermeable membrane, is _____ (G, greater than; E, equal to; L, less than) the absolute magnitude of the osmotic pressure gradient and of the _____ (S, same; O opposite) polarity.

    a. G, S
    b. E, S
    c. L, S
    d. G, O
    e. E, O
    f. L, O

28. The osmotic pressure exerted by nondiffusible solute in solution is determined by the _____ (M, mass; N, number) of particles per unit volume of fluid and approximates _____ mm. Hg for a one milliosmolar solution at body temperature.

    a. M, 7.9
    b. M, 19.3
    c. M, 3790
    d. N, 7.9
    e. N, 19.3
    f. N, 3790

29. A _____ osmolar solution of glucose, containing one gram molecular weight of glucose per liter of water, is the approximate osmotic equivalent of a _____ molar solution of sodium chloride.

    a. 180, 58.5
    b. 2, 1
    c. 1, ½
    d. 180, 117
    e. 2, 4
    f. 1, 2

30. The osmolarity of mammalian extracellular fluids is about _____ osmolar and is _____ (L, significantly less than; E, equal to; G, significantly greater than) that of intracellular fluids.

    a. 0.3, L
    b. 0.3, E
    c. 0.9, G
    d. 0.9, E
    e. 2.7, G
    f. 2.7, L

31. Bulk flow of water acts _____ (S, synergistically; A, in opposition) to net water diffusion through membrane pores in response to _____ (H, only hydrostatic; O, only osmotic; B, either hydrostatic or osmotic) pressure gradients.

    a. S, H
    b. S, O
    c. S, B
    d. A, H
    e. A, O
    f. A, B

OBJECTIVE 4–8.

Identify active transport, its operational mechanisms, and its general characteristics.

32. Active transport mechanisms generally _____ involve carrier-mediated transport, _____ exhibit saturation, and _____ transport substances against electrochemical concentration gradients. (D, do; N, do not)

    a. D, D, D
    b. D, N, D
    c. D, D, N
    d. D, N, N
    e. N, D, D
    f. N, N, N

33. The relatively _____ (N, nonspecific; S, specific) active transport mechanisms of cellular membranes perform concentration work which is proportional to the _____ (L, logarithm; A, antilogarithm) of the transmembrane concentration _____ (D, difference; R, ratio).

    a. S, L, D
    b. S, L, R
    c. S, A, D
    d. S, A, R
    e. N, L, R
    f. N, A, D

34. Active transport mechanisms of sodium ions by intestinal mucosal cells results in _____ chloride and _____ water absorption from the intestinal lumen. (E, enhanced; C, no significant influence upon; D, diminished)

  a. E, E   c. E, D   e. D, C
  b. E, C   d. D, E   f. D, D

OBJECTIVE 4-9.

Identify the characteristics and functional significance of the sodium transport mechanism or "sodium pump."

35. The sodium active transport mechanism of cell membranes more probably involves a _____ (L, lipoprotein; G, glycoprotein) carrier and sodium ion transport to the _____ (E, extracellular; I, intracellular) fluid compartment _____ (W, along with; X, in exchange for) potassium ions.

  a. L, E, W   c. L, I, X   e. G, E, W
  b. L, E, X   d. G, I, W   f. G, E, X

36. In the absence of available ATP energy, the _____ (C, continuing; L, loss of) activity of the "sodium pump" results in _____ (I, increased; N, no significant change in; D, decreased) intracellular volumes.

  a. C, I,   c. C, N   e. L, D
  b. C, D   d. L, I   f. L, N

37. The active transport mechanism for _____ (M, monosaccharides; A, amino acids; Na, sodium ions) operates more directly to prevent a tendency for increased intracellular osmolarity and _____ (I, increased; D, decreased) intracellular fluid volumes.

  a. M, I   c. Na, I   e. A, D
  b. A, I   d. M, D   f. Na, D

OBJECTIVE 4-10.

Recognize the general characteristics and significance of active transport mechanisms for other electrolytes, sugars, and amino acids.

38. The majority of mammalian cells transport monosaccharides by _____ (A, active transport; F, facilitated diffusion) mechanisms and _____ (D, do; N, do not) actively transport disaccharides.

  a. A, D   c. F, D
  b. A, N   d. F, N

39. The active transport mechanism for glucose, galactose, and a group of other sugars, _____ (E, enhanced; N, not significantly affected; D, diminished) by increased activity of the "sodium pump," appears to require an intact hydroxyl group on the _____ carbon of the molecule.

  a. E, C_2 c. D, C_2 e. N, C_5
  b. N, C_2 d. E, C_5 f. D, C_5

40. The active transport mechanism of some of the amino acids requires vitamin _____ and _____ (I, is; N, is not) dependent upon operation of the sodium metabolic pump.

  a. C, I   c. B_{12}, I   e. B_6, N
  b. B_6, I   d. C, N   f. B_{12}, N

41. _____ (I, increased; D, decreased) amino acid transport into essentially all cells of the body is more effectively accomplished with the presence of increased levels of _____ (N, insulin; G, glucocorticoids; GH, growth hormone).

  a. I, N   c. I, GH   e. D, G
  b. I, G   d. D, N   f. D, GH

5

Red Blood Cells, Anemia, and Polycythemia

OBJECTIVE 5-1.
Identify the major functions of the red blood cells.

1. The major function of the red blood cells is to _____ (S, synthesize plasma; T, transport) hemoglobin and thereby provide a major transport mechanism for _____ (O_2, oxygen; CO_2, carbon dioxide).

   a. S, O_2
   b. S, CO_2
   c. T, O_2
   d. T, CO_2

2. Carbonic anhydrase, localized within the _____ (P, plasma; RBC, red blood cell) fraction, catalyzes carbonic acid _____ (I, ionization; F, formation).

   a. P, I
   b. RBC, I
   c. P, F
   d. RBC, F

3. Carbonic anhydrase functions to _____ (I, increase; D, decrease) the rate of carbon dioxide removal from tissues by increasing the rate of reaction of carbon dioxide and water by about _____ times.

   a. I, 25
   b. I, 250
   c. I, 25,000
   d. D, 25
   e. D, 250
   f. D, 25,000

OBJECTIVE 5-2.
Identify typical values of the dimensions, cellular volume, and blood concentration of red blood cells.

4. Typical _____ (S, spherical; B, biconcave disk-like) red blood cells are closer to _____ microns in diameter.

   a. S, 2.2
   b. S, 8
   c. S, 50
   d. B, 2.2
   e. B, 8
   f. B, 50

5. The volume ratio of packed red blood cells to whole blood, divided by the red blood cell _____ (H, hematocrit; C, blood concentration; D, diameter), may be utilized to determine a typical mean red blood cell volume of _____ cubic microns.

   a. H, 87
   b. C, 87
   c. D, 87
   d. H, 870
   e. C, 870
   f. D, 870

6. A greater average number of red blood cells per cubic millimeter of blood occurs for _____ (W, women; M, men) and approximates _____ .

   a. W, 5×10^3
   b. W, 4.5×10^5
   c. W, 5.2×10^6
   d. M, 5×10^3
   e. M, 4.5×10^5
   f. M, 5.2×10^6

OBJECTIVE 5–3.

Recognize the composition of blood. Identify the hematocrit, and the hemoglobin content and oxygen-carrying capacity of red blood cells and whole blood.

7. Cellular elements, largely comprised of _____ (W, white blood cells; R, red blood cells; P, platelets), constitute about _____ % of the blood volume.

   a. W, 15 c. P, 15 e. R, 45
   b. R, 15 d. W, 45 f. P, 45

8. The hematocrit, representing a volume ratio of _____ (P, plasma; C, cells; H, hemoglobin; B, whole blood), is normally _____ (G, greater; L, less) than 50%.

   a. P/C, G c. C/B, G e. H/B, L
   b. H/B, G d. P/C, L f. C/B, L

9. The hemoglobin concentration of blood is normally closest to _____ per 100 ml., with each gram of hemoglobin capable of combining maximally with _____ ml. of O_2.

   a. 45 gm., 0.34 d. 45 gm., 1.3
   b. 20 mgm., 0.34 e. 20 mgm., 1.3
   c. 15 gm., 0.34 f. 15 gm., 1.3

10. The hemoglobin concentration in grams per 100 ml. of cells is normally about _____ (E, equal to; H, one half) a maximum value of _____ .

    a. E, 1.3 c. E, 34 e. H, 15
    b. E, 15 d. H, 1.3 f. H, 34

11. Male adult average values compared to those for women are _____ for hemoglobin content of blood and _____ for hematocrit. (H, significantly higher; E, about equal; L, significantly lower)

    a. H, H c. H, L e. L, E
    b. H, E d. L, H f. L, L

12. Low values of the hemoglobin concentration of packed red blood cells is *least likely* to result from the combination of a _____ red blood cell count and a _____ blood hemoglobin. (N, normal; L, low; H, high)

    a. N, L c. L, N
    b. H, N d. L, L

OBJECTIVE 5–4.

Identify prenatal and postnatal sites of red blood cell production, and the intermediate steps involved in their genesis.

13. The liver is the major site for embryonic red blood cell formation during the _____ trimester, and the _____ (S, spleen; M, bone marrow; L, liver) is the major formative locus at birth.

    a. 1st, S c. 1st, L e. 2nd, M
    b. 1st, M d. 2nd, S f. 2nd, L

14. The sequential order of declining erythropoietic function with increasing age following birth is that of _____ , followed by _____ , and the persistent but declining function of _____ . (M, membranous bones; P, proximal parts of long bones; D, distal parts of long bones)

    a. M, P, D c. P, D, M e. D, P, M
    b. M, D, P d. P, M, D f. D, M, P

15. The _____ , initiating hemoglobin synthesis, is derived from a unipotential stem cell called the _____ . (N, normoblast; B, basophil erythroblast; H, hemocytoblast)

    a. N, B c. B, N e. H, B
    b. N, H d. B, H f. H, N

16. _____ (N, nucleate; A, anucleate) reticulocytes, normally comprising _____ (M, more; L, less) than 1% of circulating red blood cells, are useful indicators of accelerated red blood cell _____ (S, synthesis; B, breakdown).

    a. N, M, S c. A, M, S e. A, L, S
    b. N, L, B d. A, M, B f. A, L, B

17. A loss of the remnants of the _____ (G, Golgi apparatus; N, nucleus; R, endoplasmic reticulum) or locus where hemoglobin synthesis has occurred, distinguishes reticulocytes from _____ (B, normoblasts; E, erythrocytes).

    a. G, B c. R, B e. N, E
    b. N, B d, G, E f. R, E

OBJECTIVE 5–5.

Identify the primary regulator of red blood cell production. Identify erythropoietin, its major origin, site of action, and functional significance.

18. Red blood cell production more significantly occurs in the _____ (F, femur shaft; V, vertebra) of a 35-yr.-old individual in response to altered tissue_____concentrations.

   a. F, O_2
   b. F, CO_2
   c. F, H^+
   d. V, O_2
   e. V, CO_2
   f. V, H^+

19. Erythropoietin, appearing in blood in response to _____ tissue oxygen levels, results in _____ circulating numbers of erythrocytes. (I, increased; D, decreased)

   a. I, I
   b. I, D
   c. D, I
   d. D, D

20. Erythropoietin is more effective in influencing the _____(R, release; G, genesis) of mature erythrocytes via its target action upon _____(K, renal erythropoietic factor; E, erythropoietic tissues).

   a. R, K
   b. R, E
   c. G, K
   d. G, E

21. The numbers of circulating red blood cells is _____by endurance training, _____ by high altitude exposure, and_____ by a loss of renal function. (I, increased; N, not significantly altered; D, decreased)

   a. I, I, N
   b. I, I, D
   c. I, D, D
   d. I, D, N
   e. D, D, N
   f. D, D, D

OBJECTIVE 5–6.

Recognize the significance of vitamin B_{12}, intrinsic factor, and folic acid in the formation of red blood cells.

22. Deficiencies of the _____ (I, intrinsic; E, extrinsic) factor, cyanocobalamin, result in _____ (S, microcytic; M, macrocytic) red blood cells containing _____(L, less; G, greater) than normal amounts of hemoglobin per cell.

   a. I, S, L
   b. I, S, G
   c. I, M, L
   d. E, M, G
   e. E, M, L
   f. E, S, G

23. Pernicious anemia, generally associated with _____(A, atrophy; H, hyperplasia) of the gastric mucosa, is more generally caused by _____ (D, dietary; G, gastrointestinal absorption) deficiencies of vitamin B_{12}.

   a. A, D
   b. A, G
   c. H, D
   d. H, G

24. Vitamin B_{12} is absorbed in the_____(S, stomach; I, ileum), stored in large quantities in the _____ (L, liver; M, gastric mucosa), and required for _____(DNA, desoxyribonucleic acid; H, hemoglobin) synthesis during erythrogenesis.

   a. S, L, DNA
   b. S, M, H
   c. S, M, DNA
   d. I, M, H
   e. I, L, DNA
   f. I, L, H

OBJECTIVE 5–7.

Identify the general chemical composition, formative processes, and oxygen-combining properties of hemoglobin.

25. Heme molecules, largely synthesized in _____ (M, mitochondria; R, ribosomes; G, Golgi apparati), are combined in a_____ to one ratio with globin molecules to form hemoglobin.

   a. M, 4
   b. R, 4
   c. G, 4
   d. M, 1
   e. R, 1
   f. G, 1

26. _____ (I, ionic; M, molecular) oxygen forms reversible "coordination" bonds with the _____(O, ferrous; C, ferric) iron of the hemoglobin molecule.

   a. I, O
   b. I, C
   c. M, O
   d. M, C

27. One hemoglobin molecule with a molecular weight of _____ is synthesized from _____ tetrapyrrole units and _____ globin molecules.

 a. 5,700, 1, 1   d. 68,000, 4, 4
 b. 5,700, 1, 4   e. 68,000, 4, 1
 c. 5,700, 4, 1   f. 68,000, 1, 4

28. Hemoglobin maximally transports _____ molecule(s) of oxygen per hemoglobin molecule or the equivalent of _____ ml. of oxygen per gram of hemoglobin.

 a. 1, 1.3   c. 4, 1.3   e. 2, 20
 b. 2, 1.3   d. 1, 20   f. 4, 20

OBJECTIVE 5–8.

Recognize the importance of: iron availability by cells; the available pool, dietary requirements, and daily loss of iron; and the mechanism of absorption, transport, and storage of iron.

29. Of a total quantity of iron in the body of about _____ grams, the highest percentage is found in the form of _____ (F, ferritin; H, hemoglobin; M, myoglobin).

 a. 4, F   c. 4, M   e. 40, H
 b. 4, H   d. 40, F   f. 40, M

30. In the absence of adequate dietary intake to compensate for an approximate daily excretion of _____ mg., iron is readily available from _____ (H, hemosiderin; F, ferritin) storage in the _____ (L, liver; M, bone marrow).

 a. 1, H, L   c. 1, F, L   e. 4, F, L
 b. 1, H, M   d. 4, F, M   f. 4, H, M

31. Iron, largely in the _____ (O, ferrous; C, ferric) form, is absorbed in the _____ (U, upper; L, lower) small intestine and undergoes transport in the cardiovascular system in combination with _____ (T, transferrin; H, hemosiderin).

 a. O, U, T   c. O, L, T   e. C, L, T
 b. O, U, H   d. C, L, H   f. C, U, H

32. The regulatory mechanism for total body iron during periods of dietary iron excess is dependent upon a transferrin _____ (H, hormonal regulatory; S, saturation) mechanism and _____ (D, desquamating; R, reduced levels of ferritin stores of) intestinal mucosal cells.

 a. H, D   c. S, D
 b. H, R   d. S, R

OBJECTIVE 5–9.

Identify the average life span of red blood cells, its influencing factors, and the process of red blood cell destruction.

33. An average life span of circulating red blood cells of _____ months is largely a function of their _____ (M, mitotic cycle; E, erythropoietin regulation; F, mechanical and osmotic fragility).

 a. 3, M   c. 3, F   e. 30, E
 b. 3, E   d. 30, M   f. 30, F

34. Evidence indicating the importance of the _____ (S, spleen; L, lymph nodes) in red blood cell destruction includes the observation that its removal _____ (I, increases; D, decreases) the average life span of circulating erythrocytes.

 a. S, I   c. L, I
 b. S, D   d. L, D

35. In the process of red blood cell destruction, initiated by the _____ (L, leukocytes; K, kidneys; RE, reticuloendothelial cells), hemoglobin iron is largely _____ (E, excreted as bilirubin; R, recycled).

 a. L, E   c. RE, E   e. K, R
 b. K, E   d. L, R   f. RE, R

OBJECTIVE 5–10.
Identify anemia and polycythemia and their types, causes, and consequences.

36. Anemia is generally associated with _____ red blood cell concentrations, _____ oxygen-carrying capacity of blood, and _____ blood viscosity. (H, higher than normal; L, lower than normal).

a. H, H, H   c. H, H, L   e. L, L, H
b. H, L, H   d. L, L, L   f. L, H, L

37. Major consequences of anemia generally include _____ cardiac output, _____ resistance to blood flow in peripheral vessels, and _____ work load for the heart. (I, increased; D, decreased)

a. I, I, I   c. I, I, D   e. D, D, I
b. I, D, I   d. D, D, D   f. D, I, D

38. A _____ (P, primary; S, secondary) type of polycythemia occurs as a consequence of _____ (A, bone marrow aplasia; E, excessive dietary vitamin B_{12}; H, high altitude exposure).

a. P, A   c. P, H   e. S, E
b. P, E   d. S, A   f. S, H

39. Thalassemia, sickle cell, and hereditary spherocytosis forms of anemia, associated with _____ (I, increased; D, decreased) life spans of circulating red blood cells, are types of _____ (H, hemorrhagic; L, hemolytic; M, macrocytic) anemia.

a. I, H   c. I, M   e. D, L
b. I, L   d. D, H   f. D, M

40. Cyanosis, resulting from increased tissue levels of _____ (CO_2, carbon dioxide; H, reduced hemoglobin) is a more frequent consequence of _____ (A, anemia; S, secondary polycythemia).

a. CO_2, A   c. H, A
b. CO_2, S   d. H, S

6

Resistance of the Body to Infection — The Reticuloendothelial System, Leukocytes, and Inflammation

OBJECTIVE 6–1.
Characterize leukocytes and platelets with regard to their general functional roles, origins, types, life span, and differential blood concentrations.

1. Destruction of invading agents by the process of _____ (G, phagocytosis; N, pinocytosis) is an important function of _____ (L, leukocytes; P, platelets).

a. G, L   c. N, L
b. G, P   d. N, P

2. Origins of blood cells include the _____ for polymorphonuclear leukocytes, the _____ for monocytes, and the _____ for platelets (M, bone marrow; L, lymphatic tissue).

a. M, M, M   c. M, M, L   e. L, M, M
b. M, L, M   d. M, L, L   f. L, L, M

3. _____ (P, platelets; M, monocytes; G, granulocytes) derived from _____ (T, meta-myelocytes; K, megakaryocytes) provide the more significant role in blood coagulation.

   a. P, T   c. G, T   e. M, K
   b. M, T   d. P, K   f. G, K

4. Typical adult leukocyte and platelet concentrations in blood are, respectively, _____ per cubic _____ .

   a. 1000 & 700,000, mm.
   b. 7000 & 300,000, mm.
   c. 700,000 & 10^6, mm.
   d. 1000 & 700,000, cm.
   e. 7000 & 300,000, cm.
   f. 700,000 & 10^6, cm

5. The less abundant _____ (G, granulocyte; A, agranulocyte) fraction contains the lowest concentration of _____.

   a. G, lymphocytes   d. A, lymphocytes
   b. G, basophils   e. A, eosinophils
   c. G, monocytes   f. A, monocytes

6. The most abundant leukocyte, or _____ (L, lymphocyte; N, neutrophil), normally comprises _____ % of the total leukocyte population, and is followed in frequency of occurrence by the _____ (L, lymphocyte; N, neutrophil; M, monocyte).

   a. L, 32, N   c. N, 32, M   e. N, 62, L
   b. L, 32, M   d. L, 62, N   f. N, 62, M

7. _____ (N, nucleate; A, anucleate) platelets have an average life span closest to ten _____ (D, days; W, weeks; M, months).

   a. N, D   c. N, M   e. A, W
   b. N, W   d. A, D   f. A, M

OBJECTIVE 6–2.

Identify the processes of diapedesis, ameboid motion, chemotaxis, and phagocytosis. Characterize the roles of each of these processes in differential functions of leukocytes.

8. _____ (N, pinocytosis; C, chemotaxis; D, diapedesis), or the process whereby cells move through the endothelial cell layer of blood vessels, is more significant for _____ (P, platelet; L, leukocyte) function.

   a. N, P   c. D, P   e. C, L
   b. C, P   d. N, L   f. D, L

9. The "navigational mechanism" utilized by _____ (L, ciliary; A, ameboid) locomotion of phagocytic leukocytes is that of _____ (R, random probability; C, chemotaxis; M, leukocyte memory).

   a. L, C   c. L, M   e. A, R
   b. L, R   d. A, C   f. A, M

10. The more actively motile _____ (M, monocytes; N, neutrophils) migrate through tissue spaces at maximum speeds of about three times their cellular lengths per _____ (S, second; M, minute; H, hour).

    a. M, S   c. M, H   e. N, M
    b. M, M   d. N, S   f. N, H

11. Net electronegative surface charges of _____ (D, dead; V, viable) body tissues exert a _____ (P, positive; N, negative) chemotaxic influence and serve to _____ (F, facilitate; I, inhibit) their phagocytosis.

    a. D, P, F   c. D, P, I   e. V, P, F
    b. D, N, I   d. V, N, I   f. V, N, F

12. Plasma globulins, called _____ (MIF, macrophage inhibitory factors; O, opsonins), _____ (E, enhance; P, protect against) phagocytosis through _____ (S, specific; N, nonspecific) combinations with particulate matter.

    a. MIF, E, S   c. MIF, P, S   e. O, E, S
    b. MIF, P, N   d. O, P, N   f. O, E, N

13. The more powerful phagocyte, the _____ (M, monocyte; B, basophil; N, neutrophil), plays the more important functional role in _____ (A, acute; C, chronic) inflammatory states.

    a. M, A   c. N, A   e. B, C
    b. B, A   d. M, C   f. N, C

14. Tissue macrophages, or equivalents of mature _____(N, neutrophils; B, basophils; M, monocytes), have life spans which generally _____(A, are; T, are not) limited by the magnitude of their phagocytic activity.

    a. N, A   c. M, A   e. B, T
    b. B, A   d. N, T   f. M, T

15. Digestive vesicles, containing larger amounts of _____(O, oxidative; H, hydrolytic) enzymes, are formed by the combination of phagocytic vesicles with cellular_____(M, mitochondria; L, lysosomes).

    a. O, M   c. H, M
    b. O, L   d. H, L

OBJECTIVE 6–3.

Identify the general significance of cells of the reticuloendothelial system, the differential roles of macrophages located in various body tissues, and their probable derivation.

16. The reticuloendothelial system is comprised largely of _____(M, motile; S, sessile), _____ (P, phagocytic; N, nonphagocytic) cells which have_____ (R, retained; L, lost) the capability for mitotic replication.

    a. M, P, R   c. M, N, R   e. S, P, R
    b. M, N, L   d. S, P, L   f. S, N, L

17. Interstitial particulate matter, which is not destroyed locally in peripheral tissues, is generally phagocytized by macrophages of the _____ (S, spleen; L, liver; N, lymph nodes) _____(A, after; B, before) gaining entrance to blood.

    a. S, A   c. N, A   e. L, B
    b. L, A   d. S, B   f. N, B

18. Liver sinuses are lined by tissue macrophages called _____ (K, Kupffer cells; P, Purkinje cells; M, megakaryocytes) which function more significantly to remove large numbers of bacteria transported by hepatic _____ (A, arterial; R, portal) vessels.

    a. K, A   c. M, A   e. P, R
    b. P, A   d. K, R   f. M, R

19. Macrophages of the spleen are more abundantly located in the_____(W, white; R, red) pulp, and function in the phagocytosis of _____(L, lymph; B, blood)-borne constituents of its afferent vessels.

    a. W, L   c. R, L
    b. W, B   d. R, B

OBJECTIVE 6–4.

Describe the process of inflammation and identify the functional roles contributed by leukocytes and macrophages.

20. _____ (H, histamine; N, norepinephrine), released by injured tissues, serves to _____ (I, increase; D, decrease) capillary membrane permeability and_____(C, constrict; E, dilate) local vessels.

    a. H, I, C   c. H, I, E   e. N, I, E
    b. H, D, C   d. N, D, E   f. N, D, C

21. Extracellular edema, resulting from tissue injury, is accompanied by a local dominance of _____(A, anticoagulant; C, coagulant) factors and _____ (I, increased; N, no significant change in; D, decreased) rates of extracellular fluid movement through interstitial spaces.

    a. A, I   c. A, D   e. C, N
    b. A, N   d. C, I   f. C, D

22. The inflammatory process, whose intensity is best described as_____(A, an all or none; P, a proportional) response to tissue injury, results in a local_____(I, increase; D, decrease) in pH.

    a. A, I   c. P, I
    b. A, D   d. P, D

23. _____(A, staphylococci; E, streptococci), which are locally more destructive to tissues, exhibit a correspondingly_____(G, greater; L, lesser) containment during inflammatory processes.

    a. A, G   c. E, G
    b. A, L   d. E, L

24. Inflammatory states result in a _____(L, leukopenia; N, neutrophilia) and the more appropriate temporal sequence of leukocyte _____(D, diapedesis; M, margination; C, chemotaxis).

    a. L; D, M, & C
    b. L; M, D, & C
    c. L; C, D, & M
    d. N; D, M, & C
    e. N; M, D, & C
    f. N; C, D, & M

25. Greater reserves of neutrophils, afforded by _____ (B, circulating blood; M, bone marrow stores), function more effectively during _____ (A, acute; C, chronic) stages of inflammatory states.

    a. B, A
    b. B, C
    c. M, A
    d. M, C

26. The onset of pus, generally comprised of a higher percentage of _____ (V, viable; D, dead) neutrophils, is a sign of _____ (T, the termination of; O, ongoing) inflammatory states.

    a. V, T
    b. V, O
    c. D, T
    d. D, O

OBJECTIVE 6–5.

Identify the more probable general functional roles of basophils, eosinophils, and lymphocytes.

27. Eosinophils are thought to have greater functional roles in the _____ of antigen-antibody complexes and the_____ of blood clots. (F, formation; B, breakdown)

    a. F, F
    b. F, B
    c. B, F
    d. B, B

28. Basophils appear to have functions which are more closely allied with those of _____ (PT, prothrombin; H, heparin; N, norepinephrine) secreting _____(M, mast; K, Kupffer) cells.

    a. PT, M
    b. H, M
    c. N, M
    d. PT, K
    e. H, K
    f. N, K

29. Lymphocytes, comprising a _____(J, majority; N, minority) of blood_____(G, granulocytes; A, agranulocytes), perform more significant functional roles in the process of _____ (I, immunity; C, blood coagulation).

    a. J, G, I
    b. J, A, C
    c. J, A, I
    d. N, A, C
    e. N, G, I
    f. N, G, C

OBJECTIVE 6–6.

Identify the conditions apart from inflammation which are associated with abnormal blood concentrations of leukocytes.

30. Eosinophil concentrations in circulating blood are _____ during allergic reactions and are _____ during most parasitic infections. (I, increased; D, decreased)

    a. I, I
    b. I, D
    c. D, I
    d. D, D

31. Severe exercise results in _____ (I, increased; D, decreased) blood flow velocities and a physiological_____(N, neutrophilia; L, leukopenia) resulting from altered states of leukocyte _____ (M, margination; P, diapedesis).

    a. I, N, M
    b. I, N, P
    c. I, L, M
    d. I, L, P
    e. D, N, M
    f. D, L, P

32. Agranulocytosis, resulting from irradiation injury to the _____ (L, lymphatic tissues; M, bone marrow), reflects a more substantial _____ (I, increase; D, decrease) in the rate of genesis of_____ (G, granulocytes; A, agranulocytes).

    a. L, D, A
    b. L, I, A
    c. L, D, G
    d. M, D, A
    e. M, I, G
    f. M, D, G

33. Leukemia is characterized by_____(H, higher than normal; N, normal; L, lower than normal) numbers of_____ (N, normally; A, abnormally) formed white blood cells.

    a. H, A
    b. N, A
    c. L, A
    d. H, N
    e. N, N
    f. L, N

7

Immunity and Allergy

OBJECTIVE 7-1.

Identify immunity, contrast the mechanisms of innate versus acquired forms of immunity, and distinguish humoral and cellular types of acquired immunity.

1. Phagocytic activity of the reticuloendothelial system, a form of _____ (A, acquired; I, innate; T, auto-) immunity, involves a capacity to resist injurious organisms by _____ (G, general; S, specific) processes.

   a. A, G   c. T, G   e. I, S
   b. I, G   d. A, S   f. T, S

2. The immune system is a more powerful antagonist against invading agents _____ (O, at; F, during the period following) the onset of initial exposure as a consequence of adaptive mechanisms of _____ (I, innate; A, acquired) immunity.

   a. O, I   c. F, I
   b. O, A   d. F, A

3. Humoral immunity, a type of _____ (I, innate; A, acquired) immunity, involves the combination of bloodborne _____ (P, mucolytic polysaccharides; G, globulins; T, antigens) with the invading agent.

   a. I, P   c. I, T   e. A, G
   b. I, G   d. A, P   f. A, T

4. Sensitized _____ (T, thrombocytes; L, lymphocytes) perform a more direct effector role in the destruction of infectious agents in a _____ (C, cellular; H, humoral) type of immunity.

   a. T, C   c. L, C
   b. T, H   d. L, H

OBJECTIVE 7-2.

Identify antigens, their significance in the immune response, and the nature of antigenic substances.

5. Antigens generally serve as agents which _____ (S, stimulate; I, inhibit) the formation of specifically reacting protein _____ (H, haptens; A, antibodies; G, agglutinogens).

   a. S, H   c. S, G   e. I, A
   b. S, A   d. I, H   f. I, G

6. As a general rule, for a substance to be antigenic, it must have a molecular weight _____ (L, less; G, greater) than _____ .

   a. L, 8,000   d. G, 8,000
   b. L, 80,000   e. G, 80,000
   c. L, 800,000   f. G, 800,000

7. Haptens are _____ (H, high; L, low) molecular weight substances which _____ (Q, acquire; S, lose) immunological specificity following combination with _____ (A, antigenic; N, nonantigenic) substances.

   a. H, S, A   c. H, Q, A   e. L, Q, A
   b. H, S, N   d. L, Q, N   f. L, S, N

OBJECTIVE 7-3.

Recognize the role of lymphoid tissue in acquired immunity. Identify "T" and "B" lymphocytes, their probable origins, preprocessing, and dispersal, and their differential functions in acquired immunity.

8. Lymphatic tissue, functioning more completely in mechanisms of _____ (I, innate; A, acquired) immunity, is strategically positioned to intercept antigenic materials within _____ (T, only interstitial fluid; B, only blood; TB, both interstitial fluid and blood).

    a. I, T   c. I, TB   e. A, B
    b. I, B   d. A, T   f. A, TB

9. "T" lymphocytes, so named because they are _____ (R, located; P, preprocessed) in the thymus gland, are responsible for _____ (C, cellular; H, humoral) mechanisms of immunity.

    a. R, C   c. P, C
    b. R, H   d. P, H

10. "B" lymphocytes are named for their _____ (B, bone marrow stem cell origin; P, preprocessing locus) as first discovered in _____ (D, birds; M, mammals).

    a. B, D   c. P, D
    b. B, M   d. P, M

11. Humoral antibodies are primarily produced by _____ (B, "B" lymphocytes; T, "T" lymphocytes) located in _____ (G, cortical and germinal; P, paracortical) areas of lymph nodes.

    a. B, G   c. T, G
    b. B, P   d. T, P

12. Removal of the thymus gland several months _____ (B, before; A, after) _____ (R, birth; P, puberty) results in a more complete suppression of _____ (C, cellular; H, humoral) mechanisms of immunity.

    a. B, R, C   c. A, R, H   e. B, P, H
    b. B, R, H   d. B, P, C   f. A, R, C

OBJECTIVE 7-4.

Identify the probable mechanisms for establishing specificity of sensitized lymphocytes and antibodies. Identify lymphocyte "clones;" their theoretical origin, and their manner of excitation.

13. A lymphocyte _____ (C, clone; BU, bursa) or all lymphocytes which form one specific type of antibody, is a functional unit of _____ (B, only "B"; T, only "T"; TB, both "B" and "T") lymphocytes.

    a. C, B   c. C, TB   e. BU, T
    b. C, T   d. BU, B   f. BU, TB

14. Phagocytosis of foreign material by macrophages generally _____ (C, circumvents; E, enhances) clonal stimulation of _____ (T, only "T"; B, only "B"; TB, both "B" and "T") lymphocytes.

    a. C, T   c. C, TB   e. E, B
    b. C, B   d. E, T   f. E, TB

15. Production of increased quantities of antibodies by means of excitation of _____ lymphocytes is thought to _____ (E, enhance; D, diminish) mechanisms of cellular immunity.

    a. "T," E   c. "B," E
    b. "T," D   d. "B," D

16. Repetitive excitation of a clone of lymphocytes by _____ (S, a single antigen type; M, multiple antigen groups) results in _____ (R, reduced; E, enhanced) numbers of available clonal lymphocytes.

    a. S, R   c. M, R
    b. S, E   d. M, E

OBJECTIVE 7–5.

Recognize the importance of tolerance of the system of acquired immunity to one's own tissues. Identify the probable mechanisms of "self-tolerance" and the consequences of their failure.

17. Mechanisms that are critical to the development of immunological tolerance _____ (I, occur during a specific time interval; R, reoccur continuously throughout life) and involve _____ (B, only "B"; T, only "T"; TB, both "T" and "B") lymphocytes.

    a. I, B   c. I, TB   e. R, T
    b. I, T   d. R, B   f. R, TB

18. Exposure to high concentrations of a particular group of antigens during the period of _____ (P, lymphocyte "processing"; T, involution of the thymus gland) is more likely to result in subsequent immunological tolerance of _____ (G, that particular group of; O, other) antigens.

    a. P, G   c. T, G
    b. P, O   d. T, O

19. Chronic inflammation of certain tissues with the release of sequestered _____ (A, antigens; L, lymphocytes) into the general circulation is a more probable mechanism of _____ (I, immunological paralysis; O, autoimmune diseases; P, passive immunity).

    a. A, I   c. A, P   e. L, O
    b. A, O   d. L, I   f. L, P

20. _____ (D, decreasing; I, increasing) probabilities for the occurrence of autoimmune diseases with increasing age _____ (A, are; N, are not) considered to generally result from errors occurring during lymphocyte "preprocessing."

    a. D, A   c. I, A
    b. D, N   d. I, N

OBJECTIVE 7–6.

Identify antibodies, their formative sequence of events, and the major factors which influence the magnitude and duration of their production.

21. Antibodies are specifically reacting _____ (P, immune polysaccharides; G, immunoglobulins) which are largely transported in blood within the _____ (PC, plasma cell; PP, plasma protein; L, lymphocyte) fraction.

    a. P, PC   c. P, L   e. G, PP
    b. P, PP   d. G, PC   f. G, L

22. Upon exposure to a foreign antigen, _____ differentiate to become major antibody-producing _____ . (T, "T" lymphocytes; B, "B" lymphocytes; P, plasma cells)

    a. T, B   c. B, T   e. P, B
    b. T, P   d. B, P   f. P, T

23. The _____ (P, primary; S, secondary) response of an activated clone involves an earlier onset, a _____ (G, greater; L, lesser) magnitude, and a _____ (S, shorter; L, longer) interval of antibody production.

    a. P, G, S   c. P, L, S   e. S, L, S
    b. P, G, L   d. S, L, L   f. S, G, L

24. Vaccination is generally more effective when appropriate _____ (B, antibodies; G, antigens) are administered in _____ (S, a single large dose; M, multiple smaller doses at regular intervals).

    a. B, S   c. G, S
    b. B, M   d. G, M

OBJECTIVE 7–7.

Identify the chemical nature of antibodies, the major classes of antibodies and their distinguishing chemical characteristics, and the probable mechanism of antibody specificity for a particular antigen.

25. Antibodies are generally comprised of light to heavy chains in a ratio of _____ , of which _____ (H, only the heavy; L, only the light; B, both the heavy and light) chains contribute to antigen specificity.

  a. 1:1, H   c. 1:1, B   e. 2:1, L
  b. 1:1, L   d. 2:1, H   f. 2:1, B

26. Complement is more likely to attach to the _____ (H, heavy; L, light) chains at the ends of the _____ (V, "variable"; C, "constant") portions of immunoglobulin molecules.

  a. H, V   c. L, V
  b. H, C   d. L, C

27. _____ is the most abundant blood immunoglobulin of the _____ major classes of antibodies.

  a. E, 2   c. E, 20   e. G, 5
  b. E, 5   d. G, 2   f. G, 20

OBJECTIVE 7–8.

Recognize the variety of ways in which antibodies function to protect the body against invading agents. Identify the complement and anaphylactic systems, their means of activation, and their consequences.

28. A more probable direct action of antibodies is that of _____ (E, enzymatic hydrolysis; A, agglutination) of antigen by _____ (U, univalent; B, bivalent) antibodies.

  a. E, U   c. A, U
  b. E, B   d. A, B

29. The _____ (A, anaphylactic; C, complement) system is comprised of nine separate _____ (P, plasma; L, lymphocyte; M, mast cell) components activated in a specific sequence, following its activation by certain antigen-antibody combinations.

  a. A, P   c. A, M   e. C, L
  b. A, L   d. C, P   f. C, M

30. The anaphylactic system is mediated by cytotropic antibodies, largely Ig _____ antibodies, which bind surfaces of _____ (L, lymphocytes; M, mast cells) in a more significant manner to produce the manifestations of anaphylaxis and allergy upon exposure to available antigen.

  a. A, L   c. E, L   e. G, M
  b. G, L   d. A, M   f. E, M

31. Lysis, opsonization, _____ (A, and; N, but not) chemotaxis are among the more important effects that are attributed to an activated _____ (C, complement; P, anaphylactic) system.

  a. A, C   c. N, C
  b. A, P   d. N, P

32. Activation of the _____ (P, complement; A, anaphylactic) system is more directly responsible for the release of histamine and "slow reacting substance" with subsequent local vascular _____ (C, constriction; D, dilation) and _____ (T, contraction; R, relaxation) of bronchiolar smooth muscle.

  a. P, C, T   c. P, D, T   e. A, D, T
  b. P, C, R   d. A, D, R   f. A, C, R

OBJECTIVE 7–9.

Contrast the characteristics of cellular immunity with those of humoral immunity. Identify the mechanisms of action of cellular immunity.

33. _____ (B, "B" lymphocytes; T, "T" lymphocytes; P, plasma cells) of regional lymphatic tissues mediate the effector limb of cellular immunity through the release of _____ (C, sensitized cells; S, cellular antibody secretion).

a. B, C   c. P, C   e. T, S
b. T, C   d. B, S   f. P, S

34. Combined activation of cellular and humoral immune mechanisms results in the formation of _____ (B, only "B"; T, only "T"; TB, both "T" and "B") memory cells and a _____ (G, greater; L, lesser) persistence of the humoral response.

a. B, G   c. TB, G   e. T, L
b. T, G   d. B, L   f. TB, L

35. A major function of the release of "transfer factor" by _____ (M, macrophages; L, lymphocytes) is to convert _____ (LM, lymphocytes to macrophages; NS, nonsensitized lymphocytes to sensitized lymphocytes).

a. M, LM   c. L, LM
b. M, NS   d. L, NS

36. Local release of "migration inhibition" and "chemotaxic" factors during an immune response results in the _____ (N, negative; P, positive) chemotaxis of nonsensitized _____ (L, lymphocytes; M, macrophages).

a. N, L   c. P, L
b. N, M   d. P, M

37. More significant roles are attributed to cellular immune mechanisms of _____ and humoral immune mechanisms of _____ . (R, organ transplant rejection; A, agglutination reactions; C, cancer prevention; T, bacterial toxin neutralization)

a. R & A, C & T   d. C & T, R & A
b. R & T, C & A   e. C & A, R & T
c. R & C, A & T   f. A & T, R & C

OBJECTIVE 7–10.

Contrast passive with active forms of acquired immunity. Identify interferon and its functional significance, and the process of vaccination and its rationale.

38. Vaccination is generally utilized to _____ (I, induce; S, suppress) mechanisms of _____ (N, innate; Q, acquired) immunity through the artificial administration of appropriate _____ (A, antigens; B, antibodies).

a. I, N, A   c. I, Q, A   e. S, N, A
b. I, Q, B   d. S, Q, B   f. S, N, B

39. Clinical procedures more generally include the administration of appropriate antigens to confer _____ immunity, antibodies to confer _____ immunity, and lymphocytes to confer _____ immunity. (PH, passive humoral; A, active; PC, passive cellular)

a. PH, A, PC   c. A, PH, PC   e. PC, A, PH
b. PH, PC, A   d. A, PC, PH   f. PC, PH, PC

40. Interferon, participating in a form of _____ (N, innate; Q, acquired) immunity, is thought to function as a _____ (S, specific; N, nonspecific), anti _____ (B, bacterial; V, viral) agent.

a. N, S, B   c. N, S, V   e. Q, S, V
b. N, N, B   d. Q, N, V   f. Q, N, B

41. Secretion of interferon by _____ (C, lymphocyte clone; RE, reticuloendothelial; I, infected) cells results in _____ (L, localized; W, widespread) target actions.

a. C, L   c. I, L   e. RE, W
b. RE, L   d. C, W   f. I, W

OBJECTIVE 7–11.

Identify the common and distinguishing characteristics of three different types of allergy. Identify the mechanisms involved in allergies, the basis for "allergic tendencies"; and characterize anaphylaxis, urticaria, hay fever, and asthma.

42. _____ (A, the Arthus reaction; P, a poison ivy reaction), an example of delayed-reaction allergies, results from _____ (C, cellular; H, humoral) immune mechanisms.

    a. A, C   c. P, C
    b. A, H   d. P, H

43. Severe inflammation and injury to small vessels locally, as a consequence of precipitation of antigen-antibody complexes involving complement activation, is a more likely consequence of a _____ (H, high; L, low) titer of a specific Ig _____ antibody.

    a. H, A   c. H, G   e. L, E
    b. H, E   d. L, A   f. L, G

44. Allergies in the genetically predisposed "allergic" person are more generally characterized by _____ (H, higher; L, lower) than normal Ig _____ concentrations and anaphylactoid types of immune reactions of tissues exposed to appropriate antigens called _____ (R, reagins; A, allergens).

    a. H, G, R   c. H, E, R   e. L, G, A
    b. H, G, A   d. H, E, A   f. L, E, R

45. The rapid development of circulatory shock is a more probable consequence of the peripheral vasodilation and _____ (P, plasma loss; E, erythrocyte hemolysis) accompanying _____ (U, urticaria; A, anaphylaxis).

    a. P, U   c. E, U
    b. P, A   d. E, A

46. Asthma, resulting as a consequence of _____ (C, mechanisms of cellular immunity; RA, reagin-allergin reactions), _____ (I, is; N, is not) effectively treated by the administration of antihistaminics.

    a. C, I   c. RA, I
    b. C, N   d. RA, N

8

Blood Groups; Transfusion; Tissue and Organ Transplantation

OBJECTIVE 8–1.

Recognize the antigenic nature of blood constituents and their significance in blood transfusions.

1. Blood transfusion reactions more significantly involve _____ (L, leukocyte; E, erythrocyte) agglutination and lysis and _____ (PP, plasma protein; L, leukocyte; E, erythrocyte) antigens.

    a. L, PP   c. L, E   e. E, L
    b. L, L   d. E, PP   f. E, E

2. Blood "types" are utilized to indicate major _____ (A, antigens; G, agglutinins) which are _____ (D, deficient; P, present) within stated blood samples.

    a. A, D   c. G, D
    b. A, P   d. G, P

3. Among the _____ (M, more; L, less) than 20 erythrocyte antigens known to be involved in transfusion reactions, the A-B-O system and the _____ system are more likely potential causes of major transfusion reactions.

a. L, Lewis
b. L, Kell

c. L, Rh-Hr
d. M, Lewis

e. M, Kell
f. M, Rh-Hr

OBJECTIVE 8–2.

Identify: A-B-O system of antigens, the major blood groups, blood types, and their relative prevalence; and their mode of inheritance. Identify agglutinogens and agglutinins.

4. Group-specific substances of the A-B-O system which act as _____ (N, agglutinins; G, agglutinogens) in blood transfusion reactions are found in blood groups _____, but not groups _____.

a. N; A & B; AB & O
b. N; A, B, & O; AB
c. N; A, B, & AB; O

d. G; A & B; AB & O
e. G; A, B, & O; AB
f. G; A, B, & AB; O

5. Genetic inheritance from among the _____ possible genotypes of the A-B-O system is best explained on the basis of participation of _____ separate gene(s) on each allelic member of _____ pair(s) of chromosomes.

a. 3, 3, 3
b. 3, 1, 3

c. 3, 3, 1
d. 6, 1, 3

e. 6, 3, 3
f. 6, 1, 1

6. Among the different possible genotypes for the A-B-O system, _____ is/are responsible for type O blood, _____ is/are responsible for type B blood, and _____ is/are responsible for type A blood.

a. 1, 1, 3
b. 1, 2, 2

c. 1, 2, 3
d. 3, 2, 3

e. 3, 1, 2
f. 3, 2, 2

7. Genetic principles dictate that parents with blood types A and AB produce offspring with blood types _____ but not _____.

a. B & AB; A & O
b. A & AB; AB & O
c. A; AB, B, & O

d. A & AB; B & O
e. A, B, & AB; O
f. A & O; B & AB

8. The lowest prevalence of blood types among the general population occurs for type _____ and is largely attributed to the infrequent occurrence of _____ determinant genes.

a. AB, A
b. AB, B

c. AB, O
d. O, A

e. O, B
f. O, AB

OBJECTIVE 8–3.

Identify the agglutinins for the A-B-O system of antigens, their origin and the determinant factors for their presence or absence, their titer variance with age, and their significance in potential transfusion reactions.

Match the appropriate choice of agglutinins which is normally present in association with each blood type.

9. O _____
10. A _____
11. B _____
12. AB _____

a. Only anti-A agglutinins
b. Only anti-B agglutinins
c. Both anti-A and anti-B agglutinins
d. Neither anti-A nor anti-B agglutinins

13. Agglutinins of the A-B-O system are normally _____ in the presence of corresponding agglutinogen and are normally _____ in the absence of corresponding agglutinogen. (P, present; A, absent)

a. P, P
b. P, A

c. A, P
d. A, A

14. Increased quantities of immunoglobulin type _____ agglutinins are elicited by the injection of either group _____ antigens into an appropriate recipient.

    a. G & M, O or A
    b. G & M, O or B
    c. G & M, A or B
    d. A & E, O or A
    e. A & E, O or B
    f. A & E, A or B

15. A-B-O system agglutinin titers are typically highest for _____ -year-old individuals and are typically lowest for _____ -year-old-individuals.

    a. 0, 40 to 80
    b. 8 to 10, 40 to 80
    c. 40 to 80, 8 to 10
    d. 0, 8 to 10
    d. 8 to 10, 0
    f. 40 to 80, 0

16. A-B-O system agglutinins have a direct action during transfusion reactions to _____ (C, "clump"; L, lyse) appropriate _____ (K, leukocytes; E, erythrocytes).

    a. C, K
    b. C, E
    c. L, K
    d. L, E

17. Complement activation plays a greater role in erythrocyte _____ (G, agglutination; H, hemolysis) during conditions of _____ (D, diminished; E, elevated) agglutinin titers.

    a. G, D
    b. G, E
    c. H, D
    d. H, E

OBJECTIVE 8–4.
Identify the methods of blood typing and cross-matching, and their utility and rationale.

18. Selected typing sera, containing high concentrations of A-B-O system _____ (N, agglutinins; G, agglutinogens), are utilized for blood typing _____ (A, and; B, but not) cross-matching procedures.

    a. N, A
    b. N, B
    c. G, A
    d. G, B

Match the following blood types with the appropriate combinations for the presence (+) or absence (−) of erythrocyte agglutination with anti-A and anti-B typing sera.

| | Typing Sera | | Red Blood |
	Anti-A	Anti-B	Cell Types
19. _____	(−)	(+)	a. A
20. _____	(+)	(+)	b. B
21. _____	(+)	(−)	c. AB
22. _____	(−)	(−)	d. O

23. Agglutination of the _____ (D, donors; R, recipient's) red blood cells with the transfusion of a pint of blood is the less probable event as a consequence of the dilution of an _____ (N, agglutinin; G, agglutinogen).

    a. D, N
    b. D, G
    c. R, N
    d. R, G

24. A more significant or "major" cross-match involves a test for agglutination of the _____ (D, donor's; R, recipient's) red blood cells, and is more frequently negative for type _____ red blood cells.

    a. D, AB
    b. D, O
    c. D, B
    d. R, AB
    e. R, O
    f. R, B

OBJECTIVE 8–5.
Recognize the complexity of the Rh-Hr blood group system. Identify the Rh factor, its theoretical basis of origin, its clinical means of determination, its prevalence, and the determinant factors and nature of the Rh immune response.

25. The Rh factor is thought to be a _____ (S, single; M, multiple group of) agglutinogen(s) whose presence is determined clinically by means of anti- _____ antibodies.

    a. S, C
    b. S, Hr
    c. S, Rh₀
    d. M, C
    e. M, Hr
    f. M, Rh₀

26. The highest segment of the general population is Rh _____ (N, negative; P, positive) as a consequence of _____ (I, prior immunization; L, lack of prior immunization; F, hereditary factors).

    a. N, I
    b. N, L
    c. N, F
    d. P, I
    e. P, L
    f. P, F

27. Parental blood types of A, Rh positive and A, Rh negative may produce sibling blood types of only:

   a. A+, A−, O+, & O−   d. A−
   b. A+ & A−   e. A+ & O+
   c. A+   f. A− & O−

28. In the absence of prior blood transfusions, permitted agglutinins_____ occur "spontaneously" in the A-B-O system and_____ occur "spontaneously" in the Rh-Hr system. (D, do; N, do not)

   a. D, D   c. N, D
   b. D, N   d. N, N

29. Considering the A-B-O and Rh-Hr systems, the most widespread agglutinin would be anti-_____ , whereas the least prevalent agglutinin would be anti-_____.

   a. A, Rh   c. Rh, B   e. B, A
   b. B, Rh   d. Rh, A   f. A, B

30. Anti-Rh antibodies more significantly include Ig _____ types, of which the Ig type _____ contains more "complete" antibodies for agglutination reactions.

   a. G & E, G   c. A & E, E   e. G & M, M
   b. G & E, E   d. A & E, A   f. G & M, G

31. Serological agglutination tests for the detection of Rh agglutinogens are more favorably accomplished utilizing _____ (H, high; L, low) agglutinin titers and a suspension of erythrocytes in _____ (S, isotonic saline; P, plasma protein) solutions.

   a. H, S   c. L, S
   b. H, P   d. L, P

OBJECTIVE 8–6.

Identify erythroblastosis fetalis, its primary cause, and the conditions for its existence.

32. Erythroblastosis fetalis is more frequently the result of circumstances involving the _____ system of antigens and _____ (P, passive; A, active) immunization of the fetus.

   a. ABO, P   c. Rh, P
   b. ABO, A   d. Rh, A

33. Erythroblastosis fetalis generally involves conditions of an Rh _____ fetus, an Rh _____ mother, and an Rh _____ father. (P, positive; N, negative)

   a. P, N, N   c. P, P, N   e. N, P, N
   b. P, N, P   d. N, P, P   f. N, N, P

34. The probability of occurrence for erythroblastosis fetalis in predisposed parents is less for _____ (E, heterozygous; O, homozygous) paternal conditions of Rh, and is _____ (G, greater; N, not significantly different; L, less) for subsequent as opposed to initial pregnancies.

   a. E, G   c. E, L   e. O, N
   b. E, N   d. O, G   f. O, L

35. The administration of anti-Rh agglutinins to the _____ (M, maternal; F, fetal) circulation is utilized as a _____ (P, preventative; C, curative) measure for erythroblastosis fetalis.

   a. M, P   c. F, P
   b. M, C   d. F, C

36. The clinical picture of erythroblastosis fetalis includes _____ rates of erythrocyte genesis, _____ blood hemoglobin content, and _____ bilirubin blood concentrations of the infant. (I, increased; D, decreased)

   a. I, I, I   c. I, D, I   e. I, D, D
   b. D, D, I   d. D, I, D   f. D, I, I

OBJECTIVE 8–7.
Recognize the clinical importance of blood transfusions. Identify the nature and consequences of transfusion reactions. Identify the terms "universal donor" and "universal recipient" and recognize the limitations of the term "universal."

37. A more common need for blood transfusion arises from conditions of _____ (S, hypovolemic shock; H, hemophilia) and a more common lethal consequence of transfusion reactions results from _____ (P, pyrogenic reactions; R, renal failure).

    a. S, P   c. H, P
    b. S, R   d. H, R

38. The amount of liberated hemoglobin in excess of a plasma content of about _____ mg. per 100 ml. is conveyed as _____ (F, free hemoglobin; H, hemoglobin combined with haptoglobin).

    a. 1, F   c. 100, F   e. 10, H
    b. 10, F   d. 1, H   f. 100, H

39. Extracellular fluid _____ (C, acidification; K, alkalinization) and _____ (R, reduced; E, enhanced) levels of urine output serve as therapeutic measures to reduce renal tubular hemoglobin precipitation during transfusion reactions.

    a. C, R   c. K, R
    b. C, E   d. K, E

40. Pyrogenic reactions occur _____ (I, infrequently; F, frequently) during transfusion reactions and _____ (A, are; N, are not) reliable indicators of ongoing erythrocyte agglutination.

    a. I, A   c. F, A
    b. I, N   d. F, N

41. Citrate, employed as _____ (N, a nutrient; A, an anticoagulant) for stored blood, leads to altered _____ ion concentrations and muscle _____ (P, paralysis; T, tetany) when blood is transfused too rapidly.

    a. N, K^+, P   d. A, Ca^{++}, T
    b. N, K^+, T   e. A, Ca^{++}, P
    c. N, Ca^{++}, P   f. A, K^+, T

42. Universal donor blood or type _____ blood is so named because of its lack of appropriate _____ (N, agglutinins; G, agglutinogens).

    a. Rh-, N   c. O, N   e. AB, G
    b. AB, N   d. Rh-, G   f. O, G

43. Universal recipients or type _____ individuals are _____ (L, less; M, more) frequently encountered than universal donors among the general population.

    a. Rh+, L   c. O, L   e. AB, M
    b. AB, L   d. Rh+, M   f. O, M

OBJECTIVE 8–8.
Identify the types of tissue or organ transplants, problems involved in their success, and the means of reducing rejection of transplanted grafts.

44. Transplants of tissues or organs from one species to another species are termed _____ transplants, whereas transplants of tissues or organs from one region to a different region of the same individual are termed _____ transplants. (E, heterologous; O, homologous; A, autologous).

    a. E, O   c. O, A   e. A, O
    b. E, A   d. O, E   f. A, E

45. Transplantation rejection of tissues or organs, largely mediated by mechanisms of _____ (H, humoral; C, cellular) immunity, is generally most severe for _____ and least severe for _____ transplants. (E, heterologous; O, homologous; A, autologous)

    a. H, A, O   c. H, E, O   e. C, E, A
    b. H, E., A   d. C, A, O   f. C, O, A

46. "Delaying tactics" for more severe rejection mechanisms of homologous transplants for _____ (C, cellular; A, acellular) tissues frequently include ACTH _____ (S, suppression; M, administration).

a. C, S   c. A, S
b. C, M   d. A, M

47. One of the more effective procedures for delaying the rejection of grafted tissues involves the inoculation of the tissue _____ (D, donor; R, recipient) with _____ (T, thymic hormone; C, complement; ALS, antilymphocyte serum).

a. D, T   c. D, ALS   e. R, C
b. D, C   d. R, T   f. R, ALS

48. Removal of the fetal _____ lymphocyte processing area prior to birth would more probably _____ (D, diminish; E, enhance) the chances of success for homologous transplants in the adult.

a. "T," D   c. "B," D
b. "T," E   d. "B," E

49. "Tissue typing" of _____ (E, erythrocyte agglutinogens; H, HL-A antigens) located within cellular _____ (C, cytoplasm; M, membranes) may be generally utilized to _____ (P, prevent; Z, minimize) the rejection of viable organ transplants.

a. E, C, P   c. E, M, P   e. H, M, P
b. E, C, Z   d. H, M, Z   f. H, C, M

9

Hemostasis and Blood Coagulation

OBJECTIVE 9–1.

Identify hemostasis. Characterize the significant features and functional roles of vascular constriction, platelets, clot formation and retraction, fibrous organization of the blood clot, and clot lysis in hemostatic mechanisms.

1. _____ (E, hemostasis; O, homeostasis), or the prevention of blood loss, is served by primary "plugging" mechanisms of _____ (F, only fibrin clots; P, only platelet adhesion; B, both fibrin clots and platelet adhesion).

a. E, F   c. E, B   e. O, P
b. E, P   d. O, F   f. O, B

2. Injury to blood vessels results in _____ (L, localized; W, widespread) vascular spasms as a probable primary consequence of the _____ (P, decreased vascular pressure; D, direct effects of trauma; R, indirect effects of neural reflexes).

a. L, P   c. L, R   e. W, D
b. L, D   d. W, P   f. W, R

3. Vascular spasms, occurring to a greater extent following the separation of blood vessels and surrounding tissues with _____ (S, sharp; D, rather dull) cutting instruments, involve mechanisms which _____ (C, circumvent; F, facilitate) intravascular blood coagulation.

a. S, C   c. D, C
b. S, F   d. D, F

4. The formation of platelet plugs, enhanced by the presence of _____ (NW, nonwettable surfaces; ADP, adenosine diphosphate), is _____ (R, reversed; P, rendered more permanent) by the action of thrombin.

a. NW, R   c. NW, P
b. ADP, R   d. ADP, P

5. Platelets are more directly involved in "plugging" _____ (L, the vascular lumen; E, endothelial cell defects) and _____ (A, activation; S, lysis) of blood clots.

a. L, A   c. E, A
b. L, S   d. E, S

6. Clot retraction is _____ (P, a passive; A, an active) process, which occurs in the _____ (R, presence; B, absence) of platelets following an interval of about _____ hour(s).

a. P, R, 12   c. P, B, 12   e. A, R, ½
b. P, B, ½   d. A, B, 12   f. A, R, 12

7. _____ (L, lymphocyte; F, fibroblast) invasion of a clot is _____ (C, a companion feature of; A, an alternate mechanism to) clot lysis.

a. L, C   c. F, C
b. L, A   d. F, A

OBJECTIVE 9–2.

Recognize the existence of procoagulant as well as anti-coagulant mechanisms in blood and tissues. Characterize the sequence of events leading to the formation of a fibrin clot following the formation of "prothrombin activator." Identify the participants in this sequence of events and their significant characteristics.

8. Formation of a "prothrombin activator" results in the conversion of plasma _____ (PTC, thromboplastin component; T, thrombin; P, prothrombin) to the active proteolytic enzyme _____ (S, plasmin; F, fibrin; T, thrombin).

a. PTC, S   c. P, T   e. T, F
b. T, S   d. PTC, T   f. P, F

9. Prothrombin is an alpha globulin, present in plasma in a concentration of about _____ mg./100 ml., with a molecular weight about _____ (H, one half of; E, equal to; D, double that of) hemoglobin.

a. 15, H   c. 15, D   e. 150, E
b. 15, E   d. 150, H   f. 150, D

10. Prothrombin synthesis by the _____ (M, mast cells; L, liver; RE, reticuloendothelial system) more notably requires vitamin _____ .

a. M, K   c. RE, K   e. L, A
b. L, K   d. M, A   f. RE, A

11. The formation of covalent bonding between fibrin monomer molecules, as a consequence of the presence of factor _____ follows the enzymatic action of _____ (P, plasmin; T, thrombin) upon the plasma protein _____ (F, fibrinogen; G, plasminogen; PT, prothrombin).

a. XII, T, PT   c. XIII, P, G   e. IV, P, G
b. XII, T, F   d. XIII, T, F   f. IV, T, PT

12. Calcium ions _____ required for the conversion of prothrombin to thrombin and _____ required for the conversion of fibrin monomer to a stable fibrin clot. (A, are; N, are not)

a. A, A   c. N, A
b. A, N   d. N, N

13. Fibrinogen, or factor _____ , has a normal concentration in plasma closest to _____ mg./100 ml.

a. I, 3   c. I, 300   e. III, 30
b. I, 30   d. III, 3   f. III, 300

14. Fibrinogen, with a molecular weight about _____ that of prothrombin, has a normal plasma to interstitial fluid concentration ratio _____ (L, substantially lower than; E, about equal to; H, substantially higher than) one.

a. ⅕, E   c. Equal to, H   e. 5 times, L
b. ⅕, L   d. Equal to, E   f. 5 times, H

OBJECTIVE 9–3.

Characterize blood clots and clot retraction and identify their functional significance. Identify the major factors which accelerate or retard clot formation and growth.

15. Fibrin clots are comprised of _____-dimensional networks of fibrin threads, which normally undergo clot retraction with a more significant removal of the _____ (C, cellular; S, serum) component.

 a. Two, C   c. Three, C
 b. Two, S   d. Three, S

16. A major factor in the formation of a "vicious cycle" or positive feedback mechanisms of clot formation involves the direct proteolytic action of _____ (F, fibrin; P, platelet factor 3; T, thrombin) upon _____ (PT, prothrombin; PA, prothrombin activator).

 a. F, PT   c. T, PT   e. P, PA
 b. P, PT   d. F, PA   f. T, PA

17. Extension of developing blood clots is generally _____ (F, facilitated; I, inhibited) by their contact with rapidly moving blood as a consequence of the _____ (A, fresh arrival; D, wide dissemination) of procoagulants.

 a. F, A   c. I, A
 b. F, D   d. I, D

OBJECTIVE 9–4.

Recognize the dual system for the formation of a "prothrombin activator" via both extrinsic and intrinsic pathways. Identify the extrinsic mechanism for initiating blood coagulation.

18. Activation of the _____ (I, intrinsic; E, extrinsic) pathway for the formation of prothrombin activator is thought to occur more directly in response to _____ (A, fibrinogen adsorption; T, tissue trauma) with the subsequent release of "tissue factor" and tissue _____ (PL, phospho-lipids; Ca, calcium).

 a. I, A, PL   c. I, T, PL   e. E, T, PL
 b. I, A, Ca   d. E, T, Ca   f. E, A, Ca

19. A complex formed by the proteolytic enzyme "tissue factor" and factor _____ is thought to form activated Stuart-Prower factor from factor _____ in the presence of a phospholipid component.

 a. XII, X   c. VII, X   e. V, XII
 b. V, X   d. VII, XII   f. VII, XII

20. Normal formation of prothrombin activator in the presence of suitable phospholipids is thought to require a more immediate participation of factor _____ and activated factor _____.

 a. V, XII   c. V, X   e. XII, VIII
 b. V, VII   d. XII, XIII   f. XII, VII

21. The formation of thrombin in the presence of prothrombin activator, typically requiring _____, usually _____ (I, is; N, is not) the major rate limiting factor in causing blood coagulation.

 a. 15 sec., I   c. 10 min., I   e. 2 min., N
 b. 2 min., I   d. 15 sec., N   f. 10 min., N

22. Tissue extracts by themselves are apparently not capable of forming a complete thromboplastin or _____ (PA, prothrombin activator; H, Hageman factor) but normally require the additional presence of the blood coagulation factors _____.

 a. PA; V, VII, & X   d. H; V, VII, & X
 b. PA; V, XII, & VIII   e. H; V, XII, & VIII
 c. PA; V, VI, & VII   f. H; V, VI, & VII

OBJECTIVE 9–5.

Identify the intrinsic mechanism for initiating blood coagulation.

23. Major blood coagulation factors involved in either intrinsic or extrinsic blood coagulation pathways include factors I through _____ with the exception of factor _____ .

    a. XIII, IV   c. XXIII, IV   e. XVII, VI
    b. XVII, IV   d. XIII, VI   f. XXIII, VI

24. Activation of the _____ (N, intrinsic; E, extrinsic) system of blood coagulation and a kallikrein-kinin-kininase system occurs via adsorption of factor _____ upon wettable surfaces.

    a. N, I   c. N, III   e. E, XII
    b. N, XII   d. E, I   f. E, III

25. Activation of Hageman factor by "wettable" surface contact is thought to be followed sequentially by activation of factor _____ and then factor _____ .

    a. VIII, IX   c. III, XIII   e. XI, IX
    b. IX, XI   d. IX, VIII   f. XIII, III

26. Factor _____ , or antihemophilic factor, in conjunction with a platelet factor and activated factor _____ is thought to be required for the activation of factor _____ .

    a. VIII, IX, X   d. XI, XII, XIII
    b. VIII, X, IX   e. XI, X, IX
    c. VIII, XII, XI   f. XI, VIII, XII

27. "Platelet factor," released by conditions similar to those required for the activation of factor _____ , is a _____ (PL, phospholipid; GP, glycoprotein; LP, lipoprotein).

    a. X, PL   c. X, LP   e. XII, GP
    b. X, GP   d. XII, PL   f. XII, LP

28. Prothrombin activator is formed in the intrinsic pathway by the combination of "platelet factor," activated factor _____ , and factor _____ .

    a. VII, V   c. V, III   e. X, VII
    b. VII, III   d. V, VIII   f. X, V

29. The extrinsic pathway utilizes _____ whereas the intrinsic pathway utilizes _____ for the formation of prothrombin activator. (V, only factor V; VII, only factor VII; B, both factors V and VII)

    a. V, VII   c. B, VII   e. V, B
    b. VII, B   d. B, V   f. B, B

OBJECTIVE 9–6.

Recognize the role of calcium ions in the intrinsic and extrinsic pathways for blood coagulation and its clinical significance.

30. Calcium ions, or factor _____ , are required for _____ (I, only the intrinsic pathway; E, only the extrinsic pathway; B, both the intrinsic and extrinsic pathways) of clot formation.

    a. IV, I   c. IV, B   e. VI, E
    b. IV, E   d. VI, I   f. VI, B

31. Reduction of calcium ion concentrations, through the addition of _____ (D, dicoumarin; C, citrate), is an effective and useful method of preventing blood coagulation _____ (V, only *in vivo;* T, only *in vitro;* VT, either *in vivo* or *in vitro*).

    a. D, V   c. D, VT   e. C, T
    b. D, T   d. C, V   f. C, VT

32. A "recalcification time," required for clot formation to occur following recalcification of an oxalated plasma sample, is about _____ the corresponding "clot time," and is best utilized to evaluate the _____ (I, intrinsic; E, extrinsic) pathway of coagulation.

    a. $\frac{1}{3}$, I   c. Equal to, I   e. $\frac{2}{3}$, E
    b. $\frac{2}{3}$, I   d. $\frac{1}{3}$, E   f. Equal to, E

OBJECTIVE 9-7.

Identify the intravascular anticoagulant factors and their functional significance.

33. Protein, adsorbed to the endothelial surface of the vascular lumen, is normally _____ (P, positively; N, negatively) charged and functions to _____ (E, enhance; I, inhibit) platelet adhesion and _____ (A, attract; R, repel) clotting factors.

    a. P, E, A   c. P, I, A   e. N, I, A
    b. P, E, R   d. N, I, R   f. N, E, R

34. Subendothelial collagen fibers are a powerful _____ (N, initiator; I, inhibitor) of blood coagulation as a consequence of their adsorption of _____ (T, thrombin, XII, factor XII).

    a. N, T   c. I, T
    b. N, XII   d. I, XII

35. _____ (F, fibrin; AT, antithrombin III) is a more effective agent in limiting excessive spread of clots as a consequence of its action to adsorb _____ (Ca, calcium ions; T, thrombin; H, Hageman factor).

    a. F, Ca   c. F, H   e. AT, T
    b. F, T   d. AT, Ca   f. AT, H

36. _____ (A, antithrombin III; H, heparin) is a conjugated polysaccharide _____ (P, procoagulant; N, anticoagulant) factor which is thought to be produced in larger quantities by _____ (S, mast cells; K, megakaryocytes).

    a. A, P, S   c. A, N, S   e. H, N, S
    b. A, P, K   d. H, N, K   f. H, P, K

37. Heparin _____ the action of thrombin on fibrinogen, _____ the formation of prothrombin activator via the intrinsic pathway, and _____ the inactivation of thrombin. (F, facilitates, N, does not significantly influence; I, inhibits).

    a. F, F, F   c. F, I, F   e. I, I, F
    b. F, F, I   d. I, F, I   f. I, I, I

OBJECTIVE 9-8.

Identify plasminogen (profibrinolysin), factors influencing its activation, and its functional significance.

38. Plasmin, arising from an _____ (N, endogenous; X, exogenous) protein, results in an overall pattern of enzymatic _____ (A, activation; D, destruction) of clotting factors and clot _____ (R, retraction; F, formation; L, lysis).

    a. N, A, R   c. N, D, L   e. X, A, F
    b. N, A, F   d. N, D, R   f. X, D, L

39. Activation of the plasminogen system, a more notable feature of _____ (E, streptococcal; A, staphylococcal) infections, facilitates the _____ (C, containment; S, spread) of the infectious process.

    a. E, C   c. A, C
    b. E, S   d. A, S

40. Lysis of _____ (I, intravascular; T, tissue) clots is a more probable event as a consequence of compartmental differences of _____ (P, plasminogen; A, plasminogen activator; PI, plasmin inactivator) availability.

    a. I, P   c. I, PI   e. T, A
    b. I, A   d. T, P   f. T, PI

OBJECTIVE 9–9.

Identify the potential consequences and underlying mechanisms for the effects of vitamin K deficiencies, "hemophilia," and thrombocytopenia upon hemostasis.

41. Synthesis of blood coagulation factors VII, IX, _____ in the _____ (S, spleen; L, liver) requires vitamin K.

   a. II, & X; S   c. X, & XI; S   e. II, & XI; L
   b. II, & XI; S   d. II, & X; L   f. X, & XI; L

42. Nutritional requirements of the _____ (W, water; F, fat)-soluble vitamin K are normally met by _____ (L, liver; B, intestinal bacterial) synthesis.

   a. W, L   c. F, L
   b. W, B   d. F, B

43. Existing mechanisms are such that severe liver disease results in _____ gastrointestinal absorption of vitamin K, _____ prothrombin activator formation via the intrinsic pathway, and _____ prothrombin activator formation via the entrinsic pathway. (I, impaired; N, no significant change in)

   a. I, I, I   c. N, N, I   e. I, I, N
   b. N, I, I   d. N, I, N   f. I, N, I

44. "Hemophilia" is _____ (A, an acquired; H, a hereditary) deficiency of either factor _____, resulting in "classical hemophilia" or, less frequently, factors _____.

   a. A, X, IX & XI   d. H, VIII, IX & XI
   b. A, X, IX & VII   e. H, VIII, IX & X
   c. A, VIII, IX & X   f. H, X, VIII & IX

45. A failure of the blood clot retraction mechanism and _____(J, major vascular; M, multiple minute) hemorrhages are symptoms of _____ (T, thrombocytopenia; S, splenectomy; E, streptococcal infections).

   a. J, T   c. J, E   e. M, S
   b. J, S   d. M, T   f. M, E

46. Aplasia of formative _____ (L, lymphatic tissue; M, bone marrow) cells causes hemorrhagic symptoms when the blood platelet count first declines to approximately _____ per cubic mm.

   a. L, 200,000   d. M, 200,000
   b. L, 50,000   e. M, 50,000
   c. L, 2,000   f. M, 2,000

OBJECTIVE 9–10.

Identify the causes and consequences of thromboembolic conditions in humans.

47. _____ (T, thrombi; E, emboli) originating in major systemic veins eventually plug small vessels of the _____ (S, systemic; P, pulmonary) circulation.

   a. T, S   c. E, S
   b. T, P   d. E, P

48. Intravascular blood clots tend to grow more rapidly in the _____(S, same; O, opposite) direction of _____ (L, slowly; R; rapidly) moving blood as a greater consequence of the effect of blood flow upon local _____ (A, anticoagulant; C, coagulant) factors.

   a. S, L, A   c. S, R, A   e. O, R, A
   b. S, L, C   d. O, R, C   f. O, L, C

49. The wide dissemination of small intravascular clots generally results in _____ (I, increased; D, decreased) bleeding tendencies, and is frequently caused by the action of bacterial endotoxins to _____ (F, initiate the formation of; L, partially lyse) intravascular clots.

   a. I, F   c. D, F
   b. I, L   d. D, L

OBJECTIVE 9–11.

Identify the major anticoagulants available for clinical use, their utility, and their modes of action.

50. _____ (P, protamine; H, heparin) has a _____ (R, rapid; D, delayed) onset of clinical anticoagulant activity following the injection of minimally effective doses of from 0.5 to 1 _____ (C, micrograms; M, milligrams) per kilogram of body weight.

   a. P, R, C   c. P, D, C   e. H, R, C
   b. P, D, M   d. H, D, M   f. H, R, M

51. The duration of action of clinical doses of heparin is approximately three to four _____ (H, hours; D, days) and may be rapidly terminated by the administration of _____ (F, factor II; P, protamine; Ca, calcium ions).

   a. H, F   c. H, Ca   e. D, P
   b. H, P   d. D, F   f. D, Ca

52. Dicumarol or dicoumarin-like drugs _____ (D, decrease; I, increase) blood coaguability through _____ (F, a facilatory; H, an inhibitory) influence upon vitamin K utilization by the liver.

   a. D, F   c. I, F
   b. D, H   d. I, H

53. The action of dicumarol upon blood coagulation as compared to heparin has _____ (E, an earlier; L, a later) onset of action following administration and a _____ (S, shorter; G, longer) interval of effectiveness.

   a. E, S   c. L, S
   b. E, G   d. L, G

OBJECTIVE 9–12.

Identify suitable methods for delaying or preventing blood coagulation *in vitro,* and the rationale for their use.

54. *In vitro* coagulation of withdrawn blood is impaired by the use of _____ (A, acid-cleaned; S, siliconized) glass containers or by the subsequent addition of _____ (H, only heparin; D, only dicumarol; B, either heparin or dicumarol).

   a. A, H   c. A, B   e. S, D
   b. A, D   d. S, H   f. S, B

55. The storage of blood bank blood for transfusions generally involves the use of _____ (C, citrate; O, oxalate) as an anticoagulant by virtue of its lower tissue toxicity and its formation of _____ (S, a soluble; I, an insoluble) complex with _____ (F, fibrinogen; Ca, calcium).

   a. C, S, F   c. C, I, F   e. O, I, F
   b. C, S, Ca   d. O, I, Ca   f. O, S, Ca

OBJECTIVE 9–13.

Recognize the existence of multiple blood coagulation tests that are utilized for the determination and identification of abnormalities of hemostasis. Identify typical values, methods of determination, and the significance of "bleeding times," "clotting times," and "prothrombin times."

56. The "clotting time" is normally about _____ minutes and is generally _____ (G, greater; L, less) than the "bleeding time."

   a. 1 to 2, G   d. 1 to 2, L
   b. 5 to 8, G   e. 5 to 8, L
   c. 12 to 20, G   f. 12 to 20, L

57. "Prothrombin times," typically _____ seconds, vary in a _____ (D, direct; R, reciprocal) manner with the concentration variable _____ (PT, prothrombin; PA, prothrombin activator).

   a. 12, D, PT   c. 12, R, PT   e. 52, R, PT
   b. 12, R, PA   d. 52, D, PA   f. 52, R, PA

58. Dicumarol-treated patients exhibit _____ "clotting times" and _____ "prothrombin times." (I, increased; N, no significant change in; D, decreased)

a. I, N   c. N, N   e. I, I
b. N, I   d. D, N   f. D, D

10

Membrane Potentials, Action Potentials, Excitation, and Rhythmicity

OBJECTIVE 10–1.

Recognize the specialized nature of "excitable" tissues. Identify transmembrane potentials, their magnitude and polarity, their basic means of origin, and the basic role of ionic active transport mechanisms in their origin.

1. _____ mammalian cell types are characterized by transmembrane potentials and _____ mammalian cell types are characterized by the ability to transmit electrochemical impulses along cellular membranes. (A, all; J, a slight majority of; N, a minority of)

a. A, A   c. A, N   e. J, N
b. A, J   d. J, J   f. N, N

2. Fluid in contact with the _____ (I, inner; O, outer) surface of cell membranes contains a _____ (L, slight; B, substantial) excess of anions when compared to an anion concentration of about _____ mEq. liter on the opposite side of the membrane.

a. I, L, 155   c. I, B, 155   e. O, B, 155
b. I, L, 305   d. O, B, 305   f. O, L, 305

3. Transmembrane potentials result from a(n) _____ (B, balance; I, imbalance) of mobile charges between opposite sides of the membrane as a consequence of mechanisms of ionic _____ (A, active transport only; P, passive diffusion only; AP, active transport as well as passive diffusion).

a. B, A   c. B, AP   e. I, P
b. B, P   d. I, A   f. I, AP

4. "Membrane potentials" of cells refer to _____ (A, absolute values; D, differences) of electrical potentials, the _____ (P, potential capability for; C, actual accomplishment of) net ionic movement, and the _____ (W, total electrical work; W/Q, electrical work per unit of charge) involved in ion movements across cell membranes.

a. A, P, W   c. D, P, W   e. D, C, W
b. A, C, W/Q   d. D, P, W/Q   f. D, C, W/Q

5. ATP energy is utilized in the origin of membrane potentials through the activity of a _____ (C, cation; CA, cation-anion exchange) "pump" resulting in a more effective _____ (I, intracellular; E, extracellular) transport mechanism for _____ ions.

a. C, I, Na   c. C, E, Na   e. CA, E, Na
b. C, I, K   d. CA, E, K   f. CA, I, K

6. Operation of _____ (E, an electrogenic; NE, a non-electrogenic) sodium "pump" in most nerve and muscle results in a net cation active transport which _____ (I, is; N, is not) zero.

a. E, I   c. NE, I
b. E, N   d. NE, N

7. Capacitance of cell membranes, typically closer to 1 _____ (F, farad; MF, microfarad) per square centimeter, is significant in that it reflects a relatively _____ (S, slight; B, substantial) ionic charge transfer required for a unit charge in membrane _____ (P, permeability, V, potential).

   a. F, S, P   c. F, B, P   e. MF, S, P
   b. F, B, V   d. MF, B, V   f. MF, S, V

8. Excitable tissues possess membrane potentials closer to _____ volts in which the intracellular fluid is _____ (P, positive; N, negative) with respect to the extracellular fluid.

   a. 0.009, P   c. 0.9, P   e. 0.09, N
   b. 0.09, P   d. 0.009, N   f. 0.9, N

9. Nerve cell membranes possess _____ (P, poor; E, excellent) dielectric properties and have an imposed electric field, or membrane potential divided by membrane thickness, of _____ volts per cm.

   a. P, 10^3 c. P, 10^5 e. E, 10^4
   b. P, 10^4 d. E, 10^3 f. E, 10^5

OBJECTIVE 10–2.

Identify the basic mechanisms by which membrane potentials are caused by ionic diffusion. Identify the Nernst and Goldman equations and their interpretative significance.

10. The concentration work required to establish an ion concentration gradient across a cell membrane varies in a _____ (N, linear; G, logarithmic) manner with the concentration _____ (R, ratio; D, difference).

   a. N, R   c. G, R
   b. N, D   d. G, D

11. Diffusion potentials arising from transmembrane cation diffusion are _____ (R, reduced; N, not significantly affected) by simultaneous anion diffusion, represent _____ (W, total electrical work; W/Q, electrical work per unit charge), and are predicted under ideal circumstances by _____ (F, Fick's law; S, the Nernst equation).

   a. R, W, F   c. R, W/Q, F   e. N, W, F
   b. R, W/Q, S d. N, W/Q, S   f. N, W, S

12. The maximum membrane potential in millivolts that may arise as a consequence of ion diffusion tendencies at body temperature for a particular ion species equals _____ times the _____ (L, log; AL, antilog) of its concentration _____ (D, difference; R, ratio).

   a. 0.61, L, D c. 0.61, AL, D e. 61, AL, D
   b. 0.61, L, R d. 61, AL, R   f. 61, L, R

13. The _____ (D, Donnan; G, Goldman) equation reduces to the Nernst equation when the membrane permeability to a particular _____ (C, cation only; CA, either cation or anion) is extremely _____ (L, low; H, high) relative to other membrane anion and cation permeabilities.

   a. D, C, L   c. D, CA, L   e. G, CA, L
   b. D, C, H   d. G, CA, H   f. G, C, H

14. The diffusion potential that develops as a consequence of membrane permeability to several different ions is _____ ionic valence, is _____ membrane permeability to each ion, and is _____ the concentration of respective ions on both sides of the membrane. (D, dependent upon; I, independent of)

   a. D, D, D   c. D, I, D   e. I, D, I
   b. D, D, I   d. I, D, D   f. I, I, D

15. A membrane separating two compartments, A from B, is impermeable to sodium and chloride ions, but permeable to potassium ions. Compartment A has concentrations in mEq./liter of: Na, 100; K, 10; and Cl, 110. Compartment B has corresponding concentrations of: Na, 10; K, 100; and Cl, 110. Compartment B would develop a membrane potential closest to _____ volts, _____ (N, negative; P, positive) with respect to compartment A.

   a. 6.0, N   c. 0.06, N   e. 0.6, P
   b. 0.6, N   d. 6.0, P   f. 0.06, P

OBJECTIVE 10–3.

Identify the roles that sodium and potassium "pumps" and non-diffusible anions play in the origin of cell membrane potentials.

16. A more effective active transport of _____ ions across nerve cell membranes and greater numbers of nondiffusible intracellular _____ (C, cations; A, anions) contribute to the electro _____ (P, positivity; N, negativity) of the intracellular fluid.

    a. Na, C, P   c. Na, A, P   e. K, A, P
    b. Na, A, N   d. K, A, N   f. K, C, N

17. Active extrusion of an intracellular cation to which cell membranes are relatively impermeable would tend to result in the intracellular _____ of other cations and the intracellular _____ of anions to which cell membranes are more permeable. (A, attraction; R, repulsion)

    a. A, A   c. R, A
    b. A, R   d. R, R

18. The intracellular fluid as compared to the extracellular fluid contains _____ sodium ion, _____ potassium ion, and _____ chloride ion concentrations. (H, higher; L, lower)

    a. H, H, H   c. H, L, L   e. L, H, L
    b. H, L, H   d. L, H, H   f. L, L, L

19. The sodium "pump" of cell membranes operates against an opposing _____ and the potassium "pump" operates against an opposing _____. (C, concentration gradient only; B, concentration as well as electrical gradient)

    a. C, C   c. B, C
    b. C, B   d. B, B

20. The intracellular sodium ion concentration of nerves is a more direct consequence of _____ (A, active transport of; EC, a near electrochemical equilibirum for) sodium ions and more closely approximates _____ mEq./liter.

    a. A, 10   c. A, 142   e. EC, 42
    b. A, 42   d. EC, 10   f. EC, 142

OBJECTIVE 10–4.

Identify the major determinants of potassium and chloride ion distribution across cellular membranes and their relationship to membrane potentials.

21. The intracellular potassium ion concentration of nerves is a more direct consequence of _____ (A, active transport of; EC, a near electro-chemical equilibrium for) potassium ions and a relatively _____ (I, impermeable; P, permeable) cell membrane to potassium ions.

    a. A, I   c. EC, I
    b. A, P   d. EC, P

22. A transmembrane potential of 120 millivolts (side A is positive with respect to side B) would provide an electrochemical equilibrium for a freely permeable monovalent anion at a concentration ratio (side A to side B) closest to:

    a. 1:1   c. 100:1   e. 1:100
    b. 10:1   d. 1:10   f. 1:1,000

23. An "equilibrium potential" or intracellular membrane potential required to establish an electrochemical equilibrium for potassium ion movement across nerve cell membranes approximates _____ millivolts and _____ (I, is; N, is not) predicted by the Nernst equation.

    a. +45, I   c. -88, I   e. -45, N
    b. -45, I   d. +45, N   f. -88, N

24. Chloride ion distribution across nerve cell membranes _____ involve a relatively impermeable membrane to chloride ions, _____ involve a chloride ion active transport mechanism, and _____ approximate an electrochemical equilibrium. (D, does; N, does not)

    a. D, D, D   c. D, D, N   e. N, N, D
    b. D, N, N   d. N, N, N   f. N, D, D

25. During a brief interruption of active transport mechanisms the cell membrane potential would exhibit a _____ (L, slight; B, substantial) change in value and would more closely approximate a diffusion potential of _____ as predicted by the Nernst equation.

 a. L, Na ions c. L, zero   e. B, K ions
 b. L, K ions   d. B, Na ions f. B, zero

26. Potassium ions as compared to _____ ions have similar "equilibrium potentials" as predicted by the Nernst equation and _____ (S, similar; O, opposing) concentration gradients across the membranes of nerve axons.

 a. Na, S   c. Cl, S
 b. Na, O   d. Cl, O

27. Widespread tissue injury _____ (I, increases; D, decreases) the magnitude of transmembrane potential differences of noninjured cells by _____ (E, elevating; L, lowering) the extracellular _____ ion concentration.

 a. I, E, K   c. I, E, Na   e. I, L, Na
 b. D, E, K   d. D, L, Na   f. D, L, K

28. The "Nernst potential" or "equilibrium potential" for sodium ion concentrations of nerves is closer to _____ mv, is _____ (P, positive; N, negative), and represents a more significant determinant for membrane potentials during the course of _____ (R, a "resting"; A, an "action") potential.

 a. 70, P, R   c. 70, N, R   e. 20, N, R
 b. 70, P, A   d. 20, N, A   f. 20, P, A

OBJECTIVE 10-5.

Identify action potentials, the means of eliciting action potentials, and a suitable method for recording action potentials and resting membrane potentials.

29. _____ (R, only the "resting"; A, only the "action"; B, both the "resting and "action") potential(s) of nerves are characterized by relatively stable states of higher membrane permeabilities to _____ cations.

 a. R, Na   c. B, Na   e. A, K
 b. A, Na   d. R, K   f. B, K

30. Action potentials of nerves are generally characterized by _____ (DR, depolarization-repolarization; RD, repolarization-depolarization) sequences and _____ (N, only negative; P, only positive; B, both negative and positive) intracellular potentials with respect to extracellular fluid.

 a. DR, N   c. DR, B   e. RD, P
 b. DR, P   d. RD, N   f. RD, B

31. The first event in the initiation of an action potential of nerve fibers in response to _____ (E, only electrical; M, multiple) forms of stimulation is a substantial _____ (D, decrease; I, increase) in membrane permeability to _____ ions.

 a. E, D, Na   c. E, I, Na   e. M, I, Na
 b. E, D, K   d. M, I, K   f. M, D, K

32. Reasonably accurate values of "resting" and "action" potentials have more generally been obtained by means of direct _____ (M, metal; F, fluid electrolyte) to _____ (E, extracellular; I, intracellular) fluid contacts by "sensitive" or recording electrodes.

 a. M, E   c. F, E
 b. M, I   d. F, I

OBJECTIVE 10-6.

Identify the sequence of events and probable mechanisms which characterize the generation of an action potential.

33. "Membrane activation" of excitable tissues is associated with membrane _____ (R, repolarization; D, depolarization) and enhanced electronegativity of the adjacent _____ (I, intracellular; E, extracellular) fluid with respect to an indifferent electrode.

 a. R, I   c. D, I
 b. R, E   d. D, E

34. The depolarization phase of the action potential is characterized by a _____ (N, negative; P, positive) feedback mechanism involving the effects of membrane potential upon _____ ion permeability.

 a. N, Ca   c. N, Na   e. P, K
 b. N, K   d. P, Ca   f. P, Na

35. Calcium ions, with a higher concentration in the _____ (IC, intracellular; EC, extracellular) fluid, more probably interact with nerve cell membranes to produce a more significant _____ (I, increase; D, decrease) in _____ ion permeability.

    a. IC, I, Na   c. IC, D, Na   e. EC, D, Na
    b. IC, I, K   d. EC, D, K   f. EC, I, K

36. Persistent depolarization of the squid giant axon by means of a voltage "clamp" experiment would result in an initial increase in _____ ion permeability which _____ (P, persists; I, becomes inactivated).

    a. Na, P   c. K, P
    b. Na, I   d. K, I

37. The repolarization phase of the action potential results from a greater _____ (P, passive; A, active) transport of _____ ions.

    a. P, Na   c. P, Ca   e. A, K
    b. P, K   d, A, Na   f. A, Ca

38. The "reversal potential" for nerves is characterized by a _____ (L, low; H, high) ratio of potassium conductance to sodium conductance and an extracellular to intracellular sodium ion concentration ratio _____ (G, substantially greater than; E, about equal to; L, substantially less than) one.

    a. L, G   c. L, L   e. H, E
    b. L, E   d. H, G   f. H, L

39. Membrane potentials tend to move towards the equilibrium potentials for _____ ions during the depolarization phase and to move towards the equilibrium potentials for _____ ions during the repolarization phase. (Na, only sodium; K, only potassium; NC, both sodium and chloride; KC, both potassium and chloride)

    a. K, Na   c. Na, KC
    b. K, NC   d. Na, K

40. An experimental reduction in the extracellular concentration of sodium ions would result in a greater _____ (I, increase; D, decrease) in the absolute magnitude of the _____ (R, resting membrane potential; A, action potential amplitude).

    a. I, R   c. D, R
    b. I, A   d. D, A

41. Changes in nerve membrane ionic conductances during the course of an action potential include an earlier change in _____ ion conductance and a greater percentage change in _____ ion conductance with regard to "resting" membrane values.

    a. Na, Na   c. Na, Cl   e. K, Cl
    b. Na, K   d. K, Na   f. K, K

42. During excitation of a nerve cell, the peak of potassium ion _____ (I, influx; E, efflux) occurs _____ (A, after; B, before) the peak Na _____ (I, influx; E, efflux).

    a. I, A, E   c. I, B, E   e. E, B, I
    b. I, A, I   d. E, B, E   f. E, A, I

43. The generation of an action potential _____ require the immediate activity of active transport mechanisms and _____ involve an appreciable energy expenditure by the cell. (D, does; N, does not)

    a. D, D   c. N, D
    b. D, N   d. N, N

44. Tetrodotoxin blocks _____ ion "channels" when applied to the _____ (I, inside; O, outside) membrane surface of nerve axons, while tetraethylammonium ions may be utilized to block _____ (Na, only sodium; K, only potassium; B, both sodium and potassium ion) "channels."

    a. Na, I, B   c. K, I, B   e. K, I, Na
    b. Na, O, K   d. K, O, Na   f. Na, O, Na

OBJECTIVE 10–7.
Identify the determinant mechanisms and characteristics of action potential propagation along cell membranes.

45. Depolarization of a local segment of a nerve axon during the development of its action potential involves local current flow which _____ (D, depolarizes; H, hyperpolarizes) adjacent membrane segments and _____ (N, initiates; I, inhibits) their action potential development.

   a. D, N   c. H, N
   b. D, I   d. H, I

46. An action potential initiated at a midpoint of an axon results in the propagation of an action potential in _____ (P, a preferential; B, both) direction(s) with _____ (L, a lack of; A, an appreciable) decline in action potential amplitude.

   a. P, L   c. B, L
   b. P, A   d. B, A

47. Axonal propagation of action potentials *in situ* involves a preferential direction as a consequence of their _____ (U, unidirectional propagation characteristics; L, locus of action potential origin) and a _____ (M, multiple creation; P, preferential pathway for) action potentials upon their arrival at nodal points of multiple terminal branches.

   a. U, M   c. L, M
   b. U, P   d. L, P

48. Action potentials resulting from a doubling of an effective stimulus intensity have _____ amplitudes and _____ propagation velocities as those resulting from the lesser stimuli. (T, twice the; H, half the; E, equal)

   a. T, H   c. H, T   e. E, E
   b. T, T   d. H, H   f. E, T

49. The "all-or-nothing" principle _____ apply to action potentials initiated by varying stimulus intensities and _____ normally apply to propagation mechanisms of action potentials. (D, does; N, does not)

   a. D, D   c. N, D
   b. D, N   d. N, N

50. A repolarization process is propagated along the nerve axon in the _____ (S, same; O, opposite) direction as the depolarization process and at _____ (I, almost identical; D, dissimilar) propagation velocities.

   a. S, I   c. O, I
   b. S, D   d. O, D

51. Metabolic energy utilization by a nerve axon represents _____ (I, an immediate; L, a long-term) requirement for action potential propagation along a nerve axon and _____ (S, is; N, is not) dependent upon action potential frequency.

   a. I, S   c. L, S
   b. I, N   d. L, N

OBJECTIVE 10–8
Recognize the variance of action potential form among excitable tissues. Identify the spike potentials, negative after-potentials, positive after-potentials, and action potential "plateaus" their general characteristics, and their probable mechanisms.

52. The negative after-potential is a period of after-_____ (H, hyperpolarization; D, depolarization), which is preceded by a _____ (P, positive after-potential; S, spike potential).

   a. H, P   c. D, P
   b. H, S   d. D, S

53. _____ (N, negative; P, positive) after-potentials are generally characterized by longer intervals during which the intracellular fluid is more _____ (N, negative; P, positive) than in the "resting" state.

   a. N, N   c. P, N
   b. N, P   d. P, P

54. _____ (N, negative; P, positive) after-potentials, thought to be caused by the active transport of _____ ions, are abolished by metabolic poisons.

   a. N, Na   c. P, Na
   b. N, K   d. P, K

55. The _____ (P, positive after-potentials; N, negative after-potentials; S, spike potentials) of large diameter myelinated axons have durations of about _____ milliseconds and are more likely to contain a "reversal potential."

   a. P, 5.0   c. S, 0.5   e. N, 5.0
   b. N, 0.5   d. P, 50   f. S, 5.0

56. The existence of a "plateau" is a common action potential characteristic of _____ (C, cardiac; S, skeletal) muscle fibers which coincides with a period of delayed membrane _____ (A, activation; I, inactivation) of _____ ions.

   a. C, A, Na   c. C, I, Na   e. S, I, Na
   b. C, A, K   d. S, I, K   f. S, A, K

OBJECTIVE 10–9.

Recognize the natural existence or potential capability for rhythmicity or repetitive discharge of excitable tissues. Identify the mechanisms for intrinsic rhythmicity of excitable tissues.

57. Membrane _____ (A, activation; I, inactivation) of a nerve axon occurs as a consequence of cellular _____ (D, depolarization; H, hyperpolarization) to a "critical equilibrium" or "threshold" level.

   a. A, D   c. I, D
   b. A, H   d. I, H

58. Isolated _____ (A, nerve axons; C, cardiac muscle cells) generally possess intrinsic rhythmicity and exhibit _____ (I, increasing; R, relatively constant; D, decreasing) ratios of sodium to potassium permeabilities during intervals between action potentials.

   a. A, I   c. A, D   e. C, R
   b. A, R   d. C, I   f. C, D

59. Mild forms of tissue injury and _____ (E, elevated; L, lowered) extracellular calcium ion concentrations both tend to _____ (D, depolarize; H, hyperpolarize) excitable tissues and initially _____ (F, facilitate; I, inhibit) a re-excitation mechanism for rhythmicity.

   a. E, D, F   c. L, D, F   e. L, D, I
   b. E, H, I   d. L, H, F   f. L, H, I

OBJECTIVE 10–10.

Characterize myelinated and unmyelinated nerve fibers within a peripheral nerve trunk. Contrast their manner of propagation and variance of propagation velocities.

60. Myelin is comprised of multiple layers of _____ (E, excitable; NE, non-excitable) cellular membranes of _____ (S, Schwann cells; N, neurons) which is interrupted by nodes of Ranvier at intervals approximating 1 _____.

   a. E, S, mm   c. E, N, mm   e. NE, S, mm
   b. E, N, cm   d. NE, N, cm   f. NE, S, cm

61. Myelinated axons, generally _____ (M, more; L, less) abundant in peripheral nerve trunks than unmyelinated axons, propagate action potentials in a _____ (C, continuous; S, saltatory) fashion.

   a. M, C   c. L, C
   b. M, S   d. L, S

62. Unmyelinated nerves as compared to myelinated nerve axons generally exhibit _____ propagation velocities, _____ energy expenditure, and _____ amounts of ion exchange during action potential propagation. (G, greater; N, no significant difference in their; L, lesser)

    a. L, N, G   c. L, G, G   e. G, G, G
    b. N, G, L   d. N, N, N   f. G, L, L

63. Propagation velocities vary _____ (D, directly; I, inversely) with the _____ (M, diameter; S, diameter squared; R, diameter square root) of myelinated nerve axons.

    a. D, M   c. D, R   e. I, S
    b. D, S   d. I, M   f. I, R

OBJECTIVE 10–11.

Identify the underlying mechanisms and properties of stimuli and excitable tissues in eliciting action potentials.

64. A _____ (L, local; P, propagated) excitatory state is initiated under the _____ (A, anode; C, cathode) with the application of a sub-threshold electrical stimulus as a consequence of membrane _____ (D, depolarization; H, hyperpolarization).

    a. L, A, D   c. L, C, D   e. P, C, D
    b. L, A, H   d. P, C, H   f. P, A, H

65. Under conditions which alter "resting" membrane potentials, the _____ (A, absolute value; I, increment of change; R, rate of change) of transmembrane potential represents a more stable "threshold level" for eliciting an action potential in response to cation _____ (IF, influx; EF, efflux).

    a. A, IF   c. R, IF   e. I, EF
    b. I, IF   d. A, EF   f. R, EF

66. _____ (A, accommodation; D, adaptation) is the phenomenon whereby slowly, rather than rapidly, increasing stimuli require _____ (H, higher; N, no significant difference in; L, lower) threshold values of stimulus intensity.

    a. A, H   c. A, L   e. D, N
    b. A, N   d. D, H   f. D, L

67. An "excitability curve" of nerve fibers represents _____ (D, a direct; I, an inverse) relationship between stimulus intensity and _____ (A, action potential; S, stimulus) duration required for _____ (T, threshold excitation; L, local excitatory states).

    a. D, A, T   c. D, S, T   e. I, S, T
    b. D, A, L   d. I, S, L   f. I, A, L

68. The least stimulus intensity for threshold excitation of a nerve fiber at any stimulus duration is equivalent to one _____ (H, half; F, fourth; S, sixteenth) of the threshold stimulus intensity for a stimulus duration at its _____ (R, rheobase; C, chronaxie) value.

    a. H, R   c. S, R   e. F, C
    b. F, R   d. H, C   f. S, C

69. _____ (M, myelinated; U, unmyelinated) nerve fibers generally have greater "excitabilities" and _____ (H, higher; L, lower) chronaxie values.

    a. M, H   c. U, H
    b. M, L   d. U, L

70. The maximum action potential frequency for large diameter nerve axons approximates _____ per second as a consequence of an effective limiting factor of the _____ (S, stimulus; K, spike potential) duration.

    a. 200, S   c. 20,000, S   e. 2,000, K
    b. 2,000, S   d. 200, K   f. 20,000, K

71. The relative refractory period _____ (P, precedes; F, follows) the absolute refractory period and is generally characterized by a period of _____ (H, higher; L, lower) than normal excitability.

    a. P, H   c. F, H
    b. P, L   d. F, L

OBJECTIVE 10–12.

Identify the effects of varying extracellular concentrations of calcium ions, potassium ions, and local anesthetics upon membrane excitability.

72. Local anesthetics and _____(D, decreased; I, increased) extracellular calcium concentrations act to _____ (DP, depolarize; S, stabilize; HP, hyperpolarize) membrane potentials of excitable tissues.

    a. D, DP   c. D, HP   e. I, S
    b. D, S   d. I, DP   f. I, HP

73. Familial periodic paralysis is a hereditary disease associated with _____(E, elevated; R, reduced) extracellular potassium ion concentrations and _____ (D, depolarization; H, hyperpolarization) of excitable tissues.

    a. E, D   c. R, D
    b. E, H   d. R, H

74. Local anesthetics _____(I, increase; D, decrease) the "safety factor" for action potential propagation as a consequence of their effects upon sodium ion _____ (P, permeability; C, concentration; K, chelation).

    a. I, P   c. I, K   e. D, C
    b. I, C   d. D, P   f. D, K

OBJECTIVE 10–13.

Recognize the utility of the cathode ray oscilloscope. Identify the relationship between recordings of action potentials obtained by means of extracellular electrodes and those obtained by means of intracellular electrodes.

75. Accurate characterization of the action potential form is readily accomplished by means of _____ (E, extracellular; I, intracellular) recordings of _____ (B, biphasic; M, monophasic) action potentials utilizing the _____ (O, oscilloscope; G, galvanometer).

    a. E, B, O   c. E, M, O   e. I, M, O
    b. E, B, G   d. I, M, G   f. I, B, G

76. Extracellular recordings reflect a more _____ (P, positive; N, negative) extracellular environment in the immediate vicinity of the propagated action potential and contain recorded double peaks of _____ (S, the same; O, opposite) polarities.

    a. P, S   c. N, S
    b. P, O   d. N, O

77. Recorded time intervals between the peaks of biphasic action potentials are _____ by decreasing the spacing between recording electrodes and are _____ by increasing propagation velocities. (I, increased; N, not significantly altered; D, decreased)

    a. N, N   c. D, D   e. I, D
    b. N, D   d. D, I   f. I, I

11

Contraction of Skeletal Muscle

OBJECTIVE 11–1.

Using the following diagram, identify the arrangement of myofilaments in the sarcomeres.

Directions: Match the lettered headings with the diagram and the numbered list of descriptive words and phrases.

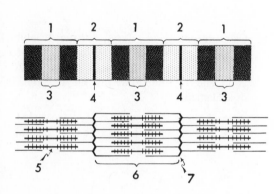

(From Guyton, A.C., Textbook of Medical Physiology, 5th ed. Philadelphia, W.B. Saunders Company, 1976.)

1. _____ Dark bands, containing thick and over-lapping ends of thin myofilaments; anisotropic to polarized light.

2. _____ Light bands, comprised of thin myofil-aments; isotropic to polarized light.

3. _____ Light regions, appearing in the center of dark bands when muscle fibers are stretched beyond their resting length.

4. _____ Membranous attachment for thin myo-filaments; used in defining the limits of a sarcomere.

5. _____ Thick myofilaments which interact with thin myofilaments during muscle contraction and relaxation; associated with dark bands.

6. _____ Repeating structural and functional unit of a myofibril, comprised of a dark band and adjacent halves of light bands.

7. _____ Thin myofilaments; associated with light bands and overlapping regions of dark bands.

a. Myosin filaments   c. Sarcomere   e. Z line   g. H zone
b. Actin filaments   d. A band   f. I band

OBJECTIVE 11–2.

Identify the structural components of skeletal muscle.

8. Skeletal muscle fibers, representing the largest _____ (N, number only; V, volume only; B, both number and volume) of cells of any tissue in the body, generally receive innervation from _____ (S, single; M, multiple) nerve ter-minals.

    a. N, S   c. B, S   e. V, M
    b. V, S   d. N, M   f. B, M

9. A "striate" appearance of skeletal muscle results from the interdigitation of _____ (A, actin; M, myoglobin; S, myosin) containing _____ (F, myofibrils; L, myofilaments).

    a. A & M, F   c. M & S, F   e. A & S, L
    b. A & S, F   d. A & M, L   f. M & S, L

10. A repeating interval of the striate appearance of skeletal muscle corresponds to the _____ (MF, myofibril; S, sarcomere) length and more closely approximates _____ microns.

 a. MF, 2   c. MF, 80   e. S, 10
 b. MF, 10   d. S, 2   f. S, 80

11. A cross-section through the _____ band of skeletal muscle is more likely to reveal the presence of a ratio of actin to myosin myofilaments of _____ .

 a. A, 6:1   c. A, 2:1   e. I, 1:6
 b. A, 1:6   d. I, 6:1   f. I, 2:1

12. _____ (S, sarcomeres; D, triads; C, cisternae) are structural arrangements of skeletal muscle involving two _____ (S, sarcomeres; D, triads; C, cisternae) per _____ tubule.

 a. S, D, T   c. C, D, T   e. D, C, L
 b. C, S, L   d. C, D, L   f. D, C, T

13. _____ (M, more; L, less) rapidly contracting types of muscle are associated with a greater development of the sarcoplasmic reticulum and T tubules containing _____ (I, intracellular; E, extracellular) fluid.

 a. M, I   c. L, I
 b. M, E   d. L, E

OBJECTIVE 11–3.

Identify the general concepts involved in the "sliding filament" model for skeletal muscle contraction.

14. Skeletal muscle is limited in its range of shortening by the observation that the A bands _____ in length while the I bands _____ in length. (D, substantially decrease; R, remain relatively constant; I, substantially increase)

 a. D, D   c. D, R   e. D, I
 b. R, D   d. I, D   f. I, R

15. The normal _____ (P, presence; A, absence) of an H zone in skeletal muscle is best validated by the observation that actin myofilaments within the same sarcomere _____ (N, do not; B, barely; S, substantially) overlap each other at their _____ (C, contracted; R, resting) length.

 a. P, S, C   c. P, N, R   e. A, B, R
 b. P, B, R   d. A, S, C   f. A, N, R

16. Calcium ions _____ (A, activate; I, inhibit) the development of attractive contractile forces between _____ (AA, actin and actin; AM, actin and myosin; MM, myosin and myosin) myofilaments.

 a. A, AA   c. A, MM   e. I, AM
 b. A, AM   d. I, AA   f. I, MM

OBJECTIVE 11–4.

Identify the components of actin and myosin myofilaments, their molecular characteristics, and their internal structural arrangement.

17. Myosin extraction from skeletal muscle is coincident with the disappearance of the _____ (N, thin; K, thick) myofilaments of the _____ (A, A band; I, I band; H, H zone).

 a. N, A   c. N, H   e. K, I
 b. N, I   d. K, A   f. K, H

18. Cross-bridge protrusions, composed of _____ (G, globular; F, double helix) protein structures, are located at the _____ (H, heavy meromyosin end; L, light meromyosin end; J, junction between heavy and light meromyosin units) of the myosin molecule.

 a. G, H   c. G, J   e. F, L
 b. G, L   d. F, H   f. F, J

19. The body of the myosin myofilament is thought to be composed of parallel strands of _____ (L, light; H, heavy) meromyosin which constitute _____ (S, a single molecule; M, multiple molecular units) of myosin.

    a. L, S   c. H, S
    b. L, M   d. H, M

20. The arms of the cross-bridges are thought to extend outward from the myosin myofilament in an oblique manner towards _____ of the myofilament leaving a deficiency of bridges at _____. (O, one particular end; B, the two ends; C, the center)

    a. O, O   c. C, C   e. B, C
    b. B, B   d. O, B   f. C, B

21. Each myosin _____ (S, myofilament; P, pair of adjacent myofilaments) represents a twisted arrangement to yield _____ (3, three; 6, three pairs of) cross-bridges per revolution.

    a. S, 3   c. P, 3
    b. S, 6   d. P, 6

22. The backbone of the actin filament is a _____ (S, single strand; H, two stranded helix) of polymerized _____ (T, troponin; TM, tropomyosin; A, G-actin) molecules.

    a. S, T   c. S, A   e. H, TM
    b. S, TM   d. H, T   f. H, A

23. _____ , having the greater affinity for calcium ions, is thought to be attached to _____ molecules within the actin myofilaments. (TM, tropomyosin; A, G-actin; T, troponin)

    a. T, A   c. A, T   e. TM, A
    b. T, TM   d. A, TM   f. TM, T

OBJECTIVE 11–5.

Identify the proposed molecular mechanisms by which myosin and actin myofilaments interact to cause contraction.

24. Calcium ions are thought to _____ an effect of troponin-tropomyosin to _____ myosin molecular binding to actin filaments. (F, facilitate; I, inhibit)

    a. F, F   c. I, F
    b. F, I   d. I, I

25. _____ availability is required as a source of energy for muscle shortening against a load and _____ (I, is; N, is not) thought to be required for its subsequent relaxation.

    a. ATP, I   c. Creatine, I   e. ADP, N
    b. ATP, N   d. Creatine, N   f. ADP, I

26. A cyclic renewal of actin and myosin interactions is thought to involve the formation of _____ -ATP and a subsequent action of _____ ATPase upon ATP. (A, an actin; M, a myosin)

    a. A, A   c. M, A
    b. A, M   d. M, M

27. Muscle contraction develops a force which is thought to be an instantaneous function of a variable _____ (R, time rate of cyclic; S, number of active sites of) interactions between actin and myosin myofilaments, and requires a greater ATP utilization when it _____ (I, is; N, is not) permitted to shorten against a load.

    a. R, I   c. S, I
    b. R, N   d. S, N

OBJECTIVE 11–6.

Identify the characteristics and probable mechanisms involved in the relationships between the degree of actin and myosin filament overlap and the tensions developed by individual sarcomeres and intact muscles.

28. A curve depicting the tension developed by an individual sarcomere as a function of its length exhibits _____ (O, only one; T, two separate) peak(s) of maximum tension occuring at sarcomere lengths of _____ microns.

a. O, 1.0 to 1.1   d. T, 0.4 & 1.1
b. O, 2.0 to 2.2   e. T, 1.1 & 2.2
c. O, 4.0 to 4.4   f. T, 1.1 & 4.4

29. An increase in muscle length above normal progressively _____ the *resting tension* and _____ the *active tension* of contraction, whereas, a decrease in muscle length below normal progressively _____ the *active tension* of contraction. (I, increases; D, decreases)

a. I, D, I   c. I, I, I   e. D, D, D
b. I, D, D   d. I, I, D   f. D, I, D

30. Overlap of _____ (M, myosin; A, actin) myofilaments with _____ (I, increased; D, decreased) muscle length beyond that which normally occurs *in vivo* results in _____ (I, increased; D, decreased) numbers of active sites available for interaction.

a. M, I, I   c. M, D, I   e. A, D, I
b. M, I, D   d. A, D, D   f. A, I, D

OBJECTIVE 11–7.

Identify the general relationship of load to velocity of contraction for skeletal muscle.

31. Contraction velocities are _____ for muscles with less extensive sarcoplasm reticulums and are _____ with greater applied loads. (G, greater; L, less)

a. G, G   c. L, G
b. G, L   d. L, L

32. The contraction velocity of a muscle is zero at _____ loads and maximal at _____ loads. (Z, zero; O, optimal; M, maximal)

a. Z, O   c. O, Z   e. M, Z
b. Z, M   d. O, M   f. M, O

OBJECTIVE 11–8.

Identify the characteristics of the action potential of skeletal muscle, its physiological origin, and its propagation along the muscle fiber membrane and T tubule system.

33. The resting membrane potential of skeletal muscle is about _____ (P, plus; M, minus) _____ millivolts with respect to an extracellular zero reference.

a. P, 8,500   c. P, 85   e. M, 850
b. P, 850   d. M, 8,500   f. M, 85

34. Action potentials of skeletal muscle have durations closer to _____ milliseconds and propagation velocities approximating _____ meters per second.

a. 0.5, 3 to 5   d. 1 to 5, 30 to 50
b. 0.5, 30 to 50   e. 10 to 15, 3 to 5
c. 1 to 5, 3 to 5   f. 10 to 15, 30 to 50

35. Action potentials are initiated at _____ (S, single; M, multiple) myoneural junctions located at the _____ (C, central portion; T, terminal ends) of the skeletal muscle fiber.

a. S, C   c. M, C
b. S, T   d. M, T

36. Electrical communication between the propagated action potential and the interior of a skeletal muscle fiber is accomplished through the _____ (L, L tubule; T, T tubule; Z, Z line), which forms a component of the _____ (S, sarcomere; TR, triad).

a. L, S   c. Z, S   e. T, TR
b. T, S   d. L, TR   f. Z, TR

37. Membranous tubes, about _____ angstroms in diameter and containing _____ (I, intracellular; E, extracellular) fluid, permit the inward spread of excitation from the skeletal muscle fiber surface.

    a. 3, I   c. 300, I   e. 30, E
    b. 30, I   d. 3, E   f. 300, E

38. The inward spread of excitation from a locus on the surface of a skeletal muscle fiber to its interior, requires _____ (S, less; M, more) than 2 milliseconds and occurs at a _____ (G, greater; L, lower) velocity than the propagation of the action potential along the sarcolemma.

    a. S, G   c. M, G
    b. S, L   d. M, L

OBJECTIVE 11–9.

Identify the roles that intracellular Ca ions and the L-tubular system play in excitation-contraction coupling of skeletal muscle.

39. Junctional feet of the _____ (T, T tubules; C, cisternae) of the sarcoplasmic reticulum of skeletal muscle are thought to play a more significant role to facilitate their _____ (U, uptake; R, release) of calcium ions.

    a. T, U   c. C, U
    b. T, R   d. C, R

40. The _____ (T, transverse; L, longitudinal) tubule system is thought to function as a "relaxing factor" for skeletal muscle through its ability to _____ (C, transport calcium ions; R, repolarize the plasma membrane).

    a. T, C   c. L, C
    b. T, R   d. L, R

41. A _____ (R, relaxation-triggered; C, contraction-triggered; K, continuously operating) calcium active transport mechanism functions to transport calcium ions _____ (I, into; O, out of) the lumen of the sarcoplasmic reticulum of skeletal muscle.

    a. R, I   c. K, I   e. C, O
    b. C, I   d. R, O   f. K, O

42. A relaxed state of skeletal muscle is a probable consequence of calcium _____ (B, binding to; D, deficient states of) troponin-tropomyosin complexes of the _____ (M, myosin; A, actin) myofilaments.

    a. B, M   c. D, M
    b. B, A   d. D, A

43. The development of an "active state" of skeletal muscle is associated with an intracellular pulse of _____ (I, increased; D, decreased) calcium ion concentration, typically about _____ milliseconds in duration, and representing _____ (M, more; L, less) than a 99% change in its resting concentration.

    a. I, 20, M   d. I, 200, L
    b. I, 20, L   e. D, 20, M
    c. I, 200, M   f. D, 200, L

44. The action potential duration for skeletal muscle _____ (A, and; N, but not) its absolute refractory period, is relatively _____ (S, short; L, long) as compared to the subsequent pulse duration of altered intracellular calcium ion concentration.

    a. A, S   c. N, S
    b. A, L   d. N, L

OBJECTIVE 11–10

Identify the energy sources for muscle contraction, the relationship of muscle energy to work performed, and the determinants for the efficiency of muscle contraction.

45. ATP generation by a predominant _____ (A, aerobic; N, anaerobic) metabolism occurs in organelles located _____ (B, between; W, within) the myofibrils.

    a. A, B   c. N, B
    b. A, W   d. N, W

46. Creatine phosphate, present in, _____ (L, lower; H, higher) concentrations than ATP in resting skeletal muscle, is utilized as a _____ (B, direct substitution; R, rephosphorylation) mechanism for ATP.

    a. L, B   c. H, B
    b. L, R   d. H, R

47. Resting skeletal muscle levels of ATP provide for sufficient energy for an interval of maximum contraction of _____ (L, less; M, more) than several seconds, whereas combined creatine phosphate and ATP levels provide for intervals of maximum contraction closer to several _____ (S, seconds; M, minutes; H, hours).

   a. L, S   c. L, H   e. M, M
   b. L, M   d. M, S   f. M, H

48. Contraction of skeletal muscle under loaded versus nonloaded conditions results in _____ (G, a greater; N, no significant change in; L, a lesser) amount of ATP utilization and the _____ (F, "Fenn"; S, staircase) effect.

   a. G, F   c. L, F   e. N, S
   b. N, F   d. G, S   f. L, S

49. The efficiency of muscle contraction, or percentage of the energy utilized by muscles that is converted into _____ (H, heat; W, work), has an upper limit of about _____%.

   a. H, 20   c. H, 77   e. W, 45
   b. H, 45   d. W, 20   f. W, 77

50. Maximal efficiency of skeletal muscle is accomplished with muscle shortening against _____ loads at _____ contraction velocities. (X, maximal; I, intermediate; Z, near zero)

   a. X, I   c. I, I   e. Z, Z
   b. Z, X   d. X, X   f. I, Z

OBJECTIVE 11–11.

Identify the characteristics of skeletal muscle twitches, and the influencing characteristics of isometric versus isotonic conditions and fast versus slow types of muscle fibers.

51. Following a brief latent period, a single effective stimulus delivered to _____ (N, only the motor nerve; M, only the muscle fibers; B, either the motor nerve or muscle fibers) results in a _____ (R, repetitive series of twitches; S, single twitch).

   a. N, R   c. B, R   e. M, S
   b. M, R   d. N, S   f. B, S

52. Contraction and relaxation phases of muscle twitches occurring under _____ (M, isometric; T, isotonic) conditions refer to varying muscle tensions at _____ (V, varying; C, constant) muscle lengths.

   a. M, V   c. T, V
   b. M, C   d. T, C

53. The performance of external work, _____ (G, greater; L, lesser) energy expenditures, and _____ (G, greater; L, lesser) twitch durations, are more frequent characterizations of _____ (M, isometric; T, isotonic) contractions.

   a. G, G, M   c. G, L, M   e. L, L, M
   b. G, G, T   d. L, L, T   f. L, G, T

54. The actual shortening between the tendonous origin and insertion of a muscle is generally slightly _____ (G, greater; L, less) than that of the equivalent contractile mechanism as a consequence of the _____ (S, series elastic component; F, "Fenn" effect; V, isovolumetric expansion) of skeletal muscle.

   a. G, S   c. G, V   e. L, F
   b. G, F   d. L, S   f. L, V

55. Slow twitch muscle fibers as compared to fast twitch muscle fibers generally possess greater _____, whereas fast twitch muscle fibers generally possess greater _____. (D, fiber diameters; M, myoglobin content; S, development of sarcoplasmic reticulum; V, vascularity; N, numbers of mitochondria)

   a. D, S, & V; M & N   d. M & V; N, D, & S
   b. D; M, S, V, & N   e. M, V, & N; D & S
   c. M, S, V, & N; D   f. M, S, & N; D & V

56. Postural muscles – e.g., the soleus – are comprised of a majority of _____ (F, fast; S, slow) twitch fibers and possess twitch durations closer to _____.

   a. F, 0.01   c. F, 1.0   e. S, 0.1
   b. F, 0.1   d. S, 0.01   f. S, 1.0

OBJECTIVE 11-12.
Identify motor units, their characteristics, and their functional significance.

57. The number of muscle fibers innervated by a single motor neuron is about _____ and along with the motor neuron constitutes the _____ (M, motor unit; G, gamma loop).

   a. 1, M
   b. 2 to 1,000, M
   c. 2 to 10, M
   d. 1, G
   e. 2 to 1,000, G
   f. 2 to 10, G

58. The number of muscle fibers in a motor unit is generally higher in muscles of _____ (S, smaller; L, larger) size which are more frequently utilized for _____ (G, gross; F, fine, discrete) movements.

   a. S, G
   b. S, F
   c. L, G
   d. L, F

59. Multiple, as opposed to single, motor neuron innervation of single skeletal muscle fibers is considered to be _____ (N, a normal; B, an abnormal) condition associated with a _____ (G, greater; L, lesser) degree of control of skeletal muscle function.

   a. N, G
   b. N, L
   c. B, G
   d. B, L

OBJECTIVE 11-13.
Identify the effects and significance of variations of motor unit action potential frequency and relative numbers of active motor units upon muscle contraction.

60. Unit action potential _____ (T, tetanic; H, twitch) response of skeletal muscle fibers are utilized physiologically to regulate the degree of skeletal muscle contraction by means of _____ (M, only multiple motor unit; W, only wave; B, both multiple motor unit and wave) summation.

   a. T, M
   b. T, W
   c. T, B
   d. H, M
   e. H, W
   f. H, B

61. Smaller motor units as compared to larger ones generally contribute _____ (S, smaller; L, larger) increments of contraction magnitude, are activated at _____ (E, earlier; T, later) stages of increased multiple motor unit summation, and afford a mechanism for greater control of muscle contraction with _____ (W, weaker; S, stronger) contractions of a skeletal muscle.

   a. S, E, W
   b. S, E, S
   c. S, T, W
   d. L, T, S
   e. L, T, W
   f. L, E, S

62. The comparatively _____ (G, greater; L, lesser) response of a skeletal muscle twitch initiated during the partially contracted state of a preceding twitch constitutes the fundamental basis of _____ (M, multiple motor unit; W, wave) summation.

   a. G, M
   b. G, W
   c. L, M
   d. L, W

63. The peak contraction of a skeletal muscle twitch generally occurs during a _____ (R, rising; P, peak; D, declining) intracellular calcium ion concentration and _____ (S, does; N, does not) generally represent a full expression of its "active state."

   a. R, S
   b. P, S
   c. D, S
   d. R, N
   e. P, N
   f. D, N

64. A *critical frequency* for the generation of _____ (W, wave summation; T, complete tetanization) is _____ (H, higher; N, not significantly different; L, lower) for slow twitch as opposed to fast twitch types of skeletal muscle.

   a. W, H
   b. W, N
   c. W, L
   d. T, H
   e. T, N
   f. T, L

65. Smooth, coordinate skeletal muscle contractions are more generally the result of physiological conditions of _____ (I, incomplete; C, complete) tetanization of individual motor units and _____ (S, synchronized; A, asynchronous) action potential firing among the different motor units.

   a. I, S
   b. I, A
   c. C, S
   d. C, A

66. Adequate insurance against damage to tendonous origins and insertion of skeletal muscles _____ (I, is; N, is not) normally vested in a maximum strength of the contractile mechanism of muscle of about _____ kg. per square centimeter.

    a. I, 3.5   c. I, 350   e. N, 35
    b. I, 35   d. N, 3.5   f. N, 350

67. A repetitive series of separate skeletal muscle twitches, following a period of inactivity, gives rise to an initial _____ (I, increase; D, decrease) in their contraction amplitudes or _____ (S, staircase; F, "Fenn"; G, fatigue) effect.

    a. I, S   c. I, G   e. D, F
    b. I, F   d. D, S   f. D, G

OBJECTIVE 11–14.

Identify skeletal muscle tone and fatigue and their basic underlying mechanisms.

68. Elevated muscle tone resulting from _____ (M, an intrinsic muscle; R, a neural reflex) mechanism would produce _____ (I, increased; N, no significant change in; D, decreased) resistance of passive muscle movements.

    a. M, I   c. M, D   e. R, N
    b. M, N   d. R, I   f. R, D

69. Destruction of the posterior roots of the spinal cord _____ (I, increases; D, decreases) corresponding muscle tone as a consequence of a loss of the influence of neural activity arising within muscle _____ (S, spindles; T, tendons).

    a. I, S   c. D, S
    b. I, T   d. D, T

70. A loss of blood supply to an actively contracting skeletal muscle results in a progressive _____ (I, increase; D, decrease) in contraction amplitudes as a major consequence of muscle _____ (DP, depolarization; H, hyperpolarization; F, fatigue).

    a. I, DP   c. I, F   e. D, H
    b. I, H   d. D, DP   f. D, F

71. Muscle fatigue would more severely _____ (D, decrease; I, increase) the duration of the _____ (C, contraction; R, relaxation) phase of isotonic twitches as a consequence of its influence upon an active transport mechanism for _____ ions.

    a. D, C, Ca   c. I, C, Ca   e. I, C, Na
    b. D, R, Na   d. I, R, Ca   f. I, R, Na

OBJECTIVE 11–15.

Identify the means by which lever systems of the body extend the functional range of utility of skeletal muscles, and the manner of accommodation of the muscle length to the length of the lever system.

72. A greater effective shortening of a _____ (M, multipennate; P, parallel) arrangement of skeletal muscle is effectively limited to about _____ % of the muscle length.

    a. M, 10   c. M, 50   e. P, 30
    b. M, 30   d. P, 10   f. P, 50

73. A third-class lever in the body has a fulcrum-applied load distance ten times the fulcrum-muscle insertion distance. The advantage of this lever system is a consequence of a potential _____ fold increase in muscle _____ (F, force; S, shortening; W, work).

    a. 10, F   c. 10, W   e. $(10)^2$, S
    b. 10, S   d. $(10)^2$, F   f. $(10)^2$, W

74. Shortened skeletal muscle, occurring as a direct consequence of broken bones which heal in place, _____ undergo further physical shortening and _____ re-establish optimal conditions for contractile mechanisms. (D, do; N, do not)

    a. D, D   c. N, D
    b. D, N   d. N, N

OBJECTIVE 11-16.

Identify the conditions and underlying causative mechanisms of muscular hypertrophy, muscular atrophy, contracture, and familial periodic paralysis.

75. Hypertrophy is an adaptive mechanism of skeletal muscle as a primary consequence of _____ (P, passive stretching; C, active contractions) of muscle with _____ (L, only smaller tension for longer periods; H, only higher tensions for shorter periods; E, either L or H).

   a. P, L   c. P, E   e. C, H
   b. P, H   d. C, L   f. C, E

76. Skeletal muscle hypertrophy involves a more significant increase in numbers of _____ (F, muscle fibers; MF, myofibrils per muscle fiber) and occurs _____ (E, at the expense of the; C, in conjunction with improved) metabolic capabilities of skeletal muscle.

   a. F, E   c. MF, E
   b. F, C   d. MF, C

77. Skeletal muscle _____ (H, hypertrophy; A, atrophy) occurs in response to muscle denervation _____ (O, or; N, but not with) reduced activity levels of innervated muscles.

   a. H, O   c. A, O
   b. H, N   d. A, N

78. A state of contracture _____ (W, as well as; N, but not) rigor is thought to result from _____ (I, injury potentials; A, action potential discharges; D, ATP depletion).

   a. W, I   c. W, D   e. N, A
   b. W, A   d. N, I   f. N, D

79. _____ (MD, muscle denervation; F, familial periodic paralysis) reduces the excitability of skeletal muscle fibers by _____ (I, increasing; D, decreasing) their intracellular to extracellular ratio of potassium ion concentrations.

   a. MD, I   c. F, I
   b. MD, D   d. F, D

OBJECTIVE 11-17.

Identify the electromyogram, its significance, and its utility in distinguishing muscle fasciculation and fibrillation.

80. Recorded electromyographic activity of skeletal muscle _____ (F, fibrillations; S, fasciculations) are generally smaller in amplitude, and when present _____ (A, are; N, are not) useful diagnostic indicators of skeletal muscle denervation.

   a. F, A   c. S, A
   b. F, N   d. S, N

81. Skeletal muscle _____ (S, fasciculations; B, fibrillations) represent a more probable physiological mechanism to enhance _____ (T, postural muscle tonus; H, metabolic heat production).

   a. S, T   c. B, T
   b. S, H   d. B, H

82. Recorded electromyographic activity of skeletal muscle by means of concentric needle electrodes implanted into the muscle record _____ (M, monophasic; B, biphasic) potentials whose amplitude is largely a function of the _____ (A, intracellular action potential amplitude; I, extracellular current flow).

   a. M, A   c. B, A
   b. M, I   d. B, I

12

Neuromuscular Transmission; Function of Smooth Muscle

OBJECTIVE 12–1.

Using the following diagrams, identify the major structural components of the neuromuscular junction.

Directions: Match the lettered headings with the diagram and the numbered list of descriptive words and phrases.

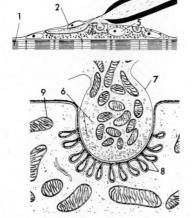

(From Guyton, A.C., Textbook of Medical Physiology, 5th ed. Philadelphia, W.B. Saunders Company, 1976.)

1. _____ Organelles containing actin and myosin myofilaments.
2. _____ Terminal Schwann cell covering the end-plate.
3. _____ Cell membrane laminate between terminal node of Ranvier and end-plate.
4. _____ Cytoplasm of motor nerve fiber.
5. _____ Multiple spherical bodies containing chromatin material.
6. _____ Small globular bodies with structureless interiors that serve as reservoirs for acetylcholine.
7. _____ Neural end-plate.
8. _____ Secondary junctional clefts that communicate with the primary junctional cleft.
9. _____ Principal locus of ATP generation.

a. Axon in junctional trough
b. Myelin sheath
c. Muscle nuclei

d. Mitochondrion
e. Myofibril
f. Synaptic vesicles

g. Teloglial cells
h. Subneural clefts
i. Axoplasm

OBJECTIVE 12–2.

Identify the general mechanisms of neuromuscular transmission of action potentials from nerve to skeletal muscle.

10. Skeletal muscles are innervated by large _____ (U, unipolar; B, bipolar; M, multipolar) neurons located in the _____ (A, anterior; P, posterior) horns of the spinal cord.

    a. U, A   c. M, A   e. B, P
    b. B, A   d. U, P   f. M, P

11. A single motoneuron transmits action potentials to _____ (S, a single; M, multiple) muscle fiber(s) by means of _____ (E, electrically; C, chemically) mediated myoneural junctions.

    a. S, E   c. M, E
    b. S, C   d. M, C

12. Secondary junctional folds at the bottom of a _____ -angstrom wide junctional gutter function primarily to _____ (E, enhance the contact surface area for; S, synthesize) a junctional transmitter agent.

    a. 7.5, E   c. 200, E   e. 75, S
    b. 75, E   d. 7.5, S   f. 200, S

OBJECTIVE 12–3.

Identify the chemical mediator of the myoneural junction and the general mechanisms involved in its synthesis, storage, release, and enzymatic breakdown.

13. Action potentials propagated into the myoneural junction result in the release of _____ (N, norepinephrine; A, acetylcholine) from _____ (P, only the prejunctional; S, only the postjunctional; B, both the prejunctional and postjunctional) membranes.

    a. N, P   c. N, B   e. A, S
    b. N, S   d. A, P   f. A, B

14. The release of chemical mediator from the end-plate by means of its _____ (D, depolarization; H, hyperpolarization) is _____ (F, facilitated; I, inhibited) by elevated extracellular calcium ion concentrations and _____ (F, facilitated; I, inhibited) by elevated extracellular magnesium ion concentrations.

    a. D, F, F   c. D, I, F   e. H, I, F
    b. D, F, I   d. H, I, I   f. H, F, I

15. Action potential propagation into the end-plate region results in the release of the contents of _____ vesicle(s) into the _____ (S, sarcoplasm; G, junctional gutter).

    a. 1, S
    b. 20 to 30, S
    c. 200 to 300, S
    d. 1, G
    e. 20 to 30, G
    f. 200 to 300, G

16. The primary function of mitochondria located in nerve terminals of myoneural junctions is to generate _____ for _____ (E, enzyme; A, acetylcholine) synthesis.

    a. ATP, E   d. ATP, A
    b. Acetyl co-A, E   e. Acetyl co-A, A
    c. RNA, E   f. RNA, A

17. _____ (CE, cholinesterase; MO, monamine oxidase; CA, choline acetylase), present on the surfaces of the folds of the junctional gutter, results in acetylcholine _____ (S, synthesis; A, activation; I, inactivation).

    a. MO, S   c. CE, A   e. CA, I
    b. CA, S   d. MO, A   f. CE, I

18. Released acetylcholine continues to function in combination with the postjunctional membranes of skeletal muscle for a period of about _____ milliseconds, which is sufficient to transmit a nerve to muscle action potential ratio of about _____.

    a. 2 to 3, 1 to 1   d. 10 to 15, 1 to 1
    b. 2 to 3, 1 to 4   e. 10 to 15, 1 to 4
    c. 2 to 3, 4 to 1   f. 10 to 15, 4 to 1

OBJECTIVE 12–4.

Identify the "end-plate potential," its characteristics, its underlying mechanisms, and its consequences. Identify the "safety factor" for transmission at the myoneural junction and the factors involved in its determination.

19. The end-plate potential is a _____ (L, local; P, propagated), _____ (A, "all or none"; G, graded) transmembrane potential.

    a. L, A   c. P, A
    b. L, G   d. P, G

20. The end-plate potential results from the more direct action of _____ (N, nerve action potentials; A, acetylcholine) upon the _____ (EJ, prejunctional; SJ, postjunctional) membrane.

    a. N, EJ   c. A, EJ
    b. N, SJ   d. A, SJ

21. The end-plate potential is typically a _____ -millivolt, _____ (H, hyperpolarizing; D, depolarizing) membrane potential resulting from _____ (I, increased; E, decreased) membrane permeability to sodium ions.

    a. 20 to 35, H, I   d. 50 to 75, D, E
    b. 20 to 35, H, E   e. 50 to 75, D, I
    c. 20 to 35, D, I   f. 50 to 75, H, E

22. A threshold level of intracellular membrane potential required for initiating action potentials of skeletal muscle typically equals _____ millivolts and is approximated with end-plate potential magnitudes of about _____ millivolts.

    a. -35, 35   c. -85, 35   e. -50, 50
    b. -50, 35   d. -35, 50   f. -85, 50

23. A "safety factor" at the myoneural junction normally involves a value _____ (S, slightly; B, substantially)_____(G, greater; L, less) than one, and which is higher at _____(H, high; L, low) frequencies of action potential transmission.

a. S, G, H   c. B, G, H   e. B, L, L
b. B, L, H   d. S, G, L   f. B, G, L

24. Fatigue is a more probable *in vivo* consequence of either a peripheral failure of the _____ (C, contractile; J, myoneural junctional transmission) mechanism or a more central nerve action potential _____(S, synaptic transmission; A, axonal propagation) mechanism.

a. C, S   c. J, S
b. C, A   d. J, A

OBJECTIVE 12–5.

Identify representative drugs and their mechanisms for the pharmacological modification of transmission at the neuromuscular junction. Identify the condition of myasthenia gravis and the rationale for its treatment.

25. Competitive inhibitory effects of neuromuscular transmission are of a depolarizing nature for _____ and a nondepolarizing nature for _____. (C, curare; D, decamethonium; N, neostigmine)

a. C, D   c. D, N   e. N, C
b. C, N   d. D, C   f. N, D

26. The mechanisms of action of _____ (P, physostigmine; N, nicotine) are similar to those of _____ (A, acetylcholine; E, acetylcholine esterase) at the myoneural junction except for a _____(G, greater; L, lesser) effective duration of action.

a. P, A, G   c. P, A, L   e. N, A, G
b. P, E, L   d. N, E, L   f. N, E, G

27. _____ (C, curare; DFP, diisopropyl fluorophosphate) results in _____ (T, transient; P, prolonged) muscular spasms as a consequence of its _____ (I, competitive inhibition of; F, facilitation of the effects of) the action of acetylcholine.

a. C, T, I   c. C, P, I   e. DFP, P, I
b. C, T, F   d. DFP, P, F   f. DFP, T, F

28. Almost normal muscular activity of a patient with myasthenia gravis can be obtained by the administration of _____ (A, acetylcholine; AC, an anticholinesterase; C, a cholinesterase) in order to compensate for an abnormally _____ (H, high; L, low) safety factor for neuromuscular transmission.

a. A, H   c. C, H   e. AC, L
b. AC, H   d. A, L   f. C, L

OBJECTIVE 12–6.

Identify the morphological characteristics of smooth muscle and contrast the properties of multiunit and visceral types of smooth muscle.

29. Smooth muscle is comprised of _____ (T, striate; NT, nonstriate), _____(M, multiple, S, single) nucleate cells, _____ (L, slightly; B, substantially) smaller in diameter than skeletal muscle cells.

a. T, M, L   c. T, S, L   e. NT, S, L
b. T, M, B   d. NT, S, B   f. NT, M, B

30. Propagation of action potentials from one smooth muscle cell to another by means of _____ (S, synaptic; E, ephaptic) transmission is largely a property of _____ (V, visceral; M, multiunit) smooth muscle.

a. S, V   c. E, V
b. S, M   d. E, M

31. _____ (V, visceral; M, multiunit) smooth muscle _____ (D, does; N, does not) exhibit spontaneous contractions and is controlled almost entirely by neural activity rather than local tissue factors.

a. V, D   c. M, D
b. V, N   d. M, N

32. Unitary smooth muscle or _____ (M, multiunit; V, visceral) smooth muscle of the _____ (C, ciliary muscle of the eye; G, wall of the gut) is innervated by the _____ (A, autonomic; S, somatic) nervous system.

a. M, C, A   c. M, G, A   e. V, G, A
b. M, C, S   d. V, G, S   f. V, C, S

OBJECTIVE 12-7.

Contrast the contractile process of smooth muscle with that occurring for skeletal muscle.

33. Smooth muscle _____ have myofibrils, _____ significantly differ from skeletal muscle in its "thin" to "thick" myofilament ratio, and _____ significantly differ in its molecular ratio of actin to myosin. (D, does; N, does not)

    a. N, D, D   c. N, N, D   e. D, N, D
    b. N, D, N   d. D, N, N   f. D, D, N

34. When compared to skeletal muscle, smooth muscle generally has a _____ economy of ATP utilization for the maintenance of prolonged tension and a _____ contraction velocity. (G, greater; L, lesser)

    a. G, G   c. L, G
    b. G, L   d. L, L

OBJECTIVE 12-8.

Identify the membrane potential and action potential characteristics, and the means of action potential origin for visceral smooth muscle.

35. Resting membrane potentials of smooth muscle are generally _____ variable and _____ negative than skeletal muscle. (M, more; L, less)

    a. M, M   c. L, M
    b. M, L   d. L, L

36. Propagated spike potentials of about _____ millisecond(s) in duration, _____ (A, and; N, but not) action potentials with plateaus, characterize certain types of _____ (M, multiunit; V, visceral) smooth muscle.

    a. 1, A, M   c. 1, N, M   e. 10, N, M
    b. 1, A, V   d. 10, N, V   f. 10, A, V

37. The *slow wave rhythm* of some _____ (M, multiunit; V, visceral) smooth muscle cells is a _____ (P, propagated; L, local) membrane potential which is more significant in that it initiates _____ (C, muscle contractions; A, action potentials).

    a. M, P, C   c. M, L, C   e. V, L, C
    b. M, P, A   d. V, L, A   f. V, P, A

38. *Pacemaker waves* of _____(M, multiunit; V, visceral) smooth muscle are _____ (A, propagated action potentials; S, slow waves) which result in rhythmical contractions of a large group of smooth muscle cells.

    a. M, A   c. V, A
    b. M, S   d. V, S

39. Action potentials with plateaus are generally associated with prolonged membrane inactivation _____ (A, and; N, but not) prolonged muscle contraction, and are generally comprised of a more rapid _____ (D, depolarization; R, repolarization) phase.

    a. A, D   c. N, D
    b. A, R   d. N, R

40. Action potentials of single visceral smooth muscle cells _____ elicited by their intrinsic rhythmicity, _____ elicited by ephaptic transmission, and _____ elicited by neurohumeral agents. (A, are; N, are not)

    a. A, A, A   c. A, A, N   e. N, N, A
    b. A, N, A   d. N, A, A   f. N, A, N

OBJECTIVE 12-9.

Identify peristalsis and the mechanism of excitation of visceral smooth muscle by stretch and action potential propagation through visceral smooth muscle.

41. Stretch is generally considered to be a means of _____ (H, inhibition; E, excitation) of visceral smooth muscle as it _____ (I, increases; D, decreases) the intracellular _____ N, negativity; P, positivity) of slow wave potentials of visceral smooth muscle.

    a. H, I, N   c. H, D, N   e. E, D, N
    b. H, I, P   d. E, D, P   f. E, I, P

42. Peristalsis, initiated by a passive _____ (I, increase; D, decrease) of the circumferential length of gut smooth muscle, is a _____ (T, tonic; P, propulsive) type of movement.

    a. I, T   c. D, T
    b. I, P   d. D, P

43. Action potential transmission from one smooth muscle cell to another is dependent upon a(n) _____ (M, mechanical; E, electrical; C, chemical) coupling mechanism mediated by _____ (T, tight junctions; A, autonomic nerves).

a. M, T   c. C, T   e. E, A
b. E, T   d. M, A   f. C, A

OBJECTIVE 12–10.

Contrast the excitation-contraction coupling and relaxation mechanisms for smooth muscle with those of skeletal muscle.

44. Smooth muscle differs from skeletal muscle in regard to a considerably greater increase in its membrane permeability to _____ ions during membrane _____ (D, depolarization; R, repolarization).

a. Na, D   c. Ca, D   e. K, R
b. K, D   d. Na, R   f. Ca, R

45. Contraction of smooth muscle as compared to skelatal muscle involves a greater participation of _____ (S, sarcoplasmic reticulum; C, extracellular calcium ions) and _____ (G, greater; E, about equal; L, longer) latent periods.

a. S, G   c. S, L   e. C, E
b. S, E   d. C, G   f. C, L

46. A relatively _____ (R, rapidly-acting; S, slow-acting) calcium pump for smooth muscle as compared to skeletal muscle is a mechanism which provides for a relatively _____ (L, longer; H, shorter) _____ (A, action potential; C, contraction phase; X, relaxation phase) duration of smooth muscle.

a. R, H, A   c. R, L, X   e. S, L, X
b. R, H, C   d. S, L, A   f. S, H, C

47. Increased extracellular calcium ion concentrations generally _____ (I, increase; D, decrease) smooth muscle tone as a consequence of their effects upon _____ (M, membrane cation permeability; C, intracellular contractile) mechanisms.

a. I, M   c. D, M
b. I, C   d. D, C

OBJECTIVE 12–11.

Identify the types of smooth muscle contraction, their characteristics, and their significance.

48. Smooth muscle _____ (I, is; N, is not) well suited for maintained tonic contractions as a consequence of its _____ (P, presence; A, absence) of wave summation, its functionally syncytial nature, and its _____ (W, well developed; R, rudimentary) sarcoplasmic reticulum.

a. I, P, W   c. I, A, W   e. N, A, W
b. I, P, R   d. N, A, R   f. N, P, R

49. Visceral smooth muscle has a useful range of shortening which is _____ 25 to 35% of its resting length and _____ that occurring for skeletal muscle. (G, greater than; E, about equal to; L, less than)

a. G, E   c. G, G   e. L, L
b. G, L   d. L, E   f. L, G

50. Smooth muscle as compared to skeletal muscle is considerably better able to maintain a relatively constant _____ (L, length; T, tension) over a wide range in muscle _____ (L, length; T, tension) as a consequence of _____ (S-R, stress-relaxation phenomena; A, autonomic reflexes; E, its elastic properties).

a. L, T, S-R   c. L, T, E   e. T, L, A
b. L, T, A   d. T, L, S-R   f. T, L, E

51. The stress-relaxation phenomenon is attributed to smooth muscle alterations in _____ (A, action potential frequency; P, "slippage" in the pattern of myofilament organization) which requires a period of several _____ (S, seconds; M, minutes; H, hours).

a. A, S   c. A, H   e. P, M
b. A, M   d. P, S   f. P, H

OBJECTIVE 12-12.

Describe the physiologic anatomy of neuromuscular junctions. Identify the chemical mediators at autonomic neuromuscular junctions, their mechanisms of action, and their functional significance.

52. _____ and _____ are the transmitter substances known to be secreted by nerves innervating smooth muscle. (A, acetylcholine; GABA, gamma aminobutyric acid; E, epinephrine; N, norepinephrine)

   a. A, GABA   c. GABA, E   e. E, N
   b. A, N   d. GABA, N   f. E, A

53. Autonomic innervation of smooth muscle involves _____ (C, only contact; N, only noncontacting; B, both C and N) types of nerve endings which exhibit a _____ (M, more; L, less) discrete neural influence than that occuring for skeletal muscle.

   a. C, M   c. B, M   e. N, L
   b. N, M   d. C, L   f. B, L

54. Acetylcholine plays _____ roles and norepinephrine plays _____ roles for certain types of smooth muscle. (I, only inhibitory; E, only excitatory; B, both excitatory and inhibitory)

   a. E, I   c. I, B   e. B, E
   b. E, B   d. I, E   f. B, B

55. The _____ (S, slow wave; J, junctional) potential of smooth muscle is mechanistically analogous to the end-plate potential except that it has a _____ (L, slightly; B, substantially) _____ (H, shorter; G, longer) duration.

   a. S, L, H   c. S, B, H   e. J, B, H
   b. S, L, G   d. J, B, G   f. J, L, G

56. A threshold potential of a smooth muscle cell is reached when the membrane potential changes from a typical resting value of _____ millivolts to a value of _____ millivolts.

   a. -75 to -85, -10 to -20
   b. -75 to -85, -30 to -40
   c. -55 to -60, -10 to -20
   d. -55 to -60, -30 to -40

57. Enhanced excitability of smooth muscle cells is associated with _____ while reduced excitability of smooth muscle cells is associated with _____. (H, only membrane hyperpolarization; D, only membrane depolarization; E, either membrane depolarization or hyperpolarization)

   a. H, D   c. D, E   e. E, D
   b. H, E   d. D, H   f. E, E

58. _____ (P, progesterone; E, estrogen) is a hormone which prevents a premature expulsion of the fetus by _____ (D, depolarizing; H, hyperpolarizing) the smooth muscle of the uterine myometrium.

   a. P, D   c. E, D
   b. P, H   d. E, H

13

Heart Muscle; the Heart as A Pump

OBJECTIVE 13–1.
Using the following diagram, identify the gross physiological anatomy of the heart.

Directions: Match the lettered headings with the diagram and the numbered list of descriptive words and phrases.

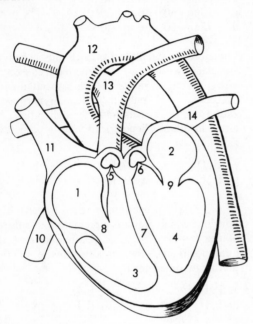

1. _____ Upper right chamber of the heart.
2. _____ Upper left chamber of the heart.
3. _____ Lower right chamber of the heart.
4. _____ Lower left chamber of the heart.
5. _____ Valve between the right ventricle and the pulmonary artery.
6. _____ Valve between the left ventricle and the aorta.
7. _____ A partition separating the lower left and right chambers of the heart.
8. _____ Right A–V valve.
9. _____ Left A–V valve.
10. _____ Vein carrying blood from the lower portions of the body to the heart.
11. _____ Vein carrying blood from the upper portions of the body to the heart.
12. _____ Main artery from which the systemic arterial system proceeds.
13. _____ Artery carrying blood from the heart to the lungs.
14. _____ Vein carrying blood from the lungs to the heart.

a. Left ventricle
b. Superior vena cava
c. Pulmonary artery
d. Aortic valve

e. Right ventricle
f. Right atrium
g. Pulmonary valve
h. Mitral valve

i. Tricuspid valve
j. Septum
k. Aorta
l. Inferior vena cava

m. Pulmonary vein
n. Left atrium

OBJECTIVE 13–2.
Recognize atrial, ventricular, and specialized excitatory and conductive muscle fibers as three major types of cardiac muscle, and identify their general functional significance.

Directions: Match the lettered answer with the numbered descriptive phrase.

15. _____ The principal contractile chamber that propels blood through the lungs.

16. _____ The principal contractile chamber that propels blood through the peripheral circulatory system.

17. _____ Tissue containing few contractile fibers and contracting only weakly.

18. _____ Weak contractile chamber for blood returning from the lungs.

19. _____ Weak contractile chamber for blood returning from the peripheral circulation.

a. Right atrium
b. Left atrium

c. Right ventricle
d. Left ventricle

e. Specialized excitatory and conductive fibers

OBJECTIVE 13-3.

Identify the functional roles of intercalated discs, the functional syncytial nature of the atrial and ventricular myocardium, and the all-or-nothing principle as it applies to the heart.

20. Intercalated discs are probable _____ (E, ephapses; S, synapses; M, myofilaments) that participate in the _____ nature of cardiac muscle.

    a. E, multiunit
    b. S, multiunit
    c. M, contractile
    d. E, functional syncytial
    e. S, functional syncytial
    f. M, all-or-nothing

21. Cardiac muscle responses to increasing stimulus intensities of fixed duration are those of:

    a. Recruitment
    b. Tetanization
    c. All-or nothing responses
    d. Increasing force of contractions

22. The range of stimulus intensities between threshold and maximal stimuli is *least* for:

    a. Ventricular muscle
    b. Skeletal muscle motor points
    c. Sensory receptors
    d. A peripheral nerve trunk

23. A graphical curve illustrating the relationship between intensity of effective stimulation plotted on the x-axis versus the amplitude of cardiac muscle contraction plotted on the y-axis would be:

    a. An exponential curve
    b. A bell-shaped curve
    c. An S-shaped or sigmoid curve
    d. A horizontal straight line

24. A threshold electrical stimulus applied to the left ventricle of a heart in momentary cardiac arrest would probably result in contraction of:

    a. Only a few local cardiac cells
    b. Only the left ventricle
    c. Only the left & right ventricles
    d. The entire heart.

OBJECTIVE 13-4.

Identify a resting membrane potential, action potential form and duration, velocity of conduction, and absolute and functional refractory periods of atrial and ventricular muscle.

25. The action potential duration for ventricular muscle is closest to _____ second and is propagated through ventricular muscle at about _____ meter(s) per second.

    a. 0.003, 0.4   c. 0.3, 0.4   e. 0.03, 4.0
    b. 0.03, 0.4   d. 0.003, 4.0   f. 0.3, 4.0

26. Ectopic ventricular stimulation is ineffective if applied during the:

    a. Atrial contraction
    b. Ventricular contraction
    c. Ventricular relaxation
    d. Entire cardiac cycle

27. The functional refractory period is about _____ second for atrial muscle, and about _____ second(s) for ventricular muscle.

    a. 0.002, 0.3   c. 0.3, 0.3   e. 0.15, 3.0
    b. 0.15, 0.3   d. 0.002, 3.0   f. 0.3, 3.0

28. A maximal rhythmical rate of atrial contraction is _____ (L, less; G, greater) than that of the ventricles as a result of differences in _____ (P, propagation velocities; R, refractory periods).

    a. G, P   c. L, P
    b. G, R   d. L, R

29. A ventricular action potential, originating from a typical intracellular membrane potential of _____ millivolts, has _____ (P, a plateau; I, an initial spike) component occupying the longest interval of its duration.

    a. −45, P   c. −150, P   e. +85, I
    b. −85, P   d. +45, I   f. +150, I

OBJECTIVE 13–5.

Describe the excitation-contraction coupling mechanism, the role of calcium ions, and the associated morphological aspects of cardiac muscle cells.

30. Transverse or T tubules of cardiac muscle are significantly different from T tubules of skeletal muscle in that they exhibit greater _____(D, diameters; N, numbers per sarcomere) and _____(G, greater; L, Lesser) roles in calcium ion transport.

    a. D, L   b. N, L   c. D, G   d. N, G

OBJECTIVE 13–6.

Identify the chronological relationship of atrial or ventricular contraction relative to its action potentials, the effect of heart rate on contraction duration, and the effect of a premature contraction on contraction strength.

31. The contraction duration for ventricular muscle is mainly a function of:

    a. Action potential duration
    b. Latent period
    c. Resting membrane potential
    d. Autonomic nerve activity
    e. Propagation velocities
    f. Intrinsic rate(s) of rhythmicity

32. The contraction phase, relaxation phase, and action potential duration of cardiac muscle are all decreased at high heart rates. (T or F)

OBJECTIVE 13–7.

Identify the cardiac cycle, and the utilization of the terms systole and diastole.

33. The average time interval in seconds of the cardiac cycle is related to the heart rate by the formula, heart rate in beats per minute equals:

    a. 60/interval   c. Interval/60
    b. Interval × 60   d. 1/(interval × 60)

34. The interval of the cardiac cycle that begins with closure of the A–V valves and ends with the opening of the A–V valves corresponds in clinical terminology to:

    a. Atrial systole   c. Ventricular systole
    b. Atrial diastole   d. Ventricular diastole

OBJECTIVE 13–8.

Recognize the S–A node action potential origin, transmission through the atria, A–V node, and A–V bundle into the ventricles, and the delay in atrial-ventricular excitation. Identify the electrocardiogram, the P, QRS complex, and T waves, their origins, and their temporal association with events of the cardiac cycle.

35. The QRS complex of the electrocardiogram results from ventricular depolarization and begins slightly prior to the onset of ventricular systole. (T or F)

36. The T wave of the electrocardiogram results from atrial depolarization and begins slightly prior to the onset of atrial systole. (T or F)

37. The P wave of the electrocardiogram results from ventricular repolarization and occurs slightly prior to the end of ventricular contraction. (T or F)

OBJECTIVE 13–9.

Identify the functional significance of atrial contraction; and the origins of the a, c, and v pressure waves of the atria and great veins.

38. The abrupt transfer of blood from the right atrium to the right ventricle with the opening of the tricuspid valve results in a recorded decrease in pressure from the peak of the:

a. c wave   c. P wave
b. a wave   d. v wave

39. Atrial contractions, associated with recorded _____ waves, _____ (N, are not; A, are) required for cardiac function adequate to sustain life.

a. a, N   c. v. N   e. c, N
b. a, A   d. v, A   f. c, A

40. Bulging of the A–V valves backward into the atria following the onset of ventricular systole is a factor in the production of the:

a. c wave   c. P wave
b. a wave   d. v wave

OBJECTIVE 13–10.

Identify the isometric contraction, ejection, protodiastole, isometric relaxation, rapid inflow, and diastasis phases of ventricular function. Recognize the significance of each phase, and identify corresponding positions of the heart valves for each phase.

41. The middle third of ventricular diastole, when the inflow of blood into the ventricles is _____(H, high; L, low) is termed the period of _____(P, protodiastole; E, ejection; D, diastasis).

a. H, P   c. H, D   e. L, E
b. H, E   d. L, P   f. L, D

42. During the isometric contraction phase of the cardiac cycle, the A–V valves are _____ and the aortic and pulmonic valves are _____ (C, closed; O, open)

a. O, O   c. C, O
b. O, C   d. C, C

43. The isometric relaxation phase of the cardiac cycle terminates with the coincident appearance of the:

a. Peak of the c wave   e. P wave
b. Beginning of the T wave   f. QRS complex
c. Peak of the v wave
d. Period of protodiastole

44. The highest percentage of blood is ejected into the pulmonary artery during the _____ (1, first; 2, second; 3, third) quarter of _____ (A, atrial; V, ventricular) systole.

a. 1, A   c. 3, A   e. 2, V
b. 2, A   d. 1, V   f. 3, V

OBJECTIVE 13–11.

Identify end-diastolic, end-systolic, and stroke volumes. Identify a typical value for stroke volume, and recognize the variability of these values.

45. Both _____ the end-diastolic volume and _____ the end-systolic volume serve to increase stroke volume output to a maximum of about double that of normal. (I, increasing; D, decreasing)

a. I, I   c. D, I
b. I, D   d. D, D

46. The stroke volume normally averages about _____ ml. and is approximately _____ times the end-diastolic volume.

a. 5, 1.0   c. 200, 0.2   e. 70, 1.0
b. 70, 0.5   d. 5, 0.2   f. 200, 0.5

OBJECTIVE 13-12.

Identify and describe the functions of the A–V (tricuspid and mitral) valves, papillary muscles and chordae tendineae, and aortic and pulmonic (semilunar) valves.

47. Mitral and tricuspid valve closure occurs largely as a result of:

   a. Papillary muscle contraction
   b. Atrial contraction
   c. A–V pressure gradients
   d. Purkinje fiber relaxation

48. Closure of the aortic valve, in *contrast* to mitral valve closure, involves:

   a. Less physical trauma
   b. Papillary muscle contraction
   c. Less rapid closure
   d. Greater backflow of blood
   e. Pressure gradient forces
   f. All of the above.

OBJECTIVE 13-13.

Recognize the aortic and left ventricular pressure curves during the cardiac cycle. Identify systolic and diastolic pressures, their typical values, and the incisura or dicrotic notch and its cause.

49. The incisura or dicrotic notch occurs in _____ (A, only the aortic; V, only the left ventricular; B, both the aortic and left ventricular) pressure curve(s) as a result of aortic valve _____ (O, opening, C, closure).

   a. A, O   c. B, O   e. V, C
   b. V, O   d. A, C   f. B, C

50. Aortic diastolic pressure, typically _____ mm. Hg., corresponds to the aortic pressure immediately prior to aortic valve _____ (O, opening; C, closure).

   a. 40, O   c. 120, O   e. 80, C
   b. 80, O   d. 40, C   f. 120, C

51. A pressure of 120 mm. Hg would be typical of the aortic _____ (D, diastolic; S, systolic) pressure and would occur during the _____ (DS, diastasis; P, protodiastole; E, ejection) phase of the cardiac cycle.

   a. D, DS   c. D, E   e. S, P
   b. D, P   d. S, DS   f. S, E

OBJECTIVE 13-14.

Graphically illustrate the temporal interrelationships of aortic, left ventricular, and atrial pressures, left ventricular volume, heart sounds, and the electrocardiogram waves during the cardiac cycle.

52. The combined periods of rapid ventricular filling and diastasis are most closely associated in time with the interval between the:

   a. P & R waves
   b. R & T waves
   c. v & P waves
   d. Third & fourth heart sounds
   e. First & second heart sounds
   f. Second & third heart sounds

53. The peak ventricular systolic pressure corresponds most closely in time with the:

   a. c wave
   b. P wave
   c. QRS complex
   d. T wave
   e. Third heart sound
   f. Fourth heart sound

54. The left ventricular volume would change least during the interval between the:

   a. c wave & second heart sound
   b. Incisura & v wave
   c. Third & first heart sound
   d. First & second heart sound

55. The ventricular ejection phase would correlate most closely in time with the interval between the:

   a. v & P waves
   b. c wave & second heart sound
   c. c wave & first heart sound
   d. P & T waves

56. The end-systolic volume represents the _____ (G, greatest; L, least) ventricular volume during the cardiac cycle and is most closely approximated at the end of the _____ wave.

   a. G, R   c. G, c   e. L, T
   b. G, T   d. L, R   f. L, c

57. An event whose onset separates protodiastole from the isometric relaxation phase of the cardiac cycle would be the:

   a. First heart sound   d. Fourth heart
   b. Second heart sound   sound
   c. Third heart sound   e. a wave
                      f. v wave

OBJECTIVE 13–15.

Recognize two components for work output by the heart — potential energy of pressure and kinetic energy of blood flow. Identify the major energy component, and contrast right versus left ventricular energy requirements.

58. Work performed by the left ventricle is substantially greater than that performed by the right ventricle largely as a result of differences in:

   a. Blood velocity   d. Arterial pressures
   b. Stroke volume   e. Atrial pressures
   c. Blood volume flow   f. Muscle efficiencies

59. A hypertensive individual with double the normal left ventricular mean ejection pressure, with other variables unchanged, would require a left ventricular work load about:

   a. Equal to normal   d. 4 times normal
   b. 2 times normal   e. ¼ normal
   c. 3 times normal   f. ½ normal

OBJECTIVE 13–16.

Identify the relationship between the amount of energy expended by the heart and its work load, and an approximate maximum efficiency of cardiac contraction.

60. The amount of energy expended by the heart is approximately proportional to the product of cardiac muscle tension and the time duration of its action. (T or F)

61. The maximum efficiency of cardiac contraction is achieved during _____ (L, low; A, average; H, high) work output by the heart, and is about _____ %.

   a. L, 20   c. H, 20   e. A, 55
   b. A, 20   d. L, 55   f. H, 55

OBJECTIVE 13–17.

Recognize the Frank-Starling law, its interpretation and significance, and the mechanisms of its existence.

62. Cardiac muscle response to increasing stretch involves _____ (D, decreasing; N, no appreciable change in; I, increasing) force of contraction, thereby constituting a _____ (HT, heterometric; HM, homeometric) mechanism of autoregulation of cardiac function.

   a. D, HT   c. N, HT   e. I, HT
   b. D, HM   d. N, HM   f. I, HM

63. The statement that "within physiological limits, the heart pumps all of the blood to it without allowing excessive damming of blood in the veins," is one way of expressing the Frank-Starling law of the heart. (T or F)

64. The Frank-Starling law primarily involves a mechanism of altered:

   a. Vagal activity   d. Action potential form
   b. Peripheral resis-   e. Propagation
      tance   velocities
   c. Cardiac   f. Sarcomere length
      metabolism

65. The Frank-Starling law of the heart implies a direct relationship between stroke volume and:

   a. End-systolic   d. Cardiac output
      volume   e. Autonomic nerve
   b. End-diastolic   activity
      volume   f. Cardiac reserve
   c. Mean arterial
      pressure

OBJECTIVE 13–18.

Identify venous return. Recognize intrinsic mechanisms of autoregulation of the heart to altered venous return and autonomic innervation as two basic means by which cardiac function is regulated.

66. Venous return is typically _____ times cardiac output, or about _____ liters per minute at rest.

    a. 0.5, 1 to 2   d. 0.5, 4 to 6
    b. 1, 1 to 2   e. 1, 4 to 6
    c. 4, 1 to 2   f. 4, 4 to 6

67. The most important determinant for intrinsic autoregulation of cardiac pumping is normally:

    a. Mean arterial pressure   d. Cardiac work
    b. Autonomic inner-   output
        vation   e. Cardiac efficiency
    c. Right atrial pressure   f. Flow resistance

68. Maximal _____ (S, sympathetic; P, parasympathetic) stimulation may _____ (I, increase; D, decrease) the strength of ventricular contraction by a factor exceeding 90% of the normal operating strength.

    a. S, I   c. P, I
    b. S, D   d. P, D

69. Autonomic activity may result in heart rates ranging from as low as 20 to 30 beats per minute with strong _____ (S, sympathetic; P, parasympathetic) tone to as high as _____ beats per minute.

    a. S, 80   c. S, 250   e. P, 125
    b. S, 125   d. P, 80   f. P, 250

70. High heart rates are especially effective in increasing the cardiac output when the:

    a. Heart rate is 350 per minute
    b. Right atrial pressure is high
    c. Sympathetic tone is low
    d. End-systolic volume is high

OBJECTIVE 13–19.

Identify ventricular function curves, their significance, and factors producing a hypereffective or hypoeffective heart.

71. Of the following, the best expression of the functional ability of the ventricles is depicted by plotting ventricular output versus:

    a. Stroke volume   d. Right atrial
    b. Stroke work   pressure
    c. Peripheral resis-   e. Mean arterial
        tance   pressure
                       f. Venous return

72. Factors resulting in a hyperdynamic heart would tend to _____ right atrial pressure and _____ the maximum permissive level of cardiac output. (I, increase; D, decrease)

    a. I, I   c. D, I
    b. I, D   d. D, D

73. Factors resulting in a hypereffective heart include conditions of:

    a. Cardiac hypoxia   d. Myocardial
    b. Valvular heart   infarction
        disease   e. Parasympathetic
    c. Hyperkalemia   stimulation
                       f. Sympathetic
                       stimulation

OBJECTIVE 13–20.

Identify the effects of temperature and variations of extracellular concentrations of potassium, sodium, and calcium ions on cardiac function.

74. Toxic effects of excess calcium ions result in sustained _____ (R, relaxation; C, contraction) of the heart, an effect which is _____ (S, similar; O, opposite) to that of potassium ion excess.

    a. R, S   c. C, S
    b. R, O   d. C, O

75. Increased body temperature _____ (I, increases; D, decreases) heart rate, presumably by _____ (R, raising; L, lowering) membrane cation permeabilities of cardiac muscle cells.

    a. I, R   c. D, R
    b. I, L   d. D, L

14

Rhythmic Excitation of the Heart

OBJECTIVE 14–1.

Locate the anatomical position of the S–A node, A–V node, and Purkinje system of the heart.

Directions: Match the lettered headings with the diagram and the numbered list of descriptive words and phrases.

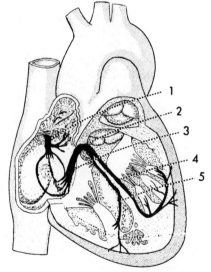

(From Guyton, A.C., Textbook of Medical Physiology, 5th ed. Philadelphia, W.B. Saunders Company, 1976.)

1. _____ A small, crescent-shaped strip of specialized muscle in the posterior wall of the right atrium, beneath the opening of the superior vena cava.

2. _____ Positioned near the opening of the coronary sinus, and embedded in the septal wall of the right atrium.

3. _____ Originates at the A–V node and follows the membranous septum toward the left atrioventricular opening for 1 to 2 cm.

4. _____ That part of the bundle of His which enters the fibrous septum and terminates under the endocardium within the left ventricle.

5. _____ That part of the bundle of His which branches and then continues under the endocardium toward the apex, where it fans out to all parts of the right ventricle.

a. Left bundle branch   b. S–A node   c. A–V bundle   d. Right bundle branch   e. A–V node

OBJECTIVE 14–2.

Identify the functional significance of the S–A node, A–V node, internodal pathways, A–V bundle, and right and left bundle branches.

Directions: Match the lettered answer with the numbered descriptive phrase.

6. _____ The locus of action potential origin of the cardiac impulse.

7. _____ The locus of a functional delay in action potential propagation between atria and ventricles.

8. _____ The locus of the most rapid action potential propagation for transmission between the atria and the ventricles.

9. _____ The locus for rapid transmission and spread of the cardiac impulse within the ventricles.

10. _____ The locus for the most rapid route of action potential propagation between the nodal tissues of the heart.

a. A–V bundle
b. A–V node

c. Right and left bundle branches
d. S–A node

e. Internodal pathways

OBJECTIVE 14–3.

Recognize the property of automatic rhythmicity of most cardiac fibers, and identify the probable mechanisms involved in self-excitation and intrinsic rates of rhythmicity of cardiac fibers.

11. Increased rates of _____ (H, hyperpolarization; DP, depolarization) immediately preceding the development of action potentials in an appropriate cardiac tissue would be associated with _____ (I, increased; D, decreased) rates of automatic rhythmicity.

a. DP, I
b. DP, D

c. H, I
d. H, D

12. That cardiac tissue having the greatest natural rate of automatic rhythmicity is the _____ , which has a comparatively _____ (L, lower; H, higher) magnitude of intracellular resting membrane potential.

a. A–V bundle, L
b. A–V node, L
c. S–A node, L

d. A–V bundle, H
e. A–V node, H
f. S–A node, H

OBJECTIVE 14–4.

Identify the magnitudes and significance of propagation velocities for action potential transmission in atrial and ventricular muscle, junctional and A–V nodal fibers, and Purkinje fibers.

13. A range of propagation velocities for inclusion of both atrial and ventricular muscle fibers would be:

a. 0.01 to 0.1 meters/second
b. 0.3 to 0.5 meters/second
c. 1.5 to 4 meters/second
d. 5 to 25 meters/second

14. Propagation velocities of 0.01 meters/second are more typical of the:

a. Left bundle branch
b. Purkinje fibers
c. Atrial muscle fibers
d. Ventricular muscle fibers

e. A–V nodal junctional fibers
f. Internodal pathway

15. That component of cardiac tissues having the slowest propagation velocity is the:

a. Purkinje system of fibers
b. A–V node
c. Atrial muscle
d. Ventricular muscle

e. Internodal pathway
f. A–V node junctional fibers

16. That component of cardiac tissues having the fastest propagation velocity is the:

a. Purkinje system of fibers
b. A–V node
c. Atrial muscle

d. Ventricular muscle
e. Internodal pathway
f. A–V node junctional fibers

17. _____ (J, junctional; A, atrial; P, Purkinje) fibers of the heart are characterized by relatively large diameters, increased numbers of nexi, few myofibrils, and have propagation velocities closest to _____ meters per second.

a. J, 0.01 to 0.1   c. P, 0.01 to 0.1   e. A, 5 to 25
b. A, 0.01 to 0.1   d. J, 5 to 25   f. P, 5 to 25

OBJECTIVE 14–5.

Describe the transmission of the cardiac impulse through the heart. Recognize approximate transit times from the S–A node to the A–V node, through the A–V node, and through the Purkinje system to the epicardial surface. Contrast the transit times with corresponding action potential durations of the cardiac tissues.

18. The propagated front of action potential depolarization enters the A–V bundle about _____ second following the origin of the cardiac impulse, an interval that is about _____ the duration of an atrial action potential.

a. 0.04, ½   d. 0.15, equal to
b. 0.04, twice   e. 0.30, twice
c. 0.15, ½   f. 0.30, equal to

19. From its arrival in the Purkinje system, the wave of excitation spreads to all parts of the ventricles in _____ seconds.

a. 0.8   c. 0.18
b. 0.06   d. 0.30

20. The sequence of activation of the ventricles in appropriate order following excitation of the Purkinje system is (P, epicardial surface; S, septum; D, endocardial surface):

a. P, S, D   c. S, D, P   e. D, P, S
b. P, D, S   d. S, P, D   f. D, S, P

OBJECTIVE 14–6.

Identify the intrinsic rates of rhythmicity of the S–A node, A–V node, and Purkinje fibers, the pacemaker concept for the heart, and the normal locus of the pacemaker.

21. The natural rate of rhythmicity of the A–V node is not normally apparent from electrocardiogram recordings because of activity of the:

a. Right vagus   d. A–V nodal junctional
b. Left vagus   fibers
c. S–A node   e. Sympathetic nerve fibers
                  f. Purkinje fibers

22. The _____ node is normally the pacemaker for the heart because of its_____ (I, innervation; L, location; R, rate of rhythmicity).

a. A–V, I   c. A–V, R   e. S–A, L
b. A–V, L   d. S–A, I   f. S–A, R

23. Intrinsic rates of rhythmicity are typically _____ beats per minute for the A–V node and _____ beats per minute for the S–A node.

a. 70 to 80, 15 to 40   d. 15 to 40, 70 to 80
b. 70 to 80, 40 to 60   e. 40 to 60, 15 to 40
c. 15 to 40, 40 to 60   f. 40 to 60, 70 to 80

OBJECTIVE 14–7.
Recognize the possibility and identify the mechanism for abnormal or ectopic pacemakers within the heart.

24. A cardiac tissue which is *not* generally considered to potentially exhibit abnormal or ectopic pacemaker activity is:

    a. The A–V node
    b. Ventricular muscle fibers
    c. The left atrium
    d. The Purkinje system
    e. The S–A node
    f. The right atrium

25. When action potential propagation between the S–A node and A–V node is prohibited, the ventricles will generally exhibit:

    a. An A–V nodal rhythm
    b. Maintained cardiac standstill
    c. An S–A nodal rhythm
    d. Ventricular fibrillation
    e. A maintained depolarization
    f. 5 to 15 beats per minute

26. Increasing the rhythmic discharge rate of some additional part of the heart to a value exceeding that of the normal pacemaker results in a(n) _____ spread of the cardiac impulse and a(n) _____ sequence of myocardial contraction. (N, normal; A, abnormal)

    a. N, N
    b. N, A
    c. A, A
    d. A, N

OBJECTIVE 14–8.
Recognize the roles and identify the significant properties of the Purkinje system in causing rapid synchronous contraction of the ventricular muscle and in preventing arrhythmias.

27. The action potential duration of Purkinje fibers is about _____ % of ventricular cardiac muscle action potential durations, and constitutes a significant factor in the function of Purkinje fibers to _____ (S, sychronize; P, prevent arrhythmic) ventricular contraction.

    a. 25, S   b. 75, S   c. 125, S   d. 25, P   e. 75, P   f. 125, P

OBJECTIVE 14–9.
Identify the distribution of the autonomic innervation to the heart, and their effects and probable mechanisms in the regulation of heart rhythmicity and conduction.

28. The action of the vagus on the heart is mediated by:

    a. Acetylcholine
    b. Norepinephrine
    c. Bradykinin
    d. Histamine
    e. GABA
    f. Epinephrine

29. Stimulation of the parasympathetic nerves to the heart would result in increased:

    a. S–A node rhythmicity
    b. Heart rate
    c. A–V conduction times
    d. Atrial contractility
    e. Ventricular excitability
    f. Ventricular contractility

30. Sympathetic stimulation to the heart results in _____ excitability and _____ force of contraction of cardiac tissues. (D, decreased; I, increased)

    a. I, I
    b. I, D
    c. D, I
    d. D, D

31. Strong _____ (S, sympathetic; P, parasympathetic) stimulation to the _____ (A, only the S–A node; V, only the A–V node; A/V, either the S–A node or the A–V node) may be utilized to demonstrate the phenomenon called "ventricular escape."

    a. S, A   c. S, A/V   e. P, V
    b. S, V   d. P, A   f. P, A/V

OBJECTIVE 14–10.

Contrast the effects of right and left vagus versus sympathetic innervation on myocardial contraction (inotrophic effect), S–A node rhythmicity and heart rate (chronotrophic effect), general myocardial excitability (bathymotrophic effect), and A-V nodal transmission (dromotrophic effect).

32. The S–A node develops from structures on one side of the embryo, as does the A–V node. In the adult, this is why the right vagus is distributed to the _____ and the left vagus mainly to the_____. (A, A–V node; S, S–A node; LV, left ventricle)

    a. A, S   c. A, LV
    b. S, A   d. S, LV

33. A negative dromotrophic effect on the heart may be demonstrated more effectively through stimulation of the _____ whereas a positive inotrophic effect on the left atrium would more likely result from stimulation of the _____. (R, right vagus; L, left vagus; S, sympathetic innervation)

    a. R, S   c. R, R   e. S, L
    b. L, S   d. L, R   f. S, R

34. Sympathetic stimulation to the heart results in a reduced or negative myocardial:

    a. Inotrophic action
    b. Chronotrophic action
    c. Dromotrophic action
    d. Bathymotrophic action
    e. All of the above
    f. None of the above

OBJECTIVE 14–11.

Recognize abnormal S–A nodal rhythmicity, ectopic pace-maker loci, regional transmissional blocks, abnormal transmission pathways, and spurious ectopic beats as causes for abnormal cardiac rhythms.

35. The possibility of an abnormal cardiac rhythm resulting from ectopic pacemaker loci would be greater with the intravenous injection of certain _____ (S, sympatheticomimetic; P, parasympatheticomimetic) agents combined with _____ (I, increased; D, decreased) vagal activity.

    a, S, I   b. S, D   c. P, I   d. P, D

OBJECTIVE 14–12.

Identify flutter, fibrillation, and cardiac arrest, their potential causes, and their potential consequences.

36. Fibrillation is not usually confined to either the atria or the ventricles alone as a consequence of the all-or-nothing behavior of the heart, but rapidly spreads to all four cardiac chambers. (T or F)

37. Mechanisms exist such that fibrillating human ventricles generally return to a rhythmic beat of their own accord. (T or F)

38. The _____ (E, ectopic focus; C, circus movement) theory of the more coordinate contractions of _____ (B, fibrillation; L, flutter) presupposes a greater dependence upon the pathway length, propagation velocity, and refractory period.

    a. E, B   c. E, L
    b. C, B   d. C, L

39. Ventricular fibrillation does not result from electrical stimuli applied to the ventricle during the _____ interval of the electrocardiogram because of the _____ (R, refractory period; D, A–V nodal delay).

   a. P–R, R   c. T–P, R   e. S–T, D
   b. S–T, R   d. P–R, D   f. T–P, D

40. Atrial fibrillation would more likely result in:

   a. Myocardial contracture
   b. Myocardial fibrillation
   c. Myocardial tetanization
   d. An irregular ventricular rhythm
   e. Cardiac arrest
   f. None of the above

OBJECTIVE 14–13.
Identify the rationale and technique of cardiac defibrillation and the rationale for internal or external cardiac massage and artificial respiration.

41. The technique of electrical defibrillation utilized for the heart is most dependent upon the process of:

   a. Maintained hyperpolarization
   b. Contracture
   c. Synchronized action potentials
   d. Coronary ischemia & fatigue
   e. Parasympathetic stimulation
   f. Sympathetic stimulation

42. A defibrillating shock timed to convert atrial flutter or fibrillation to a normal rhythm without inducing ventricular fibrillation may be accomplished by triggering the onset of the electrical shock to be delivered immediately following the:

   a. Electrocardiogram T wave
   b. Electrocardiogram P wave
   c. Electrocardiogram R wave
   d. Radial pulse
   e. Third heart sound
   f. None of the above

15

The Normal Electrocardiogram

OBJECTIVE 15–1.
Identify the component deflections and characteristics of a typical electrocardiogram record, and their chronological association with atrial and ventricular action potentials and contractions. Describe the genesis of the P, QRS, and T waves.

1. The depolarization wave of the atria, reflected in the _____ wave of the electrocardiogram, is followed within about _____ seconds by atrial repolarization.

   a. P, 0.10 to 0.20   d. P, 0.30 to 0.50
   b. R, 0.10 to 0.20   e. R, 0.30 to 0.50
   c. T, 0.10 to 0.20   f. T, 0.30 to 0.50

2. All of the principal electrocardiogram waves result from myocardial _____ (D, depolarizations; R, repolarizations) with the exception of the _____ wave.

   a. D, P   c. D, T   e. R, R
   b. D, R   d. R, Q   f. R, T

3. The atrial _____ wave is generally totally obscured in the recorded electrocardiogram by the _____ wave(s).

   a. P, T   c. T, P   e. R, T
   b. R, P   d. P, QRS   f. T, QRS

4. Electrocardiogram waves resulting from electrical activity of the ventricles normally occur only during periods of _____ and do not occur during periods of _____ . (ID, incomplete ventricular depolarization; IR, incomplete ventricular repolarization; CD, completed ventricular depolarization; A, an absence of ventricular action potentials)

   a. ID; IR, CD, & A   d. CD; ID, IR, & A
   b. ID & IR; CD & A   e. CD & ID; IR & A
   c. ID, IR, & CD; A   f. CD & IR; ID & A

5. The QRS complex, occurring immediately before the beginning of _____ (A, atrial; V, ventricular) contraction, is generally of _____ (L, lesser; S, the same; G, greater) amplitude than the T wave.

   a. A, L   c. A, G   e. V, S
   b. A, S   d. V, L   f. V, G

OBJECTIVE 15-2.

Recognize typical electrocardiogram values for the P–P, P–R (or P–Q), and Q–T intervals at a heart rate of 72 beats per minute.

6. That portion of the electrocardiogram related to the propagation of the cardiac impulse between the S–A node and the A–V node is the _____ interval, typically lasting about _____ seconds.

   a. Q-S, 0.04   c. P–R, 0.04   e. Q–T, 0.16
   b. Q–T, 0.04   d. Q–S, 0.16   f. P–R, 0.16

7. That portion of the electrocardiogram which coincides with the approximate interval of the plateau portion of a ventricular action potential is the _____ interval, lasting about _____ seconds.

   a. P–R, 0.1   c. Q–T, 0.1   e. QRS, 0.3
   b. QRS, 0.1   d. P–R, 0.3   f. Q–T, 0.3

8. At a heart rate of 72 beats per minute, the longest electrocardiogram interval would generally be the:

   a. Duration of the P wave
   b. Duration of the QRS complex
   c. P–R interval
   d. Q–T interval
   e. Duration of the T wave
   f. Q–S interval

OBJECTIVE 15-3.

Identify a typical voltage amplitude for the QRS complex in a standard lead, and the calibration standard and paper speed utilized for electrocardiogram recordings.

9. Standard electrocardiograms typically contain voltage amplitudes of the recorded waves ranging from:

   a. 0.1 to 1 millivolt   d. 0.1 to 1 microvolt
   b. 3 to 10 millivolts   e. 3 to 10 microvolts
   c. 50 to 100 millivolts   f. 50 to 100 microvolts

10. The standard for electrocardiogram recordings utilizes calibrations of _____ per major vertical line interval, and _____ second(s) per major interval on the horizontal axis.

   a. 1 microvolt, 0.2   d. 0.2 millivolt, 1
   b. 10 microvolts, 1   e. 1 millivolt, 1
   c. 100 microvolts, 1   f. 50 millivolts, 0.2

OBJECTIVE 15-4.

Identify the general factors influencing the occurrence and general direction of current flow in the volume conductor surrounding the heart during the cardiac cycle.

11. Resultant current flow in the extracellular fluid surrounding the heart occurs coincident with proximate extracellular _____ in tissues undergoing depolarization, and proximate extracellular _____ in tissues undergoing repolarization. (P, positivity; N, negativity)

   a. P, P   c. N, P
   b. P, N   d. N, N

12. During most of the interval of the spread of ventricular depolarization, the epicardial surface is _____ with respect to the endocardial surface, with the net current flow taking place primarily from the base of the heart to a more _____ apex. (N, electronegative; P, electropositive)

   a. N, N   c. P, N
   b. N, P   d. P, P

13. During ventricular _____ (D, depolarization; R, repolarization), represented by the QRS of the electrocardiogram, a recording electrode in closer proximity to the apex of the heart would record a predominantly _____ (N, negative; P, positive) potential with respect to the base of the heart.

   a. D, N   c. R, N
   b. D, P   d. R, P

OBJECTIVE 15-5.

Identify the three standard limb leads. Recognize the convention regarding electrical polarity that results in an "upright" R wave in leads I, II, and III for a typical individual.

14. The standard electrocardiogram lead I consists of electrodes recording the potential difference between the _____ (A arms; L, legs; AL, arms and legs) so arranged as to normally yield an upright deflection of the P wave when the right extremity is _____ (P, positive; N, negative) with respect to the other.

   a. A, P   c. AL, P   e. L, N
   b. L, P   d. A, N   f. AL, N

15. The standard electrocardiogram lead II consists of electrodes on the _____ (C, chest; R, right arm; L, left arm) and left leg, arranged so as to yield an upright deflection when the left leg potential is _____ (P, positive; N, negative) with respect to the other electrode.

   a. C, P   c. L, P   e. R, N
   b. R, P   d. C, N   f. L, N

16. In lead III electrocardiogram recordings, the _____ is grounded and the potential difference is recorded between _____ . (RA, right arm; LA, left arm; RL, right leg; LL, left leg)

   a. RL, LL & RA   d. LL, RL & RA
   b. RL, LL & LA   e. LL, RL & LA
   c. RL, LA & RA   f. LL, LA & RA

OBJECTIVE 15–6.

Identify the loci within the frontal plane forming Einthoven's triangle. Recognize Einthoven's law.

17. The derivation of the genesis of lead II electrocardiograms from simultaneous recordings of leads I and III is described by:

a. The Frank-Starling law
b. The law of Laplace
c. Wilson's law
d. Marey's law
e. Einthoven's law
f. None of the above

18. With standard limb lead configurations, lead I records a positive potential of positive 0.5 mv at the same instant that lead III records a positive 0.7 mv potential. The corresponding potential as recorded in lead II would be _____ mv.

a. +1.2   c. +0.2   e. +0.6
b. −1.2   d. −0.2   f. −0.6

OBJECTIVE 15–7.

Identify the chest or precordial leads, and recognize their potential utility.

19. Electrocardiogram leads designated V_1, V_2, and V_3 would refer to:

a. Bipolar standard limb leads
b. Unipolar limb leads
c. Bipolar chest leads
d. Augmented unipolar limb leads
e. Unipolar chest leads
f. Apices of Einthoven's triangle

20. Normally, in changing from leads V_1 through V_6, V_6 lying closer to the _____ (R, right, L, left) axillary region, the R wave would change progressively from an _____ (U, upright; I, inverted) deflection.

a. R, I to U   c. L, I to U
b. R, U to I   d. L, U to I

21. In leads V_4, V_5, and V_6, the QRS complexes are mainly _____ (N, negative; P, positive) because the chest electrode in these leads is near the apex, which is in the direction of _____ (EN, electronegativity; EP, electropositivity) during ventricular depolarization.

a. N, EN   c. P, EN
b. N, EP   d. P, EP

22. When a repolarization wave approaches a precordial lead electrode, the recorded potential deflection will be _____ (P, positive; N, negative) with respect to the precordial reference or _____ (L, left leg; I, indifferent; R, right arm) electrode.

a. P, L   c. P, R   e. N, I
b. P, I   d. N, L   f. N, R

OBJECTIVE 15–8.

Identify the augmented unipolar limb leads. Recognize the recording polarity convention for aV_R, aV_L, and aV_F.

23. In contrast with standard limb lead recordings, augmented limb lead recordings which are typically recorded inverted include:

a. Only aV_F
b. Only aV_F & a V_L
c. Only aV_F, a V_L, & aV_R
d. Only aV_R
e. Only aV_R & aV_F
f. Neither aV_R, aV_F, nor aV_L

24. An upright deflection in the aV_F lead signifies that the _____ of the heart is electronegative with respect to the _____ of the heart. (L, left side; R, right side; A, apex; B, base)

a. L, R   c. A, B
b. R, L   d. B, A

16

Electrocardiographic Interpretation in Cardiac Myopathies — Vectorial Analysis

OBJECTIVE 16–1.

Recognize the directional and magnitude components of a vector quantity, the transposition of vectors in space for vector addition or subtraction, the utility of representation of a summated cardiac vector in the heart as a function of time, and identify the polarity and zero reference direction conventions utilized for cardiac vector representation.

1. By convention, the length of the electrocardiagram vector is proportional to the _____ (D, direction of rotation; V, voltage generated), and the arrow points to the _____ (P, positive; R, reference zero; N, negative) direction.

    a. D, P   c. D, N   e. V, R
    b. D, R   d. V, P   f. V, N

2. A cardiac vector extending in the direction of +10 degrees is one rotated 10 degrees _____ (C, clockwise; CC, counterclockwise) from the individual's _____ (R, right horizontal; L, left horizontal; UV, upper vertical; LV, lower vertical) reference zero.

    a. C, R   c. C, UV   e. CC, L
    b. C, L   d. CC, R   f. CC, LV

3. In a normal heart the average direction of the cardiac vector during ventricular depolarization is about _____ degrees.

    a. 0   c. −10   e. −59
    b. +10   d. +59   f. −175

OBJECTIVE 16–2.

Identify the direction of the "recording axis" of the standard bipolar and augmented unipolar leads.

Directions: Match the lettered answer with the numbered descriptive phrase.

4. _____ The recording axis direction of lead I.
5. _____ The recording axis direction of lead II.
6. _____ The recording axis direction of lead III.

7. _____ The recording axis direction of aV_R.
8. _____ The recording axis direction of aV_L.
9. _____ The recording axis direction of aV_F.

a. −30 degrees   c. 60 degrees   e. 120 degrees
b. 0 degrees   d. 90 degrees   f. 210 degrees

OBJECTIVE 16–3.

Identify the graphical technique of vector component analysis. Construct a single frontal plane vector that represents instantaneous voltages obtained from the standard bipolar or augmented unipolar leads.

10. The average direction of the vector of the heart during ventricular depolarization is _____ to the long axis of the heart. A recording axis _____ to this vector would yield the correspondingly greater recorded voltage. (P, parallel; T, perpendicular)

    a. P, P   c. T, P
    b. P, T   d. T, T

11. A vector of 60 degrees would yield the largest resultant vector component in:

    a. Lead I   c. Lead III   e. Lead aV_F
    b. Lead II   d. Lead aV_L f. Lead aV_R

12. Simultaneous recordings of leads I, II, and III reveal that component vector amplitudes are negative in leads I and II and positive in lead III. The direction of the summated vector representation in the heart is therefore between _____ degrees.

    a. 0 and 60   c. 30 and 150   e. 90 and 180
    b. 0 and 120   d. 90 and 120   f. 150 and 210

13. When the R wave is absent in lead III, and both positive and of equal amplitude in leads I and II, the corresponding direction of the summated vector representation in the heart would be:

    a. +30 degrees   d. −60 degrees
    b. +120 degrees   e. −30 degrees
    c. +150 degrees   f. Absent

14. The magnitude of a component vector in lead I may be graphically obtained from a principal vector representation in the heart by drawing lines from the ends of the principal vector to the lead _____ axis. These lines are constructed to be at right angles with the _____ (L, lead axis; V, principal vector representation in the heart).

    a. I, L   c. aV_F, L
    b. I, V   d. aV_F, V

OBJECTIVE 16–4.

Recognize the sequential variance of relative amplitudes and directions of vector representation during atrial depolarization and repolarization, and ventricular depolarization and repolarization. Identify the vectorcardiogram, a means of its recording, and its constituent P, QRS, and T vectorcardiogram components.

15. Towards the end of ventricular depolarization, the vector direction typically shifts towards the _____ (R, right; L, left) side because the right ventricle depolarizes slightly more _____ (S, slowly; P, rapidly) than the left.

    a. R, S   c. L, S
    b. R, P   d. L, P

16. The maximum amplitude of the QRS vector is reached when the ventricles are about _____ % _____ (D, depolarized; R, repolarized).

    a. 25, D   c. 75, D   e. 100, R
    b. 50, D   d. 100, D   f. 50, R

17. A low amplitude vector in the apex to base or left to right direction during the spread of depolarization in the septum is best related to the existence of a standard electrocardiogram _____ (P, positive; N, negative) _____ wave.

    a. P, R   c. P, Q   e. N, S
    b. P, S   d. N, T   f. N, Q

18. The average vector direction of current flow during ventricular repolarization is normally in _____ (S, the same; O, an opposite) direction as that occurring for ventricular depolarization, as the first portions of the ventricles to depolarize are the _____ (F, first; L, last) to repolarize.

    a. S, F   c. O, F
    b. S, L   d. O, L

19. Atrial repolarization originating near the _____(A, A–V node; SA, S–A node; L, left atrium) has a vector representation which is generally in the _____(S, same; O, opposite) direction as that occurring for atrial depolarization.

   a. A, S   c. L, S   e. SA, O
   b. SA, S   d. A, O   f. L, O

20. The absence of Q and S components of the electrocardiogram in lead I recordings implies that the QRS vectorcardiogram fails to extend within the range of _____ degrees.

   a. 0 to 90   c. 90 to 270   e. 60 to 180
   b. 0 to 180   d. 90 to 360   f. 180 to 360

21. The elliptical figure generated by the _____ (U, positive; N, negative) ends of the vector sequence resulting from ventricular depolarization is called the _____ vectorcardiogram.

   a. U, QRS   c. U, T   e. N, P
   b. U, P   d. N, QRS   f. N, T

OBJECTIVE 16–5.

Identify the mean electrical axis of the ventricles and its average direction. Approximate the mean electrical axis from a standard three-lead electrocardiogram.

22. After subtracting the negative portion from the QRS complex in the standard electrocardiogram leads, the remaining net component vectors are negative in leads I and II and positive in lead III. The mean electrical axis of the ventricles therefore lies between _____degrees.

   a. 0 and 60   c. 30 and 150   e. 90 and 180
   b. 0 and 120   d. 90 and 120   f. 150 and 210

23. The predominant direction of current flow during ventricular _____(R, repolarization; D, depolarization) is the direction of the mean electrical axis and averages about _____ degrees.

   a. D, 10   c. D, 159   e. R, 59
   b. D, 59   d. R, 10   f. R, 159

OBJECTIVE 16–6.

Recognize normal variations in the direction of the mean electrical axis. Identify axis deviation. Recognize the effects of anatomical position, ventricular hypertrophy, bundle branch blocks, and muscular destruction on axis deviation.

24. Normal variations of the direction in the mean electrical axis of the ventricle are generally considered to range from about _____

   a. 0 to 20°   c. 70 to 110°   e. 100 to 210°
   b. 20 to 40°   d. 20 to 100°   f. ±90° from
                                         the average

25. The net magnitude of the QRS complexes in the electrocardiogram standard leads are: lead I = +3.0; lead II = −0.5; and lead III = −3.0. The mean electrical axis would indicate a:

   a. Right axis deviation
   b. Left axis deviation
   c. Value within the normal range
   d. Lack of an electrical axis

26. A severe right electrical axis deviation is more likely to produce an inverted R wave with the largest amplitude in lead:

   a. aV_F c. I
   b. II   d. III

27. A left axis deviation may result from a _____ (R, right; L, left) bundle branch block as a consequence of a _____ (G, greater; S, lesser) transmission velocity for ventricular muscle as compared to Purkinje fibers.

   a. R, G   c. L, G
   b. R, S   d. L, S

28. Increased pulmonary vascular resistance may cause _____ (R, right; L, left) axis deviation as a result of _____ (A, atrial; V, ventricular) muscle hypertrophy.

a. R, A
b. R, V
c. L, A
d. L, V

29. When an individual stands up there is a typical tendency of the heart to shift to the _____ and the mean electrical axis of the heart to shift to the _____ (L, left; R, right)

a. L, L
b. L, R
c. R, L
d. R, R

OBJECTIVE 16–7.

Identify the effects of increased or reduced cardiac muscle mass and alterations in surrounding fluid on recorded electrocardiogram voltage amplitudes.

30. Increased _____ (V, ventricular muscle mass only; P, pericardial fluid only; B, both V and P) tend(s) to _____ (D, decrease; I, increase) standard lead voltages by "short circuiting" potentials generated by the heart.

a. V, I
b. P, I
c. B, I
d. V, D
e. P, D
f. B, D

31. Borderline, normal–abnormally high QRS complex voltages occur when the _____ (A, average; S, sum) of the voltages of the QRS complexes of the standard leads is greater than _____ mv.

a. A, 0.5
b. A, 4
c. A, 9
d. S, 0.5
e. S, 4
f. S, 9

OBJECTIVE 16–8.

Recognize cardiac hypertrophy or dilatation, Purkinje system blocks, and destruction of cardiac muscle as potential causes of abnormal QRS complexes.

32. Block of the Purkinje fibers supplying the left ventricle would _____ (D, decrease; N, not alter; I, increase) the duration of the QRS complex and result in a _____ (R, right; L, left) axis deviation.

a. D, R
b. D, L
c. N, R
d. N, L
e. I, R
f. I, L

33. Double or even triple peaks in the QRS complex in some of the standard electrocardiographic leads might best be indicative of:

a. Atrial fibrillation
b. Increased vagal tone
c. Currents of injury
d. "Atrial T waves"
e. Localized ventricular blocks
f. Tachycardia

OBJECTIVE 16–9.

Identify "current of injury;" its potential causes, resultant electrocardiogram changes, and the method of determination of its vector.

34. Ischemia of local areas of cardiac muscle caused by coronary thrombosis results in localized partial _____ (D, depolarization; H, hyperpolarization) of cell membranes and localized extracellular _____ (EP, electropositivity; EN, electronegativity).

a. D, EP
b. D, EN
c. H, EP
d. H, EN

36. The "J point" reference, utilized in determining the electrical axis of a current of injury, corresponds to complete ventricular _____ (P, polarization; D, depolarization) and occurs at the _____ (B, beginning; E, end) of the QRS complex.

a. P, B
b. P, E
c. D, B
d. D, E

35. S–T deviations from the T–P segment of the electrocardiogram represent a classical pattern of:

a. Sympathetic stimulation
b. Parasympathetic stimulation
c. Ectopic beats
d. Electrical axis deviation
e. Myocardial injury
f. Cardiac hypertrophy

37. During the T–P interval it is determined that vector components of the current of injury are positive in V_2, negative in leads II and III, and absent in lead I. A possible explanation of these recordings would be an infarction localized near the _____ (B, base; X, apex) of the heart on the _____ (A, anterior; P, posterior) wall of the left ventricle.

a. B, A
b. B, P
c. X, A
d. X, P

OBJECTIVE 16–10.

Recognize slowed propagation and prolongation of ventricular depolarization as potential causes of T wave abnormalities.

38. When propagation of the depolarization wave through the ventricles is greatly delayed, the _____ wave generally becomes _____ (I, inverted; E, erect).

 a. P, I   c. T, I   e. R, E
 b. R, I   d. P, E   f. T, E

39. A greatly delayed propagation of the depolarization wave through the ventricles generally results in an abnormal condition, whereby the _____ wave is of the _____ (S, same; O, opposite) polarity as the QRS complex.

 a. P, S   c. P, O
 b. T, S   d. T, O

40. One method for the detection of coronary insufficiency rests upon abnormalities occurring in the _____ wave, resulting from _____ (I, increased; D, decreased) periods of depolarization of ischemic ventricular muscle following provocative exercise.

 a. P, I   c. T, I   e. R, D
 b. R, I   d. P, D   f. T, D

41. Localized cooling of the apex and epicardial surfaces of the ventricles is noted to result in inversion of the T wave in an electrocardiogram lead. This may best be attributed to _____ (I, increased; D, decreased) action potential durations at the epicardial surface causing its repolarization _____ (B, before; A, after) that of the corresponding endocardial surface.

 a. I, B   c. D, B
 b. I, A   d. D, A

17

Electrocardiographic Interpretation of Cardiac Arrhythmias

OBJECTIVE 17–1.

Identify tachycardia, bradycardia, and sinus arrhythmia. Recognize potential causes of abnormal sinus arrhythmias.

1. A fast heart rate or _____ (B, bradycardia; T, tachycardia) is generally considered to occur when the heart rate exceeds _____ beats/minute.

 a. B, 60   c. B, 150   e. T, 100
 b. B, 100   d. T, 60   f. T, 150

2. Moderate fevers generally result in a _____ (B, bradycardia; T, tachycardia) as the heart rate changes about _____ beats per minute for each degree F change in body temperature.

 a. B, 1   c. B, 20   e. T, 10
 b. B, 10   d. T, 1   f. T, 20

3. An athlete subjected to endurance training generally has a resting heart rate _____ (G, greater; L, less) than 60 beats per minute as a result of increased _____ (S, sympathetic; P, parasympathetic) tone.

 a. G, S   c. L, S
 b. G, P   d. L, P

4. The normal respiratory cycle is the most common cause of _____ (B, bradycardia; S, sinus arrhythmia), typically resulting in a _____ % variance from the average heart rate.

 a. B, 5   c. B, 50   e. S, 25
 b. B, 25   d. S, 5   f. S, 50

OBJECTIVE 17–2.

Identify sino-atrial blocks; first, second, and third degree atrioventricular blocks, the Stokes-Adams syndrome; and intraventricular blocks. Recognize potential causes and the utility of electrocardiogram recordings in their diagnosis.

5. First degree incomplete heart blocks would be associated with:

   a. S–T deviations   e. Inverted T waves
   b. Split R waves   f. a wave
   c. Right axis deviation   disappearance
   d. Increased P–R intervals

6. Cyclic heart blocks incorporating the elements of periodic fainting spells, ventricular "escape," and borderline cardiac ischemia, are known as:

   a. Cheyne-Stokes syndrome
   b. Einthoven's syndrome
   c. Stokes-Adams syndrome
   d. Sinus arrhythmia
   e. First degree A–V blocks
   f. Second degree A–V blocks

7. A P–R interval of 0.35 second would be considered:

   a. Normal   e. An intraven-
   b. A first degree block   tricular block
   c. A second degree block   f. A sino-atrial
   d. A third degree block   block

8. An electrocardiogram showing a consistent rhythmical ratio of three P waves to each QRS complex indicates:

   a. Sinus arrhythmia
   b. Ventricular escape
   c. S–A node block
   d. First degree A–V block
   e. Second degree A–V block
   f. Third degree A–V block

9. Conditions which increase the rate of propagation between the atria and ventricle include:

   a. A–V junctional fiber ischemia
   b. A–V bundle compression
   c. A–V bundle inflammation
   d. Vagal stimulation
   e. Baroreceptor stimulation
   f. Sympathetic stimulation

OBJECTIVE 17–3.

Recognize the potential existence of atrial, A–V nodal or A–V bundle, and ventricular premature beats. Identify the terms pulse deficit and bigeminal pulse. Identify the loci of ectopic premature beats from appropriate electrocardiogram recordings.

10. An electrocardiogram sequence shows a premature beat with a normal P wave, QRS complex, and T wave but the P–R interval and the interval between the preceding beat and the premature beat is shortened. The premature beat probably had its origin in the:

   a. A–V node   e. Base of the right
   b. A–V bundle   ventricle
   c. Atria   f. Ventricular apex
   d. Base of the left
      ventricle

11. An electrocardiogram sequence shows a premature beat with a QRS–T complex which is comparatively normal, except that the P wave is superimposed on the QRS–T complex. The probable locus of the premature beat is the:

   a. A–V node or   e. Base of the right
      bundle   ventricle
   b. Atria   f. Ventricular apex
   c. Base of the left
      ventricle
   d. S–A node

12. Ventricular premature beats are generally characterized by QRS complexes of _____ duration and _____ amplitude. (D, decreased; N, normal; I, increased)

a. D, D   c. N, N   e. I, N
b. D, I   d. I, D   f. I, I

13. Ectopic stimuli applied to the base of the right ventricle and the apex of the left ventricle would generally produce lead II electrocardiograms that would, for both situation, involve:

a. Upright R waves   d. Upright T waves
b. Inverted R waves   e. Inverted T waves
c. Opposing R & T   f. Large Q & S waves
     polarities

14. Premature atrial beats more readily produce _____ (E, electrical alternans; P, pulse deficits) as a consequence of _____(A, absolute refractory periods; D, decreased end-diastolic volume).

a. E, A   c. P, A
b. E, D   d. P, D

OBJECTIVE 17-4.

Identify paroxysmal tachycardia. Recognize atrial, A–V nodal, or ventricular origins of paroxysmal tachycardia and their associated electrocardiograms. Identify the electrocardiogram characteristics of atrial flutter, atrial fibrillation, and ventricular fibrillation.

15. Lead II electrocardiogram recordings would generally contain _____ P wave amplitudes in atrial flutter, and _____ P wave amplitudes in atrial fibrillation. (I, increased; D, decreased or absent)

a. I, I   c. D, I
b. I, D   d. D, D

16. An extremely bizarre electrocardiogram pattern with an absence of a rhythm of any type occurs only in _____ (P, paroxysmal tachycardia; F, atrial flutter; AF, atrial fibrillation; VF, ventricular fibrillation)

a. P & F   c. F, AF, & VF   e. P
b. AF & VF   d. P, F, AF, &   f. VF
              VF

17. Conditions which are usually serious because they more frequently initiate ventricular fibrillation include_____ (A, atrial; V, ventricular)_____ (PT, paroxysmal tachycardia; B, fibrillation; F, flutter).

a. A, F   c. V, PT
b. A, B   d. A, PT

18. Ocular pressure or pressure over the carotid sinus may be of use in terminating _____ (F, fibrillation; P, paroxysmal tachycardia) by means of _____ (I, increased; D, decreased) vagal activity.

a. F, I   c. P, I
b. F, D   d. P, D

18

Physics of Blood, Blood Flow, and Pressure: Hemodynamics

OBJECTIVE 18-1.

Recognize the two major subdivisions of the circulatory system and the implications of a closed circulatory circuit. Identify hematocrit, average values, its method of determination, and the influence of hematocrit on blood viscosity. Identify the classes of plasma proteins, their concentrations, and general functional significance.

1. The hematocrit, typically _____ %, represents the volume ratio of formed elements to _____ (B, blood; P, plasma).

   a. 20, B   c. 60, B   e. 40, P
   b. 40, B   d. 20, P   f. 60, P

2. Blood viscosity _____ as hematocrit increases, and _____ as velocity of flow increases. (I, increases; D, decreases)

   a. I, I   c. D, D
   b. I, D   d. D, I

3. Plasma contains about _____ % protein, whereas interstitial fluid contains an average of _____ % protein.

   a. 45, 4.5   c. 2.5, 0.3   e. 7, 2
   b. 7, 0.3   d. 45, 7   f. 2.5, 2

4. The major portion of the plasma colloid osmotic pressure results from plasma _____ whereas protection against infection results from the presence of plasma _____ . (F, fibrinogen; G, globulins; A, albumin)

   a. F, G   c. G, F   e. F, A
   b. A, G   d. G, G   f. A, F

5. The viscosity of blood, typically _____ times that for water, is about _____ that of plasma.

   a. ½ to 1, ½   d. ½ to 1, 4 times
   b. 3 to 4, 4 times   e. 3 to 4, 2 times
   c. 10, 2 times   f. 10, 10 times

OBJECTIVE 18–2.

Identify blood flow, its typical value, characteristics of laminar versus turbulent flow, and the interrelationships of blood flow to pressure and resistance. Recognize techniques utilized to measure blood flow.

6. The quantity of blood that passes a given point in the systemic circulation per unit of time, or blood _____ (V, velocity; F, volume flow), is typically _____ in the resting adult.

   a. V, 45 cm./min.   d. F, 45 cm./sec.
   b. V, 20 liters/sec.   e. F, 20 liters/min.
   c. V, 5 cm./sec.   f. F, 5 liters/min.

7. Blood volume flow is _____ proportional to the pressure difference and _____ proportional to the flow resistance. (I, inversely; D, directly)

   a. I, I   c. D, I
   b. I, D   d. D, D

8. The product of ventricular stroke volume, heart rate, and peripheral resistance most closely approximates the:

   a. Mean pressure gradient d. Pulse pressure
   b. Cardiac index   e. Stroke index
   c. Ventricular efficiency   f. Ventricular
                                  work

9. Increased systemic circulatory pressure differences may be associated with a decreased peripheral resistance when there is:

   a. Turbulence   e. Increased cardiac
   b. Increased hematocrit   output
   c. Vasoconstriction   f. Decreased cardiac
   d. Vasodilation   output

10. The instrument utilized for quantitating blood flow by measuring the rate of change of vascular volume during venous occlusion is the:

    a. Ultrasonic flowmeter   d. Rotameter
    b. Plethysmograph
    c. Electromagnetic flowmeter

11. At _____ (H, higher; L, lower) Reynolds' numbers, a laminar flow velocity in the axial center of a blood vessel has _____ (G, a greater; S, the same; O, a lower) velocity as compared to the vessel periphery.

    a. H, G   c. H, O   e. L, S
    b. H, S   d. L, G   f. L, O

12. The tendency for turbulent flow increases with decreased:

    a. Blood velocity   d. Blood viscosity
    b. Reynolds' numbers   e. Vessel diameter
    c. Blood density   f. Blood volume flow

13. *Eddy currents* present in _____ (T, turbulent; L, laminar) flow of blood result in _____ (I, increased; N, no appreciable change in; D, decreased) flow resistance.

    a. T, I   c. T, D   e. L, N
    b. T, N   d. L, I   f. L, D

OBJECTIVE 18-3.

Define pressure and pressure difference or gradient. Identify appropriate methods and standard units for blood pressure measurement.

14. Blood pressure, signifying force per unit of _____ (L, length; A, area) is most generally measured in units of _____ (AT, atmospheres; M, mm. Hg, LB, lbs./in.^{2}).

   a. L, AT   c. L, LB   e. A, M
   b. L, M   d. A, AT   f. A, LB

15. A pressure of 1 mm. Hg is sufficient to raise a column of water to a vertical height of:

   a. 0.68 cm.   c. 2.72 cm.   e. 13.6 cm.
   b. 1.36 cm.   d. 6.8 cm.   f. 760 mm.

OBJECTIVE 18-4.

Identify resistance and conductance, their means of determination, the peripheral resistance unit (PRU), and typical values of total peripheral resistance and total pulmonary resistance.

16. With a pressure gradient of 100 mm. Hg and a flow of 6 liters/minute, the flow resistance would approximate _____ PRU, a value _____ (H, considerably higher than; E, about equal to; L, considerably lower than) the typical total peripheral resistance of the systemic circulation at rest.

   a. 0.06, L   c. 16, H   e. 1, E
   b. 1, L   d. 0.06, E   f. 16, E

17. With a typical cardiac output and mean values of pulmonary arterial pressure of 13 mm. Hg and left atrial pressure of 4 mm. Hg, the total peripheral resistance would more closely approximate:

   a. 0.01 PRU   c. 10 PRU
   b. 0.1 PRU   d. 100 PRU

18. Blood flow per increment of pressure difference equals the _____ or the reciprocal of the _____ (C, conductance; R, resistance; Q, cardiac output; X, cardiac index)

   a. X, R   c. C, R   e. Q, C
   b. Q, R   d. X, C   f. R, C

OBJECTIVE 18-5.

Identify the relationship of blood flow to vascular cross-sectional area and the mean velocity, and the significance of this relationship for a closed circulatory system.

19. The mean velocity of blood flow is _____ proportional to the total vascular cross-sectional area, and _____ proportional to the blood flow. (D, directly; I, inversely)

   a. D, D   c. I, D
   b. D, I   d. I, I

20. An individual with a cardiac output of 6 liters/minute, a blood volume of 5.5 liters, a mean systemic arterial blood velocity of 20 cm./second, and a mean systemic capillary flow velocity of 0.05 cm./second, would have a total systemic capillary network cross-sectional area of about _____ cm^2.

   a. 300   c. 25   e. 1,100
   b. 3,000   d. 2,000   f. 2.5

21. The aortic cross-sectional area is determined to be 4.5 cm.^{2}, the mean aortic pressure 100 mm. Hg, and the mean aortic flow velocity 40 cm./second. The vena cavae, with a 10 cm.^{2} cross-sectional area and 0 mm. Hg mean pressure, consequently has a mean velocity of flow of _____ cm./second.

   a. 2.5   c. 10   e. 160
   b. 18   d. 0.04   f. 45

22. The quantity of blood that will flow through a vessel in a given period of time is equal to _____. (V = mean velocity; R = radius; L = length; P = pressure gradient)

   a. $V\pi R^2$ c. RP   e. $V(R)^4$
   b. VP   d. $V/(R)^4$ f. None of the above

OBJECTIVE 18–6.

Identify the relationship of vascular conductance to vessel diameter, vessel length, and blood viscosity. Identify Poiseuille's law and its significance.

23. When the pressure gradient, vessel length, and blood viscosity remain constant, then blood flow will vary _____ (D, directly; I, inversely) with the vessel diameter to the _____ (1, first; 2, second; 3, third; 4, fourth) power.

    a. D, 4   c. D, 2   e. I, 1
    b. D, 3   d. D, 1   f. I, 4

24. The resistance to blood flow is _____ proportional to blood viscosity, and _____ proportional to vessel length. (D, directly; I, inversely)

    a. D, D   c. I, D
    b. D, I   d. I, I

25. Poiseuille's law predicts that the blood flow varies in direct proportion to the _____ and in inverse proportion to the _____. (P, pressure difference; V, viscosity; R, vessel radius; and L, vessel length)

    a. R^4P, VL   c. VP, LR   e. VL, R^2
    b. P, R^2VL   d. VL, R^4 f. VPL, R^2

OBJECTIVE 18–7.

Calculate the total resistance to flow for "series" or "parallel" arrangements of vascular resistance.

26. The total peripheral resistance of the systemic or peripheral circulation, typically _____ PRU at rest in the adult, is _____ (L, less; M, more) than the peripheral resistance of an average organ.

    a. 0.25, L   c. 4.0, L   e. 1.0, M
    b. 1.0, L   d. 0.25, M   f. 4.0, M

27. The total resistance of vessels arranged in _____ (S, series; P, parallel) is _____ (E, equal to; I, equal to the reciprocal of) the sum of the vascular conductances.

    a. S, E   c. P, E
    b. S, I   d. P, I

28. The total peripheral _____ equals the _____ of the arteries, plus that of the arterioles, plus that of the capillaries, plus that of the venous vessels. (C, conductance; R, resistance)

    a. C, C   c. R, C
    b. C, R   d. R, R

OBJECTIVE 18–8.

Recognize the distensible nature of blood vessels, the resultant nonlinear nature of blood flow versus pressure gradients, and the corresponding influence of varied sympathetic tone. Identify critical closing pressure, its probable mechanisms, its typical value, and the corresponding influence of varied sympathetic tone.

29. The conductance of an artery is _____ as the mean arterial pressure increases as a result of _____ vascular diameter. (I, increased; C, a constant; D, decreased)

    a. I, I   c. C, C   e. D, I
    b. I, D   d. C, I   f. D, D

30. The average critical closing pressure is about _____ mm. Hg and would be significantly _____ (H, higher; L, lower) if blood were temporarily replaced with plasma.

    a. 0, H   c. 20, H   e. 20, L
    b. 7, H   d. 7, L   f. 45, L

31. Sympathetic_____ (B, inhibition; S, stimulation) generally increases the blood flow through a tissue for a given pressure gradient and tends to _____ (I, increase; D, decrease) the critical closing pressure.

   a. B, D   c. S, D
   b. B, I   d. S, I

32. Two significant factors contributing to a cessation of blood flow when the arterial pressure is decreased include the size of red blood cells and:

   a. Frank-Starling's law   e. Einthoven's law
   b. Reynold's law   f. Turbulent flow
   c. The law of Laplace
   d. Starling's hypothesis

OBJECTIVE 18–9.

Identify the law of Laplace and its physiological significance.

33. The distending force, tending to stretch the muscle fibers in the vascular wall at a given pressure, is proportional to _____ . Poiseuille's law predicts that, with a constant pressure gradient across the ends of a vessel, the flow through a vessel with laminar flow is proportional to _____ . (R = radius)

   a. R, R^2 c. R^3, R e. $1/R, 1/R^4$
   b. R, R^4 d. $1/R, 1/R^2$ f. $1/R^3, 1/R$

34. The law of_____ (P, Poiseuille; L, Laplace) implies that the circumferential force tending to stretch the vascular wall is _____ (D, directly; I, inversely) proportional to the vessel diameter.

   a. P, D   c. L, D
   b. P, I   d. L, I

35. The aortic wall strength (radius = 1 cm.) required to support a mean aortic pressure of 100 mm. Hg is compared to the wall strength of a capillary (radius = 4 microns) supporting a pressure of 25 mm. Hg. The ratio of required wall strength for the capillary to the aortic wall strength would be about:

   a. 1/4   c. 1/2,500
   b. 1/625   d. 1/10,000

36. The law of _____ (L, Laplace; S, Starling) makes clear a situation found in dilated hearts. When the radius of a cardiac chamber is increased, _____ (LS, less; G, greater) tension must be developed in the myocardium to produce a given pressure.

   a. L, LS   c. S, LS
   b. L, G   d. S, G

OBJECTIVE 18–10.

Identify vascular distensibility and compliance. Contrast the pressure-volume relationships of the arterial and venous systems and their functional significance. Identify the effects of varied sympathetic tone on these relationships.

37. The fractional increase in vessel volume for each mm. Hg rise in pressure, or vascular _____(D, distensibility; C, compliance), equals the vessel capacitance _____ (M, multiplied; R, divided) by vessel volume.

   a. D, M   c. C, M
   b. D, R   d. C, R

38. Distensibility is least for the:

   a. Pulmonary arterial segment
   b. Pulmonary venous segment
   c. Systemic arterial segment
   d. Systemic venous segment

39. The ratio of compliances of the venous system to the arterial system, about equal to _____ is _____ (G, greater; L, less) than the corresponding ratio of their distensibilities.

   a. 3, G   c. 24, G   e. 8, L
   b. 8, G   d. 3, L   f. 24, L

40. A loss of 25% of the total circulatory blood volume is compensated for through _____ (I, increased; D, decreased) sympathetic tone on the major "storage areas" of the circulation or _____(A, arterial; V, venous; C, capillary) system.

   a. I, A   c. I, C   e. D, V
   b. I, V   d. D, A   f. D, C

OBJECTIVE 18–11.

Identify mean circulatory pressure or circulatory filling pressure, its significance, its relationship to sympathetic tone, and a typical value. Contrast systemic versus pulmonary filling pressures.

41. The _____ (C, critical closing pressure; M, mean circulatory pressure) would be the pressure found in the aorta immediately following ventricular fibrillation and would approximate _____ mm. Hg.

 a. C, 0   c. C, 20   e. M, 7
 b. C, 7   d. M, 0   f. M, 20

42. Right heart failure _____ the systemic filling pressure and _____ the pulmonary filling pressure. (I, increases; D, decreases)

 a. I, I   c. D, I
 b. I, D   d. D, D

43. Increased sympathetic tone would _____ the circulatory filling pressure, and would tend to _____ venous return. (I, increase; N, not appreciably alter; D, decrease)

 a. I, I   c. D, I   e. N, D
 b. I, D   d. D, D   f. N, N

44. The mean circulatory pressure is significant in that it reflects the relationship between vascular:

 a. Venous & arterial pressures
 b. Capacity & blood volumes
 c. Systemic & pulmonary volumes
 d. Distensibility & compliance
 e. Resistance & capacitance
 f. Flow & resistance

OBJECTIVE 18–12.

Identify delayed compliance or "stress-relaxation" of vessels, and its relative significance for arterial versus venous segments of its circulation.

45. Delayed compliance is a phenomenon whereby a rapid reduction in vascular pressure, resulting from a reduction in vascular volume, results in a subsequent increase in:

 a. Smooth muscle length   d. Compliance
 b. Vascular volume
 c. Smooth muscle tonus

46. Delayed compliance is most significant and occurs to a much greater extent in:

 a. Elastic arteries   e. Precapillary
 b. Muscular arteries   sphincters
 c. Arterioles   f. Veins
 d. Arteriovenous anastomoses

19

The Systemic Circulation

OBJECTIVE 19–1.

Recognize the general structural design and functions of the systemic and pulmonary circulations. Identify the component functions of the aorta and elastic arteries, muscular arteries, arterioles, capillaries, venules, and veins.

1. _____ act as a high pressure transport system to tissues whereas _____ serve as a divergent vascular network acting as "control valves" for blood distribution. (A, arteries; V, veins; VU, venules; AO, arterioles)

 a. A, VU   c. A, AO   e. V, AO
 b. A, V   d. V, VU   f. V, A

2. _____ are the loci of fluid and nutrient exchange between blood and interstitial spaces whereas the _____ serve as a convergent vascular network for blood returning from tissues. (L, lymphatics; V, veins; C, capillaries; A, arterioles; VU, venules)

 a. L, V   c. V, VU   e. C, VU
 b. L, VU   d. V, A   f. C, A

3. A synonym for the pulmonary circulation is the:

    a. Greater circulation   c. Peripheral circulation   e. Venous system
    b. Systemic circulation   d. Arterial system   f. None of the above

OBJECTIVE 19–2.

Identify the relative percentage distribution of blood volume in various segments of the cardiovascular system.

4. Of the following cardiovascular segments, the least circulating blood volume is contained within:

    a. Capillaries   d. Small veins & venules
    b. Arteries   e. The aorta
    c. The heart   f. Arterioles

5. The cardiovascular segment containing the highest percentage of the circulating blood volume is the:

    a. Venous system   c. Heart & arteries
    b. Capillaries   d. Liver & spleen

6. About 84% of the entire circulating blood volume lies in the _____ segment of the circulation whereas 7% lies in the _____ segment. (P, pulmonary; S, systemic; V, venous reservoir; A, arterial; H, heart)

    a. P, H   c. S, A   e. V, A
    b. P, A   d. S, H   f. V, H

OBJECTIVE 19–3.

Identify the term "total cross-sectional area" and representation values for the aorta, capillaries, and venae cavae. Identify the relationship of mean velocity of blood flow to total cross-sectional area, and the form of the curves relating total cross-sectional area and mean velocity to the progression of cardiovascular segments in the systemic circulation.

7. Within the systemic circulation the total cross-sectional area is greatest in the _____ and lowest in the _____ . (C, capillaries; V, veins; A, aorta; AO, arterioles; VC, venae cavae)

    a. C, A   c. V, AO
    b. V, A   d. AO, VC

8. In the systemic circulation, the average blood velocity is highest in the _____ and lowest in the _____ as a result of their _____ . (A, aorta; C, capillaries; V, veins; R, flow resistance; X, cross-sectional area)

    a. V, A, R   c. A, C, R   e. A, C, X
    b. A, V, R   d. C, A, X   f. C, V, X

9. The mean blood velocity at progressive segments of the cardiovascular system _____ as blood traverses the aorta, its larger primary branches, smaller secondary branches, and the arterioles, and then _____ as it passes through the venules and moves centrally towards the venae cavae. (I, increases progressively; D, decreases progressively; CI, continues to increase; CD, continues to decrease)

    a. I, CI   c. D, CD
    b. D, I   d. I, D

10. The _____ (C, capillary; A, aorta) segment of the circulation has a typical mean velocity of 33 cm./second and a cross-sectional area about _____ that of the venae cavae.

    a. C, 80 times   c. C, 10 times   e. A, 4 times
    b. C, 4 times   d. A, 25%   f. A, 40 times

OBJECTIVE 19–4.

Identify the progression of average and pulse pressures through the systemic circulation and their relationship to vascular resistances. Identify systolic, diastolic, and average pressures for the aorta, average capillary and venae cavae pressures, and the locus of the highest resistance among the cardiovascular segments.

11. The magnitude of the variance in blood pressure between systolic and diastolic values is typically _____ mm. Hg for the aorta and _____ mm. Hg for the capillaries.

    a. 100, 15   c. 15, 5   e. 40, 1
    b. 40, 15   d. 100, 1   f. 15, 1

12. The mean arterial pressure decreases _____ (M, more; L, less) than 20 mm. Hg in the large and medium-sized arteries as a result of their low _____ (F, flow resistance; D, distensibility).

    a. M, F   c. L, F
    b. M, D   d. L, D

13. Compared with other segments of the systemic circulation, arterioles have the greatest:

    a. Pulse pressure   e. Distensibility
    b. Mean pressure   f. Total cross-
    c. Flow velocity   sectional area
    d. Resistance component

14. The largest incremental decrease in average pressure occurs across the:

    a. Aorta & large arteries   e. Venules & small
    b. Small arteries   veins
    c. Arterioles   f. Large veins &
    d. Capillaries   venae cavae

15. An average pressure of 20 mm. Hg is more likely to be encountered in the _____ whereas a pressure of 5.6 cm. of water is more likely to be encountered in the _____ . (A, arterioles; C, capillaries; V, veins)

    a. A, C   c. C, A   e. V, A
    b. A, V   d. C, V   f. V, C

OBJECTIVE 19–5.

Identify a normal arterial pulse contour, the incisura, and pulse pressure. Identify the influencing factors and the resultant consequences of variations of stroke volume and arterial compliance on the pulse pressure.

16. Diastolic pressure, typically _____ mm. Hg, equals the _____ (S, systolic; A, average) pressure minus the pulse pressure.

    a. 120, S   c. 40, S   e. 80, A
    b. 80, S   d. 120, A   f. 40, A

17. The incisura occurs during the _____ phase of the aortic pressure pulse contour, and the positive dicrotic wave occurs during the _____ phase. (A, ascending; D, descending)

    a. A, A   c. D, A
    b. A, D   d. D, D

18. Increased stroke volume output tends to _____ systolic pressure and _____ pulse pressure. (I, increase; D, decrease)

    a. I, I   c. D, I
    b. I, D   d. D, D

19. A characteristic of the aorta which has the greatest influence on the maintenance of a normal diastolic pressure is its:

    a. Wide lumen   d. Smooth muscle
    b. Compliance   activity
    c. Frictional resistance e. Resonant properties
                          f. Dicrotic notch

20. _____ (I, increased; D, decreased) compliance of arterial vessels, increases systolic pressure and _____ (R, raises; L, lowers) diastolic pressure.

    a. I, R   c. D, R
    b. I, L   d. D, L

21. In a person with a high arterial pressure and normal stroke volume output, the pulse pressure would be _____ slightly due to _____ compliance. (I, increased; D, decreased)

    a. D, I   c. I, I
    b. D, D   d. I, D

22. Other circulatory factors remaining constant, pulse pressure tends to increase with increased:

    a. Ejection phase durations e. Circulatory
    b. Peripheral resistance   filling pressure
    c. Arterial distensibility   f. Arterial
    d. Heart rate   compliance

OBJECTIVE 19–6.

Identify the resultant effects of arteriosclerosis, aortic regurgitation, and patent ductus arteriosus on systolic, diastolic, and pulse pressures.

23. Arteriosclerosis, with accompanying hypertension and increased peripheral resistance, would more generally be associated with _____ systolic and _____ diastolic pressures. (I, increased; D, decreased)

   a. I, D   c. D, I
   b. I, I   d. D, D

24. The absence of the incisura in an aortic pressure pulse contour may occur in the condition of _____ (P, patent ductus arteriosus; S, arteriosclerosis; A, aortic regurgitation) and is associated with _____ (I, increased; D, decreased) pulse pressures.

   a. P, I   c. A, I   e. S, D
   b. S, I   d. P, D   f. A, D

OBJECTIVE 19–7.

Identify the pressure pulse and its mode of transmission. Identify pulse wave velocity, its variance with arterial distensibility, and typical values for the aorta, large arteries, and small arteries. Contrast pulse wave velocity with velocity of blood flow.

25. When the stroke volume is ejected into the aorta during a cardiac cycle, the increase in vascular pressure occurs _____ and the increase in vascular volume occurs _____ . (A, in the proximal aorta initially; CV, in all segments of the systemic circulation simultaneously)

   a. A, CV   c. CV, CV
   b. CV, A   d. A, A

26. The velocity of pressure pulse transmission in the aorta, about _____ (1, one fifteenth; E, equal to; 15, 15 times) the velocity of blood flow, is _____ (G, greater; L, less) than the pressure pulse velocity in the more distal arteries.

   a. 1, G   c. 15, G   e. E, L
   b. E, G   d. 1, L   f. 15, L

27. Pulse wave velocity _____ (I, increases; D, decreases) primarily as a result of decreased _____ (P, pulse pressure; C, compliance; B, blood velocity) within a vascular segment.

   a. I, P   c. I, B   e. D, C
   b. I, C   d. D, P   f. D, B

28. An average arterial pulse wave velocity for arrival at the radial pulse would be closest to:

   a. The average blood velocity . d. 80 meters/
   b. 0.1 to 0.5 meters/second   second
   c. 5 to 10 meters/second

OBJECTIVE 19–8.

Recognize possible augmentation of the peripheral arterial pulse pressure, and identify the effects of vascular resistance and distensibility on damping of the pressure pulse and the existence of capillary pulsation.

29. Peripheral arterial systolic pressures 20% higher and diastolic pressures 10% lower than corresponding central aortic values would be due to:

   a. Gravitational forces   e. The law of
   b. Reflected pulse waves   Laplace
   c. Pressure pulse damping f. Experimental
   d. Arteriolar frictional   error only
      resistance

30. Pressure applied to the anterior portion of a fingernail results in a border formed between a blanched anterior portion and a red posterior portion of the nail. Pronounced back and forward oscillation of this border would more probably occur as a result of:

   a. High heart rates   e. Vasomotion
   b. Low stroke volumes f. All of the above
   c. Dilation of arterioles
   d. High arterial distensibility

31. Factors tending to increase the degree of damping of the pressure pulse include _____ flow resistance and _____ vascular distensibility. (E, elevated; L, lowered)

    a. E, E   b. E, L   c. L, E   d. L, L

OBJECTIVE 19–9.

Identify the radial pulse, its significance, the influence of altered central pulse pressure and arterial muscle tone on the peripheral pulse, pulsus paradoxus, pulse deficits, and pulsus alternans.

32. Deep breathing resulting in synchronized variations of the radial pulse amplitude is more likely to result in:

    a. Arterial muscle spasms   e. Bigeminal pulse
    b. Pulsus paradoxus   f. Electrical
    c. Pulse deficits   alternans
    d. Pulsus alternans

33. The arrhythmia of atrial fibrillation characteristically results in:

    a. Radial pulse   c. Pulsus alternans
       disappearance   d. Pulse deficits
    b. Pulsus paradoxus

34. A weak pulse in the radial artery may be indicative of _____ (I, increased; D, decreased) stroke volumes, or strong _____ (S, sympathetic; P, parasympathetic) activity to arterial smooth muscle.

    a. I, S   c. D, S
    b. I, P   d. D, P

OBJECTIVE 19–10.

Identify the general functions and related features of the arterioles, capillaries, and veins.

35. The principal cardiovascular locus for the control of blood flow to tissues is the _____ (C, capillary; V, venule; A, arteriole) which _____ (D, does; N, does not) contain a smooth muscle layer.

    a. C, N   c. V, N   e. A, N
    b. C, D   d. V, D   f. A, D

36. The two principal types of pressure gradients across the capillary membrane are the _____ (E, electrolyte; C, colloid) osmotic and the hydrostatic pressure gradients, which act in _____ (O, opposing; S, similar) directions.

    a. E, O   c. C, O
    b. E, S   d. C, S

OBJECTIVE 19–11.

Identify the "central venous pressure," its typical value and potential range, and recognize the dependence of right atrial pressure upon the balance of factors influencing venous return versus cardiac output. Identify the effects of varying blood volume, large vessel tone, and dilation of small systemic vessels on venous return and right atrial pressure.

37. The "central venous pressure" or _____ (I, intrapleural; R, right atrial) pressure is typically _____ (–5, –5 mm. Hg; A, atmospheric pressure; 20, 20 mm. Hg).

    a. I, –5   c. I, 20   e. R, A
    b. I, A   d. R, –5   f. R, 20

38. Right atrial pressure is increased by _____ pumping ability of the heart, and _____ tendency for blood to flow from peripheral tissues back to the heart. (I, increased; D, decreased)

    a. I, I   c. D, I
    b. I, D   d. D, D

39. Right atrial pressure is generally decreased by increased:

    a. Blood volume
    b. Circulatory filling pressure
    c. Venous constriction
    d. Arterial constriction
    e. Arteriolar constriction
    f. "Central venous" pressure

40. The lower limit of right atrial pressure, about _____ mm. Hg, may be approached with factors tending to _____ (I, increase; D, decrease) the flow of blood into the heart.

    a. -30, I   c. 0, I   e. -5, D
    b. -5, I   d. -30, D   f. 0, D

OBJECTIVE 19–12.

Recognize the collapsible nature of veins and its effects on peripheral venous resistance and pressures. Identify the consequences of elevated right atrial pressure and elevated abdominal pressure on peripheral venous pressures.

41. When the intra-abdominal pressure is 20 mm. Hg, which is _____ (H, higher than; L, lower than; E, about equal to) a normal average, the lowest possible pressure in the femoral vein would be _____ mm. Hg.

    a. H, 20   c. L, 20   e. E, 0
    b. E, 20   d. H, 0   f. L, 0

42. The veins within the thorax are normally _____ (C, collapsed; N, not collapsed) as a consequence of _____ pressures.

    a. C, negative venous   d. N, colloid osmotic
    b. C, gravitational   e. N, gravitational
    c. C, intrapleural   f. N, negative chest

43. Peripheral venous pressure, typically _____ mm. Hg, _____ (I, is; N, is not) generally elevated in early stages of cardiac failure.

    a. -5, I   c. 4 to 9, I   e. 0, N
    b. 0, I   d. -5, N   f. 4 to 9, N

OBJECTIVE 19–13.

Identify the effects of an intravascular hydrostatic blood column on cardiovascular pressures. Identify the function of the venous valves, the "venous pump," and the consequences of incompetent venous valves.

44. In an uninterrupted column of water, the hydrostatic pressure _____ (I, increases; D, decreases) 1 mm. Hg for each _____ cm. below the surface of the water.

    a. I, 1.36   c. I, 76   e. D, 13.6
    b. I, 13.6   d. D, 1.36   f. D, 76

45. A pressure of -10 mm. Hg would more likely be encountered in a standing individual in the _____ (N, neck veins; D, dural sinuses of the head) as a result of their _____ (C, collapsible; NC, noncollapsible) nature and gravity.

    a. N, C   c. D, C
    b. N, NC   d. D, NC

46. The hydrostatic pressure in the veins of the feet of an erect individual standing absolutely still typically approaches a maximum of about:

    a. 90 mm. of water   c. 9 mm. Hg
    b. 90 mm. Hg   d. 9 cm. of water

47. The existence of venous valves in the forearm may be readily demonstrated by occlusion of the veins on the _____ (D, distal; C, central) side of the valve and transiently massaging the blood toward the _____ (H, heart; P, periphery), thereby allowing blood to return to the valve.

    a. D, H   c. C, H
    b. D, P   d. C, P

48. A stated value of 100 mm. Hg for the mean arterial pressure in the femoral artery implies that the mean femoral arterial pressure is about _____ mm. Hg at the hydrostatic level of the _____ (F, femoral artery; H, heart) in a typical individual standing erect.

   a. 100, F   c. 60, F
   b. 100, H   d. 60, H

49. The venous pump, dependent upon the action of venous valves and rhythmical _____ (S, smooth; K, skeletal) muscle activity, tends to _____ (I, increase; D, decrease) venous pressures and augment venous flow within the lower extremities.

   a. S, I   c. K, I
   b. S, D   d. K, D

OBJECTIVE 19–14.

Identify the reference locus utilized for intravascular pressure measurement and the means for clinical estimation of venous pressure. Recognize the potential transmission of venous pulse waves to peripheral veins.

50. The reference point for pressure measurements of venous pressure is at the level of the _____ (S, sagittal sinus; T, tricuspid valve; A, ankle), and its absolute pressure _____ (I, is; N, is not) altered more than 2 mm. Hg by variations in body position with respect to gravity.

   a. S, I   c. A, I   e. T, N
   b. T, I   d. S, N   f. A, N

51. In the sitting position the lower veins of the neck _____ (A, are; N, are not) normally distended. Essentially all of the neck veins become greatly distended, with central venous pressures of about 15 mm. _____ .

   a. A, Hg   c. N, Hg
   b. A, H_2O d. N, H_2O

52. Veins, which collapse when elevated a vertical distance of 6 cm. above the right atrium, have estimated venous pressures closest to _____ mm. Hg.

   a. 2   c. 15   e. −15
   b. 4.5   d. −2   f. 60

53. Venous pulses originating from the _____ (P, arterial pulse; A, atria) are more readily recorded from neck veins when the central venous pressure is _____ (E, elevated; N, normal; D, diminished).

   a. P, E   c. P, D   e. A, N
   b. P, N   d. A, E   f. A, D

54. A recorded P–Q interval of the electrocardiogram coincides most closely in time with the venous wave _____ interval.

   a. a-c   c. v-a
   b. a-v   d. c-v

OBJECTIVE 19–15.

Recognize the potential "blood reservoir" functions of the spleen, liver sinuses, large abdominal veins, subcutaneous venous plexus, heart, and pulmonary circulation; and their influence on the circulatory filling pressure.

55. Sympathetic stimulation to the heart may _____ (I, increase; D, decrease) its volume by 100 ml. of blood and thereby participate in _____ (R, raising; L, lowering) the circulatory filling pressure.

   a. I, R   c. D, R
   b. I, L   d. D, L

56. The circulatory filling pressure, normally about _____ mm. Hg, is most sensitive to variations of the blood reservoir functions of the _____ (S, systemic; P, pulmonary) venous system.

   a. 0, S   c. 25, S   e. 7, P
   b. 7, S   d. 0, P   f. 25, P

20

Local Control of Blood Flow by the Tissues; Nervous and Humoral Regulation

OBJECTIVE 20–1.

Identify three major types of blood flow control. Contrast the blood flow to different tissues and organs under resting or basal conditions. Recognize the local control of blood flow in proportion to tissue metabolism.

1. Adjustments of blood flow to a tissue or organ in accord with its intrinsic metabolic activity is largely achieved via:

   a. Sympathetic reflexes   d. Endocrine
   b. Parasympathetic reflexes   hormones
   c. Local control

2. Approximately 50% of the cardiac output passes through the combined _____ and _____ circulations under basal conditions.

   a. Brain, coronary   e. Bronchial, brain
   b. Kidneys, liver   f. Liver, endocrine
   c. Muscle, bone   glands
   d. Skin, endocrine glands

3. A greater independence of local blood flow upon intrinsic metabolic requirements is frequently demonstrated by:

   a. Muscle and bone   d. Brain and cardiac
   b. Muscle and liver   tissue
   c. Liver and endocrine   e. Skin and kidneys
       glands   f. Brain and liver

OBJECTIVE 20–2.

Identify metarterioles, preferential channels, true capillaries, and precapillary sphincters. Contrast the significant segments and characteristics of the systemic microcirculation involved in the regulation of blood flow.

4. The _____ (A, arterioles; M, metarterioles; PS, precapillary sphincters; V, venules) of the microcirculation are supplied by extensive innervation from the _____ (P, parasympathetic; S, sympathetic) division of the autonomic nervous system.

   a. Only A & M, P   d. Only A & M, S
   b. Only A & PS, P   e. Only A & V, S
   c. A, M, PS, & V; P   f. A, M, PS, & V; S

5. The highest degree of smooth muscle investment for the microvasculature segments is found in the _____ . Precapillary sphincters are located adjacent to the _____ . (A, arterioles; M, metarterioles; V, venules)

   a. A, M   c. V, M   e. M, V
   b. M, M   d. A, V   f. V, V

OBJECTIVE 20–3.

Identify the general relationship between tissue metabolic rate and local blood flow. Contrast the "oxygen demand theory" and the "vasodilator theory" for the local regulation of blood flow.

6. A precapillary sphincter, generally _____ (E, either completely open or completely closed; P, partially open), may undergo vasomotion via _____ (N, neural; L, local) regulatory influences.

   a. E, N   c. P, N
   b. E, L   d. P, L

7. Extreme effects of tissue oxygen _____ (S, surplus; F, deficiency), associated with local cyanide poisoning, result in substantially _____ (I, increased; D, decreased) tissue blood flow.

   a. S, I   c. F, I
   b. S, D   d. F, D

8. According to the vasodilator theory, the _____ (L, less; G, greater) the rate of tissue metabolism the greater is the rate of formation of _____ (V, vasodilator; C, vasoconstrictor) substances.

   a. L, V   c. G, V
   b. L, C   d. G, C

9. Substances that have been suggested as agents active in the vasodilator theory for local blood flow in tissues include all of the following *with the notable exception of:*

   a. Adenosine   d. Carbon dioxide
      compounds   e. Histamine
   b. Potassium   f. Angiotensin
   c. Hydrogen ions

10. When theophylline is administered to an animal in an amount sufficient to block the action of adenosine, coronary blood flow is _____ (D, still decreased; I, still increased; U, no longer altered) by _____ (L, lowered; E, elevated; LE, either lowered or elevated) oxygen concentrations of cardiac muscle.

    a. I, E   c. I, L   e. I, LE
    b. D, L   d. D, LE   f. U, LE

OBJECTIVE 20–4.

Identify reactive hyperemia and functional hyperemia, and their relationship to local blood flow regulation mechanisms. Identify the effects of acute and long-term variations of arterial blood flow upon tissue autoregulation of blood flow.

11. A phenomenon which is most closely linked mechanistically to local tissue regulation of blood flow is:

    a. The Bainbridge reflex   e. The positive
    b. The Frank-Starling law   inotropic effect
    c. The Fick principle   f. Poiseuille's
    d. Reactive hyperemia   principle

12. Transient interruption of blood flow to a tissue is followed by _____ (L, lower; H, higher) than normal blood flows for an interval of time which is _____ (D, decreased; I, increased) by increased intervals of occlusion.

    a. L, D   c. H, D
    b. L, I   d. H, I

13. Enhanced secretion of gastrointestinal glands is generally accompanied by local vascular _____ (C, constriction; D, dilation) and _____ (R, reduced blood flow; K, constancy of blood flow; F, functional hyperemia).

    a. C, R   c. C, F   e. D, K
    b. C, K   d. D, R   f. D, F

14. Increased arterial pressure results in _____ (C, vasoconstriction; D, vasodilation) of vessels supplying skeletal muscle via local regulatory mechanisms, which _____ (N, largely nullifies; P, compounds) the effect of increased arterial blood pressure on blood flow.

    a. C, N   c. D, N
    b. C, P   d. D, P

15. A variance of blood flow to limits of ±10 to 15% of normal in skeletal muscle for acute changes in arterial blood pressure would occur with mean arterial pressures of _____ mm. Hg. The long-term regulation of normal tissue blood flow is _____ (L, less; M, more) effective than acute regulatory responses to altered arterial pressure.

    a. 40 & 200, L   c. 90 & 110, L   e. 75 & 175, M
    b. 75 & 175, L   d. 40 & 200, M   f. 90 & 110, M

OBJECTIVE 20–5.

Recognize the supplemental dependence of local regulatory mechanisms of cerebral blood flow upon CO_2 and H ion concentrations, and a distinctive renal autoregulatory mechanism.

16. _____ CO_2 and _____ pH locally result in cerebral vasoconstriction. (D, decreased; I, increased)

   a. D, D   c. I, D
   b. D, I   d. I, I

17. Increased cerebral blood flow tends to _____ carbon dioxide concentration and _____ hydrogen ion concentrations of cerebral tissues. (I, increase; D, decrease)

   a. D, D   c. I, D
   b. D, I   d. I, I

18. A postulated mechanism for "autoregulation" of renal blood flow involves the _____ (J, juxtaglomerular apparatus; V, vasa rectae) located at contacts of the distal tubule with the _____ (E, efferent arteriole; A, afferent arteriole; R, renal artery).

   a. J, E   c. J, R   e. V, A
   b. J, A   d. V, E   f. V, R

19. _____ blood sodium concentrations and _____ blood concentrations of end-products of protein metabolism are effective in increasing renal blood flow. (I, increased; D, decreased)

   a. D, D   c. I, D
   b. D, I   d. I, I

OBJECTIVE 20–6.

Recognize the roles of oxygen and changes in tissue vascularity in the long-term regulation of local blood flow, their significance, and the effects of age on the degree of readjustment.

20. The long-term regulation of blood flow through tissues is almost certainly related to changes in:

   a. Autonomic innervation   e. Tissue
   b. Arterial diameter   vascularity
   c. Vasopressin secretion   f. O_2 utilization
   d. Coarctation   coefficients

21. As a general rule, tissue vascularity is _____ (I, inversely; D, directly) proportional to tissue _____ (M, mass; B, metabolism).

   a. I, M   c. D, M
   b. I, B   d. D, B

22. Altered tissue oxygen availability may result in alternations in tissue vascularity that occur more rapidly in the _____ (I, infant; A, adult) and in _____ (W, well established tissues; N, newly forming tissue).

   a. I, W   c. A, W
   b. I, N   d. A, N

23. Tissue hypoxia results in _____ tissue vascularity whereas hyperoxia results in _____ tissue vascularity. (I, increased; N, no change in; D, decreased)

   a. I, N   c. D, N   e. I, D
   b. N, I   d. N, D   f. D, I

OBJECTIVE 20–7.

Identify the influence of the autonomic innervation upon the circulation. Contrast the sympathetic and parasympathetic influences on the heart and peripheral circulation.

24. The major function of the predominantly _____ (P, parasympathetic; S, sympathetic) innervation of small arteries, arterioles, venules, and small veins is to alter vascular _____ (D, distensibility; V, volume; R, resistance).

   a. P, D   c. P, R   e. S, V
   b. P, V   d. S, D   f. S, R

25. The major function of the predominantly _____ (P, parasympathetic; S, sympathetic) innervation of large veins is to alter venous _____ (V, volume; R, resistance).

   a. P, V   c. S, V
   b. P, R   d. S, R

26. The most significant effect of the parasympathetic nervous system on the general circulation is its effect upon:

   a. Cardiac contractility   b. Vascular compliance   c. Vascular resistance   d. Heart rate

OBJECTIVE 20–8.

Contrast the significance of sympathetic vasoconstrictor versus vasodilator roles, and their comparative tissue or organ distribution. Locate and identify the vasomotor and cardiac centers, their roles in vasomotor tone and cardiac performance, and the consequences of a loss of their function.

27. Sympathetic _____ (C, vasoconstrictor; D, vasodilator) fibers are widely distributed to most segments of the circulation, but have comparatively lesser influence on the _____ (KG, kidney and gut; SS, spleen and skin; BC, brain and coronary) circulations.

   a. C, KG   c. C, BC   e. D, SS
   b. C, SS   d. D, KG   f. D, BC

28. "Vasomotor centers," located bilaterally in the _____ (H, hypothalamus; P, pons and medulla; S, spinal cord; C, cerebral cortex) transmit tonic activity which _____ (A, ascends; D, descends) within the central nervous system in order to maintain vasomotor tone.

   a. H, A   c. S, A   e. P, D
   b. P, A   d. H, D   f. C, D

29. Stimulation of the _____ (L, upper and lateral; M, lower and medial) portions of the vasomotor centers results in increased vasomotor tone and _____ (I, increased; D, decreased) peripheral resistance.

   a. L, I   c. M, I
   b. L, D   d. M, D

30. A substantial reduction in arterial pressure would occur with:

   a. Adrenalectomy   e. Sympathetic
   b. Total spinal anesthesia   stimulation
   c. Midpontile transections f. None of the
   d. Decortication   above

31. Stimulation of the _____ (M, medial; L, lateral) portion of the vasomotor centers, associated with the dorsal motor nucleus of the vagus nerve, generally results in _____ (I, increased; N, no change in; D, decreased) heart rate.

   a. M, I   c. M, D   e. L, N
   b. M, N   d. L, I   f. L, D

OBJECTIVE 20–9.

Recognize the influence of higher centers upon the vasomotor centers.

32. A higher center which may exert either powerful excitatory or inhibitory influences on the vasomotor centers is the:

   a. Parietal lobe   d. Hypothalamus
   b. Occipital lobe   e. Geniculate bodies
   c. Thalamus   f. Superior colliculi

33. The more lateral and superior portions of the reticular formation of the pons, mesencephalon, and diencephalon and the _____ (A, anterior part; PL, posterolateral) portion of the hypothalamus are more significant in _____ (E, excitatory; I, inhibitory) effects on the vasomotor centers.

   a. A, E   c. PL, E
   b. A, I   d. PL, I

OBJECTIVE 20–10.

Identify the sympathetic vasoconstrictor transmitter substance at effector structures, and the relationship of the adrenal medulla to the sympathetic vasoconstrictor system.

34. The sympathetic vasoconstrictor transmitter substance at effector junctions is:

    a. Epinephrine
    b. Acetylcholine
    c. Histamine
    d. Norepinephrine
    e. GABA
    f. Serotonin

35. Coincident with _____ (I, increased; D, decreased) vasoconstrictor activity, sympathetic activity to the adrenal medulla results in the release of _____ (N, only norepinephrine; B, both epinephrine and norepinephrine).

    a. I, N
    b. I, B
    c. D, N
    d. D, B

OBJECTIVE 20–11.

Identify the sympathetic vasodilator system, its central nervous system control, and its significance.

36. Sympathetic nerves to skeletal muscles which _____ (A, also contain; L, lack) vasoconstrictor fibers carry sympathetic vasodilator fibers whose chemical transmitter agent _____ (I, is; N, is not) norepinephrine.

    a. A, I
    b. A, N
    c. L, I
    d. L, N

37. The principal area(s) of the brain controlling the sympathetic vasodilator system is (are) the:

    a. Vasomotor centers
    b. Motor cortex
    c. Anterior hypothalamus
    d. Posterolateral hypothalamus
    e. Adrenal gland
    f. Fields of Forel

38. The sympathetic vasodilator system probably functions to _____ (I, increase; D, decrease) blood flow to skeletal muscle in conjunction with _____ (C, motor cortex stimulation; V, vasomotor center inhibition).

    a. I, C
    b. I, V
    c. D, C
    d. D, V

OBJECTIVE 20–12.

Identify the "mass action," "alarm," and "motor" patterns of circulatory responses, and their component structures, resultant effects, and functional significance. Identify vasovagal syncopy, its probable cause, and its resultant effects.

39. The _____ (A, alarm; M, motor; S, mass action) pattern of circulatory responses, resulting from widespread stimulation of the vasomotor centers, prepares the circulation for _____ (D, decreased; I, increased) delivery of blood flow to the body.

    a. A, D
    b. M, D
    c. S, D
    d. A, I
    e. M, I
    f. S, I

40. Vasovagal syncopy refers to a condition of:

    a. Mass action effects
    b. Emotional fainting
    c. Myocardial infarction
    d. Stokes-Adams syndrome
    e. Ventricular premature beats
    f. Hypertension

41. "Mass action" effects on the circulation include a reduction in:

    a. Peripheral resistance
    b. Arterial blood pressure
    c. Venomotor tone
    d. Circulating epinephrine levels
    e. Sympathetic vasomotor tone
    f. Vascular capacitance

42. Fainting resulting from intense emotional experiences probably results from _____ (I, increased; D, decreased) peripheral resistance resulting from the _____ (M, muscle vasodilator system; S, "mass action" effect upon the circulation).

    a. I, M
    b. I, S
    c. D, M
    d. D, S

43. The "alarm" pattern of circulatory responses, resulting from _____ (A, anterior; D, diffuse) hypothalamic stimulation, contains the elements of _____ (M, "mass action" effects; S, sympathetic vasodilator system activation).

    a. A, only M
    b. A, neither M nor S
    c. A, both M & S
    d. D, only M
    e. D, neither M nor S
    f. D, both M & S

44. The "motor" pattern of circulatory responses _____ arterial blood pressure and _____ vasoconstrictor tone to the skin and gut. (I, increases; D, decreases)

    a. I, I
    b. I, D
    c. D, I
    d. D, D

OBJECTIVE 20–13.

Identify the general circulatory patterns associated with reflex regulation of arterial pressure, blood volume, and body temperature.

45. Increased arterial pressure acting through the baroreceptor reflex results in _____ heart rate and_____ vasomotor tone. (D, decreased; I, increased)

    a. D, D
    b. D, I
    c. I, D
    d. I, I

46. Increased stretch receptor activity from the _____ (A, arteries; V, large veins and atria) results in_____ (I, increased; D, decreased) urine output.

    a. A, I
    b. A, D
    c. V, I
    d. V, D

47. _____(I, increased; D, decreased) body temperature results in cutaneous vasodilation, primarily as a consequence of altered _____ (S, sympathetic; P, parasympathetic) activity.

    a. I, S
    b. I, P
    c. D, S
    d. D, P

OBJECTIVE 20–14.

Recognize the significance of humoral regulation. Identify the origins and resultant circulatory effects of norepinephrine, epinephrine, angiotensin, bradykinin, vasopressin, serotonin, histamine, and prostaglandins.

48. 5-hydroxytryptamine, or _____ (A, angiotensin; S, serotonin), is present in substantial concentrations in _____ (E, erythrocytes; L, leukocytes; P, platelets) and chromaffin tissue of the intestine.

    a. A, E
    b. A, L
    c. A, P
    d. S, E
    e. S, L
    f. S, P

49. Bradykinin is formed from a plasma _____ (A, albumin; G, globulin) by the enzymatic action of_____ (C, carboxypeptidase; K, kallikrein).

    a. A, C
    b. A, K
    c. G, C
    d. G, K

50. Widespread intense vasodilation and edema result from the administration of _____ whereas the administration of _____ (A, angiotensin; H, histamine; S, serotonin; N, norepinephrine) results in intense vasoconstriction.

    a. Bradykinin, A
    b. Acetylcholine, H
    c. Renin, A
    d. Prostaglandins, H
    e. Norepinephrine, S
    f. Epinephrine, N

51. _____ (R, renin; A, angiotensin; V, vaso-pressin), secreted by the kidneys, acts on a plasma protein to produce a vasoactive peptide, resulting in renal arteriolar_____ (D, dilation; C, constriction).

   a. R, D   c. V, D   e. A, C
   b. A, D   d. R, C   f. V, C

52. In comparison with norepinephrine, epinephrine has a _____ (M, more; L, less) pronounced effect on _____(I, increasing; D, decreasing) systemic peripheral resistance.

   a. M, I   c. L, I
   b. M, D   d. L, D

53. A _____ (R, renal; H, hypothalamic; P, prostate gland; C, chromaffin tissue) hormone, functioning primarily in the regulation of renal tubular reabsorption of water, has secondary cardiovascular effects on the _____ (V, venous; A, arteriolar) segment.

   a. R, V   c. P, V   e. H, A
   b. H, V   d. C, A   f. P, A

54. The class of prostaglandins, originating from _____ (P, predominantly the prostate gland; W, widespread body tissues) include hormones which cause _____(C, vasoconstriction only; D, vasodilation only; CD, both vasoconstriction and vasodilation).

   a. P, C   c. P, CD   e. W, D
   b. P, D   d. W, C   f. W, CD

OBJECTIVE 20–15.
   Identify the local effects of osmolarity and calcium, potassium, magnesium, sodium, hydrogen, acetate, and citrate ion concentrations on vascular smooth muscle. Contrast the local direct effects of carbon dioxide on the cardiovascular system with the effects resulting from carbon dioxide effects upon the vasomotor centers.

55. Both _____ (I, increased; D, decreased) osmolarity and elevated citrate content of blood result in _____ (V, vasoconstriction; L, vasodilation).

   a. I, V   c. D, V
   b. I, L   d. D, L

56. A slight increase in pH results in arteriolar _____(C, constriction; D, dilation), an effect which is_____ (I, intensified; R, reversed) as the increased pH becomes extreme.

   a. C, I   c. D, I
   b. C, R   d. D, R

21

Regulation of Mean Arterial Pressure: I. Nervous Reflex and Hormonal Mechanisms for Rapid Pressure Control

OBJECTIVE 21–1.
   Recognize the comparative constancy of arterial blood pressure versus cardiac output and peripheral resistance. Identify typical systolic, diastolic, pulse pressure, and mean pressure values of a young adult, and their increasing and decreasing tendencies as a function of younger and older ages.

1. Among the following items, the variable remaining most constant during a wide range of tissue activities would be:

   a. Heart rate   d. Peripheral resistance
   b. Stroke volume   e. Velocity of flow
   c. Cardiac output   f. Mean arterial pressure

2. An arterial blood pressure of 85/60 would lie closer to that typically occurring for a:

   a. 1-month-old infant   d. 30-year-old adult
   b. 10-year-old child   e. 40-year-old adult
   c. 20-year-old adult   f. 60-year-old adult

3. Systolic pressures of normal young adults average about _____ mm. Hg, with a corresponding pulse pressure of _____ mm. Hg.

   a. 80, 40   c. 120, 40   e. 100, 20
   b. 100, 40   d. 80, 20   f. 120, 20

4. The _____ (D, decrease; I, increase) in arterial pressures at progressively older ages in the adult includes more pronounced changes in _____ (M, mean; A, diastolic; S, systolic) values.

   a. D, M   c. D, S   e. I, A
   b. D, A   d. I, M   f. I, S

OBJECTIVE 21–2.
Identify mean arterial pressure and its significance.

5. The mean arterial pressure, averaging _____ mm. Hg in normal young adults, is equal to the average _____ (SD, of systolic and diastolic pressures; C, pressure over the pulse cycle time interval).

   a. 110, SD   c. 83, SD   e. 96, C
   b. 96, SD   d. 110, C   f. 83, C

6. The mean arterial pressure, which typically lies closer to _____ (S, systolic; D, diastolic) pressure values, approaches this value even more at _____ (H, higher; L, lower) heart rates.

   a. S, H   c. D, H
   b. S, L   d. D, L

7. The _____ (M, mean; P, pulse; DI, diastolic) pressure difference equals cardiac output _____ (T, times; D, divided by) the total peripheral resistance.

   a. M, T   c. DI, T   e. P, D
   b. P, T   d. M, D   f. DI, D

OBJECTIVE 21–3.
Identify the auscultatory method utilized clinically for the indirect determination of arterial blood pressure. Identify the Korotkow sounds, their probable cause, and their clinical utility.

8. The basis for the sounds of Korotkow is generally stated to be:

   a. A–V valve closure   d. Arterial turbulence
   b. Aortic valve closure e. Arterial resonance
   c. Arterial expansion   f. The Laplace principle

9. With the auscultatory method for arterial pressure determination, Korotkow sounds, auscultated over the commonly utilized _____ (P, popliteal; A, antecubital) artery, occur when the artery is _____ (O, open continuously; C, closed continuously; I, intermittently opened and closed) during arterial pressure cycles.

   a. P, O   c. P, I   e. A, C
   b. P, C   d. A, O   f. A, I

10. The beginning of the muffled quality of sounds, lasting throughout the next 5 to 10 mm. Hg fall in pressure before all sound disappears during the indirect determination of arterial blood pressure, is considered to be the:

    a. Second diastolic pressure
    b. Mean blood pressure
    c. Systolic pressure
    d. Diastolic or the first diastolic
    e. Second systolic pressure
    f. Average blood pressure

11. The sphygmomanometer cuff pressure at the onset of tapping sounds, occurring during cuff deflation, is an indirect determination of the arterial:

    a. Second diastolic pressure
    b. Mean blood pressure
    c. Systolic pressure
    d. Diastolic pressure
    e. Second systolic pressure
    f. Average blood pressure

OBJECTIVE 21–4.

Identify the radial pulse and oscillometric methods for the indirect determination of arterial blood pressure. Contrast the utility of these methods with the auscultatory method.

12. A mercury manometer, sphygmomanometer cuff, and stethoscope, or their functional equivalents, would all be utilized in the determination of arterial blood pressure by the:

    a. Oscillometric method
    b. Radial pulse method
    c. Palpatory method
    d. Auscultatory method
    e. Bandage method
    f. None of the above

13. The _____ (A, auscultatory; O, oscillometric) method, in which _____ (S, systolic; D, diastolic) pressure may be determined with greater accuracy, is particularly valuable for estimating arterial pressures in a shock patient where Korotkow sounds are not distinct.

    a. A, S   c. A, D
    b. O, S   d. O, D

14. With a declining cuff pressure for simultaneous radial pulse and oscillometric methods for estimating arterial pressure, a distinct radial pulse _____ (F, is first detected; D, disappears) at the point of _____ (A, appearance of the; M, maximum) oscillations of the mercury manometer.

    a. F, A   c. D, A
    b. F, M   d. D, M

OBJECTIVE 21–5.

Contrast the general characteristics of rapidly acting control mechanisms versus the long-term control mechanism for the regulation of arterial pressure.

15. The long-term _____ (P, positive; N, negative) feedback mechanism for regulation of arterial pressure operates primarily through the _____ (K, kidneys; A, arterial pressoreceptors).

    a. P, K   c. N, K
    b. P, A   d. N, A

16. A rapidly acting pressure control mechanism, serving a buffer function for transient changes in arterial pressure, involves _____ (R, arterial; A, arteriolar) receptors and significant effector structures including _____ (H, only the heart; V, only veins and arterioles; B, the heart, veins, and arterioles).

    a. R, H   c. R, B   e. A, V
    b. R, V   d. A, H   f. A, B

OBJECTIVE 21–6.

Describe the physiologic anatomy of the baroreceptors, two principle areas of their location, and their afferent nerves.

17. Arterial baroreceptors are _____ (E, encapsulated; S, spray-type) receptors, whose afferent activity is conveyed largely in _____ cranial nerves _____ .

    a. E, V & VII   c. E, V & X   e. S, IX & X
    b. E, IX & X   d. S, V & VII   f. S, V & X

18. A rise in arterial pressure _____ (D, decreases; I, increases) activity of baroreceptors, more abundantly located in the carotid and aortic _____ (B, bodies; S, sinuses).

    a. D, B   c. I, B
    b. D, S   d. I, S

OBJECTIVE 21–7.

Identify the responses of baroreceptors to pressure, their effects upon vasomotor and cardiac centers, and their resultant influence on the cardiovascular system.

19. The increment of change in action potential frequency per unit change in arterial pressure is greatest in baroreceptor afferent nerve fibers at an arterial pressure of:

    a. Normal mean values   d. 150 mm. Hg
    b. 0 mm. Hg   e. 200 mm. Hg
    c. 60 mm. Hg   f. 250 mm. Hg

20. The _____ (D, decrease; I, increase) in baroreceptor action potential firing frequency, occurring as a result of ventricular systole, is more substantial during an interval in which the mean arterial pressure exhibits a _____ (R, rising; F, falling) trend.

    a. D, R   c. I, R
    b. D, F   d. I, F

21. Baroreceptor activity _____ the vasoconstrictor center of the medulla, and _____ the vagal center. (S, stimulates; I, inhibits)

    a. S, S   c. I, S
    b. S, I   d. I, I

22. Clamping both common carotid arteries below their bifurcation results in _____ in peripheral resistance and _____ in heart rate. (I, an increase; N, no significant change; D, a decrease)

    a. I, I   c. I, D   e. D, N
    b. I, N   d. D, I   f. D, D

23. Decreased arterial baroreceptor activity results in reflexively _____ cardiac contractility and _____ P–R intervals of the electrocardiogram. (D, decreased; I, increased)

    a. D, D   c. I, D
    b. D, I   d. I, I

24. Widespread sympathetic stimulation tends to _____ (I, increase; D, decrease) S–A node rhythmicity via direct cardiac actions, which is _____ (A, augmented; O, opposed) by reflex cardiac effects resulting from the altered arterial baroreceptor activity.

    a. I, A   c. D, A
    b. I, O   d. D, O

OBJECTIVE 21–8.

Recognize the "buffer" function of the baroreceptor control system. Identify the function of baroreceptors in cardiovascular adjustments to gravitational forces, the adaptive properties of baroreceptors and their consequences, and the carotid sinus syndrome.

25. The tendency towards _____ arterial pressure in the carotid artery upon standing is compensated for by _____ baroreceptor activity. (I, increased; D, decreased)

    a. I, I   c. D, I
    b. I, D   d. D, D

26. The "buffer" system, which normally reduces daily transient changes in arterial pressure to about one half to one third that which would occur in its absence, involves the:

    a. Bainbridge reflex
    b. Juxtaglomerular apparatus
    c. Coronary ischemic reflex
    d. Carotid and aortic sinuses
    e. Carotid and aortic bodies
    f. Sympathetic vasodilator system

27. An experimental doubling of the mean arterial pressure results in an initial _____ (I, increase; D, decrease) in baroreceptor activity, which progressively _____ (CI, continues to increase; CD, continues to decrease; R, returns to control levels) during the following few days.

   a. I, CI   c. D, CD
   b. I, R   d. D, R

28. Strong pressure applied bilaterally on the neck directly over the carotid sinuses would cause the heart rate to _____ and arterial pressure to _____. (I, increase; D, decrease)

   a. I, I   c. D, I
   b. I, D   d. D, D

29. Carotid sinus syncope results from _____ (D, decreased; I, increased) baroreceptor responses to arterial pressure changes, and carotid sinus _____ (C, compression; DC, decompression).

   a. D, C   c. I, C
   b. D, DC   d. I, DC

OBJECTIVE 21–9.

Recognize the role of atrial and pulmonary reflexes in the regulation of arterial pressure. Identify the effects of atrial reflexes upon peripheral resistance, heart rate, renal arterioles, and antidiuretic hormone secretion.

30. Infusion of blood would probably result in the greatest increment of change in arterial pressure with _____ high pressure receptors and _____ low pressure receptors. (A, absent; I, intact)

   a. I, I   c. A, I
   b. I, A   d. A, A

31. Stretching the atria results in reflex arteriolar _____ (C, vasoconstriction; D, vasodilation) and a subsequent tendency to _____ (I, increase; D, decrease) arterial pressure.

   a. C, I   c. D, I
   b. C, D   d. D, D

32. Increased stretch of atrial walls results in _____ urine output resulting from _____ renal afferent arteriolar vasomotor tone. (I, increased; D, decreased)

   a. I, I   c. D, I
   b. I, D   d. D, D

33. Increased atrial volumes increase body water _____ (L, loss; R, retention) by _____ (I, increasing; D, decreasing) antidiuretic hormone secretion.

   a. L, I   c. R, I
   b. L, D   d. R, D

34. The reflex variation in heart rate occurring as a function of right atrial pressure is:

   a. Starling's law   e. The Bainbridge
   b. Marey's law   reflex
   c. The aortic sinus reflex f. The chemorecep-
   d. The Cushing reaction   tor reflex

35. Atrial distension _____ (I, increases; D, decreases) S-A node rhythmicity via direct effects, and Bainbridge reflex effects _____ (A, aid; O, oppose) these direct effects.

   a. I, A   c. D, A
   b. I, O   d. D, O

OBJECTIVE 21–10.

Identify the effects of elevated carbon dioxide and ischemia on the vasomotor and cardiac centers, the CNS ischemic response, the Cushing reaction, and the limitations of extreme ischemia on these mechanisms.

36. One of the most potent stimuli that directly effects the vasomotor centers is _____ (B, the Bainbridge response; I, ischemia), producing peripheral _____ (C, vasoconstriction; D, vasodilation).

   a. B, C   c. I, C
   b. B, D   d. I, D

37. The Cushing reaction is most similar in its resultant effects to the:

   a. CNS ischemic response e. Bainbridge
   b. Vasovagal syncope   reflex
   c. Carotid sinus syndrome f. Baroreceptor
   d. Chemoreceptor reflex   reflex

38. Elevated cerebrospinal fluid pressures causing compression of cerebral arteries result in _____ (I, increased; D, decreased) arterial pressure via the _____ (C, chemoreceptor reflex; B, baroreceptor reflex; CU, Cushing reaction).

    a. I, C   c. I, CU   e. D, B
    b. I, B   d. D, C   f. D, CU

39. Severe CNS ischemia resulting in loss of all activity of the vasomotor centers results in arterial blood pressures of _____ (H, as high as 270 mm. Hg; L, about 40 to 50 mm. Hg) as a consequence of the _____ (P, presence; A, absence) of the CNS ischemic reflex.

    a. H, P   c. L, P
    b. H, A   d. L, A

OBJECTIVE 21–11.

Identify the carotid and aortic bodies; effects of oxygen, carbon dioxide, and hydrogen ion concentrations on chemoreceptor activity; and the resultant effects of their stimulation.

40. Arterial chemoreceptors, comparatively more sensitive to _____ oxygen concentrations, also respond to _____ hydrogen ion concentration. (I, increased; D, decreased)

    a. I, I   c. D, I
    b. I, D   d. D, D

41. Elevated carbon dioxide concentrations produce peripheral _____ via their direct action on the vasomotor centers, and produce reflex peripheral _____ via their action on carotid and aortic chemoreceptors. (D, vasodilation; C, vasoconstriction)

    a. D, D   c. C, D
    b. D, C   d. C, C

42. Reduction of mean arterial pressures to extremely low levels in the range of 40 to 80 mm. Hg would result in _____ discharge of carotid and aortic sinus receptor afferent activity and _____ discharge of carotid and aortic body receptor afferent activity. (I, increased; D, decreased)

    a. I, I   c. D, I
    b. I, D   d. D, D

43. Divisions of Hering's nerves, conveying activity from the _____ (A, aortic; C, carotid) bodies and sinuses, respectively, have _____ (S, similar; O, opposing) effects on the vasomotor centers.

    a. A, S   c. C, S
    b. A, O   d. C, O

44. Denervation of the aortic and carotid chemoreceptors in an otherwise normal individual would result in _____ arterial pressure, and _____ heart rate. (I, increased; D, decreased; N, no appreciable change in)

    a. I, I   c. I, D   e. D, I
    b. I, N   d. N, N   f. D, D

OBJECTIVE 21–12.

Identify the effects of altered autonomic activity upon venous capacitance, and its consequences on venous pressure, circulatory filling pressure, arterial pressure, and pumping effectiveness of the heart. Recognize the role of these effects in the baroreceptor, ischemic, and chemoreceptor reflexes.

45. The predominantly _____ (P, parasympathetic; S, sympathetic) innervation of veins regulates a more significant venous _____ (R, resistance; C, capacity).

    a. P, R   c. S, R
    b. P, C   d. S, C

46. Increased venomotor tone results in the greatest decrease in:

    a. Venous resistance   e. Critical closing
    b. Venous capacity   pressure
    c. Venous pressure   f. Venous return
    d. Circulatory filling
       pressure

47. Variations in venomotor tone would be a significant participant in reflex circulatory patterns during activities of _____(B, the baroreceptor reflex; C, the chemoreceptor reflex; I, the CNS ischemic response; A, atrial reflexes).

    a. B only
    b. B & I only
    c. B & A only
    d. I & C only
    e. I, B, & C only
    f. B, C, I, & A

48. _____ sympathetic activity to veins due to stimulation of arterial chemoreceptors results in _____ venous capacitance. (I, increased; D, decreased)

    a. I, I
    b. I, D
    c. D, I
    d. D, D

OBJECTIVE 21-13.

Identify the association of vasomotor center activity with skeletal muscle tonus. Identify the abdominal compression reflex and its effect upon the systemic filling pressure.

49. The _____ (C, chemoreceptor; A, abdominal compression) reflex generally accompanies reflex cardiovascular alterations resulting from a _____ (D, decrease; I, increase) in carotid sinus pressure.

    a. C, D
    b. C, I
    c. A, D
    d. A, I

50. _____ skeletal muscle tonus occurs with the anticipation of exercise, and results in _____ systemic filling pressure. (I, increased; D, decreased)

    a. I, I
    b. I, D
    c. D, I
    d. D, D

51. Increased activity of the sympathetic vasoconstrictor system via the vasomotor centers will generally _____ skeletal muscle tone, an effect which tends to _____ cardiac output. (I, increase; D, decrease)

    a. I, I
    b. I, D
    c. D, I
    d. D, D

OBJECTIVE 21-14.

Identify respiratory pressure waves in arterial pressure recordings and their mechanistic origin. Identify "Mayer waves" or "Traube-Hering waves" and potential mechanisms for their existence.

52. With each cycle of respiration the arterial pressure rises and falls, with an increase in arterial pressure usually occurring during the _____ part of inspiration and _____ part of expiration and a fall in pressure occurring during the remainder of the respiratory cycle. (E, early; L, late)

    a. L, L
    b. E, E
    c. E, L
    d. L, E

53. _____ (B, baroreceptor; C, chemoreceptor; I, CNS ischemic) reflex oscillations of a typical 15 to 40 _____ (M, minute; S, second) periodicity are a probable major cause of Traube-Hering waves occurring within an arterial pressure range of 40 to 80 mm. Hg.

    a. B, M
    b. C, M
    c. I, M
    d. B, S
    e. C, S
    f. I, S

OBJECTIVE 21-15.

Recognize the norepinephrine-epinephrine, renin-angiotensin, and vasopressin (ADH) mechanisms for the rapid or moderately rapid control of arterial pressure. Identify the significant characteristics of each of these control mechanisms.

54. A hormonal system which is considered to be part of the sympathetic mechanism for the regulation of arterial pressure involves:

a. Norepinephrine-epinephrine d. Kallikrein-
b. Renin-angiotensin   bradykinin
c. Vasopressin

55. The _____ (H, hormone; E, enzyme) renin is secreted by the _____ (G, glomerulosa; P, podocyte; J, juxtaglomerular) cells of the kidney.

a. H, G   c. H, J   e. E, P
b. H, P   d. E, G   f. E, J

56. Activation of the renin-angiotensin mechanism results in a decrease in:

a. Aldosterone secretion   e. Blood volume
b. Arteriolar tone   f. Renal salt
c. Venous tone   excretion
d. Circulatory filling pressure

57. The renin-angiotensin mechanism activated as a result of _____ (I, increased; D, decreased) renal blood flow requires about 20 _____ (S, seconds; M, minutes; H, hours) to become fully active.

a. I, S   c. I, H   e. D, M
b. I, M   d. D, S   f. D, H

58. _____ arterial pressure results in posterior pituitary secretion of antidiuretic hormone, with consequent target effects of _____ renal water retention and peripheral resistance. (I, increased; D, decreased)

a. I, I   c. D, I
b. I, D   d. D, D

OBJECTIVE 21–16.

Identify the roles of capillary fluid shifts and stress-relaxation as intermediate acting feedback mechanisms for arterial pressure regulation.

59. Elevation of capillary pressure, sufficient to result in capillary fluid shifts, tends to _____ the blood volume and _____ arterial pressure. (I, increase; D, decrease)

a. I, I   c. D, I
b. I, D   d. D, D

60. Massive transfusion, resulting in an initial marked increase in arterial pressure, is followed by a _____ (C, continued rise; R, return towards normal) of the arterial pressure as a consequence of _____ (S, stress-relaxation; V, restoration of a normal blood volume).

a. C, S   c. R, S
b. C, V   d. R, V

22

Regulation of Arterial Pressure: II. The Renal-Body Fluid System for Long-Term Pressure Control. Mechanisms of Hypertension

OBJECTIVE 22–1.

Identify the short-term, intermediate-term, and long-term pressure control mechanisms, and the relative importance of these mechanisms as controllers of arterial pressure.

1. Baroreceptor, chemoreceptor, and CNS ischemic feedback mechanisms are considered _____(SH, short-; I, intermediate-; L, long-) term pressure control mechanisms, initiating action within _____(S, less than 15 seconds; M, 1 to 10 minutes; H, 1 to 24 hours) in response to a disturbance in arterial pressure.

   a. SH, S   c. L, M   e. I, M
   b. I, S   d. SH, M   f. L, H

2. A quantitative expression of the capability of an arterial pressure regulatory system to maintain arterial pressure near its normal mean value is the _____(K, damping coefficient; G, feedback gain), which has a maximum value greatest for a _____ (S, short-; L, long-) term regulatory mechanism.

   a. K, S   c. G, S
   b. K, L   d. G, L

3. The maximum feedback gains for neural regulation of arterial pressure at optimal arterial pressures are highest for the _____ mechanism, and lowest for the _____ mechanism. (C, chemoreceptor; B, baroreceptor; I, CNS ischemic response)

   a. C, B   c. C, I   e. B, I
   b. B, C   d. I, C   f. I, B

OBJECTIVE 22–2.

Identify the sequence and significant characteristics of intermediate steps involved in the renal body fluid feedback control system for the regulation of arterial pressure.

4. Increased arterial pressure results in _____ renal excretion of fluid and _____ blood volume. (I, increased; N, no significant change in; D, decreased)

   a. I, I   c. I, D   e. D, I
   b. I, N   d. N, N   f. D, D

5. Increased blood volume results in _____ cardiac output and _____ arterial blood pressure. (I, increased; N, no significant change; D, decreased)

   a. I, I   c. I, D   e. D, I
   b. I, N   d. N, N   f. D, D

6. The arterial pressure is modified by an experimental increase in cardiac output, partly through _____ (V, an inverse; P, a direct) proportionality of arterial pressure to cardiac output, and partly by local autoregulatory mechanisms of tissue blood flow involving _____(I, increased; D, decreased) peripheral resistance.

   a. V, I   c. P, I
   b. V, D   d. P, D

7. A chronic 2% increase in blood volume, via renal body fluid regulatory mechanisms, would result in the greatest percentage change among the following in:

   a. Circulatory filling   c. Blood volume
      pressure   d. Peripheral
   b. Venous return   resistance

8. The compensation for the renal body fluid pressure control system is _____% of normal, which is _____. (G, greater than; E, equal to; L, less than) that occurring for the baroreceptor system.

   a. 0, E   c. 100, G   e. 88, L
   b. 88, G   d. 0, L   f. 100, E

9. Urine output exhibits about a one-fold increase for each _____ mm. Hg increment of _____(D, decreased; I, increased) arterial pressure.

   a. 1, D   c. 100, D   e. 10, I
   b. 10, D   d. 1, I   f. 100, I

10. The long-term level of arterial pressure is primarily determined by the _____ (F, fluid intake; BB, Bainbridge reflex; BR, baroreceptor reflex) and the relationship between urinary output and _____ (C, cardiac output; A, arterial pressure).

    a. F, C   b. BB, C   c. BR, C   d. F, A   e. BB, A   f. BR, A

OBJECTIVE 22–3.

Contrast the roles of renal resistance versus total peripheral resistance as determinants of the long-term level of arterial pressure. Recognize the secondary association of increased total peripheral resistance with high blood pressure.

11. _____ (D, decreased; I, increased) total peripheral resistance associated with hypertension is more generally a _____ (C, causative; N, consequential) factor.

    a. D, C   c. D, N
    b. I, C   d. I, N

12. An increase in total peripheral resistance, secondary to mechanisms that cause hypertension, is primarily associated with mechanisms of:

    a. Tissue flow autoregulation   d. Renal
    b. Baroreceptor reflexes   resistance
    c. Chemoreceptor reflexes

13. The renal body fluid control system, responding to the opening of an arteriovenous fistula and its associated _____ (I, increase; D, decrease) in peripheral resistance, restores a steady state value of the arterial pressure to a _____ (L, hypotensive; N, normal; H, hypertensive) equilibrium level.

    a. I, L   c. I, H   e. D, N
    b. I, N   d. D, L   f. D, H

14. Increased renal resistance will usually _____ (I, increase; D, decrease) arterial blood pressure by decreasing _____ (R, renin secretion; U, urinary output).

    a. I, R   c. D, R
    b. I, U   d. D, U

OBJECTIVE 22–4.

Identify hypertension and the potential causes and characteristics of volume-loading hypertension, Goldblatt hypertension, hypertension in toxemia of pregnancy, neurogenic hypertension, hypertension caused by primary aldosteronism, and essential hypertension.

15. Volume-loading hypertension is generally caused by _____ (D, deficient; E, excess) water and salt intake in patients with _____ (N, normal; R, substantially reduced) renal mass.

    a. D, N   c. E, N
    b. D, R   d. E, R

16. Volume-loading hypertension is associated with _____ (D, decreased; I, increased) peripheral resistance as a _____ (C, causative; N, consequential) factor of the elevated arterial pressure.

    a. D, C   c. I, C
    b. D, N   d. I, N

17. Constricting devices placed on the renal arteries may result in an early rise in the _____ (G, Goldblatt; E, essential) form of hypertension as a consequence of _____ (F, fluid retention; R, the renin-angiotensin mechanism).

    a. G, F   c. E, F
    b. G, R   d. E, R

18. Chronic ischemia of localized diseased regions of the kidneys frequently results in systemic hypertension, renal arteriolar _____ (C, constriction; D, dilation) of normal renal segments, and _____ (I, increased; DE, decreased) renal excretion of water and electrolytes.

    a. C, I   c. D, I
    b. C, DE   d. D, DE

19. _____ (I, increased; D, decreased) glomerular filtration, resulting from pathological changes in the glomerular membranes with toxemia of pregnancy, renders the patient particularly susceptible to hypertension with increased _____(S, salt intake; N, neurogenic factors; U, urine output).

   a. I, S   c. I, U   e. D, N
   b. I, N   d. D, S   f. D, U

20. Neurogenic hypertension, produced for days at a time by _____ (I, increased; D, decreased) renal sympathetic tone, is _____ (R, retained; L, lost) following restoration of normal sympathetic tone.

   a. I, R   c. D, R
   b. I, L   d. D, L

21. Primary aldosteronism results in _____ (D, decreased; I, increased) rates of renal salt and water reabsorption, and alterations in arterial blood pressure which are _____ (R, reduced; A, accentuated) by increasing water and salt intake.

   a. D, A   c. I, A
   b. D, R   d. I, R

22. The patient with essential hypertension probably has _____ glomerular filtration rates at his hypertensive pressure level, and _____ glomerular filtration rates at normal arterial pressures. (N, normal; A, abnormal)

   a. N, N   c. A, N
   b. N, A   d. A, A

OBJECTIVE 22–5.

Contrast the types of renal disease processes that result in hypertension from those causing uremia.

23. A renal disease process which reduces renal mass but does not alter remaining nephron function generally _____ cause hypertension and _____ cause uremia. (D, does; N, does not)

   a. D, D   c. N, D
   b. D, N   d. N, N

24. Renal disease processes which reduce glomerular filtration of all glomeruli equally _____ cause hypertension, and consequently _____ cause uremia. (D, do; N, do not)

   a. D, D   c. N, D
   b. D, N   d. N, N

25. Factors that result in excess fluid reabsorption by renal tubules in the absence of renal injury _____ produce hypertension and _____ produce uremia. (D, do; N, do not)

   a. D, D   c. N, D
   b. D, N   d. N, N

OBJECTIVE 22–6.

Identify the effects of hypertension on the work load of the heart and vascular integrity.

26. Hypertension is accompanied by a relative _____ (H, hyperemia; I, ischemia) of the _____ (L, left; R, right) ventricle.

   a. H, L   c. I, L
   b. H, R   d. I, R

27. Sclerosis of blood vessels, a process which is _____ (E, exaggerated; A, alleviated) by hypertension, tends to _____ (I, increase; D, decrease) arterial blood pressure when it results in progressive renal deterioration.

   a. E, I   c. A, I
   b. E, D   d. A, D

23

Cardiac Output, Venous Return, and Their Regulation

OBJECTIVE 23–1.

Identify cardiac output, its significance, and its relationship to heart rate and stroke volume, venous return, and cardiac index.

1. Cardiac output in liters per minute divided by the heart rate in beats per minute equals:

   a. The cardiac index
   b. Cardiac efficiency
   c. Mean arterial pressure
   d. Mean stroke volume
   e. Mean blood velocity
   f. Zero

2. The cardiac index is equal to _____ . (HR, heart rate; SA, surface area; SV, stroke volume; CO, cardiac output)

   a. (HR) (SA)/(CO)
   b. (CO) (SA)
   c. (SV)/(SA)
   d. (HR) (SV)/(SA)

3. The relative constancy of cardiac index and arterial pressures among a group of adults with varying body sizes implies that total peripheral resistance varies _____ (D, directly; I, inversely) with the body surface area to the _____ power.

   a. D, 1st
   b. I, 1st
   c. D, 2nd
   d. D, 4th
   e. I, 2nd
   f. I, 3rd

OBJECTIVE 23–2.

Identify normal values for cardiac output and cardiac index and their variance with age, body posture, and metabolic rate.

4. Average adult values for cardiac output and body surface area are respectively _____ liters/minute and _____ square meters.

   a. 5, 1.7
   b. 10, 1.7
   c. 22, 1.7
   d. 5, 3.0
   e. 10, 3.0
   f. 22, 3.0

5. The cardiac index tends to _____ with increasing ages between 15 and 30 years, and tends to _____ with increasing ages between 40 and 80 years. (I, increase; D, decrease)

   a. I, I
   b. I, D
   c. D, I
   d. D, D

6. _____ muscle tone, occurring when a person rises from the reclining to standing position, may completely compensate for the tendency towards _____ cardiac output resulting from gravitational forces. (I, increased; D, decreased)

   a. I, I
   b. I, D
   c. D, I
   d. D, D

7. Cardiac output, exhibiting an approximate _____ (D, direct; I, inverse) proportional relationship to metabolic rate, varies as the _____ (Z, zero; F, first; S, second) power of the rate of oxygen consumption.

   a. D, Z
   b. D, F
   c. D, S
   d. I, Z
   e. I, F
   f. I, S

OBJECTIVE 23–3.

Recall the significance of the Frank-Starling law. Contrast the peripheral circulatory influences upon venous return and permissive levels of cardiac pumping as determinants of cardiac output.

8. Cardiac pumping effectiveness is a major determinant of cardiac output under conditions of _____ (N, normal circumstances; F, cardiac failure).

    a. Neither N nor F   c. F but not N
    b. N but not F   d. Both N and F

9. Under resting conditions the maximum pumping capability of the heart is about _____ times an average cardiac output of _____ liters/minute.

    a. 1, 5   c. 7, 5   e. 3, 7
    b. 3, 5   d. 1, 7   f. 7, 7

10. The volume of blood that enters the right atrium per minute is normally primarily established by:

    a. The Frank-Starling law
    b. Peripheral circulatory factors
    c. Arterial pressure
    d. Homeometric autoregulation
    e. Pulmonary resistance
    f. Cardiac permissive levels

OBJECTIVE 23–4.

Identify the effects of hypertrophy, autonomic stimulation, and heart disease upon permissive levels of cardiac pumping.

11. Under resting conditions, cardiac output approaches the cardiac "permissive level" in the _____ heart.

    a. Hypertrophied   c. Failing
    b. Normal   d. Endurance-trained

12. _____ (S, sympathetic; P, parasympathetic) stimulation to the heart, which tends to increase cardiac contractility, may increase the permissive level of heart pumping to a maximum of _____ times its nonstimulated value.

    a. S, 2   c. S, 8   e. P, 4
    b. S, 4   d. P, 2   f. P, 8

OBJECTIVE 23–5.

Recognize the normal role of peripheral resistance in determining venous return and cardiac output when arterial pressure remains about normal, and the consequences of a failure to maintain arterial pressure.

13. Venous return is _____ proportional to total peripheral resistance and _____ proportional to the difference between arterial and right atrial pressures. (D, directly; I, inversely)

    a. D, D   c. I, D
    b. D, I   d. I, I

14. Vascular dilation within the systemic circulation generally _____ venous return and _____ cardiac output. (I, increases; D, decreases)

    a. I, I   c. D, I
    b. I, D   d. D, D

15. The magnitude of cardiac output is normally established through:

    a. Mean arterial pressure   e. The heart
    b. Mean systemic pressure   f. The pulmonary
    c. Renal regulation   circulation
    d. Tissue autoregulation

16. A tendency for a _____ (R, rise; F, fall) in arterial pressure _____ (A, aids; O, opposes) the change in cardiac output resulting from widespread vasodilation of peripheral tissues.

    a. R, A   c. F, A
    b. R, O   d. F, O

17. Sudden opening of an arteriovenous fistula results in _____(D, an early decrease; N, no change; I, an early increase) in cardiac output as a result of _____ (S, sympathetic reflexes; V, altered venous return; C, a constant metabolic rate).

    a. D, V   b. D, S   c. N, S   d. N, C   e. I, V   f. I, S

OBJECTIVE 23–6.
Identify the significance of systemic filling pressure on venous return and the regulation of cardiac output.

18. Increased cardiac output is more readily achieved by _____ systemic filling pressure via _____ sympathetic venous tone. (I, increased; D, decreased)

    a. I, I   c. D, I
    b. I, D   d. D, D

19. A nonexistent pressure gradient for venous return from the peripheral tissues to the heart occurs when the systemic filling pressure equals:

    a. 25 mm. Hg   d. Capillary pressure
    b. 7 mm. Hg   e. Right atrial pressure
    c. Pulmonary filling   f. None of the above
        pressure

OBJECTIVE 23–7.
Identify the potential requirements and the peripheral and cardiac adjustments for cardiac output during strenuous exercise.

20. Just prior to the onset of exercise, the permissive level of cardiac pumping is increased while cardiac output_____(R, remains constant; I, is increased; D, is decreased) as a primary consequence of the influence of sympathetic activity upon the heart and_____ (PR, peripheral resistance; VT, venomotor tone).

    a. R, PR   c. D, PR   e. I, VT
    b. I, PR   d. R, VT   f. D, VT

21. The mechanical effects of altered skeletal muscle tonus, occurring at the very onset of exercise, is a mechanism which tends to _____ systemic filling pressure and _____ flow resistance. (I, increase; D, decrease)

    a. I, I   c. D, I
    b. I, D   d. D, D

22. At the onset of exercise the motor cortex transmits signals to_____ the sympathetic vasoconstrictor system and _____ the sympathetic vasodilator system. (A, activate; I, inhibit)

    a. A, A   c. I, A
    b. A, I   d. I, I

23. The state of the cardiovascular system at the onset of exercise, compared to resting conditions, indicates a more significant decrease in:

    a. Mean arterial pressure   d. Muscle tone
    b. Cardiac output   e. Vascular capacity
    c. Oxygen consumption   f. Sympathetic tone

24. The marked _____ (I, increase; D, decrease) in total peripheral resistance resulting from _____ (S, altered sympathetic activity; L, local auto-regulation) is the most significant factor providing for required cardiac output during strenuous muscle exercise.

    a. I, S   c. D, S
    b. I, L   d. D, L

OBJECTIVE 23-8.

Identify the conditions whereby cardiac performance becomes a limiting and determining factor for the regulation of cardiac output. Recognize abnormal factors which affect cardiac pumping ability and peripheral control of venous return.

25. _____ (C, cardiac; P, peripheral) factors _____ (I, increase; D, decrease) cardiac output when the permissive level of cardiac pumping falls below adequate levels of blood flow.

   a. C, I   c. P, I
   b. C, D   d. P, D

26. Abnormalities of the peripheral control of venous return include _____ total peripheral resistance associated with excess cardiac output, and _____ systemic filling pressure as a cause of decreased cardiac output. (I, increased; D, decreased)

   a. I, I   c. D, I
   b. I, D   d. D, D

27. A 30% reduction of blood volume with hemorrhage results in _____ (I increased; D, decreased) cardiac output as a consequence of altered _____ (O, oxygen-carrying capacity; P, systemic filling pressure; R, peripheral resistance).

   a. I, O   c. I, R   e. D, P
   b. I, P   d. D, O   f. D, R

28. A more important role of the cardiac performance in regulation of cardiac output occurs in hyperthyroid states and beriberi, in which there is abnormally _____ (L, low; H, high) cardiac output as a primary consequence of abnormal _____ (P, peripheral resistance; B, blood volume; C, cardiac performance).

   a. L, P   c. L, C   e. H, B
   b. L, B   d. H, P   f. H, C

OBJECTIVE 23-9.

Identify cardiac output curves, their significance, factors producing hypereffective and hypoeffective hearts and their influence on cardiac output curves, and the causes and consequences upon cardiac output curves of cardiac tamponade and altered intrapleural pressure.

29. A cardiac function curve, obtained by plotting cardiac output on the ordinate versus _____ (PR, peripheral resistance; A, right atrial pressure) on the abscissa, is shifted to the _____ (R, right; L, left) by opening the thoracic cage.

   a. PR, R   c. A, R
   b. PR, L   d. A, L

30. Factors producing a _____ (E, hyperdynamic; O, hypodynamic) heart include sympathetic stimulation and _____ (I, increased; D, decreased) systemic arterial pressure.

   a. E, I   c. O, I
   b. E, D   d. O, D

31. An increase in intrapleural pressure of 2 mm. Hg shifts the cardiac output to the _____ (R, right; L, left) by _____ (G, a greater; S, the same; E, a lesser) amount.

   a. R, G   c. R, E   e. L, S
   b. R, S   d. L, G   f. L, E

32. The normal extracardiac pressure is about _____ mm. Hg and becomes more _____ (P, positive; N, negative) with positive pressure forms of artificial respiration.

   a. -4, P   c. +4, P   e. 0, N
   b. 0, P   d. -4, N   f. +4, N

33. Cardiac tamponade refers to _____ (A, arrhythmia; P, pericardial fluid accumulation), and results in _____ (R, reduced; E, elevated) right atrial pressures in order to maintain a normal cardiac output.

   a. A, R   c. P, R
   b. A, E   d. P, E

34. Cardiac tamponade results in a more substantial alteration in the cardiac output curves at _____ (E, elevated; L, lowered) values of cardiac output as a consequence of the relationship between cardiac output and cardiac _____ (I, index; V, volumes).

   a. E, I   c. L, I
   b. E, V   d. L, V

OBJECTIVE 23–10.

Identify the venous return curves and their significance, and factors producing altered systemic filling pressure and peripheral resistance and their influence upon venous return curves.

35. The venous return curve obtained by plotting venous return on the ordinate versus _____ (C, capillary; A, right atrial) pressure on the abscissa exhibits a "plateau" resulting from the _____ (F, Frank-Starling law; V, venous collapse).

    a. C, F
    b. C, V
    c. A, F
    d. A, V

36. The point of intersection of the venous return curve with the abscissa equals the _____ pressure.

    a. Atomospheric
    b. Intrapleural
    c. Capillary
    d. Systemic filling
    e. Mean arterial
    f. Critical closing

37. Either a loss of sympathetic tone or _____ (I, increased; D, decreased) blood volumes shift the venous return curve and the point of its intersection with the abscissa to the _____ (L, left; R, right).

    a. I, L
    b. I, R
    c. D, L
    d. D, R

38. The systemic filling pressure, normally about _____ mm. Hg, is _____ (I, increased; D, decreased) by increased sympathetic tone.

    a. −7, D
    b. 0, D
    c. +7, D
    d. −7, I
    e. 0, I
    f. +7, I

39. Peripheral vasodilation, which _____ (I, increases; D, decreases) peripheral resistance, makes the slope of the venous return curve _____ (L, less negative; M, more negative; P, positive).

    a. I, L
    b. I, M
    c. I, P
    d. D, L
    e. D, M
    f. D, P

40. The venous return curve predicts that if both venous return and right atrial pressure maintain constant values with increasing systemic filling pressure, a simultaneous _____ (I, increase; D, decrease) must occur in _____ (P, peripheral resistance; L, left atrial pressure).

    a. I, P
    b. I, L
    c. D, P
    d. D, L

OBJECTIVE 23–11.

Recognize the equilibrium point predicted by combinations of cardiac output and venous return curves. Identify the influence of altered blood volume, sympathetic activity, and exercise upon combined graphical representations of cardiac output and venous return curves.

41. Increased blood volume shifting the venous return curve _____ (R, upward and to the right; L, downward and to the left), results in an early _____ (I, increase; D, decrease) in cardiac output.

    a. R, I
    b. R, D
    c. L, I
    d. L, D

42. The early influence of increased blood volume upon cardiac output is subsequently _____ (O, opposed; A, augmented) by _____ (I, increased; D, decreased) systemic filling pressure resulting from capillary fluid shift and stress-relaxation mechanisms.

    a. O, I
    b. O, D
    c. A, I
    d. A, D

43. Maximal sympathetic stimulation tends to _____ (I, increase; D, decrease) the equilibrium point for cardiac output and venous return for a period of time through supportive influences of altered _____ (C, cardiac output curves only; V, venous return curves only; CV, both cardiac output and venous return curves).

    a. I, C
    b. I, V
    c. I, CV
    d. D, C
    e. D, V
    f. D, CV

44. Loss of sympathetic tone results in a shift of the venous return curve to the _____ (R, right; L, left), and a slight _____ (I, increase; D, decrease) in right atrial pressure resulting from a shift in the cardiac output curve.

    a. R, I
    b. R, D
    c. L, I
    d. L, D

45. During exercise the _____ (V, venous return, O, cardiac output) curve is rotated upward to the right as a consequence of _____ (C, peripheral vasoconstriction; D, local vasodilation) and altered systemic filling pressure.

    a. V, C   b. V, D   c. O, C   d. O, D

OBJECTIVE 23–12.

Identify the mechanism providing for a steady state balance between right and left ventricular output.

46. A transient reduction in left ventricular output compared to right ventricular output is compensated for by _____ left atrial pressure and _____ pulmonary filling pressure. (I, increased; D, decreased)

    a. D, D   b. D, I   c. I, D   d. I, I

OBJECTIVE 23–13.

Identify commonly utilized methods for direct and indirect measurement of cardiac output. Identify the principle involved in the Fick, indicator dilution, and pulse pressure methods for estimation of quantitation of cardiac output.

47. The amount of a substance taken up by an organ (or by the whole body) per unit time is equal to the arterial level of the substance minus the venous level (A–V difference) _____ (M, multiplied; D, divided) by blood flow, and is known as (the) _____ (S, Starling; F, Fick; R, Reynold's) principle.

    a. M, S   c. M, F   e. M, R
    b. D, S   d. D, F   f. D, R

48. A systemic arterial-venous oxygen concentration difference of 3 volumes per cent, and an oxygen consumption rate of 180 ml./minute, would best correlate with a cardiac output closest to _____ liters/minute.

    a. 1.0   c. 4.5   e. 6.0
    b. 3.2   d. 5.4   f. 32.4

49. (X) mg. of cardiogreen injected intravenously produces an extrapolated dye dilution curve averaging (Y) mg. per liter of blood over its time interval of (Z) minutes. Cardiac output in liters per minute, therefore, equals _____ divided by _____.

    a. Z, Y   c. XZ, Y   e. XY, Z
    b. XY, X   d. Y, Z   f. X, YZ

50. The pulse pressure method for estimating cardiac output relies on the principle that increased rates of blood flow _____ (I, increase; D, decrease) the rate of decline of arterial blood pressure during _____ (S, systole; DI, diastole).

    a. I, S   c. D, S
    b. I, DI   d. D, DI

51. Continuous recordings of aortic blood flow, more readily accomplished by an _____ (F, indirect Fick method; E, electromagnetic flowmeter), indicate that aortic flow shows a transient reversal towards the end of _____ (S, systole; D, diastole).

    a. F, S   c. E, S
    b. F, D   d. E, D

24

The Pulmonary Circulation

OBJECTIVE 24–1.

Recognize the essential equality of pulmonary and systemic circulatory blood flow. Describe the physiologic anatomy of the right side of the heart, pulmonary vessels, bronchial vessels, and lymphatics.

1. A common feature of the pulmonary and systemic circulations is that they have essentially the same:

   a. Pulse pressure   d. Systolic pressure
   b. Peripheral resistance   e. Distensibility
   c. Diastolic pressure   f. Stroke volume

2. The wall thickness of the right ventricle is about _____ that of the right atrium and _____ that of the left ventricle in accord with intracardiac pressures. (E, equal to; 3, 3 times; $\frac{1}{3}$, one-third)

   a. E, E   c. E, 3   e. 3, $\frac{1}{3}$
   b. E, $\frac{1}{3}$ d. 3, E   f. 3, 3

3. Ventricular systole results in ventricular septal deviation _____ (T, towards; A, away from) the more crescent- shaped cross-sectional area of the _____ (R, right; L, left) ventricle.

   a. T, R   c. A, R
   b. T, L   d. A, L

4. Corresponding subdivisions of the pulmonary arterial network have _____ wall thicknesses, _____ diameters, and _____ compliances when compared to the systemic arterial system. (G, greater; L, lesser)

   a. G, G, G   c. L, L, G   e. G, L, L
   b. L, G, G   d. L, L, L   f. G, L, G

5. The _____ (P, pulmonary; B, bronchial) artery contains oxygenated blood and returns to the heart via the _____ (R, right; L, left) atrium.

   a. P, R   c. B, R
   b. P, L   d. B, L

6. Right ventricular output is normally _____ than left ventricular output by an amount which is _____ than 5 per cent. (G, greater; L, less)

   a. G, G   c. L, G
   b. G, L   d. L, L

OBJECTIVE 24–2.

Identify the pressures within the pulmonary circulation during a typical cardiac cycle.

7. Right ventricular pressures on the average range between a high of _____ mm. Hg and a low of _____ mm. Hg.

   a. 56, 13   c. 13, 0   e. 22, 0
   b. 22, 7   d. 56, 7   f. 13, –5

8. A pulse pressure, which normally averages about two-thirds its corresponding systolic pressure, is encountered in the:

   a. Aorta   d. Pulmonary artery
   b. Right ventricle   e. Pulmonary vein
   c. Left ventricle   f. Pulmonary capillaries

9. Of the following, the *least* mean arterial pressure would be found in the:

   a. Uterine artery   d. Cerebral arteries
   b. Left coronary artery   e. Hepatic artery
   c. Pulmonary artery   f. All would be about equal

10. Pulmonary arterial pressures approximate _____ mm. Hg for diastolic and a mean of _____ mm. Hg.

    a. 0, 13   c. 13, 22   e. 8, 22
    b. 8, 13   d. 0, 22   f. 13, 56

11. The pulmonary arterial pulse is most directly influenced by pulmonary _____ (SV, stroke volume; C, arterial compliance).

    a. SV times C   c. SV/C
    b. 1/(SV times C)   d. C/SV

12. The pulmonary capillary pressure is about halfway between the pulmonary _____ arterial pressure and _____ pressure. (S, systolic; M, mean; D, diastolic; L, left atrial)

    a. S, D   c. D, L
    b. M, L   d. M, D

13. Compared with the systemic circulation under normal conditions, the pulmonary circulation is characterized by _____ flow, _____ pressure, and _____ resistance. (H, high; L, low; E, equal)

    a. L, L, L   c. H, L, E
    b. E, L, L   d. E, L, E

14. Mean left atrial pressure normally ranges from _____ mm. Hg and contains c waves of _____ (G, greater; L, lesser) amplitude than the right atrium.

    a. -5 to 0, G   c. 5 to 10, G   e. 1 to 4, L
    b. 1 to 4, G   d. -5 to 0, L   f. 5 to 10, L

OBJECTIVE 24-3.

Identify the pulmonary blood volume, its relative distribution, its reservoir functional role, and its shift with right- or left-sided heart failure.

15. The pulmonary blood volume is about _____ % of the total blood volume, or about _____ ml.

    a. 5, 250   c. 22, 1100   e. 9, 900
    b. 9, 450   d. 5, 500   f. 22, 2200

16. Increased pulmonary air pressures may _____ (I, increase; D, decrease) pulmonary blood volume by _____ (M, more; N, no more) than 25%.

    a. I, M   c. D, M
    b. I, N   d. D, N

17. Left heart failure tends to _____ pulmonary blood volume and _____ pulmonary circulatory pressures. (I, increase; D, decrease)

    a. I, I   c. D, I
    b. I, D   d. D, D

18. Shifts of blood volume between the pulmonary and systemic circulations affect the functions of the pulmonary system to a _____ extent than the systemic circulation as a consequence of the relatively _____ blood volume of the pulmonary circulation. (G, greater; L, lesser)

    a. G, G   c. L, G
    b. G, L   d. L, L

OBJECTIVE 24-4.

Identify the regulatory mechanisms which control pulmonary blood flow. Identify the effects of hydrostatic pressure, lowered alveolar oxygen pressures, and autonomic innervation upon pulmonary flow characteristics and the regional distribution of pulmonary flow.

19. The primary determinants of pulmonary blood flow are _____ (C, cardiac; S, systemic circulatory; P, pulmonary circulatory) factors that regulate _____ (CO, cardiac output; PR, pulmonary resistance).

    a. C, CO   c. P, CO   e. S, PR
    b. S, CO   d. C, PR   f. P, PR

20. Changing body position from a lying to upright position results in _____ pulmonary vascular pressures and _____ blood flow in the region of apical lung tissues. (N, no change in; I, increased; D, decreased).

    a. N, N   c. I, N   e. D, I
    b. N, I   d. D, N   f. D, D

21. Exercise increases pulmonary _____ (R, flow resistance; P, pressures) and increases pulmonary blood volume, particularly by increasing the number of functioning capillaries at the _____ (A, apex; B, base) of the lung.

    a. R, A   c. P, A
    b. R, B   d. P, B

22. Obstruction of a bronchus results in _____ blood flow to poorly ventilated alveoli, and _____ blood flow to well ventilated alveoli. (I, increased; D, decreased; N, no change in)

    a. D, D   c. D, N   e. I, D
    b. D, I   d. I, I   f. I, N

23. Sympathetic stimulation results in _____ (O, opposing; S, similar) directional changes in vasomotor tone among the systemic and pulmonary circulations which are more significant for the _____ (SY, systemic; P, pulmonary) circulation.

    a. O, SY   c. S, SY
    b. O, P   d. S, P

OBJECTIVE 24–5.

Identify the mechanisms and significance of pulmonary capillary membrane dynamics producing its negative interstitial fluid pressure. Identify factors which normally guard against pulmonary edema, and those conditions which may result in pulmonary edema.

24. Pulmonary capillary pressures, lying between mean values of left atrial and pulmonary arterial pressures of _____ mm. Hg, respectively, are typical of systemic _____ (C, capillary; V, peripheral vein) pressures.

    a. 2 & 13, C   c. 7 & 56, C   e. 4 & 22, V
    b. 4 & 22, C   d. 2 & 13, V   f. 7 & 56, V

25. The alveolar gas to blood diffusion distance, _____ (G, greater; L, less) than one micron in length, is maintained by interstitial dehydration effects of a favorable _____ (O, osmotic; H, hydrostatic) pressure gradient.

    a. L, O   c. G, O
    b. L, H   d. G, H

26. The normal interstitial fluid pressure of the lungs, probably about _____ mm. Hg, provides the mechanism for _____ (L, low pulmonary capillary pressures; D, dry alveoli).

    a. −17, L   c. +6, L   e. −6, D
    b. −6, L   d. −17, D   f. +6, D

27. The most frequent cause of pulmonary edema is _____ (I, increased; D, decreased) pulmonary capillary pressures resulting from _____ (L, left; R, right) heart failure.

    a. I, L   c. D, L
    b. I, R   d. D, R

28. The pulmonary capillary pressure "safety factor" against pulmonary edema, about _____ mm. Hg, is _____ (I, increased; D, decreased) by chronically increased pulmonary lymphatic flows.

    a. 7, D   c. 45, D   e. 23, I
    b. 23, D   d. 7, I   f. 45, I

OBJECTIVE 24–6.

Identify the consequences of elevated cardiac output upon pulmonary capillary transit times, respiratory gas diffusion rates, pulmonary resistances, pulmonary arterial pressures, and pulmonary volumes.

29. Pulmonary capillary transit times for gas exchange, typically about _____ second(s),
_____ (S, are shortened; C, remain constant; L, are lengthened) by increased cardiac
output.

   a. 1, S   c. 1, L   e. 5, C
   b. 1, C   d. 5, S   f. 5, L

30. The rate of respiratory gas diffusion between
blood and alveoli is increased during heavy
exercise by increased _____ (T, capillary
transit times; F, blood flow) through alveoli,
_____ (A, and; B, but not) by increasing
the number of alveoli participating in gas
exchange.

   a. T, A   c. F, A
   b. T, B   d. F, B

31. Doubling a typical cardiac output results in
_____ (I, increased; D, decreased) pulmonary blood volume as a result of changes in
pulmonary _____ (V, vasomotor tone; P,
cardiovascular pressures).

   a. I, V   c. D, V
   b. I, P   d. D, P

32. A three-fold increase in cardiac output normally results in a _____ in pulmonary
vascular resistance and a _____ in pulmonary arterial pressure. (LI, large increase; SI,
slight increase; LD, large decrease; SD, slight
decrease)

   a. LD, SI   c. LD, SD   e. LI, LI
   b. SD, LI   d. SI, LI   f. SD, SD

OBJECTIVE 24-7.

Identify the consequences of increased left ventricular work load and left ventricular
failure upon left atrial pressure, pulmonary arterial pressure, and the right ventricle.

33. Pulmonary arterial pressure changes with
varied left atrial pressures are _____ in the
low left atrial pressure range and _____ in
the high left atrial pressure range. (M, marked;
S, slight)

   a. M, M   c. S, M
   b. M, S   d. S, S

34. Pulmonary resistance, which _____ (I, increases; D, decreases) with increasing left atrial
pressures, shows the greatest variance with left
atrial pressures ranging from _____ mm.
Hg.

   a. I, 2 to 7   c. I, 14 to 19   e. D, 8 to 13
   b. I, 8 to 13   d. D, 2 to 7   f. D, 14 to 19

35. Systemic hypertension, and its accompanying
increased left ventricular work load, results in
_____ pulmonary arterial pressure and consequently _____ right ventricular work
load. (I, increased; N, little or no change in; D,
decreased)

   a. I, D   c. D, D   e. N, N
   b. I, I   d. D, I   f. N, I

36. Elevated right atrial pressure as a consequence
of left heart failure is generally associated with
_____ left atrial pressures and _____
pulmonary arterial pressures. (L, abnormally
low; N, normal; H, abnormally high)

   a. L, L   c. L, H   e. H, N
   b. L, N   d. N, H   f. H, H

37. Cardiac conditions producing a left atrial pressure above 40 mm. Hg frequently result in
_____ (P, pulmonary edema, S, peripheral
edema).

   a. P but not S   c. Both P and S
   b. S but not P   d. Neither P nor S

OBJECTIVE 24-8.

Identify the pulmonary conditions associated with pulmonary embolism, emphysema,
diffuse sclerosis, atelectasis, and lung tissue removal, and their consequences upon the
pulmonary circulation.

38. Alveolar collapse of all or part of a lung, termed _____ (E, emphysema; A, atelectasis), results in increased pulmonary blood flow in the _____ (C, collapsed; N, noncollapsed) portions of the lungs.

    a. E, C   c. A, C
    b. E, N   d. A, N

40. Removal of an entire lung generally results in substantially _____ (I, increased; D, decreased) pulmonary arterial pressures at _____ (E, only elevated; N, normal as well as elevated) levels of cardiac output.

    a. I, E   c. D, E
    b. I, N   d. D, N

39. Symptoms of right heart failure associated with emphysema may be alleviated as a consequence of _____ (I, increased; D, decreased) pulmonary resistance produced by breathing air containing elevated _____ concentrations.

    a. I, O_2 c. I, N_2 e. D, CO_2
    b. I, CO_2 d. D, O_2 f. D, N_2

25

The Coronary Circulation and Ischemic Heart Disease

OBJECTIVE 25–1.

Identify the origin, distribution, and drainage of the coronary vessels. Identify normal variations in coronary flow and the consequences of ventricular systole and diastole upon the regional distribution and phasic nature of coronary flow.

1. Right coronary arterial blood flow, supplying the right ventricle _____ (O, only; P, and part of the left ventricle), is more frequently _____ (L, less; G, equal to or greater) than left coronary arterial blood flow.

    a. O, L   c. P, L
    b. O, G   d. P, G

2. Coronary venous blood, draining the _____ (R, right; L, left) ventricle, drains into the _____ (CS, coronary sinus; RA, right atrium; LA, left atrium).

    a. L, CS   c. L, LA   e. R, RA
    b. L, RA   d. R, CS   f. R, LA

3. Resting values of coronary blood flow are about _____ % of a typical cardiac output of _____ liters/minute.

    a. 0 to 1, 5   c. 12 to 14, 5   e. 4 to 5, 10
    b. 4 to 5, 5   d. 0 to 1, 10   f. 12 to 14, 10

4. Increased cardiac output results in _____ ratios of coronary blood flow to cardiac energy expenditures, and _____ ratios of cardiac work output to energy expenditure. (I, increased; D, decreased)

    a. I, I   c. D, I
    b. I, D   d. D, D

5. A pronounced reduction in coronary blood flow occurs in the _____ (R, right; L, left) coronary artery during ventricular _____ (S, systole; D, diastole).

    a. L, S   c. R, S
    b. L, D   d. R, D

6. During ventricular systole, intramyocardial compression effects are greatest for the _____ muscle layer, and least for the _____ muscle layer. (O, outer; M, middle; I, inner)

a. O, M   c. I, O   e. I, M
b. M, I   d. O, I   f. M, O

7. Subendothelial arteries are _____ (L, larger; S, smaller), and have _____ (E, lesser; G, greater) diastolic blood flows than corresponding nutrient arteries of the outer and middle cardiac muscle layers.

a. L, E   c. S, E
b. L, G   d. S, G

OBJECTIVE 25–2.

Identify the roles of oxygen demand and local metabolism in the regulation of coronary flow, factors influencing cardiac oxygen consumption, and the significance of local autoregulation of coronary blood flow.

8. The primary controller of coronary blood flow involves cardiac _____ (A, autonomic tone; M, metabolic factors), especially _____ .

a. A, sympathetic tone   c. M, O_2 lack
b. A, parasympathetic   d. M, O_2 excess
    tone

9. During resting conditions, the arteriovenous oxygen difference is greatest for:

a. Coronary vessels   c. Cutaneous vessels
b. Renal vessels   d. Skeletal muscle

10. Oxygen lack of cardiac tissues results in the local release of _____ (D, vasodilator; C, vasoconstrictor) compounds, including _____ (A, adenosine; T, theophylline; Ca, calcium ions).

a. D, A   c. D, Ca   e. C, T
b. D, T   d. C, A   f. C, Ca

11. Among the following conditions, cardiac oxygen consumption would be increased above normal to the greater extent by _____ cardiac output and _____ systemic arterial pressure. (N, normal; ½N, halving; 2N, doubling)

a. ½N, ½N   c. N, 2N   e. 2N, ½N
b. ½N, 2N   d. 2N, N   f. ½N, N

12. Cardiac oxygen consumption is _____ proportional to myocardial muscle tension and _____ proportional to the contraction duration. (D, directly; I, inversely)

a. I, I   c. D, I
b. I, D   d. D, D

13. With a constant cardiac work output, cardiac oxygen consumption _____ (C, remains constant; I, increases; D, decreases) with increasing end-diastolic volumes in the dilated heart as a consequence of _____ (P, Poiseuille's law; L, the law of Laplace).

a. C, P   c. D, P   e. I, L
b. I, P   d. C, L   f. D, L

14. Thyroxine, digitalis, and elevated cardiac temperatures _____ (I, increase; D, decrease) coronary blood flow in approximate proportion to increased cardiac _____ (W, work output; O, oxygen consumption).

a. D, W   c. I, W
b. D, O   d. I, O

15. A tendency towards _____ (A, anginal pain; H, relative hyperemia), produced by doubling the cardiac metabolic rate in the absence of a corresponding increase in coronary blood flow, _____ (I, is; N, is not) largely compensated for by increased oxygen extraction from coronary blood.

a. A, I   c. H, I
b. A, N   d. H, N

16. During a period of reactive hyperemia, coronary blood flow _____ and cardiac oxygen utilization _____ when compared to control levels. (I, increases; C, remains normal; D, decreases).

a. I, I   c. I, D   e. D, C
b. I, C   d. D, I   f. D, D

OBJECTIVE 25-3.

Identify the direct and indirect effects of autonomic nerves upon coronary flow. Distinguish the cardiac distribution and roles of alpha and beta receptors.

17. The more important _____ (D, direct; I, indirect) effects of autonomic activity upon coronary blood flow are those effects which alter activity of _____ (A, alpha receptors; C, cardiac muscle cells).

    a. D, C   c. I, C
    b. D, A   d. I, A

18. Indirect effects upon coronary blood flow include _____ blood flow with sympathetic stimulation, and _____ blood flow with parasympathetic stimulation. (I, increased; D, decreased)

    a. I, I   c. D, I
    b. I, D   d. D, D

19. Coronary vascular receptors, associated with the _____ (P, parasympathetic; S, sympathetic) division of the autonomic nervous system, include beta or _____ (C, constrictor; D, dilator) receptors.

    a. P, C   c. S, C
    b. P, D   d. S, D

20. _____ receptors predominate in the epicardial arterial distribution, and _____ receptors predominate in intramuscular arterial distribution of the heart. (A, alpha; B, beta)

    a. A, A   c. B, A
    b. A, B   d. B, B

21. The least consequential effects upon coronary blood flow occur via _____ effects.

    a. Direct sympathetic
    b. Indirect sympathetic
    c. Direct parasympathetic
    d. Indirect parasympathetic

22. The overall effects of sympathetic stimulation on coronary blood flow normally result from activity of the _____ (A, alpha; B, beta) receptors, controlling the major flow resistance of the _____ (E, epicardial; I, intramuscular) vascular network.

    a. A, E   c. B, E
    b. A, I   d. B, I

OBJECTIVE 25-4.

Identify the principle metabolic substrate utilized by cardiac muscle, and the roles of aerobic metabolism, anaerobic glycolysis, and adenosine compounds in cardiac metabolic functions.

23. Metabolic conditions associated with a cardiac index of 3.0 liters per minute per square meter are principally _____ (A, aerobic; N, anaerobic), and involve _____ (C, carbohydrate; L, lipid; P, protein) utilization as the major substrate for cardiac metabolism.

    a. A, C   c. A, P   e. N, L
    b. A, L   d. N, C   f. N, P

24. Cardiac muscle cells are more permeable to _____ , whose intracellular conversion _____ (I, increases; D, decreases) during coronary ischemia.

    a. ATP, I   d. ATP, D
    b. ADP, I   e. ADP, D
    c. Adenosine, I   f. Adenosine, D

OBJECTIVE 25-5.

Identify the processes of coronary atherosclerosis, coronary occlusion, and myocardial infarction, and their potential consequences upon cardiac function.

25. Cholesterol and lipid deposits beneath the intima, becoming fibrous and frequently calcifying, are a major cause of increased coronary arterial:

    a. Internal diameters c. Thrombus formation
    b. Distensibility   d. Coronary flow

26. Recovery from coronary occlusion is largely dependent upon:

    a. Powerful sympathetic reflexes
    b. Improved collateral circulation
    c. Increased myoglobin content
    d. Reactive hyperemia
    e. Cardiac muscle regeneration
    f. Normal coronary flow levels

27. A critical level of coronary blood flow for sustaining viability of cardiac muscle cells occurs at a percentage value of normal, approximating _____ %.

a. 1   c. 30   e. 70
b. 10   d. 50   f. 90

28. A region of the heart, which has inadequate coronary blood flow to sustain vital cardiac muscle functions, undergoes a process of:

a. Atherosclerosis   d. "Diastolic stretch"
b. Coronary occlusion e. Fibrillation
c. Myocardial   f. Myocardial rupture
   infarction

29. Decreased cardiac pumping strength, as a consequence of a regional loss of ventricular muscle function following acute coronary occlusion, is frequently made _____ (L, less; M, more) severe by the phenomenon of _____ (D, diastolic; S, systolic) stretch of the nonfunctioning muscle.

a. L, D   c. M, D
b. L, S   d. M, S

30. Inadequate cardiac output resulting from the consequences of myocardial infarction elicits powerful _____ (P, parasympathetic; S, sympathetic) reflexes which _____ (I, increase; D, decrease; N, do not appreciably alter) cardiac muscle irritability.

a. P, I   c. P, N   e. S, D
b. P, D   d. S, I   f. S, N

31. An especially dangerous period occurs within _____ following myocardial infarction as a consequence of danger from _____ (R, rupture of the infarcted area; F, fibrillation).

a. 0 to 10 minutes, R   c. ½ to 2 hours, R
b. 0 to 10 minutes, F   d. ½ to 2 hours, F

32. A patient should _____ (E, exercise; R, rest) during the first few weeks following myocardial infarction, as a consequence of the _____ (C, blood coagulation tendency; S, "coronary steal" syndrome; H, resultant myocardial hypertrophy).

a. E, C   c. E, H   e. R, S
b. E, S   d. R, C   f. R, H

OBJECTIVE 25–6.
Identify angina pectoris, its cause, and its treatment.

33. The pain of angina pectoris, generally resulting from progressive _____ (C, constriction; E, enlargement) of coronary arteries, is transmitted to the central nervous system in conjunction with _____ (V, vagal; S, sympathetic) fibers.

a. C, V   c. E, V
b. C, S   d. E, S

34. Rapid relief of the pain of angina is frequently accomplished with nitroglycerine and amyl nitrite, which result in _____ (I, increased; D, decreased) systemic filling pressure and dilation of coronary _____ (A, arteries; O, arterioles).

a. I, A   c. D, A
b. I, O   d. D, O

35. Blocking activity of beta receptors _____ cardiac work output, and _____ the ratio of myocardial oxygen requirement to oxygen availability. (I, increases; D, decreases)

a. I, I   c. D, I
b. I, D   d. D, D

36. The pain of angina is referred to _____ (S, surface areas; V, visceral structures) of the body having similar _____ (M, sensory modalities; E, embryological origin).

a. S, M   c. V, M
b. S, E   d. V, E

OBJECTIVE 25–7.
Identify the application of the Fick principle to measurements of coronary blood flow.

37. The _____ (C, coronary flow; R, reciprocal of coronary flow), determined by the Fick principle, equals the quantity of measured substance removed by cardiac tissue from each ml. of blood, _____ (M, multiplied; D, divided) by the total quantity removed per unit of time.

a. C, M   b. C, D   c. R, M   d. R, D

26

Cardiac Failure

OBJECTIVE 26–1.

Identify cardiac failure, its general causes, two primary avenues for manifestations of cardiac failure, and the relative occurrence of right, left, and bilateral cardiac failure.

1. Conditions rendering the pumping ability of the heart inadequate are also conditions resulting in:

   a. Low cardiac output
   b. High right atrial pressure
   c. High left atrial pressure
   d. Cardiac failure
   e. Coronary ischemia
   f. Low arterial pressures

2. A tendency for _____ cardiac output with cardiac failure tends to be compensated for by _____ atrial pressures. (I, increased; D, decreased)

   a. I, I
   b. D, I
   c. I, D
   d. D, D

3. Cardiac failure resulting from coronary thrombosis occurs more frequently in the _____ (R, right; L, left) ventricle, and generally _____ (D, decreases; N, does not appreciably alter; I, increases) pulmonary arterial pressure.

   a. R, D
   b. R, N
   c. R, I
   d. L, D
   e. L, N
   f. L, I

OBJECTIVE 26–2.

Identify the acute and chronic circulatory consequences, compensatory mechanisms, and recovery stages following moderate myocardial infarction and compensated cardiac failure.

4. The progression of cardiac changes occurring with a moderate heart attack involve progressive cardiac output curve changes, in which the lowest right atrial pressure at comparable cardiac output values occurs for the _____ heart.

   a. Normal
   b. Acutely damaged
   c. Damaged & compensated
   d. Partially recovered

5. Compensatory mechanisms for acute cardiac failure following a mild heart attack involve increased outflow of predominant _____ (S, sympathetic; P, parasympathetic) reflexes, resulting in a higher than normal _____ .

   a. S, Cardiac reserve
   b. S, Cardiac output
   c. S, Systemic filling pressure
   d. P, Cardiac reserve
   e. P, Cardiac output
   f. P, Systemic filling pressure

6. A transient reduction of cardiac output to 40% of normal following a heart attack is more likely to result in central nervous system _____ (F, fainting; P, permanent damage), and _____ (I, increased; D, decreased) urine formation.

   a. F, I
   b. F, D
   c. P, I
   d. P, D

7. Renal compensatory mechanisms, predominant during the _____ (A, acute; C, chronic) stage of cardiac failure, involve _____ (I, increased; D, decreased) secretion of renin and aldosterone.

   a. A, I
   b. A, D
   c. C, I
   d. C, D

8. Altered glomerular filtration, resulting from cardiac failure, occurs as a consequence of _____ (I, increased; D, decreased) arterial pressure and renal afferent arteriolar _____ (C, constriction; DI, dilation).

    a. I, C   c. D, C
    b. I, DI   d. D, DI

9. Compensatory mechanisms for cardiac failure include altered extracellular water and electrolyte content as a consequence of _____ renal sodium ion reabsorption and _____ antidiuretic hormone secretion. (I, increased; D, decreased)

    a. I, I   c. D, I
    b. I, D   d. D, D

10. Compensatory extracellular fluid volume changes in cardiac failure _____ the pressure gradient for venous return and _____ venous resistance. (I, increase; D, decrease)

    a. I, I   c. D, I
    b. I, D   d. D, D

11. Tachycardia, cold skin, sweating, and pallor are symptoms of _____ (P, parasympathetic; S, sympathetic) stimulation, prominant during the _____ (A, acute; C, chronic) stage of cardiac failure.

    a. S, A   c. P, A
    b. S, C   d. P, C

12. Renal fluid output returns to normal following a moderate heart attack, when there is a return to normal of the:

    a. Right atrial pressure   d. Cardiac pumping
    b. Extracellular fluid   ability
       volume   e. Cardiac output
    c. Systemic filling   f. Cardiac reserve
       pressure

OBJECTIVE 26–3.
Identify the circumstances that lead to decompensated heart failure and the rationale for treatment of the decompensation.

13. A cardiac output which fails to rise to a critical level required for normal renal function causes _____ (P, progressive; L, cessation of) renal body fluid retention and _____ (C, compensated; D, decompensated) heart failure.

    a. P, C   c. L, C
    b. P, D   d. L, D

14. The nontreated decompensation process of cardiac failure leads to _____ (E, static equilibrium; P, pulmonary edema) progressing to death at sustantially _____ (H, elevated; L, lowered) right atrial pressures.

    a. E, H   c. P, H
    b. E, L   d. P, L

15. The cardiac output curve of decompensated heart disease shows a secondary _____ (R, rise; D, decline) in cardiac output at elevated right atrial pressures as a partial consequence of excessively high _____ (E, end-diastolic; S, stroke) volumes.

    a. R, E   c. D, E
    b. R, S   d. D, S

16. The rationale for treatment of the decompensation process of cardiac failure is to _____ (R, raise; L, lower) cardiac output by treatment which _____ (I, increases; D, decreases) the systemic filling pressure.

    a. L, D   c. R, D
    b. L, I   d. R, I

OBJECTIVE 26–4.

Identify the sequence of events occurring with acute and chronic stages of unilateral heart failure.

17. Unilateral left heart failure _____ pulmonary filling pressure and _____ systemic filling pressure. (I, increases; D, decreases)

   a. I, I
   b. I, D
   c. D, I
   d. D, D

18. Pulmonary edema, resulting from unilateral left heart failure when pulmonary capillary pressure first rises above _____ (N, normal; O, colloid osmotic pressure), is _____ (M, more; L, less) likely to occur during the acute stage than the chronic stage.

   a. N, M
   b. N, L
   c. O, M
   d. O, L

19. A shift in blood volume from the lungs into the systemic circulation occurs with severe unilateral _____ (L, left; R, right) heart failure and _____ (D, does; N, does not) result in severe systemic congestion during the acute stage.

   a. L, D
   b. L, N
   c. R, D
   d. R, N

20. Greater depression of cardiac output results from comparable unilateral _____ (L, left; R, right) heart failure as a consequence of the greater compliance of the _____ (P, pulmonary; S, systemic) circulation.

   a. L, P
   b. L, S
   c. R, P
   d. R, S

21. The kidneys play a significant role in compensatory processes of _____ heart failure and a significant role in the decompensation process of _____ heart failure. (R, only unilateral right; L, only unilateral left; B, both unilateral right and left)

   a. R, R   c. B, B   e. L, R
   b. L, L   d. R, L   f. B, L

OBJECTIVE 26–5.

Identify "high cardiac output failure" and its causes.

22. Basic causes of "high cardiac output failure" include _____ systemic resistance and _____ systemic filling pressure. (I, increased; D, decreased)

   a. I, I
   b. I, D
   c. D, I
   d. D, D

23. Symptoms of "high cardiac output failure" include _____ (H, higher; L, lower) than normal cardiac outputs under resting conditions and _____ (E, elevated; D, depressed) atrial pressures.

   a. H, E
   b. H, D
   c. L, E
   d. L, D

24. Vitamin B deficiencies and thyroid hormone _____ (F, deficiencies; E, excess), resulting in systemic _____ (C, vasoconstriction; D, vasodilation), are potential causes of high cardiac output failure.

   a. F, C
   b. F, D
   c. E, C
   d. E, D

OBJECTIVE 26–6.

Identify cardiogenic shock, its feedback influence on cardiac performance, and its treatment rationale.

25. Mean arterial pressures, which are just sufficient to prevent progressive myocardial deterioration, are typically _____ of the normal mean arterial pressure for the healthy heart and _____ (G, greater than; E, about equal to; L, less than) this value for the infarcted heart in cardiogenic shock.

   a. ¼, G   c. ¼, L   e. ½, E
   b. ¼, E   d. ½, G   f. ½, L

26. Cardiac shock implies a reduction in _____ (P, arterial pressure; CO, cardiac output) as a consequence of reduced _____ (CF, coronary flow; CP, cardiac pumping).

   a. P, CF   c. CO, CF
   b. P, CP   d. CO, CP

27. Agents preventing systemic arterial _____ (H, hypertension; L, hypotension) are frequently employed to lessen the _____ (N, negative; P, positive) feedback influence of cardiogenic shock upon cardiac performace.

   a. H, N   c. L, N
   b. H, P   d. L, P

OBJECTIVE 26–7.

Identify the causes and consequences of systemic and pulmonary edema occurring with cardiac failure.

28. Lethal consequences of _____ (P, peripheral; Y, pulmonary) edema during the acute stages of cardiac failure is more likely to occur with _____ (U, unilateral left; B, bilateral) heart failure.

   a. P, U   c. Y, U
   b. P, B   d. Y, B

29. Acute left cardiac failure results in a _____ (R, rise; F, fall) in systemic capillary pressures and, consequently, a(n) _____ (I, increased; D, decreased) tendency for peripheral edema.

   a. R, D   c. F, D
   b. R, I   d. F, I

30. Acute pulmonary edema triggered by heavy exercise or emotional excitement during chronic cardiac failure is thought to be associated with peripheral _____ (C, vasoconstriction; D, vasodilation) and to be alleviated by administering _____ (B, blood transfusions; O, pure oxygen).

   a. C, B   c. D, B
   b. C, O   d. D, O

OBJECTIVE 26–8.

Identify the role of decreased myocardial contractility in chronic heart failure, the causative factors, and the value of cardiotonic drugs in cardiac failure.

31. Excessive stretch of cardiac muscle and diminished _____ ion transport by the sarcoplasmic reticulum are frequent _____ (I, initiating; C, secondary contributing) causes of cardiac failure.

   a. Ca, I   c. Mg, I
   b. Ca, C   d. Mg, C

32. Reduced stores of _____ (E, epinephrine; N, norepinephrine; A, acetylcholine) in cardiac sympathetic nerve endings result from _____ (I, increased; D, decreased) enzyme concentrations of tyrosine hydroxylase.

   a. E, I   c. A, I   e. N, D
   b. N, I   d. E, D   f. A, D

33. Digitalis is useful in increasing cardiac contractility of _____ (N, only normal; F, only failing; B, both normal and failing) cardiac muscle via _____ (R, reflex; I, intrinsic) mechanisms.

   a. N, R   b. F, R   c. B, R   d. N, I   e. F, I   f. B, I

OBJECTIVE 26-9.

Identify the conditions of cardiac failure that result in symptoms of low cardiac output, pulmonary congestion, and systemic congestion.

34. Generalized weakness, fainting, and increased sympathetic activity resulting from _____ (CO, low cardiac output; S, systemic congestion) are more frequently the only acute symptoms in _____ (R, right; L, left) heart failure.

   a. CO, R   c. S, R
   b. CO, L   d. S, L

35. Pulmonary congestive symptoms resulting from acute _____ (R, right; L, left) heart failure _____ (D, do; N, do not) occur if cardiac output is normal.

   a. R, D   c. L, D
   b. R, N   d. L, N

36. Systemic congestion without pulmonary congestion, and with near normal cardiac output, is more likely to occur with _____ (A, acute; C, chronic) stages of _____ (L, unilateral left; R, unilateral right; B, bilateral) heart failure.

   a. A, L   c. A, B   e. C, R
   b. A, R   d. C, L   f. C, B

OBJECTIVE 26-10.

Identify cardiac reserve, its normal, adaptive, and pathological range of values, and the consequences of its variance.

37. Cardiac _____ (I, index; R, reserve) is the percentage change of cardiac output between normal and _____ (N, minimal; X, maximal) levels.

   a. I, X   c. R, X
   b. I, N   d. R, N

38. A cardiac reserve of 500% would more likely occur in the _____ young adult, whereas a cardiac reserve of 300% would more likely occur in the _____ young adult. (E, endurance-trained; N, average normal; A, asthenic)

   a. E, N   c. E, A   e. N, E
   b. A, N   d. A, E   f. N, A

39. Dyspnea and extreme muscle fatigue as a consequence of a mild exercise test are indicative of abnormally _____ (L, low; H, high) values of _____ (O, cardiac output; R, cardiac reserve) for the individual at rest.

   a. H, O   c. H, R
   b. L, O   d. L, R

OBJECTIVE 26–11.

Identify the associated changes in cardiac output and venous return curves with cardiac failure.

40. An equilibrium point of _____ cardiac output and elevated right atrial pressure is established in compensated cardiac failure with normal renal output and _____ systemic circulatory pressure. (N, normal; D, depressed; E, elevated)

 a. N, N   c. N, E   e. D, D
 b. N, D   d. D, N   f. E, D

41. Autonomic activity resulting from a moderately severe heart attack _____ (I, increases; D, decreases) the plateau level of the cardiac output curve and shifts the venous return curve to the _____ (L, left, R, right).

 a. I, L   c. D, L
 b. I, R   d. D, R

42. Treatment of decompensated heart failure with digitalis results in a "stabilized" state when the intersection of the venous return and cardiac output curves lies _____ (A, above; B, below; O, on) the critical cardiac output level required for normal _____ (F, fluid balance; S, systemic circulatory pressure).

 a. A, F   c. O, F   e. B, S
 b. B, F   d. A, S   f. O, S

43. _____ (I, increased; D, decreased) right atrial pressures and normal cardiac output curves frequently occur with _____ (C, compensated cardiac; DC, decompensated cardiac; H, high cardiac output) failure.

 a. D, C   c. D, H   e. I, DC
 b. D, DC   d. I, C   f. I, H

27

Heart Sounds; Dynamics of Valvular and Congenital Heart Defects

OBJECTIVE 27–1.

Identify the normal heart sounds, their causes, and acoustic characteristics.

1. The second heart sound results from vibrations occurring immediately following _____ (A–V; S, semilunar) valve _____ (C, closure; O, opening).

 a. A–V, C   c. S, C
 b. A–V, O   d. S, O

2. The "lub" sound or _____ heart sound occurs as a consequence of _____ (A–V; S, semilunar) valve closure.

 a. 4th, A–V   c. 2nd, A–V   e. 1st, S
 b. 1st, A–V   d. 4th, S   f. 2nd, S

3. A possible minor contributory factor in the origin of the _____ heart sound involves contraction sounds of the _____ (A, atrial; V, ventricular) muscle.

 a. 1st, V   c. 2nd, V   e. 4th, V
 b. 1st, A   d. 2nd, A   f. 3rd, V

4. Heart sound vibrations are transmitted along the arterial system as _____ (C, acoustic compression waves; L, lateral wall vibrations) which travel at _____ (S, the speed of sound; P, pulse wave velocities; V, blood flow velocities).

 a. C, S   c. C, V   e. L, P
 b. C, P   d. L, S   f. L, V

5. _____ (H, high; L, low) pitch sounds, produced by the close proximity of arterial walls to surrounding thoracic structures, lie largely _____ (A, above; B, below) the threshold of audibility of the human ear.

    a. H, A   c. L, A
    b. H, B   d. L, B

6. Durations of the first and second heart sounds are normally within the range of _____ seconds, and consist of lower frequency components for the _____ sound.

    a. 0.0010 to 0.0015, 1st   e. 0.010 to 0.015,
    b. 0.01 to 0.015, 1st   2nd
    c. 0.10 to 0.15, 1st   f. 0.10 to 0.15,
    d. 0.0010 to 0.0015, 2nd   2nd

7. The _____ (F, frequency; D, duration; L, loudness) of the first and second heart sounds are most directly proportional to the _____ (M, magnitude; R, rate of change) of pressure gradients across the respective valves.

    a. F, M   c. L, M   e. D, R
    b. D, M   d. F, R   f. L, R

8. Increased force of ventricular contraction more notably alters the magnitude of the _____ heart sound, and hypertension more notably alters the magnitude of the _____ heart sound.

    a. 1st, 1st   c. 2nd, 1st
    b. 1st, 2nd   d. 2nd, 2nd

9. The atrial or _____ heart sound _____ (I, is; N, is not) usually audible with a stethoscope.

    a. 2nd, I   c. 4th, I   e. 3rd, N
    b. 3rd, I   d. 2nd, N   f. 4th, N

10. A usually inaudible _____ heart sound may be recorded with a phonocardiogram in many normal young adults at the beginning of the _____ (F, first; M, middle; L, last) third of diastole.

    a. 3rd, F   c. 3rd, L   e. 4th, M
    b. 3rd, M   d. 4th, F   f. 4th, L

OBJECTIVE 27–2.

Identify the areas for auscultation of normal heart sounds. Identify the phonocardiogram and its utility.

11. Vibrations originating as a result of mitral valve activity are transmitted to the chest wall through _____ (M, major vessels; V, ventricular masses), and may be maximally auscultated over the _____ (R, second right; L, fifth or sixth left) intercostal space near the sternum.

    a. M, R   c. V, R
    b. M, L   d. V, L

12. Vibrations originating as a result of semilunar valve activity are transmitted to the chest wall through _____ (M, major vessels; V, ventricular masses), and generally result in maximal auscultation of pulmonic valve sounds on the _____ (R, right; L, left) side of the chest near the sternum.

    a. M, R   c. V, R
    b. M, L   d. V, L

13. An individual with a stethoscope placed in the area of maximal auscultation of sounds resulting from activity of the tricuspid valve _____ generally be able to hear sounds resulting from the other three heart valves, and _____ generally be able to hear the third and fourth heart sounds. (W, would; N, would not)

    a. W, W   c. N, W
    b. W, N   d. N, N

14. A phonocardiogram is a _____ (M, microphone; R, recording) which reproduces the entire sound spectrum of normal and pathological heart sounds including the generally inaudible _____ (H, high; L, low) frequency components.

    a. M, H   c. R, H
    b. M, L   d. R, L

OBJECTIVE 27–3.

Identify the causes of valvular lesions, their relative occurrence, and their effects upon valvular function.

15. Valvular damage, most frequently resulting from _____ (S, vascular syphilis; R, rheumatic fever), best correlates with _____ (O, antistreptolysin O antibody titers; H, the intensity of the third heart sound).

    a. S, O   c. R, O
    b. S, H   d. R, H

16. The _____ (T, tricuspid and pulmonic; M, mitral and aortic) valves are more frequently severely damaged by rheumatic fever, with the highest incidence of valvular damage occurring in the corresponding _____ (A, A–V; S, semilunar) valve.

    a. T, A   c. M, A
    b. T, S   d. M, S

17. Aortic valvular _____ (G, regurgitation; T, stenosis), involving adherent leaflets and restriction of blood flow, occurs more frequently with _____ (S, syphilitic; R, rheumatic) valvular lesions.

    a. G, S   c. T, S
    b. G, R   d. T, R

OBJECTIVE 27–4.

Identify heart murmurs, their causes, and their diagnostic acoustic characteristics.

18. Systolic murmurs include those of aortic _____ and mitral _____. (S, stenosis; R, regurgitation)

    a. S, S   c. R, S
    b. S, R   d. R, R

19. A quality of sound described as a high frequency "blowing," "swishing" sound occurs with aortic _____ and mitral _____. (S, stenosis; R, regurgitation)

    a. S, S   c. R, S
    b. S, R   d. R, R

20. A presystolic murmur resulting from left _____ (A, atrial; V, ventricular) contraction often occurs in the early stages of mitral _____ (S, stenosis; R, regurgitation).

    a. A, S   c. V, S
    b. A, R   d. V, R

21. Diastolic murmurs which persist throughout diastole, rather than a lesser portion, frequently imply a _____ severity of mitral valve involvement and a _____ severity of aortic valve involvement. (G, greater; L, lesser)

    a. G, G   c. L, G
    b. G, L   d. L, L

22. The murmur of greatest phonocardiogram intensity is produced by aortic _____ and the lowest intensity murmur results from mitral _____. (S, stenosis; R, regurgitation)

    a. S, S   c. R, S
    b. S, R   d. R, R

OBJECTIVE 27–5.

Identify the circulatory alterations resulting from aortic stenosis and regurgitation, and mitral stenosis and regurgitation.

23. Serious valvular diseases may result in peak intraventricular pressures as high as 450 mm. Hg in aortic _____ and stroke volume outputs as high as 300 ml. in aortic _____ . (R, regurgitation; S, stenosis)

    a. R, only R
    b. R, only S
    c. R, both R & S
    d. S, only R
    e. S, only S
    f. S, both R & S

24. Left ventricular hypertrophy and _____ (I, increased; D, decreased) blood volumes are compensatory mechanisms for a reduction in cardiac output occurring with aortic _____ (R, regurgitation only; S, stenosis only; RS, regurgitation and stenosis).

    a. I, R
    b. I, S
    c. I, RS
    d. D, R
    e. D, S
    f. D, RS

25. Aortic stenosis and regurgitation, and mitral stenosis and regurgitation, are all similar in their tendency to decrease:

    a. Left atrial pressures
    b. End diastolic volume
    c. End systolic volume
    d. Arterial systolic pressure
    e. Pulse pressure
    f. Cardiac reserve

26. Adequacy of coronary blood flow to the subendocardial myocardium is particularly vulnerable to valvular pathology resulting in _____ intraventricular diastolic pressures and _____ aortic diastolic pressures. (I, increased; D, decreased)

    a. I, I
    b. I, D
    c. D, I
    d. D, D

OBJECTIVE 27–6.

Identify the three major types of congenital anomalies of the heart and its major vessels, commoner examples of each type, and their associated causes and resultant effects.

27. Patent ductus arteriosus is an example of a _____ (R, right to left; L, left to right) shunt which bypasses the _____ (S, systemic; P, pulmonary) circulation.

    a. R, S
    b. R, P
    c. L, S
    d. L, P

28. The major congenital cause of cyanosis in babies is _____ (P, patent ductus arteriosus; T, the tetralogy of Fallot), which is associated with an enlarged _____ (R, right; L, left) ventricle.

    a. P, R
    b. P, L
    c. T, R
    d. T, L

29. Pulmonary arterial blood flows through the _____ (D, ductus arteriosus; F, foramen ovale) into the aorta in the fetus as a consequence of the lower flow resistance of the _____ (L, lungs; P, placenta).

    a. D, L
    b. D, P
    c. F, L
    d. F, P

30. Closure of the fetal ductus arteriosus occurs as a consequence of a sudden _____ in aortic pressure and a _____ in pulmonary arterial pressure at birth. (R, rise; F, fall)

    a. R, R
    b. R, F
    c. F, R
    d. F, F

31. The major consequences of patent ductus arteriosus are decreased _____ (R, cardiac and respiratory reserves; CO, left ventricular output and cyanosis), which are more pronounced in the _____ (N, newborn; O, older) child.

    a. R, N
    b. CO, N
    c. R, O
    d. CO, O

32. A waxing and waning "machinery murmur," higher in intensity during ventricular _____ (D, diastole; S, systole), characterizes a _____ (O, patent foramen ovale; A, persistent patent ductus arteriosus).

    a. D, O
    b. S, O
    c. D, A
    d. S, A

33. Circulatory changes normally occurring at birth _____ the right and increase the left ventricular loads, causing a corresponding _____ in the right and increase in the left atrial pressures. (D, decrease; I, increase).

    a. D, D
    b. D, I
    c. I, D
    d. I, I

34. The presence of a systolic blowing murmur, elevated right ventricular systolic pressure, and increased oxygenation of right ventricular blood is diagnostic for:

    a. Patent foramen ovale
    b. Intraventricular septal defects
    c. Pulmonary stenosis
    d. Patent ductus arteriosus

35. Blood flow from the systemic to the pulmonary circulation occurs during _____ in patent ductus arteriosus and during _____ with intraventricular septal defects. (S, only systole; D, only diastole; SD, both systole and diastole)

    a. S, S
    b. S, D
    c. D, S
    d. SD, SD
    e. SD, S
    f. S, SD

OBJECTIVE 27-7.

Identify the incidence and adaptive role of cardiac hypertrophy in valvular and congenital heart disorders.

36. Within limits, the degree of ventricular hypertrophy is generally more directly correlated with _____ (O, ventricular output; R, opposing flow resistance; P, pressure against which the ventricle must work).

    a. O/R
    b. R/O
    c. P × R
    d. R/P
    e. O/P
    f. P × O

37. Left atrial hypertrophy in the absence of left ventricular hypertrophy is more likely to occur with:

    a. Mitral regurgitation
    b. Mitral stenosis
    c. Aortic regurgitation
    d. Aortic stenosis

38. Right ventricular hypertrophy with a smaller than normal left ventricle is more likely to occur with:

    a. Mitral regurgitation
    b. Aortic regurgitation
    c. Patent ductus arteriosus
    d. Tetralogy of Fallot

28

Circulatory Shock and Physiology of Its Treatment

OBJECTIVE 28-1.

Identify circulatory shock, its major stages, its physiological causes, and its relationship to cardiac output and arterial pressure.

1. Circulatory shock is best equated with an abnormal increase in the ratio of _____ (A, arterial pressure; O, cardiac output;, M, general metabolism).

   a. O to A   c. M to O   e. M to A
   b. A to M   d. A to O   f. O to M

2. The progressive stage of circulatory shock _____ (D, does; N, does not) include cardiovascular deterioration, and if untreated leads to _____ (C, a compensated; I, an irreversible) stage.

   a. D, C   c. N, C
   b. D, I   d. N, I

OBJECTIVE 28–2.

Identify hemorrhagic shock, the effects of hypovolemia upon cardiac output, arterial pressure, and systemic filling pressure; and the consequences of autonomic reflex compensatory mechanisms.

3. Venous return is _____ (I, increased; D, decreased) from normal during hemorrhage as a consequence of altered _____ (F, systemic filling pressure; A, autonomic reflexes).

   a. I, F   c. D, F
   b. I, A   d. D, A

4. Compensatory autonomic reflex consequences of hemorrhagic shock include reflex _____ heart rate and _____ arteriolar and venomotor tone. (I, increased; D, decreased)

   a. I, I   c. D, I
   b. I, D   d. D, D

5. Minimal blood loss of _____ % of the blood volume within a half-hour interval generally results in death. In the absence of autonomic compensatory mechanisms, this value is _____ (H, halved; U, unchanged; D, doubled).

   a. 15 to 20, H   c. 15 to 20, D   e. 30 to 40, U
   b. 15 to 20, U   d. 30 to 40, H   f. 30 to 40, D

6. During hemorrhage, _____ (S, sympathetic; P, parasympathetic) reflex activity provides comparatively better regulation for _____ (A, arterial pressure; CO, cardiac output).

   a. S, A   c. P, A
   b. S, CO   d. P, CO

7. The declining arterial pressure curve as a function of increasing blood loss exhibits an initial plateau as a consequence of _____ (B, baroreceptor; C, central nervous system ischemic) reflexes, followed by a second plateau at a mean arterial pressure of about _____ mm. Hg.

   a. B, 100   c. B, 30   e. C, 65
   b. B, 65   d. C, 100   f. C, 30

8. Blood flow to the heart and brain is _____ (M, markedly altered; R, relatively unchanged) in the arterial pressure range of 100 to 70 mm. Hg during hemorrhagic shock, as a consequence of their _____ (S, sympathetic; P, parasympathetic; A, autoregulatory) control of flow resistance.

   a. M, S   c. M, A   e. R, P
   b. M, P   d. R, S   f. R, A

OBJECTIVE 28–3.

Identify the nonprogressive stage of hemorrhagic shock, and its compensatory mechanisms.

9. _____ (P, positive; N, negative) feedback mechanisms predominate in progressive hemorrhagic shock, in which the transition from nonprogressive shock is dependent upon a rather _____ (C, critical; V, variable) amount of rapid blood loss.

   a. P, C   c. N, C
   b. P, V   d. N, V

10. The sympathetic circulatory effects are greater for the _____ (B, baroreceptor; I, central nervous system ischemic) reflex, which is maximally active at arterial pressures _____ (A, above; W, below) 60 mm. Hg.

    a. B, A   c. I, A
    b. B, W   d. I, W

11. Compensatory mechanisms for hemorrhagic shock include vascular _____ (S, stress-relaxation; R, reverse stress-relaxation) and _____ (D, diminished; E, enhanced) rates of angiotensin formation.

    a. S, D   c. R, D
    b. S, E   d. R, E

12. Massive discharge of the vasomotor centers with severe, acute hemorrhage results in a pattern of overall peripheral _____ (C, vasoconstriction; D, vasodilation), negligible effects upon _____ (R, renal; B, cerebral) resistance vessels, and _____ (C, vasoconstriction; D, vasodilation) of skin and splanchnic vessels.

    a. C, R, C   c. D, R, D
    b. C, B, C   d. D, B, D

13. In the compensated stage of hemorrhagic shock, there is generally an increase above normal of:

    a. Cutaneous temperature   e. Angiotensin
    b. Baroreceptor activity   levels
    c. Renal blood flow   f. Cardiac output
    d. Cutaneous blood flow

OBJECTIVE 28–4.

Identify the progressive stage of hemorrhagic shock, and its associated feedback mechanisms of cardiovascular deterioration.

14. Depression of cardiac function plays the more significant role in an individual's condition during _____ (E, early; L, late) stages of hemorrhagic shock as a consequence of cardiac _____ (R, reserve; MTF, formation of myocardial toxic factor).

    a. E, R   c. L, R
    b. E, MTF   d. L, MTF

15. Extreme vascular _____ (C, constriction; D, dilation) of pancreatic arterioles during shock results in the release of a peptide having toxic cardiac effects that are more similar to toxic effects of _____ (E, excess; R, reduced) blood calcium.

    a. C, E   c. D, E
    b. C, R   d. D, R

16. Metabolic alterations which generally occur in circulatory shock result in _____ (C, acidosis; K, alkalosis) and _____ (I, increased; D, decreased) intracellular volumes.

    a. C, I   c. K, I
    b. C, D   d. K, D

17. During progressive hemorrhagic shock, the altered nutritional status of the cardiovascular system results in _____ capillary permeability, and a subsequent tendency for _____ blood volume and venous return. (I, increased; D, decreased)

    a. I, I   c. D, I
    b. I, D   d. D, D

18. Vascular and vasomotor center failure during late stages of hemorrhagic shock _____ venous return, _____ peripheral resistance, and _____ vascular capacitance. (I, increase; D, decrease)

 a. I, I, I   c. I, I, D   e. D, I, D
 b. I, D, I   d. D, D, D   f. D, D, I

19. Necrotic lesions resulting from hemorrhagic shock are more pronounced in the _____ (P, periphery; C, center) of liver lobules, which is the _____ (F, first; L, last) portion of the lobule to receive blood from the liver sinusoids.

 a. P, F   c. C, F
 b. P, L   d. C, L

OBJECTIVE 28–5.
Identify the irreversible stage of hemorrhagic shock and its underlying causes.

20. The irreversible stage of hemorrhagic shock is reached when therapeutic measures fail to produce a transient reversal of the directional trend in:

 a. Cardiac output
 b. Arterial blood pressure
 c. Circulatory filling pressure
 d. Peripheral resistance
 e. Tissue deterioration
 f. Pulmonary ventilation

21. The most critical factor determining the onset of the irreversible stage of hemorrhagic shock is thought to be the magnitude of the:

 a. Arterial blood pressure
 b. Duration of shock
 c. Cardiac output
 d. Accumulated oxygen deficit
 e. Body temperature
 f. Blood glucose levels

22. Contributing factors to the irreversible stage of hemorrhagic shock include cellular transmembrane diffusion of _____ (C, creatine; A, adenosine; AMP), derived from ATP degradation, and its subsequent conversion to _____ (T, urate; U, urea).

 a. C, T   c. AMP, T   e. C, U
 b. A, T   d. AMP, U   f. A, U

OBJECTIVE 28–6.
Identify other types of hypovolemic shock in addition to hemorrhagic shock, their causes, and their resultant effects.

23. Hypovolemic shock resulting from intestinal obstruction generally involves _____ total plasma protein and _____ blood viscosity. (I, increased; N, no significant change in; D, decreased)

 a. N, I   c. D, I   e. N, N
 b. I, D   d. D, D   f. N, D

24. _____ antidiuretic hormone secretion and _____ activity of the adrenal cortex are potential causes of hypovolemic shock resulting from dehydration. (I, increased; D, decreased)

 a. I, I   c. D, I
 b. I, D   d. D, D

OBJECTIVE 28–7.
Identify neurogenic shock, its associated causes, and its resultant effects.

25. The major cause of neurogenic shock is a reduction in blood flow as a consequence of _____ vasomotor tone and _____ vascular capacitance. (I, increased; D, decreased)

 a. I, I   c. D, I
 b. I, D   d. D, D

26. The term "venous pooling" generally refers to a condition resulting from _____ systemic filling pressure and _____ venous return to the heart. (I increased; D, decreased)

 a. I, I   c. D, I
 b. I, D   d. D, D

27. The consequences of neurogenic shock, resulting from depressed _____ (P, parasympathetic; S, sympathetic) outflow of spinal anesthesia, are less severe for an individual in a _____ (V, vertical; W, sitting; T, Trendelenburg) position.

    a. P, V   c. P, T   e. S, W
    b. P, W   d. S, V   f. S, T

28. Vasovagal syncope results from _____ (I, increased; R, reduced) vagal tone to the heart and increased sympathetic _____ (C, vasoconstrictor; D, vasodilator) system activity.

    a. I, C   c. R, C
    b. I, D   d. R, D

OBJECTIVE 28–8.
Identify anaphylactic shock, its causes, and its resultant effects.

29. Anaphylactic shock results as a consequence of antigen-antibody reactions occurring _____ (I, intravascularly only; E, extravascularly only; B, both intravascularly and within surrounding tissues), with the subsequent release of _____ (C, vasoconstrictor; D, vasodilator) substances.

    a. I, C   c. B, C   e. E, D
    b. E, C   d. I, D   f. B, D

30. The _____ (I, increased; D, decreased) capillary permeability associated with anaphylactic shock would be _____ (A, augmented; O, opposed) by administered histamine, and _____ (A, augmented; O, opposed) by norepinephrine administration.

    a. I, A, A   c. I, O, A   e. D, A, O
    b. I, A, O   d. I, O, O   f. D, O, A

OBJECTIVE 28–9.
Identify septic shock and traumatic shock, their potential causes, and their characteristic effects.

31. Ischemic gut segments frequently result in septic shock as a result of the release of _____ (E, endotoxin; V, vasodilator peptides) from gram-negative bacilli of the _____ (C, colon; S, stomach).

    a. E, C   c. V, C
    b. E, S   d. V, S

32. Diagnostic features generally characteristic of septic shock are least likely to include:

    a. Infectious processes
    b. High fever
    c. Peripheral vasodilation
    d. Reduced cardiac outputs
    e. Elevated metabolic rates
    f. Sludging of blood

33. Traumatic shock is more generally thought to result as a consequence of:

    a. Histamine release
    b. Hypovolemia & pain
    c. Norepinephrine release
    d. Anaphylaxis

OBJECTIVE 28–10.
Identify the consequences of circulatory shock upon general body functions.

34. Reduction of cardiac output to 50% of normal generally results in _____ body metabolism and _____ body temperature. (I, increased; N, no appreciable change in; D, decreased)

    a. N, N   c. D, N   e. I, I
    b. N, D   d. D, D   f. I, N

35. Stages of circulatory shock are accompanied by relatively _____ symptoms of severe muscular weakness and relatively _____ symptoms of a loss of the conscious state. (E, early; L, late)

    a. E, E   c. L, E
    b. E, L   d. L, L

36. Death, occurring a week or so following apparent recovery from shock, is more probably the result of _____ (B, brain; R, renal) damage sustained during shock as a partial consequence of its characteristic _____ (H, high; L, low) metabolic rate.

   a. B, H   b. B, L   c. R, H   d. R, L

OBJECTIVE 28–11.
Identify the treatment rationale for therapeutic measures utilized to prevent or treat circulatory shock.

37. Circulatory shock requiring treatment would more probably involve _____ (B, blood; P, plasma) infusion for hypovolemic shock of intestinal obstruction, and _____ (M, sympathomimetic; L, sympatholytic) drugs for neurogenic shock of deep general anesthesia.

   a. B, M   c. P, M
   b. B, L   d. P, L

38. Dextran, a _____ (C, carbohydrate; P, protein; L, lipid) substance, is utilized in circulatory shock therapy in order to maintain _____ (M, metabolic substrate concentrations; O, colloid osmotic pressure).

   a. C, M   c. L, M   e. P, O
   b. P, M   d. C, O   f. L, O

39. The use of _____ (M, sympathomimetic; L, sympatholytic) drugs generally results in a greater alteration in circulatory patterns during most forms of circulatory shock as a result of _____ (I, inactivated; A, maximally active) circulatory reflexes.

   a. M, I   c. L, I
   b. M, A   d. L, A

40. _____ (N, norepinephrine; A, antihistamine; C, cortisol) administration is the more effective treatment for anaphylactic shock following its onset, whereas morphine is of greater general value in the treatment of _____ (G, neurogenic; T, traumatic) shock.

   a. N, G   c. C, G   e. A, T
   b. A, G   d. N, T   f. C, T

41. Administered glucocorticoids tend to _____ (E, elevate; D, depress) blood glucose concentrations and have a _____ (L, labilizing; S, stabilizing) effect upon lysosomes.

   a. E, L   c. D, L
   b. E, S   d. D, S

OBJECTIVE 28–12.
Identify circulatory arrest, its more common causes, and its consequences upon the central nervous system.

42. Circulatory arrest, commonly caused by _____ (A, cardiac arrest only; F, ventricular fibrillation only; B, both cardiac arrest and ventricular fibrillation), presents the most immediate concern for permanent injury to _____ (C, cardiac; N, brain) tissues.

   a. A, C   c. B, C   e. F, N
   b. F, C   d. A, N   f. B, N

43. Circulatory arrest for _____ minutes causes permanent brain damage in slightly over half of the patients and almost universal consequences at _____ minutes.

   a. 0.5, 2   c. 4, 10   e. 10, 45
   b. 1, 3   d. 10, 20   f. 20, 45

29

Muscle Blood Flow During Exercise; Cerebral, Splanchnic, and Skin Blood Flows

OBJECTIVE 29–1.

Identify the resting and extreme requirements for skeletal muscle blood flow, the influence of intermittent and sustained muscle contraction upon its flow, and the mechanistic roles of local regulation and autonomic innervation in the control of blood flow through skeletal muscles.

1. Skeletal muscle constitutes about _____ % of the body mass and achieves blood flows as high as _____ ml./minute/100 grams of muscle in the well trained athlete.

   a. 20, 5   c. 75, 5   e. 40, 50
   b. 40, 5   d. 20, 50   f. 75, 50

2. Blood flow to a skeletal muscle is higher _____ (D, during; B, between) contraction phases of intermittent activity, and lower in the interval _____ (F, immediately following intermittent; T, during maximal tetanic) muscle activity.

   a. D, F   c. B, F
   b. D, T   d. B, T

3. Average capillary to skeletal muscle fiber exchange of nutrients during increased muscle activity generally involves _____ diffusion distances, _____ total diffusion surface area, and _____ numbers of participating capillaries. (I, increased; N, no appreciable change in; D, decreased)

   a. N, I, I   c. N, N, N   e. D, I, N
   b. N, I, N   d. D, I, I   f. D, N, N

4. Increased blood flow during skeletal muscle activity, as a result of local _____ (D, vasodilation; C, vasoconstriction) occurs primarily via _____ (N, neural; L, local) regulatory influences.

   a. D, N   c. C, N
   b. D, L   d. C, L

5. Significant skeletal muscle effects of sympathetic vasoconstrictor activity upon _____ (A, only alpha; B, only beta; AB, both alpha and beta) receptors are mediated by the neural release of _____ (N, norepinephrine; E, epinephrine).

   a. A, N   c. AB, N   e. B, E
   b. B, N   d. A, E   f. AB, E

6. Epinephrine secreted by the adrenal _____ (M, medulla; X, cortex) results in a net _____ (C, vasoconstriction; D, vasodilation) of skeletal muscle.

   a. M, C   c. X, C
   b. M, D   d. X, D

7. Sympathetic vasodilator fibers are more directly involved in _____ (R, regulatory; A, anticipatory) mechanisms of blood flow requirements for states of skeletal muscle _____ (T, resting tonus; M, motor activity).

   a. R, T   c. A, M
   b. R, M   d. A, T

OBJECTIVE 29-2.

Identify the essential roles of the autonomic nervous system, altered cardiac output, and altered arterial pressure in providing circulatory adjustments required by exercising muscle. Recognize the cardiovascular system as the limiting factor in strenuous exercise.

8. _____ heart rates at the onset of exercise result from _____ vagal tone and _____ sympathetic tone. (I, increased; D, decreased)

a. I, I, I   c. I, I, D   e. D, I, D
b. I, D, I   d. D, D, D   f. D, D, I

9. Blood flow through the majority of body tissues is generally _____ (I, increased; D, decreased) at the onset of exercise with the exception of _____ (M, skeletal muscle; H, heart muscle; B, brain; K, kidneys).

a. I; only H & B   d. D; only M
b. I; only M, H, & B   e. D; only M, H, & B
c. I; M, H, B, & K   f. D; M, H, B, & K

10. The _____ (R, rise; F, fall) in arterial blood pressure during a difficult manual exercise is generally _____ (G, greater; L, less) than that occurring during a strenuous whole body exercise.

a. R, G   c. F, G
b. R, L   d. F, L

11. An individual lacking the function of the sympathetic nervous system would encounter a _____ in arterial pressure and a _____ in cardiac output during exercise. (R, rise; F, fall)

a. R, R   c. F, R
b. R, F   d. F, F

12. _____ (D, decreased; I, increased) systemic filling pressure at the onset of exercise tends to increase venous _____ (R, return; P, pooling).

a. D, R   c. I, R
b. D, P   d. I, P

OBJECTIVE 29-3.

Identify the magnitude of cerebral blood flow, its means of determination, and the effects and relative significance of altered oxygen, carbon dioxide, hydrogen ion concentrations, cerebral activity, arterial pressure, and autonomic nerves upon its value.

13. The average adult brain blood flow, approximating _____ % of the resting cardiac output, is closest to that occurring for _____ (R, resting; M, maximally active) skeletal muscle blood flow when compared on a unit of tissue mass basis.

a. 5, R   c. 15, R   e. 10, M
b. 10, R   d. 5, M   f. 15, M

14. Doubling the normal arterial carbon dioxide concentration results in _____ (I, increased; N, no appreciable change in; D, decreased) cerebral hydrogen ion concentration, and _____ (H, halved; B, doubled) cerebral blood flow.

a. I, B   c. D, B   e. N, H
b. N, B   d. I, H   f. D, H

15. _____ cerebral hydrogen ion concentrations decrease neuronal synaptic activity and result in _____ cerebral blood flow. (I, increased; D, decreased)

a. I, I   c. D, I
b. I, D   d. D, D

16. Interruption of normal sympathetic vasomotor tone to cerebral vessels results in _____ (I, increased; N, no appreciable change in; D, decreased) cerebral blood flow, and maximal stimulation of its parasympathetic innervation results in _____ (P, pronounced; M, mild) changes in cerebral blood flow.

a. I, P   c. D, P   e. N, M
b. N, P   d. I, M   f. D, M

17. Cerebral blood flow is more generally reduced during periods of _____ , and increased during periods of _____ . (C, convulsions; A, deep general anesthesia; H, arterial hypertension; M, increased brain metabolic rates)

a. A; C & M
b. A & C; H
c. A, C, & H; M
d. C & H; A & M

18. Significant alterations in cerebral blood flow would more probably coincide with either a 10 mm. Hg fall in cerebral _____ (A, arterial; V, venous) PO_2 or a change in mean arterial pressure to _____ mm. Hg.

a. A, 50
b. A, 150
c. V, 50
d. V, 150

19. A brain region which shows a premature decline in radioactivity following radioactive krypton equilibration of brain tissues generally exhibits _____ (I, increased; D, decreased) levels of _____ (A, anesthetic susceptibility; B, blood flow).

a. I, A
b. I, B
c. D, A
d. D, B

OBJECTIVE 29–4.
Identify two sources of blood flow to the liver, their respective magnitudes and significance, their regulatory mechanisms, their blood reservoir and cleansing functions, and methods of their measurement.

20. Liver blood flow, typically about _____ % of the cardiac output, receives the larger contribution from _____ (H, hepatic arterial; P, portal) blood.

a. 15, H
b. 30, H
c. 45, H
d. 15, P
e. 30, P
f. 45, P

21. Liver _____ (N, autonomic; L, local regulatory) mechanisms are primarily responsible for the regulation of _____ (A, only arterial; P, only portal; AP, both arterial and portal) hepatic blood flows.

a. N, A
b. N, P
c. N, AP
d. L, A
e. L, P
f. L, AP

22. Liver blood volume, typically about _____ ml., would more probably be _____ (H, halved; D, doubled) during chronic stages of severe heart failure.

a. 500, H
b. 1,000, H
c. 1,500, H
d. 500, D
e. 1,000, D
f. 1,500, D

23. Sympathetic stimulation results in a more pronounced percentage _____ (I, increase; D, decrease) in blood volume of hepatic _____ (V, veins; S, sinuses).

a. I, V
b. I, S
c. D, V
d. D, S

24. Extravascular spaces of the liver have comparatively _____ protein concentrations and _____ lymphatic outflows when compared to other body tissues. (L, low; A, average; H, high)

a. L, A
b. A, L
c. L, H
d. H, L
e. H, A
f. H, H

25. Hepatic sinusoids are lined by endothelial cells with comparatively _____ (H, higher; L, lower) permeability characteristics and _____ (B, bile-secreting; P, phagocytic; PS, parasympathetic) Kupffer cells.

a. H, B
b. H, P
c. H, PS
d. L, B
e. L, P
f. L, PS

26. Bromosulfophthalein, a substance utilized to quantitate liver blood flow via a modified _____ (I, indicator dilution; F, Fick) method, is actively secreted by hepatic cells into the _____ (S, hepatic sinuses; B, biliary system).

a. I, S
b. I, B
c. F, S
d. F, B

OBJECTIVE 29-5.

Identify the role of autoregulation of the gastrointestinal system blood flow, and its relation to glandular secretion and gut motility. Identify the effects of parasympathetic and sympathetic activity on gastrointestinal blood flow.

27. Blood flow to gastrointestinal smooth muscle is controlled _____ (S, separately from; C, conjointly with) mucosal and submucosal blood flow, and is _____ (I, increased; N, not significantly altered) by increasing smooth muscle activity.

    a. S, I   c. C, I
    b. S, N   d. C, N

28. Parasympathetic stimulation _____ (I, increases; D, decreases) gastrointestinal blood flow, with a comparatively greater effect upon blood flow to the _____ (S, stomach; G, small intestine).

    a. I, S   c. D, S
    b. I, G   d. D, G

29. "Autoregulatory escape," applied to a gastrointestinal cardiovascular situation, involves the effects of _____ (S, sympathetic; P, parasympathetic) stimulation, and the ultimate dominance of _____ (L, local; N, autonomic) regulatory influences.

    a. S, L   c. P, L
    b. S, N   d. P, N

30. Sympathetic stimulation _____ blood flow to the mucosa and submucosa, and _____ blood flow to smooth muscle layers of the gastrointestinal tract. (I, increases; D, decreases)

    a. I, I   c. D, I
    b. I, D   d. D, D

OBJECTIVE 29-6.

Identify the principal causes and consequences of blockade of the portal circulation.

31. Chronic alcoholism and carbon tetrachloride poisoning frequently result in the _____ (F, fibrous; P, parenchymal) liver tissue destruction known as liver _____ (C, cirrhosis; E, emphysema).

    a. F, C   c. P, C
    b. F, E   d. P, E

32. Slowly developing cirrhosis of the liver frequently results in esophageal varicosities as a result of developing collateral vessels between _____ (H, hepatic arteries; P, portal veins) and _____ (E, esophageal arteries; S, systemic veins).

    a. H, E   c. P, E
    b. H, S   d. P, S

33. The formation of free fluid in the peritoneal cavity, termed _____ (A, ascites; C, cirrhosis), more frequently results from slowly developing vascular blocks associated with elevated _____ (I, intestinal capillary; L, liver sinusoid) pressures.

    a. A, I   c. C, I
    b. A, L   d. C, L

OBJECTIVE 29-7.

Identify the functional roles of the spleen as a blood reservoir in erythrocyte destruction, in immunobiology, and in hemopoiesis.

34. Diminished blood volume of the spleen in humans by means of _____ (C, only capsular; V, only vascular; B, both capsular and vascular) smooth muscle contraction occurs in response to _____ (S, sympathetic; P, parasympathetic) stimulation.

    a. C, S   c. B, S   e. V, P
    b. V, S   d. C, P   f. B, P

35. Autonomic stimulation to the spleen during strenous exercise results in _____ circulating blood volume and _____ blood hematocrit. (I, increased; N, no significant change in; D, decreased)

    a. I, I   c. I, D   e. D, N
    b. I, N   d. D, I   f. D, D

36. Adult functions of the spleen are *least likely* to include:

a. Humoral antibody synthesis
b. Lymphocyte proliferation

c. Phagocytosis
d. Hemopoiesis

e. Platelet destruction
f. Erythrocyte destruction

OBJECTIVE 29–8.

Identify the physiologic anatomy of the cutaneous circulation, the nutritive and heat conduction functional roles of the cutaneous circulation, and the significance and magnitude of the range of variance of cutaneous blood flow.

37. Arteriovenous anastomoses, found in greater numbers in cutaneous regions of the _____ (T, arms, legs, and body trunk; F, hands, feet, lips, nose, and ears), are constricted by _____ (C, cholinergic; A, adrenergic) neural activity.

a. T, C
b. T, A

c. F, C
d. F, A

38. Activity of A–V anastomoses are most functionally related to cutaneous _____ (N, nutrition; H, heat conduction) as a consequence of their regulatory influence upon blood flow through _____ (D, dermal capillary loops; P, the subcutaneous venous plexus).

a. N, D
b. N, P

c. H, D
d. H, P

39. A cutaneous blood flow in a typical adult of 1.5 liters per minute would more probably occur during conditions of either _____ body metabolic rates or _____ environmental temperatures. (L, low; H, high; N, normal)

a. H, H
b. H, L

c. N, N
d. L, H

e. L, L
f. L, N

40. A patient with borderline cardiac failure is more likely to exhibit severe symptoms of cardiac failure during conditions of either _____ body metabolic rates or _____ environmental temperatures. (L, low; H, high)

a. H, H
b. H, L

c. L, H
d. L, L

OBJECTIVE 29–9.

Identify the cutaneous blood flow influences and functional significance exerted by the temperature regulatory centers, sympathetic vasoconstrictor and vasodilator mechanisms, adrenalin secretion and states of circulatory stress, local autoregulation, and cutaneous temperatures.

41. Regulation of cutaneous blood flow is largely achieved via _____ (L, local; N, neural) regulatory mechanisms which is _____ (S, similar; O, opposite) to regulatory influences in most body tissues.

a. L, S
b. L, O

c. N, S
d. N, O

42. _____ (C, cooling; W, heating) the preoptic region of the _____ (H, hypothalamus; O, occipital cortex) results in cutaneous vasodilation and sweating.

a. C, H
b. C, O

c. W, H
d. W, O

43. _____ (I, increased; D, decreased) cutaneous blood flow results from increased circulating levels of _____ (N, norepinephrine only; E, epinephrine only; B, both norepinephrine and epinephrine).

a. I, N
b. I, E

c. I, B
d. D, N

e. D, E
f. D, B

44. The incorporation of a _____ (P, parasympathomimetic; S, sympathomimetic) agent with a local anesthetic solution would best be utilized to _____ (D, enhance its duration of action; M, maintain adequate cutaneous nutrition).

a. P, D
b. P, M

c. S, D
d. S, M

45. Activation of the sympathetic _____ (D, vasodilator; C, vasoconstrictor) mechanism to cutaneous tissues occurs with rising body temperatures as a part of the regulatory mechanisms which generally _____ (P, precedes; E, coincides with) the appearance of excessive sweating.

    a. D, P   c. C, P
    b. D, E   d. C, E

46. Cutaneous vessels are maximally constricted by local cutaneous temperatures of _____ degrees C., and are maximally dilated by local cutaneous temperatures of _____ degrees C.

    a. 37, 40   c. 15, 0   e. 34, 15
    b. 15, 34   d. 0, 15   f. 40, 34

47. The dominance of _____ (L, local; A, autonomic) regulatory mechanisms in the phenomenon of "reactive hyperemia" results from inadequacy of _____ (N, nutritional; T, temperature regulatory) functions of cutaneous blood flow.

    a. L, N   c. A, N
    b. L, T   d. A, T

48. Circulatory stress results in _____ (R, reduced; I, increased) blood volume of the subcutaneous venous plexus as a consequence of sympathetic _____ (D, vasodilator; C, vasoconstrictor) activity to cutaneous regions.

    a. R, D   c. I, D
    b. R, C   d. I, C

OBJECTIVE 29-10.

Identify the contributory influences of altered patterns of cutaneous blood flow upon skin color and skin temperature.

49. Increased cutaneous blood flow, associated with _____ (D, decreased; I, increased) core body temperatures, is generally associated with a contributory _____ (B, bluish; R, reddish) skin hue.

    a. D, B   c. I, B
    b. D, R   d. I, R

50. Reduced skin temperature is more frequently associated with _____ (E, elevated; L, lowered) rates of cutaneous blood flow and a contributory _____ (R, reddish; B, bluish) hue to skin color.

    a. E, R   c. E, B
    b. L, R   d. L, B

51. Mass sympathetic discharge results in diminished cutaneous _____ (F, blood flow only; V, blood volume only; B, both blood flow and blood volume) and a subsequent cutaneous _____ (C, cyanosis; A, ashen pallor).

    a. F, C   c. V, C   e. B, C
    b. F, A   d. V, A   f. B, A

OBJECTIVE 29-11.

Identify Raynaud's disease and Buerger's disease, their consequences, and their treatment rationale. Contrast the consequences of peripheral arteriosclerosis with Raynaud's disease and Buerger's disease.

52. Raynaud's disease most frequently involves the _____ (H, hands and feet; A, arms and body trunk) of individuals living in _____ (N, northern; T, tropical) climates.

    a. H, N   c. A, N
    b. H, T   d. A, T

53. _____ (E, preganglionic; O, postganglionic) autonomic fibers are sectioned in the treatment of Raynaud's disease in order to _____ (P, produce; A, avoid) the phenomenon of denervation hypersensitivity.

    a. E, P   c. O, P
    b. E, A   d. O, A

54. Sympathectomy is of more substantial benefit for _____ (B, Buerger's disease; A, peripheral arteriosclerosis), which more frequently involves the _____ (U, upper; L, lower) extremities.

    a. B, U   b. B, L   c. A, U   d. A, L

30

Capillary Dynamics, and Exchange of Fluid Between the Blood and Interstitial Fluid

OBJECTIVE 30–1.

Describe the structural features and functional properties of a typical capillary bed and the capillary wall. Identify vasomotion, its role in flow autoregulation, and its relationship to average capillary functions.

1. Arterioles are _____ (S, smaller; G, greater) in diameter than venules and have a _____ (L, less; M, more) extensive smooth muscle layer.

    a. S, L   c. G, L
    b. S, M   d. G, M

2. Capillary diameters, approximating _____ microns, are slightly more than _____ times the erythrocyte diameter.

    a. 1 to 2, 1   c. 20 to 30, 2   e. 7 to 9, 2
    b. 7 to 9, 1   d. 1 to 2, 2   f. 20 to 30, 4

3. Size-limiting pores for substances passing from the capillary lumen to interstitial fluid are located _____ (W, within; B, between) cellular membranes, and have a width of about _____ angstroms.

    a. W, 7 to 9   d. B, 7 to 9
    b. W, 80 to 90   e. B, 80 to 90
    c. W, 200 to 300   f. B, 200 to 300

4. Smooth muscle cells at the loci of origin of true capillaries from _____ (A, arterioles; M, metarterioles) function as _____ (P, pericytes; S, sphincter valves).

    a. A, P   c. M, P
    b. A, S   d. M, S

5. Tissue flow autoregulation is generally more dependent upon intermittent activity of _____ (A, arterioles; P, precapillary sphincters) in response to altered concentrations of _____ (S, organic metabolic substrates; E, organic metabolic end-products; G, respiratory gases).

    a. A, S   c. A, G   e. P, E
    b. A, E   d. P, S   f. P, G

OBJECTIVE 30–2.

Identify the functional roles and significant properties of the capillary endothelium in the transcapillary exchange of nutrients and other subtances.

6. The most significant transcapillary exchange of substances occurs via:

   a. Pinocytosis
   b. Active transport
   c. Carrier-mediated diffusion
   d. Passive diffusion

7. Net diffusion of water across the capillary wall is greater for _____ (S, slit; M, cell membrane) pores, which have comparatively _____ (G, greater; L, lesser) cumulative pore cross-sectional areas.

   a. S, G
   b. S, L
   c. M, G
   d. M, L

8. Net diffusion of respiratory gases, involving _____ (L, lower; H, higher) transcapillary permeability coefficients than water, occurs largely _____ (T, through; B, between) endothelial cells.

   a. L, T
   b. L, B
   c. H, T
   d. H, B

9. Transcapillary exchange involves typical permeability coefficient ratios to that for water of about _____ for glucose, about _____ for sodium ions, and about _____ for hemoglobin.

   a. 0.1, 0.6, 0.01
   b. 0.1, 0.1, 0.001
   c. 0.6, 1.0, 0.001
   d. 0.6, 1.0, 0.01
   e. 0.2, 0.1, 0.001
   f. 1.0, 1.0, 1.0

10. Capillary volume flows are _____ than net transcapillary volume flows between arteriolar and venular segments of the capillary, and capillary flow velocities are _____ than transcapillary water molecule diffusion velocities. (G, greater; L, less)

    a. G, G
    b. G, L
    c. L, G
    d. L, L

11. The transcapillary concentration difference of oxygen is normally _____ (G, greater; L, less) than 5% of its capillary blood concentration as a result of _____ (V, capillary flow velocities; P, oxygen permeability coefficients).

    a. L, V
    b. L, P
    c. G, V
    d. G, P

12. Increased hydrostatic pressures are associated with _____ molecular kinetic energies of a substance and _____ molecular diffusion rates. (I, increased; N, no significant change in; D, decreased)

    a. N, N
    b. N, I
    c. I, N
    d. I, D
    e. D, D
    f. I, I

13. The phenomenon of "bulk flow" is more significant in transcapillary transport mechanisms involving _____ diameter pores, and pinocytosis is probably more significant for _____ molecular weight substances. (L, larger; S, smaller)

    a. L, L
    b. L, S
    c. S, L
    d. S, S

OBJECTIVE 30–3.

Identify the four primary pressures influencing transcapillary fluid movement, their typical magnitudes, their contributing causes, and their means of measurement.

14. Pressures tending to move fluid outward through the capillary membrane include positive values of _____ and _____ pressures. (I, interstitial fluid; C, capillary; P, plasma colloid osmotic; O, interstitial fluid colloid osmotic)

    a. I, C
    b. I, P
    c. I, O
    d. C, P
    e. C, O
    f. O, P

15. The normal functional mean capillary pressure, typically about _____ mm. Hg, is probably best measured by _____ (D, direct capillary cannulation; I, indirect isogravimetric or isovolumetric methods).

    a. 17, D
    b. 27, D
    c. 37, D
    d. 17, I
    e. 27, I
    f. 37, I

16. When precapillary sphincters are closed, the capillary pressure at its metarteriolar end would be about _____ mm. Hg, and would be _____ (H, higher than; N, not significantly different from; L, lower than) the capillary pressure at the venular end.

a. 10, H   c. 10, L   e. 20, N
b. 10, N   d. 20, H   f. 20, L

17. Direct cannulation of patent capillary loops at the base of human fingernails yield capillary pressures of _____ mm. Hg at their arterial ends and _____ mm. Hg at the venous ends.

a. 30 to 40, 20 to 30   d. 15 to 25, 10 to 15
b. 30 to 40, 10 to 15   e. 15 to 25, 5 to 7
c. 30 to 40, 5 to 7   f. 15 to 25, 0 to 3

18. A typical pressure of _____ mm. Hg, considerably _____ (H, higher; L, lower) than the midcapillary pressure as measured by direct cannulation, may be estimated experimentally by bringing the arterial and venous pressures towards each other in a manner which maintains their tissue volume constant.

a. 17, H   c. 37, H   e. 27, L
b. 27, H   d. 17, L   f. 37, L

19. A small vacuum cup is placed over the skin, connected to a vacuum, and one day later the fluid pressure is measured in the resultant subcutaneous space. This pressure would be _____ (A, above; B, below) atmospheric pressure and would approximate the _____ (I, interstitial fluid; V, vacuum cup; O, interstitial fluid colloid osmotic) pressure.

a. A, I   c. B, I
b. A, O   d. B, V

20. A typical interstitial fluid pressure of _____ mm. Hg is a pressure tending to cause fluid movement _____ (I, into; O, out of) capillaries.

a. +6 to +7, I   c. +28, I   e. −6 to −7, O
b. −6 to −7, I   d. +6 to +7, O   f. +28, O

21. The protein concentration of plasma averages _____ times that of the interstitial fluid, or about _____ gm. per 100 ml.

a. ½, 3.7   c. 4, 3.7   e. 2, 7.3
b. 2, 3.7   d. ½, 7.3   f. 4, 7.3

22. The Donnan equilibrium, resulting from _____ (P, positively; N, negatively) charged plasma proteins, causes the plasma colloid osmotic pressure to be _____ (G, greater; L, less) than that due to proteins alone.

a. P, G   c. N, G
b. P, L   d. N, L

23. The plasma colloid osmotic pressure is closest to _____ mm. Hg, a portion of which results from a surplus of plasma electrolyte _____ (C, cations; A, anions).

a. 300, C   c. 28, C   e. 79, A
b. 79, C   d. 300, A   f. 28, A

24. Each gram of plasma globulin exerts about _____ as much osmotic pressure as albumin, and the plasma concentration of plasma globulin is _____ as much as for albumin. (½, one half; T, twice)

a. ½, ½   c. T, ½
b. ½, T   d. T, T

25. The albumin to globulin concentration ratio in interstitial fluid is disproportionately _____ (H, higher; L, lower) than its plasma ratio as a consequence of capillary membrane _____ (A, active transport; P, permeability coefficients).

a. H, A   c. L, A
b. H, P   d. L, P

26. The interstitial fluid volume has a total protein content closest to _____ times that of the plasma volume, and exerts an average interstitial fluid colloid osmotic pressure closest to _____ mm. Hg.

a. 1, 5   c. 1, 15   e. 4, 10
b. 1, 10   d. 4, 5   f. 4, 15

OBJECTIVE 30–4.

Distinguish between capillary filtration and diffusion processes. Identify net filtration and reabsorption pressures, their typical magnitudes, and their contributing factors. Identify the magnitude of interstitial fluid flow and its influencing factors.

27. The outward and inward filtrations of fluid in the tissue capillaries result primarily from an arteriolar to venular capillary gradient of _____ pressures.

    a. Interstitial fluid   c. Capillary
    b. Plasma colloid   d. Interstitial colloid
       osmotic   osmotic

28. The fluid filtration rate _____ (I, into; O, out of) capillaries at their arterial ends is _____ (G, greater than; E, equivalent to; L, less than) the diffusion rate for water.

    a. I, G   c. I, L   e. O, E
    b. I, E   d. O, G   f. O, L

29. _____ (E, less; M, more) than 1% of the capillary plasma flow typically filters out of the capillaries into interstitial fluid, of which a higher percentage is returned via _____ (L, liver; S, skeletal muscle) capillaries.

    a. E, L   c. M, L
    b. E, S   d. M, S

30. With a capillary pressure at the venous capillary end of 10 mm. Hg, an interstitial fluid colloid osmotic pressure of 5 mm. Hg, a plasma colloid osmotic pressure of 28 mm. Hg, and an interstitial fluid pressure of -6.3 mm. Hg, the corresponding net reabsorption pressure would be _____ mm. Hg.

    a. 6.7   c. 26.7   e. 3.8
    b. 16.7   d. 0   f. 19.3

31. Net filtration is typically _____ (G, greater; S, less) than net reabsorption within tissue capillary networks as a consequence of _____ (V, vasomotion; P, preferential capillary channels; L, lymphatics).

    a. G, V   c. G, L   e. S, P
    b. G, P   d. S, V   f. S, L

OBJECTIVE 30–5.

Recognize the Starling near-equilibrium of capillary fluid exchange. Identify the filtration coefficient for capillary fluid filtration. Identify the consequences of an imbalance of forces acting upon the capillary membrane.

32. The filtration coefficient is typically the net rate of fluid filtration in the entire body of about _____ ml. per minute, divided by the difference between mean capillary filtration minus reabsorption pressures of about _____ mm. Hg.

    a. 1.7, 0.3   c. 1.7, 8.3   e. 7.1, 5.1
    b. 1.7, 5.1   d. 7.1, 0.3   f. 7.1, 8.3

33. The interstitial colloid osmotic pressure is generally greatest in:

    a. Subcutaneous tissue c. The liver
    b. Intestinal tissues   d. Skeletal muscle

34. A rise in mean capillary pressure results in _____ (I, increased; D, decreased; N, no significant change in) net filtration of fluid into the interstitial spaces, and would arise as a consequence of arteriolar _____ (C, vasoconstriction; E, vasodilation).

    a. I, C   c. N, C   e. D, E
    b. D, C   d. I, E   f. N, E

31

The Lymphatic System, Interstitial Fluid Dynamics, Edema, and Pulmonary Fluid

OBJECTIVE 31-1.
Identify the functional significance of the lymphatic system. Describe the physiologic anatomy of the lymphatic drainage system and the histologic features of lymphatic capillaries.

1. Principal routes of absorption from the interstitial spaces include _____ for fluid filtrate, _____ for protein, and _____ for particulate matter. (L, lymphatics; C, capillary reabsorption)

   a. L, L, L   c. C, C, L   e. L, L, C
   b. C, L, L   d. C, C, C   f. L, C, L

2. Bone and the central nervous system _____ (R, are; N, are not) included in a _____ (J, majority; M, minority) of body tissues possessing direct lymphatic drainage from the interstitial spaces.

   a. R, J   c. N, J
   b. R, M   d. N, M

3. Lymph flow from the lower portion of the body empties into the _____ (RL, right lymph; T, thoracic) duct, and is returned to the venous system at the junction of the subclavian and _____ (R, right; L, left) internal jugular veins.

   a. RL, R   c. T, R
   b. RL, L   d. T, L

4. Lymph flow from the right side of the neck and head returns to the junction of the _____ (R, right; L, left) subclavian and internal jugular veins via the _____ (RL, right lymph; T, thoracic) duct.

   a. RL, R   c. T, R
   b. RL, L   d. T, L

5. The absence of a continuous basement membrane is a functional characteristic of endothelial cells of _____ capillaries.

   a. Lymphatic   c. Renal glomerular
   b. Pulmonary   d. Coronary

6. Larger lymphatic vessels _____ (P, possess; L, lack) valves, whereas _____ (S, precapillary sphincters; E, endothelial cells) function as valves in lymphatic capillary terminals.

   a. P, S   c. L, S
   b. P, E   d. L, E

OBJECTIVE 31-2.

Identify the origin and composition of lymph, factors influencing its protein composition, and its variance with constituent tissues.

7. The protein concentration of interstitial fluid is _____ (G, greater; L, less) than that of capillary filtrate as a consequence of the properties of capillary _____ (P, pinocytosis; F, fluid reabsorption).

   a. G, P   c. L, P
   b. G, F   d. L, F

8. The protein concentration of lymph from most peripheral tissues is closer to the protein concentration of _____ (F, capillary filtrate; I, interstitial fluid) or an average of about _____ grams per cent.

   a. F, 0.2   c. F, 6   e. I, 2
   b. F, 2   d. I, 0.2   f. I, 6

9. Protein concentrations are comparatively higher in _____ (P, plasma; L, liver lymph), and comparatively lower in lymph from _____ (I, intestine; M, muscle).

   a. P, I   c. L, I
   b. P, M   d. L, M

10. Lymph within the thoracic duct has a protein concentration closest to _____ grams per cent, which is characteristically _____ (L, lower; H, higher) than that found in the right lymph duct.

    a. 1, H   c. 7, H   e. 4, L
    b. 4, H   d. 1, L   f. 7, L

OBJECTIVE 31-3.

Identify the magnitude of lymph flow, determinant factors for lymph flow, and the mechanism of action of lymphatic pumps.

11. Lymphatic flow rates at rest typically approximate _____ ml. per hour for the thoracic duct, and _____ ml. per hour for an estimated total body lymph flow.

    a. 20, 100   d. 200, 1,000
    b. 100, 120   e. 1,000, 1,200
    c. 120, 760   f. 1,200, 7,600

12. Lymphatic flow tends to be increased by factors which either _____ plasma colloid osmotic pressure, or _____ capillary permeability. (I, increase; D, decrease)

    a. I, I   c. D, I
    b. I, D   d. D, D

13. Interstitial fluid pressure tends to be increased by factors which _____ capillary pressure and _____ interstitial fluid protein concentration. (I, increase; D, decrease)

    a. I, I   c. D, I
    b. I, D   d. D, D

14. The lymphatic pump operates predominantly via _____ (I, intrinsic lymphatic smooth muscle activity only; E, extrinsic compression effects only; B, both extrinsic and intrinsic compression mechanisms), and is assisted by valves which lie closer together in _____ (L, larger; S, smaller) diameter lymph vessels.

    a. I, L   c. B, L   e. E, S
    b. E, L   d. I, S   f. B, S

15. Lymphatic terminals held in an expanded state by _____ (F, connective tissue fibers; N, a negative interstitial fluid pressure) have their endothelial "flap" valves in _____ (C, a closed; O, an open) position.

    a. F, C   c. N, C
    b. F, O   d. N, O

16. Interstitial fluid flow from interstitial spaces into lymphatic capillary terminals is accelerated as a result of transients of either _____ interstitial fluid pressure or _____ lymphatic capillary pressures. (I, increased; D, decreased)

    a. I, I   c. D, I
    b. I, D   d. D, D

17. A maximum limit of lymph flow, about _____ times normal flow rates, occurs within a _____ (P, positive; N, negative) interstitial fluid pressure range.

    a. 7, P   c. 100, P   e. 20, N
    b. 20, P   d. 7, N   f. 100, N

18. A maximum limit to lymph flow results from excessive _____ of lymphatic capillary terminals, and excessive _____ of major lymph channels. (C, compression; E, expansion)

    a. C, C   c. E, C
    b. C, E   d. E, E

OBJECTIVE 31–4.

Identify the control mechanisms of interstitial fluid protein concentration and interstitial fluid pressure, and their significance.

19. Protein accumulation in interstitial fluid tends to _____ interstitial fluid volume, _____ interstitial fluid pressure, and _____ lymphatic flow. (I, increase; N, not significantly alter; D, decrease)

    a. N, N, N   c. D, D, D   e. I, N, I
    b. I, I, I   d. N, D, I   f. D, N, D

20. Interstitial fluid protein, resulting largely from _____ (IC, intracellular; P, plasma) fluid compartment leakage, tends to be diminished by _____ lymph flows. (I, increased; D, decreased)

    a. IC, I   c. P, I
    b. IC, D   d. P, D

21. Lymphatic pump activity tends to _____ (G, augment; O, oppose) the _____ (S, subatmospheric; A, above atmospheric) interstitial fluid pressure required to balance capillary wall forces.

    a. G, S   c. O, S
    b. G, A   d. O, A

22. The major contributory factor towards maintenance of a normal interstitial fluid pressure results _____ (D, directly; I, indirectly) from lymphatic _____ (P, protein; F, fluid volume) removal.

    a. D, P   c. I, P
    b. D, F   d. I, F

23. Interstitial fluid spaces, typically _____ (C, compacted; M, containing mobile fluid), are generally maintained in this state by their _____ (A, connective tissue attachments; I, interstitial fluid pressure).

    a. C, A   c. M, A
    b. C, I   d. M, I

24. The condition referred to as _____ (D, dehydration; E, edema) occurs when the interstitial fluid pressure _____ (R, rises above; F, falls below) atmospheric pressure.

    a. D, R   c. E, R
    b. D, F   d. E, F

25. Diffusion rates of a substance between plasma and surrounding cells are greater when interstitial fluid pressures are _____ (N, negative; P, positive) as a consequence of _____ (V, greater fluid volumes; D, lesser diffusing distances).

    a. N, V   c. P, V
    b. N, D   d. P, D

OBJECTIVE 31–5.

Identify edema, its physical cause, and its characterization. Identify a typical pressure-volume curve for interstitial fluid spaces, its mechanistic causes, and its functional significance.

26. An abrupt change in interstitial fluid space compliance occurs at its pressure of _____ mm. Hg, with higher tissue compliance values occurring in the more _____ (N, negative; P, positive) pressure range.

    a. -4, N   c. +4, N   e. 0, P
    b. 0, N   d. -4, P   f. +4, P

27. A compliance of 400 ml. per mm. Hg for the entire body is more probably typical for _____ (N, normal; E, edematous) tissue conditions corresponding to _____ (G, negative; P, positive) interstitial fluid pressures.

    a. N, G   c. E, G
    b. N, P   d. E, P

28. A _____ (R, rise; D, decline) in tissue compliance at very _____ (H, high; L, low) interstitial fluid pressures results from limitations imposed by stretched tissues.

    a. R, H   c. D, H
    b. R, L   d. D, L

29. A "plus one" edema signifies that the degree of edema is _____ (B, barely detectable; M, maximal) and that the interstitial fluid volume has increased typically about _____ % above normal.

    a. B, 5   c. B, 200   e. M, 30
    b. B, 30   d. M, 5   f. M, 200

30. Chronic conditions of edema result in _____ (S, stress-relaxation; C, compensatory compaction) of tissue spaces, and a resultant _____ (I, increase; D, decrease) in interstitial fluid volumes for corresponding interstitial fluid pressures.

a. S, I　　　　　c. C, I
b. S, D　　　　　d. C, D

31. The phenomenon of "pitting" edema, _____ (S, similar to; D, different from) "brawny" edema, relates to the condition of interstitial fluid _____ (I, immobility; M, mobility).

a. S, I　　　　　c. D, I
b. S, M　　　　　d. D, M

OBJECTIVE 31–6.

Identify the "safety factors" involved in the prevention of edema, their relative magnitudes, and their functional significance.

32. A "safety factor" of _____ mm. Hg for a normally _____ (P, positive; N, negative) interstitial fluid pressure exists before edema generally develops.

a. 3.6 P　　　c. 16.3, P　　　e. 6.3, N
b. 6.3, P　　　d. 3.6, N　　　f. 16.3, N

33. Either a _____ (R, rise; F, fall) in capillary pressure or a _____ (R, rise; F, fall) in plasma colloid osmotic pressure of about _____ mm. Hg must generally occur before edema generally develops.

a. R, R, 7　　　c. F, R, 7　　　e. R, F, 17
b. R, F, 7　　　d. F, F, 17　　　f. F, R, 17

34. The consequences of _____ (I, a normal interstitial fluid pressure; L, increased lymphatic flow) generally contribute the larger component to a "total safety factor" of about _____ mm. Hg in the prevention of edema with rising capillary pressure.

a. I, 7　　　c. I, 27　　　e. L, 17
b. I, 17　　　d. L, 7　　　f. L, 27

OBJECTIVE 31–7.

Identify the conditions of altered capillary pressure, capillary permeability, plasma protein concentration, lymphatic function, and renal function resulting in edema. Recognize potential causes of these conditions.

35. A more probable primary cause of tissue edema would be a _____ (C, capillary; P, plasma colloid osmotic) pressure of _____ mm. Hg.

a. C, 9　　　　　c. P, 9
b. C, 30　　　　　d. P, 30

36. Venous obstruction by blood clots tends to _____ (I, increase; N, not significantly alter; D, decrease) corresponding tissue capillary pressures, and to _____ (R, reduce; P, promote) the tendency for tissue edema.

a. I, R　　　c. D, R　　　e. N, P
b. N, R　　　d. I, P　　　f. D, P

37. Reduced capillary pressures are likely to be a primary cause of tissue edema with:

a. Angioneurotic edema　　e. All of the
b. Urticaria　　　　　　　　above
c. Unilateral heart failure　f. None of the
d. Bilateral heart failure　　above

38. Histamine release by tissues during allergic reactions _____ (I, increases; D, decreases) venous to arteriolar resistance ratios, and produces _____ (DL, decreased lymph flow; H, "hives").

a. I, DL　　　　　c. D, DL
b. I, H　　　　　d. D, H

39. Edema of body tissues, not directly injured by localized skin burns, is a more probable consequence of a greater and more widespread alteration of _____ (I, interstitial fluid; P, plasma) protein concentrations of _____ (A, albumin; G, globulin; F, fibrinogen).

   a. I, A   c. I, F   e. P, G
   b. I, G   d. P, A   f. P, F

40. _____ (F, filiariasis; C, *Clostridium oedematiens*) is a common cause of a _____ (B, brawny; D, dependent) type of edema of the lower extremities as a consequence of lymphatic obstruction.

   a. F, B   c. C, B
   b. F, D   d. C, D

41. A tendency towards edema would more likely result from a state of:

   a. Essential hypertension
   b. Protein nutritional deficiencies
   c. Arteriolar constriction
   d. Reduced venous pressure

42. Fluid retention with renal failure tends to _____ capillary pressures, _____ interstitial fluid pressure, and _____ the tendency for edema. (I, increase; N, not appreciably alter; D, decrease)

   a. D, D, D   c. I, N, I   e. I, D, I
   b. I, N, D   d. D, N, I   f. I, I, I

OBJECTIVE 31–8.

Identify the significant characteristics of the gel phase of interstitial spaces, and their relationship to interstitial fluid volumes and edema.

43. Interstitial fluid, of which about _____ % normally exists in a freely mobile state, comprises about _____ % of average body tissues.

   a. 0, 16   c. 95, 16   e. 40, 40
   b. 40, 16   d. 0, 40   f. 95, 40

44. A maximal volume of the gel phase of interstitial spaces, about _____ % above its normal value, _____ (D, does; N, does not) occur with a +1 degree of edema.

   a. 50, D   c. 500, D   e. 100, N
   b. 100, D   d. 50, N   f. 500, N

45. The gel phase of interstitial spaces has an associated surplus of mobile _____ (A, anions; C, cations), resulting in a tendency for osmotic water movement _____ (I, into; O, out of) the gel.

   a. A, I   c. C, I
   b. A, O   d. C, O

46. A dependent form of edema results from the influence of _____ (G, gravity; S, solid tissue pressure) upon the interstitial _____ (L, gel; M, mobile liquid) phase.

   a. G, L   c. S, L
   b. G, M   d. S, M

47. The presence of a gel in the interstitial fluid normally places a more severe restriction upon interstitial _____ (D, molecular diffusion; M, fluid mobility) within the _____ (N, negative; P, positive) interstitial fluid pressure range.

   a. D, N   c. M, N
   b. D, P   d. M, P

OBJECTIVE 31-9.

Identify interstitial fluid, solid tissue, and total tissue pressures, their origins, their interrelationships, their relative magnitudes, their significance, and their methods of determination.

48. Compression of blood vessels results from _____ and interstitial fluid movement results from _____. (S, only solid tissue pressure; I, only interstitial fluid pressure; B, both solid tissue and interstitial fluid pressures)

   a. B, B   c. S, I   e. I, B
   b. I, I   d. S, B   f. B, I

49. The total tissue pressure, when the interstitial fluid volume is normal, is closest to _____ mm. Hg.

   a. -7   c. 7   e. 27
   b. 0   d. 17   f. 47

50. In the edematous state, solid tissue pressure approaches _____ (N, negative; Z, zero; P, positive) values, and total tissue pressure approaches _____ (I, interstitial fluid; S, solid tissue) pressure values.

   a. N, I   c. P, I   e. Z, S
   b. Z, I   d. N, S   f. P, S

51. The average solid tissue pressure is about _____ mm. Hg, and is normally _____ (A, aided; O, opposed) by the interstitial fluid pressure.

   a. -3.8, A   c. +8.3, A   e. +3.8, O
   b. +3.8, A   d. -3.8, O   f. +8.3, O

52. Normally, a pressure of about _____ mm. Hg is minimally required to increase the volume of a balloon inserted into the interstitial fluid spaces, and represents a measure of _____ (S, solid tissue pressure; I, interstitial fluid pressure; SI, the sum of S and I).

   a. +2, S   c. +2, SI   e. +20, I
   b. +2, I   d. +20, S   f. +20, SI

OBJECTIVE 31-10.

Contrast the significant features of pulmonary interstitial fluid dynamics with those of peripheral tissues.

53. Lymph flow from the lungs has a _____ rate of flow and a _____ protein concentration when compared to typical peripheral tissues. (H, higher; L, lower)

   a. H, H   c. L, H
   b. H, L   d. L, L

54. Pulmonary capillaries have _____ pressures and _____ protein permeabilities when compared to typical peripheral capillaries. (L, lower; E, equal; H, higher)

   a. L, H   c. E, H   e. H, H
   b. L, L   d. E, L   f. H, L

55. The pulmonary interstitial fluid space is relatively _____ and its tolerance to positive interstitial fluid pressures is relatively _____ when compared to typical peripheral tissues (G, greater; L, less)

   a. G, G   c. L, G
   b. G, L   d. L, L

56. Large slit-pores _____ present between pulmonary capillary endothelial cells and _____ present between adjacent alveolar epithelial cells. (A, are; N, are not)

   a. N, N   c. A, N
   b. N, A   d. A, A

57. Pulmonary pressures involve a significantly greater interstitial _____ (F, fluid; O, colloid osmotic) pressure and a lesser _____ (C, capillary; P, plasma colloid osmotic) pressure when compared to peripheral tissues.

   a. F, C   c. O, C
   b. F, P   d. O, P

OBJECTIVE 31-11.

Identify "interstitial fluid" and "alveolar" forms of pulmonary edema, their causes and significance, and the safety factors against chronic and acute pulmonary edema.

58. Pulmonary capillary pressures must normally _____ (R, rise; F, fall) under acute conditions to about _____ mm. Hg before pulmonary edema develops.

    a. F, 7   c. F, 45   e. R, 28
    b. F, 28   d. R, 7   f. R, 45

59. A potential, but rarer, cause of pulmonary edema occurs with conditions of:

    a. Left heart failure
    b. Elevated capillary pressures
    c. Breathing noxious substances
    d. Decreased plasma proteins

60. An "alveolar" form of pulmonary edema first occurs when the interstitial fluid volume increases _____ (M, more; L, less) than 500 ml., to about _____ % above normal.

    a. M, 50   c. L, 50
    b. M, 500   d. L, 500

61. The safety factor in _____ (A, acute; C, chronic) conditions of pulmonary edema is greater as a consequence of adaptations of pulmonary _____ (P, capillaries; L, lymphatics).

    a. A, P   c. C, P
    b. A, L   d. C, L

32

The Special Fluid Systems of the Body — Cerebrospinal, Ocular, Pleural, Pericardial, Peritoneal, and Synovial

OBJECTIVE 32-1.

Identify the cerebrospinal fluid, its locations and volume, and its cushioning function.

1. Cerebrospinal fluid volume, about _____ ml., occupies about _____ % of the entire cavity enclosing the brain and spinal cord.

    a. 150, 9   c. 150, 39   e. 1,500, 19
    b. 150, 19   d. 1,500, 9   f. 1,500, 39

2. The ability of the cerebrospinal fluid to cushion or protect the brain from injury is more dependent upon its _____ (S, specific gravity; P, fluid pressure), which is _____ (H, higher; N, not significantly different; L, lower) than that of brain tissue.

    a. S, H   c. S, L   e. P, N
    b. S, N   d. P, H   f. P, L

3. Blows to the frontal region of the skull cause more severe injury to the _____ (F, frontal; P, parietal; O, occipital) region of the brain as a consequence of _____ (C, contrecoup; PE, papilledema).

    a. F, C   c. O, C   e. P, PE
    b. P, C   d. F, PE   f. O, PE

OBJECTIVE 32–2.

Identify, for cerebrospinal fluid, its formative locations and probable mechanisms, its formation rates, its relative composition, its lining surfaces, and the function of its perivascular spaces.

4. Cerebrospinal fluid is formed as _____ (F, a simple filtrate; S, an active secretion) of the _____ (A, arachnoid granulations; C, choroid plexus).

   a. F, A   c. S, A
   b. F, C   d. S, C

5. Cerebrospinal fluid contains a higher concentration of _____ and a lower concentration of _____ than other extracellular fluids. (G, glucose; S, sodium; P, potassium)

   a. G, S & P   c. G & P, S   e. S & P, G
   b. G & S, P   d. S, G & P   f. P, G & S

6. The formation rate of a slightly _____ (H, hypertonic; L, hypotonic) cerebrospinal fluid is about _____ times that of the total cerebrospinal fluid volume per day.

   a. H, ½   c. H, 5   e. L, 1
   b. H, 1   d. L, ½   f. L, 5

7. Particulate matter from the brain is more frequently conveyed by _____ (P, perivascular spaces; L, lymphatics) into the _____ (D, subdural sinus; A, subarachnoid space).

   a. P, D   c. L, D
   b. P, A   d. L, A

8. Diffusion between the cerebrospinal fluid of the ventricles and neural tissues of the brain occurs through the _____ (N, endothelial; P, epithelial) cellular layer of the _____ (A, arachnoid; E, ependyma).

   a. N, A   c. P, A
   b. N, E   d. P, E

9. The perivascular spaces, which _____ (R, are; N, are not) separated from neural tissues by the pia, occur for _____ (A, arteries only; V, veins only; B, both arteries and veins).

   a. R, A   c. R, B   e. N, V
   b. R, V   d. N, A   f. N, B

OBJECTIVE 32–3.

Identify the location and mechanisms of cerebrospinal fluid reabsorption, the cerebrospinal fluid flow pathways, and the regulatory mechanisms for cerebrospinal fluid pressure.

10. Arachnoid villi are regions of _____ (H, high; L, low) permeability, communicating between cerebrospinal fluid and _____ (A, arterial; V, venous) blood.

    a. H, A   c. L, A
    b. H, V   d. L, V

11. Cerebrospinal fluid communications between ventricles include the foramina of _____ (L, Luschka; M, Monro) between the third and lateral ventricles, and the aqueduct of _____ (S, Sylvius; G, Magendie) between the third and fourth ventricles.

    a. L, S   c. M, S
    b. L, G   d. M, G

12. Cerebrospinal fluid flow occurs from the _____ ventricle(s) through foramina into the _____ (D, subdural; A, subarachnoid) space.

    a. 1st & 2nd, D   d. 1st & 2nd, A
    b. 3rd, D   e. 3rd, A
    c. 4th, D   f. 4th, A

OBJECTIVE 32-4.

Identify the clinical method of cerebrospinal fluid measurement, pathologic conditions elevating cerebrospinal fluid pressure, and the consequences of elevated cerebrospinal fluid upon the retina.

13. A typical cerebrospinal fluid pressure is _____ mm. Hg, or about _____ mm. water.

    a. -6, -36   c. +10, +65   e. +2, +26
    b. +2, +13   d. -6, -78   f. +10, +130

14. Cerebrospinal fluid pressure is _____ by increased rates of formation, and is _____ by increased resistance to its absorption into the cardiovascular system. (I, increased; D, decreased; N, not significantly altered)

    a. N, N   c. D, D   e. N, I
    b. I, I   d. I, N   f. N, D

15. Hemorrhage or infectious processes of the cranial vault more frequently _____ (I, increase; D, decrease) cerebrospinal fluid pressure through a loss of function of the _____ (C, choroid plexus; P, perivascular spaces; A, arachnoidal villi).

    a. I, C   c. I, A   e. D, P
    b. I, P   d. D, C   f. D, A

16. _____ cerebrospinal fluid pressures result in an edematous condition of the retina as a consequence of _____ ratios of retinal arterial to venous flow resistances. (I, increased; D, decreased)

    a. D, D   c. I, D
    b. D, I   d. I, I

17. Cerebrospinal fluid pressures are generally measured by a direct tap of the _____ (M, cisterna magna; L, lumbar cistern) with the patient in a _____ (V, vertical; H, horizontal) plane.

    a. M, V   c. L, V
    b. M, H   d. L, H

OBJECTIVE 32-5.

Identify potential causes and resultant consequences of obstructions to cerebrospinal fluid flow. Identify hydrocephalus, its classification, and its treatment rationale.

18. Blockage of the aqueduct of Sylvius results in a _____ (C, communicating; N, noncommunicating) type of hydrocephalus and _____ (I, increased; D, decreased) cerebrospinal fluid pressure in the third ventricle.

    a. C, I   c. N, I
    b. C, D   d. N, D

19. Portions of the choroid plexus of the _____ (A, subarachnoid space; V, ventricles) are frequently therapeutically destroyed in the _____ (C, communicating; N, noncommunicating) type of hydrocephalus.

    a. A, C   c. V, C
    b. A, N   d. V, N

20. Blockage of the spinal cord generally results in _____ (D, lowered; E, elevated) cerebrospinal fluid pressure below the block as a consequence of its relatively _____ (H, high; L, low) rate of fluid reabsorption into lymphatics of the spinal cord.

    a. D, H   c. E, H
    b. D, L   d. E, L

21. Bilateral compression of the internal jugular veins results in a _____ (R, rise; F, fall) in cerebrospinal fluid pressure in the lower spinal cord if the spinal canal _____ (I, is; N, is not) blocked.

    a. R, I   c. F, I
    b. R, N   d. F, N

OBJECTIVE 32–6.

Identify the barriers to diffusion between fluid compartments of the central nervous system, the permeability characteristics of these barriers, and the properties of the interstitial fluid of the central nervous system.

22. Diffusion barriers generally present between blood and _____ (B, brain tissue only; C, cerebrospinal fluid only; CB, both brain tissue and cerebrospinal fluid) are notably absent in the region of the _____ (H, hypothalamus; M, medulla).

    a. B, H   c. CB, H   e. C, M
    b. C, H   d. B, M   f. CB, M

23. Adjacent endothelial cells of brain capillaries generally have _____ (S, slit pores; T, tight junctions) and _____ (A, are; N, are not) surrounded by glial cell processes.

    a. S, A   c. T, A
    b. S, N   d. T, N

24. The central nervous system has slightly _____ interstitial fluid and _____ collagen fibers than average body tissue, when compared on a tissue weight basis. (L, less; M, more)

    a. L, L   c. M, L
    b. L, M   d. M, M

25. Blood brain barriers are _____ permeable to carbon dioxide, _____ permeable to water, and _____ permeable to sodium and chloride ions. (H, highly; S, slightly; N, relatively non)

    a. H, H, H   c. H, H, N   e. H, S, S
    b. N, N, N   d. H, H, S   f. S, H, N

OBJECTIVE 32–7.

Identify the aqueous and vitreous humors of the eye, their relative compositions, and their relative locations. Identify the mechanisms of formation, exchange, and outflow of aqueous humor.

26. The _____ (V, vitreous; A, aqueous) humor of the eye lying between the lens and the retina has a comparatively _____ (H, high; L, low) rate of fluid flow.

    a. V, H   c. A, H
    b. V, L   d. A, L

27. Aqueous humor is formed by _____ (F, filtration; A, active secretion), largely by the _____ (R, retinal capillaries; S, canal of Schlemm; C, ciliary processes).

    a. F, R   c. F, C   e. A, S
    b. F, S   d. A, R   f. A, C

28. Aqueous humor, _____ (H, hypotonic; E, hypertonic) with respect to plasma, is formed at the rate of 2 to 5 ml. per _____ (M, minute; R, hour; D, day).

    a. H, M   c. H, D   e. E, R
    b. H, R   d. E, M   f. E, D

29. More significant protein and particulate phagocytosis occurs in the _____ , and more significant fluid volume reabsorption of aqueous humor occurs in the _____ . (S, canal of Schlemm; I, iris epithelium; L, retinal lymphatics)

    a. I, L   c. L, I   e. S, S
    b. I, S   d. L, S   f. S, L

30. The canal of Schlemm, located in the _____ (A, anterior; P, posterior) chamber of the eye, is a _____ (L, lymph; V, venous) vessel generally containing _____ (H, aqueous humor; B, blood).

    a. A, V, B   c. A, L, H   e. P, L, H
    b. A, V, H   d. P, L, B   f. P, V, B

OBJECTIVE 32–8.

Identify the normal intraocular pressure, its method of measurement, its means of regulation, and the conditions and results of elevated intraocular pressure.

31. The average intraocular pressure of _____ mm. Hg is largely regulated by a formative and reabsorptive balance involving _____ (A, aqueous; V, vitreous) humor.

    a. 1 to 3, A   c. 45, A   e. 15, V
    b. 15, A   d. 1 to 3, V   f. 45, V

32. Diagnosis of the elevated intraocular pressure of _____ (P, papilledema; G, glaucoma) is more practically accomplished by _____ (S, catheterization of the canal of Schlemm; O, oculopuncture; T, tonometry).

    a. P, S   c. P, T   e. G, O
    b. P, O   d. G, S   f. G, T

33. Precise regulation of intraocular pressure is normally thought to involve more directly a _____ (N, neural; L, local) regulatory mechanism of aqueous humor _____ (F, formation; R, reabsorption).

    a. N, F   c. L, F
    b. N, R   d. L, R

34. The administration of Diamox _____ (I, increases; D, decreases) aqueous humor pressure, and is a frequent _____ (C, cause; T, treatment) of glaucoma.

    a. D, C   c. I, C
    b. D, T   d. I, T

35. Most cases of glaucoma result from increased flow resistance in the vicinity of the _____ (C, ciliary process; I, irido-corneal) junction, and are potentially capable of causing blindness with chronic threshold pressures of _____ mm. Hg.

    a. C, 10 to 20   d. I, 10 to 20
    b. C, 30 to 40   e. I, 30 to 40
    c. C, 70 to 80   f. I, 70 to 80

OBJECTIVE 32–9.

Identify the potential spaces of the body, the dynamics of their fluid exchange, their associated pressures and functions, and their conditions for transudate fluid formation.

36. Pleural, pericardial, peritoneal, and synovial cavities normally contain _____ (C, considerable; M, minimal) fluid volumes, which largely function in tissue _____ (T, movements; N, nutritional states; E, edematous states).

    a. C, T   c. C, E   e. M, N
    b. C, N   d. M, T   f. M, E

37. The tendency for transudate fluid formation in potential spaces is _____ by lymphatic blockage, and is _____ by increased capillary pressures. (I, increased; n, not significantly altered; D, decreased)

    a. N, N   c. D, D   e. N, D
    b. I, I   d. N, I   f. D, I

38. The _____ (P, positive; N, negative) fluid pressure of the intrapleural space of the lungs serves to hold the _____ (A, parietal; V, visceral) pleura tightly against the chest wall.

    a. P, A   c. N, A
    b. P, V   d. N, V

39. The _____ (PL, pleural; PT, peritoneal; PC, pericardial) cavity is generally more susceptible to transudate fluid formation than other potential spaces as a consequence of its _____ (H, higher; L, lower) capillary pressures.

    a. PL, H   c. PC, H   e. PT, L
    b. PT, H   d. PL, L   f. PC, L

40. The _____ (P, positive; N, negative) pressure of the pericardial cavity becomes more positive or less negative during _____ (E, expiration only; F, excessive filling of the heart only; B, both expiration and excessive filling of the heart).

    a. P, E   c. P, B   e. N, F
    b. P, F   d. N, E   f. N, B

41. The synovial membrane of joint cavities and bursae _____ (I, is; N, is not) a significant barrier to ion and molecular exchange with surrounding tissues, and contains a fluid comparatively high in _____ (M, mucopolysaccharides; P, plasma proteins).

    a. I, M   b. I, P   c. N, M   d. N, P

33

Partition of the Body Fluids: Osmotic Equilibria Between Extracellular and Intracellular Fluids

OBJECTIVE 33–1.

Identify typical values of total body water content among adults, its variability with age and sex, and the consequences of increased adipose tissue.

1. The total body water in an average 70-kg. adult, in liters, is closest to:

    a. 125   c. 40   e. 15
    b. 90   d. 25   f. Less than 1

2. The most constant ratio from one individual to another is the ratio of body _____ divided by body _____ . (F, fat; L, lean mass; W, water; A, weight)

    ·a. F, A   c. F, L   e. L, A
    b. F, W   d. L, W   f. A, W

3. On the average, fat-free tissue of the human body is approximately _____ % water.

    a. 10   c. 50   e. 90
    b. 30   d. 70   f. 99

4. An individual with a total body water content of 43% of the total body weight is more probably:

    a. Newborn   d. Elderly
    b. Obese   e. An average adult man
    c. Lean   f. An average adult woman

5. The intravenous injection of a pharmacological agent, distributed throughout all body water compartments, would result in its dilution with an amount of water equivalent to about _____ % of the total body mass. This estimate would have to be revised _____ (D, downward; U, upward) for individuals with greater amounts of adipose tissue.

    a. 60, D   c. 60, U
    b. 85, D   d. 85, U

OBJECTIVE 33–2.

Identify the sources of body water intake and output, their typical values, and the conditions for their variance.

6. Daily sources of body water intake average _____ liters, of which the larger percentage occurs in _____ (F, oral food intake; L, oral liquid intake; W, water of oxidation).

    a. 2 to 2½, F   c. 2 to 2½, W   e. 5 to 6, L
    b. 2 to 2½, L   d. 5 to 6, F   f. 5 to 6, W

7. Water produced by the oxidation of hydrogen during food metabolism averages _____ liters per _____ (H, hour; D, day).

    a. 0.2, H   c. 1.5, H   e. 0.5, D
    b. 0.5, H   d. 0.2, D   f. 1.5, D

8. Water lost in saturating air entering the lungs to a vapor pressure of _____ mm. Hg averages about _____ ml. per day.

   a. 27, 100   c. 27, 700   e. 47, 350
   b. 27, 350   d. 47, 100   f. 47, 700

9. Atmospheric vapor pressure tends to _____ (I, increase; D, decrease) with decreasing temperature, resulting in a greater insensible pulmonary water loss in very _____ (C, cold; W, warm) weather.

   a. I, C   c. D, C
   b. I, W   d. D, W

10. An obligatory urine output resulting from prolonged and heavy exercise is about _____ ml. per day, whereas an average urine output under average conditions approximates _____ times this value.

    a. 100, 3   c. 500, 3   e. 350, 5
    b. 350, 3   d. 100, 5   f. 500, 5

11. Cornified skin, as a consequence of a high _____ (C, cholesterol; G, collagen) content, has an insensible water loss that is _____ (H, considerably higher; E, about equal; L, considerably lower) than that for the pulmonary system under average conditions.

    a. C, H   c. C, L   e. G, E
    b. C, E   d. G, H   f. G, L

12. Contributing factors of _____ (P, insensible pulmonary loss; E, eccrine sweating rates; U, urine output) result in _____ (I, increased; D, decreased) daily loss of water with prolonged heavy exercise.

    a. E only, I   d. U only, D
    b. Both E & U, I   e. Both U & P, D
    c. Both E & P, I   f. Both E & U, D

OBJECTIVE 33–3.

Identify the intracellular and extracellular fluid compartments, their average values, and the subdivisions of the extracellular fluid compartment.

13. The higher fluid volume occurs in the _____ (E, extracellular; I, intracellular) fluid volume and averages _____ liters.

    a. E, 40   c. E, 15   e. I, 25
    b. E, 25   d. I, 40   f. I, 15

14. Within the extracellular fluid compartment, the highest volume occurs for:

    a. Interstitial fluids   e. The gastrointestinal tract
    b. Plasma
    c. Cerebrospinal fluids   f. Intraocular fluids
    d. Potential spaces

15. The extracellular fluid volume, typically about _____ liters in a 70-kg. adult, contains a plasma volume averaging _____ liters.

    a. 15, 1.5   c. 15, 5   e. 25, 3
    b. 15, 3   d. 25, 1.5   f. 25, 5

OBJECTIVE 33–4.

Identify average values of blood, plasma, and red blood cell volumes, and their variance with body weight, sex, and adipose tissue. Identify the types of measured hematocrits, their typical values, and the causes and consequences of hematocrit variance.

16. The average blood volume of normal adults is _____ liters, of which the larger volume results from the _____ (P, plasma; E, erythrocyte) component.

    a. 2, P   c. 5, P   e. 3, E
    b. 3, P   d. 2, E   f. 5, E

17. The percentage of body weight that represents the blood volume of lean adults averages _____ %, and is _____ (I, increased; D, decreased) by increasing body fat content.

    a. 5, I   c. 12, I   e. 8, D
    b. 8, I   d. 5, D   f. 12, D

18. The average adult woman has a blood volume
per unit of body weight about _____ %
_____ (L, lower; H, higher) than the
average adult man.

   a. 5, L   c. 35, L   e. 20, H
   b. 20, L   d. 5, H   f. 35, H

19. The hematocrit is the percentage of _____
(P, plasma; C, cells) in blood, and is generally
higher for adult _____ (M, men; W,
women).

   a. P, M   c. C, M
   b. P, W   d. C, W

20. The hematocrit of very small vessels of the
body is considerably _____ (G, greater; L,
less) than that in large vessels as a consequence
of _____ (A, axial streaming; M, margination).

   a. L, A   c. G, A
   b. L, M   d. G, M

21. Actual values of hematocrit from large vessels
are generally _____ than their measured
values and are _____ than actual values of
"body hematocrit." (G, greater; L, less)

   a. G, G   c. L, G
   b. G, L   d. L, L

OBJECTIVE 33–5.

Identify the diluting principle for measuring fluid volumes. Identify appropriate
methods for measuring or calculating fluid compartment volumes, their typical values, and
their measurement rationale.

22. The volume of a diluting compartment is
equivalent to _____ (Q, the quantity of test
substance injected; C, the final concentration
per ml. of dispersed fluid)

   a. (QC)   c. (C/Q)
   b. (Q/C)   d. 1/(QC)

23. Erythrocyte compartment volumes may be
more readily measured with _____,
whereas plasma volumes are more readily
measured with _____. (Cr, radioactive chro-
mium; Fe, radioactive iron; I, radioactive
iodine; EB, Evans blue dye)

   a. Cr, Fe, & I; EB   c. Fe, I, & EB; Cr
   b. Cr & Fe; I & EB   d. Fe; I, EB, & Cr

24. The "_____ spaces" are more likely to be
lower than, and the "_____ spaces" are
more likely to be higher than, the actual extracellular space. (Br, bromide; S, sucrose; I,
insulin; T, thiocyanate)

   a. Br & S; I & T   d. I & T; Br & S
   b. S & I; Br & T   e. Br & T; S & I
   c. Br, S, & I; T   f. Br, I & T; S

25. The total body water may be estimated utilizing the dilution principle and the administration of:

   a. Insulin   e. Vital dyes
   b. Antipyrine   f. Organic
   c. Sodium thiocyanate   phosphates
   d. Thiosulfate

26. The interstitial fluid volume is generally measured _____ (D, directly; I, indirectly), and
approximates _____ liters in an average
70-kg. adult.

   a. D, 8   c. D, 15   e. I, 12
   b. D, 12   d. I, 8   f. I, 15

OBJECTIVE 33–6.

Identify the constituents of extracellular and intracellular fluids and their relative concentrations.

27. Dissolved proteins and nonelectrolyte constituents compose a _____ amount for plasma, a _____ amount for interstitial fluid, and a _____ amount for intracellular fluid, when compared on a mass basis to the electrolyte fractions. (G, greater; L, lesser)

    a. G, G, G   c. L, L, G   e. G, L, L
    b. L, G, G   d. G, L, G   f. L, L, L

28. Typically, plasma concentrations include 150 mg.% of _____ , 100 mg.% of _____ and 15 mg.% of _____ . (G, glucose; B, bilirubin; U, urea; C, cholesterol)

    a. G, U, B   c. U, C, G   e. C, G, B
    b. G, B, C   d. U, G, C   f. C, G, U

29. Ratios of extracellular to intracellular fluid concentrations are greater than one for _____ (Na, sodium; Cl, chloride; Mg, magnesium; HCO_3, bicarbonate; K, potassium; SO_4, sulfate; Ca, calcium; HPO_4, phosphate).

    a. Na, Cl, Mg, & HCO_3
    b. K, Mg, SO_4, & Cl
    c. Cl, HCO_3, Ca, & Na
    d. HPO_4, HCO_3, K, & Mg
    e. Cl, Mg, Na, & HPO_4
    f. Ca, Mg, Na, & Cl

30. The protein concentration is highest in _____ fluids, and lowest in _____ fluids. (T, interstitial; I, intracellular; P, plasma)

    a. P, T   c. T, I   e. I, T
    b. I, P   d. P, I   f. T, P

31. The concentration of the major extracellular cation approximates _____ mEq./liter, whereas the major intracellular cation concentration approximates _____ mEq./liter.

    a. 40, 68   c. 300, 68   e. 140, 140
    b. 140, 68   d. 40, 140   f. 300, 140

OBJECTIVE 33–7.

Identify the basic principles underlying osmosis and osmotic pressure.

32. The movement of water through a semipermeable membrane toward the side with the _____ (G, greater; E, lesser) concentration of nondiffusible substances is termed _____ (O, osmolality; S, osmosis).

    a. G, O   c. E, O
    b. G, S   d. E, S

33. The side of a semipermeable membrane with the higher concentration of nondiffusible substances has a _____ total chemical "activity" of water, and a _____ tendency for outward water diffusion that the opposing side. (G, greater; L, lesser)

    a. G, G   c. L, G
    b. G, L   d. L, L

34. A molecule of high molecular weight exerts _____ osmotic effect as compared to a molecule of low molecular weight, and a bivalent cation exerts _____ osmotic effect as compared to a monovalent cation. (G, a greater; S, the same; L, a lesser)

    a. G, G   c. S, S   e. L, G
    b. G, S   d. S, L   f. L, S

35. The osmol is a unit of measure of the _____ (N, total number; MW, average molecular weight; O, osmotic pressure) of _____ (M, only nonionizable molecules; B, both ions and nonionizable molecules).

    a. N, M   c. O, M   e. MW, B
    b. MW, M   d. N, B   f. O, B

36. The _____ (L, osmolality; R, osmolarity) of a solution, when expressed in milliosmols per liter and _____ (M, multiplied; D, divided) by 19.3, approximates its osmotic pressure in mm. Hg at body temperature.

    a. L, M   b. L, D   c. R, M   d. R, D

OBJECTIVE 33–8.

Identify the osmolarity of body fluids, the relative magnitudes of constituent contributions, corrective factors for osmotic effects, and resultant total osmotic pressures.

37. Sodium and chloride ions account for _____ than 75% of the total osmolarity of interstitial fluid, and for _____ than 75% of the total osmolarity of plasma. (M, more; L, less)

    a. M, M   c. L, M
    b. M, L   d. L, L

38. The osmolarity of interstitial fluid is closer to _____ milliosmols per liter, and is _____ (H, higher; L, lower) than plasma.

    a. 100, H   c. 300, H   e. 200, L
    b. 200, H   d. 100, L   f. 300, L

39. The osmotic pressure difference between plasma and interstitial fluids of about _____ mm. Hg results largely from differences in _____ (H, hydrostatic pressures; P, protein concentrations).

    a. 12, H   c. 48, H   e. 24, P
    b. 24, H   d. 12, P   f. 48, P

40. The osmotic activity of interstitial fluid constituents is slightly _____ (G, greater; L, less) than that expected from the number of milliosmols present, as a result of a greater tendency for intermolecular _____ (A, attraction; R, repulsion) of ions and molecules in solution.

    a. G, A   c. L, A
    b. G, R   d. L, R

41. The osmotic pressure exerted by plasma against pure water at 37° C. lies within a pressure range of _____ atmospheres.

    a. ½ to 1   c. 4 to 5   e. 10 to 15
    b. 2 to 2½   d. 6 to 8   f. 20 to 30

OBJECTIVE 33–9.

Identify isotonic, hypertonic, and hypotonic solutions, and their influence upon cellular volumes, cellular integrity, and rate factors influencing compartmental osmotic equilibrium.

42. A cell, placed in a solution with a greater osmolarity than intracellular fluid or a _____ (H, hypotonic; E, hypertonic; IS, isotonic) solution, _____ (I, increases; D, decreases) its volume as a consequence of osmosis.

    a. H, I   c. IS, I   e. E, D
    b. E, I   d. H, D   f. IS, D

43. Either a 5% glucose solution, or a _____ % solution of sodium chloride, is _____ (H, hypertonic; O, hypotonic; I, isotonic) to body fluids.

    a. 0.9, H   c. 9.0, I   e. 4.5, H
    b. 4.5, O   d. 0.9, I   f. 9.0, H

44. 7 grams of sodium chloride (molecular weight = 58.5), dissolved in one liter of water, would result in a solution closest to _____ osmolar and would be _____ (H, hypertonic; O, hypotonic; I, isotonic) to mammalian intracellular fluids.

    a. 0.12, H   c. 0.7, H   e. 0.12, I
    b. 0.24, O   d. 0.12, O   f. 0.24, I

45. Cells neither swell nor shrink when placed in _____ (H, hypertonic; O, hypotonic; I, isotonic) solutions of a substance to which cell membranes are relatively _____ (P, permeable; M, impermeable).

    a. H, P   c. O, P   e. I, M
    b. I, P   d. H, M   f. O, M

46. A biphasic response of decreased and then increased cellular volumes occurs when cells are placed in _____ (H, hypertonic; O, hypotonic; I, isotonic) solutions of a constituent to which cell membranes are relatively _____ (P, permeable; M, impermeable).

    a. H, M   c. O, M   e. I, P
    b. I, M   d. H, P   f. O, P

47. Erythrocytes placed in tap or distilled water would _____ (H, hemolyze; C, crenate) in a time most closely approximating 10 _____ (S, seconds; M, minutes).

    a. H, S   c. C, S
    b. H, M   d. C, M

OBJECTIVE 33–10.

Identify the consequences of saline addition of varied tonicity, dehydration, and sodium chloride loss upon fluid volumes and osmolarity.

48. Intracellular and interstitial fluid compartments of the body generally have the same:

    a. Sodium ion concentrations
    b. Electrolyte osmolarity
    c. Colloid osmotic pressures
    d. Total osmotic pressures
    e. Chloride ion concentrations
    f. All of the above

49. The addition of water to body fluids result in _____ extracellular fluid volumes, _____ osmolarity, and _____ intracellular fluid volumes following osmotic equilibrium. (I, increased; U, unchanged; D, decreased)

    a. I, I, D   c. I, D, I   e. D, U, I
    b. I, D, U   d. I, I, I   f. U, D, I

50. The total osmol content of the intracellular fluid compartment is typically _____ milliosmols.

    a. 22.4   c. 1,250   e. 12,500
    b. 900   d. 7,500   f. 25,000

51. Following osmotic equilibration, the larger amount of a volume of ingested water would be added to the _____ (I, intracellular; E, extracellular) fluid compartment, which normally averages _____ liters.

    a. I, 25   c. I, 40
    b. E, 25   d. E, 40

52. The addition of a 4.5% saline solution to body fluids results in _____ extracellular fluid volumes, _____ osmolarity, and _____ intracellular fluid volumes following osmotic equilibrium. (I, increased; U, unchanged; D, decreased)

    a. I, I, U   c. I, U, U   e. D, U, I
    b. I, I, D   d. I, I, I   f. D, D, D

53. Severe sweating with subsequent water replacement results in a net loss of body sodium chloride, _____ extracellular fluid volumes, _____ osmolarity, and _____ intracellular volumes. (I, increased; U, unchanged; D, decreased)

    a. D, D, D   c. I, D, D   e. D, D, I
    b. U, D, U   d. I, D, I   f. U, I, U

54. Intracellular edema may be more effectively reduced through the administration of a _____ (E, hypertonic; O, hypotonic; I, isotonic) solution of a substance to which cell membranes have a relatively _____ (H, high; L, low) permeability.

    a. O, H   c. I, H   e. E, L
    b. E, H   d. O, L   f. I, L

OBJECTIVE 33–11.
Identify solutions utilized for physiologic and clinical replacement and nutritive purposes, and their utilization rationale.

55. _____ (T, Tyrode's; R, Ringer's; H, Hartmann's) solution contains 146 mEq./liter of sodium ions, 4 mEq./liter of potassium ions, and 5.4 mEq./liter of calcium ions and is utilized for _____ (E, extracellular; I, intracellular) fluid replacement.

a. T, E   c. H, E   e. R, I
b. R, E   d. T, I   f. H, I

56. A solution containing 167 mEq./liter of sodium ions and 167 mEq./liter of lactate would more probably be utilized for:

a. ECF replacement   d. Acidosis
b. Nutrition   e. Alkalosis
c. "Water   f. Plasma
   intoxication"   substitution

57. The more complete replacement solution for extracellular fluid is _____ solution.

a. Ringer's   c. Darrow's
b. Hartmann's   d. Tyrode's

58. A solution containing a ⅙ molar concentration of ammonium chloride would more probably be utilized for:

a. ECF replacement   d. Acidosis
b. Nutrition   e. Alkalosis
c. "Water   f. Plasma
   intoxication"   substitution

59. Isotonic saline approximates a _____ osmolar solution, and contains sodium and chloride ion concentrations of _____ mEq./liter.

a. 0.15, 75   c. 0.9, 75   e. 0.3, 155
b. 0.3, 75   d. 0.15, 155   f. 0.9, 155

34

Formation of Urine by the Kidney: Glomerular Filtration, Tubular Function and Plasma Clearance

OBJECTIVE 34–1.
Identify the physiologic anatomy of the kidney.

True

1. Complete functional units of the kidney are the _____ (N, nephrons; G, glomeruli), of which each kidney contains about _____ .

a. N, 1,000   d. G, 1,000
b. N, 120,000   e. G, 120,000
c. N, 1,200,000   f. G, 1,200,000

2. The glomerulus is a network of capillaries, derived from the _____ (A, afferent; E, efferent) arteriole, which is enclosed by the epithelial cells of _____ (D, ducts of Bellini; B, Bowman's capsule; H, Henle's loops).

a. A, D   c. A, H   e. E, B
b. A, B   d. E, D   f. E, H

3. Blood flow into the vasa recta is largely derived from _____ (A, afferent; E, efferent) arterioles of _____ (C, cortical; M, juxtamedullary) nephrons.

   a. A, C   c. E, C
   b. A, M   d. E, M

4. Glomeruli are characteristically located in the renal _____ and thin segments of loops of Henle are located in the renal _____ . (M, medulla only; C, cortex only; B, both medulla and cortex)

   a. B, M   c. M, C   e. C, B
   b. M, B   d. B, B   f. C, M

5. The chronological sequence of structures that renal tubular fluid encounters is _____ . (C, collecting duct; B, Bowman's capsule; H, loop of Henle; P, proximal tubule; D, distal tubule)

   a. P, H, D, C, B   d. C, P, H, D, B
   b. B, H, P, D, C   e. C, H, P, D, B
   c. B, P, H, D, C   f. P, H, D, B, C

6. The majority of water and solute reabsorption from renal tubules occurs into _____ (G, glomerular; V, vasa recta; P, peritubular) capillaries of the renal _____ (M, medulla; C, cortex).

   a. G, M   c. P, M   e. V, C
   b. V, M   d. G, C   f. P, C

OBJECTIVE 34–2.

Identify the basic nephron functions, and the underlying processes by which the nephron achieves these functions.

7. Functionally significant renal processes include:

   a. Active secretion   d. Filtration
   b. Active reabsorption   e. Passive secretion
   c. Passive absorption   f. All of the above

8. A basic function of the nephron is the elimination of metabolic end-products, achieved principally by filtration occurring in _____ , selective reabsorption occurring in _____ , and active secretion occurring in _____ . (G, glomeruli only; R, renal tubules only; B, both the glomeruli and renal tubules)

   a. G, G, G   c. R, R, R   e. G, B, B
   b. B, B, B   d. G, R, R   f. G, R, G

OBJECTIVE 34–3.

Identify typical values for renal blood volume flow and the renal fraction. Identify the pressures and functional significance of the "high" and "low" pressure capillary networks of the kidneys.

9. Renal blood flow in a 70-kg. male adult averages _____ ml. per minute, or about _____ % of the cardiac output.

   a. 60, 5   c. 1,200, 20   e. 600, 20
   b. 600, 10   d. 60, 10   f. 1,200, 40

10. The renal fraction is the ratio of _____ to _____ . (P, renal plasma flow; G, glomerular filtrate; R, renal blood flow; CO, cardiac output)

    a. G, P   c. G, CO   e. CO, G
    b. G, R   d. P, CO   f. R, CO

11. Compared to the majority of systemic capillary networks, the peritubular capillary network has a relatively _____ pressure, and the glomerular capillary network has a relatively _____ pressure. (H, high; L, low)

    a. H, H   c. L, H
    b. H, L   d. L, L

12. The vasa recta have a significantly _____ proportion of the total renal flow and a significantly _____ velocity of blood flow, when compared to other renal capillary systems. (H, higher; L, lower)

    a. H, H   c. L, H
    b. H, L   d. L, L

13. Glomerular capillary and peritubular capillary pressures more probably operate at about _____ % and _____ %, respectively, of the normal mean arterial pressure.

    a. 86, 60   c. 20, 6   e. 13, 60
    b. 60, 13   d. 60, 86   f. 6, 20

14. Estimates of renal pressures yield _____ values of total renal tissue pressure and _____ values of the renal interstitial fluid pressure. (N, negative; P, positive)

    a. N, N   c. P, N
    b. N, P   d. P, P

15. Peritubular capillaries which primarily function in fluid _____ (R, reabsorption; F, filtration) are involved in the net transfer of about _____ liters of renal tubular fluid per day.

a. R, 1.8
b. R, 18

c. R, 180
d. F, 1.8

e. F, 18
f. F, 180

OBJECTIVE 34–4.

Identify the properties of the glomerular membrane and the composition of glomerular filtrate.

16. Glomerular filtrate filters through _____ cellular and _____ basement membrane layer(s).

a. 1, 0
b. 2, 1

c. 2, 2
d. 1, 2

e. 1, 1
f. 2, 3

17. The glomerular filtrate to plasma concentration ratio of a substance with a molecular weight of 5,000 is closer to:

a. 0
b. 0.2

c. 0.4
d. 0.6

e. 0.8
f. 1.0

18. Capillary endothelial cells of glomerular capillaries are atypical in that they possess _____ (S, slit pores; F, fenestrae) and unusually _____ (H, high; L, low) permeabilities.

a. S, H
b. S, L

c. F, H
d. F, L

19. The chloride ion concentration of glomerular filtrate is _____ (H, higher; L, lower) than plasma as a consequence of _____ (P, protein concentration gradients; A, active transport; S, pore "sieving" effects).

a. H, P
b. H, A

c. H, S
d. L, P

e. L, A
f. L, S

20. The protein concentration of glomerular filtrate is _____ than that of plasma, and is _____ than that of interstitial fluids. (H, higher; N, not significantly different; L, lower)

a. N, H
b. L, N

c. L, H
d. L, L

OBJECTIVE 34–5.

Identify the glomerular filtration rate, the filtration fraction, and their typical values.

21. The glomerular filtration rate is expressed as the rate at which fluid enters the _____ (D, collecting ducts; C, Bowman's capsules) of _____ (O, one kidney; B, both kidneys).

a. D, O
b. D, B

c. C, O
d. C, B

22. The filtration fraction is the ratio of _____ to _____ . (P, renal plasma flow; G, glomerular filtrate; R, renal blood flow; CO, cardiac output)

a. G, P
b. G, R

c. G, CO
d. P, CO

e. CO, G
f. R, CO

23. The glomerular filtration rate is typically closest to:

a. 25 ml./minute
b. 50 ml./minute
c. 300 ml./minute
d. 1 liter/minute

e. 20% of the cardiac output
f. 125 ml./minute

24. A fluid volume equal to the total body water of an average adult is filtered into the renal tubules and reabsorbed, on the average, about _____ times daily.

a. ½
b. 2

c. 4
d. 8

e. 20
f. 150

25. The ratio of the filtration fraction to the renal fraction is typically about:

a. 1 to 5
b. 1 to 3

c. 1 to 2
d. 1 to 1

e. 2 to 1
f. 3 to 1

26. With a cardiac output of 6 liters per minute, a renal fraction of 10%, a filtration fraction of 20%, and a hematocrit of 40%, the glomerular filtration rate would be closest to _____ ml./minute.

a. 48
b. 72

c. 106
d. 125

e. 240
f. 360

OBJECTIVE 34-6.

Identify the normal net filtration pressure for glomerular filtration, and the component pressures influencing its magnitude.

27. The net filtration pressure for the formation of glomerular filtrate equals _____ . (CP, capsular pressure; OP, glomerular capillary colloid osmotic pressure; GP, glomerular capillary pressure)

    a. (GP + OP) - CP   d. (CP + OP) - GP
    b. GP - (CP + OP)   e. GP + OP + CP
    c. CP - (OP + GP)   f. (CP/GP)(OP)

28. The glomerular filtration rate divided by the normal mean net filtration pressure, or filtration _____ (F, fraction; C, coefficient), approximates _____ ml./min./mm. Hg.

    a. F, 2.5   c. F, 12.5   e. C, 7.5
    b. F, 7.5   d. C, 2.5   f. C, 12.5

29. The colloid osmotic pressure _____ (I, increases; D, decreases) about 20% from the arterial to venous ends of glomerular capillaries as a consequence of the _____ (R, renal; F, filtration) fraction.

    a. I, R   c. D, R
    b. I, F   d. D, F

30. Ureteral obstruction would _____ (I, increase; N, not significantly alter; D, decrease) Bowman's capsular pressure, normally estimated to be about _____ mm. Hg.

    a. I, 5 to 10   c. D, 5 to 10   e. N, 15 to 20
    b. N, 5 to 10   d. I, 15 to 20   f. D, 15 to 20

31. The colloid osmotic pressure of the glomerular capillaries at their arterial end approximates _____ mm. Hg, and _____ (I, is; N, is not) opposed by a significant colloid osmotic pressure of glomerular filtrate.

    a. 18, I   c. 38, I   e. 28, N
    b. 28, I   d. 18, N   f. 38, N

OBJECTIVE 34-7.

Identify the effects of afferent and efferent arteriole constriction, mild and strong sympathetic stimulation, arterial pressure, and renal blood flow upon filtration pressure and glomerular filtration rates.

32. Increased glomerular pressures tend to _____ and increased plasma colloid osmotic pressures tend to _____ the glomerular filtration rate. (I, increase; N, not significantly alter; D, decrease)

    a. I, I   c. N, I   e. N, D
    b. I, N   d. I, D   f. N, N

33. Increased rates of glomerular capillary flow, _____ the mean plasma protein concentration of glomerular capillaries, and _____ the glomerular filtration rate. (I, increase; N, does not significantly alter; D, decrease)

    a. N, N   c. N, D   e. D, D
    b. N, I   d. I, I   f. D, I

34. Constriction of renal afferent arterioles tends to _____ glomerular capillary pressures and _____ the rate of glomerular blood flow. (I, increase; N, not significantly alter; D, decrease)

    a. I, I   c. D, I   e. N, D
    b. N, I   d. I, D   f. D, D

35. Constriction of renal efferent arterioles tends to _____ glomerular capillary pressures and _____ the rate of glomerular blood flow. (I, increase; N, not significantly alter; D, decrease)

    a. I, I   c. D, I   e. N, D
    b. N, I   d. I, D   f. D, D

36. Moderate sympathetic stimulation to the kidneys results in a more pronounced _____ (C, constriction; D, dilation) of _____ (A, afferent; E, efferent) arterioles.

    a. C, A   c. D, A
    b. C, E   d. D, E

37. Elevated arterial pressures result in a more pronounced _____ (C, constriction; D, dilation) of the _____ (A, afferent; E, efferent) arterioles as a result of autoregulatory mechanisms.

    a. C, A   c. D, A
    b. C, E   d. D, E

OBJECTIVE 34–8.

Identify the conditions of tubular reabsorption and secretion, their relationship to rates of filtration versus rates of excretion of a substance and their significance.

38. The concentration of a small molecular weight substance in glomerular filtrate, which is _____ (G, greater than; E, equal to; L, less than) its plasma concentration, when multiplied by the glomerular filtration rate, equals the rate of _____ (F, filtration; X, excretion) of a substance.

    a. G, F   c. L, F   e. E, X
    b. E, F   d. G, X   f. L, X

39. The rate of urine flow times the urine concentration of a substance is equal to its rate of renal tubular:

    a. Active secretion   d. Reabsorption
    b. Passive diffusion   e. Excretion
    c. Filtration   f. Passive secretion

40. Net renal tubular _____ and _____, respectively, are conditions in which the rate of filtration of a substance exceeds its rate of excretion, and the rate of excretion exceeds its rate of filtration. (R, reabsorption; S, secretion; A, active transport; P, passive transport)

    a. R, S   c. A, P
    b. S, R   d. P, A

41. The volume percentage of the glomerular filtrate which is _____ (R, reabsorbed; S, secreted) daily by the renal tubules and ducts is normally about _____ % of the glomerular filtrate.

    a. R, 45 to 65   d. S, 45 to 65
    b. R, 80 to 85   e. S, 80 to 85
    c. R, 99 to 100   f. S, 99 to 100

OBJECTIVE 34–9.

Identify the basic mechanisms of renal tubular absorption and secretion, their transported constituents, and their significance.

42. Renal tubular active transport mechanisms exist for _____, and renal tubular passive transport mechanisms exist for _____. (R, reabsorption only; S, secretion only; B, both reabsorption and secretion)

    a. R, R   c. S, R   e. B, S
    b. R, S   d. R, B   f. B, B

43. The more significant _____ (P, passive; A, active) sodium ion transport through the brush border membrane surface of proximal renal tubular epithelial cells is _____ (O, opposed; S, assisted) by its existing electrochemical concentration gradient.

    a. P, O   c. A, O
    b. P, S   d. A, S

44. Intracellular potentials of proximal renal tubular epithelial cells, typically _____ mV., are a consequence of active transport of sodium ions along _____ (BC, basal channels; BB, brush borders; S, lateral intercellular spaces) of the epithelial cells.

    a. -30, BB only   d. -50, BB only
    b. -30, BB & BC only   e. -50, BC & S only
    c. -30, BC, BB, & S   f. -50, BC, BB, & S

45. Known renal transport mechanisms include active _____ for potassium ions, active _____ for glucose, and active _____ for urate ions. (S, secretion only; A, absorption only; B, both secretion and absorption)

    a. B, A, S   c. B, B, A   e. S, A, B
    b. B, A, B   d. S, A, S   f. S, B, S

46. The _____ (A, active; P, passive) reabsorption of water in the proximal tubule is largely linked to the active transport of _____ (N, another nonelectrolyte; C, a cation; AN, an anion).

    a. A, N   c. A, AN   e. P, C
    b. A, C   d. P, N   f. P, AN

47. Among the following, the permeability of the renal tubular membrane is highest for:

    a. Creatinine   d. Sucrose
    b. Insulin   e. Urea
    c. Mannitol   f. Polysaccharides

48. The rate of reabsorption of a nonactively transported solute from the renal tubules would be generally _____ by increased rates of water reabsorption and _____ by increased tubular membrane permeability for the solute. (I, increased; N, not significantly altered; D, decreased)

    a. N, N   c. I, D   e. I, I
    b. N, I   d. I, N   f. D, D

49. The transtubular electrical potential difference, caused primarily by active reabsorption of _____ , results in an electrical gradient which facilitates the passive reabsorption of _____ . (A, anions; C, cations; OA, other anions; OC, other cations)

    a. A, OA   c. C, A
    b. A, C   d. C, OC

50. The _____ (P, positive; N, negative) electrical potential of the renal tubular lumen, with respect to peritubular fluids, has a greater absolute magnitude in the _____ (X, proximal; D, distal) tubule.

    a. P, X   c. N, X
    b. P, D   d. N, D

51. Ammonium ions of tubular fluids, largely resulting from _____ (L, liver; RT, renal tubular) synthesis, are _____ (P, passively; A, actively) _____ (S, secreted; R, reabsorbed) by renal tubules.

    a. L, P, S   c. L, A, R   e. RT, A, S
    b. L, A, S   d. RT, P, S   f. RT, A, R

OBJECTIVE 34–10.

Identify the epithelial cellular properties and relative absorptive capabilities of the different renal tubular segments.

52. The thin segment of the loop of Henle has _____ levels of metabolic activity and _____ levels of epithelial cell permeability as compared to other renal tubular segments. (H, high; L, low)

    a. H, H   c. L, H
    b. H, L   d. L, L

53. The highest volume percentage of glomerular filtrate reabsorption occurs in the:

    a. Bowman's capsule   d. Distal tubule
    b. Proximal tubule   e. Collecting ducts
    c. Loop of Henle   f. Urinary bladder

54. The cortical portion of the collecting tubules are _____ (L, less; M, more) permeable to urea than the medullary portion, and have a permeability to water that is _____ (I, increased; N, not significantly altered; D, decreased) in the presence of antidiurectic hormone.

    a. M, I   c. M, D   e. L, N
    b. M, N   d. L, I   f. L, D

55. A well developed brush border is a characteristic of _____ renal tubular epithelial cells, and large numbers of mitochondria and a low water permeability characterize _____ renal tubular epithelial segments. (P, only proximal; D, only distal; B, both the proximal and distal)

    a. P, P   c. P, B   e. B, D
    b. P, D   d. B, P   f. B, B

56. Cells which are especially adapted for the active transport of sodium ions against high concentration and electrical gradients are located in the _____ (T, thin; K, thick) segment of the _____ (D, descending; A, ascending) limb of the loop of Henle and the _____ (P, proximal; L, distal) segment of the renal tubule.

    a. T, A, P   c. T, D, L   e. K, A, P
    b. T, D, P   d. K, A, L   f. K, D, L

OBJECTIVE 34–11.

Identify the influence of different segments of the renal tubules upon reabsorption and secretion of water, substances of nutritional value to the body, metabolic end-products, and insulin and para-aminohippuric acid.

57. Fluid flow past the end of the proximal segment of the renal tubule represents about _____ % of the original glomerular filtration rate and is _____ (H, hypertonic; I, isotonic) with respect to the peritubular fluid.

   a. 5, H   c. 65, H   e. 35, I
   b. 35, H   d. 5, I   f. 65, I

58. A fluid flow rate of 12.5 ml./minute is probably more representative of the fluid flow rate into the:

   a. Bowman's capsule   d. Distal tubule
   b. Loop of Henle   e. Collecting ducts
   c. Proximal tubule   f. Urinary bladder

59. The total concentration of glucose, proteins, amino acids, acetoacetic acid, and vitamins is typically about _____ mg.% in urine as a consequence of transport mechanisms existing in the _____ (P, proximal; D, distal) renal tubules.

   a. 0, P   c. 100, P   e. 10, D
   b. 10, P   d. 0, D   f. 100, D

60. Protein reabsorption occurs by _____ (D, passive diffusion; P, pinocytosis) in the _____ (LH, loop of Henle; PT, proximal tubule; DT, distal tubule).

   a. D, LH   c. D, DT   e. P, PT
   b. D, PT   d. P, LH   f. P, DT

61. Of the following metabolic end-products, renal tubules reabsorb the highest percentage of filtered _____ and the lowest percentage of filtered _____ . (R, urea; C, creatinine; U, urate)

   a. R, C   c. U, R   e. R, U
   b. C, R   d. U, C   f. C, U

62. The rates of renal filtration of sulfates, phosphates, and nitrates are generally _____ than their urinary excretion rates, and their urine concentrations are generally _____ than their concentrations in glomerular filtrate. (G, greater; L, less)

   a. G, G   c. L, G
   b. G, L   d. L, L

63. Inulin is _____ (R, reabsorbed; S, secreted; N, neither reabsorbed nor secreted) by renal tubules, and its rate of filtration is _____ (G, greater than; E, equal to; L, less than) its rate of urinary excretion.

   a. R, G   c. S, G   e. N, E
   b. R, L   d. S, L   f. N, G

64. Para-aminohippuric acid is largely _____ (R, reabsorbed; S, secreted) by epithelial cells of the _____ (P, proximal; D, distal) renal tubule.

   a. R, P   c. S, P
   b. R, D   d. S, D

OBJECTIVE 34–12.

Identify the influence of different segments of the renal tubules upon reabsorption and secretion of ionic constituents.

65. A net flux of ions _____ (I, into; O, out of) the descending portion of the thin loop of Henle occurs by means of _____ (P, passive; A, active) transport.

   a. I, A   c. O, A
   b. I, P   d. O, P

66. Chloride ions are largely reabsorbed from the renal tubule by _____ transport mechanisms, and bicarbonate ions are largely reabsorbed by _____ transport mechanisms. (A, active; P, passive)

   a. A, A   c. P, A
   b. A, P   d. P, P

67. Bicarbonate reabsorption from the renal tubule occurs largely in the form of _____ (C, carbon dioxide; H, HCO_3^- ions), and is generally _____ (I, increased; D, decreased; N, not altered) by deficient hydrogen ion concentrations of renal tubular fluid.

    a. C, I   c. C, D   e. H, N
    b. C, N   d. H, I   f. H, D

68. Hydrogen ions are actively _____ in the proximal, distal, and collecting tubules, and potassium ions are _____ in the distal tubules. (R, reabsorbed; S, secreted; A, actively; P, passively)

    a. R; A S   c. R; P R   e. S; A R
    b. R; A R   d. S; A S   f. S; P R

69. _____ (V, active; P, passive) renal tubular transport mechanisms for cation reabsorption, including those for calcium and magnesium ions, are _____ (A, aided; O, opposed) by existing electrical gradients.

    a. V, A   c. P, A
    b. V, O   d. P, O

OBJECTIVE 34–13.

Identify the concentration variance of constituent substances in renal tubular fluid at progressive segments of its flow.

70. Concentrations of certain essential _____ (O, organic; E, electrolyte) constituents fall to zero in renal tubular fluid by the time the fluid first reaches the end of the _____ (P, proximal; D, distal; C, collecting) tubules.

    a. O, P   c. O, C   e. E, D
    b. O, D   d. E, P   f. E, C

71. The concentration ratio of potassium ions to chloride ions in renal tubular fluid _____ in the proximal tubule and _____ in the distal and collecting tubules. (I, progressively increases; C, remains relatively constant; D, progressively decreases)

    a. C, C   c. C, I   e. D, C
    b. C, D   d. D, I   f. I, C

72. The concentration ratio of a substance in glomerular filtrate to that found in tubular fluid entering the loop of Henle would be closer to one for:

    a. Sodium & chloride ions   e. Urea & creatinine
    b. Glucose & vitamins
    c. Amino acids & proteins   f. Para-aminohippuric acid
    d. Inulin

73. The urine concentration of a substance that is neither reabsorbed nor secreted by renal tubules is _____ (H, higher than; E, equal to; L, lower than) its concentration in glomerular filtrate, and is _____ (I, increased; D, decreased; N, not significantly altered) by antidiuretic hormone.

    a. H, I   c. E, N   e. L, N
    b. H, D   d. E, I   f. L, D

74. The concentration ratio for sodium ions in glomerular filtrate to renal tubular fluid is both greater than one and less than one in the:

    a. Proximal tubule   c. Distal tubule
    b. Loop of Henle   d. Collecting tubule

OBJECTIVE 34-14.

Define plasma clearance, identify its means of calculation and its utility, and recognize relative magnitudes of plasma clearances for plasma constituents.

75. The volume of plasma required per minute to supply an equivalent amount of a substance at a rate at which it is excreted in the urine is known as:

a. The extraction ratio d. Renal fraction
b. Filtration ratio e. Plasma clearance
c. Tubular mass f. Tubular load

76. The plasma clearance for a substance may be calculated from the equation: plasma clearance, in ml./min., equals _____ . (U, urine flow in ml./min.; C, urine concentration; P, plasma concentration; G, glomerular filtration rate in ml./min.)

a. P(C/U) c. G(C/P) e. (U/G)(C/P)
b. U(P/C) d. G(P/C) f. U(C/P)

77. A plasma concentration of 0.26 mg./ml., a urine flow of 1 ml./min., and a urine concentration of 1,820 mg.%, correspond to a plasma clearance of a constituent of _____ ml./minute.

a. 4.7 c. 470 e. 70
b. 47 d. 7.0 f. 700

78. The rate of filtration of a substance is equal to the plasma concentration multiplied by the _____ , and the rate of urinary excretion of a substance is equal to the plasma concentration multiplied by the _____ . (PC, plasma clearance; GFR, glomerular filtration rate; FF, filtration fraction)

a. PC, GFR c. PC, FF e. GFR, FF
b. GFR, PC d. FF, PC f. FF, GFR

79. A plasma clearance equal to the glomerular filtration rate implies net tubular _____ , and a plasma clearance greater than the glomerular filtration rate implies net tubular _____ for a substance. (R, reabsorption only; S, secretion only; N, neither reabsorption nor secretion; B, both reabsorption and secretion)

a. N, N c. B, S e. N, R
b. B, N d. N, S f. S, R

80. At an average normal glomerular filtration rate, about _____ % of a filtered constituent with a plasma clearance of 75 ml./min. would be _____ (R, reabsorbed; S, secreted) by the kidneys.

a. 20, R c. 80, R e. 40, S
b. 40, R d. 20, S f. 80, S

81. In order, from the lowest to the highest, the plasma clearances for the following substances would be _____ . (N, Na$^+$; G, glucose; I, inulin; C, creatinine; U, urea)

a. C, I, N, U, G d. N, G, I, C, U
b. C, U, I, N, G e. G, U, N, C, I
c. N, I, C, U, G f. G, N, U, I, C

OBJECTIVE 34-15.

Identify the properties of substances utilized to measure glomerular filtration rates, the substances utilized, and the method by which glomerular filtration rate is determined.

82. A fructose polymer with a molecular weight of 5,200 most probably _____ be filtered, _____ be secreted, and _____ be reabsorbed by renal nephrons. (W, would; N, would not)

a. W, W, W c. W, N, N e. N, W, N
b. W, W, N d. N, N, N f. W, N, W

83. For a substance which is filtered but neither reabsorbed nor secreted by the renal tubules, its plasma clearance is _____ the glomerular filtration rate, and is _____ a quantity of plasma that would be completely cleared of the substance per unit time. (G, greater than; E, equal to; L, less than)

a. G, E c. G, G e. E, L
b. G, L d. E, E f. E, G

84. The plasma clearance is _____ the glomerular filtration rate for inulin, and is _____ the glomerular filtration rate for mannitol. (G, greater than; N, not significantly different from; L, less than)

a. N, N   c. N, L   e. G, G
b. N, G   d. G, N   f. L, L

85. Glomerular filtration rate times the plasma concentration equals the rate of excretion of a substance for:

a. PAH   d. Phosphate
b. Diodrast   e. Urea
c. Inulin   f. All filtered constituents

86. With a steady state concentration of plasma inulin, a free inulin plasma concentration of 2 mg./ml., a urine formation rate of 2.5 ml./min., and a urine inulin concentration of 32 mg./ml., the glomerular filtration rate would equal _____ ml./min.

a. 160   c. 80
b. 120   d. 40

OBJECTIVE 34–16.

Identify the properties of substances utilized to measure plasma flow through the kidneys, the substances utilized, and the method by which renal plasma flow is determined.

87. The renal arterial-venous concentration difference divided by its arterial concentration for a substance which is both filtered and secreted by the kidneys is equal to:

a. Zero   e. The extraction
b. One   ratio
c. The filtration   f. The plasma
    fraction   clearance
d. The renal fraction

88. The advantages of utilizing para-aminohippuric acid clinically for measurements of renal function are its _____ plasma clearance and its _____ extraction ratio. (H, high; L, low)

a. H, H   c. L, H
b. H, L   d. L, L

89. The concentration of para-aminohippuric acid in arterial and renal venous blood is measured, and the concentration difference (X) is recorded. The volume of urine formed per minute (V) is recorded and contains a para-aminohippuric acid concentration of (U). The Fick principle could be utilized to calculate the renal blood flow, equal to _____ divided by _____ .

a. UXV, 1   c. UX, V   e. XV, U
b. 1, XUV   d. UV, X   f. U, XV

90. The plasma clearance of _____ (D, Diodrast; PAH, para-aminohippuric acid) is the greater of the two substances as a result of differences in _____ (T, tubular reabsorption; E, extraction ratios).

a. PAH, T   c. D, T
b. PAH, E   d. D, E

91. The plasma clearance of unbound para-aminohippuric acid generally _____ with increasing arterial plasma concentrations for a low range of concentrations, and _____ with increasing renal plasma volume flows. (I, increases; C, remains relatively constant; D, decreases)

a. I, I   c. C, I   e. D, C
b. D, I   d. I, C   f. C, C

92. The renal plasma flow is equal to the _____ (GFR, glomerular filtration rate; PC, plasma clearance) of para-aminohippuric acid, _____ (M, multiplied by; D, divided by).

a. GFR, M 0.91   d. PC, M 0.91
b. GFR, D 0.91   e. PC, D 0.91
c. GFR, D 0.55   f. PC, M 0.45

OBJECTIVE 34–17.
Calculate the filtration fraction from measured values of plasma clearances of appropriate substances.

93. The two substances that would more probably be utilized to determine the filtration fraction would be:

   a. Inulin & mannitol
   b. PAH & Diodrast
   c. Urea & Diodrast
   d. PAH & creatinine
   e. Inulin & PAH
   f. Phenol red & PAH

94. The filtration fraction would best be calculated as the ratio of plasma clearances of _____ (M, mannitol to Diodrast; D, Diodrast to mannitol), _____ (X, multiplied; D, divided) by 0.85.

   a. M, X
   b. M, D
   c. D, X
   d. D, D

OBJECTIVE 34–18.
Identify the use of the terms "plasma load," "tubular load," "transfer maximum" (Tm), and "threshold substances." Recognize "gradient-time limited" mechanisms for sodium and potassium active transport and "transfer maximum" mechanisms for many other constituents. Identify the associated characteristics and significance of Tm mechanisms, typical Tm values for reabsorbed and secreted substances, and a means of Tm determination.

95. At a renal plasma flow of 600 ml./min., a plasma glucose concentration of 100 mg.%, and a filtration fraction of 20%, the plasma load of glucose is _____ mg./min. and the tubular load is _____ mg./min.

   a. 125, 600
   b. 20, 100
   c. 12, 60
   d. 600, 125
   e. 100, 20
   f. 60, 12

96. The maximum amount of each substance that can be transported in one minute by the kidney tubules is known as:

   a. The transfer maximum
   b. The inulin clearance
   c. The secretion rate
   d. The filtration rate
   e. The reabsorption rate
   f. None of the above

97. A substance which is actively transported by renal tubular cells, but which is not known to exhibit a Tm value, is:

   a. Ionic sodium
   b. Sulfate
   c. Glucose
   d. Hemoglobin
   e. Lactate
   f. Acetoacetic acid

98. A substance that has a "threshold" concentration in the plasma below which none of it appears in the urine, and above which progressively larger quantities appear in the urine, has a transfer _____ (M, minimum; X, maximum) and is actively _____ (S, secreted; R, reabsorbed).

   a. X, S
   b. X, R
   c. M, S
   d. M, R

99. The maximum rate at which glucose may be reabsorbed per minute by renal tubules is _____ mg./min. The plasma threshold for glucose at a typical glomerular filtration rate is _____ mg.%.

   a. 180, 220
   b. 180, 320
   c. 220, 180
   d. 180, 125
   e. 320, 180
   f. 125, 70

100. The glomerular filtration rate (GFR), the Tm, and the plasma concentration (P) are known values for a reabsorbed constituent. The amount of the substance appearing in the urine per unit of time, above a plasma threshold, would be:

   a. $P(GFR) - Tm$
   b. $Tm(GFR) - P$
   c. $Tm(GFR) + P$
   d. $Tm - P(GFR)$
   e. $Tm + P(GFR)$
   f. $P - Tm(GFR)$

101. A patient's glucose concentration is 80 mg.%, but his urine contains glucose. His glomerular filtration rate is 120 ml./min. The patient is probably a "renal diabetic" as a result of abnormal _____ (T, Tmax; F, filtered load), which apparently never equals or exceeds _____ mg./min.

   a. T, 64
   b. T, 96
   c. T, 112
   d. F, 64
   e. F, 96
   f. F, 112

102. If the filtered or _____ (P, plasma; T, tubular) load of glucose is above its Tm value, then the rate of glucose reabsorption would be _____ (I, independent of; D, directly proportional to; S, inversely proportional to) the plasma glucose concentration.

a. P, I   c. P, S   e. T, D
b. P, D   d. T, I   f. T, S

103. The administration of phlorhizin, which inhibits the glucose transport mechanism in the _____ (P, proximal; D, distal) renal tubules, would cause the renal clearance of glucose to approach _____ ml./minute.

a. P, 120   c. P, 20   e. D, 600
b. P, 600   d. D, 120   f. D, 20

104. Of the following, the substance with the greatest renal transport maximum would be:

a. Phosphate   d. Glucose
b. Urate   e. Lactate
c. Plasma protein   f. Acetoacetic acid

105. The Tm value for plasma protein _____ exceed 10 mg./min., and the Tm value for lactate _____ exceed 10 mg./min. (D, does; N, does not)

a. D, D   c. N, D
b. D, N   d. N, N

106. Creatinine, PAH, Diodrast, and phenol red _____ have transfer maximums and _____ have plasma thresholds. (D, do; N, do not)

a. D, D   c. N, D
b. D, N   d. N, N

35

Renal Mechanisms for Concentrating and Diluting the Urine; Urea, Sodium, Potassium, and Fluid Volume Excretion

OBJECTIVE 35–1.

Identify the ratio of juxtamedullary to cortical nephrons in man and its significance, and the nature and magnitude of the renal interstitial fluid osmolar gradient and its causes.

1. The kidneys utilize mechanisms of _____ (A, active; P, passive) transport of water in forming a hypertonic urine, which are made possible by _cortical_ (C, cortical; J, juxtamedullary) nephrons.

a. A, C   c. P, C
b. A, J   d. P, J

2. Nephrons lacking a long thin loop of Henle comprise _____ % of their total number in man, and function *least* in _____ (G, glucose reabsorption; C, counter-current multiplier mechanisms; K, potassium secretion).

a. 20, G   c. 20, K   e. 80, C
b. 20, C   d. 80, G   f. 80, K

3. The osmolarity of the renal tubular fluid as it enters the proximal tubule is _____ milliosmols/liter, and is _____ (G, greater; N, not significantly different; L, less) than the cortical interstitial fluid osmolarity.

a. 100, G   c. 1,200, L   e. 300, G
b. 300, N   d. 100, N   f. 1,200, G

4. The highest extracellular fluid osmolarity found in the body occurs in the renal _____ (C, cortex; MC, medullary-cortical junction; M, pelvic tip of the medulla) and is typically _____ osmols per liter.

a. C, 1.2   c. M, 1.2   e. MC, 3.0
b. MC, 1.2   d. C, 3.0   f. M, 3.0

5. The cortical-medullary osmotic gradient of renal interstitial fluid is primarily a result of mechanisms involving _____ (V, vasa recta; P, proximal renal tubules; D, distal renal tubules; H, loops of Henle).

a. V & P   c. D & V   e. H & P
b. P & D   d. H & V   f. H & D

6. The mean circulation time through the vasa recta, estimated to be about 50 seconds, is _____ (G, greater; L, less) than that for cortical peritubular capillaries, and represents about _____ % of the total renal blood flow.

a. G, 1 to 2   c. G, 90   e. L, 25 to 30
b. G, 25 to 30   d. L, 1 to 2   f. L, 90

OBJECTIVE 35–2.

Identify the counter-current "multiplier" mechanisms of the loop of Henle and vasa recta.

7. The osmotic gradient within the interstitial fluids of the kidney is dependent upon active transport of _____ (Na, sodium ions; W, water), a "hairpin" or U-tube geometry of the tubule system, and differential permeability between the ascending and descending limbs of the _____ (V, vasa recta; L, loop of Henle).

a. Na, V   c. W, V
b. Na, L   d. W, L

8. The "motor" of the counter-current mechanism involves active _____ (S, secretion; R, reabsorption) of electrolyte in the _____ (A, ascending; D, descending) limb of the loop of Henle.

a. S, A   c. R, A
b. S, D   d. R, D

9. Altered osmolar concentrations within the first half of the loop of Henle are primarily caused by _____ (A, active; P, passive) transport of ions _____ (I, into; O, out of) the renal tubule at this location.

a. A, I   c. P, I
b. A, O   d. P, O

10. The osmolarity at the tips of the vasa recta is closer to _____ milliosmols per liter as a consequence of a relatively _____ capillary permeability and a _____ flow velocity. (H, high; L, low)

a. 300, L, H   d. 300, H, H
b. 600, L, L   e. 600, H, L
c. 1,200, L, H   f. 1,200, H, L

11. The lowest permeability to water movement is encountered in the _____ (A, ascending; D, descending) limb of the _____ (V, vasa recta; L, loop of Henle).

a. A, V   c. D, V
b. A, L   d. D, L

OBJECTIVE 35–3.

Identify the renal effects of antidiuretic hormone, and the mechanisms and limits for excreting either a concentrated or a dilute urine.

12. Fluid in the ascending loop of Henle becomes progressively more _____ (D, dilute; C, concentrated) as it ascends toward the cortex, reaching an osmolarity of about _____ milliosmolar before leaving the loop of Henle.

a. D, 10   c. D, 1,000   e. C, 100
b. D, 100   d. C, 10   f. C, 1,000

13. The minimal urine osmolarity, about _____ (E, equal to; F, one fourth; T, one tenth) that of plasma, is achieved under conditions of _____ (H, high; L, low) collecting duct permeability.

a. E, H   c. T, H   e. F, L
b. F, H   d. E, L   f. T, L

14. The urine osmolarity is _____ and the urine volume flow is _____ by reduced levels of antidiuretic hormone secretion. (I, increased; N, not significantly altered; D, decreased)

a. I, I   c. I, D   e. D, N
b. I, N   d. D, I   f. D, D

15. The more significant renal effect of antidiuretic hormone involves altered _____ (A, active transport; P, permeability) of epithelial cells of _____ (D, distal tubules; C, collecting tubules; H, loops of Henle).

a. A, D   c. A, H   e. P, C
b. A, C   d. P, D   f. P, H

16. The _____ (L, loop of Henle; C, collecting duct) serves as an osmotic exchanger to translate the osmolarity of the medullary and papillary interstitium into urinary _____ (G, hypertonicity; O, hypotonicity) in the _____ (A, absence; P, presence) of antidiuretic hormone.

a. L, O, P   c. L, G, A   e. C, G, P
b. L, O, A   d. C, G, A   f. C, O, P

17. Maximal osmolarity of collecting duct fluid is _____ (G, greater; N, not significantly different; L, less) than the maximal value of renal interstitial fluid osmolarity, and _____ (I, is; X, is not) significantly influenced by active transport of collecting duct epithelial cells.

a. G, I   c. L, I   e. N, X
b. N, I   d. G, X   f. L, X

18. Net proportional movements of solute and water _____ (I, into; O, out of) the renal tubule occur in the _____ (P, proximal tubules; D, descending limb of the loops of Henle; C, collecting ducts).

a. I, P only   d. O, P only
b. I, both P & D   e. O, both P & D
c. I, C only   f. O, C only

19. The urea concentration of urine, _____ (I, increased; N, not significantly altered; D, decreased) by the effects of elevated antidiuretic hormone levels, _____ (Y, is; X, is not) normally more than twice the urea concentration of medullary interstitial fluid.

a. I, Y   c. D, Y   e. N, X
b. N, Y   d. I, X   f. D, X

OBJECTIVE 35–4.

Identify osmolar clearance and free water clearance, their means of determination, and their significance.

20. The number of osmols entering the urine per minute, divided by the _____ (U, urine; P, plasma) osmolar concentration, equals the osmolar _____ (L, load; E, excretion; C, clearance).

a. U, L   c. U, C   e. P, E
b. U, E   d. P, L   f. P, C

21. Free water clearance can be _____ (P, only positive; B, either positive or negative) in magnitude, and is _____ (I, increased; N, not significantly altered; D, decreased) by increased osmolar clearances at constant urine volume flow rates.

a. P, I   c. P, D   e. B, N
b. P, N   d. B, I   f. B, D

22. Antidiuretic hormone results in a greater alteration in the _____ (O, osmolar; F, free water) clearance, and _____ (I, increases; N, does not significantly alter; D, decreases) the urine volume flow rate.

a. O, I   c. O, D   e. F, N
b. O, N   d. F, I   f. F, D

23. The sum of free water and osmolar clearances more closely approximates:

a. Inulin clearance
b. Creatinine clearance
c. Urea clearance
d. Glomerular filtration rates
e. Renal plasma flow rates
f. Urine flow rates

OBJECTIVE 35–5.

Identify a normal plasma concentration of urea, the major factors that determine the rate of renal excretion of urea, and the influence of extreme variance of glomerular filtration rates upon urea excretion.

24. Typically, about _____ % of the filtered load of urea is _____ (R, reabsorbed; S, secreted) by the renal tubules.

   a. 90, R   c. 10, R   e. 40, S
   b. 40, R   d. 90, S   f. 10, S

25. Urea is an end-product of _____ (P, protein; N, nucleic acid) catabolism, which is synthesized in the _____ (L, liver; K, kidneys) and excreted in the kidneys.

   a. P, L   c. N, L
   b. P, K   d. N, K

26. A plasma urea concentration, normally averaging _____ mg.%, would be substantially _____ (I, increased; D, decreased) by a complete loss of renal functions.

   a. 2.6, I   c. 180, I   e. 26, D
   b. 26, I   d. 2.6, D   f. 180, D

27. Renal excretion of urea is _____ by increased urea plasma concentrations and _____ by reduced glomerular filtration rates. (I, increased; N, not significantly altered; D, decreased)

   a. I, I   c. I, D   e. N, N
   b. I, N   d. N, I   f. N, D

28. At very low glomerular filtration rates, a _____ (H, higher; L, lower) than normal percentage of the filtered load is reabsorbed as a consequence of renal tubular _____ (Tm, transfer maximum; P, permeability) mechanisms.

   a. H, Tm   c. L, Tm
   b. H, P   d. L, P

OBJECTIVE 35–6.

Identify the magnitudes and mechanisms of sodium ion transport in the proximal tubules, loops of Henle, and distal tubules. Identify the mechanism by which aldosterone influences sodium and potassium ion transport.

29. Compared to the proximal renal tubule, the distal tubule has _____ transtubular electrical potential and _____ transtubular sodium ion concentration gradient. (G, a greater; E, an equal; L, a lesser)

   a. L, L   c. E, E   e. G, E
   b. L, G   d. L, E   f. G, G

30. Reabsorption of about _____ % of the filtered load of sodium is under the influence of aldosterone, acting upon the _____ (P, proximal; D, distal) tubular epithelium.

   a. 8, P   c. 65, P   e. 35, D
   b. 35, P   d. 8, D   f. 65, D

31. Aldosterone is thought to exert its target renal effects by combining with _____ (M, cell membrane; I, intracellular) receptors to activate _____ (P, protein synthesis; AMP, the cyclic AMP mechanism).

   a. M, P   c. I, P
   b. M, AMP   d. I, AMP

32. Aldosterone is thought to activate a mechanism which actively transports sodium ions _____ , and potassium ions _____ , the interior of the renal tubular epithelial cell. (I, into; O, out of)

   a. I, I   c. O, I
   b. I, O   d. O, O

33. Aldosterone _____ net renal reabsorption of sodium ions and _____ net renal reabsorption of potassium ions. (I, increases; D, decreases)

   a. I, I   c. D, I
   b. I, D   d. D, D

OBJECTIVE 35–7.

Identify the magnitudes and mechanisms of potassium ion transport in the proximal tubules, loops of Henle, and distal tubules. Identify the regulatory mechanism for extracellular potassium ion concentration and its significance.

34. Mechanisms for both active reabsorption and active secretion, at differing segments of the renal tubule, are more significant for transport of:

a. Sodium ions
b. Para-aminohippuric acid
c. Glucose
d. Sulfate ions
e. Urea
f. Potassium ions

35. Approximately _____ % of the filtered load of potassium is reabsorbed in the proximal tubules, and _____ % in the loops of Henle.

a. 65, 10   c. 10, 65   e. 25, 0
b. 25, 25   d. 65, 0   f. 10, 0

36. A cation exchange process of active transport is involved in the _____ (R, reabsorption; S, secretion) of potassium ions in _____ (P, only the proximal; D, only the distal; B, both the proximal and distal) segments of the renal tubules.

a. R, P   c. R, B   e. S, D
b. R, D   d. S, P   f. S, B

37. The renal regulation of extracellular potassium ion concentrations is a vital function, as cardiac arrhythmias and potential cardiac fibrillation result when the extracellular potassium ion concentration first rises from a normal value of _____ mEq./liter to a toxic level of _____ mEq./liter.

a. 140, 150   c. 4, 8   e. 23, 28
b. 23, 50 to 60   d. 140, 200   f. 4, 23

38. A reduction in the transtubular electrical potential in the distal tubules, with _____ (I, increased; D, decreased) sodium transport, results in increased potassium _____ (R, reabsorption; S, secretion).

a. I, R   c. D, R
b. I, S   d. D, S

39. A comparatively greater plasma clearance of _____ (Na, sodium; K, potassium) ions occurs with average diets which is typically _____ (G, greater than; E, equal to; L, less than) the glomerular filtration rate.

a. Na, G   c. Na, L   e. K, E
b. Na, E   d. K, G   f. K, L

OBJECTIVE 35–8.

Identify glomerulotubular balance and its relationship to fluid volume excretion. Identify the causes and resultant effects upon fluid volume excretion of varied osmolar clearance, plasma colloid osmotic pressure, renal sympathetic tone, mean arterial pressure, and antidiuretic hormone secretion.

40. The concept of glomerulotubular balance implies that an increased increment of _____ (GFR, glomerular filtration rate; O, osmolar clearance) is accompanied by a corresponding equal increase in _____ (R, filtrate reabsorption rate; F, free water clearance).

a. GFR, R   c. O, R
b. GFR, F   d. O, F

41. An incremental _____ (I, increase; D, decrease) in the glomerular filtration rate of 25 ml./min. from its normal average value results in the greatest percentage increase in the rate of _____ (GF, glomerular filtration; T, tubular reabsorption; U, urine output.

a. I, GF   c. I, U   e. D, T
b. I, T   d. D, GF   f. D, U

42. Osmotic diuresis is associated with _____ osmolar clearance and _____ urine volume flow rates. (I, increased; N, no significant change in; D, decreased)

a. I, N   c. I, I   e. N, D
b. N, I   d. D, N   f. D, D

43. Diabetes mellitus refers to a condition of _____ (A, antidiuresis; D, diuresis) as a direct consequence of direct osmotic effects exerted by elevated _____ (P, plasma; R, renal tubular fluid) glucose concentrations.

a. A, P   c. D, P
b. A, R   d. D, R

44. Sucrose and mannitol, substances which are _____ (R, readily; P, poorly) reabsorbed by renal tubules, result in _____ (N, no significant; S, slight; E, extreme) osmotic diuresis when present in plasma in large quantities.

    a. R, N   c. R, E   e. P, S
    b. R, S   d. P, N   f. P, E

45. Increased plasma colloid pressure _____ glomerular filtration rates, _____ tubular reabsorption, and _____ urine volume excretion. (I, increases; N, does not significantly alter; D, decreases)

    a. I, I, I   c. I, D, I   e. D, N, D
    b. I, N, N   d. D, I, D   f. D, I, N

46. A change in plasma colloid osmotic pressure of 1 mm. Hg typically results in a change in glomerular filtration of about _____ % as a consequence of altered glomerular _____ (P, filtration pressure; V, plasma volume flow).

    a. 5, P   c. 500, P   e. 50, V
    b. 50, P   d. 5, V   f. 500, V

47. Sympathetic stimulation to the kidneys has a more powerful effect upon the _____ (A, afferent; E, efferent) arterioles, and subsequently, _____ (I, increases; D, decreases) glomerular filtration rates and _____ (I, increases; D, decreases) urine volume flow.

    a. A, D, I   c. A, I, D   e. E, I, D
    b. A, D, D   d. E, I, I   f. E, D, I

48. Acute doubling of the mean arterial pressure, when other factors remain constant, results in a _____ (C, constant; D, doubling; M, more than five-fold rise in) urine output whereas urine flow ceases when the mean arterial pressure first reaches a lowered value of _____ mm. Hg.

    a. C, 60   c. M, 60   e. D, 35
    b. D, 60   d. C, 35   f. M, 35

49. The rate of change of urinary output as a function of elevated arterial pressures under acute conditions is largely attributed to _____ (F, increased glomerular filtration; T, decreased tubular reabsorption) rates, and is _____ (G, greater than; E, equal to; L, less than) that occurring under chronic conditions.

    a. F, G   c. F, L   e. T, E
    b. F, E   d. T, G   f. T, L

50. Antidiuretic hormone results in a greater _____ (I, increase; D, decrease) in the urine output under _____ (A, acute; C, chronic) conditions.

    a. D, A   c. I, A
    b. D, C   d. I, C

51. Antidiuretic hormone _____ (I, increases; D, decreases) the cyclic AMP concentrations of target _____ (P, proximal; D, distal) tubular epithelial cells.

    a. I, P   c. D, P
    b. I, D   d. D, D

OBJECTIVE 35-9.

Identify autoregulation of renal blood flow and glomerular filtration rates, its relative sensitivity, its potential mechanisms, and its importance.

52. Acute doubling of the normal mean arterial pressure results in a percentage change from normal which is _____ 25% for renal blood flow, _____ 25% for glomerular filtration rate, and _____ 25% for urine output. (G, greater than; L, less than)

    a. G, G, G   c. L, L, G   e. L, G, L
    b. L, G, G   d. L, L, L   f. G, L, L

53. Components of the juxtaglomerular apparatus _____ thought to be directly involved in autoregulation of renal blood flow, and _____ thought to be directly involved in the autoregulation of glomerular filtration rates. (A, are; N, are not)

    a. A, A   c. N, A
    b. A, N   d. N, N

54. Renin-containing cells of the _____ are in close proximity to macula densa cells of the _____. (P, proximal tubule; D, distal tubule; A, afferent arteriole; E, efferent arteriole)

    a. A, E   c. P, E   e. D, A
    b. E, A   d. A, D   f. D, E

OBJECTIVE 35–10.

Identify the mechanisms involved in the renin-angiotensin system and their intrarenal functions.

55. Renin is released from the kidneys in response to _____ glomerular filtration rates, which results from _D_ glomerular pressure and _____ renal sympathetic tone. (I, increased; D, decreased)

   a. I, D, I   c. I, I, I   e. D, D, I
   b. I, I, D   d. D, I, D   f. D, D, D

56. Angiotensin II, formed from a decapeptide by the enzymatic action of _____ (R, renin; C, converting enzyme), results in an overall _____ (I, increase; D, decrease) in the systemic peripheral resistance.

   a. R, I   c. C, I
   b. R, D   d. C, D

57. Selective efferent arteriolar constriction results in _____ glomerular blood flow, _____ percentage reabsorption of glomerular filtrate, and _____ fluid volume excretion. (I, increased; D, decreased)

   a. D, I, I   c. D, D, I   e. I, I, I
   b. D, I, D   d. D, D, D   f. I, D, D

58. Angiotensin II effects upon renal vessels include afferent arteriolar _____ and efferent arteriolar _____. (C, constriction; D, dilation)

   a. C, C   c. D, C
   b. C, D   d. D, D

59. Intrarenal effects of angiotensin II are thought to include _____ body water retention and _____ body salt retention. (I, increased; D, decreased)

   a. I, I   c. D, I
   b. I, D   d. D, D

36

Regulation of Blood Volume, Extracellular Fluid Volume, and Extracellular Fluid Composition by the Kidneys and by the Thirst Mechanism

OBJECTIVE 36–1.

Recognize the relative constancy of the blood volume. Identify the basic mechanisms for blood volume control.

1. The blood volume, relatively _____ (V, variable; C, constant) on a day to day basis, is adequately regulated until the daily fluid intake first reaches a critical low value of about _____ liters per day.

   a. V, O   c. V, 2 to 3   e. C, 1 to 1.5
   b. V, 1 to 1.5   d. C, 0   f. C, 2 to 3

2. The daily fluid intake during average environmental conditions, normally about _____ liters, is about _____ (T, twice; E, equal to) the daily fluid loss of water.

   a. 0.5, T   c. 6.7, T   e. 2.5, E
   b. 2.5, T   d. 0.5, E   f. 6.7, E

3. Increased blood volume results in _____ cardiac output, _____ arterial pressure, and _____ urinary output. (I, increased; D, decreased; N, no significant change in)

a. N, N, N   c. I, N, I   e. I, I, N
b. I, N, N   d. I, I, I   f. I, D, D

4. Renal regulating mechanisms of body fluid volumes operate such that when the erythrocyte compartment volume is reduced, plasma volume _____ and the total blood volume _____ . (I, increases; R, remains constant; D, decreases)

a. R, D   c. I, R   e. I, I
b. D, D   d. D, R   f. I, D

5. A 5% change in blood volume results in a _____ percentage change in cardiac output, and a 5% change in mean arterial pressure results in a _____ percentage change in urinary output. (G, greater; L, lesser)

a. G, L   c. L, G
b. G, G   d. L, L

OBJECTIVE 36–2.

Identify the mechanisms of action of volume receptors and their significance in the control of blood volume.

6. Increased atrial blood volume receptor activity, with _____ (I, increased; D, decreased) blood volumes, act _____ (S, synergistically; O, in opposition) to altered arterial baroreceptor activity in the regulation of blood volume.

a. D, S   c. I, S
b. D, O   d. I, O

7. Increased atrial stretch receptor activity results in _____ sympathetic tone to the kidney and _____ urinary output. (I, increased; D, decreased)

a. I, I   c. D, I
b. I, D   d. D, D

8. Increased atrial stretch receptor activity results in _____ secretion of antidiuretic hormone and _____ free water clearance by the kidneys. (I, increased; D, decreased)

a. I, I   c. D, I
b. I, D   d. D, D

9. Enhanced atrial stretch receptor activity results in _____ peripheral arteriolar tone and resultant tendencies for _____ capillary pressures and _____ ratios of interstitial to blood compartmental volumes. (I, increased; D, decreased)

a. I, I, D   c. I, D, D   e. D, D, I
b. I, D, I   d. D, I, D   f. D, I, I

10. Volume receptor reflexes, requiring a minimal time of a few _____ (H, hours; D, days; W, weeks) for their action, are more directly involved in the _____ (S, short; L, long) -term regulation of blood volume.

a. H, S   c. W, S   e. D, L
b. D, S   d. H, L   f. W, L

OBJECTIVE 36-3.

Identify the effects and relative regulatory significance of antidiuretic hormone and aldosterone upon blood volume. Identify the influence of altered circulatory capacity upon the control of blood volume.

11. Diabetes insipidus, a condition of abnormally _____ antidiuretic hormone secretion, generally results in _____ blood volumes. (I, increased; N, no significant change in; D, decreased)

a. I, I   c. I, D   e. D, N
b. I, N   d. D, I   f. D, D

12. The development of varicose veins, which _____ (I, increases; D, decreases) the capacity of the systemic circulation system, results in the regulation of blood volumes at values _____ (H, higher than; E, equal to; L, lower than) preexisting levels.

a. I, H   c. I, L   e. D, E
b. I, E   d. D, H   f. D, L

13. Patients with primary aldosteronism have _____ (I, increased; D, decreased) blood and interstitial fluid volumes, generally by an amount _____ (G, greater; L, less) than 25% of normal.

a. I, G   c. D, G
b. I, L   d. D, L

14. The consequences of primary aldosteronism upon blood volumes, similar to that occurring with _____ (D, diabetes insipidus; I, inappropriate ADH syndrome), _____ (A, are; N, are not) significantly opposed by other mechanisms involved in the regulation of blood volume.

a. D, A   c. I, A
b. D, N   d. I, N

OBJECTIVE 36-4.

Identify the regulating mechanisms of extracellular fluid volume and their relationship to blood volume.

15. Extracellular fluid volume is controlled _____ (I, independently of; S, simultaneously with) the regulation of blood volume and _____ (Y, is; N, is not) dependent upon the physical properties of the circulation.

a. I, Y   c. S, Y
b. I, N   d. S, N

16. An incremental increase in extracellular fluid volume is more generally associated with comparatively _____ (G, greater; E, equal; L, lesser) increments of _____ (I, increased; D, decreased) blood volume.

a. G, I   c. L, I   e. E, D
b. E, I   d. G, D   f. L, D

17. The ratio of extracellular fluid volume to blood volume tends to be _____ by increased capillary pressures, and _____ by increased plasma colloid osmotic pressures. (I, increased; D, decreased)

a. I, I   c. D, I
b. I, D   d. D, D

18. Blood volume remains relatively more constant with varying interstitial fluid volumes within a range which encompasses _____ (G, greater; L, lesser) compliances of tissue spaces and _____ (N, negative; P, positive) interstitial fluid pressures.

a. G, N   c. L, N
b. G, P   d. L, P

19. Edema, associated with the administration of large volumes of fluid, functions more significantly as _____ (R, a regulatory; O, an overflow) mechanism for the regulation of the _____ (B, blood; I, interstitial fluid) compartmental volume.

a. R, B   c. O, B
b. R, I   d. O, I

OBJECTIVE 36–5.

Identify the primary regulatory mechanisms for extracellular sodium ion concentration, and their relationship to osmolarity of body fluid compartments. Identify the principal features of the osmo-sodium receptor–antidiuretic hormone feedback control system.

20. Regulatory mechanisms for the concentration of the major extracellular _____ (N, non-electrolyte; A, anion; C, cation) are the dominant influence upon the osmolarity of _____ (E, only the extracellular; B, both the extracellular and intracellular) fluid compartments.

   a. N, E   c. C, E   e. A, B
   b. A, E   d. N, B   f. C, B

21. Acid-base regulatory mechanisms create or eliminate extracellular _____ (C, cations; A, anions) to establish an extracellular _____ (CS, cation surplus; AS, anion surplus; E, cation-anion equality).

   a. C, CS   c. C, E   e. A, AS
   b. C, AS   d. A, CS   f. A, E

22. Neural osmoreceptors, located in the _____ (P, posterior pituitary; A, anterior pituitary; S, supraoptic nuclei), increase their action potential firing frequency in response to _____ (I, increased; D, decreased) osmolarity of extracellular fluids.

   a. P, I   c. S, I   e. A, D
   b. A, I   d. P, D   f. S, D

23. _____ (I, increased; D, decreased) intracellular volumes of osmoreceptors result in increased action potential discharges, which _____ (P, promotes; N, inhibit) the release of antidiuretic hormone.

   a. I, P   c. D, P
   b. I, N   d. D, N

24. Antidiuretic hormone secretion from the _____ (P, posterior pituitary; A, anterior pituitary; H, anterior hypothalamus) is more sensitive to changes in extracellular fluid concentrations of _____ (K, potassium; Na, sodium) ions.

   a. P, K   c. H, K   e. A, Na
   b. A, K   d. P, Na   f. H, Na

25. Antidiuretic hormone _____ the permeability of the collecting tubules and _____ renal water reabsorption. (I, increases; D, decreases)

   a. I, I   c. D, I
   b. I, D   d. D, D

26. Antidiuretic hormone secretion, more sensitive to _____ (V, isotonic volume; O, isovolumic osmotic) changes in the extracellular fluid compartment, _____ (D, does; N, does not) significantly increase renal electrolyte reabsorption.

   a. V, D   c. O, D
   b. V, N   d. O, N

OBJECTIVE 36–6.

Identify the characteristics and consequences of water diuresis, diabetes insipidus, and the inappropriate ADH syndrome.

27. Water diuresis, following _____ (D, deficient; E, excessive) water intake in response to thirst, is initiated following a time delay that is more probably contingent upon rates of _____ (A, gastrointestinal absorption; H, existing ADH destruction).

   a. D, A   c. E, A
   b. D, H   d. E, H

28. In diabetes insipidus, body fluid volumes generally are _____ (H, higher than normal; N, about normal; L, lower than normal) as a consequence of the _____ (T, thirst mechanism; A, aldosterone mechanism).

   a. H, T   c. L, T   e. N, A
   b. N, T   d. H, A   f. L, A

29. The inappropriate ADH syndrome, which has _____ (S, similar; O, opposite) effects upon ADH levels when compared to diabetes insipidus, results in _____ (I, increased; D, decreased; N, no significant change in) extracellular fluid concentrations of sodium.

   a. S, I   b. S, D   c. S, N   d. O, I   e. O, D   f. O, N

OBJECTIVE 36–7.

Identify "thirst," the characteristics and neural integration of the "thirst" center, stimuli leading to thirst and its relief, and the roles of thirst in regulating extracellular fluid osmolarity and sodium concentration.

30. Thirst is an important mechanism for regulating body _____ (W, only water; B, both sodium and water) concentrations, primarily by regulating fluid _____ (I, intake only; IO, both intake and output).

   a. W, I   c. B, I
   b. W, IO   d. B, IO

31. _____ (I, increased; D, decreased) intracellular neuron volumes is probably the stimulus for the _____ (H, hypothalamic; P, pituitary; M, medullary) thirst center.

   a. I, H   c. I, M   e. D, P
   b. I, P   d. D, H   f. D, M

32. Maintained electrical stimulation of the thirst center via implanted electrodes results in a _____ (B, brief; P, persistent) period of drinking by the animal following a comparatively _____ (S, short; L, long) latent period.

   a. B, S   c. P, S
   b. B, L   d. P, L

33. Excessive loss of body potassium _____ the intracellular volume of cells of the thirst center, and _____ the frequency of thirst sensation. (I, increases; N, does not significantly alter; D, decreases)

   a. I, D   c. D, I   e. N, I
   b. N, N   d. I, I   f. D, D

34. A _____ % change in the blood volume is an approximate threshold value for a sensation of thirst, and is _____ (H, significantly greater than; E, about equal to; L, significantly lower than) the threshold value for ADH secretion.

   a. 10, H   c. 10, L   e. 50, E
   b. 10, E   d. 50, H   f. 50, L

35. Normally, an individual initially receives relief from thirst with the act of _____ (F, restoration of body fluid balance; D, drinking water) as a consequence of monitoring mechanisms located in the _____ (K, kidneys; H, hypothalamus; G, gastrointestinal system).

   a. F, K   c. F, G   e. D, H
   b. F, H   d. D, K   f. D, G

36. A threshold "tripping" mechanism for drinking occurs when the extracellular fluid osmolarity rises about _____ mOsm./liter and generally results in an ingested water volume which is _____ (L, lower than; E, equal to; H, higher than) that required to return the osmolarity to normal.

   a. 4, L   c. 4, H   e. 12, E
   b. 4, E   d. 12, L   f. 12, H

OBJECTIVE 36–8.

Identify the roles of sodium intake, salt appetite, and aldosterone regulatory mechanisms upon sodium concentration of body fluids. Contrast the roles of antidiuretic hormone, thirst, sodium appetite, and aldosterone mechanisms upon regulation of extracellular fluid sodium concentration and osmolarity.

37. An extracellular sodium ion concentration which only increases 10% with a five-fold increase in sodium intake would more probably result from _____ ADH mechanisms, _____ thirst mechanisms, and _____ aldosterone mechanisms of regulation. (B, blocked; N, normal)

  a. N, N, N c. B, B, N e. N, B, N
  b. B, N, N d. N, N, B f. N, B, B

38. The major feedback mechanism for the control of extracellular fluid osmolarity, _____ (B, but not; D, and) the sodium concentration, is the _____ (ADH, ADH-thirst, AL, aldosterone) mechanism.

  a. B, ADH   c. D, ADH
  b. B, AL   d. D, AL

39. In addition to direct renal effects upon sodium active transport, aldosterone tends to indirectly _____ (I, increase; D, decrease) the glomerular filtration rate and consequently _____ (A, augment; O, oppose) its direct effects upon sodium transport.

  a. I, A   c. D, A
  b. I, O   d. D, O

40. Addison's disease, characterized by _____ (I, increased; D, decreased) aldosterone levels of secretion, results in substantially altered _____ (R, renal excretion only; C, salt appetite only; B, both renal excretion and salt appetite) of sodium chloride.

  a. I, R   c. I, B   e. D, C
  b. I, C   d. D, R   f. D, B

OBJECTIVE 36–9.

Identify the primary control mechanism for regulating extracellular fluid potassium ion concentration, its mechanisms of action and characteristics, and the circumstances and consequences of its failure. Identify the supplemental factors in addition to extracellular potassium ion concentration which influence its regulation.

41. The aldosterone mechanism, regulating a potassium ion _____ (R, reabsorptive; S, secretory) transport mechanism of the renal tubules, is of _____ (P, primary; C, secondary) importance in the control of extracellular potassium concentration.

  a. R, P   c. S, P
  b. R, C   d. S, C

42. Hydrogen ion transport has a _____ relationship, and sodium ion transport has a _____ relationship, to potassium ion transport across the renal tubular epithelium under the influence of aldosterone. (C, mutually competitive; R, reciprocal exchange)

  a. C, C   c. R, C
  b. C, R   d. R, R

43. The extracellular potassium ion concentration, normally _____ mEq./liter, is _____ (I, increased; D, decreased) in primary aldosteronism.

  a. 4, I   c. 24, I   e. 8, D
  b. 8, I   d. 4, D   f. 24, D

44. Muscular weakness and paralysis resulting from altered extracellular _____ (Na, sodium; K, potassium) ion concentrations frequently results from _____ (A, addison's disease; P, primary aldosteronism).

  a. Na, A   c. K, A
  b. Na, P   d. K, P

45. Aldosterone secretion is increased by increasing extracellular concentrations of _____ (Na, sodium; K, potassium) ions, and is _____ (I, increased; N, not significantly altered; D, decreased) by angiotensin II.

  a. Na, I   c. Na, D   e. K, N
  b. Na, N   d. K, I   f. K, D

OBJECTIVE 36–10.

Identify the regulatory mechanisms and characteristics of calcium and magnesium ions, and phospate and other anions exhibiting transfer maxima.

46. _____ (I, increased; D, decreased) extra-cellular fluid calcium ion concentrations increase parathyroid hormone secretion, thereby increasing the rate of skeletal _____ (P, deposition; R, reabsorption) of bone salts.

   a. I, P   c. D, P
   b. I, R   d. D, R

47. Parathyroid hormone _____ renal tubular calcium reabsorption and _____ renal tubular phosphate reabsorption. (I, increases; D, decreases)

   a. I, I   c. D, I
   b. I, D   d. D, D

48. The transport mechanism for magnesium by renal tubules _____ exhibit a transfer maximum, and _____ exhibit a plasma threshold. (D, does; N, does not)

   a. D, D   c. N, D
   b. D, N   d. N, N

49. The plasma concentration of _____ (L, lactate; P, phosphate) which is _____ (I, increased by; N, independent of; D, decreased by) increasing glomerular filtration rates, is normally determined by an "overflow" mechanism.

   a. L, I   c. L, D   e. P, N
   b. L, N   d. P, I   f. P, D

50. Renal regulation of plasma excesses of sulfates and nitrites is accomplished by a _____ (S, secretory; R, reabsorptive) mechanism exhibiting _____ (T, transfer maxima; P, passive diffusion).

   a. S, T   c. R, T
   b. S, P   d. R, P

37

Regulation of Acid-Base Balance

OBJECTIVE 37–1.

Identify the pH symbol and its utility, typical values of arterial, venous, and intracellular pH, acidosis and alkalosis, and the limits of pH compatible with life.

1. The pH is equal to the _____ (A, antilogarithm; L, logarithm) of the _____ (H, hydrogen; R, reciprocal of the hydrogen) ion concentration.

   a. A, H   c. L, H
   b. A, R   d. L, R

2. A solution with a pH between 7 and 8 would correspond to a solution containing a hydrogen ion concentration between _____ per liter.

   a. 10^7 to 10^8 Eq.   d. 10^{-7} to 10^{-8} Eq.
   b. 10^7 to 10^8 mEq.   e. 10^{-7} to 10^{-8} mEq.
   c. 10^6 to 10^7 mEq.   f. 10^{-6} to 10^{-7} mEq.

3. The normal pH of arterial blood, slightly more _____ (A, acidic; B, alkaline) than venous blood, is closer to _____ .

   a. A, 7.0   c. A, 7.4   e. B, 7.2
   b. A, 7.2   d. B, 7.0   f. B, 7.4

4. The intracellular pH, with an average on the _____ (A, acid; B, alkaline) side of the average blood pH, is _____ (I, increased; N, not significantly altered; D, decreased) by increased rates of cellular metabolism.

   a. A, I   c. A, D   e. B, N
   b. A, N   d. B, I   f. B, D

5. The term acidosis refers to a condition of body pH _____ (A, above; B, below) _____ (7.0; N, normal).

   a. A, 7.0   c. A, N
   b. B, 7.0   d. B, N

6. A person is more likely to die in _____ in extreme acidosis, and die in _____ in extreme alkalosis. (C, coma; T, tetany or convulsions)

   a. C, C   c. T, C
   b. C, T   d. T, T

7. The hydrogen ion concentrations of the extra-cellular fluids that are compatible with life of mammalian organisms covers a corresponding pH range of _____ to _____ .

   a. 7.35, 7.40   c. 7.0, 7.7   e. 5, 8
   b. 7.03, 7.04   d. 4, 10   f. 6.5, 9.2

OBJECTIVE 37–2.

Identify three principle mechanisms which oppose changes in body hydrogen ion concentration, their modes of action, and their required duration of action.

8. The most complete and powerful mechanism of the acid-base regulatory system is the _____ (R, respiratory; C, chemical buffer; E, renal) mechanism, requiring the _____ (L, longest; S, shortest) interval for restoration of hydrogen ion concentrations.

   a. R, L   c. E, L   e. C, S
   b. C, L   d. R, S   f. E, S

9. Acidosis _____ (S, stimulates; N, inhibits) the respiratory centers, resulting in _____ (I, increased; D, decreased) rates of carbon dioxide removal from body fluids.

   a. S, I   c. N, I
   b. S, D   d. N, D

10. Respiratory compensation for _____ (A, acidosis; L, alkalosis) by means of increased carbon dioxide concentrations of body fluids requires 1 to 3 _____ (S, seconds; M, minutes; H, hours).

   a. A, S   c. A, H   e. L, M
   b. A, M   d. L, S   f. L, H

OBJECTIVE 37–3.

Identify the bicarbonate buffer system, its quantitative dynamics, the Henderson-Hasselbalch equation and its utility, and the reaction curve of the bicarbonate buffer system.

11. The product of the hydrogen ion and _____ concentrations, divided by the _____ concentration, equals the dissociation constant of carbonic acid.

   a. CO_2, HCO_3^- c. HCO_3^-, H_2CO_3
   b. H_2CO_3, HCO_3^- d. HCO_3^-, CO_2

12. The amount of undissociated carbonic acid is _____ (D, directly; I, inversely) proportional to the amount of dissolved carbon dioxide in body fluids, and is normally about _____ (E, equal to; 1, 1/1,000th; X, 1,000 times) the dissolved carbon dioxide concentration.

   a. D, E   c. D, X   e. I, 1
   b. D, 1   d. I, E   f. I, X

13. An increased carbon dioxide content of blood, without a corresponding increase in bicarbonate ion concentration, results in _____ hydrogen ion concentration and _____ pH. (I, increased; N, no significant change in; D, decreased)

   a. N, N   c. D, D   e. D, I
   b. N, I   d. I, I   f. I, D

14. The Henderson-Hasselbalch equation, as applied to the bicarbonate buffer system, implies that the pH equals a constant _____ (P, plus; T, times) the log of the concentration ratio of _____ .

   a. P, (HCO_3^-/CO_2) c. T, (HCO_3^-/CO_2)
   b. P, (CO_2/HCO_3^-) d. T, (CO_2/HCO_3^-)

15. At a pH equal to the pK value of _____ for the bicarbonate buffer system, the concentration ratio of CO_2 to HCO_3^- would equal _____ .

    a. 6.1, 10   c. 7.4, 10   e. 6.8, 1
    b. 6.8, 10   d. 6.1, 1   f. 7.4, 1

16. A blood pH of 7.1 would best be described as _____ (A, acidosis; L, alkalosis), and would correspond to a HCO_3^-/CO_2 concentration ratio of _____ .

    a. A, 1/20   c. A, 1/10   e. L, 10/1
    b. A, 10/1   d. L, 1/20   f. L, 1/10

17. The kidneys compensate for an acidosis by _____ (I, increasing; D, decreasing) the rate of renal excretion of HCO_3 .

    a. I, HCO_3^- c. D, HCO_3^-
    b. I, CO_2 d. D, CO_2

18. Minimal rates of change of pH for the bicarbonate buffer system occur at pH values of _____ for the addition of acid, and _____ for the addition of base.

    a. 6.1, 7.8   c. 4.3, 7.8   e. 6.1, 6.1
    b. 4.3, 6.1   d. 4.3, 4.3   f. 7.8, 4.3

OBJECTIVE 37–4.

Identify the factors influencing the buffer power of a buffer system. Identify the isohydric principle. Identify three major buffer systems of the body, their pK values, relative significance, and modes of action.

19. The buffering power of a buffer system is greatest when the pH is _____ (G, greater than; E, equal to; L, less than) its pK value and is _____ (I, increased; N, not significantly altered) by increased concentrations of the buffer substances.

    a. G, I   c. L, I   e. E, N
    b. E, I   d. G, N   f. L, N

20. The higher percentage of the total buffer power of all buffer systems of the body, result from _____ (E, extracellular; I, intracellular) buffers, and is largely attributed to the _____ (B, bicarbonate; F, phosphate; P, protein) buffer system.

    a. E, B   c. E, P   e. I, F
    b. E, F   d. I, B   f. I, P

21. The highest percentage of the chemical buffering action of blood proteins is attributed to _____ . The higher blood protein concentration occurs for_____ . (P, plasma protein; H, hemoglobin)

    a. H, H   c. P, H
    b. H, P   d. P, P

22. The phosphate buffer system has a higher concentration in the _____ fluids, and has a pK value which lies closer to the average pH of _____ fluids. (E, extracellular; I, intracellular)

    a. E, E   c. I, E
    b. E, I   d. I, I

23. The bicarbonate buffer system is a more important buffer system of body fluids from the standpoint of:

    a. pK values
    b. Intracellular concentrations
    c. Extracellular concentrations
    d. Regulation of body pH
    e. Chemical buffer power
    f. All of the above

24. An increase in the ratio of HCO_3^-/CO_2 concentrations of body fluids, normally about _____ , results in _____ (I, an increase; N, no significant change in; D, a decrease) in the $HPO_4^{-2}/H_2PO_4^-$ concentration ratio.

    a. 10, I   c. 10, D   e. 20, N
    b. 10, N   d. 20, I   f. 20, D

25. _____ (E, extracellular; R, renal tubular) fluid generally has a pH closer to the pK value of _____ for the phosphate buffer system.

    a. E, 6.1   c. E, 7.4   e. R, 6.8
    b. E, 6.8   d. R, 6.1   f. R, 7.4

26. Salt to acid concentration ratios of the three major buffer systems are related to each other by their differing _____ (H, hydrogen ion concentrations; K, dissociation constants) and the _____ (I, isohydric principle; C, respective buffer capacities).

    a. H, I   c. K, I
    b. H, C   d. K, C

OBJECTIVE 37–5.

Identify the normal averages of CO_2 and HCO_3^- ion concentrations in extracellular fluid and the factors controlling the balance between CO_2 formation and removal. Identify the mechanisms and significance of the feedback relationship between hydrogen ion concentration and alveolar ventilation.

27. Increased rates of tissue metabolism tend to _____ (I, increase; D, decrease) the extracellular fluid concentration of carbon dioxide from a normal average of _____ millimols/liter.

   a. I, 1.2   c. I, 24   e. D, 4.8
   b. I, 4.8   d. D, 1.2   f. D, 24

28. The concentration of bicarbonate ions in the extracellular fluid is about _____ times that for CO_2, as its pH is _____ (G, greater; L, less) than the pK value of the bicarbonate buffer system.

   a. 10, G   c. 1/10th, L   e. 20, L
   b. 10, L   d. 20, G   f. 1/20th, G

29. Doubling the normal alveolar ventilation _____ (I, increases; D, decreases) the carbon dioxide concentration of body fluids to about _____ millimols/liter.

   a. I, 2.4   c. I, 24   e. D, 0.60
   b. I, 4.8   d. D, 0.24   f. D, 1.2

30. Doubling the normal alveolar ventilation _____ (I, increases; D, decreases) the hydrogen ion concentration of body fluids by about _____ pH unit.

   a. I, 0.23   c. D, 0.23
   b. I, 0.76   d. D, 0.76

31. Increased hydrogen ion concentration of extracellular fluids results in a negative feedback regulatory mechanism involving _____ alveolar ventilation rates and _____ rates of carbon dioxide elimination from body fluids. (I, increased; D, decreased)

   a. I, I   c. D, I
   b. I, D   d. D, D

32. The respiratory mechanism for regulation of hydrogen ion concentration via _____ (S, spinal cord; M, medullary; C, cortical) respiratory centers is about _____ % effective.

   a. S, 100   c. C, 100   e. M, 50 to 75
   b. M, 100   d. S, 50 to 75   f. C, 50 to 75

OBJECTIVE 37–6.

Recognize the renal role in regulating body pH by altering the extracellular bicarbonate concentration, and the involvement of renal hydrogen ion secretion, sodium ion reabsorption, bicarbonate ion excretion, and ammonia secretion. Identify the renal secretory mechanisms for hydrogen ions, the roles of sodium ion transport and carbonic anhydrase, differential renal tubular transport characteristics in hydrogen ion transport, and the regulation of hydrogen ion secretion by extracellular fluid carbon dioxide concentration.

33. Carbonic anhydrase in the kidney tubular cells is known to be associated with reabosrption of:

   a. Urea   d. Lactic acid
   b. Uric acid   e. Water
   c. Carbohydrate   f. Bicarbonate

34. Carbonic anhydrase localized in _____ (E, extracellular; I, intracellular) renal tubular fluids catalyzes the interconversion between carbonic acid and _____ (B, bicarbonate; C, carbon dioxide).

   a. E, B   c. I, B
   b. E, C   d. I, C

35. _____ (A, active; P, passive) hydrogen ion secretion may decrease the pH of renal tubular fluids to a minimal limit of about _____ .

   a. A, 7.4   c. A, 4.5   e. P, 6.0 to 6.5
   b. A, 6.0 to 6.5   d. P, 7.4   f. P, 4.5

36. A higher rate of hydrogen ion secretion is achieved in the _____ portion of the nephron, and a higher hydrogen ion concentration gradient is achieved across the _____ tubules. (P, proximal; D, distal)

   a. P, P   c. D, P
   b. P, D   d. D, D

37. Hydrogen ion secretion, thought to occur at the _____ (L, luminal; B, basal) membrane borders of renal tubular epithelial cells, occurs largely in exchange for _____ (Cl, chloride; Na, sodium) ion transport.

a. L, Cl
b. L, Na
c. B, Cl
d. B, Na

38. Intracellular bicarbonate ions, formed in association with hydrogen ion secretion by renal tubular epithelial cells, is largely _____ (A, actively; P, passively) transported into the _____ (T, tubular lumen; PF, peritubular fluids).

a. A, T
b. A, PF
c. P, T
d. P, PF

39. A typical rate of renal hydrogen ion secretion of 3.5 millimols per minute would be _____ by _____ extracellular fluid carbon dioxide concentrations in response to decreased alveolar ventilation. (I, increased; D, decreased)

a. I, I
b. I, D
c. D, I
d. D, D

OBJECTIVE 37–7.

Identify the renal "reabsorptive" mechanisms for bicarbonate ions, their relationship to renal hydrogen ion secretion, differential renal tubular characteristics in renal tubular bicarbonate ion transport, and normal rates and consequences of bicarbonate ion filtration and hydrogen ion secretion.

40. Bicarbonate ions are more completely reabsorbed from renal tubular fluid during _____ (E, elevated; R, reduced) rates of tubular hydrogen ion secretion as a consequence of a high tubular epithelial cell membrane permeability to _____ . (B, bicarbonate ions; C, carbon dioxide)

a. E, B
b. E, C
c. R, B
d. R, C

41. Carbonic anhydrase associated with the _____ (L, luminal; B, basal) border of the _____ (P, proximal; D, distal) tubules is associated with the larger percentage of bicarbonate ion reabsorption from renal tubular fluid.

a. L, P
b. L, D
c. B, P
d. B, D

42. Bicarbonate reabsorption by a renal tubular _____ (C, competitive; T, titration) process involving hydrogen ion secretion is _____ (I, increased; D, decreased) in acidosis.

a. C, I
b. C, D
c. T, I
d. T, D

43. Metabolic processes generally result in a _____ (S, surplus; D, deficit) of extracellular hydrogen ions, and a ratio of renal tubular hydrogen ion secretion rate (mols per minute) to filtered load of bicarbonate (mols per minute) closest to _____ .

a. S, ½
b. S, 1
c. S, 2
d. D, ½
e. D, 1
f. D, 2

OBJECTIVE 37–8.

Identify the mechanistic roles of renal function in the correction of acidosis and alkalosis.

44. The basic mechanism whereby renal correction of acidosis or alkalosis occurs, involves the incomplete titration of _____ ions against _____ ions in the renal tubule.

a. NH_3, H
b. HCO_3^-, H
c. HPO_4^-, H
d. NH_4^+, OH^-
e. HCO_3^-, Na
f. Cl^-, H

45. The renal tubular secretion of hydrogen ions is generally accompanied by renal tubular _____ of sodium ions and _____ of bicarbonate ions. (S, secretion; R, reabsorption)

a. S, S
b. S, R
c. R, S
d. R, R

46. In _____ (A, acidosis; L, alkalosis), the concentration ratio of carbon dioxide to bicarbonate ions in extracellular fluid is increased from normal, and the renal rate of hydrogen ion secretion is _____ (G, greater; S, less) than the rate of bicarbonate ion filtration.

   a. L, G   c. A, G
   b. L, S   d. A, S

47. The net effect of secreting hydrogen ion excess into renal tubular fluid is to _____ (I, increase; D, decrease) the bicarbonate concentration in the extracellular fluid, which _____ the chemical buffers of the body _____ (A, acidic; B, alkaline) di...

   a. I, A   c. D.
   b. I, B   d.

OBJECTIVE 37–9.

Identify the roles of the renal phosphate and ammonia buff [...] ms in the transport of excess hydrogen ions into urine.

48. At representative daily urine flows, a maximum hydrogen ion concentration that may be achieved in urine of _____ molar represents about _____ % of the daily excretion of excess hydrogen ions.

   a. 10^{-12}, 1   d. 10^{-12}, 50
   b. $10^{-4.5}$, 1   e. $10^{-4.5}$, 50
   c. $10^{-2.2}$, 1   f. $10^{-2.2}$, 50

49. The renal hydrogen ion secretion mechanism is _____ (G, gradient; T, transfer maximum)– limited and operates _____ (X, maximally; M, minimally) at the lower limit of urine pH values.

   a. G, X   c. T, X
   b. G, M   d. T, M

50. The phosphate buffer system concentration in glomerular filtrate has a comparatively _____ (G, greater; L, lesser) concentration than urine, and has a normal ratio of HPO_4^{-2} to $H_2PO_4^-$ ions of about _____ .

   a. G, ¼   c. G, 4   e. L, 1
   b. G, 1   d. L, ¼   f. L, 4

51. The net effect of conversion of HPO_4^{-2} ions to $H_2PO_4^-$ ions in renal tubular fluid is to _____ the renal tubular secretion of hydrogen ions and _____ the renal tubular reabsorption of sodium bicarbonate. (I, increase; N, not significantly alter; D, decrease)

   a. I, N   c. N, D   e. I, I
   b. N, I   d. D, D   f. N, N

52. Renal ammonia is largely derived from _____ (U, urea; G, glutamine; Y, glycine) catabolism in the _____ (L, liver; K, kidneys).

   a. U, L   c. Y, L   e. G, K
   b. G, L   d. U, K   f. Y, K

53. Renal tubular fluid ammonium ions owe their origin to _____ (A, active; P, passive) transport of _____ by renal tubular epithelial cells.

   a. A, NH_3 c. P, NH_3
   b. A, NH_4^+ ions   d. P, NH_4^+ ions

54. The secretion of ammonia by renal tubules is regulated largely by the renal secretion of excess _____ , and is _____ (I, increased; D, decreased) in acidosis.

   a. HCO_3^-, D   d. HCO_3^-, I
   b. Cl^-, D   e. Cl^-, I
   c. H^+, D   f. H^+, I

55. Increased urinary excretion of ammonium and chloride ions is associated with _____ bicarbonate reabsorption and _____ hydrogen ion secretion. (I, increased; N. no significant change in; D, decreased)

   a. I, D   c. D, D   e. N, I
   b. N, D   d. I, I   f. D, I

56. Increased ammonia secretion by the kidneys, in response to a persistent _____ (A, acidosis; L, alkalosis), is followed during the next few days by an adaptive _____ (I, increase; D, decline) in the rate of ammonia formation.

   a. A, I   c. L, I
   b. A, D   d. L, D

OBJECTIVE 37-10.

Identify the correction rates, degree of effectiveness, range of urinary pH, and the interrelationship between chloride and bicarbonate ion concentrations in the renal regulation of acid-base imbalance.

57. When compared to respiratory mechanisms, the renal regulatory mechanisms of acid-base balance have a _____ rapidity of action, and have a _____ degree of complete neutralization of excess acid or alkali. (G, greater; L, lesser)

    a. G, G   c. L, G
    b. G, L   d. L, L

58. The urine pH may rise from a normal average of _____ to an upper limit of _____ .

    a. 8.0, 10   c. 6.0, 8.0   e. 7.4, 12
    b. 7.4, 8.0   d. 8.0, 14   f. 6.0, 12

59. The kidneys can eliminate about _____ millimols of acid or alkali per day, thereby returning the extremes of acid-base imbalance to normal in about 1 to 3 _____ (M, minutes; H, hours; D, days).

    a. 50, M   c. 50, D   e. 500, H
    b. 50, H   d. 500, M   f. 500, D

60. The major extracellular anion is _____ , and the extracellular bicarbonate and chloride concentrations generally vary in _____ (S, similar; R, reciprocal) directions during acid-base imbalance or its correction.

    a. HCO_3^-, S   c. HCO_3^-, R
    b. Cl^-, S   d. Cl^-, R

61. The transport of hydrogen ions into urine in combination with ammonia results in _____ bicarbonate and _____ chloride ion concentrations in the extracellular fluids. (I, increased; N, no significant change in; D, decreased)

    a. I, I   c. D, N   e. D, I
    b. I, N   d. D, D   f. I, D

OBJECTIVE 37-11.

Distinguish respiratory acidosis, respiratory alkalosis, metabolic acidosis, and metabolic alkalosis, and identify their mechanistic causes.

62. Elevated concentrations of carbonic acid, termed a _____ (M, metabolic; S, respiratory) acid, is associated with _____ (E, elevated; R, reduced) alveolar ventilation.

    a. M, E   c. S, E
    b. M, R   d. S, R

63. _____ (I, increased; D, decreased) concentrations of carbon dioxide occur in the more common pathological condition of a respiratory form of _____ (A, acidosis; L, alkalosis).

    a. I, A   c. D, A
    b. I, L   d. D, L

64. Severe diarrhea is a common cause of a _____ (R, respiratory; M, metabolic) form of acidosis as a consequence of excessive loss of body _____ .

    a. R, Cl^- d. M, Cl^-
    b. R, CO_2 e. M, HCl
    c. R, HCO_3^- f. M, HCO_3^-

65. Administration of the carbonic anhydrase _____ (P, potentiator; I, inhibitor) acetazolamide (Diamox) results in a mild form of _____ (A, acidosis; L, alkalosis).

    a. P, L   c. I, L
    b. P, A   d. I, A

66. Secretion of potassium ions and _____ ions in the distal tubules are mutually competitive, resulting in _____ (A, acidosis; L, alkalosis) with elevated potassium ion concentrations in body fluids.

    a. H^+, A   c. HCO_3^-, A
    b. H^+, L   d. HCO_3^-, L

67. Excess aldosterone secretion results in a metabolic _____ (A, acidosis; L, alkalosis), and excessive ingestion of sodium bicarbonate for the treatment of gastritis or peptic ulcer results in a _____ (M, metabolic; R, respiratory) alkalosis.

    a. A, M   c. L, M
    b. A, R   d. L, R

OBJECTIVE 37–12.
Identify the mechanisms of compensation of metabolic and respiratory forms of acidosis of alkalosis, the consequences of uncompensated acidosis or alkalosis on the body, and the treatment rationale for acid-base imbalance.

68. Pulmonary ventilation is usually _____ in respiratory acidosis and is _____ in metabolic acidosis. (I, increased; N, not significantly altered; D, decreased)

   a. I, N   c. D, I   e. I, I
   b. D, N   d. I, D   f. D, D

69. A clinical method for ascertaining a tendency towards epileptic seizures utilizes the effects of _____ (A, acidosis; L, alkalosis) on the central nervous system as a consequence of _____ (O, overbreathing; B, breath-holding).

   a. A, O   c. L, O
   b. A, B   d. L, B

70. In respiratory compensation of metabolic acidosis, the pH of body fluids is close to normal, but the carbon dioxide content of body fluids is _____ as a consequence of _____ respiration. (I, increased; D, decreased; N, normal)

   a. I, I   c. D, I   e. N, D
   b. N, I   d. I, D   f. D, D

71. A state of renal compensation for a respiratory acidosis results in _____ pH values, _____ bicarbonate concentrations, and _____ carbon dioxide concentrations of extracellular fluids. (A, above normal; N, about normal; B, below normal)

   a. A, A, A   c. N, N, N   e. N, N, A
   b. B, B, B   d. N, A, A   f. B, N, A

72. The administration of sodium lactate is utilized in the treatment of _____, and sodium bicarbonate is utilized in the treatment of _____. (A, acidosis; L, alkalosis) _____

   a. L, A   c. A, A
   b. A, L   d. L, L

73. Ammonium chloride administration results in increased _____ (R, renal ammonia secretion; L, liver urea synthesis) and a tendency towards _____ (K, alkalosis; A, acidosis).

   a. R, K   c. L, K
   b. R, A   d. L, A

OBJECTIVE 37–13.
Identify the clinical measurements utilized for identification of the types and extent of acid-base abnormalities.

74. Plasma when exposed to the atmosphere, _____ its pH value and _____ its carbon dioxide concentration. (I, increases; D, decreases)

   a. I, I   c. D, I
   b. I, D   d. D, D

75. The term "buffer base" refers to the _____ (C, cation; A, anion) components of the buffer system, and _____ (D, does; N, does not) include hemoglobin at body pH values.

   a. C, D   c. A, D
   b. C, N   d. A, N

76. In the Astrup and Siggaard-Andersen nomogram, the point at which the experimental line crosses the "standard bicarbonate" curve gives the _____ concentration in the blood sample at a carbon dioxide tension of _____ mm. Hg.

   a. H^+ ion, 40   d. H^+ ion, 100
   b. HCO_3^- ion, 40   e. HCO_3^- ion, 100
   c. CO_2, 40   f. CO_2, 100

77. The Astrup and Siggaard-Andersen nomogram method _____ utilized to determine the existing types of respiratory and metabolic acidosis or alkalosis, and _____ utilized to determine the magnitude of metabolic base excess or deficiency. (I, is; N, is not)

   a. N, N   c. N, I
   b. I, N   d. I, I

38

Micturition, Renal Disease, and Diuresis

OBJECTIVE 38-1.

Describe the physiologic anatomy and neural connections of the urinary bladder. Identify micturition, the mechanisms of urine transport in the ureters, and the consequences of afferent and efferent neural activity from the ureters.

1. Micturition is a _____ (L, local intrinsic; N, neural reflex) mechanism by which the urinary bladder becomes _____ (E, emptied; F, filled).

   a. L, E   c. N, E
   b. L, F   d. N, F

2. The body of the urinary bladder contracts with _____ (P, parasympathetic; S, sympathetic) activity to the _____ (T, trigonal; D, detrusor muscle).

   a. P, T   c. S, T
   b. P, D   d. S, D

3. The external sphincter of the bladder innervated by _____ (S, sympathetic; P, parasympathetic; D, pudic) nerves is normally in a state of _____ (T, tonic contraction; R, relaxation).

   a. S, T   c. D, T   e. P, R
   b. P, T   d. S, R   f. D, R

4. Ureter peristaltic waves, traveling at a velocity of about _____ cm./sec. _____ (I, increase; D, decrease) their frequency of occurrence with parasympathetic activity to the ureters.

   a. 0.3, D   c. 30, D   e. 3, I
   b. 3, D   d. 0.3, I   f. 30, I

5. The ureterorenal reflex results in _____ (C, constriction; A, dilation) of renal arterioles and _____ (I, increased; D, decreased) urinary output from the involved kidney.

   a. C, I   c. A, I
   b. C, D   d. A, D

OBJECTIVE 38-2.

Identify the cystometrogram and micturition waves, and their mechanistic causes.

6. The pressure "plateau" for intravesical pressure occurs at _____ cm. of water for urinary bladder volumes between _____ ml. of fluid.

   a. 1, 10 & 100   d. 1, 500 & 800
   b. 10, 100 & 300   e. 10, 500 & 800
   c. 100, 100 & 300   f. 100, 10 & 100

7. The pressure "plateau" in the cystometrogram during bladder filling is largely a consequence of _____ mechanisms, and micturition waves in the cystometrogram are largely a consequence of _____ mechanisms. (I, intrinsic bladder; R, micturition reflex)

   a. I, I   c. R, I
   b. I, R   d. R, R

8. The force of micturition waves generally increases with _____ bladder volume and _____ bladder pressures. (I, increased; D, decreased)

    a. D, D   b. D, I   c. I, D   d. I, I

OBJECTIVE 38-3.

Identify the micturition reflex, its characteristics, its neural pathways, and its regulation by higher brain centers.

9. The micturition reflex involves _____ (S, sympathetic; P, parasympathetic) efferents from _____ (SC, spinal cord; M, medullary; C, cortical) reflex centers.

    a. S, SC   c. S, C   e. P, M
    b. S, M   d. P, SC   f. P, C

10. Smooth muscle contraction of the bladder wall provides a _____ (P, positive; N, negative) feedback mechanism for the micturition reflex, which becomes _____ (M, more; L, less) powerful with increasing urinary bladder volumes.

    a. N, M   c. P, M
    b. N, L   d. P, L

11. A micturition contraction cycle which does not succeed in emptying the bladder is followed by a period of micturition reflex _____ (F, facilitation; I, inhibition) and _____ (S, sustained contraction; R, restoration of the basal intrinsic tone; D, depressed basal tone) of the smooth muscle of the bladder.

    a. F, S   c. F, D   e. I, R
    b. F, R   d. I, S   f. I, D

12. Higher neural centers may _____ (F, only facilitate; I, only inhibit; E, either facilitate or inhibit) micturition by influencing _____ (M, the micturition reflex only; S, external urinary sphincter tone only; B, both the micturition reflex and external urinary sphincter tone).

    a. F, M   c. E, M   e. I, B
    b. I, S   d. F, B   f. E, B

OBJECTIVE 38-4.

Identify the causes and consequences of the "atonic bladder," the "automatic bladder," and the "uninhibited neurogenic bladder."

13. Complete transection of the spinal cord above the sacral segments _____ (D, does; N, does not) abolish micturition reflexes, and results in the _____ (A, atonic; M, automatic; U, uninhibited neurogenic) bladder.

    a. D, A   c. D, U   e. N, M
    b. D, M   d. N, A   f. N, U

14. Frequent and uncontrollable micturition reflexes of the _____ (T, atonic; U, uninhibited neurogenic) bladder results from _____ (A, peripheral afferent; E, peripheral efferent; C, central nervous system) lesions.

    a. T, A   c. T, C   e. U, E
    b. T, E   d. U, A   f. U, C

OBJECTIVE 38–5.

Identify the conditions of acute and chronic failure, their common causes, and associated consequences.

T&F

15. Renal failure caused by _____ (G, acute glomerular nephritis; P, pyelonephritis) results _____ (D, directly; I, indirectly) from infectious processes of group A beta streptococci.

    a. G, D   c. P, D
    b. G, I   d. P, I

16. Carbon tetrachloride and heavy metals, including the mercuric ion, have relatively specific toxic effects upon _____ (G, glomerular capillaries; E, tubular epithelial cells), which _____ (D, do; N, do not) have the capability of regeneration.

    a. G, D   c. E, D
    b. G, N   d. E, N

17. Elevated concentrations of free hemoglobin in plasma _____ (D, does; N, does not) result in acute renal failure during transfusion reactions as a consequence of its _____ .

    a. D, glomerular filtration
    b. D, renal secretion
    c. D, osmotic effects
    d. N, active reabsorption
    e. N, molecular size
    f. N, plasma threshold

18. Pyelonephritis usually begins as an infectious process of the renal _____ (G, glomeruli; P, pelvis) and frequently involves a rather selective inability to excrete a(n) _____ (A, acidic; D, dilute; C, concentrated) urine.

    a. G, A   c. G, C   e. P, D
    b. G, D   d. P, A   f. P, C

19. On the average, normal adults between the ages of 40 and 80 years exhibit _____ renal blood flow and _____ Diodrast renal clearance. (I, progressively increasing; N, no significant change in; D, progressively decreasing)

    a. I, I   c. D, D   e. D, I
    b. N, N   d. I, D   f. I, N

20. Thickened glomerular membranes, resulting from latter stages of _____ (G, chronic glomerulonephritis; N, benign nephrosclerosis), are associated with _____ (I, increased; D, decreased) glomerular filtration coefficients.

    a. G, I   c. N, I
    b. G, D   d. N, D

OBJECTIVE 38–6.

Identify the adaptive changes occurring in remaining functional nephrons in chronic renal failure.

21. Death resulting from chronic renal failure results when the number of nephrons are reduced to about _____ % of a normal number of about _____ million.

    a. 50 to 60, 0.5   d. 10 to 20, 0.5
    b. 50 to 60, 1   e. 10 to 20, 1
    c. 50 to 60, 2.5   f. 10 to 20, 2.5

22. In renal failure resulting from reduced numbers of nephrons, the average functional nephron remaining has _____ glomerular filtrate formation rates, _____ rates of tubular reabsorption, and _____ volume flow contributions to urine formation. (H, higher than normal; N, normal; L, lower than normal)

    a. N, H, L   c. N, L, H   e. H, H, N
    b. N, N, N   d. H, N, H   f. H, H, H

23. Reduction in the numbers of functioning nephrons in chronic renal failure results in a greater impairment of the renal urine _____ (C, concentrating; D, diluting) mechanism and the development of _____ (I, isosthenuria; A, anuria).

    a. C, I   c. D, I
    b. C, A   d. D, A

24. Progressive destruction of increasing numbers of nephrons results in a urine osmolarity which approaches a specific gravity of _____ corresponding to a fluid osmolarity of _____ milliosmols.

    a. 1.055, 100   d. 1.008, 100
    b. 1.055, 300   e. 1.008, 300
    c. 1.055, 1,200   f. 1.008, 1,200

OBJECTIVE 38-7.

Identify the effects of acute and chronic renal failure upon the body fluids, and the rationale for treatment of uremic patients with the artificial kidney.

25. The effects of chronic renal failure would more probably include a diminished extracellular fluid concentration of:

   a. Potassium ions   d. Creatinine
   b. Sodium ions   e. Bicarbonate ions
   c. Sulfate ions   f. Urea

26. A patient with uremia and renal failure, who adjusts fluid intake according to normal desire, will usually exhibit _____ (D, dehydration; E, edema) of _____ (I, only intracellular; X, only extracellular; B, both intracellular and extracellular) fluids.

   a. D, I   c. D, B   e. E, X
   b. D, X   d. E, I   f. E, B

27. Symptoms of uremic coma include _____ (K, alkalosis; C, acidosis), _____ (R, rapid; S, slow) respiratory rates, and _____ (E, elevated; D, diminished) plasma concentrations of nonprotein nitrogen compounds.

   a. K, S, E   c. K, R, E   e. C, R, D
   b. K, S, D   d. C, R, E   f. C, S, D

28. A patient with chronic renal failure of a severe degree is more likely to develop a severe _____ (P, polycythemia; A, anemia) as a consequence of _____ (D, dehydration; ED, erythropoietin deficiency; EE, erythropoietin excess).

   a. P, D   c. P, EE   e. A, ED
   b. P, ED   d. A, D   f. A, EE

29. Prolonged renal failure results in _____ (O, osteoporosis; M, osteomalacia) as a consequence of _____ (I, increased; D, decreased) availability of calcium to the skeletal system.

   a. O, I   c. M, I
   b. O, D   d. M, D

30. _____ (I, increased; D, decreased) extracellular fluid concentrations of potassium with uremia result in cardiotoxic effects at threshold values of about _____ mEq./liter.

   a. D, 2   c. D, 8   e. I, 4
   b. D, 4   d. I, 2   f. I, 8

31. The typical dialyzing fluid of an artificial kidney, when compared to normal plasma, contains _____ sodium and chloride ion concentrations, _____ phosphate and sulfate ion concentrations, and _____ potassium ion concentrations. (H, higher; N, about normal; L, lower)

   a. N, N, N   c. N, L, N   e. H, L, N
   b. N, L, L   d. H, N, L   f. L, N, H

32. The transfer of a substance across the dialyzing membrane of an artificial kidney is increased by _____ molecular concentration gradients, _____ molecular size, and _____ durations of membrane contact. (I, increased; D, decreased)

   a. D, D, D   c. I, D, I   e. D, I, D
   b. D, I, I   d. I, I, I   f. I, D, D

OBJECTIVE 38-8.

Identify the mechanistic basis of associated hypertensive, normal, or hypotensive conditions of arterial pressure for renal disease mechanisms.

33. Most renal lesions that result in either a diminished renal vascular supply or _____ (I, increased; D, decreased) glomerular filtration _____ (O, do; N, do not) result in hypertension.

   a. I, O   c. D, O
   b. I, N   d. D, N

34. The most probable cause of hypertension caused by secretion of renin is _____ (C, peripheral vasoconstriction; R, renal salt and water retention) resulting from the more direct action of _____ (A, angiotensin II; N, norepinephrine).

   a. C, A   c. R, A
   b. C, N   d. R, N

35. Renal lesions resulting in arterial hypotension are more likely to include _____ (A, renal atherosclerosis; P, medullary pyelonephritis), particularly when the individual ingests _____ (H, high; L, low) quantities of salt.

    a. A, H   c. P, H
    b. A, L   d. P, L

36. Substantial reductions of numbers of nephrons more generally results in _____ (S, substantial; T, slight) changes in arterial blood pressure, unless the dietary intake of salt is substantially _____ (I, increased; D, decreased).

    a. S, I   c. T, I
    b. S, D   d. T, D

OBJECTIVE 38–9.

Identify the characteristics and consequences of the nephrotic syndrome, and its common causes.

37. Lipoid nephrosis, more common in _____ (A, adults; C, children), is more generally associated with _____ (R, reduced; E, elevated) blood lipid concentrations.

    a. A, R   c. C, R
    b. A, E   d. C, E

38. Severe nephrosis results in _____ blood volumes and _____ interstitial fluid volumes. (I, increased; D, decreased)

    a. D, D   c. I, D
    b. D, I   d. I, I

39. The average functioning nephron in chronic glomerulonephritis has _____ (H, higher; L, lower) than normal glomerular membrane permeability and _____ (D, does; N, does not) potentially lead to the nephrotic syndrome.

    a. H, D   c. L, D
    b. H, N   d. L, N

40. The abnormal interstitial fluid volume of subcutaneous tissues as a consequence of the nephrotic syndrome is largely attributed to the _____ (I, increased; D, decreased) subcutaneous capillary _____ (P, permeability; H, hydrostatic pressure; O, colloid osmotic pressure).

    a. I, P   c. I, O   e. D, H
    b. I, H   d. D, P   f. D, O

OBJECTIVE 38–10.

Identify the tubular disorders renal glycosuria, nephrogenic diabetes insipidus, renal tubular acidosis, renal hypophosphatemia, and aminoacidurias; their causes; and their consequences.

41. Renal hypophosphatemia, resulting from abnormalities of _____ (V, vitamin D metabolism; T, a renal transfer maximum), leads to _____ (I, increased; D, decreased) calcification of the skeletal system.

    a. V, I   c. T, I
    b. V, D   d. T, D

42. Nephrogenic diabetes insipidus results from a failure of antidiuretic hormone _____ (S, secretion; T, target action), resulting in _____ (I, increased; D, decreased) urine volume flows.

    a. S, I   c. T, I
    b. S, D   d. T, D

43. Renal glycosuria is a tubular disorder involving a _____ transfer maximum, and a _____ plasma threshold for glucose. (H, higher than normal; N, normal; L, lower than normal)

    a. N, N   c. H, N   e. L, L
    b. N, H   d. H, L   f. L, N

44. Amino acid crystallization in the urine to form renal stones is a more common occurrence in _____ (G, glycinuria; C, essential cystinuria; A, beta-aminoisobutyric aciduria), as a consequence of an abnormal tubular _____ (S, secretory; R, reabsorptive) mechanism.

    a. G, S   c. A, S   e. C, R
    b. C, S   d. G, R   f. A, R

45. Renal tubular acidosis, resulting from a renal failure to _____ (A, synthesize ammonia; B, secrete bicarbonate ions; H, secrete hydrogen ions), results in a higher than normal urine concentration of _____ ions.

   a. A, HCO_3^- b. B, HCO_3^- c. H, HCO_3^- d. A, H^+ e. B, H^+ f. H, H^+

OBJECTIVE 38–11.
Identify the commonly utilized renal function tests, their measurement rationale, and their significance.

46. Clinical symptoms of renal failure generally first become apparent when the plasma clearance of urea changes from a normal of _____ ml. per minute to about _____ ml. per minute.

   a. 700, 500   c. 70, 120   e. 120, 70
   b. 120, 50   d. 700, 120   f. 70, 20

47. Phenolsulfonphthalein is an _____ (L, alkaline; A, acidic) dye which is _____ (R, reabsorbed; S, secreted; N, neither reabsorbed nor secreted) actively by the renal tubules.

   a. L, R   c. L, N   e. A, S
   b. L, S   d. A, R   f. A, N

48. Diodrast is an x-ray opaque compound containing large quantities of _____ (L, lead; B, barium; I, iodine) which is actively _____ (R, reabsorbed; S, secreted) by renal tubules.

   a. L, R   c. I, R   e. B, S
   b. B, R   d. L, S   f. I, S

49. Symptoms of renal insufficiency would more probably include _____ plasma concentrations of urea, _____ plasma concentrations of creatinine, and _____ "base excess" of blood. (I, increased; D, decreased)

   a. I, I, I   c. I, D, D   e. D, I, I
   b. I, I, D   d. I, D, I   f. D, D, I

50. The difference between the maximal and minimal specific gravities of urine for normal kidneys is about _____ and would be _____ (I, increased; D, decreased) by reductions in the numbers of functioning nephrons.

   a. 0.043, I   c. 4.3, I   e. 0.43, D
   b. 0.43, I   d. 0.043, D   f. 4.3, D

OBJECTIVE 38–12.
Identify commonly utilized diuretics, their mechanisms of action, and their resultant effects.

51. A diuretic is a substance that _____ (I, increases; D, decreases) the urine volume flow and generally _____ (O, does; N, does not) result in natriuresis.

   a. I, O   c. D, O
   b. I, N   d. D, N

52. The effects of norepinephrine to alter urinary output include _____ (I, increased; D, decreased) arterial blood pressure, and _____ (S, a synergistic; O, an opposing) effect upon renal afferent arterioles.

   a. I, S   c. D, S
   b. I, O   d. D, O

53. Theophylline and caffeine result in _____ (A, antidiuresis; D, diuresis) by means of _____ (C, constriction; I, dilation) of renal afferent arterioles.

   a. A, C   c. D, C
   b. A, I   d. D, I

54. The administration of a substance which is filtered, but neither reabsorbed nor secreted, by the renal tubules, such as _____ (D, Diodrast; M, mannitol), _____ (I, increases; N, does not significantly alter; E, decreases) urine volume flow.

   a. D, I   c. D, N   e. D, E
   b. M, I   d. M, N   f. M, E

55. The administration of spironolactone results in diuresis by blocking aldosterone _____ (S, secretion; T, target receptor effects), and results in _____ (I, increased; N, no significant alteration of; D, decreased) extracellular fluid potassium concentration.

a. S, I   c. S, D   e. T, N
b. S, N   d. T, I   f. T, D

56. _____ (C, chlorothiazide; F, furosemide), a "loop" diuretic, may increase the urine output maximally to about _____ % of the glomerular filtration rate.

a. C, 8   c. C, 95   e. F, 20
b. C, 20   d. F, 8   f. F, 95

57. Antidiuretic hormone secretion is _____ by water ingestion, and _____ by alcohol ingestion. (I, increased; N, not significantly altered; D, decreased)

a. I, I   c. I, D   e. D, N
b. I, N   d. D, D   f. D, I

39

Pulmonary Ventilation

OBJECTIVE 39-1.

Recognize the major categories of the respiratory process. Identify the basic mechanisms of lung expansion and contraction.

1. Diaphragmatic contraction pulls the lower border of the chest cavity _____ (D, downward; U, upward), thereby _____ (I, increasing; E, decreasing) thoracic cage length, and _____ (I, increasing; E, decreasing) thoracic cage volume.

a. D, E, E   c. D, I, I   e. U, E, I
b. D, E, I   d. U, E, E   f. U, I, I

2. Elevation of the anterior portion of the thorax tends to _____ (I, increase; D, decrease) the thoracic volume during _____ (N, inspiratory; E, expiratory) efforts.

a. D, N   c. I, N
b. D, E   d. I, E

3. Elevation of the sternum _____ (I, increases; D, decreases) the anteroposterior diameter of the chest about _____ %.

a. I, 20   c. I, 60   e. D, 40
b. I, 40   d. D, 20   f. D, 60

4. Expiration in a normal resting individual is accomplished _____ (P, only partially; E, almost entirely) by diaphragmatic _____ (C, contraction; R, relaxation).

a. P, C   c. E, C
b. P, R   d. E, R

5. Muscles that elevate the chest cage are generally classified as muscles of _____ (E, expiration; I, inspiration), and _____ (D, do; N, do not) include the abdominal muscles.

a. E, D   c. I, D
b. E, N   d. I, N

OBJECTIVE 39-2.

Identify the muscles of inspiration and expiration, their mode of action, and their significance.

6. Elastic structures of the lungs and thoracic cage tend to force the diaphragm _____ (U, upward; D, downward) during respiration, and abdominal muscle contraction would _____ (A, aid; O, oppose) this action.

   a. U, A   c. D, A
   b. U, O   d. D, O

7. Muscles of expiration include the:

   a. Diaphragm
   b. Internal intercostals
   c. Sternocleidomastoids
   d. Scaleni
   e. Scapular elevators
   f. External intercostals

8. Contraction of the external intercostal muscles tends to pull the upper ribs _____ (B, backward; F, forward) in relation to the lower ribs, and thereby _____ (I, increase; D, decrease) thoracic volume.

   a. B, I   c. F, I
   b. B, D   d. F, D

9. Inadequacy of the elastic recoil of the lungs and thoracic cage would require the utilization of _____ (X, external intercostals; S, scaleni; R, abdominal rectus) muscles for _____ (I, inspiration; E, expiration).

   a. X, I   c. R, I   e. S, E
   b. S, I   d. X, E   f. R, E

10. Contraction of the internal intercostal muscles, which are stretched in the _____ (I, inspiratory; X, expiratory) position, tends to _____ (E, elevate; L, lower) the chest cage.

   a. I, E   c. X, E
   b. I, L   d. X, L

OBJECTIVE 39-3.

Identify intra-alveolar, intrapleural fluid, intrapleural, and pleural surface pressures, their typical magnitudes during respiration, their mechanistic causes, and their interrelationships.

11. The intrapleural space between the lungs and the chest wall normally contains:

   a. Air at a negative pressure
   b. Air at a positive pressure
   c. Only a thin layer of fluid
   d. The residual volume
   e. Pulmonary surfactant
   f. Anchored elastic fibers

12. Intra-alveolar pressure is generally less than atmospheric pressure during quiet respiration by an amount _____ (G, greater; L, less) than 2 mm. Hg during _____ (E, only expiration; I, only inspiration; B, both expiration and inspiration).

   a. G, E   c. G, B   e. L, I
   b. G, I   d. L, E   f. L, B

13. A greater variance of intra-alveolar pressure occurs within the limits of maximal _____ (I, inspiratory; E, expiratory) efforts, and typically _____ (D, does; N, does not) exceed 500 mm. Hg.

   a. I, D   c. E, D
   b. I, N   d. E, N

14. The fluid pressure of the intrapleural space is about _____ mm. Hg as a consequence of pulmonary _____ (E, elastic properties; C, pleural capillary dynamics).

   a. -10, E   c. +10, E   e. 0, C
   b. 0, E   d. -10, C   f. +10, C

15. The total recoil tendency of the lungs, largely attributed to _____ (E, elastic fibers; T, surface tension forces), may be measured by the magnitude of _____ (P, intrapleural; A, intra-alveolar) pressure required to prevent collapse of the lungs.

    a. E, P   c. T, P
    b. E, A   d. T, A

16. _____ analogous to a solid tissue pressure, is equal to _____ . (I, intrapleural pressure; S, pleural surface pressure; F, pleural fluid pressure)

    a. S, I–F   c. I, S–F   e. F, S–I
    b. S, I+F   d. I, S+F   f. F, S+I

17. The intra-alveolar pressure is _____ the intrapleural pressure during inspiration, _____ the intrapleural pressure between respiratory efforts, and _____ the intra-pleural pressure during expiration. (G, greater than; E, equal to; L, less than)

    a. G, E, L   c. G, G, G   e. E, E, E
    b. L, E, G   d. L, L, L   f. L, L, G

18. The intrapleural pressure, normally about _____ mm. Hg between respiratory efforts, is _____ during deep inspiratory efforts.

    a. 0, positive   d. 0, negative
    b. –4, less negative   e. –4, more negative
    c. –10, less negative   f. –10, more negative

19. When air is admitted into the intrapleural space, the lung on the affected side _____ (E, expands; C, collapses) due to its _____ .

    a. E, positive pressure   c. C, elastic recoil
    b. E, surfactant   d. C, negative pressure

OBJECTIVE 39–4.

Recognize the existence of pulmonary surfactant, its chemical nature and origin, its functional significance, and the consequences of its absence.

20. Surfactant is a _____ (G, glycoprotein; L, lipoprotein) mixture, primarily secreted by surfactant-secreting cells of the _____ (T, tracheobronchiolar glands; A, alveolar epithelium).

    a. G, T   c. L, T
    b. G, A   d. L, A

21. Surfactant acts to _____ the surface tension of the fluid-air interface of alveoli, and to _____ the absolute magnitude of the expansion pressure required to open the alveoli. (I, increase; D, decrease)

    a. I, I   c. D, I
    b. I, D   d. D, D

22. Surfactant has the property of _____ (I, increasing; D, decreasing) the surface tension of the alveoli as they become smaller in diameter, which serves to _____ (A, assist; O, oppose) the collapse tendency of the alveoli as predicted by the law of Laplace.

    a. I, A   c. D, A
    b. I, O   d. D, O

OBJECTIVE 39-5.

Identify the compliances of the lungs and chest wall, their typical values and significance, their means of measurement, and factors that cause abnormal compliance.

23. The expansibility of the combined lungs and thorax, expressed as the volume increase of the lungs per unit increase in intra-alveolar pressure is the _____ (M, compliance; C, conductance), and normally averages _____ ml. per cm. of water.

   a. M, 220   c. M, 60   e. C, 130
   b. M, 130   d. C, 220   f. C, 60

24. The lungs, when removed from the thoracic cage, have a compliance of about _____ ml. per cm. of water, which is _____ (G, greater than; E, equal to; L, less than) the combined compliance of the lungs and thoracic cage.

   a. 220, G   c. 60, E   e. 130, E
   b. 130, G   d. 220, E   f. 60, L

25. The intra-esophageal pressure, more closely approximating the _____ (A, intra-alveolar; P, intrapleural) pressure, is more negative for comparable lung volumes during the _____ (E, expiratory; I, inspiratory) phase of an inspiratory-expiratory cycle.

   a. A, E   c. P, E
   b. A, I   d. P, I

26. Deformities of the chest cage, _____ (A, and; B, but not) conditions which destroy lung tissue, generally result in _____ (I, increased; D, decreased) total pulmonary compliance.

   a. A, I   c. B, I
   b. A, D   d. B, D

OBJECTIVE 39-6.

Identify nonelastic tissue resistance and airway resistance, their means of measurement, and their relationship to energy requirements for normal and abnormal respiration.

27. The work of breathing is _____ by increased nonelastic tissue resistance of the lungs, _____ by increased nonelastic tissue resistance of the thoracic cage, and _____ by increased air flow resistance. (I, increased; N, not significantly altered; D, decreased)

   a. N, N, N   c. I, D, N   e. I, N, I
   b. D, I, N   d. I, N, D   f. I, I, I

28. The hysteresis area between expiratory and inspiratory portions of "static" curves of intra-esophageal pressure versus lung volumes is _____ (D, directly; I, inversely) proportional to the energy required to overcome _____ (A, airway resistance; T, nonelastic tissue resistance).

   a. D, A   c. I, A
   b. D, T   d. I, T

29. The hysteresis area between expiratory and inspiratory portions of "dynamic" curves of intra-esophageal pressure versus lung volumes is _____ (G, greater; L, less) than that obtained from similar curves obtained under "static" conditions, as a consequence of _____ (A, airway; T, nonelastic tissue) resistance components.

   a. G, A   c. L, A
   b. G, T   d. L, T

30. The percentage of the total body energy expenditure required for pulmonary ventilation is closer to _____ % at rest, and closer to _____ % during very heavy exercise.

   a. 2 to 3, 3 to 4   d. 2 to 3, 10 to 12
   b. 10 to 12, 20 to 25   e. 10 to 12, 10 to 12
   c. 40 to 50, 60 to 70   f. 40 to 50, 20 to 30

OBJECTIVE 39-7.

Identify the resting expiratory level. Identify the pulmonary volumes and capacities, and their typical values.

31. The volume of air inspired and expired with normal breathing at rest is about _____ ml. in the normal young male adult, and constitutes about _____ % of his vital capacity.

    a. 200, 9   c. 1,100, 9   e. 500, 26
    b. 500, 9   d. 200, 26   f. 1,100, 26

32. When the inspiratory muscles are completely relaxed, the pulmonary volume equals the _____ (RV, residual volume; RC, functional residual capacity), and is usually about _____ ml. in normal young adult males.

    a. RV, 1,100   c. RV, 4,600   e. RC, 2,300
    b. RV, 2,300   d. RC, 1,100   f. RC, 4,600

33. The inspiratory reserve volume, or the additional pulmonary volume that can be inspired above a level of the normal _____ (T, tidal volume; E, resting expiratory level), is usually _____ ml. in the young male adult.

    a. T, 1,100   c. T, 4,600   e. E, 3,000
    b. T, 3,000   d. E, 1,100   f. E, 4,600

34. The total lung capacity minus the _____ (V, vital; I, inspiratory; FR, functional residual) capacity equals the _____ (E, expiratory reserve; R, residual) volume.

    a. V, E   c. FR, E   e. I, R
    b. I, E   d. V, R   f. FR, R

35. The _____ (R, residual; T, tidal) volume is the larger of the two pulmonary volumes, and averages about _____ ml. in the young male adult.

    a. R, 150   c. R, 1,200   e. T, 500
    b. R, 500   d. T, 150   f. T, 1,200

36. The expiratory reserve volume, equal to the vital capacity minus the inspiratory _____ (R, reserve volume; C, capacity), is about _____ ml. in the average young adult male.

    a. R, 500   c. R, 3,500   e. C, 1,100
    b. R, 1,100   d. C, 500   f. C, 3,500

37. A respiratory volume, (in ml.) is frequently estimated by the formula in which it equals (27.63 − (0.112 × age)) × height (in cm.) for males. It also comprises about 80% of the total lung capacity, and would be the:

    a. Functional residual volume
    b. Inspiratory reserve volume
    c. Residual volume
    d. Vital capacity
    e. Expiratory reserve volume
    f. Physiological dead space

38. Pulmonary volumes and capacities are typically _____ % _____ (G, greater; L, less) in women than men.

    a. 5 to 10, G   d. 5 to 10, L
    b. 20 to 25, G   e. 20 to 25, L
    c. 40 to 50, G   f. 40 to 50, L

OBJECTIVE 39-8.

Identify the significance, rationale for variance, and means of measurement of pulmonary volumes and capacities.

39. The total lung capacity remains, on the average, about the same for individuals from 25 through 65 years of age, whereas the vital capacity decreases with increasing age. With advancing age, residual volumes for the same age range:

    a. Increase progressively
    b. Remain constant
    c. Decrease progressively
    d. Decrease and then increase

40. Pulmonary blood volume _____ about 400 ml. in an average individual and vital capacity _____ when an individual lies down. (I, increases; D, decreases)

    a. I, I   c. D, I
    b. I, D   d. D, D

41. The diaphragm moves _____ (D, down-ward; U, upward) upon changing to the lying position, which _____ (O, opposes; S, supplements) the consequences of altered pulmonary blood volume upon vital capacity.

    a. D, O   c. U, O
    b. D, S   d. U, S

42. During normal resting respiration, marked fluctuations in alveolar gas tensions _____ (D, do; N, do not) occur as a consequence of a _____(H, high; L, low) volume ratio of tidal volume to _____ (R, functional residual; V, vital) capacity.

    a. D, L, V   c. D, H, V   e. N, L, R
    b. D, H, R   d. N, H, R   f. N, L, V

43. Vital capacity is generally_____ by diminished strength of the respiratory muscles, and is generally _____ by decreased pulmonary compliance. (I, increased; N, not significantly altered; D, decreased)

    a. N, N   c. D, N   e. D, I
    b. N, D   d. D, D   f. N, I

44. A simple method utilized for measuring the majority of pulmonary volumes and capacities is _____ (S, spirometry; N, the nitrogen washout method), which is not adequate to determine the _____(F, functional residual; V, vital; I, inspiratory) capacity.

    a. S, F   c. S, I   e. N, V
    b. S, V   d. N, F   f. N, I

OBJECTIVE 39–9.

Identify the minute respiratory volume; normal values of respiratory rates and tidal volumes, and their variance; the maximum breathing capacity and its normal average value; and the maximum expiratory flow, its normal value, and its significance.

45. The minute respiratory volume, or the volume of new air moved into the _____ (A, alveoli; U, upper respiratory tract) per minute, averages _____ liters per minute in a young adult male.

    a. A, 3   c. A, 12   e. U, 6
    b. A, 6   d. U, 3   f. U, 12

46. A constant minute respiratory volume, accompanying a reduction in the respiratory rate from a normal adult average of _____ breaths per minute, would require _____ (I, increased; N, no significant change in; D, decreased) tidal volumes.

    a. 12, I   c. 12, D   e. 24, N
    b. 12, N   d. 24, I   f. 24, D

47. A maximum minute respiratory volume in young male adults averages about _____ liters per minute during the first 15 seconds of effort, and then subsequently _____ (I, increases; D, decreases; R, remains constant) with maintained efforts.

    a. 150, I   c. 150, D   e. 250, R
    b. 150, R   d. 250, I   f. 250, D

48. Air volume flow rates during respiration are _____ proportional to the pressure gradient, and _____ proportional to air flow resistance. (D, directly; I, inversely)

    a. D, D   c. D, I
    b. I, D   d. I, I

49. The maximum expiratory air flow, normally limited by _____ (M, skeletal muscle strength; R, air flow resistance), is _____ (I, increased; D, decreased; N, not altered) by reducing lung volumes.

    a. M, I   c. M, D   e. R, N
    b. M, N   d. R, I   f. R, D

50. A loss of elastic pulmonary tissue in emphysema results in a more significant elevation of air flow resistance during _____ (I, inspiration; E, expiration), which is generally compensated for partially by increasing the _____ (F, functional residual; V, vital; N, inspiratory) capacity.

    a. I, N   c. I, F   e. E, V
    b. I, V   d. E, F   f. E, N

OBJECTIVE 39–10.

Identify the anatomical and physiological dead spaces, their means of measurement, their typical values, and their functional significance. Identify the alveolar ventilation rate, its typical value, its significance, and means of variance.

51. The difference between anatomical and physiological dead spaces is equivalent to a volume of:

    a. Functioning alveoli
    b. Non-functioning alveoli
    c. The residual volume
    d. Nasal and pharyngeal passages
    c. The trachea and bronchi
    f. The tidal volume

52. The volume of air contained within the conducting passages of the respiratory system in the average young adult is _____ ml., and tends to _____ (I, increase; D, decrease) slightly with age.

    a. 100 to 200, I   d. 100 to 200, D
    b. 500 to 600, I   e. 500 to 600, D
    c. 1,000 to 2,000, I f. 1,000 to 2,000, D

53. The volume of air that enters the alveoli with each breath, equals the _____ minus the _____, and represents about two-thirds of the _____. (V, vital capacity; T, tidal volume; R, residual volume; D, dead space)

    a. D, T, V   c. T, R, T   e. T, D, V
    b. D, T, T   d. T, R, V   f. T, D, T

54. The best criterion for the effectiveness of breathing is the:

    a. Respiratory minute volume
    b. Tidal volume
    c. Respiratory rate
    d. Alveolar ventilation
    e. Functional residual volume
    f. Physiological dead space

55. For a given, fixed respiratory minute volume, alveolar ventilation is greater if the respiratory rate is _____ (L, large; S, small) because the _____ (R, residual; D, dead space; T, tidal) volume is constant.

    a. L, R   c. L, T   e. S, D
    b. L, D   d. S, R   f. S, T

56. At a tidal volume of 500 ml., a physiological dead space of 150 ml., and a respiratory rate of 12 breaths per minute, the alveolar ventilation equals _____ ml./minute.

    a. 900   c. 3,600   e. 6,000
    b. 1,800   d. 4,200   f. 7,800

57. An emphysematous patient has a PCO_2 of 50 mm. Hg, a tidal volume of 800 ml., and a PCO_2 of expired air equal to 25 mm. Hg. The physiological dead space of the patient is therefore _____ ml.

    a. 100   c. 300   e. 500
    b. 200   d. 400   f. 600

58. The difference between the minute respiratory volume and the rate of alveolar ventilation is equal to the _____ (T, tidal volume; P, physiological dead space; R, residual volume) _____ (M, multiplied; D, divided) by the respiratory rate.

    a. T, M   c. R, M   e. P, D
    b. P, M   d. T, D   f. R, D

OBJECTIVE 39-11.

Identify the "air conditioning" functions of the upper respiratory tract, their mechanisms of action, and their degree of effectiveness.

59. 1 cubic meter of inspired air at 0° C. and 100% humidity contains 4.85 grams of water. The same air, after it has passed through a portion of the respiratory system, has a temperature of 30° C. and contains 30.4 grams of water vapor (100% humidity). Most of the water has come from the:

   a. Nasal cavity
   b. Alveoli
   c. Physiological dead space
   d. Tracheobronchiolar glands
   e. Water released in metabolism
   f. None of the above

60. Some particles in inhaled air are not normally removed completely by the nose, trachea, and bronchiolar tree. Subsequently they are found in the alveoli, but are not found in expired air. These particles are about _____ microns in size.

   a. 10   c. 0.3 to 2
   b. 2 to 10   d. Less than 0.3

61. Normal ciliary function for the removal of filtered particles from the trachea, bronchi, and bronchioles requires:

   a. A sheet of mucus
   b. Parasympathetic innervation
   c. Sympathetic innervation
   d. Gaseous irritants
   e. Surfactant
   f. All of the above

62. Nasal passages remove the majority of the _____ (L, larger; S, smaller) diameter particulate matter in inspired air by means of nasal _____ (H, hairs; T, turbulent mechanisms).

   a. L, H   c. S, H
   b. L, T   d. S, T

63. Inspired air is generally about _____ % of the way to equilibrium with body temperature, and about _____ % of the way to full saturation with water vapor, by the time it reaches the right and left bronchi.

   a. 50, 75   c. 75, 75   e. 100, 75
   b. 75, 50   d. 75, 100   f. 100, 100

OBJECTIVE 39-12.

Identify the cough reflex, the sneeze reflex, and the action of cilia of the respiratory epithelium, their mechanisms of action, and their functional significance.

64. The sneeze reflex involves irritations of the _____ (C, carina; N, nasal passages), _____ (T, trigeminal; V, vagal) afferents, and a _____ (H, hypothalamic; M, medullary) reflex center.

   a. C, T, H   c. N, T, H   e. N, V, H
   b. C, V, M   d. N, T, M   f. N, V, M

65. During the cough reflex, high air velocities occurring through a _____ (I, dilated; N, narrowed) trachea result from _____ (D, diaphragmatic; A, abdominal) muscle contraction.

   a. I, D   c. N, D
   b. I, A   d. N, A

66. It has been estimated that approximately 0.9 gm. of mucus per square meter of respiratory tract epithelial surface is secreted every 10 minutes in humans. The removal of this mucus from the respiratory passages generally occurs in association with:

   a. Mucosal absorption
   b. Enzymatic hydrolysis
   c. Phagocytosis
   d. Deglutition
   e. Cough reflexes
   f. Sneeze reflexes

OBJECTIVE 39–13.

Identify the roles of the respiratory system in vocalization, and the mechanisms of phonation, articulation, and resonance.

67. The mechanical functions of phonation is achieved by the_____ , while articulation is achieved by the structures of the_____ . (M, mouth; L, larynx; B, both the mouth and larynx)

   a. M, M   c. M, L   e. L, B
   b. M, B   d. L, M   f. L, L

68. Vocal cords vibrate_____ (A, in the direction of air flow; L, laterally) with a frequency which is dependent upon _____ (T, tension only; V, shape and mass of the vocal cord edges only; B, both the tension as well as the shape and mass of the vocal cord edges).

   a. A, T   c. A, B   e. L, V
   b. A, V   d. L, T   f. L, B

69. The _____ (L, lateral; T, transverse) arytenoid muscle pulls the arytenoid cartilages _____ (G, together; A, apart) so that they vibrate in a stream of expired air.

   a. L, G   c. T, G
   b. L, A   d. T, A

70. The entire larynx is moved upward by the external laryngeal muscles to _____ (I, increase; D, decrease) vocal cord tension in order to emit _____ (H, high; L, low) frequency sounds.

   a. I, H   c. D, H
   b. I, L   d. D, L

OBJECTIVE 39–14.

Identify the rationale for mouth to mouth resuscitation and mechanical methods of artificial respiration.

71. Normal expired air has _____ (A, adequate; I, inadequate) amounts of oxygen to sustain life if continuously inhaled, and has carbon dioxide levels which have _____ (N, inhibitory; F, facilitating) effects upon respiration when inhaled.

   a. A, N   c. I, N
   b. A, F   d. I, F

72. Positive pressure resuscitators tend to _____ (I, increase; D, decrease) cardiac output, and are generally lethal when the positive pressure first exceeds _____ mm. Hg for more than a few minutes.

   a. I, 10   c. I, 100   e. D, 30
   b. I, 30   d. D, 10   f. D, 100

40

Physical Principles of Gaseous Exchange; Diffusion of Oxygen and Carbon Dioxide Through the Respiratory Membrane

OBJECTIVE 40–1.

Identify the volume occupied by 1 gram-mol of a gas at 0° C. and 1 atmosphere of pressure. Identify Boyle's law, the law of Gay-Lussac (Charles' law), and the ideal gas law.

1. The volume occupied by 1 gram-mol of gas at standard conditions of 0° C. and 1 atmosphere of pressure is _____ liters.

   a. 6.0   c. 22.4   e. 760
   b. 11.2   d. 44.8   f. 1,120

2. According to _____ (C, Charles'; B, Boyle's) law, doubling the volume of a fixed quantity of gas at a constant temperature _____ (D, doubles; N, does not alter; H, halves) the gas pressure.

   a. C, D   c. C, H   e. B, N
   b. C, N   d. B, D   f. B, H

3. _____ (H, Henry's; C, Charles'; B, Boyle's) law states that the volume of a fixed quantity of gas varies in _____ (D, direct; I, inverse) proportion to the absolute temperature.

   a. H, D   c. B, D   e. C, I
   b. C, D   d. H, I   f. B, I

4. The ideal gas law is a combination of the laws of _____ and _____ (D, Dalton; H, Haldane; C, Charles; B, Boyle; L, Laplace).

   a. D, C   c. H, L   e. B, C
   b. D, H   d. H, B   f. B, D

5. When pressure (P) is expressed in mm. Hg, volume (V) in liters, n in gram-mols, and temperature (T) in degrees Kelvin, the ideal gas constant R equals _____ , or a numerical value of _____ .

   a. PV/nT, 36.26   d. PV/nT, 62.36
   b. PT/nV, 36.26   e. PT/nV, 62.36
   c. nV/PT, 36.26   f. nV/PT, 62.36

OBJECTIVE 40–2.

Identify water vapor pressure, its temperature dependence, and its value at body temperature.

6. The vapor pressure of water _____ (I, increases; D, decreases) with increasing temperature, and the tendency for water _____ (E, evaporation; C, condensation) increases with decreasing temperature.

   a. I, E   c. D, E
   b. I, C   d. D, C

7. The vapor pressure of water is _____ mm. Hg at a temperature of 100° C., and is _____ (G, greater than; E, equal to) 0 at 0° C.

   a. 0, G   c. 760, G   e. 100, E
   b. 100, G   d. 0, E   f. 760, E

8. The water vapor pressure for gases at a normal body temperature of _____ ° C. is _____ mm. Hg.

   a. 37, 27   b. 37, 47   c. 37, 76   d. 98.6, 27   e. 98.6, 47   f. 98.6, 76

OBJECTIVE 40–3.

Identify the factors determining the quantity of a gas dissolved in a fluid at equilibrium. Identify Henry's law and the solubility coefficients for the respiratory gases at body temperatures..

9. _____ (D, Dalton's; B, Boyle's; H, Henry's) law implies that the solubility coefficient K of a gas in a liquid, the pressure P of a gas in equilibrium with the liquid phase, and the volume V of a gas dissolved per unit volume of liquid, are related by the equation, $K =$ _____ .

   a. D, VP c. H, VP e. B, V/P
   b. B, VP d. D, V/P f. H, V/P

10. Of the respiratory gases, the solubility coefficients in body fluids are highest for _____ , intermediate for _____ and lowest for _____ . (O, oxygen; C, carbon dioxide; N, nitrogen).

   a. O, C, N   c. C, N, O   e. N, C, O
   b. O, N, C   d. C, O, N   f. N, O, C

OBJECTIVE 40–4.

Identify the concepts of partial pressure of a gas in gaseous and liquid phases. Recognize that the partial pressure exerted by a gas in a gas mixture is the same as it would be if it occupied the whole volume (Dalton's law). Identify the relationship between per cent concentrations, partial pressures, and the total pressure of constituent gases of a gaseous mixture.

11. The partial pressure of dissolved oxygen in a liquid is determined by the instantaneous sum of the forces of impact of _____ (O, oxygen; A, all dissolved) gas molecules acting upon a surface _____ (M, multiplied; D, divided) by the surface area.

   a. O, M   c. A, M
   b. O, D   d. A, D

12. The partial pressure of a gas in a liquid phase at equilibrium with a gaseous phase is _____ (G, greater than; E, equal to; L, less than) its partial pressure in the gaseous phase, and _____ (I, is; N, is not) dependent upon partial pressures of other gases in the gas mixture.

   a. G, I   c. L, I   e. E, N
   b. E, I   d. G, N   f. L, N

13. That the pressure (partial pressure) exerted by a gas in a mixture of gases is the same as it would be if it occupied the whole volume is a statement of _____ law.

   a. Gay-Lussac's   d. Henry's
   b. Boyle's   e. Dalton's
   c. Graham's   f. Avogadro's

14. A gas at 1 atmosphere of pressure containing 79 volumes % nitrogen and 21 volumes % oxygen has a P_{N_2} of _____ mm. Hg, and a P_{O_2} of _____ mm. Hg.

   a. 760, 760   c. 760, 160   e. 600, 160
   b. 160, 760   d. 160, 600   f. 80, 300

15. A few drops of water are added to a closed chamber with a 21% N_2 and 79% O_2 gas mixture at a temperature of 37° C. After equilibrium the partial pressures of nitrogen and oxygen would be relatively _____ and the total gas pressure would be relatively _____ . (H, higher; L, lower; U, unchanged)

   a. H, H   c. U, H   e. L, H
   b. H, U   d. U, U   f. L, U

OBJECTIVE 40-5.

Identify the factors which influence the rate of diffusion of a gas through body fluids and tissues, and their means of quantitation. Identify the diffusion coefficient and typical values of diffusion coefficients for gases of respiratory importance.

16. The concept of gaseous diffusion from a region of higher to lower partial pressures of a gas _____ hold true for diffusion in a gaseous mixture, _____ hold true for diffusion of dissolved gases in a solution, and _____ hold true for diffusion of a gas between gaseous and liquid phases. (D, does; N, does not)

    a. N, N, N   c. D, D, N   e. D, N, D
    b. D, N, N   d. D, D, D   f. N, D, N

17. The _____ (D, diffusion; O, solubility) coefficient, or the rate of diffusion of a gas in a liquid for a given area, distance, and pressure difference, is proportional to _____ (S, gas solubility; MW, gas molecular weight).

    a. $D, (MW)^{1/2}/S$ d. $O, (MW)^{1/2}/S$
    b. $D, S (MW)^{1/2}$ e. $O, S (MW)^{1/2}$
    c. $D, S/(MW)^{1/2}$ f. $O, S/(MW)^{1/2}$

18. The rate of diffusion of a gas in a liquid chamber, with a given partial pressure difference of a gas between the ends of the chamber, varies in _____ proportion to the chamber cross-sectional area and in _____ proportion to the length of the chamber. (D, direct; I, inverse)

    a. I, I   c. D, I
    b. I, D   d. D, D

19. Of the respiratory gases, the diffusion coefficients in body fluids are highest for _____, intermediate for _____, and lowest for _____.

    a. O_2, N_2, CO_2 d. O_2, CO_2, N_2
    b. N_2, CO_2, O_2 e. N_2, O_2, CO_2
    c. CO_2, O_2, N_2 f. CO_2, N_2, O_2

20. The diffusion coefficient of carbon monoxide in body fluids is _____ (G, greater; L, less) than that for carbon dioxide by a _____ (S, slight; C, considerable) difference.

    a. G, S   c. L, S
    b. G, C   d. L, C

21. Respiratory gases, relatively _____ (S, soluble; I, insoluble) in lipids, encounter a greater diffusion rate limiting impediment for diffusion through _____ (M, cell membranes; F, tissue fluids).

    a. S, M   c. I, M
    b. S, F   d. I, F

OBJECTIVE 40-6.

Identify typical gas compositions of atmospheric air, humidified air at 37° C., alveolar air, and expired air, and the rationale for their differences.

22. Dry air contains about _____ % oxygen, which is equivalent to a partial pressure of oxygen of _____ mm. Hg.

    a. 21, 160   c. 78, 140   e. 21, 760
    b. 78, 600   d. 40, 310   f. 79, 760

23. The sum of the partial pressures of oxygen, carbon dioxide, and nitrogen in the alveoli normally equals _____ mm. Hg as a consequence of _____ (I, intrapleural; W, water vapor) pressures.

    a. 760, I   c. 807, I   e. 713, W
    b. 713, I   d. 760, W   f. 807, W

24. The partial pressures of respiratory gases in alveolar air is highest for _____, intermediate for _____, and lowest for _____.

    a. CO_2, O_2, N_2 d. O_2, N_2, CO_2
    b. CO_2, N_2, O_2 e. N_2, O_2, CO_2
    c. O_2, CO_2, N_2 f. $N_2, CO_2 \ O_2$

25. The carbon dioxide concentration of alveolar air is about _____ %, corresponding to an alveolar P_{CO_2} of _____ mm. Hg.

    a. 0.3, 2.3   c. 5.3, 40   e. 53, 400
    b. 0.3, 23   d. 5.3, 100   f. 53, 100

26. Expired air, when compared to alveolar air, normally contains _____ PO_2 values, _____ PCO_2, values, and _____ PH_2O values. (H, higher; L, lower; E, equal)

    a. E, E, E   c. L, H, E   e. L, L, L
    b. E, E, H   d. H, L, E   f. H, H, H

27. The volume of new atmospheric air brought into the lungs per respiratory cycle is typically _____ % of the lung volume at the end of a normal expiratory effort, or the _____ (E, expiratory reserve volume; R, functional residual capacity).

    a. 15, E   c. 85, E   e. 45, R
    b. 45, E   d. 15, R   f. 85, R

28. About half the volume of a foreign gas contained within the alveoli would be removed in about _____ seconds at typical alveolar ventilation rates, and in about _____ seconds at alveolar ventilation rates 50% of normal.

    a. 18, 9   c. 18, 36   e. 84, 84
    b. 18, 18   d. 84, 42   f. 84, 168

OBJECTIVE 40-7.

Identify the factors that influence oxygen and carbon dioxide concentrations of alveolar and expired air, and their consequences.

29. The alveolar PO_2, normally _____ mm. Hg, is increased maximally to about _____ mm. Hg by _____ (I, increased; D, decreased) rates of alveolar ventilation while breathing atmospheric air.

    a. 47, 97, I   c. 40, 46, I   e. 104, 149, I
    b. 40, 46, D   d. 40, 760, I   f. 104, 149, D

30. The alveolar PCO_2, normally _____ mm. Hg, varies in _____ (D, direct; I, inverse) proportion with the rate of carbon dioxide excretion and in _____ (D, direct; I, inverse) proportion with the rate of alveolar ventilation.

    a. 40, D, I   c. 40, D, D   e. 100, I, D
    b. 40, I, D   d. 100, D, I   f. 100, I, I

31. A normal alveolar PO_2 and an oxygen consumption of 1 liter per minute, would be associated with an alveolar ventilation rate of about _____ normal.

    a. 25%   c. Equal to   e. 4 times
    b. 50%   d. 2 times   f. 8 times

32. Under resting conditions, expired air typically contains a higher percentage of _____ air, and has an average carbon dioxide concentration closer to that of _____ air. (D, dead space; A, alveolar)

    a. D, D   c. A, D
    b. D, A   d. A, A

33. If the concentration of carbon dioxide is 4% in expired air, 6% in alveolar air, and 0% in atmospheric air, and the tidal volume is 600 ml., then the volume of dead air space is _____ ml.

    a. 50   c. 200   e. 600
    b. 120   d. 400   f. 1,200

OBJECTIVE 40–8.
Identify the respiratory unit and its constituent structures, the respiratory membrane, its constituent structures, and physical characteristics, and the permeability characteristics of the respiratory membrane and its functional significance.

34. Within the respiratory unit, alveolar ducts are branch structures of _____ (T, terminal; R, respiratory) bronchioles, and terminal _____ (A, atria; L, alveoli) are the principal loci for respiratory gas exchange.

   a. T, A   c. R, A
   b. T, L   d. R, L

35. The respiratory membrane, with an estimated total surface area of about _____ square meters in the average adult, has an average thickness of about _____ microns.

   a. 7, 0.5   c. 700, 0.5   e. 70, 5
   b. 70, 0.5   d. 7, 5   f. 700, 5

36. _____ (N, endothelial; E, epithelial) cells of the respiratory membrane, with smaller effective pore sizes, are relatively _____ (P, permeable; I, impermeable) to water diffusion.

   a. N, P   c. E, P
   b. N, I   d. E, I

37. The ratio of the rates of diffusion of oxygen to carbon dioxide is about _____ in water and about _____ for the respiratory membrane.

   a. 1 to 5, 20 to 1   d. 5 to 1, 1 to 20
   b. 5 to 1, 20 to 1   e. 1 to 5, 1 to 20
   c. 20 to 1, 20 to 1   f. 1 to 20, 1 to 20

38. The respiratory membrane is _____ permeable to alcohol, _____ permeable to sodium and glucose, and _____ permeable to urea and potassium. (V, very; R, reasonably; P, poorly)

   a. V, V, V   c. R, R, R   e. V, P, P
   b. P, P, P   d. P, R, V   f. V, P, R

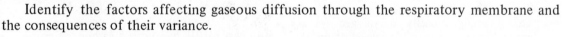

OBJECTIVE 40–9.
Identify the factors affecting gaseous diffusion through the respiratory membrane and the consequences of their variance.

39. The rate of diffusion of gases through the respiratory membrane varies in _____ proportion to its thickness and in _____ proportion to its total surface area. (D, direct; I, inverse)

   a. D, D   c. I, D
   b. D, I   d. I, I

40. Emphysema tends to _____ (E, enlarge; M, diminish) the volume of terminal chambers involved in gas exchange, and _____ (I, increase; D, decrease) the total surface area of the respiratory membrane.

   a. E, I   c. M, I
   b. E, D   d. M, D

41. The significant pressure difference across the respiratory membrane for diffusion rates of a gas involves the alveolar _____ (T, total gas pressure; G, gas partial pressure) and pulmonary capillary _____ (H, hydrostatic pressure; P, partial pressure of the dissolved gas).

   a. T, H   c. G, H
   b. T, P   d. G, P

42. The average partial pressures of carbon dioxide are higher in _____ and the average partial pressures of oxygen are higher in _____. (A, alveoli; B, pulmonary capillary blood)

   a. A, A   c. B, A
   b. A, B   d. B, B

OBJECTIVE 40–10.

Identify the diffusion capacity of the respiratory membrane, typical values for oxygen and carbon dioxide, their influencing factors, and their means of measurement. Identify the gas diffusion characteristics within the respiratory unit.

43. The volume of a gas that diffuses through the respiratory membrane per minute, divided by the driving pressure difference in mm. Hg, is the:

    a. Alveolar ventilation rate
    b. Minute respiratory volume
    c. Diffusion coefficient
    d. Diffusion capacity
    e. Ideal gas constant
    f. A–V gas difference

44. A greater pulmonary diffusing capacity for _____ largely results from differences of _____ (MW, molecular weight; S, solubility coefficients; P, alveolar partial pressures).

    a. O_2, P
    b. CO, S
    c. CO_2, MW
    d. O_2, S
    e. CO, MW
    f. CO_2, S

45. Equal rates of respiratory membrane transfer of oxygen and carbon dioxide, at a normal diffusion capacity ratio of oxygen to carbon dioxide of _____ , implies a greater mean alveolar-capillary pressure gradient for _____ .

    a. 20 to 1, O_2
    b. 2 to 1, O_2
    c. 1 to 20, O_2
    d. 20 to 1, CO_2
    e. 2 to 1, CO_2
    f. 1 to 20, CO_2

46. The diffusing capacity of oxygen for the average resting state of young adult males is _____ ml./min./mm. Hg, and corresponds to a mean oxygen pressure difference across the respiratory membrane of about _____ mm. Hg.

    a. 1, 3
    b. 20, 3
    c. 400 to 450, 3
    d. 1, 11
    e. 20, 11
    f. 400 to 450, 11

47. During exercise, the diffusing capacity in young male adults increases maximally about _____ normal, and the diffusing capacity for carbon dioxide _____ (R, remains constant; I, increases).

    a. 10% above, R
    b. 3 times, R
    c. 20 times, R
    d. 10% above, I
    e. 3 times, I
    f. 20 times, I

48. The mean carbon dioxide pressure difference across the respiratory membrane is closest to _____ mm. Hg with a normal alveolar ventilation, and is closest to _____ mm. Hg at alveolar ventilation rates of twice that of normal.

    a. 0, 0
    b. 4, 4
    c. 8, 8
    d. 0, 4
    e. 4, 2
    f. 8, 4

49. Alveolo-capillary blocks are more likely to result in impaired pulmonary _____ diffusion as a consequence of its lower _____ (D, diffusion capacity; M, mean alveolo-capillary pressure difference).

    a. O_2, D
    b. O_2, M
    c. CO_2, D
    d. CO_2, M

50. The partial pressure of carbon monoxide in blood is generally considerably _____ (I, higher; L, lower) than its alveolar partial pressure as a consequence of the high carbon monoxide _____ (D, diffusion capacity; H, hemoglobin affinity).

    a. I, D
    b. I, H
    c. L, D
    d. L, H

41

Transport of Oxygen and Carbon Dioxide in the Blood and Body Fluids

OBJECTIVE 41–1.
Identify the roles of gaseous diffusion gradients and blood volume flow upon oxygen and carbon dioxide exchange.

1. The partial pressure of oxygen is generally highest in _____ and lowest in _____ . (A, alveoli; P, pulmonary capillaries; I intracellular fluid compartments; B, arterial blood)

   a. I, A   c. A, B   e. P, B
   b. I, P   d. A, I   f. P, I

2. The partial pressure of carbon dioxide is generally highest in _____ and lowest in _____ . (A, alveoli; P, pulmonary capillaries; I, intracellular fluid compartments; B, arterial blood)

   a. I, A   c. A, B   e. P, B
   b. I, P   d. A, I   f. P, I

3. Net carbon dioxide diffusion occurs _____ (O, out of; I, into) pulmonary capillaries _____ (A, against; W, with) a favorable PCO_2 gradient.

   a. O, A   c. I, A
   b. O, W   d. I, W

4. The transport of the majority of oxygen conveyed to tissues occurs as _____ (O_2, dissolved oxygen; H, oxyhemoglobin) and the majority of carbon dioxide transport occurs as _____ (CO_2, dissolved CO_2; C, chemical substances formed from CO_2).

   a. O_2, CO_2 c. H, CO_2
   b. O_2, C   d. H, C

OBJECTIVE 41–2.
Identify typical PO_2 values for blood entering and leaving pulmonary alveoli, the effects of venous admixture upon systemic arterial oxygen content and PO_2, the average PO_2 gradient for pulmonary oxygen diffusion, and the effects of exercise upon the dynamics of alveolar oxygen diffusion.

5. The PO_2 at arterial ends of pulmonary capillaries is typically _____ mm. Hg, and the PO_2 at venous ends is typically _____ mm. Hg.

   a. 104, 46   c. 46, 11   e. 46, 64
   b. 64, 11   d. 40, 104   f. 64, 159

6. The average pressure difference of greater significance for oxygen diffusion through the pulmonary capillaries is its _____ (T, time-integrated; A, arithmetical) average PO_2 of about _____ mm. Hg.

   a. T, 11   c. T, 32   e. A, 27
   b. T, 27   d. A, 11   f. A, 32

7. During exercise, the transit time of pulmonary capillary blood flow _____ , the oxygen-diffusing capacity _____ , and the degree of oxygenation of blood _____ . (D, decreases; R, remains about constant; I, increases)

   a. D, I, D   c. D, I, R   e. I, D, I
   b. D, D, R   d. I, I, I   f. I, R, D

8. During normal resting conditions of pulmonary blood flow, about _____ % of the net oxygen exchange occurs in the first third of the pulmonary blood capillary transit time, and about _____ % during the last third.

   a. 33, 33   c. 75, 5   e. 0, 100
   b. 5, 75   d. 100, 0   f. 50, 10

9. Systemic arterial blood has a PO_2 of _____ mm. Hg as a partial result of a venous admixture which normally represents about _____ % of the total cardiac output.

a. 159, 1 to 2
b. 104, 1 to 2
c. 95, 1 to 2
d. 104, 10 to 15
e. 64, 10 to 15
f. 40, 10 to 15

10. Poorly aerated pulmonary capillaries and alveoli generally tend to _____ (I, increase; D, decrease) the _____ (Q, total quantity; P, partial pressure) of oxygen in arterial blood to a greater extent.

a. I, Q
b. I, P
c. D, Q
d. D, P

OBJECTIVE 41–3.

Identify typical values of significance in O_2 diffusion from capillaries to interstitial fluids to intracellular fluids, and the consequences of varied rates of blood flow, tissue metabolism, and hemoglobin concentration.

11. By the time blood passes through a tissue capillary, its PO_2 is typically _____ mm. Hg, which is _____ (H, considerably higher; E, about equal) to the interstitial fluid PO_2 surrounding the capillary.

a. 70, H
b. 40, H
c. 23, H
d. 70, E
e. 40, E
f. 23, E

12. _____ (I, increased; D, decreased) blood flow to a tissue tends to increase the interstitial fluid PO_2 up to an upper limit of _____ mm. Hg.

a. I, 40
b. I, 70
c. I, 95
d. D, 40
e. D, 70
f. D, 95

13. Decreased hemoglobin concentration of blood tends to have the same effect upon interstitial fluid PO_2 as does either _____ tissue blood flow or _____ rates of tissue metabolism. (I, increased; D, decreased)

a. I, I
b. I, D
c. D, I
d. D, D

14. The minimal intracellular PO_2 required for full support of metabolic processes of cells of about _____ mm. Hg is _____ (E, about equal to; B, considerably below) the average intracellular PO_2.

a. 1 to 5, E
b. 15 to 20, E
c. 35 to 40, E
d. 1 to 5, B
e. 15 to 20, B
f. 35 to 40, B

OBJECTIVE 41–4.

Identify typical PCO_2 values of significance in carbon dioxide diffusion between cells and tissue capillaries, the consequences of altered blood flow and tissue metabolism on their values, and the determinants and PCO_2 values influencing pulmonary exchange of carbon dioxide.

15. The highest partial pressure of carbon dioxide is generally found in:

a. Cells
b. Interstitial fluids
c. Arterial blood
d. Venous blood

16. Partial pressure differences between arterial and venous ends of a tissue capillary network are of _____ (G, greater; E, about equal; L, lesser) absolute magnitudes for oxygen, when compared to carbon dioxide, and are of _____ (S, similar; O, opposite) sign.

a. G, S
b. E, S
c. L, S
d. G, O
e. E, O
f. L, O

17. Partial pressure values are typically represented by _____ mm. Hg for intracellular PCO_2 and _____ mm. Hg for venous blood PCO_2.

a. 46, 40
b. 40, 46
c. 46, 45
d. 45, 46
e. 60, 46
f. 46, 60

18. Tissue PCO_2 values are increased by _____ rates of blood flow and _____ rates of tissue metabolism. (I, increased; D, decreased)

a. I, I
b. I, D
c. D, I
d. D, D

19. At the entrance to a pulmonary capillary the diffusion gradient for the transfer of carbon dioxide is about _____ mm. Hg.

   a. 1   c. 11   e. 40
   b. 5   d. 2ᴼ   f. 100

20. Mean partial pressure differences across the respiratory membrane are of greater absolute magnitudes for _____ , largely attributed to differences of _____ (R, respiratory quotients; D, diffusion coefficients).

   a. CO_2, R   c. O_2, R
   b. CO_2, D   d. O_2, D

OBJECTIVE 41–5.

Identify the modes of transport and relative percentages of oxygen transport capacity of blood, its normal value, and its significance.

21. Normally, blood contains about _____ grams of hemoglobin per 100 ml. of blood, and each gram of hemoglobin combines maximally with _____ ml. of oxygen.

   a. 0.015, 0.34   d. 0.015, 1.34
   b. 0.15, 0.34   e. 0.15, 1.34
   c. 15, 0.34   f. 15, 1.34

22. Arterial blood normally contains about _____ volumes % of oxygen of which about _____ % exists as dissolved oxygen.

   a. 10, 3   c. 40, 3   e. 20, 30
   b. 20, 3   d. 10, 30   f. 40, 30

23. Decreasing the hemoglobin concentration of a blood sample in equilibrium with air would result in _____ oxygen-carrying capacity and _____ Po_2 of the blood sample. (I, increased; N, no significant change in; D, decreased)

   a. I, N   c. N, N   e. D, D
   b. D, I   d. N, I   f. D, N

24. Increased Po_2 values result in increased _____ (B, binding; R, release) of oxygen by the _____ (H, heme; G, globin) portion of the hemoglobin molecule.

   a. B, H   c. B, G
   b. R, H   d. R, G

25. 70% saturation of hemoglobin normally represents _____ ml. of O_2/100 ml. of blood, and 30% saturation of hemoglobin normally represents _____ ml. of O_2/100 ml. of blood.

   a. 10.5, 4.5   c. 14, 6   e. 40, 30
   b. 13.4, 0.8   d. 20, 0.3   f. 100, 40

OBJECTIVE 41–6.

Diagram the oxyhemoglobin dissociation curve with appropriate axes and correlative values.

26. The oxyhemoglobin dissociation curve is a _____ (L, linear; N, nonlinear) curve relating the oxygen percentage saturation of hemoglobin versus the _____ (C, oxygen-carrying capacity; Po_2, partial pressure of O_2).

   a. L, C   c. N, C
   b. L, Po_2 d. N, Po_2

27. The percentage saturation of hemoglobin of systemic arterial blood is normally about _____ %.

   a. 97   c. 75   e. 35
   b. 87   d. 50   f. 20

28. The percentage saturation of hemoglobin of an arterial blood sample would be about _____ % at a Po_2 of 700 mm. Hg and about _____ % at a Po_2 of 70 mm. Hg.

   a. 100, 90   c. 100, 40   e. 90, 40
   b. 100, 70   d. 90, 70   f. 90, 15

29. Venous blood, normally containing a Po_2 of about _____ mm. Hg, has a percentage saturation of hemoglobin of _____ .

   a. 100, 100   c. 40, 70   e. 20, 40
   b. 100, 70   d. 40, 20   f. 20, 20

30. Decreasing the hemoglobin content of blood _____ the percentage saturation of hemoglobin, and _____ the oxygen content of blood equilibrated at a given PO_2 value. (I, increases; N, does not alter; D, decreases)

   a. I, N   c. D, N   e. N, D
   b. N, N   d. I, D   f. D, D

31. Blood containing 2 ml. of O_2 per 100 ml. and normal amounts of hemoglobin has a percentage oxyhemoglobin saturation of about _____ %.

   a. 98   c. 75   e. 20
   b. 87   d. 50   f. 10

OBJECTIVE 41–7.

Identify the utilization coefficient, its contributing components, its significance, its typical value, and its variance during strenuous exercise.

32. The fraction of hemoglobin that gives up its oxygen as it passes through tissue capillaries is normally about _____ %.

   a. 4   c. 75
   b. 25   d. 97

33. The arterial-venous difference is normally about _____ ml. of oxygen for each 100 ml. of blood. The corresponding utilization coefficient normally approximates _____ %.

   a. 5, 35 to 40   d. 0.5, 35 to 40
   b. 5, 20 to 25   e. 10, 20 to 25
   c. 0.5, 5 to 10   f. 10, 45 to 50

34. Utilization coefficients approaching 100% occur with very _____ rates of tissue blood flow and with very _____ rates of tissue metabolism. (H, high; L, low)

   a. H, H   c. L, H
   b. H, L   d. L, L

35. A utilization coefficient of 50% with a normal arterial oxygen content of about _____ ml. per 100 ml. of blood would imply a venous oxygen content of about _____ ml. per 100 ml. of blood.

   a. 20, 5   c. 20, 15   e. 10, 5
   b. 20, 10   d. 10, 20   f. 10, 2

36. The conditions of a utilization coefficient of 0.20, a cardiac output of 5 liters per minute, and a normal oxygen content of arterial blood implies a net rate of oxygen diffusion across the respiratory membrane of _____ ml./ minute.

   a. 100   c. 300   e. 500
   b. 200   d. 400   f. 600

OBJECTIVE 41–8.

Identify the "oxygen buffer" functions of hemoglobin for maintenance of a relatively constant tissue fluid PO_2 under varying conditions.

37. The "oxygen buffer" function of hemoglobin relates to its:

   a. Dissociation curve shape   e. Bohr effect
   b. pK value   f. Respiratory
   c. Utilization coefficient   exchange
   d. Haldane effect   ratio

38. Oxyhemoglobin undergoes the greater volume % reduction of bound oxygen for declining PO_2 values in the range of _____ mm.Hg.

   a. 200 to 100   c. 80 to 60   e. 40 to 20
   b. 100 to 80   d. 60 to 40   f. 20 to 0

39. Oxyhemoglobin undergoes the least volume % reduction of bound oxygen for declining PO_2 values in the range of _____ mm. Hg.

   a. 120 to 100   c. 80 to 60   e. 40 to 20
   b. 100 to 80   d. 60 to 40   f. 20 to 0

40. A change in alveolar PO_2 to a value of 400 mm. Hg _____ substantially alter the arterial PO_2, _____ substantially alter the percentage saturation of arterial hemoglobin, and _____ substantially alter the PO_2 of venous blood. (W, would; N, would not)

   a. W, W, W   c. N, N, W   e. W, W, N
   b. N, W, W   d. N, N, N   f. W, N, N

41. Decreasing the arterial PO_2 from a normal value to between 50 and 60 mm. Hg results in a decrease in arterial oxygen saturation of about _____ % and a decrease in interstitial fluid PO_2 of about _____ mm. Hg.

   a. 50, 5   b. 25, 5   c. 15, 5   d. 50, 15   e. 25, 15   f. 15, 15

OBJECTIVE 41–9.
Identify the factors that shift the hemoglobin dissociation curve to the right or to the left, and their significance.

42. The oxygen dissociation curve for fetal blood lies to the _____ (R, right; L, left) of the standard curve, indicating that the affinity for oxygen is _____ (H, higher; S, less) in fetal than in maternal arterial blood.

   a. R, S   c. L, S
   b. R, H   d. L, H

43. The curve which relates percentage saturation of hemoglobin with oxygen as a function of the PO_2 has a characteristic _____ (L, linear; S, sigmoid; B, bell) shape which is shifted to the _____ (R, right; F, left) by a decrease in temperature.

   a. L, R   c. S, R   e. B, R
   b. L, F   d. S, F   f. B, F

44. During hypoxic conditions, a(n) _____ (I, increased; D, decreased) quantity of 2,3-diphosphoglycerate in blood, shifts the oxygen dissociation curve to the _____ (L, left; R, right).

   a. I, L   c. D, L
   b. I, R   d. D, R

45. A decrease in the pH of blood shifts the oxygen dissociation curve to the _____ (R, right; L, left) and _____ (I, increases; D, decreases) the oxygen affinity of hemoglobin.

   a. R, I   c. L, I
   b. R, D   d. L, D

46. Reduced carbon dioxide concentrations in erythrocytes shift the oxygen dissociation curve to the _____ (L, left; R, right) as part of the phenomenon termed the _____ (H, Haldane; B, Bohr; Cl, Cl^- shift) effect.

   a. L, H   c. L, Cl   e. R, B
   b. L, B   d. R, H   f. R, Cl

47. A patient with both a fever and carbon dioxide retention would have a _____ (H, higher; O, lower) than normal percentage saturation of arterial hemoglobin and a shift of his oxyhemoglobin dissociation curve to the _____ (R, right; L, left)

   a. H, R   c. O, R
   b. H, L   d. O, L

48. At a tissue level, the consequences of exercise tend to shift the hemoglobin dissociation curve to the _____ (R, right; L, left), thereby tending to _____ (E, elevate; U, reduce) the tissue PO_2 for a given muscle capillary blood content of oxygen.

   a. R, E   c. L, E
   b. R, U   d. L, U

OBJECTIVE 41–10.
Identify the typical rate of oxygen transport to tissues, the reserve afforded by varied cardiac output and utilization coefficients, and the consequences of varied hematocrit upon oxygen transport maxima.

49. A typical resting rate of oxygen transport to body tissues is about _____ ml. per minute, or about _____ ml. of oxygen per 100 ml. of blood flow.

   a. 150, 5   c. 450, 5   e. 250, 15
   b. 250, 5   d. 150, 15   f. 450, 15

50. The rate of oxygen transport to tissues can be increased about _____ times normal in heavy exercise, with the greater reserve afforded by increased _____ (CO, cardiac output; UC, utilization coefficients).

   a. 3, CO   c. 15, CO   e. 5, UC
   b. 5, CO   d. 3, UC   f. 15, UC

51. The maximal rate of oxygen transport to tissues _____ with elevated hematocrits and _____ with reduced hematocrits from the normal average value. (I, increases; D, decreases)

    a. I, I   c. D, I
    b. I, D   d. D, D

52. High altitude acclimatization _____ the hematocrit and _____ the "optimal hematocrit" for oxygen transport to tissues. (I, increases; N, does not significantly alter; D, decreases)

    a. I, I   c. I, D   e. D, N
    b. I, N   d. D, I   f. D, D

53. In anemia, the oxygen-carrying capacity of blood generally _____, cardiac output generally _____ , and the rate of oxygen transport to tissues generally _____ . (I, increases; C, remains constant; D, decreases)

    a. C, D, D   c. C, I, I   e. D, I, D
    b. C, C, D   d. D, C, D   f. D, I, C

OBJECTIVE 41-11.

Identify the effects of intracellular PO_2, diffusion distances from capillaries to cells, and varied blood flow upon the metabolic utilization of oxygen by cells.

54. At an intracellular PO_2 of 15 mm. Hg the rate of oxygen utilization by cells is determined by:

    a. Blood flow
    b. Blood PO_2
    c. The respiratory quotient
    d. ADP levels
    e. The utilization coefficient
    f. Blood O_2 carrying capacity

55. Oxygen utilization by the cells becomes diffusion limited when the intracellular PO_2 falls below a critical level of _____ mm. Hg required to maintain _____ (M, minimal; X, maximal) rates of intracellular metabolism.

    a. 3 to 5, M   d. 3 to 5, X
    b. 15 to 20, M   e. 15 to 20, X
    c. 35 to 40, M   f. 35 to 40, X

56. Under typical conditions, the metabolic use of oxygen by cells is _____ by increased intracellular concentrations of ADP, _____ by increased capillary PO_2 values, and _____ by increased rates of tissue blood flow. (I, increased; N, not significantly altered; D, decreased)

    a. I, I, I   c. I, I, N   e. D, I, I
    b. I, N, I   d. I, N, N   f. D, I, N

OBJECTIVE 41-12.

Identify the relative amounts of oxygen transport to tissues as oxyhemoglobin and dissolved oxygen, the consequences of elevated alveolar PO_2 upon modes of oxygen transport and tissue PO_2, and the rationale and limitations of hyperbaric oxygen utilization.

57. Typically about _____ ml. of oxygen is transported to tissues in combination with hemoglobin and about _____ ml. of oxygen is transported as dissolved oxygen per 100 ml. of blood.

    a. 15, 0.15   c. 1.5, 0.15   e. 5, 1.5
    b. 5, 0.15   d. 15, 1.5   f. 1.5, 1.5

58. Breathing oxygen under 4 atmospheres of pressure results in arterial blood containing about _____ volumes % oxygen in the dissolved state, and about _____ volumes % oxygen in the form of oxyhemoglobin oxygen.

    a. 3, 20   c. 29, 20   e. 9, 40
    b. 9, 20   d. 3, 40   f. 29, 40

59. Strenuous exercise, resulting in a _____ (R, rise; F, fall) in the utilization coefficient, tends to _____ (I, increase; D, decrease) the percentage of oxygen transported to tissues in the dissolved state.

   a. R, I   c. F, I
   b. R, D   d. F, D

60. The interstitial fluid PO_2 first undergoes a significant increase when the alveolar PO_2 is _____ (S, slightly; C, considerably) elevated as a consequence of _____ (P, oxygen poisoning; D, altered dissolved oxygen transport; O, altered oxyhemoglobin oxygen transport).

   a. S, P   c. S, O   e. C, D
   b. S, D   d. C, P   f. C, O

OBJECTIVE 41–13.

Identify the relative affinity of hemoglobin for carbon monoxide versus oxygen, lethal PCO values, and the treatment rationale for carbon monoxide poisoning.

61. Carbon monoxide, when compared with oxygen, exhibits a _____ (G, greater; L, lesser) affinity for combination with hemoglobin, and attaches to _____ (S, the same; D, a different) locus on the hemoglobin molecule.

   a. G, S   c. L, S
   b. G, D   d. L, D

62. Lethal levels of 0.1% carbon monoxide, corresponding to a partial pressure of carbon monoxide of about _____ mm. Hg, occur as a result of decreased _____ (P, PO_2; D, dissolved O_2; O, oxyhemoglobin) of arterial blood.

   a. 0.1, P   c. 21, D   e. 0.7, O
   b. 0.7, P   d. 0.1, D   f. 21, O

63. _____ (L, low; H, high) alveolar pressures of _____ are utilized in the treatment of carbon monoxide poisoning in order to displace carbon monoxide from its combination with hemoglobin.

   a. L, O_2 c. H, O_2
   b. L, CO_2 d. H, CO_2

OBJECTIVE 41–14.

Identify the chemical forms for carbon dioxide blood transport, their transport mechanisms, and their relative transport percentages and magnitudes.

64. Carbon dioxide transport across cellular membranes largely involves the _____ (A, active; P, passive) transport of _____ .

   a. A, HCO_3^- c. A, H_2CO_3 e. P, CO_2
   b. A, CO_2 d. P, HCO_3^- f. P, H_2CO_3

65. An average of _____ ml. of carbon dioxide is transported to the _____ (T, tissues from the lungs; L, lungs from the tissues) for each 100 ml. of blood flow under normal resting conditions.

   a. 0.5, T   c. 12, T   e. 4, L
   b. 4, T   d. 0.5, L   f. 12, L

66. Typical values for PCO_2 are _____ mm.Hg for systemic venous blood and _____ mm. Hg for arterial blood.

   a. 100, 40   c. 40, 20   e. 40, 100
   b. 45, 40   d. 25, 15   f. 15, 20

67. Typically, about _____ ml. of CO_2/100 ml. of blood is transported to the lungs in the form of dissolved carbon dioxide, representing about _____ % of all the carbon dioxide transported.

   a. 0.3, 7   c. 13, 7   e. 3, 37
   b. 3, 7   d. 0.3, 37   f. 13, 37

68. The activity of carbonic anhydrase, catalyzing _____ (I, carbonic acid ionization; H, CO_2 hydration), results in a greater elevation of plasma _____ concentrations.

    a. I, HCO_3^- c. H, HCO_3^-
    b. I, H_2CO_3 d. H, H_2CO_3

69. Inhibition of carbonic anhydrase, localized in _____ (P, plasma; E, erythrocytes), would result in _____ (I, increased; N, no significant change in; D, decreased) venous PCO_2.

    a. P, I   c. P, D   e. E, N
    b. P, N   d. E, I   f. E, D

70. The "chloride shift" involves an erythrocyte exchange of chloride ions largely for _____ (B, bicarbonate; H, hydrogen; P, phosphate) ions, and a greater intracellular concentration of chloride ions in _____ (V, venous; A, arterial) erythrocytes.

    a. B, V   c. P, V   e. H, A
    b. H, V   d. B, A   f. P, A

71. As blood passes through capillaries in the lungs:

    a. Blood PCO_2 increases
    b. Blood PO_2 decreases
    c. Blood acidity increases
    d. Oxyhemoglobin decreases
    e. Plasma Cl^- increases
    f. Plasma HCO_3^- increases

72. Hydrogen ions, with a higher concentration in _____ (A, arterial; V, venous) blood, are largely buffered by _____ (I, intracellular; E, extracellular) blood proteins.

    a. A, I   c. V, I
    b. A, E   d. V, E

73. Carbaminohemoglobin refers to compounds formed between the _____ (H, heme; G, globin) portion of hemoglobin and _____.

    a. H, CO   c. H, NH_3 e. G, CO_2
    b. H, CO_2 d. G, CO   f. G, NH_3

74. Typically about _____% of carbon dioxide is transported as bicarbonate ions and _____% is transported in chemical combination with hemoglobin.

    a. 50, 50   c. 23, 7   e. 7, 70
    b. 70, 23   d. 70, 7   f. 7, 30

OBJECTIVE 41–15.

Identify the carbon dioxide dissociation curve, its significance, and typical values of PCO_2 and carbon dioxide volumes % for arterial and venous blood.

75. Normally the blood PCO_2 ranges between _____ mm. Hg in arterial blood and _____ mm. Hg in venous blood.

    a. 100, 40   c. 45, 40   e. 45, 20
    b. 40, 100   d. 40, 45   f. 20, 35

76. PCO_2 values _____ influence the dissolved carbon dioxide content of blood, and _____ influence the blood content of chemically combined forms of carbon dioxide in blood. (D, do; N, do not)

    a. D, D   c. N, D
    b. D, N   d. N, N

77. Normally the total carbon dioxide content is _____ volumes % for arterial blood and _____ volumes % for venous blood.

    a. 20, 15   c. 40, 46   e. 52, 48
    b. 15, 20   d. 46, 40   f. 48, 52

OBJECTIVE 41–16.

Identify the Haldane effect, its mechanism of action, and its significance.

78. The effect of changes in PO_2 on carbon dioxide content for blood is described by:

    a. Dalton's law   d. Boyle's law
    b. The chloride shift   e. Charles' law
    c. The Haldane effect   f. The Bohr effect

79. Combination of oxygen with hemoglobin causes hemoglobin to become a _____ (W, weaker; S, stronger) acid, and _____ (I, increases; D, decreases) the carbaminohemo-globin content of blood.

    a. W, I   c. S, I
    b. W, D   d. S, D

80. The Haldane effect results in the unloading of _____ from blood _____ (C, carbamino compounds; P, plasma proteins; H, hemo-globin; BH, bicarbonate and hydrogen ions).

    a. CO_2, C   c. O_2, P   e. CO, P
    b. O_2, C   d. CO, H   f. CO_2, BH

81. The _____ (B, Bohr; H, Haldane; O, Boyle) effect approximately doubles the amount of carbon dioxide released from the blood in the lungs as a consequence of increased _____ .

    a. B, PO_2 c. H, PO_2 e. O, PO_2
    b. B, H^+ d. H, PH f. O, H^+

82. Breathing oxygen at 3 atmospheres pressure for a short interval, so that most of the tissues needs may be met without oxyhemoglobin dissociation, results in _____ (I, increased; D, decreased) tissue PCO_2 values as a conse-quence of _____ (B, the Bohr effect; H, the Haldane effect; A, altered metabolism).

    a. I, B   c. I, A   e. D, H
    b. I, H   d. D, B   f. D, A

OBJECTIVE 41–17.

Identify the respiratory exchange ratio (respiratory quotient), its dependence upon the type of foodstuff metabolized, and its typical value.

83. The volume of carbon dioxide transported from tissues to the lungs equals the volume of oxygen transported from the lungs to the tissues _____ (T, multiplied; D, divided) by the _____ (R, respiratory; E, extraction; F, diffusion) _____ (Q, quotient; A, ratio; C, capacity).

    a. D, R, Q   c. D, E, A   e. D, F, C
    b. T, R, Q   d. T, E, A   f. T, F, C

84. The percentage of nitrogen is slightly _____ (H, higher; L, lower) in alveolar air than in humidified air, at body temperature, due to the fact that oxygen consumption is _____ (G, greater; S, less) than carbon dioxide production.

    a. H, G   c. L, G
    b. H, S   d. L, S

85. Normally the net transport of oxygen from the lungs to the tissues is about _____ ml. of O_2/100 ml. of blood, whereas the net trans-port of carbon dioxide from the tissues is typically about _____ % of this value.

    a. 20, 70   c. 20, 100   e. 5, 83
    b. 20, 83   d. 5, 70   f. 5, 100

86. Utilization of fat for metabolism yields a molecular ratio of carbon dioxide produced to oxygen utilized of _____ , and a respiratory exchange ratio of _____ .

    a. 0.7, 0.7   c. 1.4, 0.7   e. 0.8, 0.8
    b. 1.0, 0.8   d. 0.7, 1.4   f. 1.4, 1.4

87. Of the following foodstuffs, the respiratory quotient is highest for _____ and lowest for _____ . (F, fat; C, carbohydrate; P, protein)

    a. F, P   c. P, C   e. C, P
    b. C, F   d. F, C   f. P, F

42

Regulation of Respiration

OBJECTIVE 42–1.

Identify, locate, and describe the general functional roles of the three major areas of the "respiratory center."

1. Basic oscillatory neural circuits which establish the basic rhythm of respiration are located within the _____ (A, apneustic; N, pneumotaxic; R, reticular formation) of the _____ (M, medulla; P, pons).

   a. A, M   c. R, M   e. N, P
   b. N, M   d. A, P   f. R, P

2. The apneustic and pneumotaxic centers are located in the reticular substance of the _____ (S, spinal cord; M, medulla, P, pons) and _____ (A, are; N, are not) required for the maintenance of rhythmic respiration.

   a. S, A   c. P, A   e. M, N
   b. M, A   d. S, N   f. P, N

3. Breathing ceases upon destruction of the:

   a. Cerebrum
   b. Cerebellum
   c. Apneustic center
   d. Medulla oblongata
   e. Carotid & aortic chemoreceptors
   f. Hypothalamus

4. The periodicity of a typical resting respiratory cycle of about _____ seconds has an inspiratory to expiratory duration ratio of about _____ .

   a. 5, 2/3   c. 5, 3/2   e. 10, 1/1
   b. 5, 1/1   d. 10, 2/3   f. 10, 3/2

5. When the apneustic center is still connected to the medullary rhythmicity area, but the pons has been transected just above the apneustic center, the inspiratory to expiratory duration ratio becomes _____ . This effect is _____ with transection of both vagi. (I, increased; D, decreased)

   a. I, I   c. D, I
   b. I, D   d. D, D

6. Enhanced activity of the apneustic area, which is _____ (G, augmented; R, reduced) by increased afferent activity of the vagus nerve, _____ (A, as well as; B, but not) the pneumotaxic area, tends to _____ (I, increase; D, decrease) the inspiratory phase duration of the respiratory cycle.

   a. G, A, I   c. G, A, D   e. R, B, I
   b. G, B, D   d. R, A, D   f. R, A, I

7. Inspiratory neurons, transmitting action potentials to contract _____ (A, abdominal muscle; D, the diaphragm), are _____ (F, facilitated; I, inhibited) by activity of expiratory neurons of the medullary rhythmicity area.

   a. A, F   c. D, F
   b. A, I   d. D, I

OBJECTIVE 42-2.
Identify the components and consequences of the Hering-Breuer reflexes.

8. _____ (S, stretch; C, chemoreceptors) for the Hering-Breuer reflexes are located within the _____ (P, pulmonary; T, systemic) _____(A, arteries; V, veins; B, bronchi and bronchioles).

   a. S, P, A   c. S, P, B   e. C, P, A
   b. S, P, V   d. S, T, A   f. C, T, V

9. Afferent neural activity of the Hering-Breuer reflexes are transmitted through the _____ (V, fifth; X, tenth; XII, twelfth) cranial nerves into _____ (L, Lissauer's Tract; S, the tractus solitarius) before influencing the respiratory centers.

   a. V, L   c. XII, L   e. X, S
   b. X, L   d. V, S   f. XII, S

10. The Hering-Breuer inflation reflex is a _____ (P, positive; N, negative) feedback mechanism which results in _____ (H, higher; L, lower) values of tidal volume.

   a. P, H   c. N, H
   b. P, L   d. N, L

11. The major effects of the Hering-Breuer reflexes include _____ values of tidal volume _____ respiratory rates, and _____ alveolar ventilation rates. (I, increased; N, no significant change in; D, decreased)

   a. I, D, N   c. D, I, N   e. D, N, D
   b. I, N, I   d. D, I, D   f. D, D, D

OBJECTIVE 42-3.
Recognize the potential alteration of "respiratory center" activity with neural activity from other regions of the nervous system.

12. "Respiratory center" activity is _____ by cutaneous pain, and is _____ by afferent activity from joint receptors. (E, enhanced; N, not significantly altered; D, diminished)

   a. E, E   c. N, D   e. D, N
   b. E, D   d. N, E   f. D, D

13. Inhibition of the arterial baroreceptors, which _____ vasomotor activity, _____ the rate of pulmonary ventilation. (I, increases; N, does not significantly alter; D, decreases)

   a. I, I,   c. I, D   e. D, N
   b. I, N   d. D, I   f. D, D

14. Panting in animals, initiated by _____ (M, medullary; H, hypothalamic; P, pontile) thermoregulatory centers, more probably is associated with increased activity of the _____ (A, apneustic; N, pneumotaxic) areas.

   a. M, A   c. P, A   e. H, N
   b. H, A   d. M, N   f. P, N

15. Stress or anxiety generally results in _____ (H, hypoventilation; E, hyperventilation). This _____ (I, increases; D, decreases) PCO_2, which reduces cerebral blood flow, causing dizziness and producing more anxiety which may lead to a vicious cycle ending in tetany and convulsions in a few predisposed patients.

   a. H, I   c. E, I
   b. H, D   d. E, D

OBJECTIVE 42–4.

Identify the principal locus of action and resultant quantitative effects upon alveolar ventilation for varying concentrations of oxygen, carbon dioxide, and hydrogen ions in the body fluids.

16. Alveolar ventilation is influenced to a greater extent by carbon dioxide and hydrogen ion receptors located in the _____, and oxygen receptors located in the _____ . (R, respiratory centers; CB, carotid bodies; CS, carotid sinus)

    a. R, R   c. R, CS   e. CB, R
    b. R, CB   d. CS, R   f. CS, CB

17. _____ carbon dioxide concentrations, _____ hydrogen ion concentrations, and _____ oxygen concentrations are all effective stimuli for increasing alveolar ventilation. (I, increased; D, decreased)

    a. D, D, D   c. D, I, I   e. I, D, D,
    b. D, D, I   d. I, I, I   f. I, I, D

18. Alveolar ventilation is more sensitive to varied _____ (CO_2, carbon dioxide; O_2, oxygen; H, hydrogen ion) concentrations of body fluids when the two other factors are _____ (K, held at constant values; V, permitted to vary).

    a. CO_2, K   c. H, K   e. O_2, V
    b. O_2, K   d. CO_2, V   f. H, V

19. The influence of varied arterial oxygen concentration is _____ (X, maximal; M, minimal) within a _____ (H, higher than normal; N, normal; L, lower than normal) PO_2 range of 80 to 140 mm. Hg.

    a. X, H   c. X, L   e. M, N
    b. X, N   d. M, H   f. M, L

20. Alveolar ventilation increases _____ (E, exponentially; L, linearly) with increasing _____ (A, arterial; V, venous) partial pressures of _____ within a range of 40 to 63 mm. Hg.

    a. E, V, O_2 c. E, V, CO_2 e. L, V, CO_2
    b. E, A, O_2 d. L, A, CO_2 f. L, A, O_2

OBJECTIVE 42–5.

Identify the significance of carbon dioxide as the major chemical factor in the regulation of alveolar ventilation, and the consequences of breathing increased concentrations of carbon dioxide upon PCO_2 values and alveolar ventilation.

21. Alveolar ventilation is stimulated by the direct action of _____ levels of carbon dioxide, as well as through an indirect mechanism involving _____ pH. (I, increased; D, decreased)

    a. I, I   c. D, I
    b. I, D   d. D, D

22. Direct effects of substantially _____ (I, increased; D, decreased) values of PCO_2 upon alveolar ventilation are _____ (A, augmented; O, opposed) by an indirect mechanism involving resultant alterations in PO_2 values.

    a. I, A   c. D, A
    b. I, O   d. D, O

23. Increased PCO_2 levels, which _____ (I, increase; D, decrease) alveolar ventilation, are _____ (A, augmented; O, opposed) by the effects of altered alveolar ventilation upon PCO_2 values.

    a. I, A   c. D, A
    b. I, O   d. D, O

24. Individuals breathing air mixtures containing 5 to 7% carbon dioxide would have _____ PCO_2 levels of body fluids and _____ alveolar ventilation rates. (H, higher than normal; N, normal; L, lower than normal)

    a. H, H   c. N, H   e. L, L
    b. H, N   d. N, L   f. N, N

25. _____ (Z, zero; M, maximum) alveolar ventilation is reached when the inspired air contains about _____ %.

   a. Z, 0, CO_2 b. Z, 9, CO_2 c. M, 100, O_2 d. M, 0, CO_2 e. M, 9, CO_2 f. M, 20, CO_2

OBJECTIVE 42–6.

Identify the influence of hydrogen ion concentration of extracellular fluids upon alveolar ventilation and the "braking effects" exerted by alterations of carbon dioxide and oxygen concentrations.

26. Increased alveolar ventilation, resulting from _____ hydrogen ion concentrations, causes _____ carbon dioxide concentrations and _____ oxygen concentrations of body fluids. (I, increased; D, decreased)

   a. I, I, I   c. I, D, I   e. D, I, D
   b. I, D, D   d. I, I, D   f. D, D, I

27. Alveolar ventilation is inhibited by the effects of _____ levels of carbon dioxide and _____ levels of oxygen. (I, increased; D, decreased)

   a. I, I   c. D, I
   b. I, D   d. D, D

28. Increased hydrogen ion concentration has a _____ effect, and carbon dioxide has a _____ effect, upon alveolar ventilation when other humoral factors are allowed to vary than when they are held constant. (G, greater; L, lesser)

   a. G, G   c. L, G
   b. G, L   d. L, L

29. A greater percentage change of alveolar ventilation from normal results as a consequence of _____ (I, increasing; D, decreasing) the pH of body fluids 0.2 unit and involves _____ (M, more; L, less) than a 50% change from normal.

   a. I, M   c. D, M
   b. I, L   d. D, L

OBJECTIVE 42–7.

Identify the appropriate conditions and underlying rationale for the regulatory influence of oxygen levels upon alveolar ventilation.

30. Threshold values for a _____ (R, rise; F, fall) in arterial PO_2 of _____ mm. Hg from normal must occur before a marked effect upon alveolar ventilation results when PCO_2 and pH values are held constant.

   a. R, 10   c. R, 70   e. F, 30
   b. R, 30   d. F, 10   f. F, 70

31. Regulatory effects of PO_2 upon alveolar ventilation are attributed to _____ (S, slight; P, pronounced) changes in arterial oxyhemoglobin saturation occurring with normal variations of arterial PO_2, the _____ (E, existence; A, absence) of a braking effect caused by the hydrogen ion control mechanism, and the _____ (E, existence; A, absence) of a braking effect caused by the PCO_2 control mechanism.

   a. S, E, A   c. S, A, E   e. P, A, E
   b. S, E, E   d. S, A, A   f. P, E, A

32. Alveolar ventilation is more responsive to variations in _____ concentrations. Changes in alveolar ventilation result in greater variations of body _____ concentrations. (O_2, oxygen; CO_2, carbon dioxide)

   a. O_2, O_2 c. CO_2, O_2
   b. O_2, CO_2 d. CO_2, CO_2

33. Pneumonia and emphysema generally result in _____ oxygen levels, _____ levels of hydrogen ions and carbon dioxide, and _____ sensitivity of the oxygen regulatory mechanism for alveolar ventilation. (I, increased; D, decreased)

   a. I, I, I   c. I, D, I   e. D, I, D
   b. I, D, D   d. D, I, I   f. D, D, D

34. Diminished oxygen levels upon chronic high altitude exposure results in an initial _____ (I, increase; D, decrease) in alveolar ventilation, followed by a subsequent _____ (I, increase; D, decrease) over a period of a week largely as a consequence of adaptation of the _____ (CO_2, carbon dioxide and hydrogen; O_2 oxygen) control mechanism of alveolar ventilation.

a. I, I, CO_2 b. I, I, O_2 c. I, D, CO_2 d. I, D, O_2 e. D, I, CO_2 f. D, D, O_2

OBJECTIVE 42–8.
Identify the composite effects of PCO_2, pH, and PO_2 upon alveolar ventilation.

35. The effect of increased concentrations of carbon dioxide to _____ (I, increase; D, decrease) alveolar ventilation is _____ (E, enhanced; R, reduced) by decreased levels of alveolar PO_2

a. I, E   c. D, E
b. I, R   d. D, R

36. Acidosis, or _____ (I, increased; D, decreased) hydrogen ion concentrations of body fluids, results in _____ (G, greater; N, no significant change in; L, lesser) threshold concentrations of carbon dioxide required for stimulation of the respiratory centers.

a. I, G   c. I, L   e. D, N
b. I, N   d. D, G   f. D, L

OBJECTIVE 42–9.
Identify the mechanisms whereby varied arterial hydrogen ion and carbon dioxide concentrations influence the respiratory centers. Identify the mechanisms involved in chronic adaptive changes in the PCO_2 regulatory mechanism of alveolar ventilation.

37. _____ (I, increased; D, decreased) _____ (H, hydrogen ion; CO_2, carbon dioxide) concentrations of fluid in immediate contact with neurons of the respiratory center are thought to be the more direct stimulus for increased alveolar ventilation.

a. I, H   c. D, H
b. I, CO_2 d. D, CO_2

38. Altered pH of arterial blood is _____ (R, more; L, less) effective in influencing alveolar ventilation via a cerebrospinal fluid pathway involving _____ (M, medullary; H, hypothalamic) chemosensitive neurons than via a direct pathway from blood to interstitial fluid of respiratory neurons.

a. R, M   c. L, M
b. R, H   d. L, H

39. Changes in _____ (H, hydrogen ion; CO_2, carbon dioxide) concentrations of blood result in a more rapid _____ (M, and more; L, but less) intense alteration in alveolar ventilation by means of the cerebrospinal fluid pathway.

a. H, M   c. CO_2, M
b. H, L   d. CO_2, L

40. Carbon dioxide diffusion from arterial blood into cerebrospinal fluid, occurring _____ (M, more; L, less) readily than hydrogen ions, is accompanied by _____ (I, increased; N, no significant change in; D, decreased) hydrogen ion concentrations of cerebrospinal fluid.

a. M, I   c. M, D   e. L, N
b. M, N   d. L, I   f. L, D

41. A return of both respiratory drive and cerebrospinal fluid _____ concentrations towards normal with chronic elevation of carbon dioxide of arterial blood is associated with increasing levels of cerebrospinal fluid _____ concentrations; (H, hydrogen ion; HCO_3, bicarbonate ion; CO_2, carbon dioxide).

a. H, CO_2 d. HCO_3, CO_2
b. H, HCO_3 e. CO_2, H
c. HCO_3, H   f. CO_2, HCO_3

OBJECTIVE 42–10.

Identify the arterial chemoreceptors, their central connections and influence, their relative oxygen, carbon dioxide, and hydrogen ion sensitivity, their mechanism of stimulation, and their significance.

42. Chemoreceptors of the aortic and carotid _____ (S, sinuses; B, bodies) convey afferent activity within the _____ (V, fifth; VII, seventh; IX, ninth; X, tenth) cranial nerves.

    a. S, V & VII
    b. S, VII & IX
    c. S, IX & X
    d. B, V & VII
    e. B, VII & IX
    f. B, IX & X

43. Arterial chemoreceptors are more sensitive to changes in _____ concentrations in the range of _____ mm. Hg.

    a. CO_2, 10 to 30
    b. CO_2, 35 to 45
    c. O_2, 30 to 60
    d. O_2, 80 to 100

44. Minute ventilation normally increases when arterial PO_2 _____ (I, increases; D, decreases). Minute changes in ventilation are normally _____ (M, more; L, less) sensitive to changes in PO_2 than PCO_2.

    a. D, M
    b. D, L
    c. I, M
    d. I, L

45. Chronic anemia and carbon monoxide poisoning, which largely affect the _____ (PO_2, partial pressure of oxygen; CC, oxygen carrying capacity) of arterial blood, are relatively _____ (E, effective; I, ineffective) conditions for stimulating the arterial chemoreceptors.

    a. PO_2, E
    b. PO_2, I
    c. CC, E
    d. CC, I

46. The PO_2 of arterial chemoreceptors is normally closest to that of most systemic _____ (T, tissues; A, arteries; V, veins) as a consequence of comparatively _____ (H, high; L, low) blood flows and _____ (H, high; L, low) extraction coefficients.

    a. T, H, H
    b. T, L, H
    c. A, H, H
    d. A, L, L
    e. A, H, L
    f. V, L, H

47. Severe hypotension results in reflexes that _____ (E, enhance; I, inhibit) respiration and _____ (E, enhance; I, inhibit) peripheral vasomotor tone as a consequence of altered _____ (PO_2, arterial PO_2 values; EC, extraction coefficients of arterial chemoreceptors).

    a. E, E, PO_2
    b. E, E, EC
    c. E, I, EC
    d. I, I, PO_2
    e. I, I, EC
    f. I, E, PO_2

48. Pharmacological agents which stimulate the arterial chemoreceptors generally _____ (I, increase; D, decrease) alveolar ventilation and _____ (I, increase; D, decrease) blood pressure via _____ (H, hypothalamic; M, medullary) centers.

    a. I, I, H
    b. I, I, M
    c. I, D, H
    d. D, I, M
    e. D, D, M
    f. D, I, H

49. Altered arterial concentrations of _____ (H, hydrogen ions and carbon dioxide; O_2, oxygen) may directly influence both respiratory center and arterial chemoreceptor acitivty with the major direct influence occurring via the _____ (A, arterial chemoreceptors; R, respiratory centers).

    a. H, A
    b. H, R
    c. O_2, A
    d. O_2, R

50. An individual would more consistently have a low arterial PO_2 and normal or below normal arterial PCO_2 and hydrogen ion concentrations with:

    a. Asphyxia
    b. Higher altitude exposure
    c. Pernicious anemia
    d. Voluntary hyperventilation
    e. Hypoventilation
    f. Carbon monoxide poisoning

OBJECTIVE 42-11.

Identify the potential mechanisms involved in the regulation of respiration during exercise, and their consequences.

51. During moderate exercise alveolar ventilation is regulated in a manner which provides for _____ arterial values of P_{CO_2}, _____ arterial values of pH, and _____ arterial values of P_{O_2}. (H, significantly higher than normal; N, near normal; L, significantly lower than normal)

a. L, N, H c. L, H, H e. H, L, L
b. H, H, L d. N, L, H f. N, N, N

52. Alveolar ventilation is generally _____ by voluntary muscle contraction and is generally _____ by passive movements of joints. (I, increased; N, not significantly altered; D, decreased)

a. I, I c. I, D e. D, N
b. I, N d. D, I f. D, D

53. Arterial P_{CO_2} values more frequently show a transient _____ at the onset of moderate exercise and a transient _____ immediately following the termination of exercise. (I, increase; D, decrease)

a. I, I c. D, I
b. I, D d. D, D

OBJECTIVE 42-12.

Recognize the potential consequences of brain edema, pressure conus, general anesthesia, and narcotics upon activity of the respiratory centers, and the rationale for the utilization of respiratory stimulants.

54. Respiratory _____ (S, stimulation; I, inhibition), resulting from barbiturate and narcotic administration, is a consequence of direct effects upon _____ (C, arterial chemoreceptors; R, respiratory centers).

a. S, C c. I, C
b. S, R d. I, R

55. Inhalation of a _____ % CO_2 mixture in air for a few minutes may be a useful test for gauging the severity of depression of the _____ (A, arterial chemoreceptors; R, respiratory centers).

a. 1 to 2, A d. 1 to 2, R
b. 7 to 10, A e. 7 to 10, R
c. 20, A f. 20, R

56. Respiratory stimulants include Coramine, which more directly stimulates the _____, picrotoxin, which more directly stimulates the _____, and Metrazol, which more directly stimulates the _____. (R, respiratory centers; C, arterial chemoreceptors)

a. R, R, R c. R, C, R e. C, R, C
b. R, R, C d. C, C, C f. C, R, R

OBJECTIVE 42–13.

Recognize the occurrence of Cheyne-Stokes and Biot's breathing, and identify their potential mechanisms and significance.

57. Cyclic waxing and waning of tidal volume amplitudes, with a periodicity of perhaps 45 seconds to 3 minutes, characterizes:

   a. Biot's breathing
   b. Apneustic breathing
   c. Dyspnea
   d. Eupneic breathing
   e. Traube-Hering waves
   f. Cheyne-Stokes breathing

58. An abnormal form of periodic breathing characterized by periodic "runs" of several normal respirations which are separated by periods of complete cessation of breathing, characterizes:

   a. Biot's breathing
   b. Apneustic breathing
   c. Dyspnea
   d. Eupneic breathing
   e. Traube-Hering waves
   f. Cheyne-Stokes breathing

59. Abnormal damping of a humoral feedback mechanism for the regulation of alveolar ventilation is more commonly associated with _____ (B, Biot's; C, Cheyne-Stokes) breathing, and _____ (I, is; N, is not) associated with central nervous system damage.

   a. B, I   c. C, I
   b. B, N   d. C, N

60. Undamped oscillation of an arterial chemoreceptor drive mechanism for maintenance of alveolar ventilation, more frequently seen as a _____ (B, Biot's; C, Cheyne-Stokes) respiratory pattern, is generally _____ (E, enhanced; R, reduced) with the inhalation of increased oxygen levels.

   a. B, E   c. C, E
   b. B, R   d. C, R

43

Respiratory Insufficiency

OBJECTIVE 43–1.

Define eupnea, tachypnea, bradypnea, hyperpnea, hypopnea, anoxia or hypoxia, anoxemia or hypoxemia, hypocapnia or acapnia, and hypercapnia.

1. _____ (B, bradypnea; H, hypopnea) signifies reduced alveolar ventilation rates, whereas _____ (X, hypoxia; C, hypocapnia) signifies reduced levels of body carbon dioxide.

   a. B, X   c. H, X
   b. B, C   d. H, C

2. _____ (C, hypercapnia; T, tachypnea) signifies a rapid rate of breathing, whereas _____ (S, asphyxia; H, hypoxemia) is utilized to denote a reduced blood level of oxygen.

   a. C, S   c. T, S
   b. C, H   d. T, H

3. Normal quiet breathing is termed:

   a. Eupnea   d. Apnea
   b. Hyperpnea   e. Dyspnea
   c. Bradypnea   f. Acapnia

OBJECTIVE 43-2.

Identify the principal methods available for evaluating respiratory abnormalities and their significance.

4. The breathing reserve is the _____ (S, sum of; D, difference between) the _____ (M, maximum breathing; L, total lung) capacity and the normal _____ (T, tidal; RM, respiratory minute) volume.

   a. S, M, T   c. S, L, RM   e. D, L, T
   b. S, L, T   d. D, L, RM   f. D, M, RM

5. _____ (U, eupnea; D, dyspnea), or "air hunger" is frequently encountered when the ventilatory reserve falls below 100%, or a value corresponding to a maximum alveolar ventilation which is _____ (E, equal to; 2, twice; 3, three times) the normal alveolar ventilation.

   a. U, E   c. U, 3   e. D, 2
   b. U, 2   d. D, E   f. D, 3

6. _____ (VS, the Van Slyke apparatus; X, oximetry; P, polarography) is an electrochemical method for the continuous monitoring of _____ (H, oxyhemoglobin saturation; O, PO_2) by means of a platinum electrode.

   a. VS, H   c. P, H   e. X, O
   b. X, H   d. VS, O   f. P, O

7. The pH of a _____ (F, phosphate; B, bicarbonate; P, proteinate buffer) solution, placed between a glass electrode measuring pH and an enclosing membrane permeable to CO_2 _____ (A, and; N, but not) bicarbonate ions, is utilized to measure blood PCO_2 values.

   a. F, A   c. P, A   e. B, N
   b. B, A   d. F, N   f. P, N

8. The Van Slyke apparatus _____ utilized to determine the O_2 content of blood, _____ utilized to determine the CO_2 content of blood, and _____ utilized to determine the percentage oxygen saturation of blood. (I, is; N, is not)

   a. I, I, I   c. I, N, I   e. N, N, I
   b. I, I, N   d. I, N, N   f. N, N, N

OBJECTIVE 43-3.

Identify three general categories of respiratory insufficiency, their major causes, and their pathophysiological consequences.

9. Primary causes of respiratory insufficiences include abnormal mechanisms of _____ for asthma, _____ for carbon monoxide poisoning, and _____ for pulmonary edema. (B, blood oxygen transport; D, pulmonary diffusing capacity; V, alveolar ventilation)

   a. B, D, V   c. D, V, B   e. V, B, D
   b. B, V, D   d. D, B, V   f. V, D, B

10. The work of breathing is increased by diseases that increase airway resistance _____ (A, as well as; N, but not) tissue resistance, and those that _____ (D, decrease; I, increase) the compliance of the lungs _____ (A, as well as; N, but not) the thorax.

    a. A, D, A   c. A, I, A   e. N, I, A
    b. A, D, N   d. N, I, N   f. N, D, N

11. Increased airflow resistance encountered during asthma and emphysema is particularly noticeable during _____ (N, inspiration; X, expiration), and generally leads to _____ (I, increased; D, decreased) functional residual capacities.

    a. N, I   c. X, I
    b. N, D   d. X, D

12. Disease states which significantly alter pulmonary diffusing capacities _____ those which alter respiratory membrane surface area, _____ those which alter respiratory membrane thickness, and _____ those which alter pulmonary ventilation-perfusion ratios. (I, include; E, exclude)

    a. I, I, I   c. I, I, E   e. E, I, I
    b. I, E, I   d. I, E, E   f. E, E, E

13. Pulmonary edema _____ (I, increases; D, decreases) the respiratory membrane _____ (T, thickness only; A, diffusion area only; AT, thickness and diffusion area).

   a. I, T c. I, AT e. D, A
   b. I, A d. D, T f. D, AT

14. The ventilation-perfusion ratio of the lungs, normally more ideal under states of _____ (R, rest; E, heavy exercise), is generally _____ (I, increased; N, not significantly altered; D, decreased) by conditions which increase the ratio of physiological dead space to anatomical dead space.

   a. R, I c. R, D e. E, N
   b. R, N d. E, I f. E, D

OBJECTIVE 43–4.

Identify the conditions and consequences of chronic emphysema, pneumonia, atelectasis, bronchial asthma, tuberculosis, and pulmonary edema as they affect pulmonary function.

15. Chronic emphysema produces _____ (X, only hypoxia; C, only hypercapnia; B, both hypoxia and hypercapnia), and more significantly increases the work load of the _____ (R, right; L, left) side of the heart.

   a. X, R c. B, R e. C, L
   b. C, R d. X, L f. B, L

16. A state of _____ (R, remaining normal; C, concurrent reductions in) blood flow through diseased areas of hypoventilated lungs during bacterial pneumonia results in a more severe hypoxemia as a consequence of altered _____ (CW, cardiac work loads; VPR, ventilation-perfusion ratios).

   a. R, CW c. C, CW
   b. R, VPR d. C, VPR

17. Atelectasis or _____ (P, an inflammatory process in; C, a collapse of) portions of one lung generally results in greater changes in _____ (R, pulmonary cardiovascular resistance; A, aortic PO_2 values).

   a. P, R c. C, R
   b. P, A d. C, A

18. _____ (A, atelectasis; B, bronchial asthma) is a more general, direct consequence of allergic reactions which have more significant effects upon pulmonary _____ (M, smooth muscle; S, surfactant; P, parenchyma).

   a. A, M c. A, P e. B, S
   b. A, S d. B, M f. B, P

19. Pulmonary edema occurs when the pulmonary capillary pressure first exceeds a value approximating _____ mm. Hg and results in a more detrimental alteration in _____ (TR, resistance to tissue expansion; PDC, pulmonary diffusing capacity) of the lungs.

   a. 14, TR c. 56, TR e. 28, PDC
   b. 28, TR d. 14, PDC f. 56, PDC

OBJECTIVE 43–5.

Identify the types of hypoxia, their potential causes, and their consequences. Identify cyanosis, its cause, and its diagnostic significance.

20. Inadequate oxygenation of the lungs due to extrinsic factors would more consistently result in a more pronounced _____ (I, increase; D, decrease) in systemic arterial _____ (C, oxygen-carrying capacity; PCO_2, partial pressure of CO_2; PO_2, partial pressure of oxygen).

    a. I, C   c. I, PO_2 e. D, PCO_2
    b. I, PCO_2 d. D, C   f. D, PO_2

21. The arterial blood during an anemic form of hypoxia contains _____ oxygen content, _____ oxygen-carrying capacity, _____ oxygen saturation of hemoglobin, and _____ arterial PO_2. (I, increased; D, decreased; U, unchanged)

    a. I, I, I, D
    b. U, U, D, D
    c. D, D, D, D
    d. U, U, U, U
    e. D, D, U, D
    f. D, D, U, U

22. Localized tissue edema is a more significant _____ (P, compensatory; U, compounding) factor for tissue _____ (X, hypoxia; C, hypercapnia).

    a. P, X
    b. P, C
    c. U, X
    d. U, C

23. Cyanosis refers to a degree of tissue _____ (X, hypoxia; C, hypercapnia; L, color) which is more critically dependent upon the quantitative value of _____ (O, oxyhemoglobin; D, reduced hemoglobin; R, the ratio of oxyhemoglobin to reduced hemoglobin).

    a. X, O   c. L, D   e. C, D
    b. C, O   d. X, R   f. L, R

24. Definite cyanosis first occurs when the arterial blood contains 5 grams % of _____ (O, oxygenated; D, deoxygenated) blood out of a normal hemoglobin concentration of _____ grams %.

    a. O, 5   c. O, 15   e. D, 10
    b. O, 10   d. D, 5   f. D, 15

25. The degree of deoxygenation of cutaneous blood flow is generally _____ (H, high; L, low) when compared to other tissues, and is _____ (I, increased; D, decreased) by states of high cutaneous blood flow.

    a. H, I
    b. H, D
    c. L, I
    d. L, D

OBJECTIVE 43–6.

Identify hypercapnia, its causes and associated conditions, and its consequences. Identify dyspnea and its predisposing conditions.

26. _____ (E, eupnea; H, hyperpnea; D, dyspnea), or "air hunger" occurs with _____ (A, only abnormal; B, both abnormal and normal) conditions of respiratory functions.

    a. E, A   c. D, A   e. H, B
    b. H, A   d. E, B   f. D, B

27. Hypercapnia, or elevated _____ levels of the body fluids, is generally associated with _____ (E, hyperoxia; O, hypoxia) during restrictive states of _____ (H, hemoglobin; C, circulatory) deficiencies.

    a. CO_2, E, H   d. pH, O, C
    b. CO_2, O, C   e. pH, E, H
    c. CO_2, O, H   f. pH, E, C

28. Carbon dioxide when compared to oxygen has a _____ blood transport capacity and a _____ pulmonary diffusion capacity. (G, greater; L, lesser)

    a. G, G
    b. G, L
    c. L, G
    d. L, L

29. Circulatory deficiencies generally result in a more severe _____, and alveolo-capillary blocks generally result in a more severe _____. (H, hypercapnia; X, hypoxia)

    a. H, H
    b. H, X
    c. X, H
    d. X, X

OBJECTIVE 43-7.

Identify the value and the rationale for the use of oxygen therapy in different types of hypoxia. Recognize the danger of hypercapnia during oxygen therapy.

30. Permitting an individual to breath pure oxygen with states of hypoxia and hypercapnia resulting from hypoventilation _____ a corrective measure for the hypoxia, and _____ a corrective measure for the hypercapnia. (I, is; N, is not)

    a. I, I
    b. I, N
    c. N, I
    d. N, N

31. Oxygen therapy is generally of _____ value in hypoxia caused by circulatory deficiencies, of _____ value in hypoxia caused by impaired pulmonary diffusion, and of _____ value in hypoxia caused by abnormalities of hemoglobin transport. (L, little; I, intermediate; M, major)

    a. L, I, M
    b. L, M, I
    c. M, L, I
    d. M, I, L
    e. I, L, M
    f. I, M, L

32. Hypoxia resulting from _____ (CO, carbon monoxide inhalation; B, alveolo-capillary blocks) tends to _____ (I, increase; D, decrease) alveolar ventilation as a consequence of activity of _____ (A, arterial; CNS, central nervous system) receptors.

    a. CO, I, A
    b. CO, D, CNS
    c. CO, D, A
    d. B, I, CNS
    e. B, I, A
    f. B, D, CNS

33. An abrupt cessation of breathing resulting from the administration of 100% O_2 to a patient with severe states of hypercapnia and hypoxia is probably a direct consequence of the effect of altered _____ levels of body fluids upon _____ (A, arterial; M, medullary) receptors.

    a. O_2, A
    b. O_2, M
    c. CO_2, A
    d. CO_2, M

OBJECTIVE 43-8.

Identify the physical principles involved in air absorption from closed body cavities and their significance.

34. Gaseous absorption of air injected into body cavities involves the absorption of _____ (O, only oxygen; OC, only oxygen and carbon dioxide; N, nitrogen, oxygen and carbon dioxide) and continues until the air cavity volume is decreased to _____ % of its original volume.

    a. O, 0
    b. OC, 0
    c. N, 0
    d. O, 75 to 80
    e. OC, 75 to 80
    f. N, 75 to 80

35. Oxygen absorption from a body cavity containing injected air occurs in the face of _____ total cavity pressure and _____ cavity water vapor pressure. (I, an increasing; C, a relatively constant; D, a decreasing)

    a. I, C
    b. C, C,
    c. D, C
    d. I, D
    e. C, D
    f. D, D

44

Aviation, High Altitude, and Space Physiology

OBJECTIVE 44-1.
Identify the effects of varying altitudes upon barometric pressure and oxygen partial pressures.

1. The barometric pressure, approximating _____ mm. Hg at sea level, _____ (I, increases; C, remains relatively constant; D, decreases) as one ascends to higher altitudes.

   a. 47, I   c. 47, D   e. 760, C
   b. 47, C   d. 760, I   f. 760, D

2. The percentage composition of oxygen in air, approximating _____ % at sea level, _____ (I, increases; C, remains relatively constant; D, decreases) as one ascends to higher altitudes.

   a. 21, I   c. 21, D   e. 79, C
   b. 21, C   d. 79, I   f. 79, D

3. A barometric pressure of 87 mm. Hg and an associated PO_2 of _____ mm. Hg would occur at an approximate elevation of _____ feet above sea level.

   a. 47, 10,000   d. 18, 10,000
   b. 47, 30,000   e. 18, 30,000
   c. 47, 50,000   f. 18, 50,000

OBJECTIVE 44-2.
Identify the variance of alveolar PO_2 and arterial oxygen saturation values with altitude, their mechanisms of variance; and the "ceiling" for breathing air in an unpressurized airplane.

4. Upon ascending from sea level to 20,000 feet, alveolar ventilation _____, alveolar PCO_2 _____ and the alveolar water vapor pressure _____. (I, increases; R, remains about constant; D, decreases)

   a. I, D, R   c. I, I, R   e. D, D, I
   b. I, D, I   d. I, R, D   f. D, R, R

5. Altered alveolar ventilation in response to increased altitudes tends to _____ (A, augment; O, oppose) the _____ (F, fall; R, rise) in alveolar PO_2.

   a. A, F   c. O, F
   b. A, R   d. O, R

6. The percentage change in alveolar PO_2 with varying altitudes is _____ than the percentage change in atmospheric PO_2 at lower altitudes, and is _____ than the percentage change in atmospheric PO_2 at higher altitudes. (L, less; G, greater)

   a. L, L   c. G, L
   b. L, G   d. G, G

7. Increased altitudes between 20,000 and 30,000 feet, while breathing air at the barometric pressures encountered, result in relative changes in the percentage gas composition of alveoli which is _____ for water vapor, _____ for carbon dioxide, and _____ for oxygen. (I, increased; C, relatively constant; D, decreased)

   a. D, D, D   c. C, I, D   e. I, I, D
   b. C, D, D   d. D, I, C   f. I, I, I

8. A "ceiling" for nonpressurized aircraft for a conscious state while breathing air is encountered at
    _____ feet, and corresponds to minimally required arterial oxyhemoglobin saturation values of
    _____ %.

a. 3,000, 70 to 80   c. 47,000, 10 to 20   e. 23,000, 10 to 20
b. 23,000, 40 to 50   d. 3,000, 40 to 50   f. 47,000, 70 to 80

OBJECTIVE 44–3.

Identify the effects of breathing pure oxygen on the alveolar PO_2 at differing altitudes,
and the "ceiling" for breathing oxygen in an unpressurized airplane.

9. Alveolar PO_2 values of about 104 mm. Hg and   10. A "ceiling" for nonpressurized aircraft for a
    arterial oxyhemoglobin values of _____ %   conscious state while breathing pure oxygen is
    would occur while breathing pure oxygen at an   encountered at _____ feet and corresponds
    altitude between _____ feet above sea level.   to arterial oxyhemoglobin saturation values of
                                                                 _____ %.
a. 97, 10,000 and 20,000
b. 72, 10,000 and 20,000   a. 3,000, 10 to 20   d. 3,000, 40 to 50
c. 43, 10,000 and 20,000   b. 23,000, 10 to 20   e. 23,000, 40 to 50
d. 97, 30,000 and 40,000   c. 47,000, 10 to 20   f. 47,000, 40 to 50
e. 72, 30,000 and 40,000
f. 43, 30,000 and 40,000

OBJECTIVE 44–4.

Identify the effects of both mild and severe hypoxia encountered with increasing
altitudes by aviators.

11. The earliest physiological effect to occur upon   13. Mental proficiency remains relatively constant
    ascending to higher altitudes is a _____ (P,   for an aviator for altitudes up to about
    progressive; S, sudden) alteration in function   _____ feet and then progressively declines,
    of the _____ (B, arterial baroreceptors; V,   resulting in _____ (V, convulsions; M, a
    vestibular system; R, rods of the retina)   coma) above a "ceiling" level.

a. P, B   c. P, R   e. S, V   a. 9,000, V   c. 23,000, V   e. 16,000, M
b. P, V   d. S, B   f. S, R   b. 16,000, V   d. 9,000, M   f. 23,000, M

12. Between altitudes of 8,000 and 16,000 feet   14. At altitudes between 25,000 and 35,000 feet,
    alveolar ventilation _____ (I, increases; D,   the duration of consciousness following a
    decreases) as a consequence of _____ (CNS,   sudden loss of oxygen or cabin depressuriza-
    central nervous system impairment; A, altered   tion is between _____ minutes and is
    arterial chemoreceptor activity).   _____ (V, inversely related to; D, in-
                                                               dependent of) altitude.
a. I, CNS   c. D, CNS
b. I, A   d. D, A   a. 0 to 5, V   d. 0 to 5, D
                                      b. 10 to 20, V   e. 10 to 20, D
                                      c. 30 to 60, V   f. 30 to 60, D

OBJECTIVE 44-5.

Identify the principal means of acclimatization to hypoxic conditions and their significance.

15. _____ pulmonary ventilation encountered immediately following an ascent to 10,000 feet results from the combined influence of states of hypoxia, _____ carbon dioxide concentrations, and _____ extracellular pH. (I, increased; D, decreased)

a. I, I, I   c. I, I, D   e. D, D, I
b. I, D, I   d. D, I, D   f. D, D, D

16. Full acclimatization to states of hypoxia at high altitudes includes increased _____ (H, hematocrit only; V, blood volume only; B, hematocrit and blood volume) resulting in _____ (L, less; M, more) than a 50% potential increase in circulating hemoglobin content.

a. H, L   c. B, L   e. V, M
b. V, L   d. H, M   f. B, M

17. _____ (I, increased; D, decreased) erythrocyte 2,3-diphosphoglycerate levels are thought to represent a _____ (T, transient; P, prolonged) adaptive mechanism to hypoxic conditions which _____ (E, enhance; M, diminish) the affinity of hemoglobin for oxygen.

a. I, T, E   c. I, P, E   e. D, P, E
b. I, T, M   d. D, P, M   f. D, T, M

18. The pulmonary diffusing capacity for oxygen is _____ (E, elevated; R, reduced) during acclimatization to hypoxic conditions as a greater consequence of altered gas diffusion characteristics within the _____ (L, lower; U, upper) portions of the lungs.

a. E, L   c. R, L
b. E, U   d. R, U

19. Acclimatization to hypoxic conditions is accompanied by an initial _____ (D, decrease; I, increase) in cardiac output and long-term alterations in blood volume which probably reflect changes in _____ (V, vasomotor tone; T, tissue vascularity).

a. D, V   c. I, V
b. D, T   d. I, T

20. Natural acclimatization of natives living at high altitudes when compared to those living at sea level involve _____ ratios of ventilatory capacity to body mass, and _____ ratios of left to right ventricular masses. (L, lower; H, higher)

a. L, L   c. H, L
b. L, H   d. H, H

21. A greater work capacity _____ (A, and; N, but not) improved tissue efficiency in utilizing oxygen exists for individuals acclimatized to _____ (S, sea level; H, high altitude) conditions.

a. A, S   c. N, S
b. A, H   d. N, H

22. Chronic mountain sickness is generally associated with _____ (L, lower; G, greater) than normal adaptive changes in hematocrit at high elevations and events which lead to _____ (C, congestive circulatory failure; P, pulmonary vasodilation reflexes).

a. L, C   c. G, C
b. L, P   d. G, P

OBJECTIVE 44-6.

Identify the mechanisms and consequences of decompression sickness in aviation.

23. Decompression sickness occurs as a result of _____ (C, collapse of alveoli; F, formation of gas bubbles) during _____ (S, slow; R, rapid) rates of _____ (A, ascent; D, descent) in an unpressurized airplane.

a. C, S, A   c. C, R, A   e. F, R, A
b. C, R, D   d. F, S, D   f. F, R, D

24. Rates of change of altitude of about 25,000 to 30,000 feet in a few _____ (S, seconds; M, minutes; H, hours) are minimally required during ascent of an aircraft to produce _____ (T, tympanic membrane rupture; D, decompression sickness).

a. S, T   c. H, T   e. M, D
b. M, T   d. S, D   f. H, D

OBJECTIVE 44-7.

Identify the major effects of centrifugal acceleratory forces on the body.

25. The force of centrifugal acceleration is _____ proportional to the object mass, _____ proportional to the radius of rotation, and _____ proportional to the velocity of the object squared. (I, inversely; D, directly)

   a. I, D, I   c. I, I, I   e. D, I, D
   b. I, I, D   d. D, D, I   f. D, D, D

26. Hydrostatic pressures of three times normal, or about _____ mm. Hg, are encountered in ankle veins of standing individuals exposed to _____ of centrifugal force.

   a. 270, +2 G   d. 2,280, +2 G
   b. 270, -2 G   e. 2,280, -3 G
   c. 270, +3 G   f. 2,280, +3 G

27. Arterial pressures at the level of the heart decrease with higher magnitudes of _____ (P, positive; N, negative) G forces of centrifugal acceleration, while cardiac output _____ (I, increases; C, remains relatively constant; D, decreases).

   a. P, I   c. P, D   e. N, C
   b. P, C   d. N, I   f. N, D

28. Threshold values for visual "blackouts" generally occur rapidly with _____ G of centrifugal force although the human body can withstand considerably higher instantaneous forces, particularly when applied along the _____ (RC, rostral-caudal; AP, anteroposterior) axis.

   a. 1 to 2, RC   d. 1 to 2, AP
   b. 4 to 6, RC   e. 4 to 6, AP
   c. 10 to 12, RC   f. 10 to 12, AP

29. A vagal bradycardia is a more likely consequence of high values of _____ (P, positive; N, negative) G centrifugal forces, _____ (L, lowered; E, elevated) left ventricular pressures, and higher arterial pressures at the level of the _____ (A, ankles; H, head) of a sitting individual.

   a. P, L, A   c. P, E, A   e. N, E, A
   b. P, L, H   d. N, E, H   f. N, L, H

30. Visual "redouts" occur as a consequence of _____ (P, positive; N, negative) G values and intense _____ (X, hypoxia; M, hyperemia).

   a. P, X   c. N, X
   b. P, M   d. N, M

31. A maneuver by which an aviator tenses and compresses his abdominal muscles is of greater consequence in minimizing _____ (D, lung and heart tissue displacement; CO, altered cardiac output; R, "redouts") during _____ (P, positive; N, negative) G values of centrifugal acceleration.

   a. D, P   c. R, P   e. CO, N
   b. CO, P   d. D, N   f. R, N

OBJECTIVE 44-8.

Identify the nature and major effects of linear acceleratory forces on the body.

32. Maximal rates of "safe" deceleration for individuals travelling at high velocities _____ (D, does; N, does not) remain relatively constant with increasing velocities, as humans can withstand _____ (S, the same; L, far less; G, far greater) rates of deceleration when they last a long time when compared to a short time.

   a. D, S   c. D, G   e. N, L
   b. D, L   d. N, S   f. N, G

33. The force of gravity results in a constant rate of acceleration of an object of _____ feet per second per second in a vacuum _____ (A, as well as; N, but not) in air.

   a. 32, A   c. 175, A   e. 63.4, N
   b. 63.4, A   d. 32, N   f. 175, N

34. "A terminal velocity" of about _____ feet per second is reached in the lower atmosphere for a parachuting aviator prior to opening his parachute when the accelerating force of gravity equals the _____ (W, aviator's weight; F, air frictional force).

    a. 32, W   c. 980, W   e. 175, F
    b. 175, W   d. 32, F   f. 980, F

35. The usual-sized parachute is designed to slow the parachutist's fall to about _____ times the "terminal velocity" of free fall in order to permit his landing without injury with the _____ (E, legs extended; B, knees bent).

    a. 0.1, E   c. 0.001, E   e. 0.01, B
    b. 0.01, E   d. 0.1, B   f. 0.001, B

OBJECTIVE 44–9.
Identify the value and limitations of perception of equilibrium and turning in blind flying.

36. Semicircular canals are effective as perceptual organs _____ (O, only at the beginning and end; T, throughout) a turn of an airplane and have thresholds for excitation of 1 to 2 _____ (S, seconds; M, minutes; D, degrees) of arc per second per second.

    a. O, S   c. O, D   e. T, M
    b. O, M   d. T, S   f. T, D

37. During airplane ascent while flying "blind" the _____ (SC, semi-circular canals; U, utricles) function to reduce the error of ascent to a maximum of about _____ degree(s).

    a. SC, 1   d. U, 1
    b. SC, 5   e. U, 5
    c. SC, 20 to 25   f. U, 20 to 25

OBJECTIVE 44–10.
Identify the problems encountered by the range of temperatures and radiation existing at high altitudes and in space.

38. The temperature of the air generally _____ (D, decreases; I, increases) at higher altitudes, reaching a value of about _____ at an altitude of about 40,000 feet.

    a. D, absolute zero   d. D, $0°$ C.
    b. D, $-250°$ C.   e. I, $250°$ C.
    c. D, $-55°$ C.   f. I, $3000°$ C.

39. The temperature of a spacecraft in space is largely determined by its _____ (E, surrounding environmental temperature; AR, absorption-radiation properties for radient energy) as the surrounding environment consists of _____ (S, a sparsity; A, an abundance) of colliding particles with very _____ (L, low; G, great) kinetic energies.

    a. E, S, L   c. E, A, L   e. AR, S, L
    b. E, A, G   d. AR, A, G   f. AR, S, G

40. The atmosphere absorbs about _____ % of the visible wavelengths from the sun before they strike the earth's surface and a _____ (G, greater; L, lesser) percentage of the ultraviolet wavelengths.

    a. 2, G   c. 56, G   e. 18, L
    b. 18, G   d. 2, L   f. 56, L

41. The Van Allen belts are significant zones of _____ , but not _____ . (C, atmospheric condensation; UV, ultraviolet light radiation; P, high energy protons; E, high energy electrons; OS, orbiting spacecraft)

    a. UV & E; C, P, & OS
    b. C, UV, P, & E; OS
    b. OS; C, UV, P, & E
    d. C & OS; UV, P, & E
    e. P & E; C, UV, & OS
    f. UV; C, P, E, & OS

42. The widest portion of the Van Allen belts extends outward on each side of a plane through the earth's _____ (P, poles; E, equator) between altitudes above the earth of _____ miles.

    a. P, 3 & 300   d. E, 3 & 300
    b. P, 30 & 2,000   e. E, 30 & 2,000
    c. P, 300 & 20,000   f. E, 300 & 20,000

43. Shielding afforded by spacecrafts reduces the radiation hazard in travelling from the earth's surface to outer space to _____ (N, minor; J, major) levels of about 10 _____ (M, milliroentgens; R, roentgens).

    a. N, M   c. J, M
    b. N, R   d. J, R

OBJECTIVE 44-11.

Identify the problems of weightlessness and maintenance of an "artificial climate" in space.

44. Gas mixtures utilized in spacecrafts, when compared to atmospheric air, contain a _____ total pressure, a _____ percentage nitrogen composition, and a _____ percentage oxygen composition. (H, higher; L, lower).

   a. H, H, H   c. H, L, H   e. L, L, H
   b. H, H, L   d. L, L, L   f. L, H, L

45. Weightlessness in space results in _____ (I, increased; D, decreased) blood volume, maximum cardiac output and work capacity, and other effects generally experienced by individuals subjected to extended periods of _____ (B, bed rest; H, heavy exercise; X, hypoxia).

   a. I, B   c. I, X   e. D, H
   b. I, H   d. D, B   f. D, X

45

Physiology of Deep Sea Diving and Other High Pressure Operations

OBJECTIVE 45-1.

Identify the physical principles of pressure and volume changes as a function of depth below the sea surface.

1. The pressures exerted by a vertical column of sea water are directly proportional to the _____ (H, height; S, square of the height) of the column, and equals 2 atmospheres at the bottom of a fluid column _____ feet high.

   a. H, 33   c. H, 138   e. S, 64
   b. H, 64   d. S, 33   f. S, 138

2. The volume of a gas is _____ proportional to its pressure and _____ proportional to its absolute temperature. (D, directly; I, inversely)

   a. D, D   c. I, D
   b. D, I   d. I, I

3. At 4 atmospheres of pressure encountered at an ocean depth of _____ feet, a sea level volume of 1 liter of gas represents an "actual volume" of gas of _____ liter(s).

   a. 100, 1   c. 256, 4   e. 256, 1
   b. 135, 0.25   d. 100, 4   f. 135, 1

OBJECTIVE 45-2.

Identify the toxic effects of elevated partial pressures of nitrogen, oxygen, carbon dioxide, and helium.

4. Toxic effects of nitrogen are thought to result from its _____ (C, chemical reactivity; P, physical properties) at high pressures to result in a central nervous system _____ (D, depressant; E, excitatory) action.

   a. C, D   c. P, D
   b. C, E   d. P, E

5. Equilibration of nitrogen gases throughout body fluids requires _____ (S, a few seconds; M, a few minutes; H, an hour or more) and first produces severe toxic symptoms while breathing compressed air at depths approximating _____ feet.

   a. S, 100   c. H, 100   e. M, 250
   b. M, 100   d. S, 250   f. H, 250

6. Exercise causes symptoms of oxygen toxicity at high pressures to occur in a diver at _____ (E, earlier; L, later) times of onset and with _____ (G, greater; S, lesser) severity.

   a. E, G   c. L, G
   b. E, S   d. L, S

7. Toxic levels of elevated oxygen throughout body tissues are associated with _____ ability of tissues to form high energy phosphate bonds, and _____ levels of cerebral blood flow. (I, increased; D, decreased)

   a. I, I   c. D, I
   b. I, D   d. D, D

8. The *oxygen safe tolerance curve* implies that individuals breathing pure oxygen at various depths are safe for maximal times of _____ (H, 1½ hours; I, indefinite intervals) at 20 feet and _____ minutes at 40 feet.

   a. H, 23   c. H, 90   e. I, 45
   b. H, 45   d. I, 23   f. I, 90

9. Toxic levels of PO_2 _____ encountered in the lungs and _____ encountered in systemic body tissues while breathing 100% O_2 at 1 atmosphere for maintained periods of time. (A, are; N, are not)

   a. A, A   c. N, A
   b. A, N   d. N, N

10. The alveolar ventilation _____ (D, decreases; I, increases) with elevated PCO_2 values above normal up to concentrations in the alveoli of about _____ times normal.

    a. I, 2   c. I, 22   e. D, 4
    b. I, 4   d. D, 2   f. D, 22

11. Toxic levels of CO_2 at great depths result in _____ (C, convulsant; A, anesthetic) effects and are less frequently encountered with diving equipment which incorporates a _____ (X, maximum; N, minimum) of dead space air.

    a. C, X   c. A, X
    b. C, N   d. A, N

12. _____ (NO, nitrous oxide; H, helium), utilized to replace nitrogen in deep dives, has a _____ (G, greater; L, lesser) density and solubility and a _____ (G, greater; L, lesser) diffusion rate in body fluids as compared to nitrogen.

    a. NO, G, G   c. NO, L, G   e. H, L, G
    b. NO, G, L   d. H, L, L   f. H, G, L

OBJECTIVE 45-3.

Identify the mechanisms, symptoms, treatment, and available means of preventing decompression sickness.

13. The volume of nitrogen dissolved in body tissues and fluids at equilibrium _____ (I, increases; D, decreases) about _____ liters for each 100-foot increment of depth below the ocean surface.

 a. I, 1   c. I, 3   e. D, 2
 b. I, 2   d. D, 1   f. D, 3

14. Body fat contains about _____ % of dissolved nitrogen under equilibrium conditions and takes _____ (L, less; M, more) time to reach saturation with nitrogen than with the body fluids.

 a. 2, L   c. 45, L   e. 15, M
 b. 15, L   d. 2, M   f. 45, M

15. Nitrogen bubbles normally begin to form in significant amounts within body fluids when the extracellular partial pressure of nitrogen is theoretically about _____ (E, equal to; T, three times; F, one-third of) the pressure on the outside of the body as a consequence of _____ (S, supersaturation; C, a compression; L, an equilibrium) phenomenon.

 a. E, L   c. F, S   e. T, S
 b. T, C   d. E, C   f. F, C

16. _____ (PC, pulmonary cardiovascular; N, neural) symptoms resulting from the _____ (C, chemical; P, physical) properties of nitrogen are of more frequent occurrence in decompression sickness.

 a. PC, C   c. N, C
 b. PC, P   d. N, P

17. Required decompression times are _____ by increased durations of exposure to a given depth and are _____ by increased depths of descent. (I, increased; N, not significantly altered; D, decreased)

 a. I, I   c. I, N   e. D, I
 b. N, I   d. I, D   f. N, N

18. Twenty minutes of breathing compressed air at a depth of 300 feet _____ (I, is; N, is not) sufficient to reach equilibrium conditions for nitrogen saturation of body fluids, and requires a minimal total decompression time of _____ hour(s).

 a. I, ½   c. I, 8½   e. N, 2½
 b. I, 2½   d. N, ½   f. N, 8½

19. The use of _____ (H, helium-oxygen; N, nitrogen-oxygen) mixtures for breathing are favored for short dives at 50-foot depths because of comparatively _____ (G, greater; L, lesser) diffusion rates for helium and _____ (G, greater; L, lesser) tendencies of nitrogen to exist in a supersaturated dissolved state.

 a. H, G, G   c. H, L, G   e. N, G, G
 b. H, L, L   d. N, L, L   f. N, G, L

20. Individuals who exhibit symptoms of decompression sickness may be effectively treated by placing them in _____ (L, low; H, high) pressure chambers for required periods of time which are considerably _____ (S, shorter; G, longer) than the usual decompression times.

 a. L, S   c. H, S
 b. L, G   d. H, G

OBJECTIVE 45-4.

Identify the physical problems encountered during diving by alterations in pulmonary air volumes and density.

21. The actual tidal volume of divers working at increasing depths _____ (I, increases; C, remains constant; D, decreases) as a primary consequence of maintaining adequate _____ exchange.

 a. I, O_2 c. D, O_2 e. C, CO_2
 b. C, O_2 d. I, CO_2 f. D, CO_2

22. A diver working with a diving helmet requires sea level volumes of about _____ cubic feet of air per minute for adequate gas exchange at sea level, and sea level volumes of about _____ cubic feet of air per minute at a depth of 300 feet below the sea level.

 a. 1.5, 1.5   c. 1.5, 15   e. 6.5, 6.5
 b. 1.5, 6   d. 6.5, 1.5   f. 6.5, 15

23. The density of air is proportional to the _____ (P, pressure; S, square of the pressure) and the respiratory air flow resistance varies _____ (I, inversely; D, directly) with the air density.

    a. P, I   c. S, I
    b. P, D   d. S, D

24. The maximum breathing capacity _____ (A, and; N, but not) the work of breathing _____ (I, increases; D, decreases) as a function of depth to a greater extent while breathing compressed _____ (R, air; O, oxygen-helium mixtures).

    a. A, I, R   c. A, D, R   e. N, D, R
    b. A, I, O   d. N, D, O   f. N, I, O

25. A "squeeze" effect during _____ (A, ascent; D, descent) of a diver generally has more severe consequences upon the _____ (NS, nervous; P, pulmonary; V, vestibular) system.

    a. A, NS   c. A, V   e. D, P
    b. A, P   d. D, NS   f. D, V

26. The "squeeze" effect limits the depth of a dive without a compressed air source to a maximum of _____ feet below the surface when the diver _____ (F, fills; E, empties) the lungs at the onset of the dive.

    a. 30, F   c. 200, F   e. 100, E
    b. 100, F   d. 30, E   f. 200, E

27. The danger of _____ (I, increasing; D, decreasing) pulmonary air volumes during rapid ascent of a diver is reduced somewhat under conditions of _____ (C, a closed; O, an open) glottis.

    a. I, C   c. D, C
    b. I, O   d. D, O

OBJECTIVE 45-5.

Identify the underlying principles of operation of the open circuit demand system and the closed circuit system for scuba diving.

28. _____ (O, open; C, closed) circuit systems utilized for scuba diving contain soda lime to _____ (G, generate O_2 only; A, absorb CO_2 only; B, both generate O_2 and absorb CO_2).

    a. O, G   c. O, B   e. C, A
    b. O, A   d. C, G   f. C, B

29. The flow through the "demand" valve of an open circuit demand system of scuba equipment occurs during _____ (I, inspiratory; E, expiratory) efforts and is more generally _____ (D, directly proportional; N, independent of; V, inversely proportional) to the diving depth.

    a. I, D   c. I, V   e. E, N
    b. I, N   d. E, D   f. E, V

30. Most closed circuit systems of scuba gear impose a depth limit of about _____ feet for a half-hour interval as a consequence of limitations of _____ tolerance.

    a. 30, CO_2 c. 200, CO_2 e. 100, O_2
    b. 100, CO_2 d. 30, O_2 f. 200, O_2

OBJECTIVE 45–6.
Identify the consequences of inhaling fresh water or salt water and the causes of death by drowning.

31. The absence of salt water in the lungs _____ (D, does; N, does not) rule out the probability of death by drowning as a consequence of _____ (L, laryngeal reflexes; O, osmotic water movement; A, atelectasis).

    a. D, L   c. D, A   e. N, O
    b. D, O   d. N, L   f. N, A

32. Inhalation of fresh water generally causes _____ (C, crenation; H, hemolysis) of red blood cells and death by _____ (A, asphyxia; F, fibrillation).

    a. C, A   c. H, A
    b. C, F   d. H, F

33. Inhalation of sea water results in death by drowning as a consequence of _____ (H, hemolysis; A, asphyxia) accompanying a marked _____ (D, hemodilution; C, hemoconcentration).

    a. H, D   c. A, D
    b. H, C   d. A, C

46

Organization of the Nervous System; Basic Functions of Synapses

OBJECTIVE 46–1.
Identify the significance of the control functions of the nervous system, the general organization and functions of sensory and motor divisions, and the roles of information processing and storage by the nervous system.

1. In contrast to the endocrine system, the nervous system is a more important regulator of _____ (R, rapidly; S, slowly) changing _____ (M, muscular; T, metabolic) activities.

    a. R, M   c. S, M
    b. R, T   d. S, T

2. The overwhelming majority of sensory information, originating in neural _____ (E, effectors; R, receptors), _____ (D, does; N, does not) reach conscious awareness.

    a. E, D   c. R, D
    b. E, N   d. R, N

3. Sensory information is transmitted within the central nervous system in rather _____ (F, diffuse combinations with; P, separate organizational patterns from) motor information and is utilized for _____ (I, only immediate; S, only stored; B, both immediate and stored) patterns of motor functions.

    a. F, I   c. F, B   e. P, S
    b. F, S   d. P, I   f. P, B

4. The motor axis of the somatic nervous system generally exhibits increased complexity of control from its _____ (H, higher; L, lower) levels of organization, and is the primary controller for skeletal muscle _____ (A, as well as; N, but not) smooth muscle and glands.

    a. H, A   c. L, A
    b. H, N   d. L, N

5. Information processing involves _____ (U, unidirectional; B, bidirectional) synaptic transmission of _____ (F, only facilatory; I, only inhibitory; FI, both facilatory and inhibitory) influences upon postsynaptic neurons.

a. U, F   c. U, FI   e. B, I
b. U, I   d. B, F   f. B, FI

6. Neurons generally receive synaptic endings from _____ (S, single; M, multiple) neuron sources which function to _____ (A, only amplify; R, only reject; AR, amplify as well as reject) weak signals.

a. S, A   c. S, AR   e. M, R
b. S, R   d. M, A   f. M, AR

7. The function of memory is _____ (L, more highly localized in specific regions of; U, rather uniformly distributed throughout) the central nervous system, and involves processes of information storage _____ (A, as well as; N, but not) modifications of information processing.

a. L, A   c. U, A
b. L, N   d. U, N

OBJECTIVE 46–2.
Identify three major levels of nervous system function and their general functional characteristics.

8. A segmented form of neural organization is more developed in the _____ , while automatic reflexes are predominant in the _____ . (C, cerebrum; B, lower brain stem; S, spinal cord)

a. C, B   c. S, B   e. B, S
b. C, S   d. S, S   f. B, B

9. A simpler pattern of organization of the _____ (S, stretch; F, flexion withdrawal) reflex is comprised of _____ (MS, muscle spindle; P, pain) receptors, an afferent limb, and _____ (M, monosynaptic; PS, polysynaptic) excitation of an efferent or effector limb.

a. S, MS, M   c. S, P, M   e. F, P, M
b. S, MS, PS   d. F, P, PS   f. F, MS, PS

10. Basic spinal cord patterns of organization in animals _____ exist for the regulation of respiration, and _____ exist for rhythmical movements of the extremities in locomotion. (D, do; N, do not)

a. D, D   c. N, D
b. D, N   d. N, N

11. The majority of neurons within the central nervous system are located within the _____ (S, spinal cord; B, brain stem; C, cerebral cortex), and are largely associated with _____ (V, visceral function; I, information storage).

a. S, V   c. C, V   e. B, I
b. B, V   d. S, I   f. C, I

12. Wakefulness and sleep are thought to result from the degree of a _____ (S, specific thalamic; D, diffuse mesencephalic) activation of the _____ (B, cerebellar; A, cerebral) cortex.

a. S, B   c. D, B
b. S, A   d. D, A

13. The capacity to distinguish the shape, character, and cutaneous localization of objects is a more specific function of the _____ (PL, prefrontal lobes; S, somesthetic cortex; B, lower brain stem), while the seat of consciousness is a _____ (H, highly; P, poorly) localized function of the cerebral cortex.

a. PL, H   c. B, H   e. S, P
b. S, H   d. PL, P   f. B, P

14. The evolutionary process of _____ (T, telencephalization; R, rhombencephalization; M, mesencephalization) primarily involved _____ (I, only increased function of higher; D, decreased function of lower centers as well as increased function of higher) centers.

a. T, I   c. M, I   e. R, D
b. R, I   d. T, D   f. M, D

OBJECTIVE 46–3.

Identify the synapse, the physiologic anatomy of neurons and synapses, and the general effects that presynaptic terminals may have upon postsynaptic structures.

15. _____ (U, unipolar; B, bipolar; MP, multipolar) motoneurons are comprised of cell bodies, _____ (S, single; M, multiple) dendrites, and _____ (S, single; M, multiple) axons.

a. U, M, S   c. B, M, M   e. MP, S, M
b. B, S, S   d. U, S, S   f. MP, M, S

16. The larger percentage of "synaptic knobs," representing _____ (E, presynaptic; O, postsynaptic) terminals, are located upon the _____ (D, dendrites; S, soma) of motoneurons.

a. E, D   c. O, D
b. E, S   d. O, S

17. Nerve impulse transmission between differing neurons in mammals occurs by means of _____ (E, electrically; C, chemically) mediated synapses which, unlike neuromuscular junctions, are areas of _____ (U, unidirectional transmission; A, ATP utilization; I, integration).

a. E, U   c. E, I   e. C, A
b. E, A   d. C, U   f. C, I

18. Single presynaptic nerve terminals possess vesicles containing _____ (E, either an excitatory or an inhibitory; B, both excitatory and inhibitory) chemical mediator(s), and are separated from the neuronal soma by synaptic clefts of about _____ angstroms.

a. E, 75 to 100   d. B, 75 to 100
b. E, 200 to 300   e. B, 200 to 300
c. E, 1000   f. B, 1000

19. Available stores of chemical mediator are sufficient for relatively _____ (B, brief; P, prolonged) periods of high synaptic activity, and are synthesized utilizing mitochondrial ATP originating from the nerve _____ (S, soma; T, terminals).

a. B, S   c. P, S
b. B, T   d. P, T

OBJECTIVE 46–4.

Identify the mechanisms of release of chemical mediators at synapses, their means of synthesis, and their mechanisms of action upon postsynaptic neurons.

20. The number of vesicles released with each action potential entering _____ (E, presynaptic; O, postsynaptic) endings is _____ (I, increased; D, decreased) by reduced extracellular concentrations of calcium ions, and is _____ (I, increased; D, decreased) by reduced extracellular concentrations of sodium ions.

a. E, I, I   c. E, D, I   e. O, I, D
b. E, I, D   d. E, D, D   f. O, D, I

21. The release of chemical mediator at synapses during action potential transmission is _____ by prior partial depolarizations of synaptic knobs or _____ extracellular magnesium ion concentrations. (E, enhanced; R, reduced).

a. E, E   c. R, E
b. E, R   d. R, R

22. The coupling agent between the arrival of the action potential and the release of chemical mediator more probably involves synaptic nerve terminal _____ (E, efflux; I, influx) of _____ ions.

a. E, K   c. E, Mg   e. I, Ca
b. E, Ca   d. I, K   f. I, Mg

23. Compared to neuromuscular end-plates, synaptic knobs release _____ (F, considerably fewer; E, about equal; M, considerably more) vesicles of chemical mediator per action potential, and generally have _____ (G, greater; L, lesser) safety factors for action potential transmission.

a. F, G   c. M, G   e. E, L
b. E, G   d. F, L   f. M, L

24. Acetylcholine released at cholinergic synapses is largely resynthesized in the neuron _____ (S, soma; T, nerve terminal) by the enzymatic action of _____ (E, acetylcholinesterase; A, choline acetyltransferase).

a. S, E   c. T, E
b. S, A   d. T, A

25. Acetate is generated at cholinergic synapses by the action of _____ (E, acetylcholinesterase; A, choline acetyltransferase), localized mainly within the _____ (ET, presynaptic terminal; OT, postsynaptic terminal; C, synaptic cleft).

a. E, ET   c. E, C   e. A, OT
b. E, OT   d. A, ET   f. A, C

26. Intracellular contents of synaptic nerve terminals are thought to be _____ (T, returned to; P, replenished by) transport mechanisms moving axoplasm along the axon at velocities of about _____ per day.

a. T, 1 mm.   c. T, 10 cm.   e. P, 1 cm.
b. T, 1 cm.   d. P, 1 mm.   f. P, 10 cm.

27. Chemical mediators which selectively increase the _____ (E, presynaptic; O, postsynaptic) membrane permeability to potassium and chloride ions exhibit _____ (X, excitatory; I, inhibitory) synaptic effects.

a. E, X   c. O, X
b. E, I   d. O, I

28. A specific receptor protein at the _____ (E, presynaptic; O, postsynaptic) membrane is thought to react with _____ (S, either a specific excitatory or a specific inhibitory; B, both a specific excitatory as well as an inhibitory) chemical mediator.

a. E, S   c. O, S
b. E, B   d. O, B

OBJECTIVE 46–5.

Identify the synaptic chemical mediators of the peripheral nervous system and the variety of suspected chemical mediators for the central nervous system. Recognize the principle that a single neuron secretes only one type of chemical mediator at its nerve terminals.

29. The _____ (I, inhibitory; X, excitatory) transmitter at the synapses of autonomic ganglia is _____ (A, acetylcholine; N, norepinephrine; E, epinephrine).

a. I, A   c. I, E   e. X, N
b. I, N   d. X, A   f. X, E

30. _____ (A, acetylcholine; N, norepinephrine; GABA, gamma aminobutyric acid) is a known chemical mediator at motoneuron end-plates, and one of the known _____ (I, inhibitory; E, excitatory) chemical mediators of the central nervous system.

a. A, I   c. GABA, I   e. N, E
b. N, I   d. A, E   f. GABA, E

31. Gamma aminobutyric acid and glycine probably function as _____ (P, peripheral; C, central) nervous system _____ (I, inhibitory; E, excitatory) chemical mediators.

a. P, I   c. C, I
b. P, E   d. C, E

32. L-Aspartate and L-glutamate are more probable _____ (I, inhibitory; E, excitatory) transmitter agents, while _____ (G, prostaglandin; T, peptide) substances termed *P-substances* may have longer durations of excitatory or inhibitory effects upon the central nervous system.

a. I, G   c. E, G
b. I, T   d. E, T

33. A single neuron is considered to be _____ (O, only excitatory or only inhibitory; B, both excitatory and inhibitory), and to release ____ (S, the same; D, potentially differing) chemical mediator(s) at its multiple nerve terminals.

a. O, S   c. B, S
b. O, D   d. B, D

OBJECTIVE 46–6.

Identify the resting membrane potential of the neuronal soma, the significance of its variance, and the mechanisms of its origin.

34. The membrane potentials of motoneurons are typically closest to the equilibrium potential of _____ (Na, sodium; K, potassium; Cl, chloride) ion distribution, or _____ millivolts.

    a. Na, -70   c. Cl, -70   e. K, -90
    b. K, -70   d. Na, -90   f. Cl, -90

35. Active neuronal transport of sodium ions occurs against _____ (E, only an electrical; C, only a chemical; B, both electrical and chemical) concentration gradient(s), and generally exhibits a _____ (G, greater; L, lesser) backward leakage than the potassium pump mechanism.

    a. E, G   c. B, G   e. C, L
    b. C, G   d. E, L   f. B, L

36. The basis for the _____ (H, high; L, low) intracellular to extracellular chloride concentration ratio of neurons is a _____ (H, high; L, low) chloride membrane permeability and the _____ (R, resting membrane potential; K, potassium active transport).

    a. H, H, R   c. H, L, R   e. L, H, R
    b. H, L, K   d. L, L, K   f. L, H, K

37. The resting membrane potential of a motoneuron is generally poised at a value which permits enhanced excitability through membrane _____ (D, depolarization; H, hyperpolarization) mechanisms _____ (A, as well as; N, but not) reduced states of excitability.

    a. D, A   c. H, A
    b. D, N   d. H, N

38. The electrical resistance of the intracellular fluid of the neuron is _____ (C, comparable to; S, substantially lower than) that of the cell membrane so that transmembrane potential changes occurring at one segment of the soma are _____ (L, highly localized; W, widespread throughout the soma).

    a. C, L   c. S, L
    b. C, W   d. S, W

OBJECTIVE 46–7.

Distinguish between an excitatory postsynaptic potential (EPSP) and an inhibitory postsynaptic potential (IPSP). Identify the underlying mechanisms for postsynaptic excitation and postsynaptic inhibition of neurons.

39. Excitatory transmitters at neuronal synapses are thought to function by _____ (I, increasing; D, decreasing) the postsynaptic membrane permeability to _____ (Na, only sodium; K, only potassium; A, all) ions.

    a. I, Na   c. I, A   e. D, K
    b. I, K   d. D, Na   f. D, A

40. The action of excitatory chemical mediators is a greater postsynaptic transmembrane flux of _____ ions as a predominant consequence of its greater _____ (P, membrane permeability; EC, electrochemical concentration gradient).

    a. Na, P   c. Cl, P   e. K, EC
    b. K, P   d. Na, EC   f. Cl, EC

41. The _____ (I, inhibitory; E, excitatory) postsynaptic potential is a transient depolarization of the _____ (ES, presynaptic; OS, postsynaptic) membrane of synapses.

    a. I, ES   c. E, ES
    b. I, OS   d. E, OS

42. More excitable portions of the neuron membrane occur in the region of the _____ (D, dendrites; S, soma; A, initial axon segment), and elicit action potentials when the intracellular potential is reduced from typical resting membrane potentials by threshold amounts of _____ millivolts.

    a. D, 11   c. A, 11   e. S, 26
    b. S, 11   d. D, 26   f. A, 26

43. Single action potentials propagated along axons _____ (A, as well as; N, but not) the soma of neurons generally result from _____ (G, only single unit; M, only summated; E, either single unit or summated) excitatory postsynaptic potentials.

   a. A, G   c. A, E   e. N, M
   b. A, M   d. N, G   f. N, E

44. _____ (I, inhibitory; E, excitatory) postsynaptic potentials involve a selective increase in postsynaptic membrane permeability to _____ (Na, only sodium; Cl, only chloride; B, both chloride and potassium) ions.

   a. I, Na   c. I, B   e. E, Cl
   b. I, Cl   d. E, Na   f. E, B

45. The intracellular to extracellular concentration ratio of _____ (Cl, chloride; K, potassium) ions is significantly _____ (G, greater; L, less) than that predicted by the Nernst equation for an electrochemical equilibrium at typical resting membrane potentials of neurons.

   a. K, G   c. Cl, G
   b. K, L   d. Cl, L

46. _____ (I, increased; D, decreased) intracellular negativity of neurons resulting from summated inhibitory postsynaptic potentials is limited by equilibrium potentials for _____ (Na, sodium; K, potassium) ions, and result in _____ (H, hyperpolarized; D, depolarized) states.

   a. I, Na, H   c. I, K, H   e. D, K, H
   b. I, Na, D   d. D, K, D   f. D, Na, D

47. "Clamping" of the _____ (P, ionic permeability; M, resting membrane potential) refers to a physiologic mechanism of neuronal _____ (E, excitation; I, inhibition) by the action of _____ (E, excitatory; I, inhibitory) chemical mediators at synapses.

   a. P, E, I   c. P, I, I   e. M, I, I
   b. P, E, E   d. M, I, E   f. M, E, E

48. Neurons with resting membrane potentials which are in states of electrochemical equilibrium with potassium and chloride ions exhibit minimal _____ (X, excitatory; B, inhibitory) postsynaptic potentials, and possess potential mechanisms of _____ (E, excitation only; I, inhibition only; EI, excitation as well as inhibition).

   a. X, E   c. X, EI   e. B, I
   b. X, I   d. B, E   f. B, EI

OBJECTIVE 46-8.

Identify presynaptic inhibition, its probable mechanisms, and its functional significance.

49. Presynaptic inhibition is thought to involve _____ (P, an ephaptic; C, a chemically mediated) _____ (H, inhibitory; F, facilitatory) influence upon mechanisms of postsynaptic _____ (I, inhibition; E, excitation).

   a. P, H, I   c. P, F, I   e. C, H, I
   b. P, F, E   d. C, F, E   f. C, H, E

50. _____ intracellular negativity of resting membrane potentials and _____ action potential amplitudes at presynaptic terminals tend to decrease the amount of chemical mediator released per action potential at neural synapses. (I, increased; D, decreased)

   a. I, I   c. D, I
   b. I, D   d. D, D

51. The mechanism of action of presynaptic inhibition involves the modification of _____ (I, inhibitory; E, excitatory) postsynaptic potentials by means of _____ (D, depolarization; H, hyperpolarization) of their presynaptic terminals.

   a. I, D   c. E, D
   b. I, H   d. E, H

52. Picrotoxin alters central nervous system activity by a more direct _____ (F, facilitory; B, blocking) action at loci of _____ (OI, postsynaptic inhibition; EI, presynaptic inhibition; OE, postsynaptic excitation).

   a. F, OI   c. F, OE   e. B, EI
   b. F, EI   d. B, OI   f. B, OE

53. Presynaptic inhibition is a more prominent feature of _____ (I, initial; T, terminal) synapses of central nervous system pathways which function to _____ (E, enlarge; D, diminish) the spatial boundaries of transmitted signals.

   a. I, E   b. I, D   c. T, E   d. T, D

OBJECTIVE 46–9.

Identify the significant functional characteristics of postsynaptic potentials.

54. Excitatory postsynaptic potentials are typically _____ milliseconds in duration and possess longer _____ (R, "rise"; D, "decay") times.

   a. 4, R   d. 4, D
   b. 15, R   e. 15, D
   c. 200 to 300, R f. 200 to 300, D

55. Inhibitory postsynaptic potentials are _____ (G, graded; A; all-or-none) potentials whose percentage change in amplitude as a function of time more closely resembles _____ (S, spike; EI, presynaptic inhibitory; OE, postsynaptic excitatory) potentials.

   a. G, S   c. G, OE   e. A, EI
   b. G, EI   d. A, S   f. A, OE

56. The amplitude of the excitatory postsynaptic potential resulting from a single presynaptic action potential is _____ (G, considerably greater than; E, about equal to; L, considerably less than) 10 millivolts, and _____ (I, is; N, is not) generally sufficient to elicit postsynaptic action potentials in "resting" neurons.

   a. G, I   c. L, I   e. E, N
   b. E, I   d. G, N   f. L, N

57. _____ (T, temporal; S, spatial) summation is a mechanism of summation of simultaneous postsynaptic potentials created by excitation of multiple synaptic knobs on _____ (C, only contiguous; W, widely spaced as well as contiguous) regions of the neuronal soma.

   a. T, C   c. S, C
   b. T, W   d. S, W

58. _____ (T, temporal; S, spatial) summation of _____ (E, only presynaptic action; O, only postsynaptic; EO, both presynaptic action potentials and postsynaptic) potentials results as a consequence of rapidly repetitive activity at a synapse.

   a. T, E   c. T, EO   e. S, O
   b. T, O   d. S, E   f. S, EO

59. States of neuronal _____ (I, inhibition; F, facilitation; E, excitation) which _____ (A, are; N, are not) sufficient to generate action potentials, reflect intracellular potentials only slightly more negative than those at threshold levels.

   a. I, A   c. E, A   e. F, N
   b. F, A   d. I, N   f. E, N

OBJECTIVE 46–10.

Identify the significant properties and specialized functions of dendrites in the process of neuronal excitation.

60. Dendrites of anterior motoneurons extend outward from their soma for distances of _____ mm. and provide the postsynaptic surface for about _____ % of the neuronal synaptic knobs.

   a. 0.5 to 1, 10 to 20   d. 5 to 10, 10 to 20
   b. 0.5 to 1, 40 to 60   e. 5 to 10, 40 to 60
   c. 0.5 to 1, 80 to 90   f. 5 to 10, 80 to 90

61. Dendrites are _____ (M, more; S, less) prone to exhibit *decremental conduction* than their neuronal soma and exhibit comparatively _____ (H, high; L, low) safety factors for action potential propagation.

   a. M, H   c. S, H
   b. M, L   d. S, L

62. Synapses located at the _____ (P, peripheral; S, somal) ends of dendrites exert a
comparatively greater influence upon the state
of relative excitability of neurons, and are
excitatory _____ (A, as well as; N, but not)
inhibitory.

   a. P, A   c. S, A
   b. P, N   d. S, N

63. Inhibitory knobs concentrated at the _____
(P, peripheral; S, somal) ends of dendrites are
_____ (M, more; L, less) likely to be
disturbed by action potentials originating in
the postsynaptic neuron than more peripheral
dendritic knobs.

   a. P, M   c. S, M
   b. P, L   d. S, L

OBJECTIVE 46–11.

Identify the relationships of firing frequency of different types of neurons to their
central excitatory states, and altered neuronal somal states produced during and following
action potential development.

64. A "central excitatory state" of 5 millivolts
refers to _____ (R, resting; A, action
potential) conditions of a neuron during which
the membrane potential is 5 millivolts
_____ (L, less; M, more) negative than the
_____ (T, threshold level for excitation;
RM, resting membrane potential).

   a. R, L, RM   c. R, M, RM   e. A, L, RM
   b. R, L, T   d. A, M, T   f. A, L, T

65. Neuronal discharge occurs for _____ (B,
only brief; P, prolonged) intervals in which
they are maintained in conditions whereby the
central excitatory state exceeds the _____
(I, central inhibitory level; RM, resting membrane potential; T, threshold for excitation).

   a. B, I   c. B, T   e. P, RM
   b. B, RM   d. P, I   f. P, T

66. Increased activity of excitatory synapses to
"firing" neurons generally results in _____
(I, increased; N, no significant change in; D,
decreased) intervals between successive action
potentials as a consequence of the relationship
of action potentials to _____ (R, absolute
refractory periods; E, central excitatory states)
required for re-excitation.

   a. I, R   c. D, R   e. N, E
   b. N, R   d. I, E   f. D, E

67. The positive after-potential or period of neuronal _____ (D, after-depolarization; H, after-
hyperpolarization) corresponds to an interval
during which the central excitatory state
required for re-excitation is _____ (G,
greater; L, less) than normal.

   a. D, G   c. H, G
   b. D, L   d. H, L

68. Inhibitory "clamping" of the neuronal soma
results in a change in the central excitatory
state required for re-excitation which is generally _____ (G, greater than; E, equal to; L,
less than) the increment of membrane potential change during the _____ (P, positive; N,
negative) after-potential.

   a. G, P   c. L, P   e. E, N
   b. E, P   d. G, N   f. L, N

69. Comparatively different morphological types
of neurons exhibit _____ (H, relatively
homogeneous; C, considerably different)
curves of _____ (I, increasing; D, decreasing) rates of change of discharge frequency as a function of increasing central
excitatory states.

   a. H, I   c. C, I
   b. H, D   d. C, D

OBJECTIVE 46–12.

Identify the synaptic characteristics of directional transmission, transmission delay, fatigue, and post-tetanic facilitation at synapses. Identify the effects of acidosis, alkalosis, hypoxia, and drug susceptibility upon synaptic transmission.

70. _____ (U, unilateral; B, bilateral) transmission at synapses is associated with a synaptic delay of about _____ milliseconds.

   a. U, 0.5   c. U, 5   e. B, 2.5
   b. U, 2.5   d. B, 0.5   f. B, 5

71. Fatigue at synapses _____ (D, does; N, does not) generally result from exhaustion of supplies of synaptic chemical mediator, because excitatory knobs store sufficient amounts of mediator for about _____ normal synaptic transmissions.

   a. D, 10^4 c. D, 10^8 e. N, 10^6
   b. D, 10^6 d. N, 10^4 f. N, 10^8

72. Synapses are generally more responsive to a presynaptic action potential following _____ (L, a long period of inactivity; S, a short period of inactivity preceded by an interval of repetitive stimulation) as a partial consequence of _____ (T, "treppe"; M, enhanced mobilization of chemical mediator; D, summation of periods of after-depolarization).

   a. L, T   c. L, D   e. S, M
   b. L, M   d. S, T   f. S, D

73. Hyperventilation _____ (E, enhances; P, depresses) neuronal activity by _____ (I, increasing; D, decreasing) extracellular fluid pH.

   a. E, I   c. P, I
   b. E, D   d. P, D

74. A threshold _____ (F, fall; R, rise) in pH of about _____ pH unit(s) generally results in a comatose state.

   a. F, 0.2   c. F, 1.6   e. R, 0.4
   b. F, 0.4   d. R, 0.2   f. R, 1.6

75. Strychnine _____ (I, increases; D, decreases) neuronal excitability by a more direct action at _____ (H, inhibitory; E, excitatory) synapses.

   a. I, H   c. D, H
   b. I, E   d. D, E

76. Caffeine, theophylline, theobromine, _____ (A, and; N, but not) hypoxia generally act to _____ (E, enhance; D, depress) synaptic transmission.

   a. A, E   c. N, E
   b. A, D   d. N, D

47

Transmission and Processing of Information in the Central Nervous System

OBJECTIVE 47–1.

Identify the roles of impulses and signals in nervous system information transmission and processing, the mechanisms of transmission of signal strength and spatial orientation within fiber tracts, and the mechanisms of the terminal detection of signals conveyed by fiber tracts.

1. The primary function of the nervous system is the transmission _____ (A, as well as; N, but not) processing of information contained within patterns of activity called _____ (S, signals; N, neuronal pools).

a. A, S   c. N, S
b. A, N   d. N, N

2. The transduction of sensory information to neural activity occurs largely within the _____ (C, central; P, peripheral) nervous system and is transmitted in the form of _____ (T, only temporal; S, only spatial; B, both temporal and spatial) patterns.

a. C, T   c. C, B   e. P, S
b. C, S   d. P, T   f. P, B

3. Signal strength is transmitted by means of _____ (A, amplitude; F, frequency) modulation of nerve impulses _____ (W, as well as; N, but not) multiple fiber summation.

a. A, W   c. F, W
b. A, N   d. F, N

4. A cutaneous receptive field is defined by properties of the _____ (S, applied stimulus; N, sensory innervation) and exhibits greater responsiveness at its _____ (P, perimeter; C, center).

a. S, P   c. N, P
b. S, C   d. N, C

5. Recruitment of _____ (S, stimuli; F, sensory fibers) is a phenomenon of _____ (P, spatial; T, temporal) summation occurring in association with _____ (I, increasing; D, decreasing) signal strengths.

a. S, T, D   c. S, T, I   e. F, T, I
b. S, P, D   d. F, P, I   f. F, P, D

6. Threshold stimuli, as opposed to higher levels of sensory stimulus intensities, are generally transmitted by _____ firing frequencies of _____ numbers of peripheral nerve axons. (L, lower; C, relatively constant; H, higher)

a. L, C   c. L, H   e. H, C
b. L, L   d. C, L   f. H, H

7. Summating and averaging functions of information signals by synapses result in greater granularities of detected signals with _____ (H, high; L, low) action potential frequencies and _____ (S, synchronous; A, asynchronous) activity in multiple presynaptic fibers.

a. H, S   c. L, S
b. H, A   d. L, A

8. The fidelity of transmission of the temporal pattern of signals is greatest for _____ (S, small; L, large) diameter _____ (M, myelinated; U, unmyelinated) axons with comparatively _____ (H, high; W, low) firing frequencies.

a. S, M, H   c. S, U, H   e. L, M, H
b. S, U, W   d. L, U, W   f. L, M, W

9. The detection of cutaneous loci receiving sensory stimuli by higher centers is dependent upon the _____ (O, spatial orientation; T, temporally coded patterns) of fiber tract signals transmitted by _____ (F, functionally separate fibers from; S, the same fibers as) those which are utilized for the transmission of signal strength.

a. O, F   c. T, F
b. O, S   d. T, S

OBJECTIVE 47–2.

Identify the functional roles of convergence, divergence, temporal and spatial summation, and facilitation and inhibition by accessory sources upon signal transmission and processing in neuronal pools.

10. The organization of neurons within neuronal pools is generally such that each incoming fiber terminates in a greater density of synaptic connections in a _____ (C, central; P, peripheral) region of its stimulatory field, which corresponds more closely with its _____ (S, subliminal; L, liminal) zone.

a. C, S   c. P, S
b. C, L   d. P, L

11. Facilitatory signals to a neuronal pool from an accessory source generally _____ the size of the discharge zone and _____ the threshold for excitation by its primary source. (I, increases; N, does not significantly alter; D, decreases)

a. I, D   c. D, I   e. N, I
b. I, N   d. D, N   f. N, D

12. Excitation of a neuronal pool is a more likely consequence of _____ (C, convergence; D, divergence) of subthreshold stimuli and mechanisms of _____ (O, occlusion; A, accommodation; S, summation).

    a. C, O   c. C, S   e. D, A
    b. C, A   d. D, O   f. D, S

13. Terminal branches of a single excitatory fiber input to a neuronal pool more frequently evoke a _____ (E, presynaptic; O, postsynaptic) form of inhibition in addition to excitation through the release of inhibitory chemical mediators from _____ (S, the same; I, intermediate) neuron(s).

    a. E, S   c. O, S
    b. E, I   d. O, I

14. Excitation of a large number of anterior motoneurons as a consequence of stimulation of a single pyramidal cell in the motor cortex is a better illustration of _____ (A, an amplifying; M, a multiple tract) type of _____ (C, convergence; D, divergence).

    a. A, C   c. M, C
    b. A, D   d. M, D

OBJECTIVE 47–3.

Identify the general characteristics for transmission of spatial patterns through successive neuronal pools.

15. Transmission of cutaneous sensory information to the somesthetic cortex minimally involves a _____ -neuron chain, with each neuron innervating a more probable average number of postsynaptic neurons of _____ (L, 10 or less; M, 100 or more).

    a. 2, L   c. 10, L   e. 3, M
    b. 3, L   d. 2, M   f. 10, M

16. The degree of facilitation of neuronal pools of major central nervous system sensory pathways is thought to be regulated to a greater extent by _____ (L, lateral inhibitory; C, centrifugal control) mechanisms, and is associated with _____ (G, greater; S, lesser) divergence of the transmitted signals as a result of increased levels of neuronal facilitation.

    a. L, G   c. C, G
    b. L, S   d. C, S

17. Lateral inhibitory circuits of neuronal pools of the central nervous system are though to _____ (I, increase; D, decrease) the signal contrast for transmission by mechanisms of _____ (E, only presynaptic; O, only postsynaptic; EO, presynaptic as well as postsynaptic) forms of inhibition.

    a. I, E   c. I, EO   e. D, O
    b. I, O   d. D, E   f. D, EO

18. The total number of recognizably separate loci on cutaneous surfaces is generally _____ (H, higher than; E, about equal to; L, less than) their number of corresponding nerve fibers in sensory pathways of the spinal cord as a probable consequence of _____ (D, divergent pathways; C, convergent pathways; R, transmission of relative ratios of signal strengths).

    a. H, D   c. L, C   e. E, R
    b. E, C   d. H, R   f. L, D

OBJECTIVE 47-4.

Identify the significance and underlying mechanisms which provide for signal prolongation, or "after-discharge," and for the generation of continuous or rhythmic signal outputs by neuronal pools.

19. A single instantaneous signal input to presynaptic endings of a neuron _____ (I, is; N, is not) able to evoke a series of repetitive postsynaptic neuron discharges and is associated with a longer duration of _____ (D, synaptic delay; P, postsynaptic potential).

    a. I, D   c. N, D
    b. I, P   d. N, P

20. A wide variance of staggered numbers of synapses within multiple pathways is a more fundamental prerequisite for a _____ (P, parallel; R, reverberating) circuit type for _____ (A, after-discharge; S, continuous signal output) by neuronal pools.

    a. P, A   c. R, A
    b. P, S   d. R, S

21. Reverberating pathways with _____ (L, longer; S, shorter) networks of feedback pathways are thought to be comparatively less susceptible to fatigue and are utilized for mechanisms of _____ (A, only after-discharge; C, only continuous signal output; AC, after-discharge as well as continuous signal output) by neuronal pools.

    a. L, A   c. L, AC   e. S, C
    b. L, C   d. S, A   f. S, AC

22. Neurons whose intracellular membrane potentials are normally maintained at levels which are less negative than their threshold levels for firing exhibit _____ (T, transient; C, continuous) discharge durations at frequencies which _____ (A, are; N, are not) increased by facilitatory signals.

    a. T, A   c. C, A
    b. T, N   d. C, N

23. Continuous signals emitted from neuronal pools are of special importance in that they normally permit modulation by _____ (E, only excitatory; I, only inhibitory; EI, both excitatory and inhibitory) control signals and a _____ (CW, carrier wave; U, unidirectional) type of information transmission.

    a. E, CW   c. EI, CW   e. I, U
    b. I, CW   d. E, U   f. EI, U

24. Mechanisms exist to alter the _____ (AP, action potential amplitude; FC, amplitude of changes in discharge frequency) _____ (W, as well as; N, but not) the periodicity of rhythmic signal output of neuronal pools by means of facilitatory _____ (W, as well as; N, but not) inhibitory signals.

    a. AP, W, N   c. AP, N, N   e. FC, W, N
    b. AP, N, W   d. FC, N, W   f. FC, W, W

OBJECTIVE 47-5.

Identify the underlying characteristics and functional roles of inhibitory circuits and decremental conduction in the stability of neuronal circuits.

25. Sensory information generally _____ (D, does; N, does not) result in widespread excitation of the central nervous system in the absence of all inhibitory mechanisms as a consequence of _____ (DC, decremental conduction; C, convergence; V, divergence) of signal transmission.

    a. D, DC   c. D, V   e. N, C
    b. D, C   d. N, DC   f. N, V

26. Two mechanisms which more specifically function to maintain stability within the central nervous system are _____ (H, habituation; D, synaptic delay; R, reverberation; I, inhibitory circuits; S, "differential" signals)

    a. H & D   c. I & S   e. D & S
    b. D & R   d. H & I   f. I & D

27. Inhibitory circuits of known significance, function in _____ (P, positive; N, negative) feedback circuits for _____ (L, lateral inhibition; E, inhibition of initial excitatory neurons by signal terminals; W, widespread control of central nervous system inhibition).

    a. N; only L & E
    b. N; only L
    c. P; only L & E
    d. N; L, E, & W
    e. N; only W
    f. P; only L & W

28. Decremental conduction occurs to a lesser extent in _____ (DC, dorsal column; FR, flexor reflex) pathways and is a phenomenon which bears a closer resemblance to the effects of _____ (D, divergence; S, synaptic fatigue; I, lateral inhibition).

    a. DC, D
    b. DC, S
    c. DC, I
    d. FR, D
    e. FR, S
    f. FR, I

29. _____ (C, conditioned; A, automatic) mechanisms of decremental conduction serve as negative feedback mechanisms for _____ (I, only increased; D, only decreased; DI, decreased as well as increased) states of neural impulse transmission.

    a. C, I
    b. C, D
    c. C, DI
    d. A, I
    e. A, D
    f. A, DI

OBJECTIVE 47–6.

Identify "alerting signals," their probable cause, their properties, and their functional significance.

30. Alerting signals are more often relatively _____ (W, weak; S, strong) output signals which are of _____ (T, transient; P, prolonged) durations.

    a. W, T
    b. W, P
    c. S, T
    d. S, P

31. Alerting signals more closely represent a(n) _____ (D, differentiation; I, integration; M, multiplication) of the original input signal and are more responsive to _____ (R, rapidly changing; C, relatively constant) magnitudes of stimulus intensity.

    a. D, R
    b. I, R
    c. M, R
    d. D, C
    e. I, C
    f. M, C

32. Alerting responses include _____ (I, only the initiation; T, only the termination; B, both the initiation and the termination) of input signals and are generally associated with relatively _____ (H, high; L, low) levels of decremental conduction.

    a. I, H
    b. T, H
    c. B, H
    d. I, L
    e. T, L
    f. B, L

33. Alerting responses are utilized to _____ (I, increase; D, decrease) signal contrast at boundaries of _____ (S, only spatial; T, only temporal; B, both spatial and temporal) visual patterns.

    a. I, S
    b. I, T
    c. I, B
    d. D, S
    e. D, T
    f. D, B

48

Sensory Receptors and Their Basic Mechanisms of Action

OBJECTIVE 48–1.

Recognize the differential sensitivity of sensory receptors to sensory stimuli. Classify and identify the morphological correlates of the types of receptors on the basis of their differential sensitivity. Identify the law of specific nerve energies and its significance.

1. Receptors in the retina of the eye are termed _____ (C, chemoreceptors; E, electromagnetic receptors; T, thermoreceptors) on the basis of their _____ (L, location; S, differential sensitivity; M, morphological properties).

   a. C, S   c. T, L   e. E, S
   b. E, L   d. C, M   f. T, M

2. The modality of pain is served by _____ (E, encapsulated; R, Ruffini's; F, free) nerve endings classified as _____ (M, mechanoreceptors; C, chemoreceptors; N, nociceptors).

   a. E, M   c. F, C   e. R, M
   b. R, N   d. E, C   f. F, N

3. Pacinian corpuscles are _____ (E, encapsulated; N, nonencapsulated) nerve endings which function as _____ (M, mechanoreceptors; C, chemoreceptors) for skin _____ (A, as well as; B, but not) for deep tissue sensibilities.

   a. E, M, A   c. E, C, A   e. N, C, A
   b. E, M, B   d. N, C, B   f. N, M, B

4. Baroreceptors of the carotid _____ (B, bodies; S, sinuses) are _____ (C, chemoreceptors; M, mechanoreceptors) for the detection of the _____ (I, internal; E, external) environment.

   a. B, C, I   c. B, M, I   e. S, M, I
   b. B, C, E   d. S, M, E   f. S, C, E

5. Muscle spindles are rather _____ (N, nonspecialized; S, specialized) _____ (C, chemoreceptors; O, nociceptors; M, mechanoreceptors) utilized for the detection of muscle _____ (T, tension; L, length; F, fatigue).

   a. N, O, F   c. S, C, T   e. S, M, L
   b. S, M, T   d. N, C, F   f. S, M, F

6. Pain receptors are responsive to _____ (O, one; M, multiple) type(s) of stimulation at comparatively _____ (L, low; H, high) thresholds for excitation, while the majority of sensory receptors are generally responsive to _____ (O, one; M, multiple) type(s) of stimuli.

   a. O, L, O   c. O, H, O   e. M, H, O
   b. O, L, M   d. M, H, M   f. M, L, M

7. The law of _____ (NE, specific nerve energies; WF, Weber-Fechner) implies a _____ (L, lack of specificity; S, specificity) of a given nerve fiber for a specific sensory modality.

   a. NE, L   c. WF, L
   b. NE, S   d. WF, S

8. Sensory modalities of sensory pathways are established by properties of _____ loci of sensory pathways, and stimulus specificity for a particular sensory modality is determined at _____ loci of sensory pathways. (I, initial; T, terminal)

   a. I, I   c. T, I
   b. I, T   d. T, T

9. The type of sensory modality for a given sensory pathway _____ (I, is; N, is not) generally a varying function of stimulus intensity for effective stimuli, as information utilized for the determination of the type of sensory modality and stimulus intensity travels within _____ (S, the same; D, differing) pathways.

   a. I, S   c. N, S
   b. I, D   d. N, D

OBJECTIVE 48-2.

Recognize the transducer role of sensory receptors. Identify the functional properties and mechanisms for the process of transduction of sensory stimuli into nerve impulses.

10. Receptor or generator potentials probably _____ (D, do; N, do not) represent a rather universal coupling mechanism for initiating action potentials by means of receptor stimulation, and are initiated by the action of the stimulus upon _____ (R, specialized receptor cells only; T, nerve axon terminals only; RT, specialized receptor cells as well as nerve axon terminals).

    a. D, R   c. D, RT   e. N, T
    b. D, T   d. N, R   f. N, RT

11. Receptor potentials are _____ (P, propagated; L, local), _____ (G, graded; A, all-or-none) potentials which _____ (R, are; N, are not) capable of prolonged periods of continuous existence.

    a. P, G, R   c. P, A, R   e. L, G, R
    b. P, A, N   d. L, A, N   f. L, G, N

12. Deformation of the isolated _____ (C, capsule; N, nerve terminal) of the pacinian corpuscle results in a receptor potential as a consequence of its membrane _____ (D, depolarization; H, hyperpolarization).

    a. C, D   c. N, D
    b. C, H   d. N, H

13. The _____ (M, myelinated; U, nonmyelinated) terminal end of the nerve fiber supplying the pacinian corpuscle _____ (D, does; N, does not) significantly differ in its electrical properties from more remote segments of its axon, and _____ (D, does; N, does not) initiate propagated action potentials.

    a. M, D, N   c. M, N, N   e. U, D, N
    b. M, N, D   d. U, N, D   f. U, D, D

14. Receptor potentials generally exhibit _____ amplitudes and _____ rates of change in amplitude as a function of increasing stimulus intensity. (I, increasing; D, decreasing)

    a. I, I   c. D, I
    b. I, D   d. D, D

15. Receptor potentials, generated by relatively _____ (O, homogeneous; E, heterogeneous) mechanisms for different types of receptors, achieve maximal amplitudes of about _____ millivolts.

    a. O, 10   c. O, 1,000   e. E, 100
    b. O, 100   d. E, 10   f. E, 1,000

16. Action potentials initiated by pacinian corpuscles _____ (D, do; N, do not) significantly influence the relatively _____ (B, brief; P, persistent) durations of their receptor potentials.

    a. D, B   c. N, B
    b. D, P   d. N, P

17. The discharge frequency within a sensory nerve is approximately _____ (D, directly; I, inversely) proportional to the _____ (L, logarithm of the amplitude; A, amplitude) of its receptor potential.

    a. D, L   c. I, L
    b. D, A   d. I, A

OBJECTIVE 48-3.

Identify the characteristics of receptor adaptation, the mechanisms by which receptors adapt, and the distinguishing functional properties of "tonic" and "phasic" receptors.

18. The continuous stimulation of receptors at constant stimulus intensities results in action potential discharge frequencies which _____ (I, increase; C, remain constant; D, decrease) as a function of time as a consequence of mechanisms of _____ (O, accommodation; P, adaptation; R, recruitment).

    a. I, R   c. D, P   e. C, P
    b. C, O   d. I, O   f. D, R

19. _____ receptors are rapidly adapting whereas _____ receptors are examples of slowly adapting receptors. (H, hair base; S, muscle spindle; J, joint capsule; P, pacinian corpuscles)

    a. H; S, J, & P   d. P & S; H & J
    b. H & S; J & P   e. P; H, S, & J
    c. H & P; S & J   f. P, H, & J; S

20. Fluid movement from encircling _____(M, central myelin; O, outer corpuscular) layers of the pacinian corpuscle is thought to be a mechanism of receptor _____ (D, adaptation; F, fatigue).

a. M, D   c. O, D
b. M, F   d. O, F

21. Pacinian corpuscles signal the _____ mechanical compression but not the _____ mechanical compression. (O, onset of; F, offset of; M, magnitude of maintained; R, rate of change of)

a. O; F, M, & R   d. O, F, & M; R
b. O & F; M & R   e. O & R; F & M
c. O, F, & R; M   f. M; O, F, & R

22. Poorly adapting or _____(P, "phasic"; T, "tonic") receptors generally exhibit _____ (I, an increase; N, no significant change; D, a decrease) in sensory nerve discharge frequencies within the first few seconds following the onset of a constant stimulus intensity.

a. P, I   c. P, D   e. T, N
b. P, N   d. T, I   f. T, D

23. "Phasic" receptors when compared to "tonic" receptors exhibit _____ (L, less; M, more) rapid rates of initial change in discharge frequencies and adapt to _____ (E, extinction; L, low levels of maintained discharge frequencies).

a. L, E   c. M, E
b. L, L   d. M, L

24. "Predictive" signals for body movements are largely transmitted from _____(T, "tonic"; P, "phasic") receptors monitoring the _____ (A, absolute magnitudes; R, rates of change) of stimulus intensity.

a. T, A   c. P, A
b. T, R   d. P, R

OBJECTIVE 48–4.

Identify the mechanisms utilized for the psychic interpretation of stimulus strength and their functional significance.

25. A mechanism which explains the tremendous intensity range of our sensory experience is the _____ (N, linear; G, logarithmic) approximation of the relationship between receptor potentials and _____ (I, stimulus intensities; D, discharge frequencies of sensory nerves).

a. N, I   c. G, I
b. N, D   d. G, D

26. A(n) _____ (I, increase; D, decrease) in stimulus intensity results in altered psychic interpretation of signal strength by means of _____(R, recruitment of additional sensory nerves; F, increased discharge frequencies of sensory nerves).

a. I, R only   d. D, R only
b. I, F only   e. D, F only
c. I, both R & F   f. D, both R & F

27. The _____ (WF, Weber-Fechner Principle; SNE, law of specific nerve energies; B-M, law of Bell-Magendie) implies that the increment of change in stimulus intensity required for perception of the change, when _____ (D, divided; M, multiplied) by the stimulus intensity, remains relatively constant over a wide range of stimulus intensities.

a. WF, D   c. BM, D   e. SNE, M
b. SNE, D   d. WF, M   f. BM, M

28. Discrimination of gradations of signal strength occurs in approximate direct proportion to the _____ (S, stimulus strength; L, logarithm of the stimulus strength), and occurs with greater accuracy at _____ (H, high; W, low) levels of background stimulus intensities.

a. S, H   c. L, H
b. S, W   d. L, W

OBJECTIVE 48–5.

Identify the means of nerve fiber classification and the general properties of different types of mammalian nerve fibers. Recognize the two separate classification systems of nerve fibers in general use.

29. Type _____ fibers include _____ (M, myelinated; U, unmyelinated), preganglionic autonomic nerve fibers with a range of nerve fiber diameters of _____ microns.

   a. A, M, 5 to 15
   b. B, M, 1 to 4
   c. C, M, 0.2 to 1
   d. A, U, 5 to 15
   e. B, U, 1 to 4
   f. C, U, 0.2 to 1

30. Type C fibers include sympathetic _____ (E, preganglionic; O, postganglionic) fibers which _____ (D, do; N, do not) exhibit saltatory propagation.

   a. E, D
   b. E, N
   c. O, D
   d. O, N

31. Type C fibers when compared to type B fibers have _____ nerve fiber diameters, _____ spike potential durations, and _____ positive after-potential durations. (G, greater; L, lesser)

   a. G, G, G
   b. G, L, G
   c. G, L, L
   d. L, L, G
   e. L, G, G
   f. L, G, L

32. Fibers found within nerve trunks to skeletal muscle include type _____ fibers supplying the motor unit control of skeletal muscle and type _____ fibers supplying the motor excitation of muscle spindles.

   a. IA, Aα
   b. Aα, Aγ
   c. Aγ, IA
   d. IA, B
   e. Aα, C
   f. Aγ, Aδ

33. Group _____ fibers from Golgi tendon apparatuses, when compared to myelinated group _____ fibers transmitting pain information, have _____ (H, higher; L, lower) propagation velocities.

   a. IA, C, H
   b. IB, III, H
   c. IV, A, L
   d. IA, III, H
   e. IB, II, L
   f. IV, III, L

34. Sensory fibers from discrete cutaneous tactile receptors or group _____ fibers correspond to type A _____ fibers in the alternate classification system.

   a. I, α & β
   b. I, β & γ
   c. II, δ & γ
   d. II, β & γ
   e. III, α & γ
   f. III, δ & γ

35. Unmyelinated fibers carrying pain _____ (A, as well as; N, but not) crude temperature sensations correspond to group _____ fibers or type _____ fibers in the alternate classification system.

   a. A, III, C
   b. A, IV, B
   c. A, IV, C
   d. N, IV, B
   e. N, III, C
   f. N, III, B

36. The largest diameter nerve fibers have diameters of about _____ microns, propagation velocities of about _____ meters per second, and are classed as group/type _____ fibers.

   a. 22, 60, IV
   b. 22, 120, I
   c. 22, 120, Aγ
   d. 13, 60, Aγ
   e. 13, 60, IA
   f. 13, 120, IV

49

Somatic Sensations:
I. The Mechanoreceptive Sensations

OBJECTIVE 49–1.

Identify the classification systems utilized for the somatic sensations.

1. The somatic senses, which _____ (D, do; N, do not) generally include the special senses, are classified as the pain, thermoreceptive, and _____ (E, electromagnetic; C, chemoreceptive; M, mechanoreceptive) senses.

   a. D, E   c. D, M   e. N, C
   b. D, C   d. N, E   f. N, M

2. Tactile senses frequently include touch, pressure and _____ (P, pain; V, vibration; R, proprioception) senses, while the _____ (K, kinesthetic; S, visceral) senses are those which are utilized to determine relative position and rates of movement of body parts.

   a. P, K   c. R, K   e. V, S
   b. V, K   d. P, S   f. R, S

3. _____ (P, proprioceptive; S, visceral; D, deep) sensations originating from fasciae and bone include pressure, pain, and _____ (V, vibration; T, temperature).

   a. P, V   c. D, V   e. S, T
   b. S, V   d. P, T   f. D, T

OBJECTIVE 49–2.

Identify the types of receptors utilized for touch, pressure, and vibration sensations, and characteristics of transmission of tactile sensations by peripheral nerve fibers.

4. Meissner's corpuscles are _____ (E, encapsulated; N, nonencapsulated) nerve endings of _____ (M, myelinated; U, nonmyelinated) sensory nerves which are particularly abundant in _____ (H, hairy regions of the body; F, the fingertips and lips).

   a. E, M, H   c. E, U, H   e. N, M, H
   b. E, M, F   d. N, U, F   f. N, M, F

5. _____ (S, Meissner's corpuscles; K, Merkel's discs) generally adapt to extinction within a few seconds and transmit information in type A _____ peripheral nerve fibers with propagation velocities of about _____ meters per second.

   a. S, α, 10 to 15   d. K, β, 30 to 60
   b. S, β, 30 to 60   e. K, α, 10 to 15
   c. S, β, 10 to 15   f. K, α, 30 to 60

6. Free nerve endings are thought to transmit a _____ (W, well; P, poorly) localized form of tactile sensation via unmyelinated type _____ peripheral nerve fibers at propagation velocities of about _____ meter(s) per second.

   a. W, C, 1   c. W, B, 1   e. P, C, 1
   b. W, B, 10   d. P, B, 10   f. P, C, 10

7. The hair end-organ is a _____ (R, rapidly; S, slowly) adapting receptor which, _____ (L, like; U, unlike) Meissner's corpuscles, is largely utilized for the detection of the movement of objects upon cutaneous surfaces.

   a. R, L   c. S, L
   b. R, U   d. S, U

8. _____ (RE, Ruffini's endings; K, Krause's corpuscles) are nonencapsulated, expanded tip endings, which are _____ (R, rapidly; S, slowly) adapting receptors utilized to signal the degree of joint rotation _____ (A, as well as; N, but not) the magnitude of applied cutaneous pressures.

   a. RE, R, A   c. RE, S, A   e. K, R, A
   b. RE, S, N   d. K, S, N   f. K, R, N

9. _____ (C, Meissner's corpuscles; M, Merkel's discs; P, Pacinian corpuscles) are utilized to detect higher vibration frequencies which range as high as _____ (E, 80 to 100; F, 400 to 500) cycles per second.

   a. C, E   c. P, E   e. M, F
   b. M, E   d. C, F   f. P, F

10. A conscious "muscle sense" is thought to travel in class _____ peripheral nerve fibers, and probably _____ (D, does; N, does not) originate in muscle spindles and _____ (D, does; N, does not) originate in Golgi tendon organs.

   a. I, D, D   d. II & III, N, N
   b. I, D, N   e. II & III, N, D
   c. I, N, D   f. II & III, D, N

OBJECTIVE 49–3.

Identify the kinesthetic receptors, their functional significance, their mechanisms of detection of the rates of movement and degree of joint rotation, and the characteristics of their information transmission by peripheral nerves.

11. Kinesthesia, referring to _____ (C, conscious recognition of; S, largely subconscious) information involving the relative orientation and rates of change of orientation of different body parts, is largely subserved by sensory endings of _____ (M, skeletal muscles; JL, joint capsules and ligaments; MC, Meissner's corpuscles).

   a. C, M   c. C, MC   e. S, JL
   b. C, JL   d. S, M   f. S, MC

12. The most abundant kinesthetic receptor is the _____ (R, rapidly; S, slowly) adapting _____ (G, Golgi tendon organ; P, pacinian corpuscle; SR, spray type of Ruffini ending).

   a. R, G   c. R, SR   e. S, P
   b. R, P   d. S, G   f. S, SR

13. Ruffini and Golgi endings are stimulated more _____ (W, weakly; S, strongly) during initial periods following joint movements, adapt to _____ (E, extinction; L, steady firing levels), and _____ (A, are; N, are not) thought to be utilized as rate detectors for joint movements.

   a. W, E, A   c. W, L, A   e. S, L, A
   b. W, E, N   d. S, L, N   f. S, E, N

14. The firing frequency of a sensory nerve from _____ (R, rapidly; S, slowly) adapting receptors which monitor joint position _____ (I, increases progressively; D, decreases progressively; E, exhibits a specific maximum) with increasing degrees of angular rotation within the range of motion of the joint.

   a. R, I   c. R, E   e. S, D
   b. R, D   d. S, I   f. S, E

15. Kinesthetic signals are transmitted at propagation velocities which are _____ (G, greater; L, less) than 20 meters per second in predominant class/type _____ peripheral nerves.

   a. G, I   c. G, III   e. L, Aβ
   b. G, Aβ   d. L, I   f. L, Aδ

OBJECTIVE 49–4.

Recognize the dual system for the transmission of mechanoreceptive somatic sensations in the central nervous system. Contrast the general characteristics of information transmission by the dorsal column and spinothalamic systems.

16. _____ (A, all; M, most, but not all) sensory information entering the spinal cord passes through the _____ (V, ventral; D, dorsal) roots.

    a. A, V   c. M, V
    b. A, D   d. M, D

17. The dorsal column pathway is comprised largely of type _____ axons of _____ (F, first; S, second) order neurons.

    a. Aβ, F   c. C & Aδ, F   e. B, S
    b. B, F   d. Aβ, S   f. C & Aδ, S

18. The _____ (ST, spinothalamic; D, dorsal column) pathway is composed of comparatively smaller diameter _____ (F, first; S, second) order fibers which transmit impulses at propagation velocities _____ (A, above; B, below) 15 meters per second.

    a. ST, F, A   c. ST, S, A   e. D, F, A
    b. ST, S, B   d. D, S, B   f. D, F, B

19. When compared to the spinothalamic system, the dorsal column pathway has a _____ (H, higher; L, lower) degree of spatial orientation of fibers, transmits sensations that detect gradations of intensity _____ (M, more; S, less) accurately, and _____ (D, does; N, does not) transmit a broader spectrum of sensory modalities.

    a. H, M, D   c. H, S, D   e. L, S, D
    b. H, M, N   d. L, S, N   f. L, M, N

20. _____ (T, touch; P, pain; H, thermal; J, joint; S, pressure) sensations are transmitted in both the spinothalamic and dorsal column pathways, with a greater degree of temporal fidelity and spatial localization occurring for transmission within the _____ (ST, spinothalamic; DC, dorsal column) pathway.

    a. P & H, ST   c. T & S, ST   e. J & S, DC
    b. J & S, ST   d. P & H, DC   f. T & S, DC

21. _____ sensations are transmitted by the dorsal column system, while _____ sensations are transmitted by the spinothalamic system. (V, vibratory; K, kinesthetic; P, pain; T, thermal; I, tickle and itch; X, sexual)

    a. V & K; P, T, I, & X
    b. V, K, & I; P, T, & X
    c. P & T; V, K, I & X
    d. V, I, & X; K, P, & T
    e. P, I, & X; V, K, & T
    f. P, T, & I; V, K, & X

OBJECTIVE 49–5.

Identify the neuroanatomical characteristics and spatial orientation of transmitted information within the dorsal column pathway.

22. Sensory information from the _____ (S, same; O, opposite) side of the body is transmitted by axons of _____ (F, first; C, second) order neurons which terminate in the cuneate and gracile nuclei.

    a. S, F   c. O, F
    b. S, C   d. O, C

23. The medial lemniscus is a _____ order neuron pathway which terminates in the _____ (VPL, ventral posterolateral; M, midline) nuclei of the thalamus.

    a. 1st, VPL   c. 3rd, VPL   e. 2nd, M
    b. 2nd, VPL   d. 1st, M   f. 3rd, M

24. _____ order neurons of the dorsal column pathway located in the _____ (E, epithalamus; T, thalamus; H, hypothalamus) project mainly to the _____ (EC, precentral; OC, postcentral) gyrus of the cerebral cortex.

   a. 2nd, E, EC   d. 3rd, T, OC
   b. 2nd, E, OC   e. 3rd, T, EC
   c. 2nd, T, EC   f. 3rd, H, OC

25. The medial lemniscus is joined by fibers from the _____ (M, main sensory and upper descending; L, lower descending) nuclei of the _____ (V, fifth; VII seventh; X, tenth) cranial nerve which serves similar sensory functions for the head as the dorsal column system serves for the body.

   a. M, V   c. M, X   e. L, VII
   b. M, VII   d. L, V   f. L, X

26. _____ order axons of the dorsal column system decussate or cross the midline in the lower _____ (D, medulla; S, mesencephalon), so that sensory representation for the left side of the body occurs in the _____ (R, right; L, left) side of the somesthetic cortex.

   a. 1st, D, R   c. 3rd, S, R   e. 2nd, D, R
   b. 2nd, D, L   d. 1st, D, L   f. 3rd, S, L

27. Body representation of spatial loci occurs with the _____ (U, upper; W, lower) parts of the body lying more medially within the dorsal columns, and the lower parts of the body represented in the more _____ (M, medial; L, lateral) portions of the ventrobasal complex.

   a. U, M   c. W, M
   b. U, L   d. W, L

OBJECTIVE 49-6.

Locate somatic sensory areas I and II and identify their general functional significance. Identify the cortical representations of spatial projection and modality separation of sensory information, and the general pattern of excitation of vertical columns of neurons within the somesthetic cortex.

28. Somatic sensory area _____ lies posterior and inferior to the lower end of the postcentral gyrus and receives sensory input from _____ (DC, only the dorsal column; ST, only the spinothalamic; B, both the dorsal column and spinothalamic) system(s).

   a. I, DC   c. I, B   e. II, ST
   b. I, ST   d. II, DC   f. II, B

29. A greater degree of sensory representation for the left lower extremity occurs within the more _____ (M, medial; T, lateral; A, anterior; P, posterior) portion of somatic sensory area _____ in the _____ (L, left; R, right) side of the cerebral cortex.

   a. M, I, L   c. T, I, L   e. A, II, L
   b. M, I, R   d. T, II, R   f. P, II, R

30. The surface area of sensory representation on the _____ (E, precentral; O, postcentral) gyrus is proportional to the _____ (N, numbers of specialized sensory receptors; C, cutaneous surface area) of peripheral body parts and is greater for the _____ (T, body trunk; L, lips).

   a. E, N, T   c. O, N, T   e. O, C, T
   b. E, C, L   d. O, N, L   f. O, C, L

31. Tactile modalities for the oral cavity are thought to be represented in the more _____ (A, anterior; P, posterior) and _____ (M, medial; L, lateral) portions of somatic sensory area I.

   a. A, M   c. P, M
   b. A, L   d. P, L

32. The cerebral cortex is generally subdivided into _____ separate layers of neurons, with separate qualities of signals presumably arranged in corresponding columns of cells which are _____ (P, parallel; R, perpendicular) to the cortical surface.

   a. 6, P   c. 18, P   e. 11, R
   b. 11, P   d. 6, R   f. 18, R

33. Neuronal layer _____ is thought to receive initial excitation from incoming sensory signals, while layers _____ are thought to receive nonspecific input from the reticular activating system which serves to control the overall level of excitability of the cortex.

   a. I, I & II   d. II, V & VI
   b. IV, I & II   e. IV, V & VI
   c. VII, I & II   f. VII, V & VI

OBJECTIVE 49-7.

Identify the inferred functions and consequences of select or widespread damage to somatic sensory area I. Identify the somatic association areas, their location, their neural connections, and the effects of their destruction.

34. Destruction of somatic sensory area I results in a more significant decrease in the _____ (I, appreciation of pain intensity; L, degree of localization of pain) and _____ (N, no significant change in; S, a loss of) the ability to recognize the relative orientation of the extremities.

    a. I, S   c. L, S
    b. I, N   d. L, N

35. Spatial recognition, recognition of the relative intensity, _____ (A, as well as; N, but not) the recognition of similarities and differences of the size, shape, texture, and weights of objects placed in the oral cavity are thought to be more initially dependent upon the function of somatic sensory area _____.

    a. A, I   c. N, I
    b. A, II   d. N, II

36. Localization of the _____ (S, somesthetic cortex; A, somatic association areas) occurs in _____ (N, neuronal layers; B, Brodmann areas) V and VII of the cerebral cortex.

    a. S, N   c. A, N
    b. S, B   d. A, B

37. Somatic association areas possess primarily _____ (S, sensory; T, motor) functions, whose loss results in _____ (D, adiadochokinesia; M, amorphosynthesis).

    a. S, D   c. T, D
    b. S, M   d. T, M

OBJECTIVE 49-8.

Identify the functional characteristics of signal transmission and information processing within the dorsal column pathway.

38. The dorsal column pathway, which _____ (D, does; N, does not) normally transmit the majority of its signals all the way to the somesthetic cortex, represents a comparably _____ (S, susceptible; R, resistant) system to the blocking actions of moderate degrees of anesthesia.

    a. D, S   c. N, S
    b. D, R   d. N, R

39. Increased effective stimulus intensities to a peripheral receptor _____ (I, increases; N, does not significantly alter; D, decreases) the size of cortical field discharge, and _____ (A, alters; N, does not significantly alter) the cortical field position of a maximum discharge frequency for dorsal column system information.

    a. I, A   c. D, A   d. N, N
    b. N, A   d. I, N   f. D, N

40. The detection of two points as separate entities when they are minimally separated by 2 mm. occurs for stimuli applied to the _____ (F, fingertips; B, skin of the back), because a higher degree of two-point discrimination occurs in cutaneous regions with correspondingly _____ (S, small; L, large) somesthetic cortical representations.

    a. F, S   c. B, S
    b. F, L   d. B, L

41. Rapidly repetitive vibratory sensations from the left side of the body are transmitted in major spinal cord tracts on _____ (R, only the right; L, only the left; RL, both) side(s) of the spinal cord in _____ (DC, only the dorsal column; ST, only the spinothalamic; B, both the dorsal column and spinothalamic) system pathway(s).

    a. R, ST   c. RL, DC   e. L, B
    b. L, DC   d. R, B   f. RL, B

42. Maximal discharge frequencies for first order neurons responding to joint rotation generally occur at _____ (E, either minimum or maximum; I, intermediate) positions of joint rotation and at _____ (E, either minimum or maximum; I, intermediate) positions for third order neurons of the _____ (ST, spinothalamic; DC, dorsal column) pathway.

    a. E, E, ST   b. E, I, DC   c. E, I, ST   d. I, I, DC   e. I, E, ST   f. I, E, DC

OBJECTIVE 49–9.

Identify the neuroanatomical and functional characteristics of transmission within the spinothalamic system. Identify the probable characteristics of the spinoreticular tract.

43. Second order neurons of the spinothalamic tract are located within _____ (M, the medulla; S, two to six segments of the spinal cord from the point of entry of first order axons) and undergo axonal decussation within the _____ (A, anterior commissure; D, decussation of the medial lemniscus) to terminate in the _____ (T, thalamus; GC, nuclei gracilis and cuneatus).

    a. M, A, T   c. M, D, T   e. S, A, T
    b. M, D, GC   d. S, D, GC   f. S, A, GC

44. _____ order pain and temperature type C fibers within the tract of Lissauer transmit activity by means of the _____ (V, ventral; L, lateral) spinothalamic tract to a more prominent locus of the _____ (VB, ventrobasal; IL, intralaminar) group of thalamic nuclei.

    a. 1st, V, VB   d. 2nd, L, IL
    b. 1st, L, IL   e. 2nd, V, VB
    c. 1st, L, VB   f. 2nd, V, IL

45. Touch and pressure information originating from the right side of the body is transmitted _____ (U, unilaterally; B, bilaterally) in the spinal cord, as touch and pressure sensations are transmitted by the _____ (L, lateral; V, ventral) spinothalamic tract _____ (A, as well as; N, but not) the dorsal column pathways.

    a. U, L, A   c. U, V, A   e. B, V, A
    b. U, L, N   d. B, V, N   f. B, L, N

46. _____ order neurons of the spinothalamic system, located within the _____ (IL, intralaminar; VB, ventrobasal, P, pulvinar) group of thalamic nuclei, are thought to project primarily to somatic sensory area I of the cerebral cortex.

    a. 2nd, IL   c. 2nd, P   e. 3rd, VB
    b. 2nd, VB   d. 3rd, IL   f. 3rd, P

47. Type C fibers mediating pain information within the _____ (VS, ventral spinothalamic; RS, reticulospinal; SR, spinoreticular) tract originate primarily within _____ (G, dorsal root ganglia; H, dorsal horns).

    a. VS, G   c. SR, G   e. RS, H
    b. RS, G   d. VS, H   f. SR, H

48. The spinothalamic system in comparison with the dorsal column pathways transmits a _____ diversity of sensory modalities with a _____ degree of spatial localization and temporal fidelity at _____ propagation velocities. (G, greater; L, lesser)

    a. G, G, L   c. G, L, L   e. L, L, G
    b. G, L, G   d. L, G, G   f. L, L, L

OBJECTIVE 49–10.

Recognize the major functional roles of the thalamus and "corticofugal" signals for sensory functions. Identify dermatomes, the underlying basis for their existence, and their diagnostic utility in states of neurological deficits.

49. The _____ (E, epithalamus; T, thalamus; H, hypothalamus), serving as a specific relay station for almost all forms of sensory information transmission to the cerebral cortex, _____ (I, is; N, is not) thought to participate directly in the conscious awareness of pain sensibility.

    a. E, I   c. H, I   e. T, N
    b. T, I   d. E, N   f. H, N

50. "Corticofugal" signals are largely _____ (F, facilitory; I, inhibitory) signals originating in _____ (L, lower; H, higher) levels of the central nervous system.

    a. F, L   c. I, L
    b. F, H   d. I, H

51. The conscious brain directs its attention to different aspects of sensory information through mechanisms of _____ (F, only facilitation; I, only inhibition; B, both facilitation and inhibition) of cortical receptive areas, _____ (A, as well as; N, but not) by altering information transmission in the dorsal column and spinothalamic pathways at lower levels.

    a. F, A   c. B, A   e. I, N
    b. I, A   d. F, N   f. B, N

52. A dermatome is a region of cutaneous innervation by a single _____ (PN, peripheral nerve; AR, anterior root; PR, posterior root) which _____ (D, does; N, does not) overlap with adjacent dermatomes.

    a. PN, D   c. PR, D   e. AR, N
    b. AR, D   d. PN, N   f. PR, N

53. A loss of all sensory information in cutaneous regions innervated by a peripheral nerve is a more likely consequence of injury to _____ whereas a loss of all sensory information from cutaneous regions bounded by "segmental fields" is a more likely consequence of injury to _____ . (ST, spinal cord tracts; C, the cerebral cortex; PN, a peripheral nerve; DR, dorsal roots)

    a. PN, ST   c. PN, DR   e. DR, C
    b. PN, C   d. DR, ST   f. DR, PN

54. Larger cutaneous regions of the "segmental fields" of the lower extremities are represented by _____ and _____ dermatome segments. (T, thoracic; L, lumbar; US, upper sacral; LS, lower sacral)

    a. T, L   c. L, LS   e. T, US
    b. T, LS   d. L, US   f. US, LS

50

Somatic Sensations: II. Pain, Visceral Pain, Headaches, and Thermal Sensations

OBJECTIVE 50-1.

Identify pain, its functional purpose, its sensory qualities, and the methods utilized for measuring the perception of pain.

1. Pain is a general psychical adjunct to _____ (F, flexor withdrawal; S, stretch; C, clasp knife) reflexes resulting from relatively _____ (P, specific; N, nonspecific) forms of nociceptor stimulation.

   a. F, P   c. C, P   e. S, N
   b. S, P   d. F, N   f. C, N

2. Higher degrees of localization and greater propagation velocities occur for _____ types of pain, while _____ types of pain are not felt on the surface of the body but represent a deep type of pain. (A, aching; B, burning; P, pricking)

   a. P, A   c. B, P   e. A, P
   b. P, B   d. B, A   f. A, B

3. Widely differing groups of people exhibit _____ thresholds for pain by standardized techniques and _____ reactions to pain. (V, widely varying; C, rather constant).

   a. V, V   c. C, V
   b. V, C   d. C, C

4. A strength-duration curve for thermal stimulation of pain perception represents loci of _____ (B, subthreshold; T, threshold; P, suprathreshold) stimuli for which the _____ (Q, quotient; R, product) of corresponding values of stimulus intensities and durations represents a more constant set of values.

   a. B, Q   c. P, Q   e. T, R
   b. T, Q   d. B, R   f. P, R

5. Thresholds for the perception of pain from cutaneous thermal stimulation generally occur when skin temperatures reach _____ °C. as a consequence of stimulation of _____ (R, corpuscles of Ruffini; F, free nerve endings).

   a. 47, R   c. 67, R   e. 57, F
   b. 57, R   d. 47, F   f. 67, F

6. Individuals can generally barely distinguish _____ just noticeable differences (JND's) of pain intensity, with a greater capability for discriminating pain intensity occurring at _____ (L, low; H, high) levels of pain.

   a. 6, L   c. 136, L   e. 22, H
   b. 22, L   d. 6, H   f. 136, H

OBJECTIVE 50-2.

Identify the distribution and properties of receptors utilized for pain, tickling, and itch. Identify the probable mechanisms for eliciting pain and their correlative relationships to tissue injury.

7. Receptors for _____ (P, only pain; I, only itch; B, both pain and itch) are located almost exclusively in the skin and transmit information in _____ (M, myelinated; U, unmyelinated) type _____ fibers.

   a. P, M, Aδ   c. B, U, C   e. I, M, Aδ
   b. I, U, C   d. P, M, C   f. B, M, Aδ

8. The arterial walls, joint surfaces, and the falx and tentorium of the cranial vault, _____ (A, as well as; N, but not) the periosteum, represent internal tissues with _____ (B, an absence; H, higher densities) of pain receptors.

   a. A, B   c. N, B
   b. A, H   d. N, H

9. Pain receptors are characterized as _____ (R, rapidly; S, slowly) adapting receptors whose threshold frequently becomes lower and lower with maintained nociceptor stimuli as a consequence of _____ (A, receptor adaptation; H, hyperalgesia).

   a. R, A   c. S, A
   b. R, H   d. S, H

10. Threshold levels of elevated cutaneous temperatures for the perception of pain are _____ (A, considerably above; E, about equal to; L, considerably lower than) threshold levels for tissue injury, and probably result from a more direct action of _____ (H, heat; C, chemical agents) upon pain receptors.

   a. A, H   c. L, H   e. E, C
   b. E, H   d. A, C   f. L, C

11. Bradykinin is a _____ (G, globulin; LP, lipoprotein; PP, polypeptide) substance which _____ (S, stimulates; I, inhibits) pain endings.

   a. G, S   c. PP, S   e. LP, I
   b. LP, S   d. G, I   f. PP, I

12. Occlusion of blood flow to tissue, _____ (A, as well as; N, but not) states of muscular spasms, generally _____ (B, abolishes; C, causes) pain.

   a. A, B   c. N, B
   b. A, C   d. N, C

13. Muscle spasms are states characterized by _____ blood flow, _____ metabolic rates, and _____ ratios of aerobic to anaerobic metabolism of skeletal muscles. (I, increased; U, unaltered; D, decreased)

   a. I, I, U   c. I, U, I   e. D, D, D
   b. I, D, D   d. D, I, U   f. D, I, D

14. Cessation of blood flow to _____ (S, skin; SM, skeletal muscle) generally results in an earlier onset of pain as a consequence of differential _____ (N, numbers of pain endings; MR, metabolic rates; BI, rates of bradykinin inactivation).

   a. S, N   c. S, BI   e. SM, MR
   b. S, MR   d. SM, N   f. SM, BI

15. Itch and tickle sensations are mediated by _____ (S, the same; P, separate, but similar) fibers as those which mediate a _____ (R, pricking; B, burning) type of pain.

   a. S, R   c. P, R
   b. S, B   d. P, B

OBJECTIVE 50–3.

Identify properties of transmission of pain signals within the central nervous system, the functions of the thalamus and cerebral cortex in pain appreciation, and the mechanisms available for surgical interruption of the pain pathways. Contrast the characteristics of the pricking pain and burning pain pathways.

16. The more rapid transmission of the _____ (P, pricking; B, burning) sensation of a "double" pain sensation is mediated by type _____ fibers which are comparatively _____ (M, more; L, less) sensitive to low concentrations of local anesthetics.

   a. P, Aδ, M   c. P, C, M   e. B, Aδ, M
   b. P, Aδ, L   d. P, C, L   f. B, C, L

17. Pain fibers terminate on second order neurons after they pass through the _____ (L, lateral spinothalamic tract; V, ventral spinothalamic tract; S, tract of Lissauer) and _____ (B, before; A, after) they cross to the contralateral side.

   a. L, B   c. S, B   e. V, A
   b. V, B   d. L, A   f. S, A

18. A greater degree of information processing, _____ (W, as well as; N, but not) a greater degree of temporal and spatial fidelity, occurs at spinal cord levels for the _____ (DC, dorsal column; ST, spinothalamic) system.

a. W, DC   c. N, DC
b. W, ST   d. N, ST

19. A pathway for pricking types of pain is thought to project largely to somatic sensory area _____ via the _____ (IL, intralaminar; VB, ventrobasal) group of nuclei.

a. I, IL   c. II, IL
b. I, VB   d. II, VB

20. A _____ (L, localized; W, widespread) pattern of _____ (E, excitation; I, inhibition) of cerebral cortical activity results from the effects of burning and aching pain upon the reticular formation and the _____ (IL, intralaminar; VB, ventrobasal) group of thalamic nuclei.

a. L, E, IL   c. L, I, IL   e. W, E, IL
b. L, I, VB   d. W, I, VB   f. W, E, VB

21. The pathway for burning types of pain, when compared to the pathway for pricking types of pain, has a _____ (H, higher; L, lower) degree of spatial localization, a _____ (S, shorter; N, longer) latency for the perception of pain, and a comparatively _____ (G, good; P, poor) capability for discriminating gradations of pain intensity.

a. H, S, G   c. H, N, G   e. L, N, G
b. H, S, P   d. L, N, P   f. L, S, P

22. Conscious perception of pain _____ (S, is; N, is not) thought to occur below the level of the cerebral cortex and is thought to be a more important manifestation of somatic sensory area _____ .

a. S, I   c. N, I
b. S, II   d. N, II

23. A cordotomy for the relief of intractable pain involves interruption of the _____ A, anterolateral; P, posteromedial) quadrant of the _____ (S, same; O, opposite) side of the spinal cord as the side of pain origin.

a. A, S   c. P, S
b. A, O   d. P, O

OBJECTIVE 50–4.

Identify the "gating" theory for control of the reactions to pain, the phenomenon and probable mechanism of referred pain.

24. Gating mechanisms for the control of pain are thought to more significantly alter the _____ (R, reaction to; T, threshold for) pain through synaptic influences upon the soma of _____ order neurons of the pain pathway.

a. R, both 1st & 2nd   d. T, both 1st & 2nd
b. R, only 2nd   e. T, only 2nd
c. R, only 1st   f. T, only 1st

25. The transmission of pain signals is thought to be _____ (F, facilitated; I, inhibited) at spinal cord levels by activity of large-diameter sensory nerves _____ (A, as well as; N, but not) corticofugal fibers.

a. F, A   c. I, A
b. F, N   d. I, N

26. _____ order neurons of the pain pathway, located in the substantia gelatinosa, are thought to be influenced by corticofugal mechanisms of presynaptic _____ (I, inhibition; F, facilitation).

a. 1st, I   c. 3rd, I   e. 2nd, F
b. 2nd, I   d. 1st, F   f. 3rd, F

27. Referred pain, thought to result from mechanisms of _____ (A, accommodation; N, centrifugal control; C, convergence) of spinal cord activity, involves a more frequent referral of _____ (V, visceral pain to cutaneous; T, cutaneous pain to visceral) regions.

a. A, V   c. C, V   e. N, T
b. N, V   d. A, T   f. C, T

28. _____ (I, increased; D, decreased) skeletal muscle tone within referred areas of visceral pain are more frequent causes of _____ (R, reduced; E, intensified) pain magnitudes.

a. I, R   c. D, R
b. I, E   d. D, E

OBJECTIVE 50-5.

Identify the general characteristics and principal causes of true visceral pain. Recognize the potential existence and characteristics of "parietal" pain occurring in association with visceral injury, and identify the visceral structures which are relatively insensitive to pain.

29. Visceral pain is more readily elicited from highly _____ (L, localized; D, diffuse) types of visceral injury and transmission in predominantly type _____ nerve fibers.

    a. L, A   c. L, C   e. D, B
    b. L, B   d. D, A   f. D, C

30. Pain, representing _____ (M, the major sensory modality; O, one of many existing sensory modalities) for most visceral structures, is elicited by visceral _____ (S, spasm only; D, overdistention only; B, spasms as well as over distention).

    a. M, S   c. M, B   e. O, D
    b. M, D   d. O, S   f. O, B

31. Pain with a sharp and pricking quality, which occurs in association with visceral injury, is a more likely consequence of a _____ (V, true visceral; R, parietal) type of pain transmitted in conjunction with _____ (S, sympathetic; P, spinal) nerve pathways.

    a. V, S   c. R, S
    b. V, P   d. R, P

32. Rhythmic cycles of visceral pain with periodicities of several minutes, which occur in association with visceral "cramps," are generally correlated with rhythmic cycles of _____ (C, cardiac; S, smooth) muscle contraction and travel largely in association with _____ (A, autonomic; M, somatic) peripheral nerve distributions.

    a. C, A   c. S, A
    b. C, M   d. S, M

33. The liver _____ (C, capsule; P, parenchyma) and the pulmonary _____ (B, bronchi; A, alveoli; PP, parietal pleura) are included in structures which are largely insensitive to pain.

    a. C, B   c. C, PP   e. P, A
    b. C, A   d. P, B   f. P, PP

OBJECTIVE 50-6.

Contrast the characteristics of the visceral and parietal pathways for transmission of abdominal and thoracic pain.

34. The visceral pathway for pain, in contrast to the parietal pathway, is associated with a _____ (G, greater; L, lesser) degree of accurate localization, and _____ (U, only unmyelinated; B, both myelinated and unmyelinated) sensory fibers associated with _____ (S, somatic; A, autonomic) peripheral nerves.

    a. G, U, S   c. G, B, S   e. L, U, S
    b. G, B, A   d. L, B, A   f. L, U, A

35. The majority of visceral pain fibers travel in association with largely _____ (P, parasympathetic; S, sympathetic) nerves, with an exception occurring for the _____ (T, stomach; SI, small intestine; U, urinary bladder).

    a. P, T   c. P, U   e. S, SI
    b. P, SI   d. S, T   f. S, U

36. Visceral pain sensations tend to be localized _____ the injured organ, and parietal pain sensations tend to be localized _____ the injured organ. (D, directly over; E, to the dermatome of embryological origin of)

    a. D, D   c. E, D
    b. D, E   d. E, E

37. Visceral afferent fibers from the heart enter the spinal cord between levels _____ , which _____ (D, does; N, does not) correspond to the dermatome level of referral of cardiac visceral pain.

    a. C-3 & T-5, D   c. C-3 & T-5, N
    b. T-7 & T-9, D   d. T-7 & T-9, N

38. An inflamed appendix results in a sharp type of _____ (V, visceral; P, parietal) pain localized in the _____ (Q, right lower quadrant; U, umbilicus) and an aching, cramping type of pain localized in the _____ (Q, right lower quadrant; U, umbilicus).

    a. V, Q, Q   b. V, Q, U   c. V, U, Q   d. P, U, U   e. P, U, Q   f. P, Q, U

OBJECTIVE 50-7.
Identify the characteristics of visceral pain for each of the major visceral organs.

39. Cardiac pain is a more frequent result of _____ (CS, coronary sclerosis; VA, vagal afferent activity; PI, pericardial injury) which is referred more frequently to the _____ (L, left; R, right) side of the body.

    a. CS, L   c. PI, L   e. VA, R
    b. VA, L   d. CS, R   f. PI, R

40. _____ (V, visceral; P, parietal) pain from the heart is localized to the base of the neck, over the arms, pectoral muscles, and down the arms, whereas the alternate type of cardiac pain is thought to be localized _____ (N, to the neck; U, to the umbilicus; S, underneath the sternum).

    a. V, N   c. V, S   e. P, U
    b. V, U   d. P, N   f. P, S

41. The pain of peptic ulcer is more likely to be referred to the region _____ , whereas irritation of the gastric end of the esophagus is more likely to be referred to the region _____ . (D, directly over the heart; N, of the lower neck; H, halfway between the umbilicus and xyphoid process; S, at the tip of the right scapula)

    a. D, N   c. D, S   e. H, N
    b. D, H   d. H, D   f. H, S

42. The kidneys and ureters are _____ (I, intraperitoneal; R, retroperitoneal) structures for which a higher percentage of pain fibers travel via _____ (S, skeletal; C, visceral) peripheral nerves to localize the major portion of renal pains to the _____ (V, ventral; D, dorsal) body surface.

    a. I, S, V   c. I, C, V   e. R, S, V
    b. I, C, D   d. R, C, D   f. R, S, D

43. Uterine pain of dysmenorrhea is generally mediated by _____ (K, skeletal; M, sympathetic) afferent nerves, while injury of the adnexa around the uterus or lesions of the fallopian tubes and broad ligaments usually causes pain of a _____ (D, diffuse cramping; S, sharper) nature localized to the lower _____ (B, back and side; V, ventral body surface).

    a. K, D, B   c. K, S, B   e. M, S, B
    b. K, D, V   d. M, S, V   f. M, D, V

OBJECTIVE 50-8.
Identify hyperalgesia, the thalamic syndrome, herpes zoster, tic douloureux, and the Brown-Séquard syndrome, and their consequences upon pain sensations.

44. A localized _____ (P, primary; S, secondary) hyperalgesia resulting from an inflammatory process of a dental pulp is associated with _____ (I, increased; N, no significant change in; D, decreased) excitability of the local pain receptors themselves.

    a. P, I   c. P, D   e. S, N
    b. P, N   d. S, I   f. S, D

45. The thalamic syndrome is a more likely cause of a _____ (P, primary; S, secondary) type of hyperalgesia, which _____ (I, is; N, is not) generally associated with decreased thresholds of peripheral pain receptors.

    a. P, I   c. S, I
    b. P, N   d. S, N

46. Occlusion of the posterolateral branch of the posterior cerebral artery is believed to result in a greater loss of function of the _____ (M, medial; PV, posteroventral) portions of the thalamus and a greater loss of the _____ (L, degree of localization; U, unpleasant affective nature) of pain sensations.

    a. M, L   c. PV, L
    b. M, U   d. PV, U

47. Herpes zoster is a viral infection of neurons of the dorsal _____ (H, horn; RG, root ganglia) resulting in a _____ (S, segmental; R, rather random) distribution of corresponding regions of _____ (E, enhanced; D, diminished) pain appreciation.

    a. H, S, E   c. H, R, E   e. RG, S, E
    b. H, R, D   d. RG, R, D   f. RG, S, D

48. Lancinating pains of tic douloureux more generally involve the _____ (V, fifth; VII, seventh; X, tenth) cranial nerves and are more specifically treated by selective destruction of the _____ (M, mesencephalic; S, descending or spinal) tract of the corresponding cranial nerve.

    a. V, M   c. X, M   e. VII, S
    b. VII, M   d. V, S   f. X, S

49. The Brown-Séquard syndrome is associated with a loss of pain _____ (A, as well as; N, but not) thermal sensations from _____ (S, the same; O, the opposite) side of the body as the lesion for all dermatome levels _____ (I, immediately; B, two to six segments) below the level of transection.

    a. A, S, I   c. A, O, I   e. N, S, I
    b. A, O, B   d. N, O, B   f. N, S, B

50. Motor functions are lost for the _____ (S, same; O, opposite) side for segments below a spinal cord hemisection, while vibration sensations are _____ (N, normal; I, impaired, yet present; L, completely lost) on the same side as the transection.

    a. S, N   c. S, L   e. O, I
    b. S, I   d. O, N   f. O, L

OBJECTIVE 50–9.

Identify the types, causes, and characteristics of headaches of intracranial and extracranial origin.

51. The pain of headache _____ (I, is; N, is not) generally caused by damage within the brain itself, and results from stimuli arising within the cranium _____ (A, as well as; B, but not) the extracranial structures.

    a. I, A   c. N, A
    b. I, B   d. N, B

52. Pain arising within the intracranial vault above the tentorium is transmitted by the _____ (III, third; V, fifth; VII, seventh) cranial nerve and produces _____ (F, "frontal"; O, "occipital") types of headache.

    a. III, F   c. VII, F   e. V, O
    b. V, F   d. III, O   f. VII, O

53. Headache caused by constipation is thought to be mediated by _____ (ST, spinal cord tracts; CV, the cardiovascular system), while migrane headaches are thought to result from _____ (M, meningeal irritation; A, abnormal cardiovascular phenomena).

    a. ST, M   c. CV, M
    b. ST, A   d. CV, A

54. Excessive contraction of the ciliary muscles is generally associated with _____ (O, occipital; RO, retro-orbital) headaches, whereas prodromal symptoms are more generally associated with developing _____ (A, alcoholic; M, migraine) headaches.

    a. O, A   c. RO, A
    b. O, M   d. RO, M

OBJECTIVE 50–10.

Identify the types of thermal receptors, their characteristics, and the transmission characteristics of thermal signals in the nervous system.

55. Receptors for both cold and warmth modalities are stimulated at cutaneous temperatures of_____° C. Receptors for cold and pain modalities are both stimulated at cutaneous temperatures of _____° C.

   a. 33, 5   c. 33, 20   e. 20, 33
   b. 20, 10   d. 20, 50   f. 33, 50

56. _____ receptors are generally stimulated by temperatures of 5° C., whereas _____ receptors are generally stimulated at 33° C. (C, cold; W, warmth; P, pain).

   a. C, C & P   d. P, only C
   b. C, W & C   e. P, only W
   c. C & P, only C   f. P, W & C

57. When a cold receptor is subjected to a sudden fall in temperature, its corresponding discharge frequency _____ (I, increases; D, decreases) as a function of time because of receptor _____ (A, adaptation; C, centrifugal control; PT, post-tetanic potentiation).

   a. I, A   c. I, PT   e. D, C
   b. I, C   d. D, A   f. D, PT

58. Increased rates of intracellular chemical rates of reaction, of between _____ times for a 10° C. _____ (I, increase; D, decrease) in temperature are thought to be responsible for an intermediate coupling mechanism for thermoreceptor activity.

   a. 2 & 3, I   c. 40 & 50, I   e. 4 & 5, D
   b. 4 & 5, I   d. 2 & 3, D   f. 40 & 50, D

59. A maximum ability to discern temperature increments of change as low as _____° C. occurs with stimulation of relatively _____ (L, large; S, small) cutaneous regions.

   a. 0.01, L   c. 3, L   e. 1, S
   b. 1, L   d. 0.01, S   f. 3, S

60. Thermal signals are largely transmitted by type _____ peripheral nerve fibers, the _____ (DC, dorsal column; ST, spinothalamic) system pathway, and a more probable _____ (VB, ventrobasal; IL, intralaminar) thalamic relay to the somesthetic cortex.

   a. Aδ, DC, VB   d. C, ST, IL
   b. Aδ, ST, IL   e. C, DC, VB
   c. Aδ, ST, VB   f. C, DC, IL

51

Motor Functions of the Spinal Cord and the Cord Reflexes

OBJECTIVE 51–1.

Recognize the experimental preparations generally utilized to study spinal cord reflexes and their influence upon spinal cord reflexes.

1. Transection of the brain stem at a _____ (P, midpontile; C, midcollicular; R, midrhombencephalon) level in a decerebrate preparation results in _____ (I, inhibition; F, facilitation) of spinal cord reflexes.

   a. P, I   c. R, I   e. C, F
   b. C, I   d. P, F   f. R, F

2. "Spinal shock" represents a period of _____ (E, enhanced; P, depressed) spinal cord reflexes, a significant factor in _____ (S, only spinal; D, only decerebrate; B, both spinal and decerebrate) animal preparations.

   a. E, S   c. E, B   e. P, D
   b. E, D   d. P, S   f. P, B

OBJECTIVE 51–2.

Identify and locate the alpha and gamma motoneurons, Renshaw cells and interneurons, their functional and organizational characteristics, and the general organizational pattern of sensory input to interneurons and motoneurons at spinal cord levels.

3. Motoneurons, located in the _____ (D, dorsal; V, ventral) horn of the _____ (G, gray; W, white) matter of the spinal cord, consist of more numerous _____ (A, alpha; M, gamma) motoneurons.

   a. D, G, A   c. D, W, A   e. V, G, A
   b. D, W, M   d. V, W, M   f. V, G, M

4. Gamma motoneurons, in comparison with alpha motoneurons, are _____ (L, larger; S, smaller)-sized neurons with axons averaging _____ microns in diameter which innervate _____ (E, extrafusal; I, intrafusal) muscle fibers.

   a. L, 12, E   c. L, 5, E   e. S, 5, E
   b. L, 12, I   d. S, 5, I   f. S, 12, I

5. An alpha motoneuron with a single _____ (M, myelinated; U, unmyelinated) axon serves as a component of _____ (S, a single; L, multiple) motor unit(s), and receives a higher percentage of synaptic terminals from _____ (D, dorsal root ganglion cells; I, interneurons).

   a. M, S, D   c. M, L, D   e. U, S, D
   b. M, S, I   d. M, L, I   f. U, L, I

6. Interneurons, when compared to spinal cord motoneurons, are generally _____ (M, more; L, less) numerous, _____ (G, larger; S, smaller) in size, and include excitatory _____ (A, as well as; N, but not) inhibitory neurons.

   a. M, G, A   c. M, S, A   e. L, G, A
   b. M, S, N   d. L, S, N   f. L, G, N

7. Repetitive stimulation of the central end of a severed ventral root results in _____ (I, increased; N, no significant change in; D, decreased) activity of adjacent motoneuronal pools as a consequence of _____ (G, the "gamma loop"; R, Renshaw cell excitation; BM, the Bell-Magendie Law).

   a. I, G   c. D, R   e. I, BM
   b. N, BM   d. I, R   f. D, G

8. Collateral axonal branches of _____ (A, alpha; G, gamma) motoneurons result in post-synaptic _____ (E, excitation; I, inhibition) of Renshaw cells through the chemical mediator _____ (GABA, gamma aminobutyric acid; C, acetylcholine).

   a. A, E, GABA   d. G, I, C
   b. A, E, C   e. G, E, GABA
   c. A, I, GABA   f. G, E, C

9. Renshaw cells located within the _____ (V, ventral; IL, interomediolateral) horns of the spinal cord result in postsynaptic _____ (E, excitation; I, inhibition) of adjacent _____ (A, alpha; G, gamma) motoneurons.

   a. V, E, A   c. V, I, A   e. IL, E, A
   b. V, I, G   d. IL, I, G   f. IL, E, G

10. Motoneurons of motor units generally utilize _____ (H, higher; L, lower) firing frequencies than interneurons and utilize Renshaw cells as a mechanism for _____ (I, increasing; D, decreasing) spatial contrast of their transmitted signals.

    a. H, I   c. L, I
    b. H, D   d. L, D

11. Afferent nerves originating from cell bodies of origin in the _____ (DR, dorsal root; PV, prevertebral) ganglia transmit information in ascending sensory spinal cord tracts _____ (A, as well as; N, but not) for multisegmental cord reflexes.

    a. DR, A   c. PV, A
    b. DR, N   d. PV, N

12. Propriospinal fibers are ascending _____ (A, as well as; N, but not) descending spinal cord fibers, which originate in the spinal cord and terminate in the _____ (T, thalamus; SC spinal cord) and comprise about 50% of the _____ (G, gray; W, white) matter of the spinal cord.

    a. A, T, W   c. A, SC, W   e. N, SC, W
    b. A, T, G   d. N, SC, G   f. N, T, G

OBJECTIVE 51–3.

Identify the principal function of the muscle spindle, its structural components, and the characteristics of its afferent and efferent innervation.

13. Muscle spindles serve as receptors for changes in muscle _____ (T, tension only; L, length only; B, both length and tension) and are connected in _____ (S, series; P, parallel) with extrafusal skeletal muscle fibers.

   a. T, S   c. B, S   e. L, P
   b. L, S   d. T, P   f. B, P

14. Muscle spindles _____ (A, as well as; N, but not) Golgi tendon organs operate at a subconscious _____ (A, as well as; N, but not) a conscious level of perception.

   a. A, A   c. N, A
   b. A, N   d. N, N

15. Shortening of the more contractile _____ (C, central region; E, ends) of the nuclear bag fibers occurs in response to direct excitation of class/type _____ fibers.

   a. C, Aα c. C, Ia   e. E, Aγ
   b. C, Aγ d. E, Aα f. E, Ia

16. Primary endings are largely _____ (FS, flower spray; AS, annulospiral) endings innervating _____ (B, only nuclear bag; C, only nuclear chain; BC, both nuclear bag and nuclear chain) fibers.

   a. FS, B   c. FS, BC   e. AS, C
   b. FS, C   d. AS, B   f. AS, BC

17. Secondary endings of type _____ fibers innervate the nuclear _____ (B, bag; C, chain) type of intrafusal fibers of the muscle spindles.

   a. Ia, B   c. II, B   e. Ib, C
   b. Ib, B   d. Ia, C   f. II, C

18. Gamma efferent fibers to muscle spindles are _____ (L, larger; S, smaller) in diameter than the corresponding sensory nerves and innervate _____ (B, only nuclear bag; C, only nuclear chain; BC, both nuclear bag and nuclear chain) fibers.

   a. L, B   c. L, BC   e. S, C
   b. L, C   d. S, B   f. S, BC

19. Excitation of gamma efferent fibers _____ (I, increases; D, decreases) the length of the receptor portion of intrafusal fibers and _____ (E, enhances; R, reduces) the discharge frequencies of their sensory fibers.

   a. I, E   c. D, E
   b. I, R   d. D, R

20. Contraction of the intrafusal fibers of the muscle spindles in response to _____ (E, ephaptic. EP, miniature end-plate) transmission at gamma efferent terminals results in _____ (S, only a static; D, only a dynamic; SD, both static and dynamic) response(s) of sensory nerves.

   a. E, S   c. E, SD   e. EP, D
   b. E, D   d. EP, S   f. EP, SD

21. Stretch of skeletal muscles _____ (A, as well as; N, but not) contraction of intrafusal fibers _____ (I, increases; D, decreases) afferent activity in corresponding type _____ fibers primary endings.

   a. A, I, Ia   c. A, D, Ia   e. N, D, Ia
   b. A, I, II   d. N, D, II   f. N, I, II

22. Gamma-s fibers are believed to innervate largely nuclear _____ (B, bag; C, chain) types of intrafusal fibers and to elicit a _____ (M, more; L, less) rapidly adapting receptor response than gamma-d fibers.

   a. B, M   c. C, M
   b. B, L   d. C, L

23. _____ (P, primary; S, secondary) endings of the muscle spindle are thought to function more effectively as rate detectors and exhibit a strong dynamic response and _____ (N, no; W, a weaker) static response.

   a. P, N   c. S, N
   b. P, W   d. S, W

OBJECTIVE 51–4.

Identify the stretch or myotatic reflex, its components and neural organization, and its stimulus-response characteristics.

24. The stretch, myotatic, or _____ (M, mono-synaptic; P, polysynaptic) reflex is initiated by _____ (C, only contraction; S, only stretch; E, either stretch or contraction) of the muscle involved.

    a. M, C   c. M, E   e. P, S
    b. M, S   d. P, C   f. P, E

25. Monosynaptic _____ (E, excitation, I, inhibition) of alpha motoneurons by _____ afferent fibers results in the _____ (S, stretch; NS, negative stretch) reflex.

    a. E, Ia, S   c. E, II, S   e. I, II, S
    b. E, Ia, NS   d. I, II, NS   f. I, Ia, NS

26. A more potent _____ (D, dynamic; C, static) component of the stretch reflex from an agonist muscle results in a greater reflex contraction of the _____ (A, same; S, synergist; T, antagonist) muscle(s).

    a. D, A   c. D, T   e. C, S
    b. D, S   d. C, A   f. C, T

27. Static components of the stretch reflex function as a _____ (P, positive; N, negative) feedback mechanism for muscle length, and the negative stretch reflex functions as a _____ (P, positive; N, negative) feedback mechanism for muscle length.

    a. P, P   c. N, P
    b. P, N   d. N, N

OBJECTIVE 51–5.

Identify the functional roles of the stretch reflex and gamma efferent system during muscle contraction, and the major brain areas for control of the gamma efferent system.

28. The "load reflex" is viewed as a _____ (M, monosynaptic; P, polysynaptic) reflex whose effectiveness is _____ (I, increased; N, not significantly altered; D, decreased) by diminished background states of gamma efferent tone.

    a. M, I   c. M, D   e. P, N
    b. M, N   d. P, I   f. P, D

29. The muscle spindle serves as a comparator for the difference in _____ (T, tension; L, length) between _____ (I, the two types of intrafusal; EI, extrafusal and intrafusal) fibers.

    a. T, I   c. L, I
    b. T, EI   d. L, EI

30. The temporal sequence of events for excitation of skeletal muscle following selective stimulation of gamma efferent fibers would be _____ . (A, alpha motoneuron excitation; E, extrafusal muscle fiber contraction; D, excitation of dorsal root ganglion cells; I, intrafusal muscle fiber contraction; P, excitation of primary endings of Ia fibers)

    a. E, D, I, P, A   d. I, P, D, A, E
    b. E, P, A, D, I   e. D, P, I, A, E
    c. D, I, P, A, E   f. I, D, A, P, E

31. Severing the dorsal roots of the spinal cord _____ muscle tone and _____ the degree of damping of muscular contractions. (I, increases; N, does not significantly alter; D, decreases)

    a. I, D   c. D, D   e. N, I
    b. N, D   d. I, I   f. D, I

32. Proper damping and load responsiveness functions of the muscle spindle is accomplished by a _____ (R, reciprocal decrease; S, simultaneous increase) in gamma efferent activity with excitation of the _____ (L, less; M, more) abundant A alpha motor fibers.

    a. R, L   c. S, L
    b. R, M   d. S, M

33. Muscle spindle activity is transmitted largely to the _____ (S, somesthetic cortex; C, cerebellum), while activity of the gamma efferent system is _____ (I, increased; N, not significantly altered; D, decreased) by excitation of the bulboreticular facilitatory region of the brain stem.

    a. S, I   c. S, D   e. C, N
    b. S, N   d. C, I   f. C, D

OBJECTIVE 51–6.

Identify the significance of altered states of the stretch reflex and their clinical means of testing.

34. Simply striking the patellar tendon with a percussion hammer results in a _____ (T, transient; P, persistent) contraction of the quadriceps muscle as a consequence of its _____ (D, direct; M, monosynaptic reflex; PR, polysynaptic reflex) excitation.

    a. T, D   c. T, PR   e. P, M
    b. T, M   d. P, D   f. P, PR

35. Diffuse lesions of the _____ (I, ipsilateral; C, contralateral) motor area of the cerebral cortex frequently _____ (E, enhance; D, diminish) the _____ (M, dynamic; S, static) stretch reflexes referred to as "muscle jerks."

    a. I, E, M   c. I, D, M   e. C, E, M
    b. I, D, S   d. C, D, S   f. C, E, S

36. Clonus involves neural patterns of organization of the _____ (F, flexion; S, stretch; E, crossed extension) reflex and occurs _____ (L, less; M, more) readily with states of facilitation of spinal cord reflexes.

    a. F, L   c. E, L   e. S, M
    b. S, L   d. F, M   f. E, M

OBJECTIVE 51–7.

Identify the tendon reflex, its receptors, its afferents and central connections, its effector responses, and its functional characteristics.

37. _____ (GTO, Golgi tendon organ; MS, muscle spindle) receptors of the tendon or "clasp knife" reflex are stimulated by _____ (S, only passive stretch; C, only active contraction; E, either passive stretch or active contraction) of skeletal muscles.

    a. GTO, S   c. GTO, E   e. MS, C
    b. GTO, C   d. MS, S   f. MS, E

38. Tendon reflex receptors are primarily _____ (L, length; T, tension) detectors arranged in _____ (S, series; P, parallel) with skeletal muscle fibers.

    a. L, S   c. T, S
    b. L, P   d. T, P

39. Increased activity in class _____ afferent fibers from muscle tendons results in _____ (E, excitation; I, inhibition) of alpha motoneurons supplying the same muscle through central _____ (M, monosynaptic; IN, interneuron) connections.

    a. Ia, E, M   c. Ia, I, M   e. Ib, I, M
    b. Ia, E, IN   d. Ib, I, IN   f. Ib, E, IN

40. Golgi tendon receptor activity _____ (D, does; N, does not) reach conscious awareness and is transmitted primarily through ascending _____ (DC, dorsal column; SC, spinocerebellar; CS, corticospinal) pathways.

    a. D, DC   c. D, CS   e. N, SC
    b. D, SC   d. N, DC   f. N, CS

41. The _____ (S, shortening; L, lengthening) reaction of the tendon reflex of a skeletal muscle is considered to have a relatively _____ (W, lower; H, higher) threshold for excitation than the stretch reflex.

    a. S, W   c. L, W
    b. S, H   d. L, H

42. If the extended limb of a decerebrate animal is forcibly flexed, considerable resistance is met until a point is reached at which the limb suddenly gives way to the force. Afferent activity involved in this sudden loss of resistance arises in _____ (M, muscle; T, tendon) receptors associated with _____ (F, flexor; E, extensor) muscles.

    a. M, F   c. T, F
    b. M, E   d. T, E

43. The tendon reflex possibly operates as a tension _____ (SC, servocontroller; SA, servo-assist) mechanism, while maximal levels of voluntary muscle contractions are more probably limited by _____ (I, intrinsic; S, stretch reflex; T, tendon reflex) mechanisms of skeletal muscle.

    a. SC, I   b. SC, S   c. SC, T   d. SA, I   e. SA, S   f. SA, T

OBJECTIVE 51-8.

Identify the flexor group of reflexes, characteristics of its afferents, central connections, and effector responses, and its functional characteristics.

44. Increasing stimulus intensities to nociceptors generally results in patterns of excitation of predominant physiological _____ (F, flexor; E, extensor) muscles and _____ (I, increasing; N, no significant change in; D, decreasing) duration of effector response.

    a. F, I   c. F, D   e. E, N
    b. F, N   d. E, I   f. E, D

45. The flexor reflex or _____ (CK, "clasp knife"; W, withdrawal) reflex is mediated by only type(s) _____ afferent nerve fiber activity.

    a. CK; I   d. W; II
    b. CK; I & II   e. W; III & IV
    c. CK; II & III   f. W; II, III & IV

46. Prolonged after-discharges of alpha motoneurons is a characteristic of the _____ (M, monosynaptic; D, disynaptic; P, polysynaptic) influence of flexor reflex activity in type _____ afferent fibers.

    a. M, Aα c. P, Aα e. D, C
    b. D, Aα d. M, C   f. P, C

47. Flexor reflexes are especially highly developed for the _____ (T, trunk; E, extremities) of the body, and possess a pattern of withdrawal which is relatively _____ (I, independent of; C, unique and characteristic for) differing cutaneous regions of stimulation.

    a. T, I   c. E, I
    b. T, C   d. E, C

48. Increased stimulus intensities to a cutaneous nerve generally results in _____ (I, increased; N, no significant change in; D, decreased) latencies of onset of flexion withdrawal reflexes as a consequence of mechanisms of _____ (S, temporal and spatial summation; R, reciprocal inhibition).

    a. I, S   c. D, S   e. N, R
    b. N, S   d. I, R   f. D, R

OBJECTIVE 51-9.

Identify the principle of reciprocal innervation and its significance. Identify the crossed extensor reflex and the general properties of flexor reflexes.

49. Reflex contraction of agonist muscles results in patterns of _____ (E, excitation; I, inhibition) of their antagonist muscles through _____ (CE, crossed extensor reflexes; RI, reciprocal innervation; RP, rebound phenomena).

    a. E, CE   c. E, RI   e. I, RP
    b. E, RP   d. I, CE   f. I, RI

50. The pattern of reciprocal innervation for the lower extremities is such that flexor reflex excitation of _____ (X, flexor; E, extensor) muscles in the same limb results in a pattern of _____ (F, facilitation; I, inhibition) of contralateral flexor muscles and _____ (F, facilitation; I, inhibition) of contralateral extensor muscles.

    a. X, F, F   c. X, I, F   e. E, F, I
    b. X, F, I   d. X, I, I   f. E, I, F

51. Class Ia afferent fiber activity from an agonist muscle results in _____ (F, facilitation; I, inhibition) of alpha motoneurons to antagonist muscles through _____ (M, monosynaptic connections; IN, interneurons).

   a. F, M   c. I, M
   b. F, IN   d. I, IN

52. The crossed extensor reflex is a companion reflex to the _____ (S, stretch; T, tendon; F, flexor) reflex and results in a _____ (M, similar; O, opposing) pattern of movement between contralateral extremities.

   a. S, M   c. F. M   e. T, O
   b. T, M   d. S, O   f. F, O

53. The _____ reflex generally has an earlier onset, and the _____ reflex is characterized by a relatively longer period of after-discharge. (F, flexor; CE, crossed extensor)

   a. F, F   c. CE, F
   b. F, CE   d. CE, CE

54. The _____ (LS, "local sign"; R, rebound; RI, reciprocal inhibition) phenomenon refers to the observation that a subsequent reflex of the same type is _____ (M, more; L, less) readily elicited in the interval immediately following a flexor reflex.

   a. LS, M   c. RI, M   e. R, L
   b. R, M   d. LS, L   f. RI, L

OBJECTIVE 51–10.

Identify the reflex patterns and functional characteristics of the positive supporting reaction, the magnet reaction, cord "righting" reflexes, rhythmic stepping of a single limb, reciprocal stepping of opposite limbs, diagonal stepping of all four limbs, the galloping reflex, and the scratch reflex.

55. The bottoms of the feet of decerebrate animals tend to move _____ (T, towards; A, away from) the point of application of applied pressure as a consequence of neuronal patterns giving rise to the _____ (R, righting reflexes; M, magnet reaction; CK, clasp knife reflex).

   a. T, R   c. T, CK   e. A, M
   b. T, M   d. A, R   f. A, CK

56. The righting reflex is considered to be a form of a _____ (L, locomotive; R, postural; M, monosynaptic) reflex which is _____ (A, absent; P, present) in spinal animal preparations.

   a. L, A   c. M, A   e. R, P
   b. R, A   d. L, P   f. M, P

57. The rhythmic stepping reflex of a single limb in a spinal animal more probably results from a combination of the phenomena of _____ (RI, reciprocal inhibition; M, magnet reactions; LI, lateral inhibition) and _____ (PSR, the positive supporting reaction; R, rebound).

   a. RI, PSR   c. LI, PSR   e. M, R
   b. M, PSR   d. RI, R   f. LI, R

58. Stepping reflexes in the spinal animal more generally involve similar responses between _____ (D, diagonally opposite; I, ipsilateral) pairs of upper and lower extremities in a pattern characteristic of the _____ (G, galloping; M, "mark time") reflex.

   a. D, G   c. I, G
   b. D, M   d. I, M

59. The scratch reflex involves a _____ (P, poorly; H, highly) developed "position sense" and involves rhythmic to-and-fro scratching movements in animals which _____ (A, are; N, are not) abolished by sectioning the sensory roots from the oscillating limb.

   a. P, A   c. H, A
   b. P, N   d. H, N

OBJECTIVE 51-11.

Identify the probable causes and consequences of muscle spasms or cramps. Recognize a potential "guarding" function of muscle responses to certain nociceptive stimuli.

60. Muscle spasms accompanying bone fractures generally occur _____ (C, in conjunction with; A, in the absence of) pain, _____ (R, are; N, are not) generally abolished by local anesthetics, and _____ (R, are; N, are not) generally abolished by general anesthetics.

    a. C, R, N   c. C, N, N   e. A, R, N
    b. C, R, R   d. A, N, R   f. A, R, R

61. Muscle spasms generally result from _____ (D, direct; R, reflex) excitation of skeletal muscle and generally produce a relative degree of muscle _____ (H, hyperemia; I, ischemia).

    a. D, H   c. R, H
    b. D, I   d. R, I

62. Muscle cramps generally result from _____ (D, direct; R, reflex) excitation of skeletal muscles and are more frequently relieved by voluntary contractions of the _____ (S, same; A, antagonist) muscles.

    a. D, S   c. R, S
    b. D, A   d. R, A

OBJECTIVE 51-12.

Recognize the existence of segmental autonomic reflexes at spinal cord levels. Identify the general effects of spinal cord transection upon autonomic and somatic functions.

63. The term _____ (SS, spinal shock; M, mass reflexes) refers to widespread excitation of _____ (A, only autonomic; S, only somatic; B, both autonomic and somatic) reflexes by nociceptive stimuli in the spinal animal.

    a. SS, A   c. SS, B   e. M, S
    b. SS, S   d. M, A   f. M, B

64. Spinal cord transection in man is followed by a period of a few _____ (D, days; W, weeks; M, months) during which the spinal cord reflexes are _____ (F, facilitated; I, inhibited).

    a. D, F   c. M, F   e. W, I
    b. W, F   d. D, I   f. M, I

65. Spinal cord transection generally results in a _____ (T, transient; P, permanent) _____ (D, decrease; I, increase) in arterial blood pressure.

    a. T, D   c. P, D
    b. T, I   d. P, I

52

Motor Functions of the Brain Stem and Basal Ganglia — Reticular Formation, Vestibular Apparatus, Equilibrium, and Brain Stem Reflexes

OBJECTIVE 52–1.

Characterize the reticular formation. Identify the general location and general functions of the bulboreticular facilitatory and inhibitory areas.

1. The reticular formation is a diffuse collection of _____ (S, only sensory; M, only motor; SM, both sensory and motor) neurons which is generally regarded as an upward extension of the _____ (V, ventral; D, dorsal; I, intermediate) column of neurons of the gray matter of the spinal cord.

   a. S, D   c. SM, I   e. M, D
   b. M, V   d. S, I   f. SM, V

2. _____ (S, smaller; L, larger) neurons which are largely motor in function are generally located within the more _____ (M, medial; T, lateral) portions of the reticular formation.

   a. S, M   c. L, M
   b. S, T   d. L, T

3. Specific nuclei within the reticular formation, such as the subthalamic, interstitial, and prestitial nuclei, are loci of "preprogrammed" control of _____ (A, acquired motor skills; S, stereotyped movements) and operate _____ (I, relatively independently of; C, in close association with) adjacent areas of the reticular formation.

   a. A, I   c. S, I
   b. A, C   d. S, C

4. The reticular formation receives ascending activity from the spinoreticular tract _____ (A, as well as; N, but not) the spinothalamic system, and descending activity from higher centers including more notably the _____ (S, sensory; M, motor) regions of the cerebral cortex.

   a. A, S   c. N, S
   b. A, M   d. N, M

5. The bulboreticular facilitatory area, representing a _____ (L, larger; S, smaller) region of the reticular formation than the bulboreticular inhibitory area, is localized in the more _____ (CM, caudal and medial; RL, rostral and lateral) regions of the reticular formation.

   a. L, CM   c. S, CM
   b. L, RL   d. S, RL

6. A transection of the brain stem at a level slightly below the vestibular nuclei results in _____ (I, increased; N, no significant change in; D, decreased) peripheral muscle tone as a consequence of a loss of the influence of motor functions of the reticular _____ (F, facilitatory; B, inhibitory) areas.

   a. I, F   c. D, F   e. N, B
   b. N, F   d. I, B   f. D, B

7. The bulboreticular inhibitory area located within the boundaries of the _____ (D, diencephalon; S, mesencephalon; M, medulla) possesses a _____ (L, low; H, high) degree of intrinsic excitability.

   a. D, L   c. M, L   e. S, H
   b. S, L   d. D, H   f. M, H

8. Removal of the influence of the basal ganglia, cerebral cortex, and cerebellum _____ (I, increases; D, decreases) muscle tone throughout the body as a consequence of a loss of their excitatory influence upon the bulboreticular _____ (F, facilitatory; B, inhibitory) area.

   a. I, F   c. D, F
   b. I, B   d. D, B

9. The bulboreticular _____ (I, inhibitory; F, facilitatory) area and the _____ cranial nucleus are particularly important areas for the maintenance of antigravity muscle tone, particularly for the _____ (E, extensor; L, flexor) group of muscles.

    a. I, 6th, E   b. I, 7th, L   c. F, 8th, E   d. F, 8th, L   e. F, 9th, E   f. F, 9th, L

OBJECTIVE 52-2.

Identify the physiologic anatomy of the vestibular apparatus, the mode of excitation of vestibular receptors, and their neuronal connections with the central nervous system.

10. The _____ (V, vestibular; C, cochlear) apparatus detects sensations associated with equilibrium by means of receptors localized _____ (W, within; O, outside) the membranous labyrinth.

    a. V, W   c. C, W
    b. V, O   d. C, O

11. Otoconia are _____ (H, hydroxyapatite; CC, calcium carbonate) crystals found in association with hair cells of the _____ (M, macula; C, cristae) of the _____ (US, utricle and saccule; A, ampullae) of man.

    a. H, M, US   c. H, C, US   e. CC, M, US
    b. H, C, A   d. CC, C, A   f. CC, M, A

12. Otoconia are utilized in the detection of _____ (A, angular; L, linear) accelerating movements of the head by virtue of their _____ (P, piezoelectric properties; M, relative mass).

    a. A, P   c. L, P
    b. A, M   d. L, M

13. Hair cells of the _____ (C, cristae; M, maculae), utilized to detect the orientation of the head with respect to gravity, exhibit a _____ (P, preferential; N, nonpreferential) directional sensitivity to bending movements of their stereocilia.

    a. C, P   c. M, P
    b. C, N   d. M, N

14. When the head is _____ (V, vertical; BF, bent forward about 30 degrees; BB, bent backward about 30 degrees), the external semicircular canals lie parallel to the surface of the earth and the posterior vertical canal on one side is in a plane parallel to the _____ (P, posterior; S, superior) canal on the other side.

    a. V, P   c. BB, P   e. BF, S
    b. BF, P   d. V, S   f. BB, S

15. A gelatinous mass called the _____ (P, cupula; A, ampulla) is located within the _____ (A, ampulla; U, utricles; S, saccules) of the semicircular canals and surrounds the projections of receptor cells of the _____ (C, cristae; M, maculae).

    a. P, A, C   c. P, U, C   e. A, S, C
    b. P, A, M   d. A, U, M   f. A, S, M

16. Right semicircular canal receptor excitation associated with fluid movement into its ampulla occurs in association with left semicircular canal receptor _____ (E, excitation; H, inhibition) and fluid movement _____ (I, into; O, out of) the left ampulla.

    a. E, I   c. H, I
    b. E, O   d. H, O

17. The vestibular system maintains significant central connections with the _____ (G, globose and emboliform; D, dentate; F, fastigial) nuclei and the _____ (N, neocerebellum; FN, flocculonodular lobe).

    a. G, N   c. F, N   e. D, FN
    b. D, N   d. G, FN   f. F, FN

18. Vestibular signals are transmitted to the motor nuclei of the extraocular eye muscles through the _____ (L, lateral lemniscus; MLF, medial longitudinal fasciculus; M, medial lemniscus) and into the spinal cord by means of the vestibulospinal tract _____ (A, as well as; N, but not) the reticulospinal tract.

    a. L, A   c. M, A   e. MLF, N
    b. MLF, A   d. L, N   f. M, N

19. The uvula of the _____ (SC, semicircular canals; CC, cerebral cortex; C, cerebellum) is thought to play a more important functional role in regard to _____ (R, rapid changes in the direction of motion; E, static equilibrium).

    a. SC, R   c. C, R   e. CC, E
    b. CC, R   d. SC, E   f. C, E

OBJECTIVE 52-3.

Identify the functions of the utricle, saccule, and semicircular canals, and their underlying mechanisms of action.

20. A threshold capability for the detection of as little as _____ degree(s) occurs for the perception of an angular change in head position from a near _____ (V, vertical; H, horizontal) position.

    a. 0.5, V   c. 10, V   e. 5, H
    b. 5, V   d. 0.5, H   f. 10, H

21. A forward acceleration _____ (A, as well as; N, but not) a forward uniform velocity of the body results in a _____ (F, forward; B, backward) relative displacement of the otoconia and body reflexes which cause the body to lean _____ (F, forward; B, backward).

    a. A, F, F   c. A, B, F   e. N, B, F
    b. A, F, B   d. N, B, B   f. N, F, B

22. Threshold excitation of the semicircular canals occurs with _____ (A, angular; L, linear) accelerations of about _____ (H, 0.001; T, 0.1; O, 1) _____ (M, mm.; D, degree) per second per second.

    a. L, H, M   c. L, O, M   e. A, T, D
    b. L, T, M   d. A, H, D   f. A, O, D

23. The direction of endolymph movement relative to the semicircular canals, _____ (A, as well as; N, but not) the relative direction of cupula movement, is generally in the _____ (S, same; O, opposite) direction as the rotation of the head during angular acceleration and the _____ (S, same; O, opposite) direction as the rotation of the head during angular deceleration.

    a. A, S, O   c. A, O, O   e. N, S, O
    b. A, O, S   d. N, O, S   f. N, S, S

24. Pairs of semicircular canals which lie more _____ (N, perpendicular; L, parallel) to the plane of head rotation exhibit changes in their afferent discharge frequency which corresponds more closely with relative displacements of their _____ (E, endolymph; C, cupula).

    a. N, E   c. L, E
    b. N, C   d. L, C

25. A rapid angular acceleration of the head and body to a uniform rotational velocity lasting 5 minutes results in a longer interval of relative displacement of the _____ (C, cupula; E, endolymph) of the semicircular canals which outlasts the period of rapid angular acceleration by about _____ (O, 1 to 2 seconds; T, 20 seconds; F, 5 minutes).

    a. C, O   c. C, F   e. E, T
    b. C, T   d. E, O   f. E, F

26. The semicircular canals function largely as _____ (P, a "predictive" organ for; E, an "evaluator" of the extent of existing) malequilibrium, particularly during _____ (S, slow and deliberate; R, rapid and intricate) body movements.

    a. P, S   c. E, S
    b. P, R   d. E, R

OBJECTIVE 52-4.

Identify the vestibular reflexes involved in phasic postural and vestibular "righting" reflexes, the vestibular mechanisms for stabilization of retinal images, and the clinical tests utilized to assess the integrity of vestibular function.

27. Vestibular phasic postural reflexes generally act _____ (S, synergistically with; O, in opposition to) sudden movements of the body axis from a vertical position; e.g., sudden movements of the head and body to the right result in _____ (E, extension; F, flexion) of the right leg.

    a. S, E   c. O, E
    b. S, F   d. O, F

28. Stimulation of the right horizontal semicircular canals by rotation of the head to the right results in extension of the _____ leg during acceleration, extension of the _____ leg during deceleration, and a postrotational tendency to fall to the _____ side. (L, left; R, right)

    a. L, L, R   c. L, R, R   e. R, L, R
    b. L, R, L   d. R, R, L   f. R, L, L

29. Vestibular reflexes, utilizing signals originating predominantly from the _____ (SC, semicircular canals; US, utricle and saccule), act _____ (S, synergistically with; O, in opposition to) spinal cord "righting" reflexes.

    a. SC, S   c. US, S
    b. SC, O   d. US, O

30. Vestibular nystagmus, named for the direction of its fast component of movement, occurs in the _____ (S, same; O, opposite) direction as the angular acceleration of the head and functions to _____ (I, increase; D, decrease) the duration of stationary retinal images during rotation.

    a. S, I   c. O, I
    b. S, D   d. O, D

31. The vestibular system is the primary controller of the rate of movement of the _____ (S, slow; F, fast) component of nystagmus during prerotational _____ (A, as well as; N, but not) postrotational forms of nystagmus.

    a. S, A   c. F, A
    b. S, N   d. F, N

32. Sudden termination of angular rotation to the left in a "Barany chair" normally results in postrotational effects including past pointing and a tendency to fall to the _____ (L, left; R, right), and a relative endolymph movement and _____ (F, fast; S, slow) component of nystagmus to the _____ (L, left; R, right).

    a. L, F, L   c. L, S, L   e. R, F, L
    b. L, S, R   d. R, S, R   f. R, F, R

33. Vestibular disease processes more frequently result from a _____ (S, rather specific; C, combined) alteration of function of the utricles, saccules, and semicircular canals, and result in states of mal-equilibrium which are _____ (I, increased; N, not significantly altered by; D, decreased) by closing the eyes.

    a. S, I   c. S, D   e. C, N
    b. S, N   d. C, I   f. C, D

34. The ice water test is a means of stimulating the _____ (US, utricle and saccule; SC, semicircular canals) by means of _____ (CC, centrifugal control mechanisms; TC, thermal convection currents; PE, pyroelectric phenomena).

    a. US, CC   c. US, PE   e. SC, TC
    b. US, TC   d. SC, CC   f. SC, PE

OBJECTIVE 52–5.

Identify the roles of neck proprioceptors and reflexes, visual information, and general proprioceptive and exteroceptive information from other parts of the body upon equilibrium. Locate and identify the significance of a center for conscious perception of equilibrium.

35. Changing the position of the _____ (H, head upon the body; E, entire body axis) results in a lesser sense of mal-equilibrium as a primary consequence of the opposing action of _____ (V, visual; MS, neck muscle spindle; JR, neck joint receptor) activity upon vestibular information.

    a. H, V   c. H, JR   e. E, MS
    b. H, MS   d. E, V   f. E, JR

36. Ventriflexion of the neck in the decerebrate labyrinthectomized animal results in _____ (I, increased; D, decreased) extensor tone of the forelimbs which is _____ (S, similar; O, opposite) to the influence of vestibular phasic reflexes upon the upper extremities.

    a. I, S   c. D, S
    b. I, O   d. D, O

37. The center for conscious perception of the state of equilibrium is thought to lie in close proximity to the primary receiving area for _____ (V, vision; H, hearing) in the _____ (O, occipital; P, parietal; T, temporal) lobes of the cerebral cortex.

    a. V, O   c. V, T   e. H, P
    b. V, P   d. H, O   f. H, T

OBJECTIVE 52–6.

Identify the functions of the reticular formation and specific brain stem nuclei in controlling subconscious, stereotyped movements.

38. Specific nuclei located within the mesencephalic and lower diencephalic regions function to control largely _____ (I, involuntary; V, voluntary) types of rather _____ (S, simple; P, complex) forms of motor movement.

    a. I, S   c. V, S
    b. I, P   d. V, P

39. The _____ (NP, nucleus precommissuralis; PN, prestitial nucleus; IN, interstitial nucleus) transmits a higher percentage of its activity through the medial longitudinal fasciculus for the control of turning movements of the _____ (B, entire body; E, head and eyes).

    a. NP, B   c. IN, B   e. PN, E
    b. PN, B   d. NP, E   f. IN, E

40. The _____ (M, "magnocellular"; S, "small cellular") portion of the red nucleus primarily _____ (T, transmits information to; R, receives information from) the cerebellum.

    a. M, T   c. S, T
    b. M, R   d. S, R

41. Stimulation of the _____ (M, "magnocellular"; S, "small cellular") portion of the red nucleus more generally results in stereotyped movements of the _____ (E, extremities; A, body axis) through descending activity transmitted by the _____ (CS, corticospinal; RS, rubrospinal) tracts.

    a. M, E, CS   c. M, A, CS   e. S, E, CS
    b. M, A, RS   d. S, A, RS   f. S, E, RS

42. Decerebrate animals with their brain stem sectioned at a level _____ (A, above; B, below) the subthalamus _____ (D, do; N, do not) generally exhibit normal walking patterns of forward progression.

    a. A, D   c. B, D
    b. A, N   d. B, N

43. The _____ (PN, prestitial nucleus; NP, nucleus precommissuralis; SN, subthalamic nucleus) is thought to be an area which controls forward progression in a manner which _____ (P, possesses; L, lacks) purposefulness of locomotion.

    a. PN, P   c. SN, P   e. NP, L
    b. NP, P   d. PN, L   f. SN, L

OBJECTIVE 52–7.

Identify the basal ganglia, their general connections, and their known motor functions.

44. The basal ganglia are largely _____ (S, single; P, paired) structures of the _____ (D, diencephalon; T, telencephalon; M, mesencephalon).

    a. S, D   c. S, M   e. P, T
    b. S, T   d. P, D   f. P, M

45. Anatomically the basal ganglia include predominantly _____ (S, sensory; M, motor) structures of the caudate nucleus, globus pallidus, and _____ (SN, substantia nigra; P, putamen), as well as the claustrum and _____ (A, amygdaloid nucleus; CC, corpus callosum).

    a. S, SN, A   c. M, SN, A   e. M, SN, CC
    b. S, P, CC   d. M, P, A   f. M, P, CC

46. Decortication, which _____ (D, destroys; S, spares) the basal ganglia, results in a more pronounced loss of motor function of the _____ (E, peripheral extremities; T, body trunk).

    a. D, E   c. S, E
    b. D, T   d. S, T

47. The _____ (AC, amygdaloid nucleus and claustrum; CP, caudate nucleus and putamen) function to initiate and regulate gross intentional movements of motor activity that is transmitted by extracorticospinal _____ (A, as well as; N, but not) corticospinal pathways.

    a. AC, A   c. CP, A
    b. AC, N   d. CP, N

48. Electrical stimulation within the _____ (GP, globus pallidus; CP, caudate nucleus and putamen; A, amygdaloid nucleus) frequently locks different body parts into specific positions during gross movements, while widespread stimulation of all basal ganglia generally results in _____ (I, increased; D, decreased) muscle tone throughout the body.

   a. GP, I   c. A, I   e. CP, D
   b. CP, I   d. GP, D   f. A, D

49. The principal function of the globus pallidus is thought to be to _____ (I, initiate; B, provide background muscle tone for) intended movements via connections with the caudate nucleus and putamen _____ (A, as well as; N, but not) the cerebral cortex.

   a. I, A   c. B, A
   b. I, N   d. B, N

50. Feedback pathways from the premotor cortex are thought to involve a more significant sequence of signal transmission of _____ . (PM, premotor cortex; T, thalamus; GP, globus pallidus; CP, caudate nucelus and putamen)

   a. PM, T, GP, CP, & PM
   b. PM, T, CP, GP, & PM
   c. PM, GP, T, CP, & PM
   d. PM, GP, CP, T, & PM
   e. PM, CP, GP, T, & PM
   f. PM, CP, T; GP, & PM

51. Signals transmitted from the motor cortex to the cerebellum via the _____ are transmitted back to the motor cortex via a more significant relay station in the _____ . (R, red nucleus; C, claustrum; P, pons; T, thalamus)

   a. R, T   c. P, T   e. T, C
   b. R, C   d. P, R   f. T, R

OBJECTIVE 52–8.
Identify the clinical syndromes resulting from damage to the basal ganglia.

52. Hemiballismus occurring in one of the left extremities generally results from lesions of the _____ (C, caudate; ST, subthalamic; A, amygdaloid) nucleus on the _____ (R, right; L, left) side.

   a. C, R   c. A, R   e. ST, L
   b. ST, R   d. C, L   f. A, L

53. _____ (C, chorea; P, Parkinson's disease) is more consistently associated with widespread destruction of the substantia nigra and is frequently characterized by muscle rigidity, akinesia, and _____ (I, an intentional tremor; R, a tremor at rest).

   a. C, I   c. P, I
   b. C, R   d. P, R

54. Neurons which originate in the substantia nigra and project to the globus pallidus are thought to be largely _____ (E, excitatory; I, inhibitory) neurons which probably secrete _____ (A, acetylcholine; GABA, gamma aminobutyric acid; D, dopamine).

   a. E, A   c. E, D   e. I, GABA
   b. E, GABA   d. I, A   f. I, D

55. The rigidity of Parkinsonism, _____ (L, like; U, unlike) that of decerebrate rigidity, is primarily _____ (A, an alpha; G, a gamma) type of rigidity.

   a. L, A   c. U, A
   b. L, G   d. U, G

56. Continuous wormlike movements of the upper extremities, which occur more frequently in _____ (P, Parkinsonism; A, athetosis; H, hemiballismus) generally _____ (R, are; N, are not) associated with seriously impaired voluntary motor movements of the affected limbs.

   a. P, R   c. H, R   e. A, N
   b. A, R   d. P, N   f. H, N

OBJECTIVE 52–9.

Contrast the general functions of varying levels of the central nervous system in posture and locomotion.

57. The cerebral cortex is more integrally required for fuller expressions of the _____ (PS, positive supporting; P, placing) and _____ (H, hopping; S, scratching; M, magnet) reactions.

a. PS, H   c. PS, M   e. P, S
b. PS, S   d. P, H   f. P, M

58. Cerebral cortical function is considered to be required for _____ (B, only vestibular; S, only visual; BS, both vestibular and visual) righting reflexes, and _____ (P, purposefulness of; M, mechanisms of forward progression for) locomotion.

a. B, P   c. BS, P   e. S, M
b. S, P   d. B, M   f. BS, M

59. Transection of the medulla at a level below the vestibular nuclei generally results in a state of spinal cord reflex _____ (F, facilitation; H, inhibition), an _____ (A, ability; I, inability) of the animal to stand, and an _____ (A, ability; I, inability) of the animal to exhibit forward progression.

a. F, A, A   c. F, I, I   e. H, A, I
b. F, A, I   d. H, I, I   f. H, A, A

60. The subthalamic animal when compared to a decerebrate animal exhibits a _____ degree of muscle rigidity, a _____ ability to stand, and a _____ capability for forward progression. (G, greater; L, lesser)

a. G, L, G   c. G, G, G   e. L, G, G
b. G, L, L   d. L, G, L   f. L, L, L

61. Different types of animals generally exhibit greater variance of posture and locomotion for comparable _____ (E, decerebrate; O, decorticate) preparations for which brain stem connections with the thalamus and basal ganglia _____ (S, are severed; I, remain intact).

a. E, S   c. O, S
b. E, I   d. O, I

62. Postural muscle contractions play a more important role to _____ (E, maintain equilibrium; G, support the body against gravity) in man when compared to lower animals as a consequence of _____ (T, telencephalization; M, his greater mass; L, the relative location of his center of gravity).

a. E, T   c. E, L   e. G, M
b. E, M   d. G, T   f. G, L

63. The _____ (P, alpha; G, gamma) motoneuron represents the final common pathway for spinal cord reflexes _____ (A, as well as; N, but not) for reflexes of posture and locomotion initiated at higher levels.

a. P, A   c. G, A
b. P, N   d. G, N

53

Cortical and Cerebellar Control of Motor Functions

OBJECTIVE 53–1.

Identify the "area pyramidalis," the pyramidal and extrapyramidal pathways, and their functional significance.

1. The motor cortex _____ (W, as well as; N, but not) the somesthetic cortex lies _____ (A, anterior; P, posterior) to the central sulcus, separating the parietal lobes from the _____ (T, temporal; F, frontal) lobes.

   a. W, A, T   c. W, P, T   e. N, A, T
   b. W, P, F   d. N, P, F   f. N, A, F

2. The area pyramidalis, _____ (C, containing; L, lacking) giant Betz cells, occupies the _____ (E, precentral; P, postcentral) gyrus of the cerebral cortex.

   a. C, E   c. L, E
   b. C, P   d. L, P

3. The area pyramidalis of the right cerebral cortex is considered a _____ (P, primary; S, secondary) _____ (M, motor; R, sensory) area for _____ (I, ipsilateral; C, contralateral) peripheral muscles.

   a. P, M, I   c. P, R, I   e. S, R, I
   b. P, M, C   d. S, R, C   f. S, M, C

4. Large myelinated fibers of the giant Betz cells of the motor cortex, comprising _____ (M, more; L, less) than 10% of the pyramidal tract fibers, generally _____ (D, do; N, do not) decussate before reaching upper spinal cord levels.

   a. M, D   c. L, D
   b. M, N   d. L, N

5. The _____ (V, ventral; L, lateral) corticospinal tract of the spinal cord is comprised largely of fibers whose cell bodies of origin are located on the ipsilateral side of the _____ (M, medulla; X, cerebral cortex; S, spinal cord).

   a. V, M   c. V, S   e. L, X
   b. V, X   d. L, M   f. L, S

6. _____ (A, only alpha motoneurons; B, only giant Betz cells; AB, both alpha motoneurons and giant Betz cells) are thought to "sharpen" the boundaries of their excitatory signals through mechanisms of recurrent _____ (F, facilitation; I, inhibition).

   a. A, F   c. AB, F   e. B, I
   b. B, F   d. A, I   f. AB, I

7. Collateral branches of the pyramidal tract are transmitted to the cerebellum predominately via the _____ (R, red; P, pontile; H, hypothalamic) nuclei and olivocerebellar fibers from the _____ (S, superior; I, inferior) olivary nuclei.

   a. R, S   c. H, S   e. P, I
   b. P, S   d. R, I   f. H, I

8. The tectospinal and vestibulospinal tracts, _____ (A, as well as; N, but not) the rubrospinal and _____ (C, corticospinal; R, reticulospinal) tracts, are parts of the _____ (P, pyramidal; E, extrapyramidal) pathway to the spinal cord.

   a. A, C, P   c. A, R, P   e. N, C, P
   b. A, R, E   d. N, R, E   f. N, C, E

9. Localized threshold stimuli applied within the region of the primary motor area of the cortex are more likely to result in relatively _____ (F, diffuse; C, discrete) patterns of motor excitation of muscles via predominantly _____ (P, pyramidal; E, extrapyramidal) pathway transmission.

   a. F, P   c. C, P
   b. F, E   d. C, E

OBJECTIVE 53–2.

Distinguish the different motor areas of the sensorimotor cortex. Identify the primary motor cortex, its neuroanatomical organization, its spatial representation of muscle groups, and its functional significance.

10. The somatotopical map of the somatic sensory area _____ more closely approximates a mirror image of the map of the _____ (P, primary; S, supplementary) motor area located in the _____ (R, parietal; F, frontal; T, temporal) lobes.

    a. I, P, R   c. I, S, R   e. II, S, T
    b. I, P, F   d. II, S, R   f. II, P, F

11. Avoidance movements frequently result from stimulation of the _____ (P, primary; S, supplementary) motor area in the vicinity of the _____ (L, lateral; M, medial) wall of the hemisphere in the _____ (G, longitudinal; V, sylvian) fissure.

    a. P, L, G   c. P, M, G   e. S, M, G
    b. P, M, V   d. S, L, V   f. S, M, V

12. Discrete muscle movements relating to swallowing, chewing, and facial movements are elicited more readily from stimulation of the _____ (M, medial; L, lateral) portions of the _____ (P, primary; S, supplemental) motor areas.

    a. M, P   c. L, P
    b. M, S   d. L, S

13. The _____ (CC, corpus callosum; IC, internal capsule; EC, external capsule) is a major structure connecting the two sides of the cerebral hemispheres for transmission of motor _____ (A, as well as; N, but not) sensory information.

    a. CC, A   c. EC, A   e. IC, N
    b. IC, A   d. CC, N   f. EC, N

14. The primary motor cortex receives afferent signals from nonspecific _____ (A, as well as; N, but not) specific thalamic nuclei, the basal ganglia and cerebellum via the _____ (M, medial; L, lateral) nuclei of the thalamus, and subcortical fibers from adjacent somatic sensory areas.

    a. A, M   c. N, M
    b. A, L   d. N, L

15. Layers _____ of the motor cortex contain neurons which give rise to efferent or output fibers, with the large Betz cells located in the _____ (S, more superficial; D, deeper) of the two layers.

    a. I & IV, S   c. V & VI, S   e. II & III, D
    b. II & III, S   d. I & IV, D   f. V & VI, D

16. Different types of afferent signal result in patterns of excitation of _____ (V, vertical; H, horizontal) columns of cells within the _____ primary layers of the motor cortex.

    a. V, 6   c. V, 10   e. H, 8
    b. V, 8   d. H, 6   f. H, 10

17. Finger muscles generally have a _____ surface area of representation within the motor cortex and a _____ motor unit size when compared to muscles of the upper arm. (G, greater; L, lesser)

    a. G, G   c. L, G
    b. G, L   d. L, L

OBJECTIVE 53–3.

Identify the functional significance and location of the "premotor cortex" and associated Broca's area, voluntary eye movement area, head rotation area, and area for hand skills.

18. The premotor cortex is located within the _____ (F, frontal; R, parietal) lobe immediately _____ (A, anterior; P, posterior) to the primary _____ (M, motor cortex; SS, somatosensory area).

    a. F, A, M   c. F, P, M   e. R, P, SS
    b. F, A, SS   d. R, P, M   f. R, A, SS

19. _____ (H, the area of Heschl; B, Broca's area) is located within the _____ (M, medial; L, lateral) aspect of the _____ (T, primary motor; P, premotor) cortex.

    a. H, M, T   c. H, L, T   e. B, L, T
    b. H, M, P   d. B, L, P   f. B, M, P

20. Damage to the _____ (V, voluntary; I, involuntary) eye movement fields of the frontal lobes results in a loss of the ability to _____ (X, maintain visual fixation; M, move the eyes toward different ojbects).

    a. V, X   c. I, X
    b. V, M   d. I, M

21. Motor _____ (P, paralysis; X, apraxia) results in damage to the area for hand skills which lies in close association with the primary _____ (SS, somatosensory; M, motor) cortex for the hands and fingers.

    a. P, SS   c. X, SS
    b. P, M   d. X, M

OBJECTIVE 53–4.

Identify the consequences of damage to the motor and premotor cortical regions.

22. Destruction of the primary motor cortex results in a loss of _____ (D, predominantly discrete; A, all) types of voluntary muscle movement, whereas subcortical motor areas, particularly the _____ (T, thalamic; B, basal ganglia; C, cerebellar) structures, are responsible for involuntary body movements.

    a. D, T   c. D, C   e. A, B
    b. D, B   d. A, T   f. A, C

23. Muscle tone is normally continuously _____ by pyramidal tract activity and _____ by extrapyramidal tract activity transmitted through the basal ganglia and bulboreticular system. (F, facilitated; I, inhibited)

    a. F, F   c. I, F
    b. F, I   d. I, I

24. Pathologic lesions restricted to the _____ (M, motor cortex; S, larger portion of the sensorimotor cortex) are more likely to result in muscle spasms as a consequence of a greater alteration of _____ (P, pyramidal; E, extrapyramidal) tract activity.

    a. M, P   c. S, P
    b. M, E   d. S, E

25. Strokes which involve the motor cortex generally _____ (D, do; N, do not) inflict damage to the basal ganglia, and result in muscle _____ (S, spasticity; F, flaccidity) on the _____ (I, ipsilateral; C, contralateral) side of the body.

    a. D, S, I   c. D, F, I   e. N, F, I
    b. D, S, C   d. N, F, C   f. N, S, C

26. The "Babinski sign" results from destruction of either the _____ (H, hand; F, foot) region of the motor cortex or the corresponding region of the _____ (P, pyramidal; E, extrapyramidal) tract.

    a. H, P   c. F, P
    b. H, E   d. F, E

27. Severe muscle spasms resulting from damage to the _____ (P, pyramidal; EP, extrapyramidal) system involve _____ (E, primarily the extensors; B, both extensor and flexor muscles) of the body.

    a. P, E   c. EP, E
    b. P, B   d. EP, B

28. Drawing a blunt instrument along the inner aspect of a subject's foot with moderate pressure normally results in the toes bending _____ (U, upward; D, downward) and constitutes a _____ (P, positive; N, negative) "Babinski sign."

    a. U, P   c. D, P
    b. U, N   d. D, N

OBJECTIVE 53–5.

Identify the general organizational patterns of innervation of spinal motoneurons by pyramidal and extrapyramidal tract terminals.

29. The pyramidal tract when compared to the extrapyramidal tract generally terminates more _____ (A, anteriorly; P, posteriorly) within the spinal cord gray matter and has a comparatively _____ (G, greater; L, lesser) percentage of direct synaptic contacts with anterior motoneurons.

    a. A, G   c. P, G
    b. A, L   d. P, L

30. The corticospinal tract, containing a _____ (J, majority; N, minority) of fibers which make direct synaptic contact with anterior motoneurons, exhibits a _____ (M, more; L, less) specific control of muscle contractions than the extrapyramidal system.

    a. J, M   c. N, M
    b. J, L   d. N, L

31. The extrapyramidal tract has an overall _____ (F, facilitatory; I, inhibitory) influence upon anterior motoneurons through its _____ (D, direct synaptic; IN, interneuronal) connections.

    a. F, D   c. I, D
    b. F, IN   d. I, IN

32. Excitation of an agonist muscle via the corticospinal tract results in an intrinsic pattern of _____ (F, facilitation; I, inhibition) of antagonist muscles as a consequence of _____ (SC, spinal cord organizational patterns; EP, reciprocal extrapyramidal tract activity).

    a. F, SC   c. I, SC
    b. F, EP   d. I, EP

OBJECTIVE 53–6.

Identify the major subdivisions and general functions of the cerebellum.

33. Stimulation of areas of the cerebellum generally _____ result in motor movements of skeletal muscle, _____ result in motor movements of visceral structures, and _____ result in sensory experiences. (D, does; N, does not)

    a. N, D, D   c. N, N, N   e. D, D, N
    b. N, D, N   d. D, N, D   f. D, D, D

34. The cerebellum functions largely as a _____ (R, specific relay station; U, memory unit; C, comparator) for _____ (M, predominantly motor; S, predominantly sensory; D, differences between sensory and motor) types of information.

    a. R, M   c. C, D   e. R, S
    b. U, S   d. C, M   f. U, D

35. Removal of the cerebellum results in a greater abnormality of _____ (S, sensory modalities; M, motor responses) during _____ (R, rapidly changing; P, static postural) patterns of muscle contractions.

    a. S, R   c. M, R
    b. S, P   d. M; P

36. The neocerebellum occupies a _____ (J, major; N, minor) portion of the cerebellar hemispheres in the larger _____ (A, anterior; P, posterior) lobe of the cerebellum.

    a. J, A   c. N, A
    b. J, P   d. N, P

37. The _____ (A, archicerebellum; N, neocerebellum; P, paleocerebellum) includes the anterior cerebellum and the _____ (L, lateral; M, midline) section(s) of the posterior cerebellum.

    a. A, L   c. P, L   e. N, M
    b. N, L   d. A, M   f. P, M

OBJECTIVE 53-7.

Identify the major afferent pathways to the cerebellum and their spatial projection within the cerebellum.

38. The corticopontocerebellar tract, originating mainly in the _____ (N, sensory; M, motor) cortex, projects largely to the _____ (I, ipsilateral; C, contralateral) cerebellar regions.

    a. N, I
    b. N, C
    c. M, I
    d. M, C

39. The olivocerebellar tract from the _____ (S, superior; I, inferior) olives transmits information to the cerebellum from the motor cortex, basal ganglia, and reticular formation, _____ (A, as well as; N, but not) the spinal cord.

    a. S, A
    b. S, N
    c. I, A
    d. I, N

40. Vestibulocerebellar fibers project to the _____ (L, lingula; F, flocculonodular lobe), located in the more _____ (R, rostral; C, caudal) region of the _____ (A, anterior; P, posterior) lobe.

    a. L, R, A
    b. L, R, P
    c. L, C, A
    d. F, C, P
    e. F, C, A
    f. F, R, P

41. The spinocerebellar tracts transmit signals to the _____ (I, ipsilateral; C, contralateral) side of the cerebellum at relatively _____ (H, higher; L, lower) propagation velocities when compared to transmission in the dorsal column pathways.

    a. I, H
    b. I, L
    c. C, H
    d. C, L

42. Clarke's column, located in the _____ (D, dorsal; V, ventral) region of the spinal cord, is primarily a locus for _____ order neurons of the _____ (SR, spinoreticular; SC, spinocerebellar) tract pathways to the cerebellum.

    a. D, 2nd, SR
    b. D, 2nd, SC
    c. D, 3rd, SR
    d. V, 3rd, SC
    e. V, 3rd, SR
    f. V, 2nd, SC

43. Spatial localization of sensory input within the cerebellum occurs in a _____ (S, single; D, dual) representation, grouped primarily according to peripheral _____ (L, location; T, types) of sensory receptors.

    a. S, L
    b. S, T
    c. D, L
    d. D, T

44. Homuncular representations of the left cerebellar hemispheres correspond to homuncular representations of the _____ cerebral hemispheres, patterns of sensory information originating in the _____ side of the body and motor information for muscles on the _____ side of the body. (L, left; R, right)

    a. L, L, L
    b. L, R, R
    c. L, L, R
    d. R, R, R
    e. R, L, L
    f. R, L, R

OBJECTIVE 53-8.

Identify the deep cerebellar nuclei, their general functions and neural patterns of organization, and the efferent pathways from the cerebellum.

45. Input signals generally project directly to the cerebellar cortex _____ (W, as well as; N, but not) the deep cerebellar nuclei, while efferent tracts originate from _____ (CN, only the deep cerebellar nuclei; CC, only the cerebellar cortex; B, both the deep cerebellar nuclei and the cerebellar cortex).

    a. W, CN
    b. W, CC
    c. W, B
    d. N, CN
    e. N, CC
    f. N, B

46. The majority of fibers for a major pathway originating in the _____ (F, fastigial; N, dentate) nuclei transmit their activity _____ (D, directly; T, via thalamic second order neurons) to the motor cortex.

    a. F, D
    b. F, T
    c. N, D
    d. N, T

47. The red nucleus is a major relay station between cerebellar _____ (A, afferent; E, efferent) fibers and _____ (S, spinal cord levels; N, deep cerebellar nuclei; CC, the cerebral cortex).

    a. A, S   c. A, CC   e. E, N
    b. A, N   d. E, S   f. E, CC

48. The _____ (G, globose and emboliform; F, fastigial; D, dentate) nuclei function primarily in conjunction with the flocculonodular lobe and the _____ (N, neocerebellum; V, vestibular nuclei).

    a. G, N   c. D, N   e. F, V
    b. F, N   d. G, V   f. D, V

OBJECTIVE 53–9.

Identify the cytoarchitectural characteristics of the cerebellar cortex, the basic patterns of neuronal organization of the cerebellum, and the probable functional characteristics of its neuronal circuitry.

49. A cross-section of a single fold or _____ (G, gyrus; S, sulcus; F, folium) of the cerebellar cortex exhibits _____ major layers.

    a. G, 3   c. F, 3   e. S, 6
    b. S, 3   d. G, 6   f. F, 6

50. The functional unit of the cerebellum centers around the _____ (G, granular; P, Purkinje; L, Golgi) cell whose cell body is located in the _____ (I, innermost; M, intermediate; O, outermost) layer of the cerebral cortex.

    a. G, I   c. L, O   e. P, I
    b. P, M   d. G, M   f. L, O

51. The deep cerebellar nuclei receive a predominantly _____ influence from direct connections from afferent fibers entering the cerebellum and a predominantly _____ influence from cerebellar Purkinje cells. (E, excitatory; I, inhibitory)

    a. E, E   c. I, E
    b. E, I   d. I, I

52. Climbing fibers originate in the _____ (C, cerebral cortex; O, inferior olives), result in _____ (E, excitation; I, inhibition) of the deep cerebellar nuclei via collateral branches, and _____ (E, excite; I, inhibit) Purkinje cells.

    a. C, E, I   c. C, I, I   e. O, E, I
    b. C, I, E   d. O, I, E   f. O, E, E

53. Mossy fibers _____ (E, excite; I, inhibit) neurons of the deep cerebellar nuclei via collateral branches, and _____ (E, excite; I, inhibit) Purkinje cells _____ (D, directly; G, via intermediate granular cells).

    a. E, E, D   c. E, I, D   e. I, I, D
    b. E, E, G   d. I, I, G   f. I, E, G

54. Climbing fibers, when compared with mossy fibers, are _____ (M, more; L, less) numerous, originate from _____ (M, more; L, less) diverse loci, and exhibit a _____ (G, greater; S, lesser) fidelity of signal transmission to Purkinje cells.

    a. M, M, G   c. M, L, G   e. L, L, G
    b. M, M, S   d. L, L, S   f. L, M, S

55. Corresponding inhibition of deep cerebellar nuclear cells by means of _____ (G, Golgi; P, Purkinje; C, cerebellar afferent) axons generally _____ (E, precedes; F, follows) deep cerebellar excitation.

    a. G, E   c. C, E   e. P, F
    b. P, E   d. G, F   f. C, F

56. Feedback pathways involving cerebellar Purkinje cells are thought to function largely as a _____ (R, reverberating; D, delay-line) type of _____ (P, positive; N, negative) feedback.

    a. R, P   c. D, P
    b. R, N   d. D, N

57. Basket cells and stellate cells, _____ (W, as well as; B, but not) Golgi cells are _____ (E, excitatory; I, inhibitory) cells located within the _____ (N, deep cerebellar nuclei; C, cortex) of the cerebellum.

    a. W, E, N   c. W, I, N   e. B, E, N
    b. W, I, C   d. B, I, C   f. B, E, C

OBJECTIVE 53–10.

Identify the functional roles of the cerebellum in voluntary movements and extramotor predictive functions.

58. The cerebellum receives afferent signals from _____ (E, only the extremities; EM, both the extremities and the motor cortex) during voluntary movements of the extremities, and transmits efferent activity to the motor cortex primarily via the _____ (G, globose and emboliform; D, dentate) nucleus and a relay station in the _____ (T, thalamus; R, red nucleus).

    a. E, G, T   c. E, D, T   e. EM, D, T
    b. E, G, R   d. EM, D, R   f. EM, G, R

59. "Error signals," _____ (R, received; P, produced) by the cerebellum during voluntary muscle contractions, reflect comparisons of the "intentions" of the _____ (B, cerebellum; X, cerebral cortex) with "performance" of the _____ (B, cerebellum; MS, musculoskeletal system; X, cerebral cortex).

    a. R, B, X   c. R, B, MS   e. P, X, MS
    b. R, X, MS   d. P, X, B   f. P, B, X

60. Learned knowledge of the _____ (S, sequence of initiation of motor patterns; I, inertia of the musculoskeletal system), stored in the _____ (B, cerebellar; C, cerebral) cortex, is a more important feature of cerebellar function.

    a. S, B   c. I, B
    b. S, C   d. I, C

61. The cerebellum functions largely at a _____ (C, conscious; S, subconscious) level to _____ (E, excite; I, inhibit) the agonist muscle and to _____ (E, excite; I, inhibit) the antagonist muscles during rapid voluntary movements of the extremities.

    a. C, E, E   c. C, I, E   e. S, I, E
    b. C, E, I   d. S, I, I   f. S, E, I

62. _____ (I, intentional tremors; R, tremors at rest) are more characteristic of _____ (F, excessive facilitation; L, loss of function) of the cerebellum.

    a. I, F   c. R, F
    b. I, L   d. R, L

63. The cerebellum is thought to provide "damping" functions through the transmission of signals into the spinal cord which _____ (F, only facilitate; FI, facilitate as well as inhibit) _____ (G, only gamma; A, only alpha; AG, gamma as well as alpha) motoneurons.

    a. F, G   c. F, AG   e. FI, A
    b. F, A   d. FI, G   f. FI, AG

64. Disturbances of cerebellar function result in disturbances of motor _____ (A, and; N, but not) extramotor predictive functions and an inability to regulate the distance of movement of body parts, _____ (W, as well as; N, but not) the normal progression sequence of rapid motor movements.

    a. A, W   c. N, W
    b. A, N   d. N, N

OBJECTIVE 53–11.

Identify the functional roles of the cerebellum in involuntary movements and equilibrium.

65. Involuntary types of motor movements transmit associated activity to the _____ (N, neocerebellum; PC, paleocerebellum) largely via the _____ (P, pyramidal; E, extrapyramidal) tracts and a relay station in the _____ (S, superior; I, inferior) olives.

    a. N, P, S   c. N, E, S   e. PC, E, S
    b. N, P, I   d. PC, E, I   f. PC, P, I

66. Efferent signals from the paleocerebellum are thought to be transmitted to the basal ganglia largely via the _____ (D, dentate; E, emboliform; F, fastigial) nucleus and the _____ (O, inferior olives; T, thalamus).

    a. D, O   c. F, O   e. E, T
    b. E, O   d. D, T   f. F, T

67. The cerebellum transmits corrective signals to the spinal cord during involuntary movements largely via the _____ (CS, corticospinal; RS, reticulospinal) tracts, which are thought to exert a greater direct influence upon _____ (A, alpha; G, gamma) motoneurons.

    a. CS, A   c. RS, A
    b. CS, G   d. RS, G

68. Destruction of the cerebellum _____ (I, increases; D, decreases) the resistance to sudden movements of the extremities as a consequence of _____ (E, enhanced; R, reduced) sensitivity of stretch reflexes.

    a. I, E   c. D, E
    b. I, R   d. D, R

69. Removal of the _____ (N, nodulus; F, flocculus) results in a more complete and permanent loss of function of the vestibular system in _____ (S, static; D, dynamic) equilibrium.

    a. N, S   c. F, S
    b. N, D   d. F, D

OBJECTIVE 53–12.

Identify the major clinical symptoms resulting from cerebellar abnormalities.

70. The _____ (R, rapid; S, slow) performance of motor movements appear almost normal with destruction of comparatively large areas of the cerebellar cortex, particularly if the deep cerebellar nuclei _____ (I, remain intact; D, are also destroyed).

    a. R, I   c. S, I
    b. R, D   d. S, D

71. Cerebellar disease symptoms include _____ (N, nystagmus; R, dysarthria; D, dysmetria), a condition in which motor movements of the extremities generally _____ (S, fall short; O, overshoot) their intended target.

    a. N, S   c. D, S   e. R, O
    b. R, S   d. N, O   f. D, O

72. Destruction of the deep cerebellar nuclei, particularly the dentate nucleus, results in _____ (I, increased; D, decreased) muscle tone on the _____ (P, ipsilateral; C, contralateral) side as the lesion.

    a. I, P   c. D, P
    b. I, C   d. D, C

73. A failure of "progression of motor movements" occurring with cerebellar abnormalities is termed _____ (D, dysdiadochokinesia; A, aphasia; X, alexia), of which dysarthria _____ (I, is; N, is not) an example.

    a. D, I   c. X, I   e. A, N
    b. A, I   d. D, N   f. X, N

74. A cerebellar nystagmus resulting more frequently from damage to the _____ (L, lingula; F, flocculonodular lobe), past pointing, rebound, and _____ (R, a tremor of rest; I, an intentional tremor) are manifestations of cerebellar deficits.

    a. L, R   b. L, I   c. F, R   d. F, I

OBJECTIVE 53-13.

Identify the mechanisms of utilization of feedback control in motor functions and the relative role of the motor cortex in the initiation of motor activity.

75. Primary somatosensory areas receive sensory signals _____ transmit motor signals, and primary motor cortices transmit motor signals _____ receive sensory signals. (W, as well as; N, but does not)

    a. W, W   c. N, W
    b. W, N   d. N, N

76. Sensory engrams for "learned" motor performance are thought to be largely stored in association with the _____ (S, sensory; M, motor) cortical region and are generally able to be repeated more rapidly when _____ (V, visual; O, somatic proprioceptor) feedback is utilized for their execution.

    a. S, V   c. M, V
    b. S, O   d. M, O

77. "Skilled" motor patterns are thought to be stored in association with predominantly _____ (S, sensory; M, motor) cortical regions and _____ (A, are; N, are not) considered to be limited by sensory feedback control characteristics in the rapidity of their execution.

    a. S, A   c. M, A
    b. S, N   d. M, N

78. The major component of feedback mechanisms for dissociating muscle load from muscle performance during skilled motor activities is hypothetically considered to be the:

    a. Cerebellum
    b. Somatosensory area I
    c. Basal ganglia
    d. Hypothalamus
    e. Vestibular nuclei
    f. Geniculate bodies

54

Activation of the Brain — the Reticular Activating System; the Diffuse Thalamocortical System; Brain Waves; Epilepsy; Wakefulness and Sleep

OBJECTIVE 54–1.
Identify the reticular activating system, its major pathways and their respective functions, and the mechanisms of their excitation.

1. Stimulation of the _____ (CM, caudal and medial; RL, rostral and lateral) portions of the reticular formation enhances mental alertness, whereas interruption of the ascending reticular activating system results in a _____ (C, less attentive but conscious; M, comatose) state.

   a. CM, C   c. CM, M
   b. RL, C   d. RL, M

2. Normal wakefulness is attributed largely to activity of the _____ (T, thalamic; B, brain stem) portion of the reticular activating system, because stimulation of the thalamic portion of the reticular activating system results in a more _____ (G, general; S, specific) activation of regions of the cerebral cortex.

   a. T, G   c. B, G
   b. T, S   d. B, S

3. The pathway from the brain stem portion of the reticular formation through the intralaminar, midline, and reticular nuclei of the thalamus results in a relatively _____ (S, specific; D, diffuse) activation of regions of the cerebral cortex _____ (W, as well as; N, but not) the basal ganglia.

   a. S, W   c. D, W
   b. S, N   d. D, N

4. The reticular activating system is effectively stimulated by comparatively _____ (S, specific; D, diverse) types of relatively _____ (H, high; L, low) magnitudes of sensory stimuli.

   a. S, H   c. D, H
   b. S, L   d. D, L

5. _____ (T, tactile; P, pain) receptors, _____ (W, as well as; N, but not) proprioceptive somatic afferent activity, are included as potent activators of the reticular activating system.

   a. T, W   c. P, W
   b. T, N   d. P, N

6. Reticular activating system activity is _____ (U, not significantly altered; D, decreased) during states of sleep, and is influenced by ascending spinal cord tracts _____ (W, as well as; N, but not) by activity originating within the sensorimotor cortex.

   a. U, W   c. D, W
   b. U, N   d. D, N

OBJECTIVE 54–2.

Identify and contrast the characteristics of the diffuse and specific thalamocortical systems.

7. The _____ (E, epithalamus; T, thalamus) is a specific relay station for the transmission of all sensory modalities to the cerebral cortex with the exception of _____ (V, vision; G, gustation; O, olfaction).

   a. E, V   c. E, O   e. T, G
   b. E, G   d. T, V   f. T, O

8. The paleospinothalamic pathway for pain terminates in association with a thalamocortical system having a relatively _____ thalamic neuronal association and a relatively _____ projection to regions of the cerebral cortex. (S, specific; D, diffuse)

   a. S, S   c. D, S
   b. S, D   d. D, D

9. Thalamic stimulations of the _____ (P, specific; D, diffuse) thalamocortical system results in a shorter latency period for cortical activation and a comparatively _____ (L, longer; S, shorter) duration of response.

   a. P, L   c. D, L
   b. P, S   d. D, S

10. The specific thalamic nuclei project largely to layer(s) _____ of the cerebral cortex, whereas the diffuse thalamic system projects mainly to layer(s) _____ .

    a. I & II, V & VI   d. IV, V & VI
    b. I & II, IV   e. V & VI, IV
    c. IV, I & II   f. V & VI, I & II

11. Stimulation of the _____ (S, specific; D, diffuse) thalamocortical system at a repetitive frequency of 10 to 12 stimuli per second is more likely to result in _____ (F, fatigue; R, a recruiting response) of the cerebral cortex.

    a. S, F   c. D, F
    b. S, R   d. D, R

OBJECTIVE 54–3.

Identify the probable roles of the reticular activating system in directing attention to various aspects of the mental environment, its possible "scanning" and "programming" functions, and mechanisms of action of barbiturate anesthesia.

12. More probable candidates for structures which regulate the overall degree of mental "attentiveness" and direct attention to specific aspects of the mental environment are the _____ (T, thalamic; B, brain stem) portions of the reticular activating system and specific nuclei of the _____ (C, cerebellum; H, thalamus; E, epithalamus), respectively.

    a. T, C   c. T, E   e. B, H
    b. T, H   d. B, C   f. B, E

13. Centrifugal control mechanisms for _____ (S, only the special; G, only the general somatic; B, both the special as well as general somatic) sensations, which _____ (O, originate; T, terminate) in the cerebral cortex, may be a mechanism whereby the brain is able to direct its attention to specific aspects of its mental environment.

    a. S, O   c. B, O   e. G, T
    b. G, O   d. S, T   f. B, T

14. "Scanning" mechanisms for information retrieval more probably involve the _____ (C, cerebellar; T, thalamic; ST, subthalamic) nuclei, while "programming units" controlling the orderly sequence of thoughts _____ (A, are; N, are not) thought to involve structures below the level of the cerebral cortex.

    a. C, A   c. ST, A   e. T, N
    b. T, A   d. C, N   f. ST, N

15. Barbiturates are considered a central nervous system _____ (S, stimulant; D, depressant) by virtue of their major action upon _____ (SD, spinothalamic and dorsal column; RA, reticular activating) system pathways.

    a. S, SD   c. D, SD
    b. S, RA   d. D, RA

OBJECTIVE 54–4.

Identify the "brain waves" of the electroencephalogram, their characteristic types, and their associated conditions of occurrence.

16. The sequence of types of brain waves, arranged in ascending order from lowest to highest frequencies, involves _____ (A, alpha; B, beta; D, delta; T, theta) waves.

    a. A, B, T, & D
    b. A, B, D, & T
    c. B, A, T, & D
    d. T, D, B, & A
    e. D, T, A, & B
    f. D, T, B, & A

17. Brain waves are recorded on the surface of the scalp during states of wakefulness _____ (W, as well as; N, but not) sleep, and their intensities generally range from 0 to 0.3 _____ (M, millivolts; U, microvolts; V, volts).

    a. W, M
    b. W, U
    c. W, V
    d. N, M
    e. N, U
    f. N, V

18. _____ (A, alpha; T, theta; D, delta) waves range in frequency from 8 to 13 cycles per second, and characterize the awake, mentally _____ (AT, attentive; I, inattentive) individual with the eyes _____ (C, closed; O, open).

    a. D, AT, C
    b. T, I, C
    c. A, I, C
    d. D, I, O
    e. T, I, O
    f. A, AT, O

19. The frequencies of _____ (D, delta; T, theta) waves range from 4 to 7 cycles per second, and are more frequently recorded in the _____ (O, occipital; PT, parietal and temporal) region(s) of _____ (A, adults; C, children).

    a. D, O, A
    b. D, PT, C
    c. D, PT, A
    d. T, PT, C
    e. T, O, A
    f. T, O, C

20. _____ (B, beta; T, theta; D, delta) waves occur at frequencies below 3½ cycles per second, and are more frequently associated with states of _____ (S, deep sleep; W, wakefulness).

    a. B, S
    b. T, S
    c. D, S
    d. B, W
    e. T, W
    f. D, W

OBJECTIVE 54–5.

Identify the probable origin of the brain waves, the effects of varying levels of cerebral activity upon electroencephalogram rhythm, and the clinical utility of the electroencephalogram.

21. Delta wave patterns arising _____ (W, within; B, below the level of the) cerebral cortex occur with _____ (S, stimulation; I, interruption) of the reticular activating system.

    a. W, S
    b. W, I
    c. B, S
    d. B, I

22. Alpha waves are thought to result from _____ (A, activated states; SA, spontaneous activity) of the _____ (S, specific; D, diffuse) thalamocortical system.

    a. A, S
    b. SA, S
    c. A, D
    d. SA, D

23. Electroencephalogram recordings are largely thought to result from summated responses of _____ (A, action potentials; DH, altered states of excitability) of neurons which are significantly altered by _____ (B, only the brain stem; T, only the thalamic; TB, both the brain stem and thalamic) portions of the reticular activating system.

    a. A, B
    b. A, T
    c. A, TB
    d. DH, B
    e. DH, T
    f. DH, TB

24. During periods of increased mental activity, the electroencephalogram _____ (I, increases; D, decreases) in frequency and becomes more _____ (S, synchronous; A, asynchronous).

    a. I, S
    b. I, A
    c. D, S
    d. D, A

25. The outer surface of the cerebral cortex is largely comprised of _____ (T, fiber tracts; D, dentritic processes of neurons), resulting in a more electronegative surface with summated _____ (E, excitatory; I, inhibitory) postsynaptic potentials.

   a. T, E   c. D, E
   b. T, I   d. D, I

26. Brain tumors are more frequently localized by the existence of abnormally _____ (H, high; L, low) voltage waves of the electroencephalogram which generally yield surface recordings of _____ (S, similar; O, opposite) polarity on opposing sides of the affected region.

   a. H, S   c. L, S
   b. H, O   d. L, O

27. _____ waves occur more often in the adult in association with brain disorders or severe emotional stress accompanying disappointment and frustration, whereas _____ waves are associated with intense activation of the central nervous system or during tension. (A, alpha; B, beta; D, delta; T, theta; I, beta I; II, beta II)

   a. A, G   c. D, T   e. A, I
   b. T, II   d. B, T   f. T, D

OBJECTIVE 54–6.

Identify epilepsy, its major types, its associated characteristics, and its underlying mechanisms.

28. Epilepsy is a state generally characterized by an uncontrolled _____ (A, absence; E, excess) of activity of _____ (F, only focal; W, only widespread; FW, either focal or widespread) areas of the central nervous system.

   a. A, F   c. A, FW   e. E, W
   b. A, W   d. E, F   f. E, FW

29. Grand mal seizures are thought to originate in the _____ (CC, cerebral cortex; B, brain stem), and generally change from a pattern of _____ (C, clonic to tonic; T, tonic to clonic; P, preseizure depression to clonic) convulsions during the attack.

   a. B, P   c. B, C   e. CC, T
   b. B, T   d. CC, P   f. CC, C

30. The spike and dome pattern of the electroencephalogram characterizing _____ (G, grand; P, petit) mal seizures may be recorded over _____ (F, only focal areas of the; E, the entire) cerebral cortex.

   a. G, F   c. P, F
   b. G, E   d. P, E

31. _____ (P, psychomotor; J, jacksonian) epilepsy is generally classified as a _____ (G, generalized; F, focal) type of epilepsy which more frequently results in a progressive "march" of muscular contractions originating in the _____ (M, mouth region; L, lower extremities).

   a. P, G, M   c. P, F, M   e. J, F, M
   b. P, G, L   d. J, F, L   f. J, G, L

32. Psychomotor seizures are _____ (F, focal; G, generalized) types of seizures involving involuntary sensory _____ (W, as well as; N, but not) motor acts for which the individual _____ (R, retains; MN, may or may not retain) awareness.

   a. F, W, R   c. F, N, R   e. G, N, R
   b. F, W, MN   d. G, N, MN   f. G, W, MN

OBJECTIVE 54–7.

Identify the general conditions associated with sleep, the different types of sleep, and their differential characteristics.

33. Sleep is a state of _____ (C, consciousness; U, unconsciousness), for which the majority of sleep during the night occurs in association with _____ (I, increased; N, no significant alteration in; D, decreased) activity of the reticular activating system as compared to the waking state.

    a. C, I   c. C, D   e. U, N
    b. C, N   d. U, I   f. U, D

34. Spindles of _____ (A, alpha; D, delta; T, theta) waves called "sleep spindles" are characteristic of _____ (L, light; DS, desynchronized; S, deep slow wave) sleep.

    a. A, L   c. T, DS   e. D, S
    b. D, DS   d. A, S   f. T, L

35. Mean arterial pressures _____ (W, as well as; N, but not) respiration and metabolic rates are _____ (I, increased; NA, not significantly altered; D, decreased) during "slow wave" sleep.

    a. W, I   c. W, D   e. N, NA
    b. W, NA   d. N, I   f. N, D

36. Rapid eye movements and a comparatively _____ (G, greater; L, lesser) occurrence of dreaming is associated with a _____ (P, "paradoxical"; S, "slow wave") type of sleep.

    a. G, P   c. L, P
    b. G, S   d. L, S

37. Periods of "REM" sleep, associated with considerably _____ muscle tone throughout the body, are generally _____ in duration when individuals are extremely tired. (I, increased; D, decreased)

    a. I, I   c. D, I
    b. I, D   d. D, D

OBJECTIVE 54–8.

Identify the basic theories and the physiological effects of sleep.

38. A theory for _____ (A, an active; S, a passive) process of sleep involves largely _____ (P, positive; N, negative) feedback mechanisms for the maintenance of the "waking" state between the reticular activating system and the cerebral cortex, _____ (W, as well as; B, but not) the peripheral musculature.

    a. A, P, W   c. A, N, W   e. S, P, W
    b. A, N, B   d. S, N, B   f. S, P, B

39. Destruction of probable _____ (A, acetylcholine; GABA, gamma aminobutyric acid; S, serotonin) secreting system of neurons of the _____ (L, locus ceruleus; R, nuclei of the raphe) in the brain stem reportedly results in insomnia.

    a. A, L   c. S, L   e. GABA, R
    b. GABA, L d. A, R   f. S, R

40. Paradoxical sleep appears to occur coincident with _____ (E, enhanced; D, depressed) periods of rhythmic alterations of reticular activating system activity occurring at about _____ minute intervals.

    a. E, 15   c. E, 400   e. D, 90
    b. E, 90   d. D, 15   f. D, 400

41. A _____ (C, cerebral cortical; BS, lower brain stem; T, thalamic) locus is more frequently considered as a sleep producing center by virtue of its active _____ (F, facilitation; I, inhibition) of the reticular activating system.

    a. C, F   c. T, F   e. BS, I
    b. BS, F   d. C, I   f. T, I

42. Sleep and wakefulness cycles are considered more necessary for maintained functions of the _____ (H, higher; S, spinal cord) levels of central nervous system function and _____ (A, are; N, are not) considered essential for somatic functions of the body.

a. H, A   b. H,N   c. S, A   d. S, N

55

The Cerebral Cortex and Intellectual Functions of the Brain

OBJECTIVE 55–1.

Identify the physiologic anatomy of the cerebral cortex and its anatomical relationship to the thalamus and lower centers.

1. The human cerebral cortex is about _____ (Z, 0.2 to 0.5; T, 2 to 5) mm. in thickness, covers a convoluted surface area of _____ (O, more; S, less) than 1 square meter, and contains an estimated 10 _____ (M, million; B, billion) neurons.

a. Z, O, M   d. T, S, B
b. Z, O, B   e. T, S, M
c. Z, S, M   f. T, O, B

2. Large numbers of pyramidal cells are found in the _____ (G, granular; A, agranular) cortex, characterized by the _____ (M, motor; S, sensory) cortex.

a. G, M   c. A, M
b. G, S   d. A, S

3. Brodmann's classification of areas of the cerebral cortex is based primarily upon _____ (F, functional; C, cytoarchitectural) differences of the _____ major layers of the _____ (G, gray; W, white) matter of the cerebral cortex.

a. F, 3, G   c. F, 6, G   e. C, 6, G
b. F, 3, W   d. C, 6, W   f. C, 3, W

4. The cerebral cortex is generally considered to be a functional and anatomical outgrowth of the _____ (H, hypothalamus; T, thalamus; C, cerebellum) by virtue of the widespread _____ (U, unidirectional; B, bidirectional) connecting pathways between them.

a. H, U   c. C, U   e. T, B
b. T, U   d. H, B   f. C, B

5. The nuclei ventralis posterolateralis and ventralis posteromedialis project largely to the region of the _____ , while the lateral geniculate body projects largely to the region of the _____ . (P, prefrontal lobe; C, central sulcus; O, occipital lobe; H, hypothalamus)

a. H, O   c. C, P   e. O, H
b. H, P   d. C, O   f. O, C

OBJECTIVE 55–2.

Identify the functions and significance of the primary sensory areas and their association areas.

6. The primary sensory areas are associated with a _____ degree of spatial localization and a _____ capacity than corresponding sensory association areas to analyze the content of signals from peripheral receptors. (G, greater; L, lesser)

    a. G, G
    b. G, L
    c. L, G
    d. L, L

7. Stimulation of primary somatic sensory areas, localized within the _____ (F, frontal; T, temporal; P, parietal) lobes, results in rather _____ (C, complicated; U, uncomplicated) sensory experiences.

    a. F, C
    b. T, C
    c. P, C
    d. F, U
    e. T, U
    f. P, U

8. Destruction of the postcentral gyrus results in _____ (D, depression; L, a total loss) of somatic sensory sensations, while loss of the primary visual cortex of one hemisphere results in blindness of the _____ (R, contralateral; I, ipsilateral halves of the retinas of both, C, contralateral halves of the retinas of both) eye(s).

    a. D, R
    b. D, I
    c. D, C
    d. L, R
    e. L, I
    f. L, C

9. The thalamus plays a _____ role in the spread of signals from primary sensory areas to adjacent sensory association areas, and a _____ role in the transmission of auditory and visual information to their primary sensory areas. (J, major; N, minor)

    a. J, J
    b. J, N
    c. N, J
    d. N, N

10. Brodmann areas 18 and 19 are _____ (P, primary sensory; A, sensory association) areas whose destruction results in _____ (X, alexia; B, blindness; W, Wernicke's aphasia).

    a. P, X
    b. P, B
    c. P, W
    d. A, X
    e. A, B
    f. A, W

OBJECTIVE 55–3.

Locate and identify the functions of the "general interpretative" or "gnostic" area. Identify the significance of a "dominant hemisphere."

11. The "general interpretative" or "gnostic" area is located in the _____ (A, anterior; P, posterior) portion of the _____ (C, calcarine; H, horizontal; L, lateral) fissure.

    a. A, C
    b. A, H
    c. A, L
    d. P, C
    e. P, H
    f. P, L

12. The confluence of somatic, visual, and auditory _____ (P, primary sensory; A, association) areas represents _____ (M, an area of motor skills; I, a general interpretative area).

    a. P, M
    b. P, I
    c. A, M
    d. A, I

13. Highly complex auditory or visual memory patterns are more readily elicited with electrical stimulation of the _____ (S, more superficial; D, deeper) thalamocortical regions of _____ (A, the angular gyrus; P, the postcentral gyrus; B, Broca's area).

    a. S, A
    b. S, P
    c. S, B
    d. D, A
    e. D, P
    f. D, B

14. A greater and more persistent deficit of the intellect results following damage to the _____ (P, prefrontal lobes; G, gnostic area; C, calcarine cortex) in the _____ (N, newborn infant; A, adult).

    a. P, N
    b. G, N
    c. C, N
    d. P, A
    e. G, A
    f. C, A

15. Most individuals possess more highly developed interpretative functions within the _____ cerebral hemisphere and a more highly developed premotor speech area within the _____ cerebral hemisphere. (R, right; L, left)

   a. R, R   c. L, R
   b. R, L   d. L, L

16. A "dominant hemisphere," usually on the _____ (R, right; L, left) side, is thought to develop primarily _____ (B, before; A, after) birth, and transmits signals to the contralateral hemisphere via the _____ (C, corpus callosum; M, massa intermedia of the thalamus).

   a. R, B, M   c. R, A, M   e. L, A, M
   b. R, B, C   d. L, A, C   f. L, B, C

17. Language interpretation is more closely allied to _____ (A, auditory; V, visual) association areas, while interpretation of printed material generally occurs _____ (D, directly from visual images of words; L, following conversion to language form).

   a. A, D   c. V, D
   b. A, L   d. V, L

18. "Hemispheric dominance" generally refers to the _____ (S, somatic; L, language; V, visual)-related intellectual functions of corresponding _____ (A, anterior; P, posterior) portions of the superior temporal lobe and angular gyrus regions.

   a. S, A   c. V, A   e. L, P
   b. L, A   d. S, P   f. V, P

OBJECTIVE 55–4.

Identify the prefrontal areas, their probable functions, and the consequences of their destruction.

19. The prefrontal lobes lie just anterior to the _____ (S, somatosensory; M, motor) areas within the _____ (F, frontal; P, parietal; O, occipital) lobes.

   a. S, F   c. S, O   e. M, P
   b. S, P   d. M, F   f. M, O

20. A comparatively greater expansion of the _____ (PF, prefrontal; C, confluence of the parietal, occipital, and temporal) lobes of man when compared to lower primates is associated with a correlative functional expansion of _____ (S, motor skills; T, elaboration of thought; M, memory storage and retrieval).

   a. PF, S   c. PF, M   e. C, T
   b. PF, T   d. C, S   f. C, M

21. Destruction of the prefrontal areas results in an _____ (I, increased likelihood; N, inability) of individuals to be distracted from a sequence of thoughts and a more _____ (H, inhibited; U, uninhibited) type of social behavior.

   a. I, H   c. N, H
   b. I, U   d. N, U

22. Prefrontal lobotomy is thought to _____ (E, enhance; D, diminish) learning via alteration in the ability to _____ (C, classify and code; S, store) memory patterns of incoming information.

   a. E, C   c. D, C
   b. E, S   d. D, S

OBJECTIVE 55–5.

Identify the functions of the central nervous system in thought patterns, learning, consciousness, memory, and the different types of memory.

23. Thoughts probably originate from neuronal activity occurring from _____ (L, well localized centers; W, widespread areas) of the cerebral cortex _____ (A, as well as; N, but not) deeper structures.

   a. L, A   c. W, A
   b. L, N   d. W, N

24. "Storage units" for individual short-term _____ (A, as well as; N, but not) long-term memories are comprised of altered states of unit _____ (S, single; P, patterns of multiple) neurons.

   a. A, S   c. N, S
   b. A, P   d. N, P

25. The rhinencephalon, _____ (W, as well as; N, but not) the thalamus, is thought to play a more influential role in contributing attributes regarding the _____(A, affective nature; D, discrete characteristics) of thoughts.

    a. W, A   c. N, A
    b. W, D   d. N, D

26. A _____ (P, primary; T, tertiary) memory, with a comparatively longer persistence, generally has a _____ (L, longer; S, shorter) latency for recall than secondary types of memory.

    a. P, L   c. T, L
    b. P, S   d. T, S

27. The addition of excess information to the lesser storage capacity of mechanisms of _____ (G, long; S, short)-term memory results in a more probable lack of retention of its _____ (M, more; L, less) recently acquired information.

    a. G, M   c. S, M
    b. G, L   d. S, L

OBJECTIVE 55–6.

Identify the potential mechanisms and roles of differing parts of the brain in different types of memory.

28. Mechanisms of _____ (S, short-term; L, long-term) memory are thought to include those which result in the generation of electrotonic potentials within the dentritic layers of the cortex, _____ (W, as well as; N, but not) mechanisms of posttetanic potentiation.

    a. S, W   c. L, W
    b. S, N   d. L, N

29. Long-term memory is thought to result from _____ (R, reverberating circuits; S, chemical or physical alterations of synapses) and requires a minimal time of _____ (L, less; M, more) than 5 to 10 minutes for maximal consolidation.

    a. R, L   c. S, L
    b. R, M   d. S, M

30. Long-term _____ (W, as well as; N, but not) short-term stored memories are _____ (A, disrupted by; I, relatively independent of) factors which disrupt central nervous system activity, such as hypoxia.

    a. W, A   c. N, A
    b. W, I   d. N, I

31. Reported physical changes in cerebral cortical synaptic layers include a "disuse" _____ in thickness, as well as a general _____ in the average number of synaptic terminals per neuron with increasing age. (I, increase; D, decrease)

    a. I, I   c. D, I
    b. I, D   d. D, D

32. A reported _____ (I, increase; D, decrease) in the rate of RNA synthesis occurring in response to increased neuronal activity is probably associated with a more direct _____ (R, DNA-RNA template; S, synaptic modification) mechanism underlying long-term memory.

    a. I, R   c. D, R
    b. I, S   d. D, S

33. Consolidation of long-term memory is thought to be greater when learning involves _____ (S, small; L, large) amounts of information studied _____ (P, superficially; D, in depth) during states of _____ (RS, REM sleep; DS, deep sleep; W, wakefulness).

    a. S, D, RS   c. S, D, W   e. L, P, DS
    b. S, D, DS   d. L, P, RS   f. L, P, W

34. The transference of short-term to long-term memory generally involves conversion of _____ (E, the entire; S, selected features of the) sensory experience through a _____ (R, rapid; H, rehearsal) process of consolidation.

    a. E, R   c. S, R
    b. E, H   d. S, H

35. Rehearsal plays an important role in the conversion of _____ (S, secondary to tertiary; T, tertiary to secondary) memories with an associated _____ (I, increase; D, decrease) in latency for recall.

    a. S, I   c. T, I
    b. S, D   d. T, D

36. Destruction of the hippocampal gyri results in _____ (R, retrograde; A, anterograde) amnesia, or a greatly reduced ability to _____ (E, establish new; C, recall old) memories.

    a. R, E   c. A, E
    b. R, C   d. A, C

37. The thalamus _____ considered a major structure for the maintenance of the conscious state, and _____ considered to play a major role in the recall of long-term memories. (I, is; N, is not)

    a. I, I   c. N, I
    b. I, N   d. N, N

OBJECTIVE 55–7.

Identify the general characteristics of the intellectual operations of the brain in the analysis of incoming information.

38. Differing qualities of a set of information signals are distinguished within the central signal relatively _____ (E, early; L, late) in its analysis by _____ (S, subcomponents of a single; M, multiple) region(s) of the brain.

    a. E, S   c. L, S
    b. E, M   d. L, M

39. The _____ (P, peripheral localization; A, affective nature) of pain information, transmitted via the _____ (ST, spinothalamic; DC, dorsal column) system, is largely established at thalamic and hypothalamic, as opposed to cerebral cortical, levels.

    a. P, ST   c. A, ST
    b. P, DC   d. A, DC

40. Analysis of incoming sensory information is thought to generally involve comparisons of _____ (V, discrete values; P, patterns) of sensory information with _____ (A, all; S, largely similar) memory traces.

    a. V, A   c. P, A
    b. V, S   d. P, S

41. The _____ (L, lateral geniculate body; V, visual cortex) detects _____ (D, deemphasized; E, emphasized) patterns of visual information which are _____ (C, critically dependent; I, relatively independent) of their position or angle of rotation within the visual field.

    a. L, D, C   c. L, E, C   e. V, E, C
    b. L, D, I   d. V, E, I   f. V, D, I

42. For every primary sensory area, _____ (A, as well as; N, but not) for every association area in the cerebral cortex, there exists reciprocally connected areas within the _____ (H, hypothalamus; T, thalamus; C, cerebellum).

    a. A, H   c. A, C   e. N, T
    b. A, T   d. N, H   f. N, C

OBJECTIVE 55–8.
Identify the intellectual aspects of motor control, the sensory and motor functions of the brain in communication, and the functional significance of the corpus callosum and anterior commissure.

43. Intellectual operations of the brain are localized _____ (E, about equally between; P, predominantly in one of) the cerebral hemispheres, and generally occur _____ (I, independently of; A, in association with) altered states of motor activity.

    a. E, I   c. P, I
    b. E, A   d. P, A

44. The general interpretative area of the brain plays a more major role in _____ (O, the conception; S, establishing the sequence of movements; C, control of muscle movements) of _____ (V, voluntary; I, involuntary) motor activities.

    a. O, V   c. C, V   e. S, I
    b. S, V   d. O, I   f. C, I

45. _____ (S, syntactical; M, motor) aphasia almost always results from damage to Broca's area located within the _____ (F, frontal; P, parietal; T, temporal) lobe generally on the _____ (L, left; R, right) side.

    a. S, P, L   c. S, T, L   e. M, F, L
    b. S, P, R   d. M, T, R   f. M, F, R

46. The capability of understanding spoken or written words, while being unable to interpret properly the meanings of words when they are used to formulate a thought, would more probably result from a combination of _____ primary sensory areas for vision and hearing, _____ visual and auditory association areas, and _____ regions of the angular gyri and posterior part of the superior temporal gyri. (T, intact; J, injured)

    a. T, J, J   c. T, T, J   e. J, J, J
    b. T, J, T   d. T, T, T   f. J, T, T

47. Visual _____ (W, as well as; N, but not) somatic information, originating from the _____ (R, right; L, left) side of the body, more frequently fails to reach the "gnostic" or general interpretative area of the brain following transection of the _____ (P, posterior commissure; C, corpus callosum).

    a. W, R, P   c. W, L, P   e. N, R, P
    b. W, L, C   d. N, L, C   f. N, R, C

56

Behavioral Functions of the Brain: The Limbic System, Role of the Hypothalamus, and Control of Vegetative Functions of the Body

OBJECTIVE 56–1.

Identify the limbic system, its general location, its component structures, and its general functions.

1. Behavior is a manifestation of relatively _____ (S, specific; G, general) regions of the nervous system, with emotional aspects of behavior occurring in closer association with the _____ (P, paleocortex; N, neocortex).

   a. S, P   c. G, P
   b. S, N   d. G, N

2. Behavior associated with emotions, subconscious motor and sensory drives, and the affective nature of sensations are largely associated with _____ (C, cortical; S, subcortical) structures of the _____ (PF, prefrontal regions; L, limbic system; G, gnostic area).

   a. C, PF   c. C, G   e. S, L
   b. C, L   d. S, PF   f. S, G

3. The limbic cortex is comprised of _____ (D, rather discontinuous regions; C, a rather continuous ring) of cerebral cortex located along its more _____ (MV, medial and ventral; PL, posterior and lateral) hemispheric surfaces.

   a. D, MV   c. C, MV
   b. D, PL   d. C, PL

4. _____ (S, singular; P, paired) hypothalamic structures of the _____ (T, telencephalon; D, diencephalon; M, mesencephalon) are located within _____ (C, more central; L, lateral; O, posterior) positions relative to the limbic cortex.

   a. S, T, L   c. S, D, C   e. P, M, L
   b. S, M, O   d. P, T, C   f. P, D, O

5. Subcortical limbic structures more notably include the _____ (M, metathalamus; E, epithalamus), the _____ (A, anterior; P, posterior) group of thalamic nuclei, and the _____ (G, amygdaloid nuclei; C, superior and inferior colliculi).

   a. M, A, G   c. M, P, G   e. E, A, G
   b. M, P, C   d. E, P, C   f. E, A, C

OBJECTIVE 56–2.

Identify the vegetative functions of various regions of the hypothalamus.

6. The most important efferent pathway by which the limbic system controls particularly the _____ (S, somatic; V, vegetative) functions of the body occurs from the _____ (C, cingulate gyrus; E, epithalamus; H, hypothalamus).

   a. S, C   c. S, H   e. V, E
   b. S, E   d. V, C   f. V, H

7. Stimulation of the _____ (PL, posterior and lateral; AM, anterior and medial) portions of the hypothalamus more generally increases blood pressure and heart rate via its _____ (N, neurosecretion of releasing factors; M, influence upon centers located within the brain stem reticular formation).

   a. PL, N   c. AM, N
   b. PL, M   d. AM, M

8. A heat loss center is localized in the _____ (SO, supraoptic; PO, preoptic) area of the _____ (A, anterior; P, posterior) region of the _____ (H, hypothalamus; T, thalamus).

   a. SO, A, H   c. SO, P, H   e. PO, A, H
   b. SO, P, T   d. PO, P, T   f. PO, A, T

9. The hypothalamus functions in the regulation of body water by means of a thirst center, located in the more _____ (M, medial; L, lateral) hypothalamus, _____ (W, as well as; N, but not) by regulation of renal excretion of water.

   a. M, W   c. L, W
   b. M, N   d. L, N

10. Stimulation of the supraoptic nucleus results in a _____ (D, direct; RF, releasing factor mediated) secretion of _____ (O, oxytocin; ADH, antidiuretic hormone).

   a. D, O   c. RF, O
   b. D, ADH   d. RF, ADH

11. Hormonal secretion resulting from increased activity of the _____ (M, mammillary bodies; PV, paraventricular nuclei) of the hypothalamic region functions to _____ (I, increase; D, decrease) uterine contractility and _____ (F, facilitate; H, inhibit) lactation.

   a. M, I, F   c. M, D, F   e. PV, I, F
   b. M, D, H   d. PV, D, H   f. PV, I, H

12. The lateral hypothalamic area and the _____ (VM, ventromedial; PF, perifornical) nucleus are hypothalamic areas associated with the development of hunger, whereas _____ (H, another hypothalamic area; M, the medulla; C, the cerebral cortex) is the locus of a "satiety center."

   a. VM, H   c. VM, C   e. PF, M
   b. VM, M   d. PF, H   f. PF, C

13. Stimulation of particularly the _____ (DM, dorsal medial nucleus; I, infundibulum; PO, preoptic nucleus) of the hypothalamus results in increased gastrointestinal secretion _____ (W, as well as; N, but not) motility.

   a. DM, W   c. PO, W   e. I, N
   b. I, W   d. DM, N   f. PO, N

14. The hypothalamus functions to regulate the secretion of _____ (A, only the adenohypophysis; N, only the neurohypophysis; AN, the neurohypophysis and the adenohypophysis) by means of its neurosecretion of "releasing factors" and their transport to the pituitary by means of _____ (X, axoplasmic flow; P, a cardiovascular portal system).

   a. A, X   c. AN, X   e. N, P
   b. N, X   d. A, P   f. AN, P

15. The median eminence of the infundibulum is thought to function more specifically in the regulation of _____ (T, thyrotropin; C, corticotropin) _____ (W, as well as; N, but not) follicle-stimulating hormone secretion.

   a. T, W   c. C, W
   b. T, N   d. C, N

OBJECTIVE 56–3.

Recognize the role of the reticular formation in behavioral functions of the limbic system. Identify the behavioral functions of the limbic system, their localization, and their significance.

16. The reticular formation, associated with centers of _____ activity, serves as a major system for descending activity originating within the limbic system, which influences _____ structures. (S, only somatic; A, only autonomic; B, both somatic and autonomic)

    a. S, A   c. A, S   e. B, S
    b. S, B   d. A, B   f. B, B

17. The _____ (MT, mamillothalamic bundle; MFB, medial forebrain bundle; AC, anterior commissure) is a specialized structure which communicates information _____ (U, unidirectionally; B, bidirectionally) through the hypothalamus between the septal and orbitofrontal regions and the reticular formation.

    a. MT, U   c. AC, U   e. MFB, B
    b. MFB, U   d. MT, B   f. AC, B

18. The affective nature of reward and punishment, characterization of an affective-defense pattern, and subconscious control of many of the body's internal activities are thought to be functions of the:

    a. Occipital lobe
    b. Cerebellum
    c. Hypothalamus
    d. Geniculate bodies
    e. Superior & inferior colliculi
    f. Medullary centers

19. Principal centers for _____ (R, reward; P, pain, punishment, and escape tendencies) are localized more significantly within the _____ (C, central gray; V, ventral) region of the mesencephalon and the perifornical nucleus of the hypothalamus.

    a. R, C   c. P, C
    b. R, V   d. P, V

20. More potent reward centers occur in the _____ (VM, ventromedial; PV, paraventricular; PF, perifornical) nuclei of the _____ (A, amygdala; T, thalamus; H, hypothalamus).

    a. VM, A   c. PF, H   e. PV, A
    b. PV, T   d. VM, H   f. PF, T

21. Stimulation of the _____ (R, reward; P, punishment) center is more effective in a mutual _____ (E, excitation; I, inhibition) of the other center.

    a. R, E   c. P, E
    b. R, I   d. P, I

22. Repetitive experiences which stimulate the reward _____ (A, as well as; N, but not the) punishment centers result in _____ (H, habituated; R, reinforced) responses with comparatively _____ (W, weak; S, strong) memory traces.

    a. A, H, W   c. A, R, W   e. N, H, W
    b. A, R, S   d. N, R, S   f. N, H, S

23. Tranquilizers such as chlorpromazine _____ (S, stimulate; H, inhibit) the reward centers, _____ (S, stimulate; H, inhibit) the punishment centers, and _____ (D, decrease; I, increase) the affective reactivity of animals.

    a. S, S, I   c. S, H, I   e. H, S, I
    b. S, H, D   d. H, S, D   f. H, H, D

24. The development of a pattern of spontaneous "rage" is more likely to occur in the _____ (P, presence; A, absence) of thalamic structures and _____ (D, does; N, does not) require the presence of the cerebral cortex.

    a. P, D   c. A, D
    b. P, N   d. A, N

OBJECTIVE 56–4.

Recognize the influence of the hypothalamus and adjacent areas upon the reticular activating system. Identify the potential mechanisms of action and psychosomatic effects of the hypothalamus and reticular activating system.

25. Psychosomatic effects of the hypothalamus and reticular activating system are mediated via _____. (A, autonomic nerves; S, somatic nerves; E, the endocrine system)

a. Only A
b. Only A & S
c. Only A & E
d. Only S
e. Only S & E
f. A, S, & E

26. States of anxiety generally _____ muscle tone, _____ activity of the reticular activating system and _____ activity of the sympathetic nervous system. (I, increase; N, do not significantly alter; D, decrease)

a. I, I, I
b. I, D, D
c. I, I, N
d. N, D, I
e. D, I, I
f. D, D, I

27. Increased arterial pressure resulting from psychosomatic abnormalities generally occurs in association with a relatively _____ (L, selective; W, widespread) excitation of the _____ (S, sympathetic; P, parasympathetic) nervous system.

a. L, S
b. L, P
c. W, S
d. W, P

28. Increased activity of the posterior hypothalamus increases the secretion of corticotropin and results in gastric _____ (O, hypoacidity; H, hyperacidity) via _____ (E, enhanced; D, diminished) adrenocortical hormone secretion.

a. O, E
b. O, D
c. H, E
d. H, D

29. Peptic ulcers _____ (W, as well as; N, but not) thyrotoxicosis are manifestations of potential psychosomatic disorders mediated via abnormal control of the _____ (A, anterior; P, posterior) pituitary.

a. W, A
b. W, P
c. N, A
d. N, P

OBJECTIVE 56–5.

Identify the probable functions of the amygdala and hippocampus.

30. The _____ (CM, corticomedial; BL, basolateral) nuclear portion of the amygdala in the human are more directly associated with a predominantly _____ (G, gustatory; A, auditory; O, olfactory) function of the amygdaloid complex of lower animals.

a. CM, G
b. CM, A
c. CM, O
d. BL, G
e. BL, A
f. BL, O

31. The amygdaloid nuclei are thought to function in _____ (R, relatively restricted; W, widespread) patterns of behavior which include _____ (S, only sexual activities; O, only oral movements associated with eating; B, both sexual activities and oral movements associated with eating).

a. R, S
b. R, O
c. R, B
d. W, S
e. W, O
f. W, B

32. The _____ (A, amygdala; H, hippocampus) consists of a _____ -layered portion of the cortex.

a. A, 3
b. A, 6
c. A, 12
d. H, 3
e. H, 6
f. H, 12

33. An "arrest reaction" of motor movements is a more likely consequence of stimulation of the _____ , and a loss of attentive contact is a more likely consequence of stimulation of the _____ . (A, amygdala; H, hippocampus; PO, posterior orbital cortex; SA, supraoptic nuclei)

a. A, PO
b. PO, SA
c. A, H
d. SA, H
e. H, SA
f. H, A

34. Destruction of the _____ (H, hippocampus; G, amygdala; M, mammillary bodies), probably functioning as a major structure in the transmission of sensory information to the major reward and punishment centers, results in a type of _____ (R, retrograde; A, anterograde) amnesia.

     a. H, R   b. G, R   c. M, R   d. H, A   e. G, A   f. M, A

OBJECTIVE 56–6.

Identify the probably functions of the hippocampal, cingulate, and orbitofrontal cortex.

35. The limbic cortex of man is thought to function in a comparatively _____ (J, major; N, minor) role in gustatory, olfactory, and feeding phenomena, and in a _____ (J, major; N, minor) role as a cerebral association area for behavior.

     a. J, J   c. N, J
     b. J, N   d. N, N

36. Stimulation of the various regions of the limbic cortex results in _____ (F, very few; D, diverse) motor responses which are more similar to those obtained by means of stimulation of the _____ (S, primary sensorimotor cortex; M, primary motor cortex; H, hypothalamus, amygdala, or hippocampus).

     a. F, S   c. F, H   e. D, M
     b. F, M   d. D, S   f. D, H

37. The _____ (MFB, medial forebrain bundle; F, fornix; UF, uncinate fasciculus) is a more direct pathway for the transmission of information between the _____ (T, thalamus; H, hippocampus; O, orbitofrontal cortex) and the hypothalamus and associated regions.

     a. MFB, H   c. UF, O   e. F, H
     b. F, T   d. MFB, O   f. UF, T

38. Destruction of the anterior portions of the cingulate gyri results in increased _____ (H, hostility; D, docility), whereas destruction of the posterior orbital cortex more frequently results in an increased tendency towards _____ (S, somnolence; I, insomnia).

     a. H, S   c. D, S
     b. H, I   d. D, I

39. Excessive tendency to examine objects, changes in dietary habits, and a general _____ (I, increase; D, decrease) in aggressiveness are part of the Klüver-Bucy syndrome, resulting particularly from damage to portions of the _____ (F, frontal; T, temporal; O, occipital) lobes.

     a. I, F   c. I, O   e. D, T
     b. I, T   d. D, F   f. D, O

57

The Autonomic Nervous System;
The Adrenal Medulla

OBJECTIVE 57-1.
Identify the general function and organizational pattern of the autonomic nervous system.

1. _____ (S, somatic; V, visceral) structures controlled by the autonomic nervous system usually _____ (L, lack; P, possess) intrinsic activity in the absence of their innervation.

   a. S, L   c. V, L
   b. S, P   d. V, P

2. A more distinctive separation of the _____ (SP, sympathetic and parasympathetic; IV, involuntary and visceral) divisions of the _____ (A, autonomic, M, somatic) nervous system occurs in its _____ (S, sensory; E, efferent) division.

   a. SP, A, S   c. SP, M, S   e. IV, M, S
   b. SP, A, E   d. IV, M, E   f. IV, A, E

3. _____ (T, thalamic; H, hypothalamic; C, calcarine cortical) and brain stem structures are more significant regions for the regulation of parasympathetic _____ (W, as well as; N, but not) sympathetic outflow.

   a. T, W   c. C, W   e. H, N
   b. H, W   d. T, N   f. C, N

OBJECTIVE 57-2.
Identify the physiologic anatomy of the sympathetic nervous system.

4. Sympathetic efferents are comprised of a _____ (S, single; T, two-; H, three-) neuron chain(s) whose cell bodies of origin are located within the _____ (TL, thoracolumbar; CS, craniosacral) regions of the central nervous system.

   a. S, TL   c. H, TL   e. T, CS
   b. T, TL   d. S, CS   f. H, CS

5. Cell bodies of origin of _____ (E, preganglionic; O postganglionic) autonomic neurons are located within the _____ (L, interomediolateral; V, ventral) horns of the thoracolumbar segments of the spinal cord between _____ .

   a. E, L, T-5 & L-5   d. E, V, T-1 & L-5
   b. E, L, T-1 & L-2   e. O, L, T-5 & L-5
   c. E, V, T-5 & L-5   f. O, V, T-1 & L-2

6. _____ (M, myelinated; U, unmyelinated) fibers of autonomic neurons emerge from the _____ (D, dorsal; V, ventral) roots of the thoracolumbar segments of the spinal cord and pass to or through the sympathetic chain via the _____ (G, gray; W, white) rami communicantes.

   a. M, D, W   c. M, V, W   e. U, D, W
   b. M, V, G   d. U, V, G   f. U, D, G

7. Type _____ fibers of the gray rami communicantes are largely _____ (E, preganglionic; O, postganglionic) _____ (S, sympathetic; P, parasympathetic) fibers.

   a. B, E, S   c. B, O, S   e. C, O, S
   b. B, E, P   d. C, O, P   f. C, E, P

8. The head receives sympathetic outflow originating from _____ , with synaptic relays in the _____ (S, superior cervical; T, stellate; C, celiac) ganglia.

a. T-1, S   d. T-3 to T-6, S
b. T-1, T   e. T-3 to T-6, T
c. T-1, C   f. T-3 to T-6, C

9. Cells of the adrenal medulla secrete _____ (S, steroids; N, norepinephrine and epinephrine), and are innervated by _____ (E, preganglionic; O, postganglionic) fibers of _____ (S, only the sympathetic; B, both the sympathetic and parasympathetic) division(s) of the autonomic nervous system.

a. S, E, S   c. S, O, S   e. N, E, S
b. S, O, B   d. N, O, B   f. N, E, B

10. Abdominal organs receive the majority of their sympathetic innervation from the _____ (T, upper thoracic; L, lower thoracic; M, lumbar) segments because of the embryological migration of the _____ (C, sympathetic chain; G, gut).

a. T, C   c. M, C   e. L, G
b. L, C   d. T, G   f. M, G

OBJECTIVE 57–3.
Identify the physiologic anatomy of the parasympathetic nervous system.

11. The craniosacral or _____ (S, sympathetic; P, parasympathetic) division of the autonomic nervous system transmits activity via cranial nerves _____ and the sacral spinal nerves.

a. S; IV, VI, X, & XII   d. P; IV, VI, X, & XII
b. S; III, VII, IX, & X   e. P; III, VII, IX, & X
c. S; III, IV, X, & XII   f. P; III, IV, X, & XII

12. Compared to the sympathetic nervous system, the preganglionic parasympathetic fibers are relatively _____ and the postganglionic fibers are relatively _____ . (S, shorter; L, longer)

a. S, S   c. L, S
b. S, L   d. L, L

13. About 75% of all parasympathetic _____ (E, preganglionic; O, postganglionic) fibers are conveyed by the _____ (NE, nervi erigentes; V, vagus nerves; G, gray rami communicantes).

a. E, NE   c. E, G   e. O, V
b. E, V   d. O, NE   f. O, G

14. Parasympathetic fibers originating from the _____ (L, lumbar; S, sacral) segments of the spinal cord innervate the bladder _____ (W, as well as; N, but not) the terminal portions of the large intestine.

a. L, W   c. S, W
b. L, N   d. S, N

15. Parasympathetic fibers in cranial nerve _____ supply the pupillary sphincters and ciliary muscle of the eye, whereas those of cranial nerve _____ pass to the lacrimal, nasal, and _____ (P, parotid; S, submandibular) glands.

a. II, VII, P   d. III, VII, S
b. II, IX, P   e. III, IX, S
c. II, X, P   f. III, X, S

OBJECTIVE 57–4.

Identify the chemical mediators of autonomic nerve endings and their respective localization.

16. Chemical mediators of the peripheral autonomic nervous system are principally _____ (E, epinephrine; N, norepinephrine) and _____ (GABA, gamma aminobutyric acid; D, dopamine; A, acetylcholine).

    a. E, GABA   c. E, A   e. N, D
    b. E, D   d. N, GABA   f. N, A

17. Preganglionic sympathetic fibers result in excitation _____ (W, as well as; N, but not) inhibition of ganglionic neurons through the release of _____ (A, only acetylcholine; N, only norepinephrine; E, either acetylcholine or norepinephrine).

    a. W, A   c. W, E   e. N, N
    b. W, N   d. N, A   f. N, E

18. Autonomic neurons generally include cholinergic _____ neurons and adrenergic _____ neurons. (SP, sympathetic preganglionic; SG, sympathetic ganglionic; PP, parasympathetic preganglionic; PG, parasympathetic ganglionic)

    a. SP, SG, & PP; PG   d. SP & SG; PP & PG
    b. SG, PP, & PG; SP   e. SP & PP; PG & SG
    c. SP, PP, & PG; SG   f. SP & PG; PP & SG

19. Exceptions to the general rule for the adrenergic and cholinergic nature of autonomic neurons are known to occur for _____ (E, preganglionic; O, postganglionic) neurons of the _____ (S, sympathetic; P, parasympathetic) division.

    a. E, S   c. O, S
    b. E, P   d. O, P

OBJECTIVE 57–5.

Identify the general mechanisms of synthesis, secretion, and removal of chemical mediators for the autonomic nervous system.

20. Acetylcholine-containing vesicles of cholinergic nerve terminals are about _____ angstroms in diameter and are _____ (C, clear; G, granular) in appearance in electronmicrographs.

    a. 2 to 3, C   d. 2 to 3, G
    b. 20 to 30, C   e. 20 to 30, G
    c. 200 to 300, C   f. 200 to 300, G

21. Visceral smooth muscle is largely innervated by autonomic nerve endings containing vesicles of chemical mediator _____ (W, as well as; N, but not) mitochondria localized within _____ (E, specialized end-plates; V, varicosities).

    a. W, E   c. N, E
    b. W, V   d. N, V

22. Vesicles of chemical mediator of autonomic nerve endings are thought to release their contents to the extracellular fluid due to an _____ (E, efflux; I, influx) of _____ ions in response to depolarization, _____ (W, as well as; N, but not) hyperpolarization of the nerve terminals.

    a. E, Mg, W   c. E, Ca, W   e. I, Ca, W
    b. E, Mg, N   d. I, Ca, N   f. I, Mg, N

23. The majority of acetylcholine released from cholinergic endings _____ (P, usually persists for; D, is usually destroyed within) several seconds as a result of the action of _____ (A, choline acetylase; E, cholinesterase; MAO, monoamine oxidase).

    a. P, A   c. P, MAO   e. D, E
    b. P, E   d. D, A   f. D, MAO

24. Epinephrine is synthesized more directly from _____ (T, tyrosine; N, norepinephrine) by a process of _____ (M, methylation; H, hydroxylation) occurring largely within _____ (R, autonomic nerve terminals; A, the adrenal medulla).

    a. T, M, R   c. T, H, R   e. N, M, R
    b. T, H, A   d. N, H, A   f. N, M, A

25. Norepinephrine removal at autonomic nerve terminals following its secretion normally requires an interval of _____ seconds, and is accomplished largely by _____ (E, local enzymatic destruction; R, reuptake by nerve terminals; D, diffusion into blood).

    a. 1 to 3, E   d. 10 to 30, E
    b. 1 to 3, R   e. 10 to 30; R
    c. 1 to 3, D   f. 10 to 30; D

26. Enzymes which inactivate norepinephrine include monoamine oxidase, _____ (L, more highly localized within nerve endings; D, present diffusely in most tissues), _____ (W, as well as; N, but not) catechol-O-methyl transferase.

   a. L, W   c. D, W
   b. L, N   d. D, N

27. Epinephrine _____ (W, as well as; N, but not) norepinephrine, when secreted by the adrenal medulla, remain(s) active for a _____ (L, longer; S, shorter) time as compared to norepinephrine secretion by autonomic nerve terminals.

   a. W, L   c. N, L
   b. W, S   d. N, S

OBJECTIVE 57-6.

Identify the types, characteristics, and significance of differing receptor substances of ganglionic neurons and effector organs of the autonomic nervous system.

28. The combination of autonomic transmitting agents with _____ (M, cell membrane; C, cytoplasmic) receptors results in membrane _____ (D, depolarization only; H, hyperpolarization only; B, depolarization or hyperpolarization).

   a. M, D   c. M, B   e. C, H
   b. M, H   d. C, D   f. C, B

29. _____ (A, acetylcholine; N, norepinephrine; E, epinephrine) is known to alter adenyl cyclase activity in some cells, and thereby _____ (I, increases; D, decreases) the rate of intracellular formation of cyclic AMP.

   a. A, I   c. E, I   e. N, D
   b. N, I   d. A, D   f. E, D

30. _____ (M, muscarinic; N, nicotinic) receptors are located in the membranes of parasympathetic _____ (W, as well as; B, but not) sympathetic ganglionic neurons.

   a. M, W   c. N, W
   b. M, B   d. N, B

31. _____ (M, muscarine; N, nicotine) is a pharmacological agent which mimics the action of _____ (A, acetylcholine; E, epinephrine; NE, norepinephrine) upon effector structures innervated by autonomic ganglionic neurons.

   a. M, A   c. M, NE   e. N, E
   b. M, E   d. N, A   f. N, NE

32. _____ (M, muscarinic; N, nicotinic) receptors are found in postjunctional membranes of skeletal muscle and in autonomic _____ (G, ganglionic; E, effector) cells.

   a. M, G   c. N, G
   b. M, E   d. N, E

33. The "alpha" and "beta" types of receptors are utilized to explain differential pharmacological actions of _____ agents.

   a. Muscarinic   c. Cholinergic
   b. Nicotinic   d. Adrenergic

34. Norepinephrine excites _____ , while epinephrine excites _____ . (A, primarily alpha receptors; B, primarily beta receptors; AB, alpha and beta receptors about equally)

   a. A, B   c. B, A   e. AB, A
   b. A, AB   d. B, AB   f. AB, B

35. The pharmocological agent _____ (A, atropine; IN, isopropyl norepinephrine; P, propanolol) has a specific excitatory action upon _____ (H, alpha; B, beta) receptors.

   a. A, H   d. IN, B
   b. IN, H   e. P, B
   c. P, H   f. A, B

36. Stimulation of alpha receptors results in _____ (D, vasodilation; C, vasoconstriction) and intestinal _____ (N, contraction; R, relaxation).

   a. D, N   c. C, N
   b. D, R   d. C, R

37. Alterations of cardiac contractility and heart rate occur primarily with altered states of _____ (A, alpha; B, beta) receptors, whereas _____ (D, bronchodilation; C, bronchoconstriction) and enhanced glycogenolysis and lipolysis are largely functions of _____ (A, alpha; B, beta) receptors.

   a. A, C, A   c. A, D, A   e. B, D, A
   b. A, C, B   d. B, D, B   f. B, C, B

OBJECTIVE 57-7.

Identify the effects of sympathetic and parasympathetic stimula? of the body.

38. Focusing the lens of the eye and control of the _____ (M, meridional; C, circular) muscle fibers of the iris is largely a function of _____ (P, parasympathetic; S, sympathetic) efferent activity.

a. M, P   c. C, P
b. M, S   d. C, S

39. Eccrine sweat glands in man are innervated directly by _____ (C, cholinergic; A, adrenergic) _____ (E, preganglionic; O, postganglionic) _____ (S, sympathetic; P, parasympathetic) fibers.

a. C, E, S   c. C, O, S   e. A, O, S
b. C, E, P   d. A, O, P   f. A, E, P

40. Secretion of gastrointestinal, salivary, nasal, and lacrimal glands, _____ (W, as well as; N, but not) apocrine odoriferous glands, is enhanced by increased _____ (P, parasympathetic; S, sympathetic) efferent activity.

a. W, P   c. N, P
b. W, S   d. N, S

41. The _____ (P, parasympathetic; S, sympathetic) nervous system performs a more direct role in regulating secretion of particularly the _____ (U, upper; L, lower) gastrointestinal system in a manner which acts _____ (G, synergistically; R, reciprocally) to its influence upon gastrointestinal propulsive movements.

a. P, U, G   c. P, L, G   e. S, L, G
b. P, U, R   d. S, L, R   f. S, U, R

42. _____ (S, only sympathetic; B, b_ pathetic) stimulation to increases; D, decreases) cardiac increases its pumping effectiveness.

a. S, I   c. B, I   e. P, D
b. P, I   d. S, D   f. B, D

43. Sympathetic stimulation tends to _____ (I, increase; D, decrease) arterial blood pressure via its greater influence upon _____ (M, skeletal muscle; SC, splanchnic and cutaneous) _____ (A, alpha; B, beta) vascular receptors.

a. I, M, A   c. I, SC, A   e. D, M, A
b. I, SC, B   d. D, SC, B   f. D, M, B

44. Parasympathetic effects include a _____ (M, mild; P, pronounced) _____ (C, constriction; D, dilation) of bronchioles and largely _____ (E, excitatory; I, inhibitory) effects upon the gallbladder, ureter, and bladder.

a. M, C, E   c. M, D, E   e. P, D, E
b. M, C, I   d. P, D, I   f. P, C, I

OBJECTIVE 57-8.

Identify the secretions of the adrenal medulla, their physiological effects, and their functional significance.

45. _____ (N, norepinephrine; E, epinephrine) released by sympathetic nerve terminals generally has a _____ (L, longer; S, shorter) duration of action than that resulting from secretion by the adrenal glands.

a. N, L   c. E, L
b. N, S   d. E, S

46. Norepinephrine is _____ (J, the major; N, a minor) constituent of adrenal _____ (M, medullary; C, cortical) secretion.

a. J, M   c. J, C
b. N, M   d. N, C

adrenal medulla provides _____ (S, a
...ergistic; O, an opposing) limb within the
_____ (M, sympathetic; P, parasympathetic) nervous system which functions in a
_____ (R, reciprocally; U, simultaneously)
active manner.

a. S, M, R   c. S, P, R   e. O, P, R
b. S, M, U   d. O, P, U   f. O, M, U

48. Greater effects upon cardiac activity occur
with _____ , while greater effects upon
peripheral resistance occur in response to
_____ . (N, norepinephrine; E, epinephrine)

a. N, N   c. E, N
b. N, E   d. E, E

49. _____ (D, vasodilation; C, vasoconstriction), muscle glycogenolysis, lipolysis, and
bronchodilation are largely attributed to the
action of _____ (E, epinephrine; N, norepinephrine) upon _____ (A, alpha; B, beta)
receptors.

a. D, E, A   c. D, N, A   e. C, E, A
b. D, E, B   d. C, N, B   f. C, E, B

OBJECTIVE 57–9.

Characterize sympathetic and parasympathetic tone and the consequences of denervation of sympathetic and parasympathetic organs.

50. Tonic discharge frequencies for sympathetic
_____ (W, as well as; N, but not) parasympathetic efferents, typically _____ (G,
greater; L, less) than 5 per second, are generally _____ (G, greater; L, less) than somatic
efferent discharge frequencies for comparable
levels of activity.

a. W, G, L   c. W, L, L   e. N, L, L
b. W, G, G   d. N, L, G   f. N, G, G

51. Vasodilation more generally results from
_____ (E, excitatory; I, inhibitory) influences upon _____ (S, sympathetic; P, parasympathetic) tonic activity to vascular smooth
muscle.

a. E, S   c. I, S
b. E, P   d. I, P

52. A more pronounced gastric and intestinal
"atony" results from _____ (A, an absence
of; E, strongly enhanced) tonic activity of its
_____ (S, sympathetic; P, parasympathetic)
innervation.

a. A, S   c. E, S
b. A, P   d. E, P

53. "Resting" rates of adrenal medullary secretion
contain predominantly _____ (N, norepinephrine; E, epinephrine) and, in comparison
with "resting" tonic activity by ganglionic
sympathetic neurons, are thought to constitute
a _____ (J, major; M, minor) _____ (S,
supporting; O, opposing) influence upon vasomotor tonus.

a. N, J, S   c. N, M, S   e. E, J, S
b. N, M, O   d. E, M, O   f. E, J, O

54. Denervation supersensitivity is associated with
interruption of _____ (E, preganglionic; O,
postganglionic) axons of _____ (S, only
sympathetic; P, only parasympathetic; PS,
either sympathetic or parasympathetic) neurons.

a. E, S   c. E, PS   e. O, P
b. E, P   d. O, S   f. O, PS

55. Vascular denervation generally results in immediate _____ (C, vasoconstriction; D,
vasodilation) followed by a progressive
_____ (R, rise; F, fall) in intrinsic tonus
and _____ (E, enhanced; M, diminished)
chemical mediator sensitivity of the vascular
smooth muscle.

a. C, F, E   c. C, R, E   e. D, R, E
b. C, F, M   d. D, R, M   f. D, F, M

56. A loss of enzymatic synthesis by degenerating autonomic nerve terminals at denervated effector structures would tend to _____ (A, assist; O, oppose) the phenomenon of denervation super-sensitivity for sympathetic _____ (W, as well as; N, but not) parasympathetic structures.

   a. A, W   b. A, N   c. O, W   d. O, N

OBJECTIVE 57–10.

Identify the characteristics of parasympathetic and sympathetic reflexes, their differential functions, and the influence of central nervous system centers upon the control of autonomic function.

57. Autonomic reflexes are generally comprised of a _____ (S, single; T, two-) neuron peripheral afferent limb, _____ (M, monosynaptic; P, polysynaptic) central connections, and a _____ (S, single; T, two-) neuron efferent limb.

   a. S, M, T   c. S, P, S   e. T, M, S
   b. S, P, T   d. T, P, S   f. T, M, T

58. The baroreceptor reflex is an example of _____ reflex while salivation in response to the smell and sight of appetizing food is considered _____ reflex. (S, a somatic; A, an autonomic)

   a. S, S   c. A, S
   b. S, A   d. A, A

59. Eccrine sweating and localized alterations in blood flow during processes of heat regulation represent _____ (X, examples; C, exceptions) to a generally _____ (S, specific; D, diffuse) nature of _____ (M, sympathetic; P, parasympathetic) function.

   a. X, S, M   c. X, D, M   e. C, D, M
   b. X, S, P   d. C, D, P   f. C, S, P

60. "Mass discharge" is a phenomenon of the _____ (S, sympathetic; P, parasympathetic) system which results in _____ (E, extra; R, reduced) activation of the body in states of physical stress.

   a. S, E   c. P, E
   b. S, R   d. P, R

61. The phenomenon of "mass discharge" results in _____ arterial pressure, _____ blood glucose concentrations, and _____ muscle strength and mental activity. (I, increased; N, no significant change in; D, decreased) _____

   a. D, D, D   c. N, I, N   e. I, I, N
   b. N, D, I   d. I, N, N   f. I, I, I

62. "Rage," elicited by stimulation of _____ (M, medullary; H, hypothalamic) centers, results in a _____ (D, discrete; W, widespread) mode of excitation of the _____ (S, sympathetic; P, parasympathetic) nervous system.

   a. M, D, S   c. M, W, S   e. H, W, S
   b. M, D, P   d. H, W, P   f. H, D, P

63. Arterial blood pressure, heart rate, _____ (W, as well as; N, but not) respiration, are regulated predominantly via _____ (H, higher; L, lower) brain stem centers.

   a. W, H   c. N, H
   b. W, L   d. N, L

64. A more significant _____ (T, thalamic; H, hypothalamic) control center for the autonomic nervous system largely _____ (B, bypasses; A, acts through) the lower brain stem centers.

   a. T, B   c. H, B
   b. T, A   d. H, A

OBJECTIVE 57–11.

Identify the representative pharmacological agents which act upon the peripheral autonomic nervous system and their mechanisms of action.

65. _____ (S, sympathomimetic; P, parasympathomimetic) pharmacological agents, including phenylephrine and methoxamine, have comparatively _____ (L, longer; H, shorter) durations of action than the corresponding chemical mediator.

a. S, L   c. P, L
b. S, H   d. P, H

66. Ephedrine, tyramine, and amphetamine act as _____ (S, sympathomimetic; P, parasympathomimetic) agents by virtue of their _____ (D, direct; I, indirect) action upon effector organs.

a. S, D   c. P, D
b. S, I   d. P, I

67. Synthesis and storage of norepinephrine is prevented by _____ whereas _____ is thought to block the release of norepinephrine. (P, pilocarpine; G, guanethidine; S, scopolamine; R, reserpine)

a. S, P   c. G, R   e. R, G
b. G, P   d. S, R   f. R, S

68. Phenoxybenzamine and phentolamine block _____ , whereas propanolol blocks transmission at _____ . (G, all autonomic ganglia; A, alpha receptors; B, beta receptors)

a. A, G   c. A, B   e. B, A
b. A, A   d. B, G   f. B, B

69. Neostigmine, physostigmine, and diisopropyl fluorophosphate _____ (P, potentiate; I, inhibit) the action of _____ (A, acetylcholine; N, norepinephrine) by their actions upon a corresponding _____ (R, receptor protein; E, enzyme).

a. P, A, R   c. P, N, R   e. I, N, R
b. P, A, E   d. I, N, E   f. I, A, E

70. _____ (P, parasympathomimetic; S, sympathomimetic) agents such as methacholine, carbamylcholine, _____ (W, as well as; N, but not) pilocarpine, are more effective in stimulating sympathetic _____ (A, adrenergic; C, cholinergic) fiber effector structures.

a. P, W, A   c. P, N, A   e. S, N, A
b. P, W, C   d. S, N, C   f. S, W, C

71. Atropine blocks the _____ (N, nicotinic; M, muscarinic) action of _____ (E, norepinephrine; A, acetylcholine) at autonomic _____ (G, ganglia; F, effector organs).

a. N, E, G   c. N, A, G   e. M, A, G
b. N, E, F   d. M, A, F   f. M, E, F

72. Nicotine stimulates _____ (S, only sympathetic; P, only parasympathetic; B, both sympathetic and parasympathetic) receptors of _____ (G, ganglia only; E, effector structures only; GE, autonomic ganglia as well as effector structures).

a. S, GE   c. B, G   e. P, E
b. P, G   d. S, E   f. B, GE

73. Tetraethyl ammonium ions, hexamethonium ions, and pentolinium are utilized in the treatment for _____ (O, hypotensive; E, hypertensive) states of arterial pressure by virtue of their blocking action at receptors of autonomic _____ (F, effector structures; G, ganglia).

a. O, F   c. E, F
b. O, G   d. E, G

58

The Eye: I. Optics of Vision

OBJECTIVE 58–1.

Using the following diagram, identify the various structural components of the eye.

Directions: Match the lettered headings with the diagram and numbered list of descriptive words and phrases.

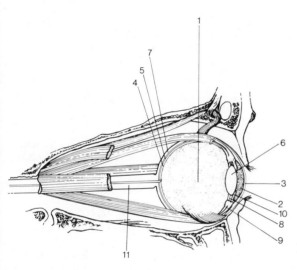

1. _____ Jelly-like substance filling the large chamber behind the lens.
2. _____ Watery fluid filling the small chamber in front of the lens.
3. _____ Transparent extension of the sclera.
4. _____ Inner layer of the eye, containing rods and cones.
5. _____ Intermediate layer of the eye, containing pigment and blood vessels.
6. _____ Crystalline, biconvex structure, capable of magnifying objects.
7. _____ Outer layer of the eye, composed of tough fibrous tissue.
8. _____ Ligaments that hold the lens in place.
9. _____ Fibers arranged circularly around the eye, producing a sphincter-like action.
10. _____ Pigmented extension of the choroid, surrounding the pupillary aperture.
11. _____ Nerve pathway leading from the retina to the brain.

a. Iris
b. Ciliary muscle fibers
c. Cornea
d. Aqueous humor

e. Choroid
f. Optic nerve
g. Vitreous humor
h. Suspensory ligaments

i. Lens
j. Sclera
k. Retina

OBJECTIVE 58–2.

Identify the physical principles underlying light refraction and their application to the physical properties of lenses.

12. Refraction, or the degree of _____ (B, bending; V, change in velocity; S, splitting) of light rays at an interface, is _____ (I, increased; N, not significantly altered; D, decreased) by increasing the degree of angulation between the entering wave front of the beam and the interface.

    a. B, N   c. S, D   e. V, D
    b. V, N   d. B, I   f. S, I

13. The refractive index of a material is the _____ (R, ratio; RR, reciprocal of the ratio) of its propagation velocity for light to the propagation velocity of light in _____ (G, glass; V, a vacuum).

    a. R, G   c. RR, G
    b. R, V   d. RR, V

14. Light travels at a velocity of 200,000 _____ (M, meters; K, kilometers) per second in a material with a refractive index of _____ .

    a. M, 0.67   c. M, 1.50   e. K, 1.00
    b. M, 1.00   d. K, 0.67   f. K, 1.50

15. Light rays entering a medium with an interface perpendicular to the beam of light and a higher refractive index travel at _____ (H, a higher; S, the same; L, a lower) velocity and _____ (A, are; N, are not) bent as they traverse the interface.

    a. H, A   c. L, A   e. S, N
    b. S, A   d. H, N   f. L, N

16. Light rays entering transparent biological materials obliquely from air encounter _____ (D, a decrease; I, an increase) in refractive index and are bent _____ (T, towards; A, away) from a direction perpendicular to the interface.

    a. D, T   c. I, T
    b. D, A   d. I, A

17. Parallel light rays are brought to a focal point on the opposite side of the lens by a _____ (V, concave; X, convex) form of _____ (C, only cylindrical; S, only spherical; B, both cylindrical as well as spherical) lenses.

    a. V, C   c. V, B   e. X, S
    b. V, S   d. X, C   f. X, B

OBJECTIVE 58–3.

Identify focal length and dioptric strength, and the principles underlying the formation of images by lens systems.

18. The distance from a convex lens at which parallel light rays converge to a common point is called the _____ (N, nodal point; F, focal length; R, refractive power), which is _____ (I, inversely; D, directly) related to the dioptric strength.

    a. N, I   c. F, I   e. R, I
    b. N, D   d. F, D   f. R, D

19. A lens that brings parallel light rays to a focus 2 cm. from its optical center has a focal length of _____ meters, and a dioptric strength of _____ diopters.

    a. 50, 0.05   d. 0.02, more than 40
    b. 50, 5   e. 0.02, less than 1
    c. 50, more than 40 f. 0.02, 5

20. Light rays which pass through the optical center or nodal point of a convex lens are those which enter the lens at _____ (O, an oblique; P, a 90°) angle to the interface and are _____ (R, largely refracted; N, not refracted).

    a. O, R   c. P, R
    b. O, N   d. P, N

21. The reciprocal of the object distance from a convex lens equals the _____ (D, dioptic strength; F, focal length) of the lens _____ (P, plus; M, minus) the _____ (I, image; R, reciprocal of the image) distance from the lens.

    a. D, P, I   c. D, M, I   e. F, P, I
    b. D, M, R   d. F, M, R   f. F, P, R

22. Maintenance of a fixed image distance, while decreasing the object distance, requires a lens system of _____ convexity and _____ dioptric strengths. (I, increasing; D, decreasing)

    a. I, I   c. D, D
    b. I, D   d. D, I

23. The image produced by a simple convex lens, when compared to the original object, is _____ (R, upright; U, upside down), with the right side of the image corresponding to the _____ (G, right; F, left) side of the object.

    a. R, G   c. U, G
    b. R, F   d. U, F

24. The _____ (C, convergent; D, divergent) effect upon light rays of a _____ (V, concave; X, convex) lens of –2 diopters may be cancelled by the series addition of a lens of _____ diopters.

    a. C, V, –½   c. C, X, +½   e. D, V, +2
    b. C, X, +2   d. D, X, –½   f. D, V, +½

25. A series of three lenses with dioptric strengths of +1 diopter, –1 diopter, and +2 diopters have the same combined _____ (D, diverging; C, converging) effect upon light rays as a lens system with a dioptric strength of _____ diopter(s).

    a. D, ½   c. D, 2   e. C, ⅓
    b. D, ⅓   d. C, ½   f. C, 2

OBJECTIVE 58–4.

Identify the optical properties of the lens system of the "reduced eye" and its major components.

26. The dioptric strength of the eye for focusing parallel light rays upon the retina is nearly _____ diopters, corresponding to a focal length of approximately _____ mm.

    a. +10, 10   c. −6, 170   e. −60, 50
    b. +6, 17   d. +60, 17   f. −10, 100

27. A higher percentage of the dioptric strength of the optical system of the eye is provided by the anterior surface of the _____ (C, cornea; A, aqueous humor; L, crystalline lens) with a refractive index of _____ .

    a. C, 1.38   c. L, 1.38   e. A, 3.81
    b. A, 1.38   d. C, 3.81   f. L, 3.81

28. The posterior surface of the cornea provides about _____ (P, plus; M, minus) 4 diopters to the dioptric strength of the reduced eye as a consequence of _____ (I, an increase; D, a decrease) in refractive index for light traversing its interface and its _____ (X, convex; V, concave) inner surface.

    a. P, I, X   c. P, D, X   e. M, D, X
    b. P, I, V   d. M, D, V   f. M, I, V

29. A dioptric strength of _____ diopters, attributed to the crystalline lens when accommodated for far vision, would be _____ (I, increased; D, decreased) by its removal and suspension in air.

    a. 2.9, I   c. 48, I   e. 15, D
    b. 15, I   d. 2.9, D   f. 48, D

30. The perception of an upright object is associated with an _____ retinal image, _____ visual information traversing the aqueous humor, and _____ visual information traversing the vitreous humor. (U, upright; I, inverted)

    a. U, U, U   c. U, U, I   e. I, U, I
    b. U, I, U   d. I, I, U   f. I, I, I

31. When holding a lighted candle in a darkened room in front of and slightly to one side of the subject's eyes, the observer should expect to see an _____ image reflected from the corneal surface, an _____ image reflected from the anterior lens surface, and an _____ image reflected from the posterior lens surface. (I, inverted; E, erect)

    a. I, I, E   c. E, I, E   e. I, I, I
    b. E, E, E   d. E, E, I   f. I, E, I

OBJECTIVE 58–5.

Identify the mechanisms underlying accommodation of the "refractive power" of the eye, their range of action, their significance, and the consequences of presbyopia upon their action.

32. The ciliary muscle, innervated by _____ (S, sympathetic; P, parasympathetic) efferents, functions to adjust the _____ (PS, pupil size; R, refractive power; I, image distance) of the eye.

    a. S, PS   c. S, I   e. P, R
    b. S, R   d. P, PS   f. P, I

33. The crystalline lens is normally _____ (N, a nonelastic; E, an elastic) structure with a _____ (BN, biconcave; BX, biconvex; NX, concavoconvex) shape.

    a. N, BN   c. N, NX   e. E, BX
    b. N, BX   d. E, BN   f. E, NX

34. The _____ (C, corneal; L, crystalline lens) tension is _____ (I, increased; D, decreased) through contraction of circular _____ (W, as well as; N, but not) meridional fibers of the ciliary muscle.

    a. C, I, W   c. L, I, W   e. L, D, W
    b. L, I, N   d. C, D, N   f. L, D, N

35. The excised crystalline lens has surfaces comparable to its _____ (H, higher; L, lower) dioptric strength in the normal eye when accommodated for _____ (N, near; F, far) vision.

    a. H, N   c. L, N
    b. H, F   d. L, F

36. A range of accommodation of about _____ diopters shortly after birth more normally approximates the _____ (V, value; R, the reciprocal of the value) of the _____ (F, far point; N, near point) expressed in meters.

    a. 4 to 5, V, F   d. 14, R, N
    b. 4 to 5, V, N   e. 14, R, F
    c. 4 to 5, R, F   f. 14, V, N

37. The reciprocal of the near point _____ (P, plus; M, minus) the _____ (F, far point; RF, reciprocal of the far point) equals the _____ (A, range of accommodation; RA, reciprocal of the range of accommodation) of the eye in diopters.

    a. P, F, A   c. P, RF, A   e. M, RF, A
    b. P, F, RA   d. M, RF, RA   f. M, F, RA

38. Assuming a normal _____ mm. distance between the nodal point of a reduced eye and the retina, the dioptric strength of a subject's eye accommodated for a near point of 20 cm. is about _____ diopters.

    a. 17, 45   c. 17, 78   e. 59, 64
    b. 17, 64   d. 59, 45   f. 59, 78

39. Individuals who require differing lens systems of different dioptric strengths in order to focus far and near objects clearly—i.e., bifocals or trifocals—probably have a defect of the _____ (L, crystalline lens; R, retina; C, cornea) termed _____ (M, myopia; A, astigmatism; P, presbyopia).

    a. L, M   c. C, M   e. R, P
    b. R, A   d. L, P   f. C, A

40. Presbyopia is a condition of altered states of _____ (O, lens opacity; E, lens elasticity; M, smooth muscle contractility) primarily affecting _____ (C, children; A, adults).

    a. O, C   c. M, C   e. E, A
    b. E, C   d. O, A   f. M, A

OBJECTIVE 58-6.

Identify the major functions of the iris and the normal "aberrations" of vision; and their significance.

41. The amount of light entering the eye is _____ (D, directly; I, inversely) proportional to the _____ (R, radius; S, square of the radius) of the pupillary aperture.

    a. D, R   c. I, R
    b. D, S   d. I, S

42. The pupil diameter ranges between minimum and maximal values of _____ mm. respectively, and is associated with a greater "depth of focus" at its _____ (G, greater; L, lesser) diameters.

    a. 0.5 & 4, G   d. 0.5 & 4, L
    b. 1.5 & 8, G   e. 1.5 & 8, L
    c. 4 & 12, G   f. 4 & 12, L

43. Increasing the size of the pupil aperture generally tends to _____ spherical aberration, _____ chromatic aberration, and _____ the sharpness of focus of vision. (I, increase; D, decrease)

    a. I, I, I   c. I, D, D   e. D, D, I
    b. I, I, D   d. D, I, D   f. D, D, F

44. Chromatic aberration results from a _____ (G, greater; L, lesser) optical refraction of the _____ (N, longer; S, shorter) wavelengths of the red region of the spectrum when compared to the blue region.

    a. G, N   c. L, N
    b. G, S   d. L, S

45. _____ (S, spherical aberrations; D, diffractive errors) of the eye result in "interference" patterns which tend to reduce the sharpness of vision, particularly at _____ (N, minimal; X, maximal) pupil diameters.

    a. S, N   c. D, N
    b. S, X   d. D, X

OBJECTIVE 58-7.

Identify emmetropia, hypermetropia, myopia, and astigmatism, the underlying mechanisms resulting in errors of refraction, and their means of correction.

46. When the emmetropic eye observes distant objects, it does so with a _____ (C, contracted; R, relaxed) state of the ciliary muscle, and _____ (I, is; N, is not) generally able to see distant objects in sharp focus.

    a. C, I   c. R, I
    b. C, N   d. R, N

47. "Far-sightedness" is associated with a more pronounced _____ (I, increase; D, decrease) in the magnitude of the _____ (F, "far point"; N, "near point") of vision.

    a. I, F   c. D, F
    b. I, N   d. D, N

48. "Near-sightedness," or _____ (E, emmetropia; HO, hyperopia; M, myopia), is associated with too _____ (L, low; H, high) a dioptric strength of the optical system relative to the length of the eyeball.

    a. E, L   c. M, L   e. HO, H
    b. HO, L   d. E, H   f. M, H

49. Myopia generally results in a greater _____ (I, increase; D, decrease) in the magnitude of _____ (N, "near point"; F, "far point") values from normal.

    a. I, N   c. D, N
    b. I, F   d. D, F

50. The presbyopic and _____ (HO, hyperopic; E, emmetropic; M, myopic) individual is less likely to focus clearly on objects at any distance from the eye as a consequence of an image plane which lies _____ (F, in front of; B, behind) the retina.

    a. HO, F   c. M, F   e. E, B
    b. E, F   d. HO, B   f. M, B

51. Swimming underwater with the eyes open, so that light enters the cornea from water rather than air, would produce an underwater:

    a. Myopia   d. Diplopia
    b. Emmetropia   e. Presbyopia
    c. Hyperopia   f. Astigmatism

52. In most cases of _____ (M, myopia; P, presbyopia; H, hyperopia) the subject's distant visual acuity is excellent because he can compensate for this condition by _____ (C, contracting; R, relaxing) his ciliary muscles.

    a. M, C   c. P, C   e. H, C
    b. M, R   d. P, R   f. H, R

53. An eye has a dioptric strength of 67 diopters with ciliary muscle relaxation. The distance from the nodal point to the retina is 17 mm. The condition would best be described as _____ (H, hyperopia; M, myopia) and could be corrected with an appropriate _____ (C, convex; V, concave) lens.

    a. H, C   c. M, C
    b. H, V   d. M, V

54. The condition of _____ (G, glaucoma; A, astigmatism; C, cataract) is a more frequent result of an oblong shape of _____ (R, corneal; L, crystalline lens) surfaces.

    a. G, R   c. C, R   e. A, L
    b. A, R   d. G, L   f. C, L

55. A prescription for corrective eyeglasses reads -4.5, $+1.5$ axis $180°$, referring to a -4.5 diopter lens necessary to correct for _____ (H, hyperopia; M, myopia) and a $+1.5$ diopter cylindrical lens at an axis of $180°$ to correct for _____ (P, presbyopia; A, astigmatism; N, protanopia).

    a. H, P   c. M, N
    b. H, A   d. M, A

Directions: Select the appropriate lettered answer from the right-hand column. Use an answer only once.

56. Strabismus
57. Hyperopia
58. Myopia
59. Astigmatism
60. Emmetropia

a. Correct with convex lens
b. Correct with concave lens
c. Correct with cylindrical lens
d. Correct with prism lens
e. No correction

OBJECTIVE 58-8.

Identify visual acuity, its significance, and the clinical method for its determination and statement.

61. The maximum visual acuity occurs within the _____ (R, rod; C, cone)-containing, _____ (P, peripheral; F, foveal) regions of the retina.

a. R, P   c. C, P
b. R, F   d. C, F

62. The fovea, _____ (M, more; L, less) than 1 mm. in diameter, receives a thousand-fold reduction in object size when the distance from the object to the eye approximates _____ meters.

a. M, 1.7   c. M, 170   e. L, 17
b. M, 17   d. L, 1.7   f. L, 170

63. The maximal visual acuity for discriminating between two point sources of light corresponds to an angle subtended at the nodal point of the eye of about ½ _____ (S, second; M, minute; D, degree), and a retinal separation of their centers of about _____ microns.

a. S, 2   c. D, 2   e. M, 20
b. M, 2   d. S, 20   f. D, 20

64. An individual with 20/40 vision in both eyes has a visual acuity of _____ , which is considered _____ (H, higher than a normal; N, normal; L, lower than a normal) average.

a. ½, H   c. ½, L   e. 2, N
b. ½, N   d. 2, H   f. 2, L

65. The Snellen test charts are generally read by subjects at a distance of 20 _____ (F, feet; M, meters) and reflect a measure of visual acuity which is _____ (D, directly; I, inversely) proportional to the minimum separable visual angle subtended at the nodal point of the eye.

a. F, D   c. F, I
b. M, D   d. M, I

OBJECTIVE 58-9.

Identify the major mechanisms utilized for the determination of distance of an object from the eye and their relative importance.

66. The retinal size of a corresponding object within the visual field is _____ (D, directly; I, inversely) proportional to the _____ (T, distance; S, square of the distance; R, square root of the distance) of the object from the eye.

a. D, T   c. D, R   e. I, S
b. D, S   d. I, T   f. I, R

67. Slight movements of the eyes result in a comparatively greater retinal displacement formed by _____ (N, near; I, intermediate; D, distant) objects, resulting in a means of determination of _____ (V, vernier acuity; P, depth perception).

a. N, V   c. D, V   e. I, P
b. I, V   d. N, P   f. D, P

68. The determination of visual distances by _____ (S, relative sizes; P, moving parallax; T, stereopsis) is a major mechanism for objects located _____ (C, closer; F, further) than 200 feet which is lost by closing one eye.

a. S, C   c. T, C   e. P, F
b. P, C   d. S, F   f. T, F

69. When the eyes are focused on a distant object, near objects within the visual field tend to project more _____ (M, medially; L, laterally) within the retina and to produce a _____ (V, moving; B, binocular) type of parallax.

a. M, V   c. L, V
b. M, B   d. L, B

OBJECTIVE 58-10.

Identify the ophthalmoscope and retinoscope, and their utility.

70. The _____ (O, ophthalmoscope; R, retino-scope) is an instrument which is designed so that _____ (S, the subject; B, an observer) can see the retina of the subject with clarity.

a. O, S   c. R, S
b. O, B   d. R, B

71. While using the retinoscope, a first glow appearing on the side of a subject's pupil from which the light beam is being moved is an indication of _____ , whereas a first glow appearing on the opposite side of the pupil is an indication of _____ . (M, myopia; E, emmetropia; H, hyperopia)

a. E, H   c. H, M   e. M, H
b. H, E   d. M, E   f. E, M

72. A natural tendency for ciliary muscle _____ (R, relaxation; C, contraction) on the part of an emmetropic subject and an emmetropic observer while using the ophthalmoscope requires the insertion of a corrective lens of about _____ diopters.

a. R, +4   c. R, +12   e. C, -4
b. R, -4   d. C, +4   f. C, +12

59

The Eye: II. Receptor Functions
of the Retina

OBJECTIVE 59–1.

Locate the various layers and membranes of the retina.

Directions: Match the lettered headings with the diagram of the various layers and membranes, and place the answers in the numbered column.

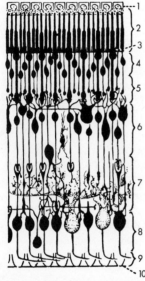

(From Guyton, A.C., Textbook of Medical Physiology, 5th ed. Philadelphia, W.B. Saunders Company, 1976.)

1. _____

2. _____

3. _____

4. _____

5. _____

6. _____

7. _____

8. _____

9. _____

10. _____

a. Ganglionic layer
b. Outer plexiform layer
c. Rod and cone layer
d. Inner limiting membrane

e. Inner plexiform layer
f. Outer limiting membrane
g. Optic nerve fiber layer

h. Inner nuclear layer
i. Pigmented layer
j. Outer nuclear layer

OBJECTIVE 59-2.

Identify the physiologic anatomy of the retina, the macula and foveal regions, and their functional significance.

11. Cones are _____ (C, chemoreceptors; E, electromagnetic receptors) located within the _____ (H, choroid; R, retina) of the eye, which are responsible for _____ (CV, color; DV, dim light) vision.

a. C, H, CV   c. C, R, CV   e. E, R, CV
b. C, H, DV   d. E, R, DV   f. E, H, DV

12. Light passes through the layer of rods and cones _____ (B, before; A, after) passing through the ganglionic layer, whose axons exit from the eye in the region of the _____ (F, fovea; M, macula; D, optic disc).

a. B, F   c. B, D   e. A, M
b. B, M   d. A, F   f. A, D

13. A smaller region of the _____ (M, macula; F, fovea) is an area of a relatively _____ (H, high; L, low) visual acuity, which functions primarily at _____ (H, high; L, low) levels of illumination.

a. M, H, H   c. M, L, H   e. F, H, H
b. M, L, L   d. F, L, L   f. F, H, L

14. The _____ portion of the retina contains a higher density of rods and the _____ portion of the retina contains a higher density of cones. (O, optic disc; F, foveal; P, peripheral)

a. O, F   c. F, O   e. P, F
b. O, P   d. F, P   f. P, O

OBJECTIVE 59-3.

Identify the rods and cones, their retinal distribution, and their morphological and functional characteristics.

15. Foveal receptors, compared to those of the peripheral regions of the retina, are _____ (L, larger; S, smaller) in diameter, possess a _____ (Y, cylindrical; C, conical) shaped outer segment, and _____ (D, do; N, do not) possess rhodopsin.

a. L, Y, D   c. L, C, D   e. S, C, D
b. L, Y, N   d. S, C, N   f. S, Y, N

16. Photochemicals of retinal receptors are largely localized along shelves formed by infoldings of the _____ (C, cell membrane; ER, endoplasmic reticulum; S, synaptic body) located within the _____ (I, inner; O, outer) segments of the cells.

a. C, I   c. S, I   e. ER, O
b. ER, I   d. C, O   f. S, O

17. A continuous replacement of receptor_____ (D, discs; C, cells) is a more notable feature of _____ (O, only cones; R, only rods; B, both rods and cones).

a. D, O   c. D, B   e. C, R
b. D, R   d. C, O   f. C, B

18. The _____ (O, outer segment; I, inner segment; S, synaptic body) is that portion of rods or cones which generally connects more directly with _____ (G, ganglion; HP, horizontal and bipolar; AM, amacrine and Müller) cells.

a. O, G   c. I, AM   e. S, HP
b. O, AM   d. I, HP   f. S, G

OBJECTIVE 59-4.

Identify the pigment layer of the retina and its function, the vascular supply of the retina, and potential causes and consequences of retinal detachment.

19. The pigment layer of the human eye primarily refers to a _____ (R, rhodopsin-containing; M, melanin-containing; T, tapetum) layer which functions to _____ (E, enhance; D, reduce) light reflection.

a. R, E   c. T, E   e. M, D
b. M, E   d. R, D   f. T, D

20. The pigment layer functions to store large quantities of vitamin _____ , cause dissolution of discs of _____ (R, rods; C, cones), and enhance visual acuity, particularly during _____ (D, daylight; M, dim light) conditions.

a. A, R, D   c. A, C, D   e. E, C, D
b. A, R, M   d. E, C, M   f. E, R, M

21. Albinos lack _____ (P, only photopsin; M, only melanin; A, melanin and rhodopsin) and have _____ (B, complete blindness; V, poor visual acuity; E, enhanced night vision).

a. P, E   c. A, B   e. M, B
b. M, V   d. P, V   f. A, E

22. The central retinal artery provides a nutrient blood supply largely for the _____ (O, outer; I, inner) layers of the retina, whereas the remainder of the retina is largely supplied by diffusion from _____ (C, choroid vessels; S, vessels of the sclera; V, vitreous humor).

a. O, C   c. O, V   e. I, S
b. O, S   d. I, C   f. I, V

23. Retinal detachment more frequently results from separation at the _____ (RV, retina-vitreous humor; RC, retina-choroid; CS, choroid-sclera) interface as a consequence of _____ (E, localized edema; V, alterations of collagenous fibrils of the vitreous humor).

a. RV, E   c. CS, E   e. RC, V
b. RC, E   d. RV, V   f. CS, V

24. Detached retinae generally _____ (D, do; N, do not) resist degenerative changes for prolonged periods of time as a consequence of their _____ (S, nutrient supply; A, disuse atrophy; G, separation from ganglion cells).

a. D, S   c. N, A
b. N, S   d. N, G

OBJECTIVE 59–5.

Identify the rhodopsin-retinene visual cycle, the significance of vitamin A, and the mechanisms of excitation of the rods.

25. The combination of the protein scotopsin with _____ (R, retinene; V, vitamin A) yields the light-sensitive pigment of the _____ (D, rods; C, cones) called _____ (P, visual purple; Y, cyanolabe).

a. R, D, P   c. R, C, P   e. V, C, P
b. R, D, Y   d. V, C, Y   f. V, D, Y

26. A chemical sequence of conversions, involving prelumirhodopsin, lumirhodopsin, and metarhodopsins, generates an all- _____ (C, cis; T, trans) retinene and _____ (PT, photopsin; ST, scotopsin) in the _____ (P, presence; A, absence) of light.

a. C, PT, P   c. C, ST, P   e. T, ST, P
b. C, PT, A   d. T, ST, A   f. T, PT, A

27. An 11-*cis* retinene _____ (I, is; N, is not) formed directly from an all-*trans* retinene, and is formed via 11-*cis* vitamin A by the process of _____ (M, methylation; D, dehydrogenation; H, hydrogenation).

a. I, M   c. I, H   e. N, D
b. I, D   d. N, M   f. N, H

28. The absence of light favors the conversion of scotopsin and _____ (P, prelumirhodopsin; T, all-*trans* vitamin A; S, 11-*cis* retinene) to _____ (A, vitamin A; D, vitamin A aldehyde; R, rhodopsin).

a. P, R   c. S, R   e. T, D
b. T, A   d. P, A   f. S, D

29. Absorption of light photons by _____ (D, rhodopsin; E, retinene), located within the _____ (I, inner; O, outer) rod segment, is thought to result in a locally enhanced cell membrane _____ (C, conductance; R, resistance) to sodium ion movement.

a. D, I, C   c. D, O, C   e. E, I, C
b. D, O, R   d. E, O, R   f. E, I, R

30. A process of receptor _____ (D, depolarization; H, hyperpolarization), thought to occur in rods during their excitation, is _____ (S, similar; O, opposite) to that occurring within most types of receptors.

a. D, S   c. H, S
b. D, O   d. H, O

31. The eye is just able to detect an increase of light intensity of about one _____ (P, single photon; PC, per cent) above most preexisting light intensities as a consequence of the _____ (N, linear; G, logarithmic) nature of the retinal response.

a. P, N   c. PC, N
b. P, G   d. PC, G

32. Night blindness is associated with reduced rhodopsin concentrations, _____ (W, as well as; N, but not) color photosensitive chemicals of cones, and occurs as a consequence of _____ (D, deficiencies of; X, excess) vitamin _____ .

a. W, D, A   c. W, X, A   e. N, X, A
b. W, D, E   d. N, X, E   f. N, D, E

OBJECTIVE 59–6.
Identify the photochemicals and receptors responsible for color vision, and their spectral absorption characteristics.

33. Color vision is thought to involve _____ basic types of _____ (R, rods; C, cones).

    a. 2, R   c. 5, R   e. 3, C
    b. 3, R   d. 2, C   f. 5, C

34. Differing types of cones have been characterized largely on the basis of their _____ (A, spectral absorption; M, morphological) characteristics, and differ chemically in the _____ (O, opsin; R, retinene) portion of their _____ (S, scotopsins; P, photopsins).

    a. A, O, S   c. A, R, S   e. M, R, S
    b. A, O, P   d. M, R, P   f. M, O, P

35. Color-sensitive pigments of retinal receptors exhibit absorption maxima at 535, 575, and _____ millimicrons in contrast to a peak spectral absorption maximum of _____ millimicrons for rhodopsin.

    a. 430, 590   c. 430, 630   e. 630, 505
    b. 430, 505   d. 630, 430   f. 630, 590

36. A spectral wavelength corresponding to a peak absorption by rhodopsin is absorbed to the greatest extent by the _____ cone and to the least extent by the _____ cone. (B, blue; R, red; G, green)

    a. B, R   c. R, B   e. G, R
    b. B, G   d. R, G   f. G, B

OBJECTIVE 59–7.
Identify scotopic and photopic vision. Identify dark and light adaptation, their influence upon retinal sensitivity, their underlying and associated mechanisms, and their significance.

37. Scotopic and photopic vision are to a great extent _____ (S, separate; O, simultaneous) retinal functions at varying levels of light intensity with _____ (C, scotopic; P, photopic) vision providing a mechanism for distinguishing color.

    a. S, C
    b. S, P
    c. O, C
    d. O, P

38. Scotopic vision is _____ (M, most; S, least) sensitive when the image is confined to the fovea centralis or when the eye is adapted to _____ (L, light; D, dark).

    a. M, L
    b. M, D
    c. S, L
    d. S, D

39. The sensitivity of the rods _____ (W, as well as; N, but not) the cones is thought to be directly proportional to the _____ (L, logarithm; A, antilogarithm; M, magnitude) of their light-sensitive photochemical concentrations.

    a. W, L   c. W, M   e. N, A
    b. W, A   d. N, L   f. N, M

40. Decreasing levels of light-sensitive photochemical concentrations of retinal receptors are associated with _____ (I, increasing; D, decreasing) sensitivity of the eye to light during the process of _____ (L, light; K, dark) adaptation.

    a. I, L
    b. I, K
    c. D, L
    d. D, K

41. _____ (L, light; D, dark) adaptation requires a longer time interval for maximal adaptation, particularly for adaptation involving the _____ (R, rods; C, cones).

    a. L, R
    b. L, C
    c. D, R
    d. D, C

42. Comparatively _____ (S, shorter; G, longer) prior exposures to bright light permit a more rapid dark adaptation of an eye as a consequence of its _____ (H, higher; W, lower) concentration of vitamin A.

    a. S, H
    b. S, W
    c. G, H
    d. G, W

43. The inflection point of dark adaptation curves, representing a change between _____ (NP, neural and photochemical; CR, cone and rod) adaptive mechanisms, is more pronounced for dark adaptation following _____ (B, brief; P, prolonged) exposures to bright light.

    a. NP, B   c. CR, B
    b. NP, P   d. CR, P

44. Maximal states of retinal sensitivity, requiring adaptive time intervals of about 10 _____ (M, minutes; H, hours) in the dark, are largely achieved via _____ (P, photochemical; N, neural) adaptive changes of a _____ (C, cone; R, rod) type of vision.

    a. M, P, C   c. M, N, C   e. H, P, C
    b. M, N, R   d. H, N, R   f. H, P, R

45. Registration of images by the retina are possible up to about one _____ (T, thousand; H, hundred thousand; M, million) times minimal levels as a consequence of light adaptation and its associated pupillary _____ (C, constriction; D, dilation).

    a. T, C   c. M, C   e. H, D
    b. H, C   d. T, D   f. M, D

46. Light adaptation, _____ (W, as well as; N, but not) dark adaptation, is a process which is largely dependent upon _____ (E, the entire retina as a; I, individual rods and cones as) functioning unit(s).

    a. W, E   c. N, E
    b. W, I   d. N, I

47. Negative after-images are produced by transferring the gaze to a _____ (B, brightly illuminated; D, dark) background and result from _____ (A, dark and light adaptation; P, visual persistence).

    a. B, A   c. D, A
    b. B, P   d. D, P

OBJECTIVE 59–8.

Identify the mechanism underlying the fusion of flickering lights by the retina, its properties, and its significance.

48. The mechanism for the existence of the "critical fusion frequency" for vision is also the mechanism for:

    a. Perimetry   d. Dark adaptation
    b. Negative   e. Positive
       after-images   after-images
    c. Purkinje shifts   f. Presbyopia

49. Critical frequencies for fusion are _____ (R, directly; V, inversely) related to durations of visual persistence and _____ (I, increase; K, remain relatively constant; D, decrease) as a function of increasing light intensity.

    a. R, I   c. R, D   e. V, K
    b. R, K   d. V, I   f. V, D

50. The detection of flicker in a repetitively flashing light at a frequency of 40 times per second more probably involves _____ (C, cone; R, rod) stimulation and relatively _____ (H, high; L, low) levels of illumination.

    a. C, H   c. R, H
    b. C, L   d. R, L

OBJECTIVE 59-9.

Identify the basic mechanisms underlying the functions of the retina in the perception of color.

51. The _____ (Y, Young-Helmholtz theory; G, Gibbs-Donnan theory; B, law of Bell-Megendie) implies that a full range of color hues may be achieved through the mixing of a minimum of _____ primary colors.

   a. Y, 3   c. B, 3   e. G, 5
   b. G, 3   d. Y, 5   f. B, 5

52. Hue discrimination is _____ (X, maximal; N, minimal) at spectral wavelengths about midway between absorption peaks of _____ different primary types of _____ (C, cones; R, rods).

   a. X, 3, C   c. X, 5, C   e. N, 5, C
   b. X, 3, R   d. N, 5, R   f. N, 3, R

53. Fixation of the eyes on a bright yellow object for a time, followed by transference of the gaze to a well illuminated background, results in a _____ (Y, yellow; B, blue) _____ (E, phosphene; N, negative after-image; P, positive after-image).

   a. Y, E   c. Y, P   e. B, N
   b. Y, N   d. B, E   f. B, P

54. A sensation of white from a wide visual field is elicited by stimulation of a minimum of _____ type(s) of cone(s) by light containing a minimum of _____ spectral wavelength(s). (S, a single; T, two differing; M, more than two differing)

   a. S, S   c. M, S   e. T, M
   b. T, S   d. S, M   f. M, T

55. Yellow spectral wavelengths normally stimulate the _____ (BG, blue and green; GR, green and red; BR, red and blue) cones about equally and result in a sensation of _____ (B, blue; Y, yellow).

   a. BG, B   c. BR, B   e. GR, Y
   b. GR, B   d. BG, Y   f. BR, Y

56. Yellow and _____ (G, green; R, red; B, blue) are complementary colors, because their spectral illumination of the retina results in a sensation of _____ (K, black; W, white) when they stimulate all cones equally.

   a. G, K   c. B, K   e. R, W
   b. R, K   d. G, W   f. B, W

57. Pigment (A) is found to absorb blue and reflect green, yellow, and red spectral wavelengths. Pigment (B) absorbs red and yellow wavelengths. Consequently, pigment (A) produces a visual sensation of _____ , pigment (B) produces a visual sensation of _____ , and a mixture of pigments (A) and (B) produces a visual sensation of _____ . (R, red; G, green; K, black; Y, yellow; B, blue; W, white; O, orange; N, brown; P, purple)

   a. R, G, G   c. Y, B, G   e. B, O, N
   b. R, G, K   d. Y, B, W   f. B, O, P

58. A greater excitation of the _____ (B, blue; R, red; G, green) type of cone with lesser excitation of the other primary types of cones would occur with a single monochromatic spectral wavelength of _____ millimicrons.

   a. B, 505   c. G, 505   e. R, 610
   b. R, 505   d. B, 610   f. G, 610

OBJECTIVE 59-10.

Identify the types of color blindness, their probable mechanistic basis, and their means of clinical determination.

59. Red-green color blindness is generally associated with an absence of appropriate genes located in the _____ chromosome and results in a confusion of colors within the _____ (BG, blue-green; RG, red-green) region of the color spectrum.

   a. X, BG   c. Y, BG
   b. X, RG   d. Y, RG

60. About 6% of the _____ (M, male; F, female) population have the condition termed _____ (P, protanopia; D, deuteranopia; T, tritanopia).

   a. M, P   c. M, T   e. F, D
   b. M, D   d. F, P   f. F, T

61. _____ (P, protanopia; D, deuteranopia) is associated with a reduction in the visual spectral width by a defect of spectral sensitivity occurring at the _____ (S, shorter; I, intermediate; L, longer) wavelengths of the visible spectrum.

   a. P, S   c. P, L   e. D, I
   b. P, I   d. D, S   f. D, L

62. A deuteranomalous individual with a deficiency of _____ (G, green; R, red) cones, when asked to match spectral combinations of red and green to produce a sensation of _____ (B, blue; Y, yellow), would require greater amounts of _____ (R, red; G, green) than normal.

   a. G, B, R   c. G, Y, R   e. R, B, R
   b. G, Y, G   d. R, Y, G   f. R, B, G

63. The _____ (S, Snellen; I, Ishihara) test chart is a test for color blindness based upon _____ (E, relative emphasis of differing patterns of colored dots; W, grouping of different colored tufts of wool).

   a. S, E   c. I, E
   b. S, W   d. I, W

60

The Eye: III. Neurophysiology of Vision

OBJECTIVE 60–1.

Identify the components of the visual pathway.

1. The optic chiasm _____ (I, is; N, is not) a synaptic relay station for visual information which separates the optic tract from the _____ (G, lateral geniculate body; R, optic radiation; O, optic nerve).

   a. I, G   c. I, O   e. N, R
   b. I, R   d. N, G   f. N, O

2. The _____ (OC, optic chiasma; LG, lateral geniculate body) represents a locus of decussating fibers from the_____(N, nasal; T, temporal) halves of the retinae.

   a. OC, N   c. OC, T
   b. LG, N   d. LG, T

3. Visual information from the left visual field of the right eye and the _____ visual field of the left eye project to the _____ side of the visual cortex. (R, right; L, left)

   a. R, R   c. L, R
   b. R, L   d. L, L

4. The _____ (L, left; R, right) _____ (SC, superior colliculus; PT, pretectal nucleus; LG, lateral geniculate body) is a specific relay station for projection of visual information from the left visual fields to the visual cortex.

   a. L, SC   c. L, LG   e. R, PT
   b. L, PT   d. R, SC   f. R, LG

5. The pretectal nuclei _____ (W, as well as; N, but not) the superior colliculi receive visual information from the lateral geniculate bodies _____ (W, as well as; N, but not) directly from the optic tracts.

   a. W, W   c. N, W
   b. W, N   d. N, N

OBJECTIVE 60-2.

Identify the principal neural organizational features of the retina.

6. The fovea is a region devoid of _____ (R, rods; C, cones) which, when compared to the peripheral regions of the retina, has a comparably _____ (H, high; L, low) light sensitivity and a _____ (H, high; L, low) visual acuity.

   a. R, H, H   c. R, L, H   e. C, H, H
   b. R, L, L   d. C, L, L   f. C, H, L

7. Each retina contains a higher number of _____ , an intermediate number of _____ , and a smaller number of _____ . (F, optic nerve fibers; R, rods; C, cones)

   a. F, R, C   c. R, F, C   e. C, R, F
   b. F, C, R   d. R, C, F   f. C, F, R

8. The _____ (C, cones; R, rods) exhibit a higher degree of convergence upon ganglion cells, particularly for those located within the _____ (N, central; P, peripheral) portion of the retina.

   a. C, N   c. R, N
   b. C, P   d. R, P

9. _____ (A, amacrine; B, bipolar; H, horizontal) cells, providing the major connecting linkage between rods and cones and ganglion cells, receive _____ (C, convergent; D, divergent) signals from the rods and cones of the peripheral regions of the retina.

   a. A, C   c. H, C   e. B, D
   b. B, C   d. A, D   f. H, D

10. The horizontal and _____ (G, ganglion; B, bipolar; A, amacrine) cells are special types of cells of the _____ (I, inner; O, outer) nuclear layer which function to transmit signals laterally within the retina.

    a. G, I   c. A, I   e. B, O
    b. B, I   d. G, O   f. A, O

OBJECTIVE 60-3.

Identify the general characteristics of the rods and cones, bipolar cells, horizontal cells, amacrine cells, and ganglion cells in the transmission of visual signals.

11. _____ (A, action potentials; E, electrotonic current flow) generated by the rods and cones is/are transmitted to _____ (X, excitatory; H, inhibitory) bipolar cells and _____ (X, excitatory; H, inhibitory) horizontal cells.

    a. A, X, X   c. A, H, X   e. E, X, X
    b. A, H, H   d. E, H, H   f. E, X, H

12. Horizontal cells transmit information from a region of excitation of rods and cones through predominantly direct connections with _____ (C, centrally placed; L, laterally placed) _____ (B, bipolar; G, ganglion) cells.

    a. C, B   c. L, B
    b. C, G   d. L, G

13. Amacrine cells are excited largely by _____ (H, horizontal cells; B, bipolar cells; R, rods and cones) and tend to _____ (F, facilitate; I, inhibit) ganglion cells.

    a. H, F   c. R, F   e. B, I
    b. B, F   d. H, I   f. R, I

14. Amacrine cells transmit signals either as a simple electrotonic _____ (D, depolarization; H, hyperpolarization) or as occasional action potentials which occur more frequently at the _____ (O, onset; T, termination) of the electrotonic potentials.

    a. D, O   c. H, O
    b. D, T   d. H, T

15. Ganglion cells transmit their signals through _____ (O, optic nerve fibers; A, amacrine cells; H, horizontal cells) and exhibit _____ (B, an absence of; L, a low frequence) tonic discharge in the absence of light stimulation of the rods and cones.

    a. O, B   c. H, B   e. A, L
    b. A, B   d. O, L   f. H, L

OBJECTIVE 60–4.

Identify the different types of signals transmitted by ganglion cells through the optic nerve, and their probable mechanistic basis.

16. Visual scenes consisting of _____ (U, uniformly illuminated fields; C, contrasting checkerboard patterns of light and dark regions) result in greater activity within the optic nerve as a consequence of _____ (L, mechanisms of lateral inhibition; A, average luminosity; M, myopia).

    a. U, L   c. U, M   e. C, A
    b. U, A   d. C, L   f. C, M

17. Ganglion cells which transmit largely _____ (C, contrast border; L, luminosity) signals represent a _____ (J, majority; N, minority) of the retinal ganglion cells, and more frequently transmit "on-off" and "off-on" responses.

    a. C, J   c. L, J
    b. C, N   d. L, N

18. Detection of transient changes in light intensity is thought to be more directly related to the relatively _____ (R, rapid; S, slow) adaptation characteristics of _____ (H, horizontal; A, amacrine; B, bipolar) cells.

    a. R, H   c. R, B   e. S, A
    b. R, A   d. S, H   f. S, B

19. The capability for the detection of rapid changes in light intensity is more highly developed within the _____ portion of the retina, which corresponds to the _____ portion of the visual field. (C, central; P, peripheral)

    a. C, C   c. P, P
    b. C, P   d. P, C

20. Response characteristics of_____(C, color-contrast; L, luminosity) ganglion cells, stimulated by all three types of cones, are _____ (S, strongly influenced by; I, relatively independent of) varying wavelength within the visible spectrum.

    a. C, S   c. L, S
    b. C, I   d. L, I

21. A "color-contrast" type of ganglion cell is largely excited directly via _____ cells and inhibited by opposing colors through the indirect action of _____ cells. (H, horizontal; A, amacrine; B, bipolar)

    a. H, A   c. B, H   e. A, H
    b. A, B   d. H, B   f. B, A

OBJECTIVE 60–5.

Recognize the existence of centrifugal control mechanisms of retinal function. Identify the neuroanatomical organization and functional characteristics of the lateral geniculate bodies.

22. Centrifugal fibers are thought to function in the regulation of visual _____ (P, perimetry; S, sensitivity) via direct synaptic connections with _____ (R, rods and cones; B, bipolar cells; G, ganglion cells) of the retina.

    a. P, R   c. P, G   e. S, B
    b. P, B   d. S, R   f. S, G

23. _____ (N, only alternating; A, all) layers of the _____ nuclear layers of the lateral geniculate body relay visual information to the visual cortex.

    a. N, 4   c. N, 8   e. A, 6
    b. N, 6   d. A, 4   f. A, 8

24. Layers _____ from the lateral geniculate body surface inward receive synaptic connections from the _____ (T, temporal; N, nasal) retina of the contralateral eye.

    a. 1, 3, & 5; T   d. 1, 3, & 5; N
    b. 1, 4, & 6; T   e. 1, 4, & 6; N
    c. 2, 4, & 6; T   f. 2, 4, & 6; N

25. Corresponding loci of a visual scene from opposite eyes project to _____ (S, similar layers of opposing sides; A, alternate layers of the same side) of the lateral geniculate bodies with a higher number of geniculate neurons responding to _____ (L, luminosity; C, contrast borders) of the visual image.

    a. S, L   c. A, L
    b. S, C   d. A, C

26. So called "red-green cells," located primarily within the _____ (O, outer surface; N, inner) layers of the geniculate bodies, are _____ (E, excited by both; EI, excited by one and inhibited by the other; I, inhibited by both) color(s).

    a. O, E   b. O, EI   c. O, I   d. N, E   e. N, EI   f. N, I

OBJECTIVE 60–6.

Locate the primary visual cortex and identify its major functional roles in the processing of visual information.

27. The primary visual cortex lies adjacent to the _____ (S, sylvian; C, calcarine) fissure on the _____ (M, medial; L, lateral) aspect of each _____ (O, occipital; T, temporal; P, parietal) lobe.

    a. S, L, O   c. S, M, P   e. C, M, O
    b. S, L, T   d. C, M, P   f. C, L, T

28. Concentric circles which lie _____ (F, furthest from; C, closest to) the occipital pole represent visual signals from the macula, with the upper half of the retina projecting to a more _____ (S, superior; I, inferior) location within the primary visual cortex.

    a. F, S   c. C, S
    b. F, I   d. C, I

29. _____ (S, similar; D, considerably distorted) patterns of neuronal excitation of the primary visual cortex, when compared to the visual scene, are formed largely by _____ (B, enhanced border contrast; F, fidelity of transmission of luminosity signals).

    a. S, B   c. D, B
    b. S, F   d. D, F

30. The primary visual cortex, containing _____ (M, many times more; E, about equal numbers of; L, considerably fewer) neurons as compared to the number of optic nerve fibers, receives terminal projections of geniculate neurons to layer _____ .

    a. M, 4   c. L, 4   e. E, 1
    b. E, 4   d. M, 1   f. L, 1

31. The primary visual cortex functions in the detection of the length _____ (W, as well as; N, but not) the orientation of lines and borders of visual patterns of _____ (P, only photopic; S, only scotopic; B, both photopic and scotopic) vision.

    a. W, P   c. W, B   e. N, S
    b. W, S   d. N, P   f. N, B

OBJECTIVE 60–7.

Identify the principal cerebral cortical regions for processing and storing of visual information, their locations, and their probable significance.

32. The striate area of the cerebral cortex is a _____ (A, visual association; P, primary visual cortical) area corresponding to Brodmann area _____ .

    a. A, 8   c. A, 41   e. P, 17
    b. A, 17   d. P, 8   f. P, 41

33. A topographical representation of the visual field and neurons which are stimulated by simple geometric patterns are characteristic features of visual association areas _____ located within the _____ (PS, prestriate area; T, temporal lobes; G, geniculate bodies).

    a. 18 & 19, PS   d. 20 & 21, PS
    b. 18 & 19, T   e. 20 & 21, T
    c. 18 & 19, G   f. 20 & 21, G

34. Complicated visual perceptions of past visual scenes are more readily elicited by means of electrical stimulation of Brodmann areas _____ located within the _____ (O, occipital; P, parietal; T, temporal) lobes.

a. 18 & 19, O   d. 20 & 21, O
b. 18 & 19, P   e. 20 & 21, P
c. 18 & 19, T   f. 20 & 21, T

35. Selective destruction of Brodmann area(s) _____ results in total blindness in man, whereas destruction of Brodmann area(s) _____ is thought to result in word blindness or dyslexia.

a. 17, 44   d. 18 & 19, 17
b. 17, 18 & 19   e. 20 & 21, 18 & 19
c. 18 & 19, 20 & 21   f. 20 & 21, 44

OBJECTIVE 60–8.

Identify the normal fields of vision and "perimetry," retinal abnormalities affecting the fields of vision, and the effects of lesions in the optic pathway upon the fields of vision.

36. A nasal field of vision refers to information received by the _____ (T, temporal; N, nasal) portion of the retina, and the process of charting the fields of vision is known as _____ (H, hemianopsia; P, perimetry; O, opticokinesia).

a. T, H   c. T, O   e. N, P
b. T, P   d. N, H   f. N, O

37. The greatest visual field for photopic _____ (W, as well as; B, but not) scotopic vision occurs in the _____ (S, superior; I, inferior) quadrant of the _____ (T, temporal; N, nasal) visual field.

a. W, S, T   c. W, I, T   e. B, S, T
b. W, I, N   d. B, I, N   f. B, S, N

38. The blind spot lies in the _____ (T, temporal; N, nasal) visual field at an angle of about _____° with respect to the central point of vision.

a. T, 45°   c. T, 15°   e. T, 75°
b. N, 45°   d. N, 15°   f. N, 75°

39. The _____ (O, optic disc; M, macula), lying _____ (L, laterally; D, medially) to the fovea centralis and lacking cones _____ (W, as well as; N, but not) rods, is the cause of the blind spot within the visual field.

a. O, L, W   c. O, D, W   e. M, L, W
b. O, D, N   d. M, D, N   f. M, L, N

40. Retinitis pigmentosa more frequently causes blindness in the _____ (C, central; P, peripheral) portions of the visual field initially, and is associated with excessive deposition of _____ (V, visual pigments; M, melanin) in the affected areas of the retina.

a. C, V   c. P, V
b. C, M   d. P, M

41. Total blindness in the _____ (R, right; L, left) eye would occur with destruction of the right optic _____ (N, nerve; T, tract).

a. R, N   c. L, N
b. R, T   d. L, T

42. Organization of the nerve fibers from retinal ganglionic cells is such that loss of both temporal visual fields—i.e., "tunnel vision"—would result from sectioning the:

a. Optic chiasma   c. Right optic tract
b. Right optic nerve   d. Left optic tract

43. The patient with a total transection of the right optic tract would have an ipsilateral loss of the _____ field of vision and a contralateral loss of the _____ field of vision. (N, nasal; T, temporal)

a. N, N   c. T, N
b. N, T   d. T, T

44. A unilateral destruction of the _____ (O, optic; G, geniculocalcarine) tract results in a _____ (B, bitemporal; H, homonymous) hemianopsia, with an associated loss of the pupillary light reflex from corresponding regions of the retina.

a. O, B   c. G, B
b. O, H   d. G, H

45. A more common cause of destruction of regions of the visual cortex is thrombosis of the _____ (A, anterior; M, middle; O, posterior) cerebral artery, which results in a more frequent sparing of _____ (C, central; P, peripheral) vision.

a. A, C   c. O, C   e. M, P
b. M, C   d. A, P   f. O, P

OBJECTIVE 60-9.

Recognize the conjugate movements of the eyes. Identify the muscular mechanisms and neural pathways for the control of eye movements.

46. Conjugate or _____ (P, independent; S, simultaneous) movements of the two eyes occur by means of final common pathways from cranial nuclei _____ .

    a. P; I, II, & III   d. S; I, II, & III
    b. P; III, IV, & VI   e. S; III, IV, & VI
    c. P; II, VI, & VIII   f. S; II, VI, & VIII

47. The superior rectus and the _____ (M, medial; L, lateral; I, inferior) rectus muscles generally function together in a _____ (S, synchronous; R, reciprocal) fashion to move the eyes vertically.

    a. M, S   c. I, S   e. L, R
    b. L, S   d. M, R   f. I, R

48. Oculomotor nuclei innervate all of the extra-ocular eye muscles with the exception of the _____ and _____ muscles. (MR, medial rectus; LR, lateral rectus; SO, superior oblique; IO, inferior oblique)

    a. MR, SO   c. IO, MR   e. LR, IO
    b. LR, SO   d. MR, LR   f. IO, SO

49. The occipital visual areas, the pretectal region, and the _____ (S, superior colliculi; M, medial geniculate body), _____ (W, as well as; N, but not) the vestibular nuclei, are more notable centers which control eye movements.

    a. S, W   c. M, W
    b. S, N   d. M, N

50. A pathway providing interconnections between the cranial motor nuclei for extraocular eye movements is the:

    a. Medial lemniscus
    b. Lateral lemniscus
    c. Trigeminal lemniscus
    d. Fasciculus gracilis
    e. Median longitudinal fasciculus
    f. Fasciculus proprius

OBJECTIVE 60-10.

Identify the types of fixation movements of the eyes, their characteristic features, their areas of control, and their significance.

51. An involuntary fixation mechanism, causing a "locking" of the eyes upon an object of attention, is found in eye fields of the _____ , whereas a voluntary fixation control center is associated with a region of the _____ . (F, frontal lobes; S, somatosensory cortex; O, occipital cortex; T, anterior temporal lobe).

    a. T, S   c. F, O   e. O, F
    b. S, F   d. S, O   f. O, S

52. Visual fixation mechanisms involve particularly the _____ (S, superior; I, inferior) colliculi, and act to maintain an object of interest _____ (L, at a stationary locus on the retina; F, within the boundaries of the fovea; O, within the boundaries of the optic disc).

    a. S, L   c. S, O   e. I, F
    b. S, F   d. I, L   f. I, O

53. As an object of attention approaches the foveal border, a _____ (S, slow drift; F, flicking) type of movement occurs during visual fixation to move the object image _____ (T, toward the center of; A, away from) the fovea.

    a. S, T   c. F, T
    b. S, A   d. F, A

54. Saccades are rather _____ (S, slow; R, rapid) types of movements associated with _____ (I, independent; C, conjugate) ocular movements and _____ (B, "blanking" of the visual image; V, vestibular functions).

    a. S, I, B   c. S, C, B   e. R, C, B
    b. S, I, V   d. R, C, V   f. R, I, V

55. A per-rotational form of vestibular nystagmus, acting _____ (S, synergistically; O, in opposition) to a coincident opticokinetic nystagmus, has a _____ (H, shorter; L, longer) latency for action.

   a. S, H   b. S, L   c. O, H   d. O, L

OBJECTIVE 60–11.

Identify the probable basis for fusion of images of the visual pathway, its significance, and its abnormal consequences.

56. Fusion of the images of the two eyes is believed to coincide with the _____ (P, presence; B, absence) of "interference patterns" of stimulation in specific cells of the visual cortex resulting from comparisons of neural activity between _____ (A, adjacent; I, identical) layers of _____ (O, opposing; S, the same) lateral geniculate bodies.

   a. P, I, O   c. B, I, O
   b. P, A, S   d. B, A, S

57. _____ (F, fusion of retinal images; S, stereopsis) is dependent upon a mechanism whereby the closer an object is to the eyes, the _____ (L, less; G, greater) is the disparity between the two corresponding retinal images.

   a. F, L   c. S, L
   b. F, G   d. S, G

58. _____ (S, strabismus; D, dyslexia; M, myopia), or "cross-eyedness," is associated with an abnormal "set" of the fusion mechanism of the visual system and _____ (N, a normal; A, an abnormal) pattern of conjugate movement of the eyes.

   a. S, N   c. M, N   e. D, A
   b. D, N   d. S, A   f. M, A

59. A more severe depression of visual acuity in one eye is associated with a _____ (J, majority; N, minority) of patients with strabismus who _____ (A, alternate the eyes; D, develop a dominant eye) in fixating on objects of attention.

   a. J, A   c. N, A
   b. J, D   d. N, D

OBJECTIVE 60–12.

Identify the autonomic innervation of the eyes.

60. _____ (P, parasympathetic; S, sympathetic) _____ (E, preganglionic; O, postganglionic) neurons, supplying the ocular structures, are located within the Edinger-Westphal nucleus in association with cranial nucleus _____ .

   a. P, E, III   c. P, O, III   e. S, O, III
   b. P, E, VII   d. S, O, VII   f. S, E, VII

61. _____ (E, preganglionic; O, postganglionic) parasympathetic fibers originate within the _____ (S, superior cervical; C, ciliary) ganglion.

   a. E, S   c. O, S
   b. E, C   d. O, C

62. _____ (E, preganglionic; O, postganglionic) _____ (P, parasympathetic; S, sympathetic) fibers transmit autonomic efferent activity to the eyes via the carotid plexus.

   a. E, P   c. O, P
   b. E, S   d. O, S

63. _____ (P, parasympathetic; S, sympathetic) autonomic efferents are largely responsible for regulation of ciliary muscle contraction, and have _____ (E, excitatory; I, inhibitory) effects upon the sphincter muscle of the iris.

   a. P, E   c. S, E
   b. P, I   d. S, I

OBJECTIVE 60-13.

Identify the control mechanisms for visual accommodation.

64. Visual accommodation operates via a feedback control mechanism that adjusts the _____ (I, image distance; D, dioptric strength) in a manner which maximizes _____ (S, image size; A, visual acuity; P, diplopia).

a. I, S   c. I, P   e. D, A
b. I, A   d. D, S   f. D, P

65. Near accommodation is associated with _____ (V, divergence; C, convergence) of the eyes and _____ (I, increased; D, decreased) dioptric strength of the visual system.

a. V, I   c. C, I
b. V, D   d. C, D

66. Accommodation of the eyes to look at a close object would include _____ (N, contraction; X, relaxation) of the ciliary muscle as well as contraction of the _____ (C, circular; R, radial) muscle of the iris to decrese the pupil size.

a. N, C   c. X, C
b. N, R   d. X, R

67. The proper directional change of accommodation during voluntary fixation movement is _____ (A, initially accurate; E, established by a "hunting" mechanism) and involves _____ (I, increased; D, decreased) parasympathetic activity to the eyes during near accommodation.

a. A, I   c. E, I
b. A, D   d. E, D

68. _____ (S, spherical aberrations; C, chromatic aberrations; D, diffraction errors) of the optical system of the eye are associated with a greater focal length for spectral wavelengths at the _____ (R, red; B, blue) end of the spectrum.

a. S, R   c. D, R   e. C, B
b. C, R   d. S, B   f. D, B

OBJECTIVE 60-14.

Identify the control mechanisms for the pupillary aperture and their functional and clinical significance.

69. Pupillary _____ (Y, mydriasis; I, miosis), or dilation, occurs in response to increased _____ (S, sympathetic; P, parasympathetic) activity to the eye.

a. Y, S   c. I, S
b. Y, P   d. I, P

70. Pupillary _____ (Y, mydriasis; I, miosis), accompanying near accommodation, functions to increase _____ (D, depth of focus; L, incident light; F, field of vision).

a. I, D   c. I, F   e. Y, L
b. I, L   d. Y, D   f. Y, F

71. Shining a bright light onto the retina of one eye results in a _____ (U, unilateral; B, bilateral) pupil _____ (D, dilation; C, constriction).

a. U, D   c. B, D
b. U, C   d. B, C

72. The pupillary light reflex is thought to involve activity transmitted from the optic tract to the _____ (T, pretectal region; C, superior colliculus) and predominant alterations in _____ (S, sympathetic; P, parasympathetic) outflow.

a. T, S   c. C, S
b. T, P   d. C, P

73. The Argyll Robertson pupil, utilized as a diagnostic sign of central nervous system syphilis, involves a tonic _____ (C, constriction; D, dilation) of the pupil, a _____ (R, retention; L, loss) of the pupillary light reflex, and a _____ (R, retention; L, loss) of the pupillary reflex accompanying visual accommodation.

a. C, R, L   c. C, L, L   e. D, L, L
b. C, L, R   d. D, L, R   f. D, R, R

74. Horner's syndrome, caused by interruption of _____ (S, sympathetic; P, parasympathetic) efferents on one side of the body, results in _____ (I, ipsilateral; C, contralateral; B, bilateral) symptoms.

    a. S, I   c. S, B   e. P, C
    b. S, C   d. P, I   f. P, B

75. Horner's syndrome includes symptoms of vascular _____ (C, constriction; D, dilation), pupillary _____ (C, constriction; D, dilation), and _____ (P, persistent; A, an absence of) sweating in the affected regions.

    a. C, C, P   c. C, D, P   e. D, C, P
    b. C, D, A   d. D, D, A   f. D, C, A

76. A loss of _____ (P, parasympathetic; S, sympathetic; M, somatic) efferent activity supplying the superior palpebral muscle results in _____ (C, an inability to close the; D, a drooping) eyelid.

    a. P, C   c. M, C   e. S, D
    b. S, C   d. P, D   f. M, D

61

The Sense of Hearing

OBJECTIVE 61–1.
Using the following diagram, identify the various structural components of the human ear.

Directions: Match the lettered headings with the diagram and numbered list of descriptive words and phrases.

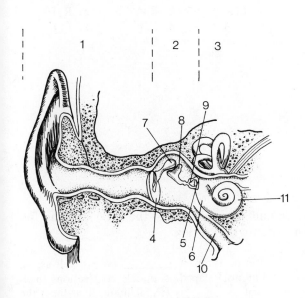

1. _____ Includes the pinna and external auditory meatus.
2. _____ An air-filled chamber that contains the ossicles.
3. _____ Includes the cochlea and cochlear nerves.
4. _____ Conical membrane, commonly called the eardrum.
5. _____ Integument between the stapes and the cochlea.
6. _____ Membrane at the end of the scala tympani and on the medial wall of the middle ear.
7. _____ The ossicle attached to the tympanic membrane.
8. _____ The ossicle connecting the malleus to the stapes.
9. _____ The ossicle attached to the oval window.
10. _____ Fleshy tube connecting the middle ear to the pharynx.
11. _____ Partitioned membranous chamber containing auditory receptors.

a. Tympanic membrane
b. Round window
c. Malleus
d. Outer ear
e. Oval window
f. Incus
g. Cochlea
h. Inner ear
i. Middle ear
j. Stapes
k. Eustachian tube

OBJECTIVE 61–2.
Identify the mechanisms and impedance matching characteristics for the transmission of sound from the outer ear to the inner ear.

12. The articulation of the _____ (M, malleus; C, incus) with the stapes is such that the stapes moves inward when the tympanic membrane moves _____ (I, inward; O, outward).

a. M, I
b. M, O
c. C, I
d. C, O

13. The ossicular lever system functions to amplify the _____ (D, distance; F, force) of movement with the combined _____ (IS, incus and stapes; MS, malleus and stapes; MI, malleus and incus) functioning as a single lever.

a. D, IS
b. D, MS
c. D, MI
d. F, IS
e. F, MS
f. F, MI

14. A greater _____ (R, reduction; A, amplification) of acoustic _____ (F, force; P, pressure) occurs in the transmission of sound between the outer and inner ears.

a. R, F
b. R, P
c. A, F
d. A, P

15. Transmission of acoustic energy from the tympanic membrane to the _____ (O, oval; R, round) window is associated with a _____ _____ (I, increase; D, reduction) of surface area for force application.

a. O, 17-fold, D
b. O, 17-fold, I
c. O, 50%, D
d. R, 50%, I
e. R, 50%, D
f. R, 17-fold, I

16. Impedance matching for improved acoustic energy transfer between air and a fluid medium of the _____ (M, middle; I, inner) ear is a function of the _____ (OT, otoliths; OS, ossicles; OW, oval window; RW, round window) and the tympanic membrane.

a. M, OT & RW
b. M, OT & OW
c. M, OS & RW
d. I, OS & OW
e. I, OS & RW
f. I, OT & OW

17. An impedance mismatch of 25 to 50 _____ (B, bels; D, decibels; P, per cent) within a frequency range of 300 to 3,000 cycles per second normally occurs for acoustic energy transfer between the _____ (L, lower; H, higher) inertia of air and perilymph.

a. B, L
b. D, L
c. P, L
d. B, H
e. D, H
f. P, H

18. Fixation of the ossicular chain results in _____. (L, a complete loss of; R, reduction but not a complete loss of; E, enhanced) transmission of acoustic energy from the tympanic membrane to the _____ (A, air; F, fluid)-filled inner ear.

a. L, A
b. R, A
c. E, A
d. L, F
e. R, F
f. E, F

OBJECTIVE 61–3.

Identify the transmission characteristics of sound transmission to the inner ear, and the characteristics and functional significance of the attenuation reflex.

19. Acoustic transmission is generally greatest through the _____ (A, air; O, osseous; C, ossicular) route of the _____ (M, middle; I, inner) ear.

a. A, M
b. O, M
c. C, M
d. A, I
e. O, I
f. C, I

20. The external auditory canal has _____ (E, enhanced; R, reduced) acoustic transmission at a natural resonating frequency of about _____ cycles per second.

a. E, 1,200
b. E, 3,000
c. E, 8,600
d. R, 1,200
e. R, 3,000
f. R, 8,600

21. The ossicular system possesses a _____ (S, sharply; B, broadly) tuned natural resonant peak at a _____ (H, higher; L, lower) frequency than that occurring for the external auditory meatus.

a. S, H
b. S, L
c. B, H
d. B, L

22. Tensor tympani _____ (W, as well as; N, but not) stapedius muscle contractions function to _____ (F, facilitate; R, reduce) the acoustic transmission characteristics through the _____ (M, middle; I, inner) ear.

a. W, F, M
b. W, R, I
c. W, R, M
d. N, R, I
e. N, F, M
f. N, F, I

23. The attenuation reflex is generally _____ by an individual's own speech and _____ by loud sounds emanating from the external environment. (A, activated; S, suppressed)

a. A, A
b. A, S
c. S, A
d. S, S

24. Decreased pressure within the middle ear, more frequently resulting from blockage of the _____ (D, cochlear duct; E, eustachian tube), has _____ (S, a similar; O, an opposing) action on sound transmission when compared with the attenuation reflex.

a. D, S
b. D, O
c. E, S
d. E, O

25. The attenuation reflex has a greater protective action against potentially injurious effects of loud sounds of _____ (E, an explosive; P, a persistent) nature, and provides a more effective "masking" function for sounds _____ (A, above; B, below) 1,000 cycles per second.

a. E, A   b. E, B   c. P, A   d. P, B

OBJECTIVE 61–4.

Using the following diagram, identify the functional anatomy of the cochlea.

Directions: Match the lettered headings with the diagrams and numbered list of descriptive words and phrases.

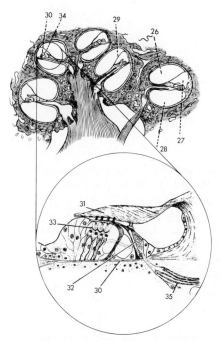

26. _____ Perilymph-containing chamber of the cochlea, communicating with the oval window.

27. _____ Endolymph-containing chamber of the cochlea.

28. _____ Perilymph-containing chamber of the cochlea, communicating with the round window.

29. _____ Membranous partition between the scala media and scala vestibuli (Reissner's membrane).

30. _____ Membranous partition between the scala media and scala tympani.

31. _____ A membrane within the scala media in contact with ciliary projections of the hair cells.

32. _____ A less abundant type of sensory hair cells.

33. _____ A more abundant type of sensory hair cells.

34. _____ A receptor organ for hearing.

35. _____ Pathway for auditory information transmission from the Organ of Corti.

(From Guyton, A.C., Textbook of Medical Physiology, 5th ed. Philadelphia, W.B. Saunders Company, 1976.)

a. Internal hair cells
b. Spiral organ of Corti
c. Scala vestibuli
d. Cochlear nerve

e. External hair cells
f. Scala media
g. Basilar membrane

h. Scala tympani
i. Tectorial membrane
j. Vestibular membrane

OBJECTIVE 61–5.

Identify the acoustic transmission and resonant characteristics of the cochlea.

36. An inward movement of the stapes results in an _____ (I, inward; O, outward) movement of the round window, _____ (E, an enhanced; R, a reduced) pressure within the scala vestibuli, and _____ (E, an enhanced; R, a reduced) pressure within the scala tympani.

a. I, E, E   c. I, R, E   e. O, E, E
b. I, E, R   d. O, R, R   f. O, E, R

37. The separating membrane between the scala media and the scala _____ (V, vestibuli; T, tympani) performs a lesser partitioning role regarding its acoustic properties, and functions primarily to _____ (P, secrete perilymph; EP, separate endolymph from perilymph; S, physically support receptor cells).

a. V, P   c. V, S   e. T, EP
b. V, EP   d. T, P   f. T, S

38. The _____ (E, Eustachian tube; C, cochlear duct; H, helicotrema) provides for steady state pressure equalization between the scala tympani and scala _____ (M, media; V, vestibuli).

a. E, M   c. C, M   e. H, M
b. E, V   d. C, V   f. H, V

39. Basilar fibers of the cochlea, projecting from the modiolus towards the _____ (C, center; O, outer wall) of the cochlea, exhibit _____ (I, an increase; D, a decrease) in length from the base of the cochlea to its apex.

a. C, I   c. O, I
b. C, D   d. O, D

40. The effects of "loading" by the fluid mass of the cochlea favor a higher frequency resonance at the _____ end of the cochlea, while differences in the stiffness of basilar fibers favor a higher frequency resonance at the _____ end. (B, basal; A, apical)

a. B, A   c. A, B
b. B, B   d. A, A

OBJECTIVE 61–6.

Identify the "traveling wave" characteristics and underlying mechanisms for the transmission of sound waves in the cochlea.

41. Cochlear "traveling waves" are propagated at velocities which _____ (V, vary; C, remain constant) along the length of the cochlea, and _____ (E, equal; D, are substantially different from) the local velocity of sound in a liquid.

a. V, E   c. C, E
b. V, D   d. C, D

42. "Traveling waves" of the cochlea for most sounds are associated with higher propagation velocities at the _____ of the cochlea and maximal amplitudes at the _____ of the cochlea. (B, basal ends; I, intermediate regions; A, apical ends)

a. B, I   c. I, B   e. A, I
b. B, A   d. I, A   f. A, B

43. A decline in the amplitude in the "traveling wave" is associated with _____ (I, an increase; D, a decrease) in its energy through fluid displacements occurring at the _____ (H, helicotrema; P, place of resonance).

a. I, H   c. D, H
b. I, P   d. D, P

44. The basal end of the cochlea, when compared to the apical end, is associated with a place of resonance for _____ (H, higher; L, lower) auditory frequencies and a comparatively _____ (H, higher; L, lower) "traveling wave" propagation velocity for high _____ (W, as well as; N, but not) low auditory frequencies.

a. H, H, W   c. H, L, W   e. L, L, W
b. H, H, N   d. L, L, N   f. L, H, N

45. The amplitude pattern of vibration of the basilar membrane as a function of cochlear distance exhibits a more pronounced rate of change along its _____ (A, ascending; D, descending) portion with a locus of its _____ (X, maximum displacement; C, "cut off") corresponding to a "place" for frequency discrimination.

a. A, X   c. D, X
b. A, C   d. D, C

OBJECTIVE 61–7.

Identify the organ of Corti, the means of excitation and functional characteristics of the hair cells, and the characteristics and origin of the endocochlear potential.

46. Receptors for hearing are found within the _____ (M, modiolus; C, organ of Corti), located upon the _____ (V, vestibular; B, basilar) membrane of the _____ (D, middle; I, inner) ear.

a. M, V, D   c. M, B, D   e. C, B, D
b. M, V, I   d. C, B, I   f. C, B, I

47. More numerous _____ (I, internal; E, external) hair cells, located on the _____ (N, inner; O, outer) aspect of the "rods of Corti," are comparatively _____ (L, larger; S, smaller) in diameter.

a. I, N, L   c. I, O, L   e. E, O, L
b. I, N, S   d. E, O, S   f. E, N, S

48. Ciliary contact of the hair cells with the _____ (B, basilar fibers; T, tectorial membrane; V, vestibular membrane) and movement of the reticular lamina during auditory stimulation results in a _____ (S, shearing; C, compressional) action upon the cilia.

a. B, S   c. V, S   e. T, C
b. T, S   d. B, C   f. V, C

49. Increased discharge frequencies of innervating _____ (CNS, central nervous system; SG, spiral ganglion) neurons occur in response to ciliary movement of the hair cells in _____ (O, only one major; B, both) directions(s).

a. CNS, O   c. SG, O
b. CNS, B   d. SG, B

50. Endolymph is thought to be secreted largely by the _____ (V, stria vascularis; E, endolymphatic sac) of the _____ (C, cochlea; S, semicircular canals).

a. V, C   c. E, C
b. V, S   d. E, S

51. The endolymph potential within the cochlea is _____ (P, positive; N, negative) with respect to perilymph by about _____ millivolts.

a. P, 20   c. P, 150   e. N, 80
b. P, 80   d. N, 20   f. N, 150

52. Perilymph contains a higher concentration of _____ , which is _____ (S, similar; O, opposite) to the relative concentrations of Na and K for endolymph.

a. Na, O   c. Na, S
b. K, O   d. K, S

53. Hair cells have a _____ (N, negative; P, positive) intracellular potential of about _____ millivolts with respect to endolymph of the scala _____ (V, vestibuli; M, media).

a. N, 70, M   c. N, 150, M   e. P, 70, M
b. N, 150, V   d. P, 150, V   f. P, 70, V

OBJECTIVE 61–8.

Contrast pitch and frequency, and describe the "place principle" as a basis for pitch discrimination.

54. The wide range of pitch perception is established largely upon the basis of _____ (F, action potential frequency; P, the place of maximal stimulation of the cochlea), and is a frequency-dependent and intensity- _____ (I, independent; D, dependent) function of absolute frequency.

a. F, I
b. F, D
c. P, I
d. P, D

55. Pitch is more likely to deviate from true sound frequencies within the _____ (H, high; M, middle) portion of the frequency spectrum, corresponding to loci of selective activation of the _____ (B, basal; M, middle; A, apical) region of the basilar membrane.

a. H, B   c. H, A   e. M, M
b. H, M   d. M, B   f. M, A

56. Select damage to a short segment of the hair cells along the cochlea near the helicotrema would affect _____ (P, only pitch; L, only loudness; PL, both pitch and loudness) of _____ (G, high; L, low) frequency sounds.

 a. P, G   b. L, G   c. PL, G   d. P, L   e. L, L   f. PL, L

OBJECTIVE 61–9.

Identify sound intensity and the decibel unit. Contrast loudness and intensity, and identify the physiological mechanisms for the determination of loudness.

57. Loudness is a frequency- _____ (I, independent; D, dependent) function of sound intensity, detected by means of major changes in _____ (F, only action potential frequency; S, only spatial summation; B, action potential frequency as well as spatial summation) of nerve fibers stimulated by a given acoustic frequency.

 a. I, F   c. I, B   e. D, S
 b. I, S   d. D, F   f. D, B

58. The scale of actual sound intensity is greatly _____ (C, compressed; E, expanded) by a perception mechanism which interprets sensation changes in approximate proportion to the _____ (I, actual; S, square of the actual; R, cube root of the actual) sound intensity.

 a. C, I   c. C, R   e. E, S
 b. C, S   d. E, I   f. E, R

59. Measurements of sound intensity are generally expressed in _____ (P, phon; M, mel; B, bel) units, representing _____ (A, absolute; R, relative) levels of sound intensity.

 a. P, A   c. B, A   e. M, R
 b. M, A   d. P, R   f. B, R

60. Decibel units represent _____ (F, one-tenth of; T, ten times) the _____ (L, logarithm; A, antilogarithm) of ratios of sound _____ (P, pressures; E, energies).

 a. F, L, P   c. F, A, E   e. T, L, E
 b. F, A, P   d. T, A, E   f. T, L, P

61. One decibel represents an actual increase in sound intensity of _____ times.

 a. 0.01   c. 1.26   e. 20
 b. 0.13   d. 10   f. 10^2

62. When the energy of sound is increased by 10^6 times its reference value, the intensity is increased by _____ db, or an equivalent of a _____ -fold increase in the acoustic pressure.

 a. 60, 10^3 c. 30, 10^3
 b. 6, 10^6 d. 3, 10^6

63. The _____ (E, energy; P, pressure) reference level of _____ dyne(s) per square centimeter is frequently utilized to represent a zero decibel level.

 a. E, 0.13   c. E, 40   e. P, 1
 b. E, 1   d. P, 0.13   f. P, 40

OBJECTIVE 61–10

Identify the threshold intensity curve for audibility as a function of frequency, its characteristics, its significance, and its normal variance.

64. A 60-db sound intensity, when compared to a 20-db sound intensity, yields _____ (G, a greater; S, the same; L, a lesser) frequency range of audibility as a consequence of the _____ (T, threshold curve of audibility; R, acoustic attenuation reflex).

 a. G, T   c. L, T   e. S, R
 b. S, T   d. G, R   f. L, R

65. The limits of audibility at 30 and 20,000 cycles per second is characteristic of varying frequencies of _____ (C, conversational; I, very intense) sound intensities perceived through the ears of _____ (Y, youth; O, old age).

 a. C, Y   c. I, Y
 b. C, O   d. I, O

66. Intensities of _____ (T, tactual thresholds; H, thresholds for hearing) for the auditory system exhibit a more pronounced _____ (X, maximum; N, minimum) as a function of sound frequency.

   a. T, X
   b. T, N
   c. H, X
   d. H, N

67. The auditory system is more sensitive to sound intensities at _____ cycles per second and is more significantly influenced by the effects of age upon the _____ (L, low; H, high) end of the audible frequency spectrum.

   a. 2,000, L
   b. 5,000, L
   c. 10,000, L
   d. 2,000, H
   e. 5,000, H
   f. 10,000, H

OBJECTIVE 61–11.

Identify the components of the auditory pathway and their functional characteristics.

68. Nerve fibers from the spiral ganglion terminate largely within the _____ (L, nucleus of the lateral lemniscus; O, superior olivary nucleus; C, cochlear nucleus) on the _____ (I, ipsilateral; N, contralateral) side.

   a. L, I
   b. O, I
   c. C, I
   d. L, N
   e. O, N
   f. C, N

69. _____ (S, second; T, third) order auditory fibers of the trapezoid bodies, largely project to the _____ (IC, inferior colliculus; SO, superior olivary nucleus) of the _____ (I, ipsilateral; C, contralateral) side of their origin.

   a. S, IC, I
   b. S, SO, C
   c. S, SO, I
   d. T, SO, C
   e. T, IC, I
   f. T, IC, C

70. The auditory pathway for the right cochlea is comprised of _____ (T, three; F, four or more) neurons with significant projections to _____ (R, only the right; L, only the left; B, both) cerebral hemisphere(s).

   a. T, R
   b. T, L
   c. T, B
   d. F, R
   e. F, L
   f. F, B

71. The auditory pathway contains major synaptic connections within neurons of the _____ (M, medial; L, lateral) lemniscus, the _____ (I, inferior; S, superior) colliculus, and the _____ (M, medial; L, lateral) geniculate bodies.

   a. M, S, M
   b. M, S, L
   c. M, I, M
   d. L, I, L
   e. L, I, M
   f. L, S, L

72. Increased activity within central auditory pathway fibers, generally associated with basilar membrane displacements towards the scala _____ (V, vestibuli; T, tympani), is limited by maximum neuronal firing frequencies of about _____ action potentials per second.

   a. V, 1,000
   b. V, 10,000
   c. V, 20,000
   d. T, 1,000
   e. T, 10,000
   f. T, 20,000

73. A high degree of spatial orientation of auditory neurons for sounds of differing _____ (I, intensity; F, frequency) is found in the cerebral cortex and cochlear nuclei _____ (W, as well as; N, but not) in the inferior colliculi.

   a. I, W
   b. I, N
   c. F, W
   d. F, N

OBJECTIVE 61–12.

Identify the functions of auditory regions of the cerebral cortex.

74. The _____ (P, primary auditory; A, auditory association) cortex, located in the _____ (T, temporal; O, occipital) lobes, receives more direct projections from the _____ (L, lateral; M, medial) geniculate bodies.

   a. P, T, L
   b. P, T, M
   c. P, O, L
   d. A, O, M
   e. A, O, L
   f. A, T, M

75. Selective lesions of the auditory association areas, which spare the primary auditory cortex, result in _____ (L, a loss; R, retention) of the capability to differentiate sound tones, and _____ (L, a loss; R, retention) of the ability to interpret the "meaning" of sounds.

   a. L, L
   b. L, R
   c. R, L
   d. R, R

76. Neurons of the auditory cortex are generally responsive to a _____ (B, broader; N, narrower) range of auditory frequencies than second order neurons of the _____ (C, cochlear nuclei; G, geniculate bodies; S, spiral ganglia).

a. B, C   b. B, G   c. B, S   d. N, C   e. N, G   f. N, S

OBJECTIVE 61-13.

Identify the mechanisms and characteristic features associated with the discrimination of the direction from which sound emanates. Recognize the existence of centrifugal control mechanisms for cochlear receptors.

77. Sound localization by differences in auditory intensity, _____ (W, as well as; N, but not) by differences in times of sound arrival, operates more favorably at frequencies _____ (B, below; A, above) 3,000 cycles per second.

a. W, B   c. N, B
b. W, A   d. N, A

78. Sound localization mechanisms are more accurate for discriminating sounds between _____ (AP, anterior and posterior; RL, right and left) auditory fields and have the greatest degree of resolution utilizing differences in auditory _____ (I, intensity; T, times of arrival) between the two ears.

a. AP, I   c. RL, I
b. AP, T   d. RL, T

79. An earlier auditory response in one ear is thought to result in _____ (F, facilitation; H, inhibition) of the contralateral auditory pathway, particularly at the level of the _____ (I, inferior colliculi; O, superior olivary nuclei; C, cerebral cortex).

a. F, I   c. F, C   e. H, O
b. F, O   d. H, I   f. H, C

80. Separation of the quality of sound direction from other qualities of sound tones is thought to occur in the _____ , while the perception of the direction of origin of sounds is thought to occur in the _____ . (G, geniculate bodies; C, cerebral cortex; SC, superior colliculi; SO, superior olivary nuclei)

a. SO, SO   c. SO, SC   e. C, C
b. SO, C   d. SC, G   f. G, C

81. A final _____ (F, facilitatory; I, inhibitory) pathway of centrifugal control of the organ of Corti originates within the _____ (V, vestibular nuclei; SO, superior olivary nuclei; L, nuclei of the lateral lemniscus).

a. F, V   c. F, L   e. I, SO
b. F, SO   d. I, V   f. I, L

OBJECTIVE 61-14.

Identify the types of hearing abnormalities and their clinical means of detection.

82. Fixation of the ossicular chain results in a _____ type of deafness, while a _____ type of deafness is the rarest form of deafness. (C, "central"; D, "conduction"; N, "neural")

a. C, D   c. D, C   e. N, C
b. C, N   d. D, N   f. N, D

83. Lower auditory thresholds for an osseous route, when compared to an ossicular route of sound transmission, characteristically occur with:

a. Central deafness   e. Low frequency
b. Neural deafness   sounds
c. Conduction deafness   f. High frequency
d. Midfrequency ranges   sounds

84. Nerve deafness results in a hearing loss when tested for _____ conduction, and "conduction" deafness results in a hearing loss when tested for _____ conduction. (A, only air; B, only bone; E, either air or bone)

a. A, B   c. B, A   e. E, A
b. A, E   d. B, E   f. E, B

85. Vibrations from a tuning fork placed on the vertex of the skull are referred to the right ear in individuals with "conduction" deafness in the _____ ear or "nerve" deafness in the _____ ear. (R, right; L, left)

a. R, R   c. L, R
b. R, L   d. L, L

86. An old-age type of nerve deafness is associated with _____ (R, reduced; I, increased) hearing loss, particularly at the _____ (H, high; L, low) frequency end of the auditory spectrum.

a. R, H   c. I, H
b. R, L   d. I, L

87. Audiograms are graphical plots of _____ (L, loudness; H, hearing loss) in _____ (B, decibels; S, sones) as a function of _____ (F, frequency; I, sound intensity).

a. L, B, F   c. L, S, F   e. H, B, F
b. L, S, I   d. H, S, I   f. H, B, I

62

The Chemical Senses —
Taste and Smell

OBJECTIVE 62-1.

Using the following diagrams, locate the approximate positions of the primary taste sensations on the tongue, and identify the histological structures of a typical taste bud.

Directions: Match the lettered headings with the diagrams and the numbered list of descriptive words and phrases.

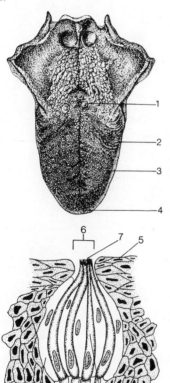

1. _____ Taste modality located on the posterior portion of the tongue.
2. _____ Taste modality located on the lateral edges of the tongue.
3. _____ Taste modality located on the lateral edges of the tongue, but usually more anterior.
4. _____ Taste modality located on the anterior portion of the tongue.
5. _____ Lingual epithelial cells forming the oral mucosa.
6. _____ Openings on the surface of the tongue, communicating with taste cells.
7. _____ Hairlike, taste cell extensions that protrude outward through the taste pores.
8. _____ Receptor cells.
9. _____ Sustentacular cells.
10. _____ Terminal nerve network, stimulated by taste cells.
11. _____ Subepithelial tissue component.

a. Taste pore
b. Taste cell
c. Sweet
d. Nerve fibers
e. Supporting cells
f. Sour

g. Stratified squamous epithelium
h. Bitter
i. Connective tissue
j. Salty
k. Microvilli

OBJECTIVE 62-2.

Identify the primary sensations of taste and their relative specificity for various types of chemical agents.

12. Lingual receptors mediate _____ primary sensations of _____ (G, gustation; O, olfaction).

   a. 3, G   d. 3, O
   b. 4, G   e. 4, O
   c. Hundreds of, G   f. Hundreds of, O

13. The intensity of a sour taste is proportional to the _____ (M, magnitude; L, logarithm; A, antilogarithm) of the _____ (H, hydrogen ion; K, alkaloid) concentration.

   a. M, H   c. A, H   e. L, K
   b. L, H   d. M, K   f. A, K

14. Chlorides of K, Mg, NH$_4$, Li, certain sulfides, bromides, iodides and sodium and potassium nitrites produce a _____ taste, whereas hydrogen ions produce a _____ taste. (B, bitter; M, metallic; S, salty; O, sour; W, sweet)

    a. B, S   c. W, O   e. S, O
    b. O, M   d. B, O   f. S, B

15. Alkaloids (quinine, strychnine, morphine, etc.), certain Mg, NH$_4$, and calcium salts, and a derivative of saccharine are found to cause a similar taste when applied to a certain region of the tongue. The taste would best be described as:

    a. Astringent   c. Metallic   e. Alkaline
    b. Sweet   d. Bitter   f. Acid

16. The sweet taste is elicited by _____ (S, a single; M, multiple) class(es) of organic chemicals and certain inorganic salts of _____ (LB, lead and beryllium; FC, iron and copper; LA, lead and antimony).

    a. S, LB   c. S, LA   e. M, FC
    b. S, FC   d. M, LB   f. M, LA

17. Sodium fluoride elicits a _____ taste, chloroform elicits a _____ taste, and caffeine elicits a _____ taste. (S, sour; B, bitter; W, sweet; T, salty)

    a. S, B, W   c. T, W, B   e. B, S, W
    b. B, W, S   d. T, T, B   f. T, B, B

18. Some substances such as saccharine taste _____ initially but have a _____ aftertaste. (S, sour; L, salty; B, bitter; W, sweet)

    a. S, W   c. L, W   e. W, B
    b. B, W   d. W, S   f. W, L

OBJECTIVE 62–3.

Contrast the relative taste thresholds for representative substances eliciting each of the primary taste sensations.

19. Relative taste indices are based upon _____ (D, direct; I, inverse) relationships to the taste _____ (A, adaptation; P, preference; T, threshold).

    a. D, A   c. D, T   e. I, P
    b. D, P   d. I, A   f. I, T

20. When stimulated by appropriate solutions of sucrose, sodium chloride, quinine, and hydrochloric acid, the lowest threshold concentration generally occurs for the _____ taste, followed by the _____ taste. (S, sour; W, sweet; B, bitter; L, salty)

    a. B, W   c. W, B   e. L, B
    b. B, S   d. W, S   f. L, W

21. Comparison of the relative taste indices among fructose, sucrose, glucose, galactose, and saccharin yields the highest taste index for _____ (S, saccharin; U, sucrose) and the lowest taste index for _____ (F, fructose; G, galactose; L, glucose).

    a. U, F   c. U, L   e. S, L
    b. U, G   d. S, G   f. S, F

22. _____ (C, citric; H, hydrochloric) acid generally has a higher taste index and a _____ (G, higher; L, lower) taste threshold for _____ (S, sour; B, bitter) tastes.

    a. C, G, S   c. C, L, S   e. H, L, S
    b. C, G, B   d. H, L, B   f. H, G, B

23. "Taste blindness" is a more frequent occurrence for _____ (H, halogenated acids; S, saccharin derivatives; T, thiourea compounds) which generally elicit a _____ (W, sweet; B, bitter; O, sour) taste.

    a. H, O   c. T, B   e. S, O
    b. S, W   d. H, B   f. T, W

OBJECTIVE 62–4.

Identify the characteristic features of the taste buds, their types and relative abundance, and their locations.

24. _____ (S, a single; M, multiple) taste cell(s) of a single taste bud have an average life span of _____ (D, 10 days; O, 6 months; L, the individual's life span).

    a. S, D   c. S, L   e. M, O
    b. S, O   d. M, D   f. M, L

25. A loss of function incurred by degeneration and regeneration of nerve fibers to taste buds _____ associated with degeneration of taste buds and _____ associated with a restoration of function following nerve regeneration. (I, is; N, is not)

    a. I, I   c. N, I
    b. I, N   d. N, N

26. Taste buds are not found in association with _____ (F, fungiform; L, filiform; O, foliate) papillae and _____ (A, are; N, are not) restricted to mucosal surfaces of the tongue.

    a. F, A   c. O, A   e. L, N
    b. L, A   d. F, N   f. O, N

27. A lower number of taste buds found in individuals of _____ years of age _____ (I, is; N, is not) generally sufficient to produce significant alterations in taste perception from a normal average.

    a. 10, I   c. 65, I   e. 30, N
    b. 30, I   d. 10, N   f. 65, N

28. An area of the tongue which is least sensitive to any of the gustatory modalities is the _____ of the tongue.

    a. Tip   c. Dorsal midline
    b. Lateral sides   d. Posterior surface

29. Substances such as sodium salicylate which may give rise to either sweet or bitter tastes are more likely to give rise to a sweet taste when applied to the _____ surface of the tongue and a bitter taste when applied to the _____ surface of the tongue. (A, anterior; L, lateral; P, posterior)

    a. A, L   c. L, A   e. P, A
    b. A, P   d. L, P   f. P, L

OBJECTIVE 62–5.

Identify the mechanisms of stimulation of the taste buds, their relative specificity for primary taste stimuli, and their probable means of detection of different sensations of taste.

30. Taste buds respond to _____ (S, a single; M, multiple) class(es) of primary taste substances by means of receptor _____ (D, depolarization; H, hyperpolarization) of taste cells.

    a. S, D   c. M, D
    b. S, H   d. M, H

31. The detection of different types of primary tastes is dependent upon their solubility in _____ (L, lipid; W, water) solvents and is more probably accomplished through _____ (R, ratios of stimulation; A, absolute specificity) of different types of taste buds.

    a. L, R   c. W, R
    b. L, A   d. W, A

32. The receptor potential of taste cells is approximately proportional to the _____ (L, logarithm; A, antilogarithm) of the _____ (P, action potential frequency; C, concentration of the stimulating substance).

    a. L, P   c. A, P
    b. L, C   d. A, C

33. A single nerve fiber, responsive to _____ (S, only a single; M, potentially multiple) taste cell(s) by means of _____ (I, intimate receptor contact; R, regional receptor association), exhibits a _____ (W, slow; P, rapid) rate of initial decline in firing frequency.

    a. S, I, W   c. S, R, W   e. M, I, W
    b. S, R, P   d. M, R, P   f. M, I, P

OBJECTIVE 62–6.

Identify the peripheral innervation, central connections, and reflex responses of gustatory afferent nerve fibers. Identify the rate of adaptation of taste perception.

34. The perception of sweet-tasting substances is accomplished by receptors on the _____ (A, anterior two thirds; P, posterior third) of the tongue, which is innervated by nerve fibers of the _____ nerve.

    a. A, glossopharyngeal   c. A, chorda tympani
    b. P, glossopharyngeal   d. P, vagus

35. Taste buds from the circumvallate papillae are innervated by the cranial nerve _____, while those located within the pharyngeal region are innervated by the cranial nerve _____.

    a. V, VII   c. V, IX   e. VII, X
    b. IX, X   d. VII, IX   f. X, X

36. The nuclei of the _____ (L, lateral lemnisci; S, tractus solitarius) are the primary locus for _____ order neurons of the taste pathway.

    a. L, 1st   c. L, 3rd   e. S, 2nd
    b. L, 2nd   d. S, 1st   f. S, 3rd

37. The taste pathway is thought to involve the projection of _____ (T, thalamic; C, collicular) axons to the _____ (I, opercular-insular; G, cingulate; S, striate) region of the cerebral cortex.

    a. T, I   c. T, S   e. C, G
    b. T, G   d. C, I   f. C, S

38. Taste fibers from the _____ (MLF, median longitudinal fasciculus; TS, tractus solitarius) mediate salivary reflexes via _____ (S, only the superior; I, only the inferior; B, both the superior and inferior) salivatory nuclei.

    a. MLF, S   c. MLF, B   e. TS, I
    b. MLF, I   d. TS, S   f. TS, B

39. Decreased firing frequencies in the chorda tympani nerve as a function of time with the application of a sapid substance would more likely involve _____ (A, accommodation; D, adaptation) of receptors located in _____ (C, circumvallate; F, fungiform; I, filiform) papillae.

    a. A, C   c. A, I   e. D, F
    b. A, F   d. D, C   f. D, I

40. A relatively _____ (S, slow; R, rapid) rate of adaptation of the taste mechanism is thought to occur for _____ (P, only peripheral receptor; C, only central neural; B, both peripheral receptor and central neural) mechanisms.

    a. S, P   c. S, B   e. R, C
    b. S, C   d. R, P   f. R, B

OBJECTIVE 62–7.

Recognize the the affective nature of taste, the close association of olfaction with taste, and the roles of taste in food preference and the regulation of dietary intake.

41. The sweet taste is more likely to be associated with an unpleasant affective nature at very _____ (L, low; H, high) stimulus concentrations, while salt _____ (W, as well as; N, but not) sour tastes exhibit concentration-dependent pleasant as well as unpleasant attributes.

    a. L, W   c. H, W
    b. L, N   d. H, N

42. Alterations of food appreciation with severe colds are more generally associated with depression of a _____ (M, more; L, less) sensitive _____ (G, gustatory; O, olfactory) mechanism.

    a. M, G   c. L, G
    b. M, O   d. L, O

43. Adrenalectomized animals _____ (D, do; N, do not) exhibit a special dietary preference for sodium chloride, while animals administered insulin in excess of their body needs have _____ (E, an enhanced; R, a reduced) preference for _____ (C, calcium salts; G, glucose).

    a. D, E, C   c. D, R, C   e. N, R, C
    b. D, E, G   d. N, R, G   f. N, E, G

OBJECTIVE 62–8.

Using the accompanying diagrams, identify the following structures of the nasal cavity and olfactory apparatus.

Directions: Match the lettered headings with the diagrams and numbered list of descriptive words and phrases.

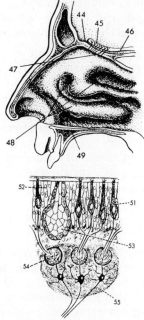

44. _____ Perforated bone through which axons pass to the olfactory bulb.

45. _____ Enlarged synaptic area above the cribriform plate.

46. _____ Composed of nerve fibers leading from the olfactory bulb to the brain.

47. _____ Specialized region of the nasal mucosa.

48. _____ Partitioning structures for the nasal cavity.

49. _____ Roof of the mouth.

50. _____ Cilia that project into the mucous coating of the inner surface of the nasal cavity.

51. _____ Receptor cells, bipolar neurons.

52. _____ Supporting cells.

53. _____ Secretes a solvent for odorous substances.

54. _____ Richly synaptic area in the olfactory bulb.

55. _____ Neurons located between the glomeruli and the olfactory tract.

(Bottom figure from Guyton, A.C., Textbook of Medical Physiology, 5th ed. Philadelphia, W.B. Saunders Company, 1976.)

a. Olfactory mucous membrane
b. Olfactory tract
c. Hard palate
d. Olfactory bulb

e. Nasal conchae
f. Cribriform plate of ethmoid
g. Bowman's gland
h. Olfactory hair

i. Glomerulus
j. Mitral cell
k. Olfactory cell
l. Sustentacular cell

OBJECTIVE 62–9.

Identify the olfactory cells, their means of exitation, and their response characteristics.

56. Olfactory cells are _____ (E, epithelial cells; B, bipolar neurons) which _____ (A, are; N, are not) in a continuous mode of replacement.

a. E, A   c. B, A
b. E, N   d. B, N

57. Olfactory substances are _____ (V, only volatile; E, either volatile or nonvolatile) substances requiring physical attributes of _____ (W, only water; L, only lipid; B, both water and lipid) solubility.

a. V, W   c. V, B   e. E, L
b. V, L   d. E, W   f. E, B

58. Olfaction is a more effective means of detecting odorous substances, particularly during _____ (I, inspiratory; E, expiratory) efforts and relatively _____ (L, low; H, high) nasal air flow velocities.

a. I, L   c. E, L
b. I, H   d. E, H

59. The amplitude of the electro-olfactogram _____ (W, as well as; N, but not) the firing frequency of olfactory nerve action potentials is proportional to the _____ (M, magnitude; L, logarithm; A, antilogarithm) of the stimulus strength.

a. W, M   c. W, A   e. N, L
b. W, L   d. N, M   f. N, A

60. The rate of adaptation of olfactory perception _____ (E, equals; X, exceeds) the rate of adaptation of olfactory receptors, and leads to near extinction within a minimum interval of several _____ (M, minutes; H, hours; D, days).

    a. E, M   b. E, H   c. E, D   d. X, M   e. X, H   f. X, D

OBJECTIVE 62–10.

Recognize the complex nature of olfactory sensations. Characterize olfactory perception in regard to olfactory thresholds, judgments of olfactory intensities, and the affective nature of olfaction.

61. The number of classes of primary receptor sensations for color vision (CV), gustation (G), and olfaction (O), is greatest for _____ and least for _____ .

    a. CV, G   c. G, O   e. O, CV
    b. CV, O   d. G, CV   f. O, G

62. _____ chemoreceptors have the lowest thresholds, while the narrowest range of intensity discrimination is associated with _____ chemoreceptors. (G, gustatory; O, olfactory; V, visual)

    a. G, O   c. V, V   e. V, O
    b. O, G   d. O, O   f. G, G

63. Affective qualities of pleasantness _____ unpleasantness are associated with gustation _____ olfaction. (W, as well as; N, but not)

    a. W, W   c. N, W
    b. W, N   d. N, N

64. A _____ (L, less acute; M, more developed) olfactory sense in humans, when compared to lower animals, is primarily concerned with the _____ (Q, quantitative detection of odor intensities; O, presence or absence of odors).

    a. L, Q   c. M, Q
    b. L, O   d. M, O

OBJECTIVE 62–11.

Identify the major features of the pathways for transmission of olfactory sensations into the central nervous system.

65. _____ (G, granule; M, mitral) cells, receiving direct connections from axons of olfactory cells, transmit activity via axons which terminate in major olfactory centers of _____ (D, only the medial; L, only the lateral; B, both the medial and lateral) olfactory area(s).

    a. G, D   c. G, B   e. M, L
    b. G, L   d. M, D   f. M, B

66. Complete removal of the _____ (L, lateral; M, medial) olfactory areas, including structures of the uncus, prepyriform area, and amygdaloid nuclei, results in a more severe loss of _____ (P, "primative" responses; C, complicated conditioned reflexes) to olfactory stimuli.

    a. L, P   c. M, P
    b. L, C   d. M, C

67. A "primary olfactory cortex" is considered by many investigators to lie within the:

    a. Medial olfactory   d. Lateral olfactory
       area   area
    b. Prefrontal cortex   e. Hypothalamus
    c. Occipital cortex   f. Olfactory bulb

68. A centrifugal control pathway terminates upon _____ (P, Purkinje; Y, pyramidal; G, granule) cells of the olfactory bulb and exerts _____ (F, a facilatatory; I, an inhibitory) influence upon mitral cells.

    a. P, F   c. G, F   e. Y, I
    b. Y, F   d. P, I   f. G, I

63

Movement of Food Through the Alimentary Tract

OBJECTIVE 63–1.
Identify the primary function of the alimentary tract.

1. The primary function of the alimentary tract is to provide the internal environment with a continual supply of:

   a. Water
   b. Nutrients
   c. Electrolytes
   d. All of the above

2. A necessary chronological sequence of events, underlying the primary function of the alimentary system, is that of _____ . (D, digestion; A, absorption; M, movement; S, secretion)

   a. D, A, M, & S
   b. M, S, D, & A
   c. M, S, A, & D
   d. S, M, A, & D

OBJECTIVE 63–2.
Using the following diagram, identify the various parts of the alimentary tract.

Directions: Match the lettered headings with the diagram and the numbered list of descriptive words and phrases.

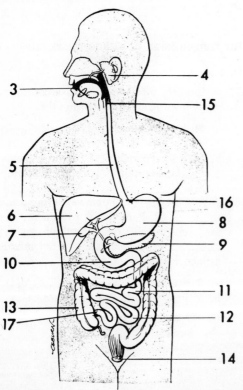

3. _____ Oral cavity.
4. _____ Secretes a viscous fluid that contains ptyalin.
5. _____ Elongate tube that transports food to the stomach.
6. _____ "Bell-shaped" organ that manufactures bile.
7. _____ Stores bile.
8. _____ Storage chamber for chyme before it enters the duodenum.
9. _____ "Leaf-shaped" organ that produces enzymes capable of digesting proteins, carbohydrates, and fats.
10. _____ Uppermost portion of the small intestine, nearest the stomach.
11. _____ Middle portion of the small intestine.
12. _____ Terminal portion of the small intestine, which empties into the colon.
13. _____ That portion of the large intestine nearest the ileum, which contains the cecum.
14. _____ Terminal portion of the alimentary canal.
15. _____ Sphincter, located at the upper end of the esophagus, just below the pharynx.

(From Guyton, A.C., Textbook of Medical Physiology, 5th ed. Philadelphia, W.B. Saunders Company, 1976.

16. _____ Sphincter, located at the lower end of the esophagus, just before entrance into the stomach.

17. _____ Sphincter, located at the terminal portion of the ileum.

a. Gallbladder
b. Jejunum
c. Ileum
d. Stomach

e. Anus
f. Pancreas
g. Hypoharyngeal
h. Mouth

i. Gastroesophageal
j. Esophagus
k. Duodenum
l. Liver

m. Salivary glands
n. Ileocecal
o. Ascending colon

OBJECTIVE 63–3.

Using the following diagram, identify the layers of a typical cross-section of the gut.

Directions: Match the lettered headings with the diagram and numbered list of descriptive words and phrases.

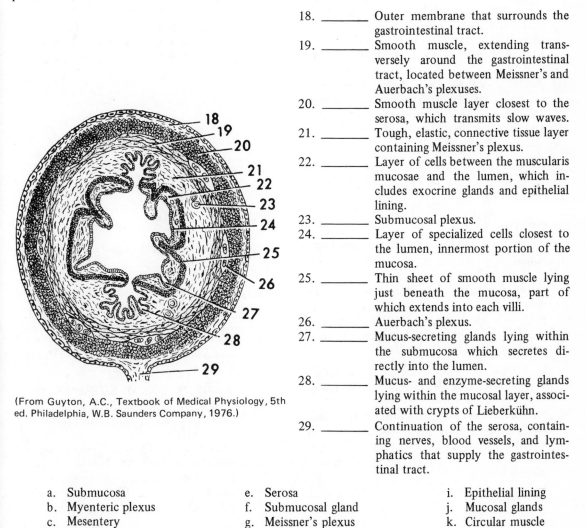

(From Guyton, A.C., Textbook of Medical Physiology, 5th ed. Philadelphia, W.B. Saunders Company, 1976.)

18. _____ Outer membrane that surrounds the gastrointestinal tract.

19. _____ Smooth muscle, extending transversely around the gastrointestinal tract, located between Meissner's and Auerbach's plexuses.

20. _____ Smooth muscle layer closest to the serosa, which transmits slow waves.

21. _____ Tough, elastic, connective tissue layer containing Meissner's plexus.

22. _____ Layer of cells between the muscularis mucosae and the lumen, which includes exocrine glands and epithelial lining.

23. _____ Submucosal plexus.

24. _____ Layer of specialized cells closest to the lumen, innermost portion of the mucosa.

25. _____ Thin sheet of smooth muscle lying just beneath the mucosa, part of which extends into each villi.

26. _____ Auerbach's plexus.

27. _____ Mucus-secreting glands lying within the submucosa which secretes directly into the lumen.

28. _____ Mucus- and enzyme-secreting glands lying within the mucosal layer, associated with crypts of Lieberkühn.

29. _____ Continuation of the serosa, containing nerves, blood vessels, and lymphatics that supply the gastrointestinal tract.

a. Submucosa
b. Myenteric plexus
c. Mesentery
d. Mucosa

e. Serosa
f. Submucosal gland
g. Meissner's plexus
h. Longitudinal muscle

i. Epithelial lining
j. Mucosal glands
k. Circular muscle
l. Muscularis mucosae

OBJECTIVE 63–4.

Characterize the electrical properties of intestinal smooth muscle.

30. The _____ (M, multiunit; F, functionally syncytial) nature of intestinal smooth muscle results from specialized regions of contact between muscle fibers and _____ (N, nerve; L, muscle) fibers in the form of _____ (S, synapses; X, a nexus).

   a. M, N, S   c. M, L, S   e. F, L, S
   b. M, N, X   d. F, L, X   f. F, N, X

31. Slow waves of intestinal smooth muscle are _____ (L, local; P, propagated), _____ (G, graded; A, all-or-none) oscillations in membrane potential occurring at periodicities of about _____ seconds.

   a. L, G, 5 to 20   d. P, A, 60
   b. P, G, 5 to 20   e. L, A, 60
   c. L, A, 5 to 20   f. P, G, 60

32. Action potentials are initiated by slow wave _____ (D, depolarizations; H, hyperpolarizations) of intestinal smooth muscle cells in excess of membrane potential levels of minus _____ millivolts.

   a. D, 15   c. D, 70   e. H, 40
   b. D, 40   d. H, 15   f. H, 70

33. Action potentials are propagated from smooth muscle cell to smooth muscle cell by mechanisms of _____ (S, synaptic; E, ephaptic) transmission at regions of low electrical _____ (C, conductance; R, resistance).

   a. S, C   c. E, C
   b. S, R   d. E, R

34. Depolarization of intestinal smooth muscle cells is generally effected by the actions of _____ , whereas _____ generally result(s) in membrane hyperpolarizations. (Y, sympathetics; P, parasympathetics; S, stretch; A, acetylcholine; N, norepinephrine)

   a. Y & A; P, S, & N   d. P, S, & N; Y & A
   b. Y, S, & N; P & A   e. P & A; Y, S, & N
   c. P, S, & A; Y & N   f. Y & N; P, S, & A

OBJECTIVE 63–5.

Characterize the contractile properties of intestinal smooth muscle and their underlying mechanisms.

35. Smooth muscle contractions are generally coupled to _____ (W, slow wave; S, spike) potentials by a predominant mechanism of _____ (I, intracellular release; T, transmembrane flux) of _____ ions.

   a. S, I, Ca   c. S, T, Ca   e. W, I, Ca
   b. S, T, Mg   d. W, T, Mg   f. W, I, Mg

36. Smooth muscle of the gastrointestinal tract exhibits rhythmic _____ (W, as well as; N, but not) tonic contractions, which is _____ (A, atypical; T, typical) for most types of smooth muscle.

   a. W, A   c. N, T
   b. W, T   d. N, A

37. Rhythmic contractions of gastrointestinal smooth muscle, largely responsible for _____ (P, phasic; S, sphincter) functions of the gastrointestinal system, occur at rates of _____ contractions per _____ (M, minute; H, hour).

   a. P, 3 to 12, M   d. S, 30 to 120, H
   b. P, 3 to 12, H   e. S, 30 to 120, M
   c. P, 30 to 120, M   f. S, 3 to 12, H

38. The frequency of rhythmic contractions of gastrointestinal smooth muscle is largely determined by the _____ (A, amplitude; F, frequency) of _____ (P, action potentials; S, slow waves).

   a. A, P   c. F, P
   b. A, S   d. F, S

OBJECTIVE 63–6.
Identify the characteristic features and functions of the intramural plexus.

39. _____ (M, the myenteric; A, Auerbach's; S, Meissner's) plexus forms the innermost plexus of the _____ distinctive layers of the intramural plexus of the intestine.

   a. M, 2   c. S, 2   e. A, 3
   b. A, 2   d. M, 3   f. S, 3

40. Auerbach's plexus is largely _____ in function, and Meissner's plexus is largely _____ in function. (M, motor; S, sensory)

   a. M, M   c. S, M
   b. M, S   d. S, S

41. Excitatory fibers of the myenteric plexus are _____ (A, adrenergic; C, cholinergic) fibers which comprise _____ (J, a predominant majority; H, about half; N, a predominant minority) of the total number.

   a. A, J   c. A, N   e. C, H
   b. A, H   d. C, J   f. C, N

42. Stimulation of the myenteric plexus results in _____ (I, increased; D, decreased) tone and rhythmical contraction intensity and _____ (I, increased; N, no significant change in; D, decreased) rate of rhythmic contraction and velocity of excitatory propagation of intestinal smooth muscle.

   a. I, I   c. I, D   e. D, N
   b. I, N   d. D, I   f. D, D

43. Reflex alterations of contractile activity of the gut muscle, _____ (W, as well as; N, but not) reflex alterations in glandular secretion by the gut, are integrated at the level of _____ (P, only the intramural plexus; C, only the central nervous system; B, the intramural plexus as well as the central nervous system)

   a. W, P   c. W, B   e. N, C
   b. W, C   d. N, P   f. N, B

OBJECTIVE 63–7.
Identify the characteristic features and functional significance of the autonomic innervation of the gastrointestinal tract.

44. The _____ (M myenteric; S, submucosal) plexus contains abundant _____ (E, preganglionic; O, ganglionic) neurons of the _____ (Y, sympathetic; P, parasympathetic) nervous system.

   a. M, E, Y   c. M, O, Y   e. S, E, Y
   b. M, O, P   d. S, O, P   f. S, E, P

45. Stimulation of the _____ (P, parasympathetic; S, sympathetic) nervous system increases gastric secretion and _____ (T, stimulates; I, inhibits) gut motility.

   a. P, T   c. S, T
   b. P, I   d. S, I

46. The celiac ganglia and mesenteric ganglia are more important loci of _____ (S, sympathetic; P, parasympathetic) _____ (E, preganglionic; O, ganglionic) neurons, whose stimulation results in enhanced tone of the _____ (IW, intestinal wall; IS, ileocecal sphincter).

   a. S, E, IW   c. S, O, IW   e. P, E, IW
   b. S, O, IS   d. P, O, IS   f. P, E, IS

47. Destruction of preganglionic _____ (S, sympathetic; P, parasympathetic) efferents to the gut results in more severe consequences, including _____ (E, enhanced; D, depressed) gastrointestinal tone _____ (W, as well as; N, but not) peristalsis.

   a. S, E, W   c. S, D, N   e. P, D, W
   b. S, E, N   d. P, D, N   f. P, E, W

48. Afferent fibers responsive to chemical stimulation, irritation, or distention of the gut have cell bodies located in the intramural plexus, particularly the _____ (S, submucosal; M, myenteric) plexus, _____ (W, as well as; N, but not) dorsal root ganglia.

   a. S, W   c. M, W
   b. S, N   d. M, N

49. Pain fibers from the gastrointestinal system are largely transmitted via the _____ (P, parasympathetic; S, sympathetic) pathway, while the majority of vagal nerve fibers are _____ (E, efferent; A, afferent) fibers.

   a. P, E   c. S, E
   b. P, A   d. S, A

OBJECTIVE 63–8.

Identify the basic types of movements in the gastrointestinal tract, their characteristics, and their significance.

50. Peristalsis, an intrinsic property of the _____ (S, syncytial smooth muscle; M, myenteric plexus; SM, submucosal plexus), is utilized for _____ (X, only mixing; P, only propulsive; B, both mixing and propulsive) types of gastrointestinal movements.

a. S, P   c. SM, B   e. M, X
b. M, P   d. S, B   f. SM, X

51. Peristalsis occurs _____ (N, only; B, bidirectionally, with a preferential direction occurring) in the _____ (O, orad; A, analward) direction.

a. N, O   c. B, O
b. N, A   d. B, A

52. Effective peristalsis _____ (S, is; X, is not) dependent upon the myenteric plexus, and is _____ (I, increased; N, not significantly altered; D, decreased) by atropine administration.

a. S, I   c. S, D   e. X, N
b. S, N   d. X, I   f. X, D

53. Contractile rings, associated with _____ (E, only the leading; G, only the lagging; B, both the leading and lagging) phase(s) of peristaltic waves, are normally initiated in response to gut _____ (C, constriction; D, distention).

a. E, C   c. B, C   e. G, D
b. G, C   d. E, D   f. B, D

54. Electrical stimulation of regions of the gastrointestinal tract results in _____ (M, myenteric; SI, somatointestinal; I, intestinointestinal) reflexes associated with _____ (D, leading; G, lagging) waves of receptive relaxation.

a. M, D   c. I, D   e. SI, G
b. SI, D   d. M, G   f. I, G

55. A basic contractile rhythm of the duodenum of about _____ contractions per minute is _____ (H, higher; L, lower) than that of the ileum and _____ (H, higher; L, lower) than that occurring for the jejunum.

a. 3, H, L   c. 3, L, L   e. 11, H, L
b. 3, L, H   d. 11, L, H   f. 11, H, H

OBJECTIVE 63–9.

Identify mastication, its mechanistic basis, and its functional significance.

56. Incisors, or _____ (A, anterior; P, posterior) teeth, function in primarily a _____ (G, grinding; C, cutting) mode and are capable of exerting maximal forces of about _____ kg.

a. A, G, 250   c. A, C, 25   e. P, G, 100
b. A, C, 100   d. P, C, 25   f. P, G, 250

57. Mastication involves afferent activity mediated largely by the _____ cranial nucleus and "muscles of mastication" innervated by the _____ cranial nuclei. (F, fifth; S, seventh; N, ninth)

a. F, S   c. N, F   e. S, F
b. S, S   d. F, F   f. N, S

58. Areas which elicit rhythmic chewing movements in response to stimulation include the _____ (B, cerebellar; C, cerebral) cortex, the amygdaloid nuclei, _____ (W, as well as; N, but not) the hypothalamus.

a. B, W   c. C, W
b. B, N   d. C, N

59. Application of occlusal forces to the teeth of a decerebrate animal results in _____ (E, excitation; I, inhibition) of jaw-closing muscles as a consequence of central _____ (A, amygdaloid; T, trigeminal; F, facial) nuclear centers.

a. E, A   c. E, F   e. I, T
b. E, T   d. I, A   f. I, F

60. Mastication in animals performs a more critical functional role in the subsequent digestion of raw foods of _____ (A, animal; P, plant) origin as a consequence of their _____ (H, higher; L, lower) _____ (G, glycogen; S, starch; C, cellulose) content.

a. A, H, G   c. A, L, S   e. P, H, C
b. A, L, G   d. P, H, S   f. P, L, C

OBJECTIVE 63–10.

Identify the stages of deglutition, their underlying mechanisms and characteristic features, and their functional significance.

61. Deglutition, or _____ (M, vomiting; S, swallowing), involves _____ (V, a voluntary; I, an involuntary) oral stage and _____ (V, a voluntary; I, an involuntary) pharyngeal stage.

    a. M, V, V   c. M, I, V   e. S, V, V
    b. M, I, I   d. S, I, I   f. S, V, I

62. A voluntary stage of deglutition is _____ (I, initiated; T, terminated) by stimulation of deglutition receptors, more abundantly located at the entrance to the _____ (O, oral cavity; P, pharynx; E, esophagus).

    a. I, O   c. I, E   e. T, P
    b. I, P   d. T, O   f. T, E

63. Food movement from the oral cavity into the pharynx is initiated by _____ (U, upward; D, downward; F, forward; B, backward) movements of the tongue and _____ (S, a squeezing action of the tongue; G, gravity) as a primary mover.

    a. U & F, S   c. D & B, S   e. U & B, G
    b. U & B, S   d. D & F, G   f. D & B, G

64. Entry of food material into the upper and lower respiratory passages is prevented during the _____ stage of deglutition by _____ (D, depression; E, elevation) of the soft palate and _____ (D, depression; E, elevation) of the larynx.

    a. 2nd, D, E   c. 2nd, E, E   e. 3rd, D, E
    b. 2nd, E, D   d. 3rd, E, D   f. 3rd, D, D

65. Palatopharyngeal folds on either side of the pharynx move _____ (M, medially; L, laterally) during the pharyngeal stage of deglutition in order to _____ (F, facilitate; I, inhibit) the movement of _____ (A, air; P, larger food particles).

    a. M, F, A   c. M, I, A   e. L, F, A
    b. M, I, P   d. L, I, P   f. L, F, P

66. Contraction of the superior constrictor muscle of the pharynx during deglutition occurs in association with _____ (D, a depressed; N, a resting position of the; E, an elevated) larynx and a _____ (C, tonically constricted; T, transient constriction of the; R, relaxed) pharyngoesophageal sphincter.

    a. D, T   c. N, C   e. E, R
    b. D, R   d. N, T   f. E, C

67. Removal of the epiglottis usually _____ (D, does; N, does not) result in food aspiration into the trachea during deglutition as a consequence of _____ (P, pulmonary-pharyngeal pressure gradients; V, the position of the vocal cords).

    a. D, P   c. N, P
    b. D, V   d. N, V

68. Receptors which initiate the pharyngeal stage of deglutition transmit activity to the deglutition center of the _____ (R, rhombencephalon; M, mesencephalon; T, telencephalon) via the _____ (N, ninth; W, twelfth) cranial nerve.

    a. R, N   c. T, N   e. M, W
    b. M, N   d. R, W   f. T, W

69. An overriding cycle of events of _____ (D, deglutition; R, respiration) interrupts the other cycle _____ (B, only at its beginning; T, at any time during its progress).

    a. D, B   c. R, B
    b. D, T   d. R, T

70. The esophagus is characterized in a resting state as a _____ (C, collapsed; D, distended) tube with closed valves at _____ (U, only its upper; L, only its lower; B, both its upper and lower) end(s).

    a. C, U   c. C, B   e. D, L
    b. C, L   d. D, U   f. D, B

71. Most of the esophageal muscle is _____ (K, skeletal; S, smooth) muscle, except for its _____ (U, upper; L, lower) end.

    a. K, U   c. S, U
    b. K, L   d. S, L

72. A primary peristaltic wave passing from the pharynx to the stomach, requiring _____ (M, more; L, less) time than the pharyngeal stage of deglutition, is completed in _____ seconds.

    a. M, 1 to 2   c. M, 15 to 30   e. L, 1 to 2
    b. M, 5 to 10   d. L, 0.1 to 0.2   f. L, 5 to 10

73. Secondary peristaltic contractions of the esophagus generally arise as a consequence of _____ (C, continuous pacemaker activity; D, local distention) and _____ (A, are; N, are not) propagated to the cardiac junction.

   a. C, A
   b. C, N
   c. D, A
   d. D, N

74. Maintained loss of the _____ (N, ninth; T, tenth) cranial nerve supplying most of the esophageal muscle is accompanied by an immediate loss of _____ (P, only primary; S, only secondary; B, both primary and secondary) peristaltic waves, followed by a partial recovery of _____ (P, only primary; S, only secondary) peristaltic waves.

   a. N, B, S
   b. N, S, S
   c. N, B, P
   d. T, B, S
   e. T, P, P
   f. T, B, P

OBJECTIVE 63–11.

Identify the properties and functional characteristics of the gastrointestinal sphincter.

75. A "flutter-valve" arrangement of the gastroesophageal sphincter, _____ (W, as well as; N, but not) the existing pressure gradient, acts as a more favorable barrier to movement of substances in the _____ (EG, esophagogastric; GE, gastroesophageal) direction.

   a. W, EG
   b. W, GE
   c. N, EG
   d. N, GE

76. Movements of substances from the esophagus into the stomach are associated with _____ (C, chronically patent gastroesophageal junctions; W, waves of "receptive relaxation") and _____ (R, "receptive relaxation" of; N, no significant change in; E, enhanced tonus of) gastric smooth muscle.

   a. C, R
   b. C, N
   c. C, E
   d. W, R
   e. W, N
   f. W, E

77. The principal function of the gastroesophageal sphincter is to regulate gastric _____ (F, filling; R, reflux). Increased intragastric pressure is associated with _____ (I, increased; N, no significant change in; D, decreased) gastroesophageal sphincter tone.

   a. F, I
   b. F, N
   c. F, D
   d. R, I
   e. R, N
   f. R, D

OBJECTIVE 63–12.

Using the following diagram, identify the various anatomical features of the stomach.

Directions: Match the lettered headings with the diagram and the numbered list of descriptive words and phrases.

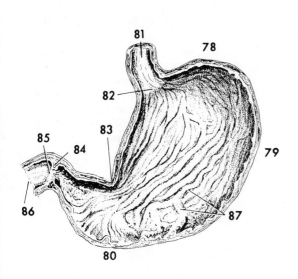

(From Guyton, A.C., Textbook of Medical Physiology, 5th ed. Philadelphia, W.B. Saunders Company, 1976.)

78. _____ The upper portion of the stomach nearest the esophagus.

79. _____ The main portion or body of the stomach.

80. _____ The lowermost portion of the stomach nearest the duodenum.

81. _____ Fleshy tube leading from the pharynx to the stomach.

82. _____ Opening at the gastroesophageal junction.

83. _____ Notch on the lesser curvature of the stomach between the body and the antrum.

84. _____ Muscular valve between the antrum and duodenum.

85. _____ Opening at the gastrointestinal junction.

86. _____ The first part of the small intestine, leading from the stomach to the jejunum.

87. _____ Wrinkles or folds on the interior stomach wall.

a. Rugae
b. Esophagus
c. Antrum
d. Pylorus
e. Incisura angularis
f. Fundus
g. Duodenum
h. Pyloric sphincter
i. Cardia
j. Corpus

OBJECTIVE 63–13.

Identify the major motor functions of the stomach, their underlying mechanisms, and their functional characteristics.

88. A higher percentage of the storage capacity of the stomach of about _____ liter(s) is attributable to the _____ (P, pylorus; C, corpus; A, antrum; F, fundus).

a. 1, C   c. 1, F   e. 5, F
b. 1, A   d. 5, P   f. 5, C

89. A moderate increase in gastric stretch with its filling results in a _____ (S, slight; B, substantial) _____ (I, increase; D, decrease) in intragastric pressure and _____ (E, enhanced; M, diminished) smooth muscle tonic activity to the body of the stomach.

a. S, I, E   c. S, D, E   e. B, D, E
b. S, I, M   d. B, D, M   f. B, I, M

90. For a relatively constant intragastric pressure, the stretching force or wall tension of the stomach would be approximately _____ (D, directly; I, inversely) proportional to the _____ (Q, square of the radius; R, radius) of curvature as predicted by _____ (S, starling's law; L, the law of Laplace).

a. D, Q, L   c. D, R, L   e. I, Q, L
b. D, R, S   d. I, R, S   f. I, Q, S

91. Propulsive peristaltic waves in the stomach, _____ (L, like; U, unlike) mixing waves, occur at _____ -second intervals.

a. L, 5   c. L, 80   e. U, 20
b. L, 20   d. U, 5   f. U, 80

92. Mixing waves beginning most frequently near the _____ (C, cardia; D, midpoint of the stomach) generally become _____ (L, less; M, more) intense as they approach the antral region.

    a. C, L   c. D, L
    b. C, M   d. D, M

93. Intense mixing of gastric contents by peristaltic waves which may obliterate the lumen is associated with the _____ (F, fundus; A, antrum; C, corpus) and a relatively _____ (L, low; H, high) degree of gastric fluidity.

    a. F, L   c. C, L   e. A, H
    b. A, L   d. F, H   f. C, H

94. Gastric filling through food ingestion generally results in _____ gastric tone to its storage regions, _____ intensity of mixing and perstaltic waves, and _____ gastric secretion. (I, increased; D, decreased)

    a. I, D, I   c. I, I, I   e. D, I, D
    b. I, D, D   d. D, I, I   f. D, D, I

95. Hunger "pangs" occur more frequently _____ hours after the last ingestion of food, occur more frequently in association with _____ (L, low; H, high) levels of gastrointestinal tonus, and are _____ (F, facilitated; N, not significantly affected) by reduced blood glucose levels.

    a. 6 to 8, L, F   d. 24, H, N
    b. 6 to 8, H, N   e. 24, H, F
    c. 6 to 8, H, F   f. 24, L, N

OBJECTIVE 63–14.

Identify the determinant factors regulating gastric emptying, their associated mechanisms, and their functional significance.

96. The rate of exit of chyme from the stomach, determined principally by _____ (P, antral peristaltic activity; T, pyloric sphincter tone), is _____ (E, enhanced; R, reduced) by the distention of the stomach with food.

    a. P, E   c. T, E
    b. P, R   d. T, R

97. The hormone _____ (P, pepsin; G, gastrin) _____ (D, decreases; I, increases) pyloric pump activity.

    a. P, D   c. G, D
    b. P, I   d. G, I

98. Gastrin secretion by the _____ (D, duodenal; A, antral) mucosa occurs in response to the presence of particularly _____ (L, lipid; P, protein) foodstuffs _____ (W, as well as; N, but not) stretch.

    a. D, L, W   c. D, P, W   e. A, P, W
    b. D, L, N   d. A, P, N   f. A, L, N

99. Gastrin _____ (H, inhibits; S, stimulates) gastric secretion, and _____ (I, increases; D, decreases) the tonus of the gastroesophageal sphincter _____ (W, as well as; N, but not) the pyloris.

    a. H, I, W   c. H, D, W   e. S, D, W
    b. H, I, N   d. S, D, N   f. S, I, N

100. The enterogastric reflex, largely initiated by afferent activity originating within the _____ (S, stomach; D, duodenum), exerts _____ (E, excitatory; I, inhibitory) effects upon pyloric activity.

    a. S, E   c. D, E
    b. S, I   d. D, I

101. The enterogastric reflex is more effectively elicited by a _____ (G, gastric; D, duodenal) pH range of _____ and the presence of particularly _____ (O, hypotonic; E, hypertonic) fluids.

    a. G, 4.5 to 6.5, O   d. D, below 4, E
    b. G, below 4, O   e. D, 4.5 to 6.5, E
    c. G, 4.5 to 6.5, E   f. D, below 4, O

102. A(n) _____ (F, facilitatory; I, inhibitory) effect of the presence of fat in the duodenum upon pyloric pump activity is _____ (R, retained; L, lost) following adequate blockage of the enterogastric reflex.

    a. F, R   c. I, R
    b. F, L   d. I, L

103. Enterogastrone is thought to be_____ (H, a hormone; E, an enzyme) released from the _____ (G, gastric; O, duodenal) mucosa which functions to _____ (D, decrease; I, increase) the duodenal content of lipid.

    a. H, G, D   c. H, O, D   e. E, G, D
    b. H, O, I   d. E, O, I   f. E, G, I

104. Secretin, released from the _____ (P, pancreas; D, duodenal mucosa) acts _____ (S, synergistically; A, antagonistically) to the action of enterogastrone upon the pyloric pump.

    a. P, S   c. D, S
    b. P, A   d. D, A

105. Pyloric sphincter tonus is more effectively increased by variations in duodenal _____ (O, osmolarity; A, acidity; P, protein concentration) and factors which _____ (I, increase; D, decrease) gastric motility.

    a. O, I   c. P, I   e. A, D
    b. A, I   d. O, D   f. P, D

106. The rate of stomach emptying is largely determined by _____ (G, gastric; D, duodenal) factors acting through mechanisms regulating _____ (S, stomach food intake; P, pyloric sphincter tonus; I, intragastric pressure; M, pyloric pump activity).

    a. G, S   c. G, I   e. D, I
    b. G, M   d. D, P   f. D, M

OBJECTIVE 63–15.

Identify the types and characteristic features of the mixing and propulsive movements of the small intestine.

107. Ring-like contractions appearing at fairly regular intervals along the gut, which then disappear and are replaced by another set of ring-like contractions in the segments between the previous contractions, are termed:

    a. Mass movements
    b. Pendular movements
    c. Antiperistaltic movements
    d. Segmentation contractions
    e. Haustral contractions
    f. Peristaltic rushes

108. _____ (H, haustral; S, segmenting) contractions are the most common type of smooth muscle activity of the small intestine, occurring most frequently in the _____ (D, duodenum; J, jejunum; I, ileum).

    a. H, D   c. H, I   e. S, J
    b. H, J   d. S, D   f. S, I

109. Segmenting movements are integrated largely within the _____ (C, central nervous system; M, myenteric plexus) and are enhanced by increased activity of _____ (P, only parasympathetic; S, only sympathetic; B, both sympathetic and parasympathetic) efferents.

    a. C, P   c. C, B   e. M, S
    b. C, S   d. M, P   f. M, B

110. Peristaltic waves of the small intestine more frequently travel _____ (S, short; L, long) distances _____ (O, orad; A, analward) at velocities of _____ cm. per second.

    a. S, O, ½ to 5   d. L, A, 10 to 20
    b. S, A, 10 to 20   e. L, O, ½ to 5
    c. S, A, ½ to 5   f. L, O, 10 to 20

111. The gastroenteric reflex _____ (I, increases; D, decreases) peristaltic activity, _____ (W, as well as; N, but not) secretion of the small intestine, in response to _____ (S, gastric stretch; R, reduced gastric volume).

    a. I, W, S   c. I, W, R   e. D, W, R
    b. I, N, S   d. D, N, R   f. D, N, S

112. Passage of chyme through the ileocecal valve is _____ (R, reduced; A, augmented) at mealtime as a consequence of _____ (EG, the enterogastric; GI, the gastroileal; SI, the somatointestinal) reflex.

    a. R, EG   c. R, SI   e. A, GI
    b. R, GI   d. A, EG   f. A, SI

113. The major determinant of the rhythmical contraction frequency of the small intestine is the_____ (B, basic rhythm; A, action potential frequency) of slow waves of the _____ (L, longitudinal; C, circular) smooth muscle component.

    a. B, L   c. A, L
    b. B, C   d. A, C

114. A higher "pacemaker" frequency of the region of the small intestine near the _____ (O, sphincter of Oddi; I, ileocecal sphincter) is associated with propagated _____ (C, ciliary motion; M, movements of the intestinal villi; S, slow waves).

    a. O, C   c. O, S   e. I, M
    b. O, M   d. I, C   f. I, S

115. Movements of the muscularis mucosae, enhanced by _____ (P, parasympathetic; S, sympathetic) activity, result in _____ (V, contractions of the villi; F, the appearance of folds in the intestinal mucosa).

    a. P, only V   d. S, only V
    b. P, only F   e. S, only F
    c. P, both V & F   f. S, both V & F

116. The hormone _____ (V, villikinin; C, cholecystokinin), thought to be released from the _____ (P, pancreas; S, small intestine), _____ (E, enhances; R, reduces) the degree of movement of intestinal villi.

    a. V, P, E   c. V, P, R   e. C, P, R
    b. V, S, E   d. C, S, R   f. C, S, E

OBJECTIVE 63–16.

Identify the underlying mechanisms and characteristics of bile storage and transport.

117. Bile is synthesized in the _____ (G, gallbladder; P, pancreas; L, liver), and stored _____ (W, as well as; N, but not) concentrated in the _____ (G, gallbladder; P, pancreas; L, liver).

    a. G, W. L   c. L, W, G   e. P, N, L
    b. P, W, G   d. G, N, P   f. L, N, G

118. Duodenal lipids _____ (I, increase; D, decrease) gastric motility and provide a major _____ (I, increase; D, decrease) in _____ (C, cholecystokinin; S, secretin) secretion.

    a. I, I, C   c. I, D, C   e. D, I, C
    b. I, D, S   d. D, D, S   f. D, I, S

119. Cholecystokinin is released from the _____ (C, pancreas; D, duodenum) in response to digested products of _____ (F, only fat; P, only protein; B, both protein and fat).

    a. C, F   c. C, B   e. D, P
    b. C, P   d. D, F   f. D, B

120. _____ (C, cholecystokinin; S, secretin) results in _____ (T, contraction; R, relaxation) of the gallbladder and _____ (T, contraction; N, no significant change in tonus; R, relaxation) of the sphincter of Oddi.

    a. C, R, T   c. C, T, R   e. S, R, N
    b. C, T, N   d. S, R, T   f. S, T, R

121. Movement of bile through the sphincter of Oddi occurs via the action(s) of cholecystokinin _____ (W, as well as; N, but not) intestinal peristalsis and performs a greater functional role regarding _____ (F, fat; P, protein; C, carbohydrate) digestion and absorption.

    a. W, F   c. W, C   e. N, P
    b. W, P   d. N, F   f. N, C

OBJECTIVE 63–17.

Identify the regulatory factors and principal function of the ileocecal valve.

122. When constricted, the ileocecal valve is better suited to resist movements of chyme in _____ (O, oral; A, anal) directions, and most of the time provides _____ (I, an insignificant; S, a significant) resistance to the rate of emptying of ileal contents.

    a. O, I
    b. O, S
    c. A, I
    d. A, S

123. The hormone gastrin, _____ (W, as well as; N, but not) the gastroileal reflex, function(s) to _____ (E, enhance; R, retard) the rate of movement of chyme into the cecum.

    a. W, E
    b. W, R
    c. N, E
    d. N, R

124. Pressure or chemical irritation of the cecum _____ the ileocecal sphincter, whereas pressure or chemical irritation of the ileum _____ the ileocecal sphincter. (E, excites; R, relaxes)

    a. E, E
    b. E, R
    c. R, E
    d. R, R

125. Pressure or chemical irritation of the cecum _____ peristalsis of the ileum, whereas pressure or chemical irritation of the ileum _____ ileal peristalsis. (F, facilitates; I, inhibits)

    a. F, F
    b. F, I
    c. I, F
    d. I, I

126. When food leaves the stomach, the passage of chyme through the ileocecal valve _____. Sympathetic stimulation _____ the tonic contraction of the valve. (I, increases; D, decreases)

    a. I, I
    b. I, D
    c. D, I
    d. D, D

OBJECTIVE 63–18.

Identify the types of movements of the colon, their underlying mechanisms and characteristics, and their functional significance.

127. A larger number of known hormones regulating gastrointestinal motility and secretion are released from the:

    a. Stomach
    b. Pancreas
    c. Colon
    d. Duodenum
    e. Jejunum
    f. Ileum

128. The colon functions primarily for storage, occurring primarily in its _____ (P, proximal; D, distal) half, and absorption of _____ (E, water and electrolytes; C, caloric substances).

    a. P, E
    b. P, C
    c. D, E
    d. D, C

129. Contractions of aggregate arrangements of the _____ (C, circular; L, longitudinal) muscle, or tineae coli, of the _____ (S, small; G, large) intestine gives rise to _____ (H, haustrations; V, vomiting; P, primarily propulsive movements).

    a. C, S, P
    b. C, G, P
    c. C, S, V
    d. L, G, H
    e. L, S, H
    f. L, G, P

130. "Mass movements" of the colon, occurring _____ (C, rather continuously; F, only a few times) during the day, involve _____ (P, only peristaltic; H, only haustral; B, combined peristaltic and haustral) contractions.

    a. C, P
    b. C, H
    c. C, B
    d. F, P
    e. F, H
    f. F, B

131. Mass movements occur most frequently within the _____ (A, ascending; D, descending) and transverse colon, and are primarily a _____ (M, mixing; P, propulsive) type of movement.

    a. A, M
    b. A, P
    c. D, M
    d. D, P

132. Gastrin, secreted by the _____ (G, gastric; L, intestinal) mucosa, results in _____ (I, increased; D, decreased) tone of the ileocecal valve and _____ (I, increased; D, decreased) activity of the colon.

a. G, I, I   c. G, D, I   e. L, I, I
b. G, D, D   d. L, D, D   f. L, I, D

133. Mass movements of the colon are _____ by parasympathetic stimulation, _____ by gastrocolic and duodenocolic reflexes, and _____ by irritative processes of the colon. (F, facilitated; I, inhibited)

a. F, F, F   c. F, I, F   e. I, F, F
b. F, F, I   d. I, I, I   f. I, F, I

OBJECTIVE 63–19.

Identify the defecation reflex, its underlying mechanisms, and its means of regulation.

134. The greatest straining, as measured by both duration and intra-abdominal pressure, occurs with the _____ (S, smallest; L, largest) stool as _____ (T, rectal wall tension; P, colon peristalsis) is the normal stimulus for defecation.

a. S, T   c. L, T
b. S, P   d. L, P

135. The rectum is a _____ (C, contractile; N, noncontractile) chamber which _____ (E, is empty of; F, contains) feces most of the time.

a. C, E   c. N, E
b. C, F   d. N, F

136. Voluntary control over defecation is largely exerted via the _____ (I, internal; E, external) anal sphincter comprised of _____ (S, striated; N, nonstriated) muscle.

a. I, S   c. E, S
b. I, N   d. E, N

137. Destruction of both vagi, _____ (I, increases; N, does not seriously alter; D, decreases) the tone and degree of peristaltic activity of the terminal esophagus, stomach, and intestine, and _____ (S, does; SN, does not) seriously impair defecation.

a. I, S   c. D, S   e. N, SN
b. N, S   d. I, SN   f. D, SN

138. Parasympathetic fibers of the _____ (N, nervi erigentes; V, vagus nerves) provide a mechanism for _____ (R, reinforcing; I, inhibiting) a relatively _____ (W, weak; P, powerful) defecation reflex involving the myenteric plexus.

a. N, R, W   c. N, I, W   e. V, I, W
b. N, R, P   d. V, I, P   f. V, R, P

OBJECTIVE 63–20.

Identify the characteristics and significance of intestinointestinal, peritoneointestinal, renointestinal, vesicointestinal, and somatointestinal reflexes.

139. The renointestinal and vesicointestinal reflexes _____ (I, inhibit; F, facilitate) intestinal activity as a result of _____ (N, normal; A, abnormal) conditions of kidney and bladder function.

a. I, N   c. F, N
b. I, A   d. F, A

140. The intestinointestinal reflex resulting from intestinal mucosal irritation is associated with a localized _____ of intestinal activity and _____ of more remote segments of the intestine. (F, facilitation; I, inhibition)

a. F, F   c. I, F
b. F, I   d. I, I

141. The peritoneointestinal, renointestinal, and somatointestinal reflexes are largely _____ (F, facilitatory; I, inhibitory) reflexes acting upon the intestinal system via enhanced _____ (P, parasympathetic; S, sympathetic) outflow.

a. F, P   c. I, P
b. F, S   d. I, S

64

Secretory Functions of the Alimentary Tract

OBJECTIVE 64–1.
Identify the general functions and anatomical types of glands associated with the gastrointestinal tract.

1. _____ (E, enzymes; M, mucus) secreted by glands present along the full extent of the gastrointestinal tract subserve _____ (L, lubricative; P, protective; D, digestive) functions.

   a. E, only D
   b. E, only P
   c. E, both D & P
   d. M, only P
   e. M, only L
   f. M, both P & L

2. Digestive secretions of the gastrointestinal tract are generally _____ the amount of food in the alimentary tract, and are generally _____ the compositional nature of ingested food. (D, dependent upon; I, independent of)

   a. D, D
   b. D, I
   c. I, D
   d. I, I

3. Glands associated with the gastrointestinal tract include the pancreas, _____ (W, as well as; B, but not) the liver, and _____ (T, tubular; C, compound acinous) glands of the stomach, _____ (A, and; N, but not) the crypts of Lieberkühn.

   a. W, T, A
   b. W, T, N
   c. W, C, A
   d. B, C, N
   e. B, C, A
   f. B, T, N

OBJECTIVE 64–2.
Identify the basic mechanisms of stimulation of the gastrointestinal glands.

4. Secretions of the gastrointestinal tract are generally _____ by chemical irritation of the mucosa, _____ by distention of the gut segments, and _____ by decreased gut motility. (F, facilitated; I, inhibited)

   a. F, F, F
   b. F, F, I
   c. F, I, F
   d. I, I, I
   e. I, F, I
   f. I, I, F

5. Stimulation of the _____ (P, parasympathetic; S, sympathetic) nerves increases secretion rates of gastrointestinal glands, particularly for the _____ (U, upper; L, lower) part of the alimentary tract.

   a. P, U
   b. P, L
   c. S, U
   d. S, L

6. Stimulation of the salivary glands, esophageal glands, and the pancreas, _____ (W, as well as; N, but not) the gastric and Brunner's glands, is largely mediated by _____ (A, automatic nerves; L, direct, local stimuli).

   a. W, A
   b. W, L
   c. N, A
   d. N, L

7. The effect of sympathetic nerve activity to _____ (D, decrease; I, increase) glandular secretion via its action upon glandular blood flow is greatest when the parasympathetic activity to the glands is _____ (H, high; L, low).

   a. D, H
   b. D, L
   c. I, H
   d. I, L

8. The _____ (C, acidic; K, alkaline) mucoid secretion of Brunner's glands is important in the etiology of peptic ulcers, since its secretion is generally _____ (F, facilitated; I, inhibited) by sympathetic stimulation.

   a. C, F   c. K, F
   b. C, I   d. K, I

9. Gastrointestinal hormones which are _____ (S, steroids; P, polypeptides) or their derivatives significantly regulate _____ (C, only the relative composition; V, only the volume; B, both the volume and relative composition) of gastrointestinal secretions.

   a. S, C   c. S, B   e. P, V
   b. S, V   d. P, C   f. P, B

OBJECTIVE 64–3.

Identify the basic mechanism of secretion by glandular cells, and the properties and functional significance of mucus secretion by the gastrointestinal tract.

10. An extensive endoplasmic reticulum, secretory granules, _____ (W, as well as; N, but not) the presence of a Golgi apparatus, are distinguishing features of glandular cells which secrete primarily _____ (O, organic substances; E, water and electrolytes).

   a. W, O   c. N, O
   b. W, E   d. N, E

11. Parasympathetic stimulation to gastrointestinal glandular-cells is thought to _____ (I, increase; D, decrease) the intracellular negativity of their typical nonstimulated membrane potentials of about _____ millivolts.

   a. I, 10 to 20   d. D, 10 to 20
   b. I, 30 to 40   e. D, 30 to 40
   c. I, 70 to 90   f. D, 70 to 90

12. Mucus functions as _____ (L, only a lubricant; P, only a protectant; B, both a lubricant and protectant) for the gastrointestinal tract and is relatively _____ (D, digestible; N, nondigestible) by gastrointestinal enzymes.

   a. L, D   c. B, D   e. P, N
   b. P, D   d. L, N   f. B, N

13. Mucus contains a mixture of _____ (L, lipoproteins; M, mucopolysaccharides), and forms a relatively _____ (D, adherent; N, nonadherent) substance with _____ (C, choletic; A, amphoteric) properties.

   a. L, D, C   c. L, N, C   e. M, D, C
   b. L, N, A   d. M, N, A   f. M, D, A

OBJECTIVE 64–4.

Identify the salivary glands, and the composition and characteristics of their secretions.

14. The _____ major types of salivary glands in conjunction with the minor salivary glands secrete an average daily secretion of _____ liters of saliva.

   a. 2, 5 to 7   d. 2, 1 to 1½
   b. 3, 5 to 7   e. 3, 1 to 1½
   c. 5, 5 to 7   f. 5, 1 to 1½

15. The parotid glands secrete a _____ secretion, while the submaxillary glands secrete a _____ secretion. (S, predominantly serous; M, predominantly mucous; X, mixed serous and mucous)

   a. S, M   c. X, S   e. M, X
   b. M, S   d. S, X   f. X, M

16. The sublingual _____ (W, as well as; N, but not) the buccal glands secrete a _____ (S, predominantly serous; X, mixed serous and mucous; M, predominantly mucous) fluid.

   a. W, S   c. W, M   e. N, X
   b. W, X   d. N, S   f. N, M

17. Saliva becomes more _____ (C, acidic; K, alkaline) upon exposure to air and has _____ (F, a favorable; U, an unfavorable) pH range for the digestive action of ptyalin of _____ .

   a. C, F, 4.0 to 6.0   d. K, U, 7.4 to 8.7
   b. C, U, 4.0 to 6.0   e. K, F, 7.4 to 8.7
   c. C, U, 6.0 to 7.4   f. K, F, 6.0 to 7.4

18. Saliva when compared to plasma, generally has _____ concentrations of sodium, _____ concentrations of chloride, and _____ concentrations of potassium. (H, higher; L, lower)

    a. H, H, H   c. L, L, H   e. H, H, L
    b. L, H, H   d. L, L, L   f. H, L, L

19. Saliva, formed by _____ (F, passive filtration; S, active secretion), is more hypotonic with respect to plasma during _____ (H, high; L, low) rates of salivary flow.

    a. F, H   c. S, H
    b. F, L   d. S, L

20. The sodium, _____ (W, as well as; N, but not) the potassium, concentration of saliva is lowest in the _____ (P, presence; A, absence) of aldosterone and _____ (H, high; L, low) salivary flow rates.

    a. W, P, H   c. W, A, H   e. N, P, H
    b. W, A, L   d. N, A, L   f. N, P, L

21. Potassium ions, _____ (W, as well as; N, but not) bicarbonate ions, are actively _____ (R, reabsorbed; S, secreted) largely by the salivary _____ (A, acini; D, ductal epithelium).

    a. W, R, D   c. W, S, D   e. N, R, D
    b. W, S, A   d. N, S, A   f. N, R, A

OBJECTIVE 64–5.

Identify the major functions of saliva and the neural mechanisms regulating salivary secretion.

22. Ptyalin is a salivary _____ (A, amylase; L, lipase; Y, lysosyme) which is secreted largely in response to _____ (N, neural; H, hormonal) mechanisms.

    a. A, N   c. Y, N   e. L, H
    b. L, N   d. A, H   f. Y, H

23. "Basal" secretions of saliva occur at rates of about _____ ml./minute and generally have a _____ (L, lower; H, higher) viscosity and a _____ (L, lower; H, higher) amylase content than saliva secreted during mastication.

    a. 1, L, L   c. 1, H, L   e. 5, L, L
    b. 1, H, H   d. 5, H, H   f. 5, L, H

24. The oral cavity is normally a relatively _____ (A, aseptic; P, septic) environment in which saliva performs major roles in preventing destructive lesions of _____ (M, only the mucosa; T, only the teeth; B, both the teeth and soft tissues) of the oral cavity.

    a. A, M   c. A, B   e. P, T
    b. A, T   d. P, M   f. P, B

25. The _____ (X, submaxillary; P, parotid; L, sublingual) glands are regulated primarily from the inferior portion of the salivatory nuclei located within the _____ (T, telencephalon; M, mesencephalon; R, rhombencephalon).

    a. X, T   c. L, M   e. P, T
    b. P, R   d. X, M   f. L, R

26. Stimulation of the primary gustatory modalities, _____ (E, except; P, particularly) the sour taste, elicit _____ (C, copious; S, scanty, mucous) secretions of saliva.

    a. E, C   c. P, C
    b. E, S   d. P, S

27. Salivation occurs reflexly in response to gustatory _____ (W, as well as; N, but not) olfactory stimuli and nongustatory receptors of the oral cavity _____ (W, as well as; N, but not) the stomach and small intestine.

    a. W, W   c. N, W
    b. W, N   d. N, N

28. A _____ (W, thin, watery; K, thick, viscid) saliva from the parotid glands is elicited by stimulation of _____ (P, parasympathetic; S, sympathetic) efferents found within the _____ (F, facial; G, glossopharyngeal) nerves.

    a. W, P, F   c. W, P, G   e. K, P, F
    b. W, S, G   d. K, S, G   f. K, S, F

29. Salivation is _____ (S, strongly influenced; N, not significantly affected) by the appetite area of the _____ (H, anterior hypothalamus; T, posterior thalamus; C, somatosensory cortex).

    a. S, H   c. S, C   e. N, T
    b. S, T   d. N, H   f. N, C

OBJECTIVE 64–6.

Using the following diagram, locate the types of tubular glands in the stomach.

Directions: Match the lettered headings with the diagram and numbered list of descriptive words and phrases.

30. _____ Mucous-secreting glands that help protect the gastroesophageal junction from excoriation and attack by HCl.

31. _____ Glands that secrete digestive juices, including HCl and pepsinogen. Also called the gastric or oxyntic glandular mucosa.

32. _____ Glands that secrete mucus for the protection of the pyloric mucosa.

a. Fundic glands   b. Pyloric glands   c. Cardiac glands

OBJECTIVE 64–7.

Identify the characteristics of esophageal and gastric secretions.

33. _____ (S, only simple; C, only compound; B, both simple and compound) glands of the esophagus secrete predominantly a _____ (U, serous; M, mucoid) secretion.

a. S, U   c. B, U   e. C, M
b. C, U   d. S, M   f. B, M

34. Mucous neck cells and chief cells, _____ (W, as well as; N, but not) parietal cells, are components of the _____ (C, cardiac; F, fundic; P, pyloric) glands.

a. W, C   c. W, P   e. N, F
b. W, F   d. N, C   f. N, P

35. The relatively _____ (L, low; H, high) concentration of hydrogen ions in gastric juice is attributed to the secretion of _____ (C, chief; P, parietal; A, argentaffin) cells.

a. L, C   c. L, A   e. H, P
b. L, P   d. H, C   f. H, A

36. Acetazolamide, which inhibits _____ (ATP, ATP formation; CA, carbonic anhydrase), _____ (I, increases; D, decreases) the secretion of the principal _____ (O, organic; I, inorganic) acid of gastric secretions.

a. ATP, I, O   c. ATP, D, O   e. CA, D, O
b. ATP, I, I   d. CA, D, I   f. CA, I, I

37. A _____ (R, rise; F, fall) in pH of portal blood is associated with secretions of the intracellular system of canaliculi of the parietal cells into the _____ (G, gastric lumen; P, portal circulation; L, lymphatic system).

a. R, G   c. R, L   e. F, P
b. R, P   d. F, G   f. F, L

38. _____ (G, gastrin; P, pepsin), the principal proteolytic enzyme of gastric juice, is secreted in an _____ (A, active; I, inactive) form by _____ (R, parietal; C, chief) cells.

a. G, A, R   c. G, I, R   e. P, I, R
b. G, A, C   d. P, I, C   f. P, A, C

39. Pepsinogen is converted to an active proteolytic enzyme with an optimal pH of _____ by the direct action of _____ (P, pepsin; G, gastrin; E, enterogastrone).

a. 2, P   c. 2, E   e. 5, G
b. 2, G   d. 5, P   f. 5, E

40. Gastric enzymes which play _____ (J, major; N, minor) roles in gastrointestinal digestion include a gastric amylase _____ (W, as well as; B, but not) a lipase.

a. J, W   c. N, W
b. J, B   d. N, B

41. A more viscid and alkaline mucus is secreted by mucous cells located _____ (W, within; B, between) the gastric glands largely in response to direct _____ (A, autonomic innervation; G, circulating gastrin; M, mechanical irritation).

    a. W, A   b. W, G   c. W, M   d. B, A   e. B, G   f. B, M

OBJECTIVE 64–8.

Identify the nervous and hormonal mechanisms of regulation of gastric secretion.

42. Gastric secretion, _____ (L, like; U, unlike) salivary secretion, is regulated by _____ (N, only neural; H, only hormonal; B, both neural and hormonal) mechanisms.

    a. L, N   c. L, B   e. U, H
    b. L, H   d. U, N   f. U, B

43. _____ (I, increased; D, decreased) vagal tone to the stomach is associated with enhanced gastric secretion with a _____ (G, greater; L, lesser) concentration of pepsin than that elicited by the gastrin mechanism.

    a. I, G   c. D, G
    b. I, L   d. D, L

44. _____ (P, parasympathetic; S, sympathetic) stimulation increases gastric secretion via _____ (N, only direct neural; G, only gastrin; B, both direct neural and gastrin) mechanisms.

    a. P, N   c. P, B   e. S, G
    b. P, G   d. S, N   f. S, B

45. Gastric distention _____ (W, as well as; N, but not) secretagogues _____ (I, increase(s); D, decrease(s)) gastrin secretion by means of _____ (M, myenteric reflexes; A, a direct action).

    a. W, I, M   c. W, D, M   e. N, D, M
    b. W, I, A   d. N, D, A   f. N, I, A

46. Gastrin is secreted largely by the mucosa of the _____ (CF, corpus and fundus; A, antrum) and results in a greater secretion of _____ (C, chief; M, mucous neck; P, parietal) cells.

    a. CF, C   c. CF, P   e. A, M
    b. CF, M   d. A, C   f. A, P

47. Vagal stimulation, when compared with the gastrin mechanism, results in a _____ volume rate of gastric secretion which continues for a _____ duration. (G, greater; L, lesser)

    a. G, G   c. L, G
    b. G, L   d. L, L

48. Histamine _____ (I, increases; D, decreases) gastric secretion in a manner similar to that of _____ (E, enterogastrone; G, gastrin; P, pepsin).

    a. I, E   c. I, P   e. D, G
    b. I, G   d. D, E   f. D, P

49. The maximal secretory response to histamine would correlate well with the total number of secreting _____ (C, chief; A, argentaffin; P, parietal) cells in the _____ (G, gastric; I, intestinal) mucosa.

    a. C, G   c. A, G   e. P, G
    b. C, I   d. A, I   f. P, I

50. Gastrin and _____ (H, histamine; C, cholecystokinin) are hormones with a similar _____ (PP, polypeptide; S, steroid; GP, glycoprotein) structure.

    a. H, PP   c. H, GP   e. C, S
    b. H, S   d. C, PP   f. C, GP

51. A _____ (F, fall; R, rise) in gastric pH beyond a level of about _____ results in _____ (I, increased; D, decreased) gastric secretion and inhibition of gastric secretion.

    a. F, 2, I   c. F, 5, D   e. R, 5, I
    b. F, 2, D   d. R, 5, D   f. R, 2, I

OBJECTIVE 64-9.

Identify the principal phases of gastric secretion, their underlying mechanisms, their characteristics, and their functional significance.

52. The cephalic phase of gastric secretion generally begins _____ (P, prior to; O, at the onset of) food entry into the stomach and accounts for _____ (L, significantly less than; A, about; S, significantly more than) one-third of the total gastric secretion associated with food ingestion.

    a. P, L   c. P, S   e. O, A
    b. P, A   d. O, L   f. O, S

53. The cephalic phase of gastric secretion _____ significantly altered by varied conditions of appetite and _____ blocked by bilateral section of the vagi. (I, is; N, is not)

    a. I, I   c. N, I
    b. I, N   d. N, N

54. A higher percentage of a total daily gastric secretion of about _____ liter(s) is contributed during the _____ (I, interdigestive period; G, gastric phase; T, intestinal phase).

    a. 2, I   c. 2, T   e. 5, G
    b. 2, G   d. 5, I   f. 5, T

55. The gastric phase of gastric secretion is mediated by _____ (M, local myenteric; C, central nervous system) centers _____ (W, as well as; N, but not) gastrin.

    a. Only M, W   d. Only M, N
    b. Only C, W   e. Only C, N
    c. Both M & C, W   f. Both M & C, N

56. A phase of gastric secretion which probably accounts for _____ (M, more; S, less) than 10% of the total acid secretion by the stomach and which persists for a longer time interval than any other phase is the _____ (C, cephalic; G, gastric; I, intestinal) phase.

    a. M, C   c. M, I   e. S, G
    b. M, G   d. S, C   f. S, I

57. The enterogastric reflex _____ (W, as well as; N, but not) enteric hormones act to _____ (I, increase; D, decrease) the magnitude of the gastric phase of secretion and to _____ (I, increase; D, decrease) gastric motility.

    a. W, I, D   c. W, D, I   e. N, I, I
    b. W, D, D   d. N, D, I   f. N, I, D

58. Cholecystokinin, _____ (W, as well as; N, but not) secretin, exert(s) an inhibitory effect upon gastric secretion when gastrin secretion is at a relatively _____ (H, high; L, low) level.

    a. W, H   c. N, H
    b. W, L   d. N, L

59. Gastric secretion during the interdigestive period is characterized by a relatively higher _____ (P, pepsin; A, acid; M, mucus) content and a secretion rate typically _____ (L, less; G, greater) than 10 ml. per hour.

    a. P, L   c. M, L   e. A, G
    b. A, L   d. P, G   f. M, G

OBJECTIVE 64-10.

Identify the components of pancreatic secretion, their characteristics, and their functional significance.

60. A secretion described as a colorless, odorless, alkaline fluid of low viscosity, tasting strongly of sodium bicarbonate, and having a pH between 8.0 and 8.3, would more probably be that of the:

    a. Large intestine   d. Gallbladder
    b. Gastric glands   e. Submandibular gland
    c. Pancreas   f. Brunner's glands

61. Pancreatic secretions contain enzymes for digesting _____ of the major classes of food stuffs, for which enzymes hydrolyzing _____ of these major classes of foodstuffs are secreted largely as inactive proenzymes. (O, only one; T, only two; A, all three)

    a. O, O   c. T, T   e. A, T
    b. T, O   d. A, O   f. A, A

62. Pancreatic enzymes include enzymes which hydrolyze partially digested _____ (W, as well as; N, but not) intact protein molecules, _____ (A, and; N, but not) enzymes which hydrolyze nucleic acids.

    a. W, A   c. N, A
    b. W, N   d. N, N

63. An enzyme present in secretions of the small intestine but absent from pancreatic secretions is:

    a. Carboxypolypeptidase   d. Amylase
    b. Cholesterol esterase   e. Enterokinase
    c. Lipase   f. Ribonuclease

64. The secretion of _____ (T, trypsin; C, chymotrypsin; X, carboxypolypeptidase), largely responsible for the activation of other pancreatic proteolytic enzymes, is accompanied by the pancreatic secretion of a specific enzyme _____ (A, activator; I, inhibitor).

    a. T, A   c. X, A   e. C, I
    b. C, A   d. T, I   f. X, I

65. _____ (E, enterokinase; V, villikinin), secreted by the intestinal mucosa, is a more specific _____ (A, activating; I, inactivating) enzyme for _____ (T, trypsin; C, chymotrypsin).

    a. E, A, T   c. E, I, T   e. V, I, T
    b. E, A, C   d. V, I, C   f. V, A, C

66. Pancreatic secretion contains higher bicarbonate ion concentrations at _____ rates of flow, and salivary secretions contain higher bicarbonate ion concentrations at _____ rates of flow. (H, high; L, low)

    a. H, H   c. L, H
    b. H, L   d. L, L

67. Bicarbonate ion concentrations of pancreatic juice _____ (D, do; N, do not) exceed plasma concentrations and reflect transport characteristics of the pancreatic _____ (U, ductules; A, acini).

    a. D, U   c. N, U
    b. D, A   d. N, A

OBJECTIVE 64–11.

Identify the regulatory mechanisms of pancreatic secretion, their characteristics, and their significance.

68. Regulatory mechanisms for pancreatic secretion, _____ (L, like; U, unlike) those for gastric secretion, involve _____ (N, only neural; H, only hormonal; B, both neural and hormonal) mechanisms.

    a. L, N   c. L, B   e. U, H
    b. L, H   d. U, N   f. U, B

69. The cephalic phase of gastric secretions is associated with a _____ (S, scanty; C, copious) secretion of pancreatic juices containing a relatively _____ (H, high; L, low) concentration of pancreatic enzymes.

    a. S, H   c. C, H
    b. S, L   d. C, L

70. Vagal stimulation, like the action of _____ (S, secretin; C, cholecystokinin), results in _____ (H, a "hydrelatic"; E, an "ecbolic") secretion of the pancreas.

    a. S, H   c. C, H
    b. S, E   d. C, E

71. Secretin is _____ (H, a hormone; E, an enzyme) secreted in an inactive form by the _____ (S, stomach; P, pancreas; D, duodenum).

    a. H, S   c. H, D   e. E, P
    b. H, P   d. E, S   f. E, D

72. The most effective stimulus for secretion of the _____ (S, steroid; P, polypeptide) secretin is the presence of high concentrations of _____ (N, pepsin; H, hydrogen ions; L, lipids).

    a. S, N   c. S, L   e. P, H
    b. S, H   d. P, N   f. P, L

73. Secretin results in a _____ (S, scanty; C, copious) flow of pancreatic secretion with a high concentration of _____ (E, enzymes; Cl⁻ chloride ions; HCO₃⁻, bicarbonate ions).

    a. S, E   c. S, HCO_3^- e. C, Cl^-
    b. S, Cl^- d. C, E   f. C, HCO_3^-

74. _____ (C, cholecystokinin; S, secretin) functions to maintain an optimal pH for pancreatic _____ (W, as well as; N, but not) gastric proteolytic enzymes within the small intestine.

    a. C, W   c. S, W
    b. C, N   d. S, N

75. Cholecystokinin, secreted by the _____ (S, small intestine; L, liver), functions to _____ (I, increase; D, decrease) pancreatic _____ (W, as well as; N, but not) gallbladder secretion.

    a. S, I, W   c. S, D, W   e. L, D, W
    b. S, I, N   d. L, D, N   f. L, I, N

76. A(n) _____ (H, hydrelatic; E, ecbolic) secretion of the pancreas, involving the hormone _____ (S, secretin; C, cholecystokinin), occurs in response to the presence of proteoses and peptones _____ (W, as well as; N, but not) intestinal lipids.

    a. H, S, W   c. H, C, N   e. E, C, W
    b. H, S, N   d. E, C, N   f. E, S, W

77. The highest percentage of a daily pancreatic secretion of about _____ ml. is attributable to the action of _____ (V, vagal stimulation; C, cholecystokinin; S, secretin).

    a. 200, V   c. 200, S   e. 1200, C
    b. 200, C   d. 1200, V   f. 1200, S

OBJECTIVE 64–12.

Identify the components of biliary secretion, their characteristics, and their functional significance.

78. Bile is secreted _____ (I, intermittently; C, continually) by the liver and contains significant amounts of _____ (S, only bile salts; E, only digestive enzymes; B, both bile salts and digestive enzymes) which contribute to the gastrointestinal digestive processes.

    a. I, S   c. I, B   e. C, E
    b. I, E   d. C, S   f. C, B

79. _____ (C, cholecystokinin; S, secretin) has a more notable effect of increasing the liver _____ (BA, formation of bile acids; BS, rate of bile secretion).

    a. C, BA   c. S, BA
    b. C, BS   d. S, BS

80. Bile salts are largely _____ (E, excreted; R, reabsorbed) by the gastrointestinal tract, and their secretion by the liver is _____ (N, not significantly affected; I, increased) by elevated plasma concentrations of bile salts.

    a. E, N   c. R, N
    b. E, I   d. R, I

81. Vagal stimulation _____ (I, increases; N, does not significantly alter; D, decreases) the rate of liver secretion of an average daily volume of bile production of about _____ ml.

    a. I, 60 to 70   d. I, 600 to 700
    b. N, 60 to 70   e. N, 600 to 700
    c. D, 60 to 70   f. D, 600 to 700

OBJECTIVE 64–13.

Identify the components of the secretions of the small intestine, their characteristics, their means of regulation, and their functional significance.

82. Brunner's glands secrete _____ (M, largely mucous; E, digestive enzymes; H, enteric hormones) into the _____ (D, duodenum; I, ileum).

    a. M, D   c. H, D   e. E, I
    b. E, D   d. M, I   f. H, I

83. Enhanced secretion of Brunner's glands occurs in response to _____ (S, sympathetic; P, parasympathetic) stimulation when the rate of gastric secretion is _____ (H, high; L, low).

    a. S, H   c. P, H
    b. S, L   d. P, L

84. Intestinal secretions by the crypts of Lieberkühn occur at a rate of about _____ ml. per day, have a pH of about _____ , and are largely _____ (R, reabsorbed; E, excreted in feces).

   a. 200, 8.5, R   d. 2000, 7, E
   b. 200, 8.5, E   e. 2000, 7, R
   c. 200, 7, E   f. 2000, 8.5, R

85. Disaccharide intolerance, more commonly occurring as a consequence of the congenital absence of lactase, involves defective synthesis by:

   a. The exocrine pancreas   e. The salivary
   b. The endocrine pancreas   glands
   c. The liver   f. The duodenal
   d. Brunner's gland   mucosa

86. Sucrase, maltase, and isomaltase, _____ (W, as well as; N, but not) an intestinal lipase, are associated with the _____ (D, cellular debris; F, pure fluid) fraction of intestinal secretions.

   a. W, D   c. N, D
   b. W, F   d. N, F

87. The migration of epithelial cells towards the _____ (T, tips of the villi; I, inner aspect of the crypts of Lieberkühn) is associated with an approximate _____ -day life cycle of intestinal epithelial cells.

   a. T, 1   c. T, 25   e. I, 5
   b. T, 5   d. I, 1   f. I, 25

88. The greater the volume of chyme in the small intestine, the _____ (L, less; G, greater) is the secretion of the small intestine as a consequence of reflexes integrated largely within the _____ (H, hypothalamus; S, spinal cord; I, intramural plexus).

   a. L, H   c. L, I   e. G, S
   b. L, S   d. G, H   f. G, I

89. Small intestinal secretions are largely attributed to _____ (M, mechanical distention; C, chemical secretagogues) and mechanisms of _____ (L, local; C, central neural; H, hormonal) regulation.

   a. M; L   c. M, H   e. C, C
   b. M, C   d. C, L   f. C, H

90. _____ (V, villikinin; E, enterocrinin; K, enterokinase) is a hormone which reportedly _____ (I, increases; D, decreases) secretions of the small intestine.

   a. V, I   c. K, I   e. E, D
   b. E, I   d. V, D   f. K, D

OBJECTIVE 64–14.

Characterize the secretions of the large intestine and their means of regulation.

91. Crypts of Lieberkühn of the large intestine, containing _____ (L, lesser; S, about the same; G, greater) numbers of goblet cells when compared to those of the small intestine, secrete a fluid which _____ (C, contains; K, lacks) significant digestive enzymes.

   a. L, C   c. G, C   e. S, K
   b. S, C   d. L, K   f. G, K

92. Secretions of the large intestine are largely a _____ (M, mucous; S, serous), _____ (C, acidic; K, alkaline) fluid serving _____ (D, digestive; P, protective) functions.

   a. M, C, P   c. M, K, P   e. S, C, P
   b. M, K, D   d. S, K, D   f. S, C, D

93. Stimulation of the nervi erigentes results in _____ (I, increased; D, decreased) motility and _____ (I, increased; N, no significant change in; D, decreased) mucus secretion of the large intestine.

   a. I, I   c. I, D   e. D, N
   b. I, N   d. D, I   f. D, D

94. A copious secretion of a serous fluid by the large intestine is associated with _____ (P, parasympathetic stimulation; S, sympathetic stimulation; I, intense irritation) and _____ (D, diminished; E, enhanced) motility of the colon.

   a. P, D   c. I, D   e. S, E
   b. S, D   d. P, E   f. I, E

65

Digestion and Absorption in the Gastrointestinal Tract

OBJECTIVE 65–1.

Using the following diagrams, locate the various parts of the villus and mucosal epithelial cells.

Directions: Match the lettered headings with the diagrams and numbered list of descriptive words and phrases.

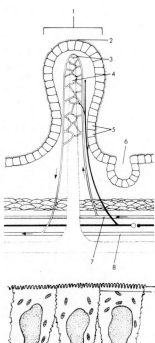

1. _____ One of the many finger-like projections on the intestinal mucosa.

2. _____ Small projections on the villi, about 1 micron in length and 0.1 micron in diameter.

3. _____ Minute, thin-walled blood vessels between arteries and veins.

4. _____ A single blind-ended lymph vessel occupying the center of each villus.

5. _____ Cells making up the intestinal mucosa.

6. _____ Depressions in the intestinal mucosa associated with epithelial cell production.

7. _____ Type of blood vessel that takes blood away from the villus.

8. _____ Type of blood vessel that moves blood toward the villus.

9. _____ Name given to the fuzzy appearance of numerous microvilli on the luminal border of epithelial cells.

10. _____ A junction between the epithelial cells that is impermeable to fluid movement.

11. _____ Area between the epithelial cells of the intestinal mucosa.

12. _____ Membrane towards the serosal border of epithelial cells of the intestinal mucosa.

a. Intercellular space
b. Tight junction
c. Crypts of Lieberkühn
d. Central lacteal

e. Blood capillaries
f. Artery
g. Vein
h. Brush border

i. Basement membrane
j. Villus
k. Microvilli
l. Epithelial cells

OBJECTIVE 65–2.

Recognize the importance of digestive and absorptive processes of the gastrointestinal tract, and identify the basic process of digestion.

13. Carbohydrates _____ (W, as well as; BN, but not) fats and proteins _____ (A, are; AN, are not) generally absorbed from the gastrointestinal tract in their natural forms.

   a. W, A   c. BN, A
   b. W, AN   d. BN, AN

14. The basic process of digestion of fats and proteins _____ (W, as well as; N, but not) carbohydrates is that of _____ (H, hydrogenation; S, hydrolysis; C, chelation).

   a. W, H   c. W, C   e. N, S
   b. W, S   d. N, H   f. N, C

15. Digestion of neutral fat by digestive enzymses of the gastrointestinal tract involves the formation of only _____ (CO_2, carbon dioxide; W, water; G, glycerol; F, fatty acids; P, peptides).

   a. CO_2 & W   c. F, P, & W   e. G & F
   b. F & P   d. F & CO_2 f. G, F, & W

16. _____ (A, all; M, most, but not all) of the digestive enzymes of the gastrointestinal tract are _____ (S, steroids; P, proteins; C, conjugated lipids).

   a. A, S   c. A, C   e. M, P
   b. A, P   d. M, S   f. M, C

OBJECTIVE 65–3.

Identify the characteristics of carbohydrate digestion by the gastrointestinal system.

17. The three major dietary forms of carbohydrate are the polysaccharide(s) _____ and the disaccharide(s) _____ . (G, glycogen; C, cellulose; T, starch; S, sucrose; L, lactose; A, galactose)

   a. G, S & L   c. T, S & A   e. T, S & L
   b. G & T, S   d. C & T, L   f. G, S & A

18. Carbohydrates which are not digested within the human digestive system include:

   a. Alcohol   c. Dextins   e. Glycogen
   b. Pectins   d. Cellulose   f. Starches

19. Chewing a piece of bread for several minutes makes the bread taste sweet due to _____ (G, glucose; M, maltose) produced by the enzyme _____ (A, maltose; AA, alpha-amylase; BA, beta-amylase).

   a. G, A   c. G, BA
   b. M, AA   d. M, A

20. Upon entry of the bolus of food into the stomach, digestion of starch by salivary amylase:

   a. Ceases immediately
   b. Continues for a few minutes
   c. Continues for an hour or two
   d. Increases because of the lower pH

21. Salivary amylase is secreted primarily by the _____ (P, parotid; SM, submaxillary; SL, sublingual) gland, and may account for up to _____ % of all polysaccharide hydrolysis in the gastrointestinal tract.

   a. P, 40   c. SL, 10   e. SM, 40
   b. SM, 30   d. P, 100   f. SL, 40

22. The _____ (L, low; H, high) pH of gastric juices is responsible for a _____ (N, minor; J, major) extent of hydrolysis of starches _____ (W, as well as; B, but not) disaccharides.

   a. L, N, W   c. L, J, W   e. H, J, W
   b. L, N, B   d. H, J. B   f. H, N, B

23. A major conversion of starches into _____ (G, glucose; M, maltose and isomaltose) occurs via the specific action of a _____ (P, pancreatic; L, liver) enzyme.

   a. G, P   c. M, P
   b. G, L   d. M, L

24. The major source of maltase and isomaltase _____ (W, as well as; N, but not) lactase and sucrase is from _____ (P, the pancreas; E, epithelial cells of the small intestine).

   a. W, P   c. N, P
   b. W, E   d. N, E

25. Carbohydrate absorption is largely in the form of _____ (M, monosaccharides; D, disaccharides), of which the most abundant form for the average diet is _____ (T, maltose; S, sucrose; G, glucose; F, fructose; L, lactose).

   a. M, T   b. M, G   c. M, F   d. D, T   e. D, S   f. D, L

OBJECTIVE 65-4.
Identify the characteristics of fat digestion by the gastrointestinal system.

26. _____ (P, phospholipids; C, cholesterol derivatives; N, neutral fats) are the most common type of dietary fat, digested primarily within the _____ (S, stomach; I, small intestine).

   a. P, S   c. N, S   e. C, I
   b. C, S   d. P, I   f. N, I

27. Regulation of gallbladder emptying, through the action of _____ (C, cholecystokinin; V, villikinin; S, secretin), is functionally significant as bile _____ (P, pigments; T, salts) assist in the digestion and absorption of fats.

   a. C, P   c. V, P   e. S, P
   b. C, T   d. V, T   f. S, T

28. A _____ (L, lipid; W, water)-soluble lipase is found in _____ (B, bile; P, pancreatic secretions).

   a. L, B   c. W, B
   b. L, P   d. W, P

29. Bile salts are soluble in _____ (W, only water; L, only lipid; B, both water and lipid) solvents and function to _____ (A, activate; S, increase the surface area for the action of) lipase.

   a. W, A   c. B, A   e. L, S
   b. L, A   d. W, S   f. B, S

30. A loss of _____ (G, gastric; E, enteric; P, pancreatic) lipase, normally the major enzyme for lipid digestion, _____ (D, does; N, does not) significantly alter the total extent of lipid digestion and absorption.

   a. G, D   c. P, D   e. E, N
   b. E, D   d. G, N   f. P, N

31. _____ (C, cholymicra; M, micelles) containing _____ (P, phospholipids and cholesterol; S, bile salts) facilitate lipid absorption by serving as a transport mechanism to the brush border of intestinal mucosal cells.

   a. C, P   c. M, P
   b. C, S   d. M, S

32. Absorption of dietary forms of cholesterol is dependent upon a _____ (B, biliary; P, pancreatic) cholesterol esterase, _____ (W, as well as; N, but not) the presence of bile salts.

   a. B, W   c. P, W
   b. B, N   d. P, N

OBJECTIVE 65-5.
Identify the characteristics of protein digestion by the gastrointestinal system.

33. One of the most important features of protein digestion in the stomach involves the digestion of:

   a. Small polypeptides
   b. Dipeptides
   c. Collagen
   d. Proteoses
   e. Peptones
   f. No protein or products

34. _____ (G, gastrin; T, trypsin; P, pepsin), secreted by the stomach, has optimal enzymatic activity _____ (A, above; B, below) a pH of 5.

   a. G, A   c. P, A   e. T, B
   b. T, A   d. G, B   f. P, B

35. Protein digestion in the gastrointestinal tract occurs through the combined actions of the enzymes _____ . (T, trypsin; A, amylase; CT, chymotrypsin; CP, carboxypolypeptidase; G, gastrin; P, pepsin; AP, aminopolypeptidase; DP, dipeptidases; TA, transaminase)

a. A, CT, CP, G, AP & DP
b. T, CP, P, AP, DP & TA
c. CT, G, P, AP & DP
d. T, CT, CP, G, P, AP, DP & TA
e. T, CT, CP, P, AP & DP
f. CT, CP, G, DP & TA

36. The end-products of protein digestion by trypsin, chymotrypsin, and carboxypolypeptidase, secreted by the _____ (S, stomach; P, pancreas; I, small intestine), are largely _____ (PP, dipeptides and small polypeptides; AA, free amino acids).

a. S, PP   c. I, PP   e. P, AA
b. P, PP   d. S, AA   f. I, AA

37. Aminopolypeptidases and dipeptidases, largely secreted by the _____ (S, stomach; P, pancreas; T, small intestine), are necessary for the _____ (I, initial; F, final) stages of protein digestion to amino acids.

a. S, I   c. T, I   e. P, F
b. P, I   d. S, F   f. T, F

OBJECTIVE 65–6.
Identify the basic characteristics of gastrointestinal absorption.

38. A volume of fluid which represents about _____ % of the extracellular fluid volume is absorbed daily, largely from the _____ (S, stomach; SI, small intestine; LA, large intestine).

a. 20, S   c. 100, LA   e. 200, SI
b. 60, SI   d. 5, S   f. 20, LA

39. Valvulae conniventes are intestinal mucosal _____ (V, valves; F, folds) which are more prominent in the _____ (D, duodenum; I, ileum) and function to increase the absorptive _____ (S, surface area; T, time interval).

a. V, D, S   c. V, I, S   e. F, D, S
b. V, I, T   d. F, I, T   f. F, D, T

40. Functions of the stomach include a comparatively _____ (G, good; P, poor) absorptive area for water and electrolytes and a more significant role in its direct absorption of _____ (V, vitamin B_{12}; A, alcohol and some drugs; CP, carbohydrate and protein).

a. G, V   c. G, CP   e. P, A
b. G, A   d. P, V   f. P, CP

41. Villi, more abundantly located in the _____ (U, upper; L, lower) portion of the _____ (S, small; G, large) intestine, have a greater function to provide _____ (C, ciliary motility; A, altered surface area).

a. U, S, C   c. U, G, C   e. L, G, C
b. U, S, A   d. L, G, A   f. L, S, A

42. Absorptive mechanisms of the gastrointestinal system of the adult include passive _____ (W, as well as; B, but not) active molecular transport mechanisms, and _____ (N, pinocytosis; G, phagocytosis).

a. W, N   c. B, N
b. W, G   d. B, G

43. The _____ (S, small; L, large) intestine has the greatest absorptive capacity for _____ (C, carbohydrates; D, lipids; P, proteins).

a. S, C   c. S, P   e. L, D
b. S, D   d. L, C   f. L, P

OBJECTIVE 65–7.

Identify the mechanisms and general characteristics of the absorption of water and electrolytes by the small intestine.

44. Water moves through the intestinal mucosa primarily by processes of _____ (S, simple diffusion; F, facilitated diffusion; A, active transport) to maintain _____ (O, hypotonic; I, isotonic; E, hypertonic) osmotic conditions within the succus entericus.

   a. S, E   c. A, O   e. F, O
   b. F, I   d. S, I   f. A, E

45. Water transport by segments of the intestinal mucosa is a _____ (U, unidirectional; B, bidirectional) mechanism which is _____ (I, increased; N, not significantly altered; D, decreased) by accompanying ion and nutrient absorption.

   a. U, I   c. U, D   e. B, N
   b. U, N   d. B, I   f. B, D

46. Sodium moves from the lumen of the gut into the epithelial cells by _____ and into the blood via _____. (A, active transport; F, facilitated diffusion; S, simple diffusion)

   a. A, S   c. F, A   e. S, A
   b. A, F   d. F, S   f. S, F

47. An active transport _____ (S, secretory; A, absorptive) mechanism for chloride is particularly well developed within epithelial cells lining the _____ (D, duodenum; J, jejunum; I, terminal ileum and large intestine).

   a. S, D   c. S, I   e. A, J
   b. S, J   d. A, D   f. A, I

48. Chloride absorption, occurring predominantly by means of _____ (A, active; P, passive) transport of chloride from the intestinal lumen, is associated with _____ (F, a favorable; O, an opposing) electrical gradient between the intestinal lumen and interstitial fluids.

   a. A, F   c. P, F
   b. A, O   d. P, O

49. Diarrhea-producing toxins of cholera, certain colon bacilli, and staphylococci appear to _____ (I, inhibit; S, stimulate) a tightly coupled active transport mechanism involving bicarbonate _____ (A, absorption; E, secretion) and _____ (C, chloride; N, sodium) transport.

   a. I, A, C   c. I, E, C   e. S, E, C
   b. I, A, N   d. S, E, N   f. S, A, N

50. Regulating mechanisms which adjust intestinal uptake in order to compensate for dietary deficiencies exist for:

   a. Neither Fe nor Ca   c. Ca but not Fe
   b. Fe but not Ca   d. Both Ca and Fe

51. Calcium ion absorption, particularly from the _____ (D, duodenum; L, ileum), is _____ (I, inhibited; E, enhanced) by parathyroid hormone.

   a. D, I   c. L, I
   b. D, E   d. L, E

52. Vitamin _____ plays a major regulatory role in calcium _____ (W, as well as; N, but not) iron absorption.

   a. A, W   c. D, W   e. C, N
   b. C, W   d. A, N   f. D, N

53. Magnesium ions _____ (W, as well as; B, but not) potassium ions _____ (A, are; N, are not) known to be actively transported by the intestinal mucosa.

   a. W, A   c. B, A
   b. W, N   d. B, N

OBJECTIVE 65–8.

Identify the mechanisms and general characteristics of the absorption of carbohydrates by the small intestine.

54. Intestinal transport of carbohydrates occurs largely by _____ (PD, passive diffusion; C, carrier mechanisms) involving _____ (M, monosaccharides; D, disaccharides; P, polysaccharides).

   a. PD, M   c. PD, P   e. C, D
   b. PD, D   d. C, M   f. C, P

55. The transport mechanism for carbohydrate is considered relatively _____ (N, nonspecific; S, specific), with more rapid rates of transport occurring for _____ (F, fructose; C, sucrose; G, galactose).

   a. N, F   c. N, G   e. S, C
   b. N, C   d. S, F   f. S, G

56. A common transport mechanism is thought to exist for glucose and _____ (F, fructose; G, galactose), which _____ (C, continues to operate; B, is blocked) in the absence of sodium transport mechanisms.

   a. F, C   c. G, C
   b. F, B   d. G, B

57. A "carrier drag" mechanism for glucose, _____ (W, as well as; N, but not) amino acid transport, has been suggested to explain its dependence upon _____ (F, fructose; G, galactose; Na, Na ion; Cl, Cl ion) transport.

   a. W, F   c. W, G   e. N, Na
   b. W, Na   d. N, F   f. N, Cl

58. Phlorhizin _____ (F, facilitates; I, inhibits) glucose _____ (W, as well as; N, but not) fructose transport by the intestinal mucosa.

   a. F, W   c. I, W
   b. F, N   d. I, N

59. Fructose absorption from the gastrointestinal lumen is associated with _____ absorption into portal blood and an active transport mechanism for _____ . (G, glucose; F, fructose)

   a. G, G   c. F, F
   b. G, F   d. F, G

OBJECTIVE 65–9.

Identify the mechanisms and general characteristics of the absorption of proteins by the small intestine.

60. Most proteins are absorbed as _____ (D, dipeptides; AA, amino acids) at a rate which is primarily dependent upon _____ (A, active transport; P, passive transport; G, digestion) mechanisms.

   a. D, A   c. D, G   e. AA, P
   b. D, P   d. AA, A   f. AA, G

61. Protein absorption, _____ (W, as well as; N, but not) digestion, occurs primarily within the _____ (S, small; G, large) intestine and is associated with comparatively _____ (L, low; H, high) intestinal concentrations of free amino acids.

   a. W, S, L   c. W, G, L   e. N, G, L
   b. W, S, H   d. N, G, H   f. N, S, H

62. Intestinal transport mechanisms have a greater affinity for _____ -stereoisomers of amino acids and possess a unique requirement for _____ . (B, vitamin B_{12} ; E, vitamin E; P, pyridoxal phosphate)

   a. L, B   c. L, P   e. D, E
   b. L, E   d. D, B   f. D, P

63. Operation of amino acid active transport mechanisms within the _____ (B, brush; F, intestitial fluid) border of intestinal epithelial cells _____ (I, is; N, is not) dependent upon active transport mechanisms for sodium ions.

   a. B, I   c. F, I
   b. B, N   d. F, N

64. Proline _____ (W, as well as; N, but not) hydroxyproline is transported by a carrier utilized for _____ (U, neutral amino; B, basic amino; A, acidic amino; I, imino) acids.

   a. W, B   b. W, I   c. W, U   d. N, U   e. N, A   f. N, I

OBJECTIVE 65–10.

Identify the mechanisms and characteristics of the absorption of fats by the small intestine.

65. Triglycerides and diglycerides, which are more miscible with _____ (B, bile salt micelles; M, cell membranes of the brush border), represent _____ (J, the major; N, a minor) percentage of lipid absorption by the small intestine.

   a. B, J   c. B, N

   b. M, J   d. M, N

66. Chylomicra, _____ (W, as well as; N, but not) micelles, exhibit surface _____ (O, hydrophobia; I, hydrophilia) and _____ (L, very low; M, moderate) degrees of suspension stability in aqueous solutions.

   a. W, O, L   c. W, I, L   e. N, O, L

   b. W, I, M   d. N, I, M   f. N, O, M

67. Monoglyceride and fatty acid absorption is dependent upon a "ferrying" mechanism of bile salt _____ (M, micelles; C, chylomicra) and a brush border transport mechanism which is dependent upon their _____ (L, lipid solubility characteristics; P, polar groups).

   a. M, L   c. C, L

   b. M, P   d. C, P

68. Chylomicra are synthesized within the _____ (C, central lacteal of; E, epithelial cells lining) the villi and are transported to the general circulatory system via _____ (P, portal veins; L, the lymphatic system).

   a. C, P   c. E, P

   b. C, L   d. E, L

69. A _____ (J, major; N, minor) fraction of absorbed fat, transported from the intestinal mucosa via portal veins, is composed primarily of _____ (S, shorter; L, longer) chain fatty acids.

   a. J. S   c. N, S

   b. J, L   d. N, L

70. Synthesis of_____ (B, bile salts; P, protein),_____(W, as well as; N, but not) glycerol, by mucosal epithelial cells is an important requirement for the transport of particularly _____ (S, short; L, long) chain fatty acids from the intestinal tract.

   a. B, W, S   c. B, N, S   e. P, W, S

   b. B, N, L   d. P, N, L   f. P, W, L

71. Bile salts are absorbed _____ (I, only in the initial; A, throughout all; T, only in the terminal) segments of the small intestine by means of _____ (AT, active; PT, passive) transport mechanisms.

   a. I, AT   c. T, AT   e. A, PT

   b. A, AT   d. I, PT   f. T, PT

72. The absorptive mechanism for the enterohepatic circulation of bile salts is _____ (G, considerably greater than; S, about the same as; L, considerably less than) the normal efficiency for conditions of ideal fat absorption, and approximates _____ % recovery.

   a. G, 50   c. L, 50   e. S, 95

   b. S, 50   d. G, 95   f. L, 25

OBJECTIVE 65–11.

Identify the mechanisms and characteristics of the absorption and storage functions of the large intestine.

73. The "absorbing" function of the _____ (S, small; L, large) intestine, or colon, is largely that of _____ (N, nutrient; E, electrolyte) absorption and _____ (A, active; P, passive) transport of water.

a. S, N, A   c. S, E, A   e. L, E, A
b. S, N, P   d. L, E, P   f. L, N, P

74. The proximal half of the human colon, or _____ (S, "storage"; A, "absorbing") colon, is associated with the formation of _____ (G, significant; NG, insignificant) levels of cellulose digestion.

a. S, G   c. A, G
b. S, NG   d. A, NG

75. Bacteria present in large amounts in the absorbing _____ (F, fundus; I, ileum; C, colon) constitute a significant _____ (D, drain on; S, supplement to) vitamin absorption.

a. F, D   c. C, D   e. I, S
b. I, D   d. F, S   f. C, S

76. Vitamin _____ , fulfilling special requirements for the synthesis of some of the blood coagulation factors, is normally present in ingested foods in _____ (Q, adequate; I, inadequate) amounts.

a. A, Q   c. K, Q   e. E, I
b. E, Q   d. A, I   f. K, I

77. Indole, skatole, and mercaptans are significant _____ (N, nutritional; V, volatile) components of the colon, whereas stereobilin and urobilin, derived from _____ (S, bile salts; R, bilirubin), impart a _____ (B, brown; G, green; C, clay-colored) pigment to the color of feces.

a. N, S, C   c. N, R, C   e. V, R, G
b. N, S, G   d. V, R, B   f. V, S, B

78. Feces, normally comprised of 25% _____ (W, water; S, solid matter), contains a higher percentage of _____ (F, fat; I, inorganic matter; B, dead bacteria).

a. W, F   c. W, B   e. S, I
b. W, I   d. S, F   f. S, B

66

Physiology of Gastrointestinal Disorders

OBJECTIVE 66–1.

Recognize the physiological basis and consequences of common disorders of swallowing.

1. Passage of _____ (A, air into the stomach; S, stomach contents into the lungs) is a serious danger of general anesthesia as a result of a more direct depression of the _____ (SC, swallowing center; MN, myoneural junction; MC, muscle contractile apparatus) by the action of general anesthetics.

a. A, SC   c. A, MC   e. S, MN
b. A, MN   d. S, SC   f. S, MC

2. Failure of the _____ (C, cricopharyngeal; R, cardiac; P, palatopharyngeal) sphincter to remain tonically closed during normal respiration permits air to enter the normally _____ (O, open; L, collapsed) esophagus.

a. C, O   c. P, O   e. R, L
b. R, O   d. C, L   f. P, L

3. Achalasia is a condition whereby the _____ (HP, hypopharyngeal; GE, gastroesophageal) sphincter fails to undergo _____ (C, tonic constriction; R, receptive relaxation).

    a. HP, C   c. HP, R
    b. GE, C   d. GE, R

4. Damage to the myenteric plexus of the _____ (U, upper; L, lower) region of the esophagus occurs more frequently in association with _____ (A, achalasia; P, poliomyelitis; H, achlorhydria).

    a. U, A   c. U, H   e. L, P
    b. U, P   d. L, A   f. L, H

OBJECTIVE 66–2.

Recognize the physiological basis and consequences of common disorders of the stomach.

5. Gastritis is more frequently associated with _____ (B, bacterial infections; P, peptic excoriation) of the stomach and a more significant penetration of the gastric mucosa by _____ ions.

    a. B, K   c. B, H   e. P, Ca
    b. B, Ca   d. P, K   f. P, H

6. Gastritis results more frequently in _____ (I, intensified; S, suppressed) salivation and a diffuse burning pain referred to the _____ (U, umbilicus; E, high epigastrium).

    a. I, U   c. S, U
    b. I, E   d. S, E

7. Achlorhydria is generally associated with _____ stomach acid secretion, _____ pepsin secretion, and _____ digestion of food by the overall gastrointestinal tract. (R, substantially reduced; N, almost normal)

    a. N, N, N   c. N, N, R   e. R, N, R
    b. N, R, N   d. R, R, R   f. R, R, N

8. The amount of _____ (F, free; C, combined) acid from gastric contents, as determined by titration to a pH of 3.5 using dimethylaminoazobenzene as an indicator, is normally relatively _____ (H, high; L, low) in the presence of gastric contents mixed with food.

    a. F, H   c. C, H
    b. F, L   d. C, L

9. Pernicious anemia of gastric atrophy results from abnormal gastric _____ (A, absorption; S, secretion) of the _____ (I, intrinsic; E, extrinsic) factor.

    a. A, I   c. S, I
    b. A, E   d. S, E

10. Intrinsic factor is a _____ (M, mucopolypeptide; L, lipoprotein) secreted by _____ (P, parietal cells; I, the terminal ileum; B, Brunner's glands).

    a. M, P   c. M, B   e. L, I
    b. M, I   d. L, P   f. L, B

11. Pernicious anemia, associated with gastric atrophy, is a consequence of _____ (E, excess; D, deficient) vitamin _____ absorption by the mucosa of the _____ (S, stomach; I, ileum).

    a. E, K, S   c. E, B_{12}, S   e. D, B_6, S
    b. E, B_6, I   d. D, K, I   f. D, B_{12}, I

OBJECTIVE 66–3.

Identify peptic ulcers, their locations, their underlying causes, and their treatment rationale.

12. Peptic ulcers occur more frequently along the _____ (L, lesser; G, greater) curvature of the stomach near the _____ (C, cardia; B, corpus; A, antral end).

    a. L, C   c. L, A   e. G, B
    b. L, B   d. G, C   f. G, A

13. Peptic ulcers, occurring more frequently in the _____ (P, esophagus; D, duodenum), are more generally associated with patients having _____ (N, normal or paradoxically low; E, elevated) interdigestive gastric rates of secretion.

    a. P, N   c. D, N
    b. P, E   d. D, E

14. Peptic ulceration is _____ (F, facilitated; T, treated) by mechanisms which inhibit Brunner's glands secretion or _____ (E, enhance; I, inhibit) secretion of gastric juice.

    a. F, E   c. T, E
    b. F, I   d. T, I

15. Protective mechanisms against injurious effects of gastric acids _____ peptic digestion are afforded by mucus secretion of the pyloric glands _____ the glands of the lower esophagus and upper duodenum. (W, as well as; N, but not)

    a. W, W   c. N, W
    b. W, N   d. N, N

16. Mechanisms whereby excess acid in the duodenum results in _____ (E, enhanced; R, reduced) activity of the pyloric pump afford a greater protection for the _____ (G, gastric; D, duodenal) mucosa.

    a. E, G   c. R, G
    b. E, D   d. R, D

17. Reflux of duodenal contents into the stomach would more probably serve as a _____ (S, cause of; C, cure for) gastric ulcers as a consequence of the _____ (B, bicarbonate; L, bile acid; H, hormonal; E, enzyme) content of the duodenal contents.

    a. S, L   c. S, E   e. C, H
    b. S, H   d. C, B   f. C, E

18. Neutralization of acidic contents of the duodenum involves alkaline secretion of the pancreas, _____ (W, as well as; N, but not) Brunner's glands, and the hormone _____ (P, pancreozymin; S, secretin).

    a. W, P   c. N, P
    b. W, S   d. N, S

19. Aspirin, _____ (W, as well as; N, but not) alcohol, are generally _____ (T, therapeutic agents; C, contraindicated) for patients with peptic ulcers.

    a. W, T   c. N, T
    b. W, C   d. N, C

20. More frequent and smaller meals with _____ (E, elevated; L, lowered) fat intake are generally _____ (R, recommended; C, contraindicated) for peptic ulcer patients.

    a. E, R   c. L, R
    b. E, C   d. L, C

OBJECTIVE 66-4.

Recognize the physiological bases and consequences of common disorders of the small intestine.

21. Disorders of the small intestine occur _____ (L, less; M, more) frequently than disorders of either the stomach or colon, and _____ (R, rarely; F, frequently) cause malnutrition.

a. L, R   c. M, R
b. L, F   d. M, F

22. Pancreatitis is generally associated with _____ (E, elevated; C, no significant change in; R, reduced) pancreatic secretion and more severe abnormalities of _____ (N, endocrine; X, exocrine) pancreatic function.

a. E, N c. R, N e. C, X
b. C, N d. E, X f. R, X

23. Chronic pancreatitis, _____ (L, like; U, unlike) acute pancreatitis, is more commonly the result of _____ (G, gallstone blockage; I, inflammation of the islets of Langerhans; D, deficiencies of enterokinase).

a. L, G c. L, D e. U, I
b. L, I d. U, G f. U, D

24. Wheat and rye grains have been implicated in disorders of the _____ (S, stomach; I, small intestine; L, large intestine) as a consequence of their high _____ (G, gluten; T, starch; P, phytate) content.

a. S, G c. L, T e. I, P
b. I, G d. S, T f. L, P

25. Early stages of sprue are associated with defects of _____ (D, digestive; A, absorptive) mechanisms, particularly for _____ (C, carbohydrates; L, lipids; P, proteins).

a. D, C c. D, P e. A, L
b. D, L d. A, C f. A, P

26. Tropical sprue, _____ (W, as well as; N, but not) nontropical sprue, is generally associated with _____ (D, a dietary; I, an infectious) etiology and treatment.

a. W, D   c. N, D
b. W, I   d. N, I

27. Appendicitis is an infectious process of the _____ (S, small; L, large) intestine which generally _____ (I, inhibits; E, enhances) motility of the small intestine.

a. S, I   c. L, I
b. S, E   d. L, E

OBJECTIVE 66-5.

Recognize the physiological bases and consequences of common disorders of the large intestine.

28. Constipation is associated with a _____ (R, more rapid; S, slower) movement of a _____ (D, dry, hard; L, more liquid) feces.

a. R, D   c. S, D
b. R, L   d. S, L

29. Constipation occurs more frequently in the _____ (N, newborn; D, adult) as a consequence of _____ (A, an acquired; C, a congenital) condition of _____ (AT, an atonic; HT, a hypertonic) colon.

a. N, A, AT c. N, C, AT e. D, A, AT
b. N, C, HT d. D, C, HT f. D, A, HT

30. Cyclic alterations of constipation and diarrhea occur in _____ (H, Hirschsprung's disease; I, the irritable colon syndrome; C, ulcerative colitis), believed to result from _____ (P, psychogenic states; E, enteritis).

a. H, P c. C, P e. I, E
b. I, P d. H, E f. C, E

31. The most frequent cause of "megacolon" is _____ (A, an acquired; C, a congenital) absence of normal function of the _____ (B, "absorbing"; S, "storage") colon.

a. A, B   c. C, B
b. A, S   d. C, S

32. Infectious irritation of the large intestine generally results in _____ (I, increased; D, decreased) secretion of the colon and _____ (C, constipation; H, diarrhea).

   a. I, C
   b. I, H
   c. D, C
   d. D, H

33. Psychogenic diarrhea is associated with _____ (E, enhanced; D, diminished) parasympathetic activity to the large intestine and a more significant loss of _____ (C, caloric substances; W, water and electrolytes).

   a. E, C
   b. E, W
   c. D, C
   d. D, W

34. Destruction of the _____ (T, thoracic; L, lumbar; S, sacral) spinal cord center of the defecation reflex spares a persistent and _____ (A, adequate; I, inadequate) intrinsic defecation reflex of the myenteric plexus.

   a. T, A
   b. L, A
   c. S, A
   d. T, I
   e. L, I
   f. S, I

OBJECTIVE 66–6.
Identify the vomiting reflex, its characteristics and significance, and its common causes.

35. Vomiting is the primary means by which the _____ (U, upper; E, entire) gastrointestinal tract rids itself of its contents in response to irritation or overdistention arising particularly within the _____ (S, stomach and duodenum; I, jejunum and ileum; C, colon).

   a. U, S
   b. U, I
   c. U, C
   d. E, S
   e. E, I
   f. E, C

36. The vomiting reflex involves a "vomiting center" of the _____ (C, spinal cord; M, medulla; SM, somatosensory cortex) and significant efferent activity transmitted via _____ (A, only autonomic; S, only somatic; B, both autonomic and somatic) efferents.

   a. C, S
   b. M, B
   c. SM, S
   d. C, B
   e. M, A
   f. SM, B

37. The esophagus is relaxed throughout, the lower esophageal sphincter is relaxed, the glottis is closed, respiration is inhibited, and the larynx and hyoid bone are drawn upward and forward and held rigidly in this position during:

   a. The 1st stage of deglutition
   b. The 2nd stage of deglutition
   c. The 3rd stage of deglutition
   d. Vomiting
   e. The Valsalva maneuver
   f. All of the above

38. The propelling force for the vomiting act arises largely from contraction of _____ (M, smooth; K, skeletal) muscles innervated largely by _____ (C, cranial; S, spinal) nerves.

   a. M, C
   b. M, S
   c. K, C
   d. K, S

39. Apomorphine and certain digitalis derivatives are included among drugs which act to _____ (I, inhibit; N, initiate) vomiting through their action upon _____ (P, peripheral; C, central) chemoreceptors.

   a. I, P
   b. I, C
   c. N, P
   d. N, C

40. Vomiting associated with motion sickness is primarily the result of _____ (GI, gastrointestinal tract; V, vestibular) receptors and _____ (R, cerebral; B, cerebellar) connections with the vomiting center.

   a. GI, R
   b. GI, B
   c. V, R
   d. V, B

41. A prodromal sensation of nausea, resulting from duodenal irritation or distention, is frequently associated with simultaneous _____ of the duodenum and _____ of the stomach. (C, contraction; R, relaxation)

   a. C, C
   b. C, R
   c. R, C
   d. R, R

42. Obstruction of the gut results in a local _____ (I, increase; D, decrease) in fluid secretion to absorption ratios, and a higher degree of severe vomiting as well as a metabolic _____ (C, acidosis; K, alkalosis) with obstruction of the _____ (P, pylorus; S, storage colon).

   a. I, C, P
   b. I, K, S
   c. I, K, P
   d. D, K, S
   e. D, C, P
   f. D, C, S

OBJECTIVE 66–7.
Identify the origins and characteristics of gases in the gastrointestinal tract.

43. Gas which is discharged from the stomach into the esophagus and mouth following a meal — i.e., belching — normally consists of:

    a. Stomach fermentative gases
    b. Small intestine gases
    c. Large intestine gases
    d. Swallowed air
    e. Trapped esophageal air
    f. Gastric air displaced by foods

44. The highest percentage of methane and hydrogen in intestinal gases is found in conjunction with _____ (S, slow; R, rapid) rates of movements of gases within the _____ (L, small; G, large) intestine.

    a. S, L   c. R, L
    b. S, G   d. R, G

45. The majority of gases which enter or are formed within the intestinal tract are _____ (A, absorbed; E, expelled as flatus). A higher percentage of gas absorption within the gastrointestinal tract occurs within the _____ (S, stomach; M, small intestine; L, large intestine).

    a. A, S   c. A, L   e. E, M
    b. A, M   d. E, S   f. E, L

67

Metabolism of Carbohydrates and Formation of Adenosine Triphosphate

OBJECTIVE 67–1.
Recognize the significance of the release of energy from foodstuffs, the concept of free "energy," and the role of adenosine triphosphate (ATP) in metabolism.

1. The release of large amounts of energy for biological functions is associated with the process of _____ (O, oxidation; R, reduction) of carbohydrates and proteins, _____ (W, as well as; N, but not) fats.

    a. O, W   c. R, W
    b. O, N   d. R, N

2. Energy is made available for physiological processes of cells largely in the form of _____ (H, heat energy; B, high energy bonds) resulting from _____ (C, a "coupling" of chemical reactions; W, water hydrolysis; D, the direct hydrogenation of foodstuffs).

    a. H, C   c. H, D   e. B, W
    b. H, W   d. B, C   f. B, D

3. The amount of energy liberated by the complete oxidation of 1 mol of glucose or_____ grams is_____kilocalories.

    a. 18, 180   c. 18, 8000   e. 180, 686
    b. 18, 686   d. 180, 7   f. 180, 8000

4. About _____ % of the energy required for physiological mechanisms is obtained directly from _____ (P, phosphorylation; D, dephosphorylation) of _____ (G, glucose; ADP, adenosine diphosphate; ATP, adenosine triphosphate).

    a. 100, P, ADP   d. 50, D, ATP
    b. 100, D, G   e. 50, P, G
    c. 100, D, ATP   f. 50, P, ADP

5. ATP is comprised of adenine, _____ (D, deoxyribose; R, ribose), and three phosphate radicals; _____ of these radicals has/have so-called high energy bonds containing _____ kilocalories per mol.

    a. D, only 1, 8   b. D, only 1, 7000   c. D, 2, 7000   d. R, 2, 7000   e. R, 2, 8   f. R, only 1, 8

OBJECTIVE 67–2.
Identify the mechanisms of monosaccharide membrane transport and phosphorylation by cells, and their significance.

6. Monosaccharides, _____ (W, as well as; N, but not) disaccharides, are transported through the _____ (P, aqueous pores; M, matrix) of the membrane largely by _____ (S, simple; F, facilitated) diffusion.

    a. W, P, S   c. W, M, S   e. N, M, S
    b. W, P, F   d. N, M, F   f. N, P, F

7. The rate of monosaccharide transport by cell membranes is _____ (I, independent of; P, approximately proportional to) the transmembrane concentration difference of monosaccharides and involves a carrier which transports monosaccharides _____ (O, in only one direction; B, in both directions).

    a. I, O   c. P, O
    b. I, B   d. P, B

8. The intracellular transport of monosaccharides, mainly _____ (L, galactose; G, glucose; F, fructose), is regulated largely by the hormone _____ (C, glucagon; H, hexokinase; I, insulin).

    a. L, H   c. F, I   e. G, I
    b. G, C   d. L, C   f. F, H

9. _____ (M, minute; S, substantial) amounts of disaccharides, which are absorbed from the gastrointestinal tract, are largely _____ (C, converted to glucose in the liver; E, excreted in the kidneys; T, transported by most cells by pinocytosis).

    a. M, C   c. M, T   e. S, E
    b. M, E   d. S, C   f. S, T

10. Insulin produced by the _____ (L, liver; P, pancreas) provides a regulatory mechanism for _____ (A, all; S, most, but not all) cells of the body by _____ (I, increasing; D, decreasing) glucose uptake by cells.

    a. L, A, I   c. L, S, D   e. P, S, I
    b. L, A, D   d. P, S, D   f. P, A, I

11. Hexokinases are _____ (S, specific; N, nonspecific) enzymes for glucose, fructose, and galactose _____ (D, only dephosphorylation; P, only phosphorylation; B, both D and P).

    a. S, D   c. S, B   e. N, P
    b. S, P   d. N, D   f. N, B

12. Liver cells, the renal tubular epithelium, and intestinal epithelial cells are among a _____ (N, minority; J, majority) of cell types which _____ (P, possess; L, lack) specific phosphatases and _____ (P, possess; L, lack) hexokinases.

    a. N, P, P   c. N, L, P   e. J, L, P
    b. N, P, L   d. J, L, L   f. J, P, L

13. The pores of cell membranes _____ (W, as well as; N, but not) the membrane matrix are relatively _____ (P, permeable; I, impermeable) to phosphorylated intermediates of metabolism.

    a. W, P   c. N, P
    b. W, I   d. N, I

14. An 18-hour fast depletes mammalian liver of glycogen but reduces the concentration of glycogen in muscle only slightly. This may be partially explained by the fact that muscle, unlike liver, lacks:

    a. Phosphorylase
    b. Creatine phosphate
    c. Creatinine
    d. Glucose 6-phosphatase
    e. Anaerobic glycolysis
    f. Glucokinase

15. _____ (G, galactose; F, fructose) usually must be converted to glucose by the _____ (P, pancreas; L, liver; E, enzymes of plasma) before its general utilization by body tissues.

    a. G, P   c. G, E   e. F, L
    b. G, L   d. F, P   f. F, E

OBJECTIVE 67–3.

Identify glycogen, glycogenesis, and glycogenolysis, their significance, and the regulatory role of phosphorylase.

16. Carbohydrate storage of mammalian cells occurs largely as _____ (D, dissolved; S, solid granules of) polymerized monosaccharide in the form of _____ (H, starch; G, glycogen; C, cellulose).

    a. D, H   c. D, C   e. S, G
    b. D, G   d. S, H   f. S, C

17. _____ (E, glycogenesis; Y, glycogenolysis), or the breakdown of glycogen, is accomplished largely by splitting _____ (R, ribose; G, glucose) units from the glycogen polymer through the process of _____ (P, phosphorylation; D, dephosphorylation).

    a. E, R, P   c. E, G, P   e. Y, G, P
    b. E, R, D   d. Y, G, D   f. Y, R, D

18. Uridine diphosphate glucose is formed more directly from _____ (G, glycogen; O, glucose 1-phosphate; S, glucose 6-phosphate) during _____ (Y, glycolysis; E, glycogenesis; NY, glycogenolysis).

    a. G, Y   c. S, Y   e. O, NY
    b. O, E   d. G, NY   f. S, E

19. _____ (A, activation; I, inactivation) of phosphorylase by epinephrine _____ (W, as well as; N, but not) glucagon tends to favor the process of _____ (E, glycogenesis; Y, glycogenolysis).

    a. A, W, E   c. A, N, Y   e. I, N, E
    b. A, W, Y   d. I, N, Y   f. I, W, E

20. Glucagon, secreted by the _____ (P, pancreas; L, liver), _____ (I, increases; D, decreases) blood glucose concentrations by a mechanism involving _____ (S, stimulation; H, inhibition) of the formation of cyclic adenylate.

    a. P, I, S   c. P, D, S   e. L, D, S
    b. P, I, H   d. L, D, H   f. L, I, H

21. Stimulation of the _____ (S, sympathetic; P, parasympathetic) innervation of the adrenal medulla tends to _____ (E, elevate; R, reduce) blood glucose concentrations as a consequence of alterations of glycogen stores of _____ (M, skeletal muscle; L, the liver).

    a. S, E, M   c. S, R, L   e. P, R, M
    b. S, E, L   d. P, R, L   f. P, E, M

OBJECTIVE 67–4.

Identify the general mechanisms and characteristics of the release of energy from glucose by the glycolytic pathway, the conversion of pyruvate to acetyl coenzyme A, and the citric acid cycle.

22. Complete oxidation of 1 gram-mol of glucose to carbon dioxide and _____ (U, urea; W, water) involves the release of _____ kilocalories of energy, most of which is released during _____ (G, glycolysis; C, the citric acid cycle).

    a. U, 180, G   c. U, 686, C   e. W, 686, G
    b. U, 180, C   d. W, 686, C   f. W, 180, G

23. Glycolysis involves the formation of _____ (G, glycogen; P, pyruvate; C, complete oxidation products) from glucose during which most of the released energy is _____ (R, retained as ATP; L, lost as heat) energy.

    a. G, R   c. C, R   e. P, L
    b. P, R   d. G, L   f. C, L

24. The conversion of 1 mol of glucose to fructose 1,6-phosphate _____ (R, requires; P, produces) _____ mols of ATP, while the conversion of 1 mol of fructose 1,6-phosphate to pyruvate produces _____ mols of ATP plus 4 mols of hydrogen atoms.

    a. R, 2, 2   c. R, 4, 6   e. P, 4, 4
    b. R, 2, 4   d. P, 4, 6   f. P, 2, 2

25. The conversion of 2 mols of pyruvic acid to acetyl Co-A involves the release of _____ mols of carbon dioxide, _____ mols of hydrogen atoms, and the direct formation of _____ mols of ATP.

    a. 2, 0, 0   c. 2, 4, 0   e. 4, 2, 0
    b. 2, 2, 2   d. 4, 0, 2   f. 4, 4, 2

26. Enzymes of the citric acid cycle, or the _____ (K, Krebs; P, phosphogluconate) cycle, are localized within the _____ (ER, endoplasmic reticulum; M, mitochondria; N, nucleoplasm).

    a. K, ER   c. K, N   e. P, M
    b. K, M   d. P, ER   f. P, N

27. The citric acid cycle is initiated by the combination of _____ to produce _____ . (C, citric acid; A, acetyl Co-A; K, alpha-keto-glutarate; O, oxaloacetate)

    a. C & K, O   c. C & O, K   e. K & O, C
    b. C & A, K   d. K & A, C   f. A & O, C

28. The catabolism of two molecules of acetyl Co-A in the citric acid cycle produces two molecules of coenzyme A, _____ hydrogen atoms, four carbon dioxide molecules, and the direct production of _____ molecules of ATP.

    a. 8, 6   c. 16, 0   e. 38, 2
    b. 8, 0   d. 16, 2   f. 38, 6

29. A hydrogen carrier derived from _____ (T, thiamine; N, niacin; F, flavin) functions to transport a _____ (J, majority; M, minority) of hydrogen pairs released during glycolysis _____ (W, as well as; B, but not) the citric acid cycle.

    a. T, J, W   c. F, J, B   e. N, M, B
    b. N, J, W   d. T, M, B   f. F, J, W

30. Enzymes of oxidative phosphorylation located within the _____ (M, mitochondria; F, cytoplasmic fluid) catalyze the oxidation of _____ (H, only hydrogen; C, only carbon; B, both hydrogen and carbon).

    a. M, H   c. M, B   e. F, C
    b. M, C   d. F, H   f. F, B

31. Oxidative phosphorylation, producing a _____ ratio of ATP molecules per pair of hydrogen atoms, results in the formation of _____ (M, more; L, less) than 50% of the ATP produced during the complete oxidation of glucose.

    a. 1:1, M   c. 6:1, M   e. 3:1, L
    b. 3:1, M   d. 1:1, L   f. 6:1, L

32. The complete oxidation of one molecule of glucose results in the formation of _____ molecules of ATP, representing an efficiency of about _____ %.

    a. 14, 29   c. 38, 44   e. 24, 71
    b. 24, 29   d. 14, 44   f. 38, 71

OBJECTIVE 67–5.

Identify the role of ADP availability in the regulation of glucose metabolism and the characteristics of ATP production by "anaerobic glycolysis." Recognize the conversion of excess glucose to glycogen or fat stores.

33. The regulatory factor for ATP production by glucose catabolism primarily involves the availability of:

    a. Carbon dioxide
    b. Oxaloacetate
    c. Oxygen
    d. Adenosine diphosphate
    e. NADH
    f. Glycogen stores

34. The majority of ATP production occurs within the _____ , and the majority of ATP utilization occurs within the _____ . (N, nucleus; C, cytoplasm; M, mitochondria)

    a. C, N   c. C, M   e. M, C
    b. C, C   d. M, N   f. M, M

35. The preferential saturation of cellular _____ (G, glycogen; L, lipid) stores by glucose which is not immediately required for energy purposes represents an energy reserve sufficient to supply the needs of the body for a few _____ (H, hours; D, days; W, weeks).

    a. G, H   c. G, W   e. L, D
    b. G, D   d. L, H   f. L, W

36. The formation of lactic acid from _____ (O, oxaloacetate; R, pyruvate; K, alpha-ketoglutarate) is an NADH and H^+ _____ (P, -producing; Q, -requiring) step.

    a. O, P   c. K, P   e. R, Q
    b. R, P   d. O, Q   f. K, Q

37. The _____ (A, aerobic; N, anaerobic) conversion of glucose to lactic acid produces _____ ATP molecules per glucose molecule and involves a mechanism which can continue _____ (S, for only a short period; I, indefinitely).

    a. A, 2, S   c. A, 18, S   e. N, 6, I
    b. A, 6, I   d. N, 2, S   f. N, 18, S

38. _____ (P, pyruvate; L, lactate), the metabolic _____ (S, substrate; E, accumulating end-product) of "anaerobic glycolysis," _____ (D, does; N, does not) pass readily through cell membranes.

    a. P, S, D   c. P, E, D   e. L, E, D
    b. P, S, N   d. L, E, N   f. L, S, N

39. Accumulating metabolic end-products of "anaerobic glycolysis" associated with a metabolic _____ (C, acidosis; K, alkalosis) are largely _____ (E, excreted by the kidneys; M, metabolized by cardiac tissues; R, reconverted to glucose by the liver).

    a. C, E   c. C, R   e. K, M
    b. C, M   d. K, E   f. K, R

OBJECTIVE 67–6.

Identify the mechanisms and characteristics of the release of energy from glucose by the phosphogluconate pathway.

40. Hydrogen released during the operation of the phosphogluconate pathway largely combines with _____ and provides a higher percentage of ATP availability for _____ (L, liver; C, cardiac muscle; F, fat) cells.

    a. NAD, L   c. NAD, F   e. NADP, C
    b. NAD, C   d. NADP, L   f. NADP, F

41. The phosphogluconate pathway provides three-carbon _____ (W, as well as; B, but not) five- and seven-carbon sugars, and _____ (I, is; N, is not) able to catabolize glucose completely to carbon dioxide and water.

    a. W, I   c. B, I
    b. W, N   d. B, N

42. The phosphogluconate pathway is a cyclic process for catabolizing _____ molecule(s) of glucose _____ (A, aerobically; N, anaerobically) for each revolution of the cycle.

    a. 1, A   c. 6, A   e. 2, N
    b. 2, A   d. 1, N   f. 6, N

43. The percentage of ATP produced by the phosphogluconate pathway of liver cells is generally _____ (I, increased; D, decreased) by diminished rates of glycolysis as a consequence of requirements for _____ (G, glycogen; F, fatty acid) synthesis.

    a. I, G   c. D, G
    b. I, F   d. D, F

OBJECTIVE 67–7.

Identify "gluconeogenesis," its mechanisms, its means of regulation, and its significance.

44. Gluconeogenesis involves a process of synthesizing _____ (L, lipids; P, proteins; C, carbohydrate), largely from _____ (C, carbohydrate; N, noncarbohydrate) sources.

    a. L, C   c. C, C   e. P, N
    b. P, C   d. L, N   f. C, N

45. Gluconeogenesis is enhanced by _____ (I, increased; D, decreased) availability of cellular _____ (W, as well as; N, but not) blood levels of carbohydrate substrate.

    a. I, W   c. D, W
    b. I, N   d. D, N

46. _____ (M, more; L, less) than 50% of the types of amino acids are converted into glucose by the process of _____ (G, glycolysis; N, glycogenesis; Y, glycogenolysis; O, gluconeogenesis)

   a. M, N   c. M, O   e. L, Y
   b. M, G   d. L, N   f. L, O

47. Gluconeogenesis is _____ (I, increased; D, decreased) by the action of glucocorticoid hormones secreted by the _____ (L, liver; A, adrenal glands; N, neurohypophysis).

   a. I, L   c. I, N   e. D, A
   b. I, A   d. D, L   f. D, N

48. The secretion of the principal glucocorticoid, _____ (A, aldosterone; C, cortisol; S, stilbestrol) is _____ (E, enhanced; I, inhibited) by corticotropin secretion from the _____ (N, anterior; P, posterior) pituitary.

   a. A, E, N   c. S, E, P   e. C, I, P
   b. C, E, N   d. A, I, P   f. S, I, N

49. Thyroxine is thought to _____ (S, spare; M, mobilize) cellular proteins and to _____ (I, increase; D, decrease) the overall rate of gluconeogenesis.

   a. S, I   c. M, I
   b. S, D   d. M, D

OBJECTIVE 67–8.
Identify typical concentrations of blood glucose and its influencing factors.

50. Dietary fructose is largely converted to _____ (G, glucose; T, galactose) by _____ (L, the liver; I, intestinal epithelial cells), while dietary galactose is largely converted to glucose by _____ (L, the liver; I, intestinal epithelial cells).

   a. G, L, I   c. G, I, I   e. T, L, I
   b. G, I, L   d. T, I, L   f. T, L, L

51. A typical blood glucose concentration four hours after the last meal is typically _____ mg. per cent, and reflects _____ (S, simple absorption and utilization rates; E, endocrine regulation) of glucose.

   a. 20, S   c. 180, S   e. 90, E
   b. 90, S   d. 20, E   f. 180, E

68

Lipid Metabolism

OBJECTIVE 68–1.
Identify "lipids," their principal means of storage, and the transport characteristics and functions of chylomicra.

1. Lipids include neutral fats, or _____ (P, phospholipids; T, triglycerides), _____ (W, as well as; N, but not) cholesterol.

   a. P, W   c. T, W
   b. P, N   d. T, N

2. _____ (S, short; L, long) chain fatty acids and a _____ (N, minor; J, major) fraction of absorbed cholesterol and phospholipids are transported in blood as chylomicra.

   a. S, N   c. L, N
   b. S, J   d. L, J

3. Chylomicra are synthesized in the _____ (L, liver; I, intestinal mucosa; A, adipose tissue) and have a half-life in blood of about _____ (M, 10 to 15 minutes; H, 2 to 3 hours).

   a. L, M   c. A, M   e. I, H
   b. I, M   d. L, H   f. A, H

4. Chylomicra, comprised largely of _____ (F, free fatty acids; T, triglycerides), are removed from the circulation largely by the _____ (L, liver; RE, reticuloendothelial system; A, adipose tissue).

     a. F, L   c. F, A   e. T, RE
     b. F, RE   d. T, L   f. T, A

5. Protein serves as a _____ (S, surface coating; C, central core) of chylomicra and comprises about _____ % of their mass.

     a. S, 1   c. S, 85   e. C, 7
     b. S, 7   d. C, 1   f. C, 85

6. The activity of lipoprotein lipase of adipose cellular membranes is _____ (R, reduced; A, augmented) by insulin, and normally functions to hydrolyze _____ (F, free fatty acids; N, neutral fats) _____ (W, as well as; B, but not) phospholipids.

     a. R, F, W   c. R, N, B   e. A, N, W
     b. R, F, B   d. A, N, B   f. A, F, W

7. Insulin _____ (I, increases; D, decreases) glucose transport into fat cells, _____ (I, increases; D, decreases) the rate of alpha-glycerophosphate formation by fat cells, and promotes fat _____ (B, breakdown; S, synthesis).

     a. I, I, B   c. I, D, S   e. D, D, B
     b. I, I, S   d. D, D, S   f. D, I, B

OBJECTIVE 68-2.

Identify "free fatty acids" of plasma, their transport characteristics, and conditions which influence their concentration and "turnover."

8. Fat cells provide a _____ (J, major; N, minor) fraction of body lipid storage largely in the form of _____ (FFA, free fatty acids; NF, neutral fats), and a _____ (J, major; N, minor) fraction of the body catabolism of free fatty acids.

     a. J, FFA, N   c. J, NF, J   e. N, FFA, J
     b. J, NF, N   d. N, NF, J   f. N, FFA, N

9. Activation of a cellular lipase of adipose tissue occurs in response to _____ (I, increased; D, decreased) cellular availability of glucose and _____ (H, higher; L, lower) than normal average levels of cellular alpha-glycerophosphate.

     a. I, H   c. D, H
     b. I, L   d. D, L

10. "Free fatty acids," or " _____ (E, esterified; N, nonesterified) fatty acids," of plasma are largely bound to _____ (G, globulins; A, albumins; C, cholesterol and glycerol).

     a. E, G   c. E, C   e. N, A
     b. E, A   d. N, G   f. N, C

11. A free fatty acid blood concentration at rest is about _____ mg. %, reflects a relatively _____ (L, low; H, high) rate of turnover, and is _____ (G, greater; S, less) than the corresponding glucose concentration.

     a. 15, L, G   c. 15, L, S   e. 150, L, G
     b. 15, H, S   d. 150, H, S   f. 150, H, G

12. Starvation and diabetes are conditions which _____ carbohydrate utilization for energy, _____ lipid utilization for energy, and _____ the free fatty acid concentration of plasma. (I, increase; D, decrease)

     a. I, I, D   c. D, I, D   e. D, D, D
     b. I, D, I   d. D, I, I   f. D, D, I

OBJECTIVE 68-3.

Identify the lipoproteins of plasma, their transport characteristics, and their major locus of synthesis.

13. Lipoproteins of plasma average about _____ mg. % and are involved in a higher percentage of the _____ (M, total mass; T, rate of transport) of plasma lipids.

 a. 7, M   c. 700, M   e. 70, T
 b. 70, M   d. 7, T   f. 700, T

14. Plasma lipoprotein particles in the postabsorptive state are _____ (S, smaller; L, larger) than 0.5 microns in diameter and comprise about _____ % of the total lipid fraction of plasma.

 a. S, 7   c. S, 97   e. L, 67
 b. S, 67   d. L, 7   f. L, 97

15. The lipoprotein fraction of plasma includes phospholipids, _____ (W, as well as; N, but not) triglycerides and cholesterol, which are largely synthesized by _____ (A, adipose tissue; L, the liver).

 a. W, A   c. N, A
 b. W, L   d. N, L

16. Of the total concentration of lipoproteins in plasma, plasma contains about _____ mg. % cholesterol and _____ mg. % of lipoprotein protein.

 a. 60, 180   c. 180, 200   e. 35, 35
 b. 120, 35   d. 6, 60   f. 120, 6

17. A higher percentage of cholesterol _____ (W, as well as; N, but not) triglycerides are associated with the _____ (VL, very low; L, low) density lipoprotein fraction of plasma.

 a. W, VL   c. N, VL
 b. W, L   d. N, L

18. Atherosclerosis is frequently associated with particularly _____ concentrations of the _____ density lipoproteins. (VL, very low; L, low; H, high)

 a. L, VL   c. L, H   e. H, L
 b. L, L   d. H, VL   f. H, H

OBJECTIVE 68-4.

Identify the major functional roles of adipose tissue and the liver in lipid metabolism.

19. Significant functions of adipose tissue include _____ , but not _____ . (I, heat insulation; K, ketone body production; M, mechanical protection and esthetic value; P, phospholipid synthesis; T, triglyceride storage)

 a. I, M, & T; K & P   d. M, P, & T; I & K
 b. I, M, T, & K; P   e. I, P, & T; K & M
 c. I, M, T, & P; K   f. P; I, K, M & T

20. Triglycerides, generally in a _____ (L, liquid; S, solid) form, are stored in fat cells up to a maximum of about _____ % of the cellular volume.

 a. L, 20 to 25   d. S, 20 to 25
 b. L, 45 to 60   e. S, 45 to 60
 c. L, 80 to 95   f. S, 80 to 95

21. Fat cells _____ (L, lack; P, possess) the enzymes which synthesize fatty acids and triglycerides from carbohydrate sources, a function which is largely accomplished in the _____ (V, liver; M, intestinal mucosa).

 a. L, V   c. P, V
 b. L, M   d. P, M

22. Lipases are utilized for the release _____ (W, as well as; N, but not) the uptake of triglyceride by fat cells, and participate in a renewal of stored fat every several _____ (D, days; K, weeks; Y, years).

 a. W, D   c. W, Y   e. N, K
 b. W, K   d. N, D   f. N, Y

23. Liver triglycerides generally contain a higher percentage of _____ (S, saturated; U, unsaturated) fatty acids than adipose tissue, as the major locus for their interconversion occurs in _____ (A, adipose tissue; L, the liver).

 a. S, A   c. U, A
 b. S, L   d. U, L

24. _____ (L, larger; S, smaller) amounts of liver triglycerides occur in the liver during starvation, diabetes mellitus, and conditions in which fat _____ (I, is; N, is not) the major source of metabolic energy by body tissues.

 a. L, I   c. S, I
 b. L, N   d. S, N

25. Phospholipids of plasma _____ (W, as well as; N, but not) endogenous cholesterol are largely synthesized by the _____ (A, adipose tissues; M, intestinal mucosa; L, liver).

    a. W, A   b. W, M   c. W, L   d. N, A   e. N, M   f. N, L

OBJECTIVE 68-5.

Identify the mechanisms and characteristics of the utilization of triglycerides for energy purposes by body tissues.

26. Triglycerides provide _____ (M, more; S, less) than 50% of the direct energy utilization by body tissues, since dietary fat comprises about _____ % of the calories of the average American adult, and 40% of the dietary _____ (C, carbohydrate is converted to triglyceride; L, lipid is converted to glycogen) stores.

    a. M, 15 to 20, C   d. S, 15 to 20, C
    b. M, 40 to 45, C   e. S, 40 to 45, L
    c. M, 75 to 80, L   f. S, 75 to 80, L

27. Fatty acids are utilized almost interchangeably with glucose for energy purposes by most tissues with the exception of the _____ (H, heart; L, liver; B, brain), since fatty acid and carbohydrate metabolic pathways converge at _____ (P, pyruvate; A, acetyl coenzyme A; G, alpha-glycerophosphate).

    a. H, G   c. B, P   e. L, G
    b. L, A   d. H, P   f. B, A

28. The degradation of fatty acids occurs in the _____ (C, cytoplasm; M, mitochondria) of cells by the progressive release of _____ carbon segments through the process of _____ (A, alpha; B, beta) oxidation.

    a. C, 2, A   c. C, 4, B   e. M, 3, B
    b. C, 3, A   d. M, 2, B   f. M, 4, A

29. The _____ (G, glycerol; L, lecithin; C, cholesterol) component of neutral fat _____ (I, is; IN, is not) catabolized by cells, while intracellular membrane transport mechanisms for fatty acids are thought to utilize _____ (N, carnitine; A, arginine) as a carrier substance.

    a. G, I, N   c. C, I, N   e. L, IN, N
    b. L, I, A   d. G, IN, A   f. C, IN, A

30. The complete oxidation of stearic acid, an 18-carbon chain which is _____ (S, a fully saturated; U, an unsaturated) fatty acid, produces a net gain of _____ molecules of ATP.

    a. S, 38   c. S, 146   e. U, 104
    b. S, 104   d. U, 38   f. U, 146

OBJECTIVE 68-6.

Identify the mechanisms, characteristics, and functional significance of the synthesis of ketone bodies.

31. Acetoacetic acid and acetone, _____ (W, as well as; N, but not) beta-hydroxybutyric acid, are produced largely by _____ (A, adipose; L, liver) tissue during predominantly _____ (P, lipid; C, carbohydrate) catabolism.

    a. W, A, P   c. W, L, P   e. N, A, P
    b. W, L, C   d. N, L, C   f. N, A, C

32. Ketone body formation is _____ during starvation and diabetes mellitus, _____ by increased glucocorticoid hormone secretion, and _____ by increased insulin secretion. (I, increased; D, decreased)

    a. I, I, I   c. I, D, I   e. D, D, I
    b. I, I, D   d. D, D, D   f. D, I, D

33. Ketone bodies are formed as a consequence of _____ (A, predominantly anaerobic glycolysis; F, accumulation of acetyl coenzyme A fragments), and are produced in larger quantities by the _____ (D, adapted; N, non-adapted) individual subsisting on a mainly _____ (C, carbohydrate; L, lipid) diet.

    a. A, D, C   c. A, N, C   e. F, N, C
    b. A, D, L   d. F, N, L   f. F, D, L

34. The rate of degradation of available acetoacetic acid by cells is _____ by low levels of carbohydrate intermediates and _____ by increased rates of cellular utilization of ATP. (I, increased; N, not significantly affected; D, decreased)

a. I, I   c. I, D   e. D, N
b. I, N   d. D, I   f. D, D

35. An odor of _____ (A, acetoacetate; T, acetone) in the breath and _____ (R, rapid, shallow; S, slow, deep) breathing accompanying a metabolic _____ (C, acidosis; K, alkalosis) are diagnostic symptoms of ketosis.

a. A, R, C   c. A, S, K   e. T, R, K
b. A, S, C   d. T, S, K   f. T, R, C

OBJECTIVE 68–7.

Identify the mechanisms and characteristics of triglyceride synthesis from carbohydrates. Recognize the existence of triglyceride synthesis from proteins.

36. _____ (A, adipose tissue; L, the liver) is the major locus for gluconeogenesis, _____ (W, as well as; N, but not) the major locus for the conversion of protein and carbohydrate to lipid.

a. A, W   c. L, W
b. A, N   d. L, N

37. Increased intracellular concentrations of alpha-glycerophosphate, and particularly _____, facilitate triglyceride _____ (C, catabolism; S, synthesis).

a. NADPH, C   c. ADP, C   e. NADH, S
b. NADH, C   d. NADPH, S   f. ADP, S

38. NADPH formation from glucose, largely by mechanisms of the _____ (G, glycolytic; P, phosphogluconate; K, Krebs cycle) pathway, is _____ (I, increased; N, not significantly altered; D, decreased) by insulin lack.

a. G, N   c. K, I   e. P, I
b. P, D   d. G, D   f. K, N

39. The synthesis of a molecule of stearic acid involves _____ acetyl coenzyme A molecule(s), _____ malonyl coenzyme A molecule(s), 16 molecules of _____, and 16 hydrogen atoms.

a. 16, 1, NADH   d. 1, 8, NADPH
b. 8, 1, NADH   e. 16, 8, NADPH
c. 1, 8, NADH   f. 8, 1, NADPH

40. Triglyceride synthesis from glucose generally involves a _____ % loss of the original energy of glucose as heat energy, and fatty acid chains which are generally _____ (S, shorter; L, longer) than 14 carbon atoms.

a. 15, S   c. 63, S   e. 36, L
b. 36, S   d. 15, L   f. 63, L

41. Storage fat has about _____ as much caloric energy per gram as glycogen and about _____ as much available mass in an average individual. (O, 50%; T, 2¼ times; N, 9 times; Y, 90 times)

a. O, N   c. N, Y   e. T, N
b. T, Y   d. O, Y   f. N, N

OBJECTIVE 68–8

Identify the principal mechanisms which regulate energy release from triglycerides.

42. The primary factor initiating energy release from carbohydrate _____ (W, as well as; N, but not) lipid stores is increased intracellular availability of _____ (O, oxygen; ADP, adenosine diphosphate; R, intracellular stores of glycogen and lipid).

a. W, O   c. W, R   e. N, ADP
b. W, ADP   d. N, O   f. N, R

43. The rate of lipid catabolism is _____ by saturation of NADH levels and _____ by increased concentrations of acetyl Co-A. (I, increased; N, not significantly affected; D, decreased)

a. I, I   c. I, D   e. D, N
b. I, N   d. D, I   f. D, D

44. Glycogen catabolism tends to _____ (I, increase; D, decrease) cellular alpha-glycerophosphate and NADPH levels, _____ (I, increase; D, decrease) cellular ratios of free fatty acids to triglycerides, and _____ (S, spare; R, reduce) cellular lipid stores.

    a. I, D, S   c. I, I, S   e. D, I, S
    b. I, D, R   d. D, D, R   f. D, I, R

45. The activity of acetyl Co-A carboxylase, an enzyme which promotes the carboxylation of _____ (A, acetyl; M, malonyl) Co-A during fatty acid _____ (B, anabolism; C, catabolism), is _____ (I, inhibited; F, facilitated) by intermediates of the Krebs cycle.

    a. A, B, I   c. A, C, I   e. M, C, I
    b. A, B, F   d. M, C, F   f. M, B, F

46. Insulin lack _____ (I, increases; D, decreases) cellular glucose utilization, increases fat _____ (S, synthesis; M, mobilization), and _____ (I, increases; D, decreases) the rate of fat utilization.

    a. I, S, I   c. I, M, D   e. D, M, I
    b. I, S, D   d. D, M, D   f. D, S, I

47. Elevation of _____ (T, thyroid hormone; G, glucocorticoid; E, epinephrine and norepinephrine) levels _____ (I, increases; D, decreases) the rate of fat mobilization and is more likely to cause ketosis.

    a. T, I   c. E, I   e. G, D
    b. G, I   d. T, D   f. E, D

48. _____ (G, growth hormone; C, corticotropin; L, luteotropin) has a greater effect in promoting fatty acid _____ (U, uptake; M, mobilization) of adipose tissue.

    a. G, U   c. L, U   e. C, M
    b. C, U   d. G, M   f. L, M

49. Sympathetic stimulation _____ (W, as well as; N, but not) adrenal medullary secretions tend to _____ (I, increase; D, decrease) the free fatty acid concentration of plasma.

    a. W, I   c. N, I
    b. W, D   d. N, D

OBJECTIVE 68-9.

Identify the formative characteristics and functions of phospholipids, and the formative characteristics, regulatory factors, and functions of cholesterol.

50. Lecithins and sphingomyelins, _____ (W, as well as; N, but not) cephalins, are classified as _____ (P, phospholipids; C, cholesterol derivatives; L, lipoproteins).

    a. W, P   c. W, L   e. N, C
    b. W, C   d. N, P   f. N, L

51. Phospholipids, synthesized by _____ (A, all; L, only liver) tissues of the body, largely function for _____ (H, hemostatic; S, general structural; E, metabolic energy) purposes.

    a. A, H   c. A, E   e. L, S
    b. A, S   d. L, H   f. L, E

52. _____ (A, arginine; C, carnitine; H, choline) is a nitrogenous base which _____ (I, inhibits the liver synthesis; T, is an essential constituent) of some of the phospholipids.

    a. A, I   c. H, I   e. C, T
    b. C, I   d. A, T   f. H, T

53. _____ performs a major functional role in the formation of thromboplastin, whereas _____ is utilized in insulating layers which permit saltatory conduction of action potentials. (L, lecithins; C, cephalins; S, sphingomyelins)

    a. L, C   c. C, L   e. S, L
    b. L, S   d. C, S   f. S, C

54. Endogenous cholesterol of plasma, largely in _____ (F, a free; E, an esterified) form, is formed _____ (L, almost entirely by the liver; M, by multiple tissues of the body).

    a. F, L   c. E, L
    b. F, M   d. E, M

55. Cholesterol is comprised of a basic _____ (S, sterol; G, glycerol) structural nucleus which is similar to hormonal structures of the adrenal _____ (C, cortex; M, medulla) and _____ (GI, gastrointestinal tract; N, gonads).

    a. S, C, GI   c. S, M, N   e. G, M, GI
    b. S, C, N   d. G, M, N   f. G, C, GI

56. Increased dietary levels of cholesterol or diets high in _____ (U, unsaturated; S, saturated) fats normally increase the plasma cholesterol by maximal amounts of _____ %.

    a. U, 15 to 30   d. S, 15 to 30
    b. U, 65 to 70   e. S, 65 to 70
    c. U, 150 to 300 f. S, 150 to 300

57. Blood cholesterol levels are _____ (I, increased; D, decreased) in diabetes mellitus and in _____ (O, hypothyroid; R, hyperthyroid) states.

    a. I, O   c. D, O
    b. I, R   d. D, R

58. Androgens _____ (W, as well as; N, but not) estrogens tend to _____ (I, increase; D, decrease) blood cholesterol levels.

    a. W, I   c. N, I
    b. W, D   d. N, D

59. Cholesterol, synthesized in the liver largely from _____ (P, phospholipids; A, acetyl Co-A), is largely utilized by the body to _____ (W, reduce insensible water loss; H, synthesize hormones; C, synthesize cholic acid).

    a. P, W   c. P, C   e. A, H
    b. P, H   d. A, W   f. A, C

OBJECTIVE 68–10.

Identify atherosclerosis and arteriosclerosis, their causative factors, and their consequences.

60. Atherosclerosis is primarily a disease process of _____ (L, large; S, small) vessels which is characterized by the deposition of _____ (R, calcium carbonate; A, calcium apatite; C, cholesterol; P, phospholipid) plaques.

    a. L, R   c. L, C   e. S, C
    b. L, A   d. S, A   f. S, P

61. Individuals with diabetes and _____ (P, hyperthyroidism; O, hypothyroidism), _____ (W, as well as; N, but not) hypertension, have a _____ (L, lower; H, higher) than normal incidence of atherosclerotic conditions.

    a. P, W, L   c. P, N, H   e. O, W, H
    b. P, N, L   d. O, N, H   f. O, W, L

62. _____ (H, atherosclerosis; R, arteriosclerosis), or "hardening of the arteries," occurs when lipids along with _____ (B, blood clots; C, calcified plaques) are formed in the vascular walls.

    a. H, B   c. R, B
    b. H, C   d. R, C

63. Estrogens tend to _____ plasma cholesterol levels and _____ the incidence of atherosclerosis. (I, increase; N, not significantly influence; D, decrease)

    a. I, I   c. I, D   e. D, N
    b. I, N   d. D, I   f. D, D

69

Protein Metabolism

OBJECTIVE 69-1.
Identify the basic types, chemical composition, and physical characteristics of proteins.

1. Proteins, comprising _____ (L, less; M, more) than 50% of the body solids, are aggregates of _____ (A, amino acids; N, nucleic acids; S, monosaccharides) formed into long chains.

   a. L, A   c. L, S   e. M, N
   b. L, N   d. M, A   f. M, S

2. Peptide bonds are formed between _____ (H, hydroxyl; A, amino; C, carboxyl) radicals of about _____ differing types of amino acids found in the body.

   a. H & A, 11   c. A & C, 11   e. H & C, 21
   b. H & C, 11   d. H & A, 21   f. A & C, 21

3. The major protein fraction of plasma is a _____ (L, lower; H, higher) molecular weight _____ (B, globular; F, fibrillar) protein fraction called _____ (A, albumin; G, globulin).

   a. L, B, A   c. L, F, A   e. H, F, A
   b. L, B, G   d. H, F, G   f. H, B, G

4. Cellular enzymes are largely conjugated forms of _____, whereas _____ are primarily found in cellular nucleoproteins. (A, albumins; G, globulins; H, histones; P, protamines)

   a. A, G   c. A, H & P   e. G, H & P
   b. A & H, P   d. H & P, A   f. G, A

5. Collagens, elastins, and keratins, _____ (W, as well as; N, but not) actins and myosin, are classified as _____ (C, conjugated; G, globular; F, fibrillar) proteins.

   a. W, C   c. W, F   e. N, G
   b. W, G   d. N, C   f. N, F

6. Fibrillar_____(W, as well as; N, but not) globular proteins are relatively_____(S, soluble; I, insoluble) in water or physiologic salt solutions.

   a. W, S   c. N, S
   b. W, I   d. N, I

7. The principal structural proteins of the body are _____ (F, fibrillar; G, globular) proteins which _____ (L, lack; P, possess) "creep" characteristics and possess greater _____ (T, tensile; C, compressive) strengths.

   a. F, L, T   c. F, P, T   e. G, L, T
   b. F, P, C   d. G, P, C   f. G, L, C

8. Nucleoproteins _____ (W, as well as; N, but not) mucoproteins are classified as _____ (S, simple; C, conjugated) proteins.

   a. W, S   c. N, S
   b. W, C   d. N, C

OBJECTIVE 69-2.
Identify the normal blood concentration and mechanism of transport and storage for amino acids.

9. Amino acids, largely present in_____(N, nonionized; I, ionized) form, represent a normal blood concentration of_____mg. %.

   a. N, 3 to 6   d. I, 3 to 6
   b. N, 35 to 65   e. I, 35 to 65
   c. N, 350 to 650   d. I, 350 to 650

10. A _____ (L, slight; B, substantial) rise in amino acid concentrations in blood immediately following a high protein meal is largely attributed to a relatively _____ (W, slow; R, rapid) rate-limiting mechanism of _____ (A, gut absorption; D, digestion; U, liver uptake).

    a. L, W, D   c. L, R, D   e. B, R, D
    b. L, W, A   d. B, W, U   f. B, R, A

11. Molecules of "essential" _____ (W, as well as; B, but not) "nonessential" amino acids are transported through cell membranes by _____ (S, simple diffusion; F, facilitated diffusion; A, active transport) mechanisms.

 a. W, S   c. W, A   e. B, F
 b. W, F   d. B, S   f. B, A

12. Amino acids _____ (L, lack; P, possess) renal plasma thresholds and are actively _____ (S, secreted; R, reabsorbed) by the _____ (X, proximal; D, distal) renal tubules.

 a. L, S, X   c. L, R, X   e. P, R, X
 b. L, S, D   d. P, R, D   f. P, S, D

13. Proteins digested largely by _____ (E, extracellular; I, intracellular) fluid enzymes constitute a _____ (N, minor; J, major) fraction of an available storage of amino acids.

 a. E, N   c. I, N
 b. E, J   d. I, J

14. _____ (F, fibrillar; G, globular) proteins which comprise a more available storage form of amino acids include _____ (N, nucleoproteins; E, cellular enzymes; C, collagen; M, muscle contractile proteins).

 a. F, N   c. F, M   e. G, E
 b. F, C   d. G, N   f. G, M

15. Glucocorticoid hormones _____ the protein stores of body tissues, _____ the rate of gluconeogenesis, and _____ the amino acid concentration of blood. (I, increase; D, decrease)

 a. I, I, I   c. I, D, I   e. D, I, D
 b. I, I, D   d. D, I, I   f. D, D, I

OBJECTIVE 69–3.

Identify the types, formative loci, and general functions of plasma proteins, and their relationship to exchangeable tissue protein stores.

16. The plasma _____ (A, albumin; G, globulin; F, fibrinogen) fraction functions in mechanisms of acquired immunity _____ (W, as well as; N, but not) in blood coagulation.

 a. A, W   c. F, W   e. G, N
 b. G, W   d. A, N   f. F, N

17. The plasma _____ fraction provides for a larger component of the plasma colloid osmotic pressure, whereas the _____ fraction functions to polymerize into a clot during blood coagulation. (F, fibrinogen; H, histone; G, globulin; A, albumin)

 a. H, G   c. G, A   e. A, H
 b. F, G   d. G, F   f. A, F

18. All of the _____ (F, fibrinogen; A, albumin; G, globulin) proteins of plasma are formed in the _____ (RES, reticuloendothelial system; L, liver) at a maximum rate which is considerably _____(M, greater; S, less) than 8 grams per day.

 a. G, RES, M   d. F & A, L, S
 b. G, RES, S   e. F & A, L, M
 c. G, L, M   f. F & A, RES, S

19. Plasma proteins _____ (A, are; N, are not) a readily available storage form of amino acids for tissues, and are largely _____ (E, excreted; D, degraded) by the _____ (K, kidneys; RES, reticuloendothelial system; M, intestinal mucosa).

 a. A, E, K   d. N, D, RES
 b. A, D, RES   e. N, E, K
 c. A, D, M   f. N, E, M

20. Plasma amino acid concentrations reflect _____ exchange with cellular proteins, and _____ exchange with plasma proteins. (R, a reversible; I, an irreversible)

 a. R, R   c. I, R
 b. R, I   d. I, I

21. The ratio of total tissue protein of the body to total plasma proteins is _____ (V, highly variable; C, relatively constant) and generally _____ (G, greater; L, less) than 1:1.

 a. V, G   c. C, G
 b. V, L   d. C, L

OBJECTIVE 69-4.

Recognize the basic chemical processes that form proteins, and the "essential" and "nonessential" nature of different types of amino acids.

22. Intracellular proteins of most tissues of the body arise from significant mechanisms of intracellular synthesis _____ (W, as well as; N, but not) membrane transport of exogenous proteins, and generally include _____ (E, only "essential"; NE, only "nonessential"; B, both E and NE) amino acids.

   a. W, E   c. W, B   e. N, NE
   b. W, NE   d. N, E   f. N, B

23. Phenylalanine, tryptophan, and histidine are among the _____ (E, essential; N, nonessential) amino acids, which comprise about _____ % of the different types of amino acids.

   a. E, 25   c. E, 75   e. N, 50
   b. E, 50   d. N, 25   f. N, 75

24. Alanine is formed by the _____ (D, deamination; T, transamination; M, methylation) of _____ (A, asparagine; P, pyruvate; K, alpha-ketoglutarate).

   a. D, A   c. M, A   e. T, P
   b. T, K   d. D, K   f. M, P

25. _____ (P, pyridoxine; C, vitamin C; D, vitamin D) is essential for the synthesis of nonessential amino acids as a consequence of its role in _____ (T, transamination reactions: S, keto acid synthesis).

   a. P, T   c. D, T   e. C, S
   b. C, T   d. P, S   f. D, S

OBJECTIVE 69-5.

Identify the principal characteristics of protein degradation for caloric value.

26. Transamination reactions, when compared to direct oxidative deamination, occur to a _____ (G, greater; L, lesser) extent in the deamination of amino acids, and involve the transfer of the amino group largely to _____ (P, pyruvate; K, alpha-ketoglutarate).

   a. G, P   c. L, P
   b. G, K   d. L, K

27. Amino acid nitrogen is eliminated from the body largely in the form of _____ (A, ammonia; U, urea; R, uric acid) produced by the _____ (L, liver; K, kidneys).

   a. A, L   c. R, L   e. U, K
   b. U, L   d. A, K   f. R, K

28. The sequential conversion of _____ (C, citrulline; A, arginine; O, ornithine) utilizes a _____ ratio of ammonia to carbon dioxide molecules.

   a. O to C to A, 2:1   d. C to O to A, 2:1
   b. O to C to A, 3:1   e. C to O to A, 3:1
   c. O to C to A, 1:2   f. C to O to A, 1:2

29. _____ (G, gluconeogenesis; K, ketogenesis), involving the conversion of deaminated amino acids to glucose, is possible for _____ (A, all; M, most, but not all; F, only a few) of the different types of amino acids.

   a. G, A   c. G, F   e. K, M
   b. G, M   d. K, A   f. K, F

30. An "obligatory loss" of protein refers to an amount of protein which is _____ (E, dietary excess; D, degraded in the absence of dietary protein intake), and is typically _____ grams per day.

   a. E, 3   c. E, 3000   e. D, 30
   b. E, 30   d. D, 3   f. D, 3000

31. Partial or incomplete proteins are inefficient proteins for their _____ (C, caloric value; S, synthesis of body proteins), and are generally characterized by a significantly low percentage of certain _____ (E, essential; N, nonessential) amino acids.

   a. C, E   c. S, E
   b. C, N   d. S, N

32. In starvation, the body utilizes its food supplies for metabolic energy in the order of_____. (F, fat; C, carbohydrate; P, protein)

a. F, C, P   b. F, P, C   c. C, F, P   d. C, P, F

OBJECTIVE 69–6.
Identify the regulatory influences of growth hormone, insulin, glucocorticoids, testosterone, and thyroxine upon protein metabolism.

33. Growth hormone results in a net increase of fat _____ and a net increase of cellular protein _____ . (S, synthesis; B, breakdown)

a. S, S   c. B, S
b. S, B   d. B, B

34. Insulin deficiencies result in _____ glucose availability for cellular metabolism and _____ protein synthesis. (I, increased; N, no significant change in; D, decreased)

a. I, I   c. I, D   e. D, N
b. I, N   d. D, I   f. D, D

35. Glucocorticoids _____ cellular protein stores of most body tissues and _____ plasma concentrations of amino acids. (I, increase; D, decrease)

a. I, I   c. D, I
b. I, D   d. D, D

36. Glucocorticoids_____ (I, increase; D, decrease) liver proteins_____(W, as well as; N, but not) plasma proteins.

a. I, W   c. D, W
b. I, N   d. D, N

37. _____ (E, estrogen; T, testosterone) results in a greater _____ (I, increase; D, decrease) in body protein.

a. E, I   c. T, I
b. E, D   d. T, D

38. Thyroxine has a more important _____ (D, direct; I, indirect) effect upon increased rates of protein deposition during growth when carbohydrates and fats _____ (A, are; N, are not) readily available.

a. D, A   c. I, A
b. D, N   d. I, N

70

The Liver and Biliary System

OBJECTIVE 70–1.
Using the following diagram, identify the basic structure of a liver lobule.

Directions: Match the lettered headings with the diagram and numbered list of descriptive words and phrases.

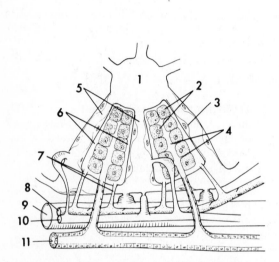

(From Guyton, A.C., Textbook of Medical Physiology, 5th ed. Philadelphia, W.B. Saunders Company, 1976.)

1. _____ Vein around which the liver lobule is constructed.
2. _____ Cells making up the hepatic plates.
3. _____ Reticuloendothelial cell which phagocytizes foreign material in the blood.
4. _____ Small passageways between liver cells which empty bile into the bile ducts.
5. _____ Flat, branching spaces lying between hepatic plates.
6. _____ Narrow space between the endothelial lining of venous sinusoids and the liver cells.
7. _____ Small lymphatic vessels in the interlobular spaces connecting the space of Disse with larger lymphatic vessels.
8. _____ Vessel that receives excess fluids from the terminal lymphatics.
9. _____ Large vein that carries blood from the digestive organs and spleen to the liver.
10. _____ Blood vessel entering the liver at the porta and ramifying through the sinusoids.
11. _____ Vessel that receives bile from the bile canaliculi.

a. Terminal lymphatics
b. Portal vein
c. Hepatic artery
d. Central vein

e. Bile duct
f. Space of Disse
g. Kupffer cell
h. Sinusoids

i. Bile canaliculi
j. Liver cells
k. Lymphatic vessel

OBJECTIVE 70–2.
Identify the characteristic features of the hepatic vascular supply and their functional significance.

12. Vascular functions of the adult liver include blood_____but not_____. (E, erythropoiesis; F, filtration; S, storage; P, plasma protein synthesis)

a. E, F, & S; P
b. E & S; F & P
c. F & P; E & S

d. S & P; E & F
e. F, S, & P; E
f. F & S; P & E

13. The total blood flow to the liver is typically closest to _____ liters per minute of which the higher percentage is derived from the _____ (A, hepatic artery; P, portal vein).

a. 0.5, A
b. 1.5, A

c. 2.8, A
d. 0.5, P

e. 1.5, P
f. 2.8, P

14. The pressure difference between blood _____ (E, entering; X, exiting from) the liver in the portal vein and blood of the hepatic vein is typically _____ mm. Hg, and is _____ (I, increased; D, decreased) by the disease process of liver cirrhosis.

   a. E, 8, I   c. E, 65, D   e. X, 22, D
   b. E, 22, I   d. X, 8, I   f. X, 65, D

15. The liver is a relatively _____ (S, scanty; P, prolific) producer of lymph, which contains a relatively _____ (H, high; L, low) concentration of protein as compared to other tissues.

   a. S, H   c. P, H
   b. S, L   d. P, L

16. Ascitic fluid, largely _____ (B, liver bile; P, plasma) in composition, is generally associated with _____ (D, diminished; E, elevated) hepatic _____ (A, arterial; V, venous) pressures.

   a. B, D, A   c. B, E, A   e. P, E, A
   b. B, D, V   d. P, E, V   f. P, D, V

17. Kupffer cells line the liver _____ (S, sinusoids; D, biliary system of ducts) and function as relatively _____ (E, efficient; I, inefficient) _____ (P, phagocytes; B, bile producers; V, lymphatic valves).

   a. S, E, P   c. S, E, V   e. D, E, B
   b. S, I, P   d. D, I, V   f. D, I, B

OBJECTIVE 70–3.

Using the following diagram, identify the physiologic anatomy of the biliary system and associated organs.

Directions: Match the lettered headings with the diagram and numbered list of descriptive words and phrases.

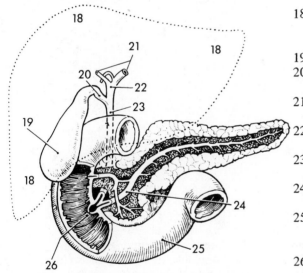

(From Bell, G.H., et al.: Textbook of Physiology and Biochemistry. 9th Ed. Churchill Livingstone, Edinburgh, 1976.)

18. _____ Bell-shaped organ in the upper right-hand portion of the abdomen, used for numerous metabolic functions.
19. _____ Stores bile.
20. _____ The duct connecting the gallbladder and the common bile duct.
21. _____ The ducts carrying bile from the right and left portions of the liver.
22. _____ The duct formed by the union of the right and left hepatic ducts.
23. _____ The duct formed by the union of the cystic and hepatic ducts.
24. _____ Main excretory duct from the pancreas.
25. _____ The proximal portion of the small intestine, located between the pylorus and the jejunum.
26. _____ Muscular sheath in the terminal portions of the bile and pancreatic ducts, which controls the movements of these fluids into the duodenum.

   a. Cystic duct   d. Common bile duct   g. Sphincter of Oddi
   b. Duodenum   e. Gallbladder   h. Pancreatic duct
   c. Liver   f. Hepatic duct   i. Hepatic ducts

OBJECTIVE 70-4.

Identify the composition of bile and the functions of the gallbladder in bile secretion and storage.

27. Bile is formed _____ (C, continuously; I, intermittently) by a _____ (J, majority; N, minority) of parenchyma cells of the _____ (P, pancreas; L, liver).

   a. C, J, P   c. C, N, L   e. I, N, P
   b. C, J, L   d. I, N, L   f. I, J, P

28. A total daily secretion of liver bile of about _____ ml. is _____ (C, concentrated; D, diluted) by the transport mechanism of the gallbladder.

   a. 90, C   c. 9000, C   e. 90, D
   b. 900, C   d. 9, D   f. 900, D

29. The propelling force for biliary secretion into the intestinal tract is derived from contraction of smooth muscle of _____ (D, bile ducts; G, the gallbladder) in response to _____ (C, cholecystokinin; S, secretin) secretion from the _____ (I, intestinal mucosa; L, liver).

   a. D, C, I   c. D, S, I   e. G, C, I
   b. D, S, L   d. G, S, L   f. G, C, L

30. The action of intestinal fat products to _____ (I, increase; D, decrease) biliary flow into the duodenum is _____ (A, augmented; O, opposed) by vagal stimulation.

   a. I, A   c. D, A
   b. I, O   d. D, O

31. The movement of fluid through the sphincter of Oddi is _____ (I, increased; D, decreased) by increased states of cholecystokinin secretion _____ (W, as well as; N, but not) intestinal peristalsis.

   a. I, W   c. D, W
   b. I, N   d. D, N

32. The highest concentration of biliary constituents occurs for _____ (C, cholesterol; F, fatty acids; S, bile salts; R, bilirubin) in _____ (L, liver; G, gallbladder) bile.

   a. R, L   c. S, L   e. S, G
   b. F, L   d. C, G   f. F, G

33. The gallbladder performs a more significant function of biliary _____ (A, absorption; S, secretion) of water and electrolytes _____ (W, as well as; N, but not) lipid-soluble substances.

   a. A, W   c. S, W
   b. A, N   d. S, N

OBJECTIVE 70-5.

Identify the bile salts and their functional characteristics.

34. The _____ (L, liver; B, gallbladder) forms bile salts from taurine and _____ (M, methionine; G, glycine)-conjugated derivatives of _____ (P, phospholipids; C, cholesterol; PP, polypeptides).

   a. L, M, P   c. L, G, C   e. B, M, P
   b. L, G, P   d. B, G, C   f. B, M, PP

35. _____ (R, bilirubin; S, bile salts) function to form micelles within the _____ (I, intestinal tract; L, lacteals) and function to _____ (C, consolidate; E, emulsify) undigested intestinal lipids.

   a. R, I, C   c. R, L, C   e. S, I, C
   b. R, L, E   d. S, L, E   f. S, I, E

36. The intestinal absorption of vitamin _____ , the fat-soluble vitamin which is stored to the least extent by body tissues, _____ (I, is; N, is not) significantly influenced by biliary obstruction.

   a. D, I   c. K, I   e. E, N
   b. A, I   d. D, N   f. K, N

37. Most of the bile _____ (S, salts; P, pigments) are reabsorbed in the distal _____ (D, duodenum; I, ileum; C, colon).

   a. S, D   c. S, C   e. P, I
   b. S, I   d. P, D   f. P, C

38. _____normal intestinal absorption of bile salts to 50% would result in transient_____secretion of bile salts by the liver and_____volume rates of secretion of liver bile. (I, increased; U, unaltered; D, decreased).

    a. I, I, U   b. I, D, I   c. I, D, D   d. D, D, D   e. D, I, I   f. D, D, U

OBJECTIVE 70–6.

Identify the sequence of events leading to the secretion of bile pigments and their subsequent fate. Identify the types of jaundice, their underlying causes, and their diagnostic distinction.

39. Bile _____ (S, salts; P, pigments) are formed from the _____ (H, heme; G, globin) portion of hemoglobin.

    a. S, H   c. P, H
    b. S, G   d. P, G

40. The principal bile pigment _____ (S, stercobilin; R, bilirubin) is largely conjugated with _____ (A, albumin; G, glucuronide) by the _____ (L, liver; P, spleen).

    a. S, A, L   c. S, G, P   e. R, G, L
    b. S, A, P   d. R, G, P   f. R, A, L

41. Plasma _____ (W, as well as; N, but not) urine generally contains a higher percentage of _____ (C, "conjugated"; F, "free") bilirubin, and bile contains a higher percentage of _____ (C, "conjugated"; F, "free") bilirubin.

    a. W, C, F   c. W, F, F   e. N, F, F
    b. W, C, C   d. N, F, C   f. N, C, C

42. A relatively _____ (I, insoluble; S, soluble) urobilinogen is produced within the _____ (K, kidneys; G, gastrointestinal tract) and largely excreted by the _____ (K, kidneys; G, gastrointestinal tract).

    a. I, K, G   c. I, G, G   e. S, G, G
    b. I, K, K   d. S, G, K   f. S, K, K

43. Jaundice is caused by elevated _____ (F, only "free"; C, only "conjugated"; E, either "free" or "conjugated") bilirubin concentrations of plasma resulting from enhanced hemolysis _____ (W, as well as; N, but not) biliary obstruction.

    a. F, W   c. E, W   e. C, N
    b. C, W   d. F, N   f. E, N

44. A "direct van den Bergh reaction" implies a high concentration of _____ (F, "free"; C, "conjugated") bilirubin in plasma and _____ (H, a hemolytic; O, an obstructive) jaundice.

    a. F, H   c. C, H
    b. F, O   d. C, O

45. A total obstructive jaundice is associated with the _____ (P, presence; A, absence) of urobilinogen in urine and a _____ (B, dark brown; C, clay; G, green) -colored stool.

    a. P, B   c. P, G   e. A, C
    b. P, C   d. A, B   f. A, G

OBJECTIVE 70–7.

Recognize the presence of cholesterol in bile, and identify the compositional characteristics and mode of occurrence of gallstones.

46. Cholesterol, _____ (S, relatively soluble; I, almost insoluble) in water, is largely _____ (A, absorbed from; C, secreted into) bile by _____ (L, liver; G, gallbladder) cells.

    a. S, A, G   c. S, C, L   e. I, A, L
    b. S, C, G   d. I, C, L   f. I, A, G

47. Cholesterol, more soluble in bile containing _____ concentrations of bile salts, has a _____ concentration in liver bile and a _____ concentration in gallbladder bile than the corresponding bile salt concentrations. (H, higher; L, lower)

    a. H, L, L   c. H, H, L   e. L, H, L
    b. H, L, H   d. L, H, H   f. L, L, H

48. Causes of gallstones include gallbladder inflammation, too _____ (L, low; H, high) a concentration of bile acids in bile, too _____ (L, low; H, high) a concentration of cholesterol in bile, and too _____ (T, little; M, much) absorption of water from bile.

 a. L, H, T   c. L, L, M   e. H, L, T
 b. L, H, M   d. H, L, M   f. H, H, T

49. The x-ray opacity of gallstones, resulting from their _____ (C, cholesterol; S, calcium salt) content, is sufficient to detect about _____ % of all cases of gallstones.

 a. C, 25   c. C, 99   e. S, 75
 b. C, 75   d. S, 25   f. S, 99

OBJECTIVE 70–8.
Identify the principal metabolic functions of the liver.

50. The "glucose buffer function" of the _____ (S, skeletal muscle tissues; L, liver) includes mechanisms of _____ (G, only glycogenolysis; N, only gluconeogenesis; B, both gluconeogenesis and glycogenolysis).

 a. S, G   c. S, B   e. L, N
 b. S, N   d. L, G   f. L, B

51. Beta oxidation of _____ (F, fats; C, carbohydrates; P, proteins) occurs _____ (M, more; L, less) rapidly in the liver than in other tissues.

 a. F, M   c. P, M   e. C, L
 b. C, M   d. F, L   f. P, L

52. Acetoacetic acid and ketone bodies, largely formed by _____ (A, adipose tissue; L, the liver) in association with _____ (I, increased; R, reduced) availability of acetyl Co-A, are normally _____ (X, excreted from; T, utilized by other tissues of) the body.

 a. A, I, X   c. A, R, X   e. L, I, X
 b. A, R, T   d. L, R, T   f. L, I, T

53. Fat synthesis from carbohydrates _____ (W, as well as; N, but not) from proteins is largely accomplished by _____ (A, adipose tissue; L, the liver; I, the intestinal mucosa).

 a. W, A   c. W, I   e. N, L
 b. W, L   d. N, A   f. N, I

54. The liver is a major producer of endogenous cholesterol, _____ (W, as well as; N, but not) phospholipids; about _____ % of its cholesterol is converted to bile _____ (P, pigments; S, salts).

 a. W, 20, S   c. W, 80, P   e. N, 20, S
 b. W, 80, S   d. N, 80, S   f. N, 20, P

55. All of the following are known functions of the liver with the exception of:

 a. Carbohydrate storage
 b. Vitamin storage
 c. Drug detoxification
 d. Intestinal enzyme synthesis
 e. Urea synthesis
 f. Amino acid deamination

56. A loss of liver metabolic functions, considered more critical regarding its role in _____ (L, lipid; C, carbohydrate; P, protein) metabolism, is associated with _____ (D, decreased; I, increased) ammonia concentrations of body fluids.

 a. L, D   c. P, D   e. C, I
 b. C, D   d. L, I   f. P, I

57. Vitamins _____ are exceptions to the general case, whereby _____ (W, water; F, fat) -soluble vitamins are stored to a greater extent in the liver.

 a. B_{12} & A, W   d. E & D, F
 b. B_{12} & D, W   e. B_{12} & K, F
 c. D & C, F   f. B_{12} & C, W

58. Vitamin _____ is required by the _____ (B, bone marrow; L, liver) to synthesize the blood coagulation factors VII, IX, X, and _____ (F, fibrinogen; P, prothrombin).

 a. K, B, F   c. K, L, P   e. E, L, F
 b. K, B, P   d. E, L, P   f. E, B, F

59. The greater proportion of iron in the body is normally stored in the _____ (S, spleen; M, red marrow; L, liver) in the form of _____ (F, ferritin; H, hemosiderin).

 a. S, F   c. L, F   e. M, H
 b. M, F   d. S, H   f. L, H

71

Energetics and Metabolic Rate

OBJECTIVE 71-1.
Recognize the importance of ATP in metabolism.

1. The intracellular formation of _____ (G, glycogen; P, protein; L, lipid) molecules has a greater energy requirement in the form of _____ (H, heat; C, chemical bond) energy sources.

   a. G, H   c. L, H   e. P, C
   b. P, H   d. G, C   f. L, C

2. _____ , termed the "energy currency" of cells, is produced by the catabolism of carbohydrates, _____ (W, as well as; N, but not) the catabolism of fatty acids.

   a. NAD, W   c. DNA, W   e. ATP, N
   b. ATP, W   d. NAD, N   f. DNA, N

3. ATP makes available about _____ kilocalories per mol in each of its _____ "high energy" phosphate bonds.

   a. 4, 2   c. 16, 2   e. 8, 3
   b. 8, 2   d. 4, 3   f. 16, 3

4. ATP is utilized in the synthesis of glucose from lactate, _____ (W, as well as; N, but not) in the synthesis of fatty acids from acetyl Co-A, and _____ (I, is; O, is not) required to form urea.

   a. W, I   c. N, I
   b. W, O   d. N, O

5. ATP, utilized _____ (D, directly; I, indirectly) to provide energy for the propagation of action potentials, is consumed at a rate which is relatively _____ (IP, independent of; DP, dependent upon) action potential frequency.

   a. D, IP   c. I, IP
   b. D, DP   d. I, DP

OBJECTIVE 72-2.
Identify the functional characteristics of creatine phosphate stores, ATP production under aerobic and anaerobic conditions, and the "oxygen debt."

6. The _____ (H, higher; L, lower) skeletal muscle concentration of creatine phosphate, when compared to ATP, functions as _____ (S, only a substitute; R, only an energy reserve; B, both a substitute and an energy reserve) for intracellular functions of ATP.

   a. H, S   c. H, B   e. L, R
   b. H, R   d. L, S   f. L, B

7. A higher energy of the "high energy" bonds of _____ (C, creatine phosphate; A, adenosine triphosphate) favors the saturation of _____ (C, creatine phosphate; A, adenosine triphosphate) stores by the transfer of a _____ (P, phosphate; PP, pyrophosphate) radical.

   a. C, A, P   c. C, C, P   e. A, C, P
   b. C, A, PP   d. A, C, PP   f. A, A, PP

8. The catabolism of carbohydrate, _____ lipid, provides cellular energy during anaerobic _____ aerobic conditions. (W, as well as; N, but not)

   a. W, W   c. N, W
   b. W, N   d. N, N

9. The lowest efficiency for the cellular production of ATP involves the utilization of _____ (P, plasma glucose; G, glycogen stores) during _____ (A, aerobic; N, anaerobic) conditions.

   a. P, A   c. G, A
   b. P, N   d. G, N

10. Maximal intracellular concentrations of ATP in muscle of about _____ millimolar are sufficient to maintain maximal skeletal muscle contraction for _____ (L, less; C, considerably more) than 10 seconds.

    a. 0.005, L   c. 5, L   e. 0.05, C
    b. 0.05, L   d. 0.005, C f. 5, C

11. ATP, available from _____ (A, preexisting ATP and creatine phosphate stores; G, anaerobic glycolysis), provides the major source of continued vital body processes for only a few _____ (S, seconds; M, minutes; H, hours) following cessation of oxygen availability to tissues.

    a. A, S   c. A, H   e. G, M
    b. A, M   d. G, S   f. G, H

12. An "oxygen debt" refers to a _____ (L, lower; H, higher) than normal rate of oxygen utilization following exercise as a consequence of _____ (G, continued anaerobic glycolysis; A, additional energy required to restore normal conditions; E, elevated cardiac output).

    a. L, G   c. L, E   e. H, A
    b. L, A   d. H, G   f. H, E

13. Following strenuous exercise, about _____ % of the lactate is reconverted to glucose primarily by _____ (M, muscle; L, liver) tissues.

    a. 20, M   c. 80, M   e. 50, L
    b. 50, M   d. 20, L   f. 80, L

14. Skeletal muscle ratios of ATP to creatine phosphate concentrations are highest _____ periods of severe exercise, while ratios of ATP production to ATP utilization are highest _____ periods of severe exercise. (B, before; D, during; F, immediately following)

    a. B, B   c. B, F   e. D, D
    b. B, D   d. D, B   f. D, F

15. The _____ (A, aerobic; N, anaerobic) mechanisms of generation of ATP, localized within _____ (M, mitochondria; C, the cytoplasm), operate more rapidly and with _____ (G, greater; L, lesser) energy efficiency.

    a. A, M, G   c. A, C, G   e. N, C, G
    b. A, M, L   d. N, C, L   f. N, M, L

OBJECTIVE 71–3.

Identify the general mechanisms which control the rate of energy release by cells.

16. The _____ (H, Henderson-Hasselbalch; M, Michaelis-Menten) equation predicts that the rate of a chemical reaction is determined by the substrate concentration _____ (W, as well as; N, but not) the concentration of the enzyme.

    a. H, W   c. M, W
    b. H, N   d. M, N

17. The rate of chemical reactions in biological systems is generally more directly influenced in a(n) _____ (D, direct; I, inverse) manner by variations of the _____ (E, enzyme; S, substrate) concentration when the substrate concentration is exceptionally high.

    a. D, E   c. I, E
    b. D, S   d. I, S

18. Functionally related chemical reactions are generally arranged in a _____ (S, series; P, parallel) fashion, so that the overall reaction time is largely determined by the _____ (W, slowest; A, average; F, fastest) of the reaction steps.

    a. S, W   c. S, F   e. P, A
    b. S, A   d. P, W   f. P, F

19. _____ (G, glucose; O, oxygen; ADP, adenosine diphosphate) is generally the major rate-limiting factor for energy metabolism, as its concentration is comparatively _____ (H, high; L, low) under resting conditions of activity.

    a. G, H   c. ADP, H   e. O, L
    b. O, H   d. G, L   f. ADP, L

20. The rate of energy release by the cell, controlled by factors which are largely _____ (U, dependent upon; O, independent of) the degree of cellular activity, is _____ (D, decreased; N, not significantly affected; I, increased) by increased cellular levels of ADP.

    a. U, D   b. U, N   c. U, I   d. O, D   e. O, N   f. O, I

OBJECTIVE 71–4.
Identify the metabolic rate, its representative origin, its means of measurement, and the principles of direct and indirect calorimetry.

21. The metabolic rate reflects _____ (A, only anabolic; C, only catabolic; S, the sum total of anabolic and catabolic) processes and is generally expressed in terms of the rate of _____ (W, work performance; H, heat production).

    a. A, W   c. S, W   e. C, H
    b. C, W   d. A, H   f. S, H

22. The proportion of energy obtained from food that reaches the functional systems of cells is about _____ %, while the proportion of food energy that is ultimately converted into heat energy is about _____ %.

    a. 25, 25   c. 25, 100   e. 75, 75
    b. 25, 75   d. 75, 25   f. 75, 100

23. The large Calorie, spelled with a capital "C," is the amount of heat energy required to increase the temperature of 1 _____ (G, gram; K, kilogram) of water _____ ° (1; 10) _____ (F.; C.)

    a. G, 1, F.   c. G, 10, F.   e. K, 1, F.
    b. G, 10, C.   d. K, 10, C.   f. K, 1, C.

24. Direct calorimetry involves the direct measurement of the total _____ (O, oxygen consumed; H, heat energy lost; F, amount of ingested food) by the body and is relatively _____ (E, easy; D, difficult) to accomplish.

    a. O, E   c. F, E   e. H, D
    b. H, E   d. O, D   f. F, D

25. A slightly greater energy yield of _____ (S, starch; F, fat; P, protein) catabolism, per liter of oxygen utilized, approximates _____ Calories per liter.

    a. S, 5   c. P, 5   e. F, 8.45
    b. F, 5   d. S, 8.45   f. P, 8.45

26. An individual utilizing 100 Calories of energy per hour from average dietary sources would utilize 100 liters of oxygen in a period of about _____ hour(s).

    a. ½   c. 10   e. 20
    b. 5   d. 15   f. 48

27. The metabolator consists of a(n) _____ (E, energy; V, volume; D, dietary intake) measuring device which is generally utilized for the _____ (R, direct; I, indirect) measurement of metabolic rates.

    a. E, R   c. D, R   e. V, I
    b. V, R   d. E, I   f. D, I

OBJECTIVE 71–5.
Identify the factors that affect the metabolic rate and their significance.

28. Daily energy requirements for simply existing, that is, performing essential functions only, for a 70-kg. adult most closely approximates _____ Calories.

    a. 7,000   c. 700   e. 40
    b. 2,000   d. 80   f. 20

29. The energy expenditure of a 70-kg. man sitting at rest is about _____ Calories per hour, whereas the same individual walking slowly would utilize about _____ times this amount.

    a. 40, 2   c. 400, 2   e. 100, 10
    b. 100, 2   d. 40, 10   f. 400, 10

30. Eating a high _____ (L, lipid; C, carbohydrate; P, protein) meal is more effective in producing a _____ (F, fall; R, rise) in the metabolic rate several hours after a meal.

    a. L, F   c. P, F   e. C, R
    b. C, F   d. L, R   f. P, R

31. The "specific dynamic action" of protein produces a change in metabolic rate which is maximally about _____ % and is largely attributed to _____ (P, protein digestion and absorption; A, the action of amino acids to directly influence cell metabolism).

    a. 2 to 4, P   d. 2 to 4, A
    b. 20 to 30, P   e. 20 to 30, A
    c. 100 to 150, P   f. 100 to 150, A

32. The metabolic rate, expressed in Calories per hour per unit of body surface area, is generally higher for _____ (W, women; M, men), lowest _____ (B, at birth; O, in old age) and highest _____ (B, at birth; A, in the young adult; O, in old age).

    a. W, B, A   c. W, O, B   e. M, O, A
    b. W, B, O   d. M, O, B   f. M, B, A

33. Metabolic rates are increased to a greater extent by _____ (M, male; F, female) sex hormones, _____ (I, increased; D, decreased) by sympathetic stimulation, and _____ (I, increased; N, not significantly influenced) by growth hormone.

    a. M, I, I   c. M, D, I   e. F, D, I
    b. M, I, N   d. F, D, N   f. F, I, N

34. _____ (M, maximal; L, a loss of) secretion of the thyroid gland decreases the metabolic rate to about _____ % of normal.

    a. M, 5 to 10   d. L, 5 to 10
    b. M, 50 to 60   e. L, 50 to 60
    c. M, 85 to 90   f. L, 85 to 90

35. People living in _____ (T, tropical; A, arctic; P, temperate) climates generally have higher levels of thyroid gland secretion and _____ (L, lower; H, higher) metabolic rates.

    a. T, L   c. P, L   e. A, H
    b. A, L   d. T, H   f. P, H

36. Sleep is generally associated with _____ muscle tone, _____ sympathetic activity, and _____ metabolic rates. (D, decreased; N, no significant change in; I, increased).

    a. D, D, D   c. D, N, N   e. N, I, N
    b. D, D, N   d. N, N, I   f. N, I, D

OBJECTIVE 71–6.

Recognize the significance of the basal metabolic rate, its mode of expression, and the customary technique for its measurement.

37. Comparisons of metabolic rates between differing individuals are generally expressed as the rate of energy utilized per unit of _____ (M, body mass; S, body surface area; C, caloric intake) during _____ (L, sleeping; B, basal; E, standard exercise) conditions.

    a. M, L   c. C, E   e. S, E
    b. S, B   d. M, B   f. C, L

38. Basal conditions for measuring the metabolic rate include ambient temperatures of _____ ° F., 30 minutes of rest in a _____ (R, reclining; S, standing) position, and measurements made _____ (I, immediately; F, at least 12 hours) following a meal.

    a. 50 to 68, R, I   d. 68 to 80, S, F
    b. 50 to 68, S, F   e. 68 to 80, R, I
    c. 50 to 68, S, I   f. 68 to 80, R, F

39. The surface area of the average 70-kg. adult is _____ square meters and varies in approximate proportion to the body weight raised to the _____ power.

    a. 1.73, $\frac{1}{3}$ c. 1.73, $\frac{3}{2}$ e. 3.71, $\frac{2}{3}$
    b. 1.73, $\frac{2}{3}$ d. 3.71, $\frac{1}{3}$ f. 3.71, $\frac{3}{2}$

40. The product of the oxygen consumption rate in liters per hour and _____ , when _____ (M, multiplied; D, divided) by a factor of weight$^{0.425}$ × height$^{0.725}$ × 0.007184, equals the _____ (T, metabolic rate; R, reciprocal of the metabolic rate).

    a. 4.825, M, T   d. 8.425, D, R
    b. 4.825, D, R   e. 8.425, M, T
    c. 4.825, D, T   f. 8.425, M, R

41. A basal metabolic rate of +25% implies a _____ (G, greater; L, lesser) _____ (C, caloric equivalent of oxygen; O, oxygen consumption rate; B, body surface area) than normal.

    a. G, C   c. G, B   e. L, O
    b. G, O   d. L, C   f. L, B

42. An abnormally low basal metabolic rate, when expressed in percentages above or below normal, more probably reflects a _____ (H, higher; L, lower) than normal _____ (G, age; S, specific dynamic action of foodstuffs; T, thyroxine level).

    a. H, G   c. H, T   e. L, S
    b. H, S   d. L, G   f. L, T

43. A basal metabolic rate of 48 Calories per hour per square meter for a 20-year-old male reflects an oxygen consumption rate of about _____ liters per hour and a metabolic rate of about _____ %.

    a. 1.5, −75   c. 1.5, +25   e. 15, +25
    b. 1.5, −25   d. 15, −25   f. 15, +75

72

Body Temperature, Temperature Regulation, and Fever

OBJECTIVE 72–1.

Identify core, surface, and average body temperatures. Identify the normal body temperature, its normal range of variance and influencing factors, and the specific heat of body tissues and its significance.

1. The term "body temperature" is utilized to designate the _____ temperature, while the _____ temperature provides a means of calculation of the body heat content. (S, surface; C, core; A, average)

    a. S, C   c. C, S   e. A, S
    b. S, A   d. C, A   f. A, C

2. The _____ (A, average; C, core) body temperature remains more constant, while the surface temperature generally changes in the _____ (S, same; O, opposite) direction as the environmental temperature.

    a. A, S   c. C, S
    b. A, O   d. C, O

3. The _____ (S, surface; C, core body) temperature lies closer to the average body temperature as a consequence of its associated _____ (G, greater; L, lesser) _____ (M, mass; R, role in body heat dissipation).

    a. S, G, M   c. S, G, R   e. C, L, M
    b. S, L, M   d. C, L, R   f. C, G, M

4. The _____ (O, oral; R, rectal) temperature, about _____° F. higher than the other, is _____ (D, decreased; I, increased) by exercise.

    a. O, 1, D   c. O, 3, I   e. R, 1, I
    b. O, 3, D   d. R, 3, I   f. R, 1, D

5. A normal body temperature of about _____° C. is equivalent to a temperature of _____° F.

    a. 20, 37   c. 98.6, 37   e. 37, 20
    b. 37, 98.6   d. 20, 98.6   f. 98.6, 20

6. The amount of heat energy of an object is directly proportional to _____ (T, its temperature; S, the square of its temperature) and _____ (D, directly proportional to; I, independent of; V, inversely proportional to) its heat capacity.

   a. T, D c. T, V e. S, I
   b. T, I d. S, D f. S, V

7. A net increase of _____ Calories of body heat for a 1° C. rise in average temperature of a 70-kg. man reflects a specific heat of body tissues of _____ Cal./kg./° C.

   a. 84, 0.83 c. 22.6, 3.1 e. 121, 1.73
   b. 40, 1.73 d. 58, 0.83 f. 217, 3.1

OBJECTIVE 72–2.

Identify the relationship between heat production, heat loss, specific heat, and body temperature. Identify the major factors that influence heat production and the major mechanisms of heat loss.

8. A relatively constant core body temperature, varying about ± _____ ° F. under normal daily conditions, reflects a ratio of body heat production to body heat loss which is _____ (G, greater than; E, equal to; L, less than) one.

   a. 1, G c. 1, L e. 5, E
   b. 1, E d. 5, G f. 5, L

9. The energy expenditure of an average sleeping adult is just sufficient in the absence of heat loss to increase the body temperature 1° C. in an interval closest to:

   a. 1 second d. 1 hour
   b. 1 minute e. 2 hours
   c. 30 minutes f. 6 hours

10. The rate of body heat production is _____ by sympathetic stimulation, and _____ by the direct effects of elevated temperatures upon cellular metabolic processes. (I, increased; N, not significantly influenced; D, decreased)

    a. I, I c. I, D e. D, N
    b. I, N d. D, I f. D, D

11. Chronic exposures to _____ environmental temperatures result in _____ rates of thyroxine secretion and increased rates of heat production. (I, increased; D, decreased)

    a. I, I   c. D, I
    b. I, D   d. D, D

12. Evaporation is a mechanism of body heat _____ (G, gain from; L, loss to) the environment; body heat is generally lost to the environment through mechanisms of conduction _____ (W, as well as; N, but not) radiation.

    a. G, W   c. L, W
    b. G, N   d. L, N

OBJECTIVE 72–3.

Identify the processes of radiation, conduction, and convection, their influencing factors, and their contributing roles in dissipation of body heat.

13. Heat loss by radiation varies _____ (D, directly; I, inversely) with the difference between the _____ (S, second; F, fourth) powers of the temperature of the body surface and the _____ (C, core body; A, average surrounding) temperature.

    a. D, S, C c. D, S, A e. I, S, C
    b. D, F, A d. I, F, A f. I, F, C

14. Radiation is a mechanism which provides for _____ (L, only body heat loss; G, potential heat gain as well as heat loss) and about _____ % of the total heat loss of the nude body at normal room temperatures.

    a. L, 25 c. L, 95 e. G, 60
    b. L, 60 d. G, 25 f. G, 95

15. Radiant loss of heat energy from the body involves largely wavelengths between 5 to 20 _____ (MC, millimicrons; C, microns) in the _____ (U, ultraviolet; V, visible; I, infrared) portion of the electromagnetic spectrum.

    a. MC, U   c. MC, I   e. C, V
    b. MC, V   d. C, U   f. C, I

16. Enhanced melanin pigmentation of skin results in decreased cholecalciferol formation, increased _____ (A, absorption; R, reflection) of visible wavelengths by the skin, and _____ (I, significantly increased; N, no significant change in; D, significantly decreased) infrared absorption and emission by the skin.

    a. A, I   c. A, D   e. R, N
    b. A, N   d. R, I   f. R, D

17. Conduction is a mechanism for heat loss, _____ (W, as well as potential; N, but not) heat gain, of the body by means of _____ (D, direct transference of molecular kinetic energy; E, emission of electromagnetic waves; L, an evaporative loss of body water).

    a. W, D   c. W, L   e. N, E
    b. W, E   d. N, D   f. N, L

18. A greater body heat loss generally occurs through heat conduction to surrounding _____ (A, air; S, solid objects), largely as a consequence of _____ (D, differences of heat capacity; V, convection; T, thermal conductivity).

    a. A, D   c. A, T   e. S, V
    b. A, V   d. S, D   f. S, T

19. The consequences of _____ (D, conduction; V, convection; R, radiation), or movement of air, are such that a wind velocity across the skin of 4 miles per hour is about _____ (T, twice; S, 16 times) as effective for cooling the skin as a velocity of 1 mile per hour.

    a. D, T   c. R, T   e. V, S
    b. V, T   d. D, S   f. R, S

20. Convection is a more significant variable in body heat loss to surrounding _____ (A, air; W, water) as a consequence of differences of specific heat and _____ (E, evaporation; T, thermal conductivity; R, radiation).

    a. A, E   c. A, R   e. W, T
    b. A, T   d. W, E   f. W, R

OBJECTIVE 72–4.

Identify the effects of water evaporation upon body heat loss, its significance, and conditions which alter evaporative heat loss.

21. An insensible rate of water evaporation from the skin and lungs of about _____ ml. per day is associated with a net _____ (G, gain; L, loss) of body heat of _____ Calorie(s) for each ml. of water evaporated.

    a. 60, G, 80   d. 600, L, 0.58
    b. 60, G, 0.58   e. 600, L, 80
    c. 60, L, 80   f. 600, G, 0.58

22. The rate of body heat loss by _____ mechanisms is substantially decreased by increased humidity, whereas the rate of body heat loss by _____ mechanisms is substantially decreased by environmental temperatures which approach or exceed core body temperature. (C, conduction; R, radiation; E, evaporation)

    a. Only C, both E & R   e. Both E & R,
    b. Only E, only R   only C
    c. Both E & C, only R   f. Only E, both
    d. Only E, only C   C & R

23. A greater _____ (E, evaporative; C, conductive) heat loss from the body on most hot days is _____ (L, only slightly; B, substantially) _____ (I, increased; D, decreased) by the presence of convection air currents.

    a. E, L, I   c. E, L, D   e. C, L, D
    b. E, B, I   d. C, B, D   f. C, B, I

24. _____ (L, light; D, dark) -colored _____ (P, plastic; C, cotton) clothing permits a greater net loss of body heat in hot climates, which _____ (I, is; N, is not) substantially different from that permitted by the nude body.

    a. L, P, I   c. D, P, I   e. L, C, I
    b. D, P, N   d. L, C, N   f. D, C, N

OBJECTIVE 72–5.
Identify the characteristics of sweating, its regulatory mechanisms, and its adaptive mechanisms.

25. The usual stimulus for _____ (A, apocrine; E, eccrine) sweating is excitation of the _____ (N, anterior; P, posterior) hypothalamus by _____ (I, increased; D, decreased) body temperature.

    a. A, N, I   c. A, N, D   e. E, N, I
    b. A, P, D   d. E, P, D   f. E, P, I

26. The sweat glands receive innervation by predominantly _____ (A, adrenergic; C, cholinergic) _____ (P, parasympathetic; S, sympathetic) fibers.

    a. A, P   c. C, P
    b. A, S   d. C, S

27. Adrenal medullary secretion, generally _____ (H, inhibited; A, augmented) by low body temperature and muscular exercise, _____ (I, increases; D, decreases) sweat gland secretion of particularly the _____ (T, body trunk and upper extremities; F, hands and feet).

    a. H, I, T   c. H, D, T   e. A, I, T
    b. H, D, F   d. A, D, F   f. A, I, F

28. Sweat is formed from a precursor fluid formed by _____ (S, active secretion; F, passive filtration) in the _____ (C, coiled; D, duct) portion of the sweat gland.

    a. S, C   c. F, C
    b. S, D   d. F, D

29. Maximal rates of sweating of _____ liters per hour, when compared to lesser rates, have _____ (H, higher; L, lower) concentrations of urea, lactate, and potassium, and _____ (H, higher; L, lower) concentrations of sodium and chloride.

    a. 1.5 to 4, H, L   d. 4.5 to 8, L, H
    b. 1.5 to 4, L, H   e. 4.5 to 8, H, L
    c. 1.5 to 4, L, L   f. 4.5 to 8, H, H

30. Sodium _____ (W, as well as; N, but not) chloride ion _____ (S, secretion; R, reabsorption) by the sweat gland duct is _____ (I, increased; D, decreased) by aldosterone.

    a. W, S, I   c. W, R, I   e. N, S, I
    b. W, R, D   d. N, R, D   f. N, S, D

31. An average individual living in the tropics since birth, when compared to individuals of temperate zones, has _____ (S, about the same; G, significantly greater) total numbers of sweat glands, a _____ (W, lower; H, higher) average maximum rate of sweat production, and a _____ (W, lower; H, higher) daily salt loss in sweat under comparable environmental conditions.

    a. S, W, W   c. S, H, H   e. G, H, W
    b. S, W, H   d. G, H, H   f. G, W, W

OBJECTIVE 72–6.
Recognize the role of the skin and subcutaneous tissues as a heat insulator for the body. Identify the functions of the cutaneous vascular system for the regulation of body heat loss.

32. A greater insulating quality of the _____ (S, skin; F, subcutaneous fat) results from the fact that it conducts heat about _____ (A, 1/30; B, 1/3; C, 30 times) as readily as most body tissues.

    a. S, A   c. S, C   e. F, B
    b. S, B   d. F, A   f. F, C

33. Arteriovenous anastomoses, particularly abundant in cutaneous regions of the _____ (A, arms, legs, and body trunk; H, hands, feet, and ears), are open during _____ (C, cold; W, warm) environmental conditions owing to _____ (D, decreased; I, increased) sympathetic tone.

    a. A, C, D   c. A, W, D   e. H, W, D
    b. A, C, I   d. H, W, I   f. H, C, I

34. The temperature difference between _____ (C, the body core; E, the environment; B, both C and E) and the cutaneous "radiator" system is increased by a fall in body temperature and a resultant _____ (I, increase; D, decrease) in cutaneous blood flow.

    a. only C, I   c. B, I   e. only E, D
    b. only E, I   d. only C, D   f. B, D

35. Stimulation of a heat _____ (C, conservation; S, loss) center of the anterior hypothalamus results in _____ (I, increased; D, decreased) sympathetic tone and cutaneous _____ (R, vasoconstriction; L, vasodilation).

    a. C, I, R   c. C, D, R   e. S, D, R
    b. C, I, L   d. S, D, L   f. S, I, L

OBJECTIVE 72–7.

Identify the variance of body temperature as a function of environmental temperature, and the roles of heat- and cold-sensitive hypothalamic neurons and cutaneous receptors in the regulation of body temperature.

36. The nude body is capable of maintaining essentially a normal body temperature between a minimal lower limit and a maximal upper limit of _____° F. for air of very _____ (H, high; L, low) humidity.

    a. 60 & 211, H   d. 60 & 211, L
    b. 60 & 130, H   e. 60 & 130, L
    c. 33 & 211, H   f. 33 & 211, L

37. Temperature receptors of the _____ (S, skin; L, liver; H, hypothalamus) have a greater functional significance in body temperature regulation by centers located in the _____ (M, medulla; H, hypothalamus).

    a. S, M   c. H, M   e. L, H
    b. L, M   d. S, H   f. H, H

38. The preoptic area of the _____ (A, anterior; P, posterior) hypothalamus contains a larger number of _____ (C, "cold-sensitive"; H, "heat-sensitive") neurons and functions as a _____ (R, primary; S, secondary) center for body temperature regulation.

    a. A, C, R   c. A, H, R   e. P, C, R
    b. A, H, S   d. P, H, S   f. P, C, S

39. Stimulation of _____ (A, anterior; P, posterior) hypothalamic neurons to produce shivering is attributed to a greater _____ (D, direct; I, indirect) stimulation of these neurons by _____ (E, elevated; R, reduced) body temperatures.

    a. A, D, E   c. A, I, E   e. P, I, E
    b. A, D, R   d. P, I, R   f. P, D, R

40. The temperature of the body's _____ (M, medullary; C, cutaneous; H, hypothalamic) "thermostat" refers to the temperature of its _____ (B, blood supply; S, "set point" for temperature regulatory mechanisms).

    a. M, B   c. H, B   e. C, S
    b. C, B   d. M, S   f. H, S

41. Hypothalamic temperatures above a critical value of _____° C. result in sweating and _____ (E, substantially elevated; C, relatively constant; D, substantially decreased) rates of body heat production.

    a. 37, E   c. 37, D   e. 40, C
    b. 37, C   d. 40, E   f. 40, D

42. The "hypothalamic thermostat" is more effective in _____ (I, increasing; D, decreasing) the rate of body heat production when the activity of the temperature-sensitive neurons of the preoptic area is _____ (H, high; L, low).

    a. I, H   c. D, H
    b. I, L   d. D, L

OBJECTIVE 72–8.

Identify the mechanisms of regulating body heat loss and conservation, their means of regulation, and their associated characteristics.

43. _____ (O, overheating; C, cooling) the preoptic area of the hypothalamus is more effective in _____ (D, decreasing body heat production; I, increasing evaporative heat loss) and _____ (L, dilating; S, constricting) cutaneous vessels.

    a. O, D, L   c. O, I, L   e. C, I, L
    b. O, D, S   d. C, I, S   f. C, D, S

44. Cutaneous vasoconstriction promotes body heat _____ (L, loss; C, conservation) and _____ (I, increases; D, decreases) the effectiveness of the insulating layer of the skin.

    a. L, I   c. C, I
    b. L, D   d. C, D

45. "Goose bumps," analogous to pilo-erection in animals, occur during predominant mechanisms of enhanced body heat _____ (L, loss; C, conservation) and are _____ (E, effective; I, ineffective) in altering the insulating thickness of skin _____ (W, as well as; N, but not) the adjacent air layer in humans.

    a. L, E, W   c. L, I, N   e. C, E, W
    b. L, E, N   d. C, I, W   f. C, E, N

46. Shivering is associated with _____ (S, stimulation; H, inhibition) of heat-sensitive neurons of the preoptic area of the hypothalamus, _____ (I, increased; D, decreased) muscle tone, and an overall _____ (I, increase; D, decrease) in muscle metabolic heat production.

    a. S, I, I   c. S, D, D   e. H, D, I
    b. S, D, I   d. H, I, I   f. H, D, D

47. _____ (I, increased; D, decreased) rates of adrenalin secretion during exposure to cold climatic conditions are associated with _____ (I, increased; D, decreased) rates of chemical thermogenesis and cutaneous _____ (C, vasoconstriction; L, vasodilation).

    a. I, I, C   c. I, D, C   e. D, D, C
    b. I, I, L   d. D, D, L   f. D, I, L

48. A higher degree of chemical thermogenesis, achieved by the _____ (I, newborn infant; A, adult), is attributed to the action of the _____ (P, parasympathetic; S, sympathetic) nervous system upon body stores of _____ (B, brown; Y, yellow) fat.

    a. I, P, B   c. I, S, B   e. A, S, B
    b. I, P, Y   d. A, S, Y   f. A, P, Y

49. A higher incidence of toxic thyroid goiter among individuals living in _____ (C, colder; W, warmer) climates is associated with _____ (H, higher; L, lower) than normal rates of secretion of thyrotropin-releasing factor.

    a. C, H   c. W, H
    b. C, L   d. W, L

50. Increased thyroid gland rates of secretion, _____ (W, as well as; N, but not) thyroid gland size, are related to _____ (A, acute; C, chronic) conditions of _____ (D, decreased; I, increased) body temperatures.

    a. W, A, I   c. W, C, D   e. N, A, D
    b. W, C, I   d. N, C, I   f. N, A, I

OBJECTIVE 72–9.

Identify the effects of cutaneous thermoreceptor activity in altering the "set point" of the hypothalamic thermostat, and the role of behavior and local skin reflexes in the control of body temperature.

51. A(n)_____ in cutaneous temperatures and fever-producing agents tends to_____ the hypothalamic "set point." (I, increase; D, decrease)

    a. I, I   c. D, I
    b. I, D   d. D, D

52. A psychic sensation of _____ (C, chill; O, being overheated) occurs when the "set point" for the hypothalamic thermostat is less than the core body temperature as a consequence of activity arising largely from the _____ (R, cutaneous receptors; PA, preoptic area; PH, posterior hypothalamus).

    a. C, R   c. C, PH   e. O, PA
    b. C, PA   d. O, R   f. O, PH

53. Cutaneous receptors responding to _____ (C, cold; W, warm) environments are thought to perform a greater role in corresponding mechanisms of body heat regulation, _____ (A, as well as; N, but not) in psychic sensations.

    a. C, A   c. W, A
    b. C, N   d. W, N

54. Localized cutaneous heating results in localized sweating and _____ (L, localized; W, widespread body) _____ (C, vasoconstriction; D, vasodilation) as a consequence of spinal cord reflexes operating _____ (I, independently of; J, in conjunction with) the hypothalamic temperature regulatory centers.

    a. L, C, I   c. L, D, J   e. W, D, I
    b. L, C, J   d. W, D, J   f. W, C, I

OBJECTIVE 72–10.

Identify the causes, underlying mechanisms, associated characteristics, and potentially harmful consequences of febrile conditions.

55. Pyrogens are exogenous _____ (W, as well as; N, but not) endogenous substances which _____ (I, increase; D, decrease) body temperature through their more direct actions upon _____ (M, medullary centers; H, hypothalamic centers; P, peripheral receptors).

    a. W, I, M   c. W, D, P   e. N, D, H
    b. W, I, H   d. N, D, M   f. N, I, P

56. Polymorphonuclear leukocytes, monocytes, macrophages, and reticuloendothelial cells are thought to _____ (F, facilitate; H, inhibit) the effects of exogenous pyrogens to _____ (I, increase; D, decrease) the "set point" of the hypothalamic thermostat.

    a. F, I   c. H, I
    b. F, D   d. H, D

57. Dehydration probably _____ the "set point" of the hypothalamic thermostat, _____ the heat capacity of the body, and _____ body temperature. (I, increases; N, does not significantly alter; D, decreases)

    a. I, I, N   c. D, I, D   e. N, D, I
    b. N, I, D   d. I, D, I   f. D, D, N

58. A blood temperature of 102° F. and a normal "set point" of the hypothalamic thermostat is associated with cutaneous _____ (C, vasoconstriction; D, vasodilation), _____ (W, sweating; H, shivering), _____ (A, and; N, but not) chills.

    a. C, H, A   c. C, H, N   e. D, H, A
    b. C, W, N   d. D, W, N   f. D, W, A

59. The "crisis" in febrile conditions refers to the onset of states of _____ (H, higher than normal; N, normal; L, lower than normal) body temperature, _____ (H, higher than normal; N, normal) "set points" of the hypothalamic thermostat, and a _____ (C, cold and dry; W, warm and moist) skin.

    a. H, H, C   c. L, N, C   e. N, H, C
    b. N, H, W   d. H, N, W   f. L, H, W

60. Chills occurring during febrile conditions are generally associated with _____ (H, higher; L, lower) than normal core body temperatures, _____ (C, cold; W, warm) skin, and predominant mechanisms of body heat _____ (V, conservation; D, dissipation).

    a. H, C, V   c. H, W, V   e. L, C, V
    b. H, C, D   d. L, W, D   f. L, C, D

61. The level of basal metabolism of body tissues _____ (I, increases; D, decreases) about _____ % per each degree C. rise in body temperature above normal.

    a. I, 0.6   c. I, 60   e. D, 10
    b. I, 10   d. D, 0.6   f. D, 60

62. A vicious cycle of increased body heat _____ (P, production; L, loss) is associated with the _____ (I, initiation; C, cessation) of sweating at body temperatures which exceed critical levels of about _____° F.

    a. P, I, 101 to 104   d. L, C, 101 to 104
    b. P, C, 107 to 110   e. L, I, 107 to 110
    c. P, C, 116 to 120   f. L, I, 116 to 120

63. Antipyrine and aminopyrine, _____ (W, as well as; N, but not) aspirin, act as _____ (P, pyretic; A, antipyretic) agents.

 a. W, P
 b. W, A
 c. N, P
 d. N, A

64. Maintained high fevers, which are more likely to result in permanent _____ (R, respiratory; N, neural; H, hematopoietic) tissue damage, may be effectively reduced by administering _____ (A, alcohol sponge baths; P, pyrexins; D, diuretics).

 a. R, P
 b. N, D
 c. H, A
 d. R, D
 e. N, A
 f. H, P

OBJECTIVE 72–11.
Identify the physiologic and pathologic consequences of exposure of the body to extremely cold conditions.

65. Ice water immersion for _____ (A, 2 to 3; B, 20 to 30) _____ (M, minutes; H, hours) generally results in a fall of the body temperature to _____° F. and cardiac fibrillation or standstill.

 a. A, M, 92
 b. B, M, 77
 c. A, H, 32
 d. A, M, 77
 e. B, M, 92
 f. A, H, 77

66. Minimal tissue injury and maximum effectiveness in warming an individual whose body temperature has fallen to critically low values are afforded by water immersion at:

 a. 37° C.
 b. 65° C.
 c. 99° C.
 d. 110° F.
 e. 165° F.
 f. 200° F.

67. Maximal _____ (C, constriction; D, dilation) of cutaneous vessels and _____ (E, enhanced; R, reduced) oxyhemoglobin dissociation are concomitant with skin temperatures approaching freezing.

 a. C, E
 b. C, R
 c. D, E
 d. D, R

73

Dietary Balances, Regulation of Feeding; Obesity and Starvation

OBJECTIVE 73-1.
Recognize the physiologically available energy (Calories/gram) for the three different foodstuffs, and the necessity for their inclusion in the diet.

1. The physiologically available energy per gram of each of the three types of foodstuffs is _____ Calories for carbohydrate, _____ Calories for fat, and _____ Calories for protein.

 a. 4, 9, 4
 b. 4, 40, 9
 c. 40, 9, 4
 d. 9, 4, 9
 e. 9, 4, 4
 f. 90, 90, 40

2. A higher percentage of _____ (C, carbohydrate; L, lipid; P, protein) absorption from the gastrointestinal tract contributes to an average for the three types of foodstuffs which is generally _____ (G, greater; S, less) than 90%.

 a. C, G
 b. L, G
 c. P, G
 d. C, S
 e. L, S
 f. P, S

3. A diet prescription is intended to yield a total of 2000 Calories. The diet includes 70 grams of protein and 85 grams of fat. The amount of carbohydrate required would be closest to _____ grams.

   a. 15   c. 150   e. 240
   b. 85   d. 175   f. 280

4. The average American diet provides the highest percentage of energy from _____ and the lowest percentage of energy from _____ . (C, carbohydrate; P, protein; F, fat)

   a. C, F   c. P, C   e. F, P
   b. C, P   d. P, F   f. F, C

5. A minimum of _____ grams of protein, largely complete proteins of particularly _____(A, animal; V, vegetable) origin, are required daily to maintain body protein levels.

   a. 5 to 10, A   d. 5 to 10, V
   b. 30 to 45, A   e. 30 to 45, V
   c. 200 to 300, A   f. 200 to 300, V

6. Kwashiorkor is _____ (U, an unsaturated fatty acid; V, a vitamin; A, an amino acid) deficiency state affecting especially _____ (D, adults; H, children).

   a. U, D   c. A, D   e. V, H
   b. V, D   d. U, H   f. A, H

7. Linoleic, linolenic, and arachidonic acids are _____ (U, unsaturated fatty acids; S, saturated fatty acids; A, amino acids) which are _____(T, toxic; E, essential) to body tissues.

   a. U, T   c. A, T   e. S, E
   b. S, T   d. U, E   f. A, E

8. Deficiencies of _____(C, carbohydrate; L, lipid) in the diet are frequently associated with muscular weakness, _____ (D, decreased; I, increased) rates of protein metabolism, and _____ (K, ketosis; O, obesity).

   a. C, D, K   c. C, I, K   e. L, D, K
   b. C, I, O   d. L, I, O   f. L, D, O

9. Caloric substances of foods containing largely carbohydrate, _____ (W, as well as; N, but not) those containing largely fat, comprise _____ (L, less; M, more) than 50% of the weight of the ingested foodstuffs.

   a. W, L   c. N, L
   b. W, M   d. N, M

OBJECTIVE 73–2.

Identify "nitrogen balance" in the body, its means of determination, and its significance.

10. Nitrogen comprises about _____ % of the weight of dietary protein and is largely excreted in the _____ (U, urine; F, feces).

    a. 6, U   c. 34, U   e. 16, F
    b. 16, U   d. 6, F   f. 34, F

11. The total nitrogen excretion, or _____ % of the urinary nitrogen excretion, when multiplied by _____ is a calculated estimate of the total protein catabolized.

    a. 110, 16   d. 110, 6.25
    b. 135, 0.16   e. 135, 16
    c. 190, 6.25   f. 190, 0.16

12. A _____ (P, positive; N, negative) nitrogen balance, or a condition of greater protein utilization than protein intake, occurs in response to _____ (G, glucocorticoid hormone; W, growth hormone; T, testosterone).

    a. P, G   c. P, T   e. N, W
    b. P, W   d. N, G   f. N, T

OBJECTIVE 73–3.

Identify the respiratory quotient, its variance, and its utility for estimating relative rates of carbohydrate and fat metabolism.

13. The respiratory quotient, defined as a volume ratio of metabolic _____ (O, oxygen used; C, carbon dioxide produced), is generally _____ (G, greater; L, less) than 1.00.

   a. O/C, G   c. C/O, G
   b. O/C, L   d. C/O, L

14. A higher respiratory quotient for _____ (C, carbohydrates; D, lipids; P, proteins) is _____ (G, greater than; E, equal to; L, less than) 1.00.

   a. C, E   c. P, G   e. D, L
   b. D, G   d. C, L   f. P, E

15. A nonprotein respiratory quotient of _____ implies approximately equal utilization of carbohydrate and fat, whereas a nonprotein respiratory quotient of _____ implies almost 100% metabolic utilization of fat.

   a. 0.71, 0.85   d. 0.85, 1.00
   b. 0.71, 1.00   e. 1.00, 0.71
   c. 0.85, 0.71   f. 1.00, 0.85

16. A lower metabolic utilization of _____ for metabolic energy brings the average respiratory quotient closer to the respiratory quotient for _____. (C, carbohydrate; L, lipid; P, protein)

   a. C, P   c. P, L   e. L, L
   b. L, C   d. C, C   f. P, P

17. The respiratory quotient shortly after a typical meal approaches a value of _____ , whereas 8 to 10 hours after a meal it approaches _____ .

   a. 0.71, 0.85   c. 0.85, 0.71   e. 1.00, 0.71
   b. 0.71, 1.00   d. 0.85, 1.00   f. 1.00, 0.85

18. A respiratory quotient which is greater than 1.00 occurs _____ (Q, frequently; I, infrequently) in man and results from the conversion of _____ (F, fat into carbohydrate; C, carbohydrate into fat; P, protein into carbohydrate).

   a. Q, F   c. Q, P   e. I, C
   b. Q, C   d. I, F   f. I, P

OBJECTIVE 73–4.

Identify the factors and neural centers involved in the regulation of food intake.

19. Satiety is the _____ (S, same; O, opposite) as hunger, and is elicited by stimulation of the _____ (L, lateral; V, ventromedial) nuclei of the _____ (H, hypothalamus; T, thalamus).

   a. S, L, H   c. S, V, H   e. O, V, T
   b. S, L, T   d. O, V, H   f. O, L, H

20. "Psychic blindness" in the choice of foodstuffs is a major consequence of a loss of function of the _____ (G, "gnostic area"; A, amygdala; E, Edinger-Westphal nuclei), whereas the actual mechanisms of feeding are largely controlled by more immediate centers located within the _____ (C, cerebral cortex; B, brainstem).

   a. G, C   c. E, C   e. A, B
   b. A, C   d. G, B   f. E, B

21. Mechanisms which regulate the long-term quantity of food intake are thought to _____ (D, directly involve; I, be largely independent of) the "feeding center" of the _____ (L, lateral; V, ventromedial) nuclei of the hypothalamus.

   a. D, L   c. I, L
   b. D, V   d. I, V

22. Activity of the feeding center is increased by _____ (I, increased; D, decreased) blood glucose levels and _____ (I, increased; N, not significantly influenced; D, decreased) by gastric distention.

   a. I, I   c. I, D   e. D, N
   b. I, N   d. D, I   f. D, D

23. A _____ (M, more; L, less) powerful mechanism than the glucostatic mechanism involves increased activity of the hunger center by _____ (E, elevated; D, diminished) blood concentrations of amino acids.

    a. M, E   c. L, E
    b. M, D   d. L, D

24. The long-term average concentration of free fatty acids in blood varies _____ (D, in direct proportion to; I, independently of; V, in inverse proportion to) the quantity of adipose tissue of the body, and is thought to be a _____ (N, minor; J, major) factor in the regulation of food intake.

    a. D, N   c. V, N   e. I, J
    b. I, N   d. D, J   f. V, J

25. Forced starvation in animals tends to foster voluntary states of _____ (U, undereating; O, overeating), while cold climates tend to _____ (I, increase; D, decrease) the dietary intake.

    a. U, I   c. O, I
    b. U, D   d. O, D

26. Receptors of the gastrointestinal tract, _____ (W, as well as; N, but not) feeding habits, are generally considered to be _____ (S, short-term; L, long-term) regulatory mechanisms of food intake.

    a. W, S   c. N, S
    b. W, L   d. N, L

27. Short-term regulatory mechanisms of food intake, involving peripheral gastric _____ (W, as well as; B, but not) oral and pharyngeal receptors, are primarily related to _____ (N, mechanisms of nutritional regulation; G, gastrointestinal levels of function).

    a. W, N   c. B, N
    b. W, G   d. B, G

OBJECTIVE 73–5.
Identify obesity, its causes, and its treatment.

28. Extremely obese individuals when compared to leaner individuals are thought to have _____ numbers of fat cells in the body and _____ settings of the hypothalamic feedback mechanism for control of adipose tissues. (S, the same; H, substantially higher)

    a. S, S   c. H, S
    b. S, H   d. H, H

29. An excess of energy input over energy expenditure is required for the development _____ (W, as well as; N, but not) the maintenance of obesity, since each gram of stored fat represents _____ Calories of excess energy.

    a. W, 0.83   c. W, 9.3   e. N, 0.48
    b. W, 4.8   d. N, 0.83   f. N, 9.3

30. Psychogenic factors are probably _____ (I, insignificant; S, highly significant) in the development of obesity. Obese individuals reducing to normal weights generally exhibit _____ (N, a normal; G, a far greater than normal) level of hunger.

    a. I, N   c. S, N
    b. I, G   d. S, G

31. Obesity is frequently associated with _____ (P, pituitary; A, pancreatic; T, thyroid) tumors that result in a loss of function of the _____ (L, lateral; VM, ventromedial) nuclei of the hypothalamus.

    a. P, L   c. T, L   e. A, VM
    b. A, L   d. P, VM   f. T, VM

32. A genetic lack of lipoprotein lipase of adipose tissue in rats _____ the quantity of stored fat and _____ rates of insulin secretion. (I, increases; N, does not significantly alter; D, decreases)

    a. I, I   c. I, D   e. D, N
    b. I, N   d. D, I   f. D, D

OBJECTIVE 73–6.

Identify inanition and its causes. Identify the consequences of starvation.

33. Anorexia nervosa is a condition of _____ (O, obesity; I, inanition) resulting from _____ (H, hypothalamic lesions; P, psychogenic abnormalities).

   a. O, H   c. I, H
   b. O, P   d. I, P

34. During starvation _____ stores in the body are depleted first, followed by _____ and then _____ depletion. (F, fat; C, carbohydrate; P, protein)

   a. F, C, P   c. C, P, F   e. P, F, C
   b. F, P, C   d. C, F, P   f. P, C, F

35. Body stores of glycogen, totaling several _____ (G, grams; H, hundred grams; K, kilograms), are present in major amounts in liver _____ (W, as well as; N, but not) muscle tissues.

   a. G, W   c. K, W   e. H, N
   b. H, W   d. G, N   f. K, N

36. The rate of utilization of protein stores is more rapid during the _____ week of a period of starvation and provides glucose which is largely utilized by the _____ (H, heart; B, brain; L, liver).

   a. 1st, H   c. 1st, L   e. 3rd, B
   b. 1st, B   d. 3rd, H   f. 3rd, L

37. Ketone bodies, largely produced by the _____ (L, liver; B, brain) during high rates of _____ (P, protein; D, lipid) catabolism, _____ (A, are; N, are not) utilized by the brain during starvation.

   a. L, P, A   c. L, D, A   e. B, D, A
   b. L, P, N   d. B, D, N   f. B, P, N

38. Death as a result of starvation occurs when the _____ (G, glycogen; F, neutral fat; P, protein) of the body is reduced to about _____ % of its normal level.

   a. G, 10 to 20   d. G, 50
   b. F, 10 to 20   e. F, 50
   c. P, 10 to 20   f. P, 50

39. Mild vitamin deficiencies, particularly of the _____ (W, water; F, fat) -soluble vitamin _____ groups, first become manifest at the end of several _____ (K, weeks; M, months) following a loss of their dietary intake.

   a. W, B & C, K   d. F, A & D, M
   b. W, B & C, M   e. F, A & D, K
   c. W, A & D, K   f. F, B & C, M

74

Vitamin and Mineral Metabolism

OBJECTIVE 74–1.

Identify vitamins, the relative storage of vitamins in the body, and relative magnitudes of the daily requirements of the vitamins.

1. Vitamins are _____ (O, organic; I, inorganic) compounds required in _____ (S, small; J, major) quantities for body metabolism which _____ (A, are; N, are not) synthesized by the body.

   a. O, S, A   c. O, J, N   e. I, J, A
   b. O, S, N   d. I, J, N   f. I, S, A

2. Daily vitamin requirements are _____ by growth, _____ by exercise, and _____ by disease and fever. (I, increased; N, not significantly altered; D, decreased)

   a. I, N, N   c. I, I, I   e. N, D, D
   b. I, I, N   d. I, I, D   f. N, D, N

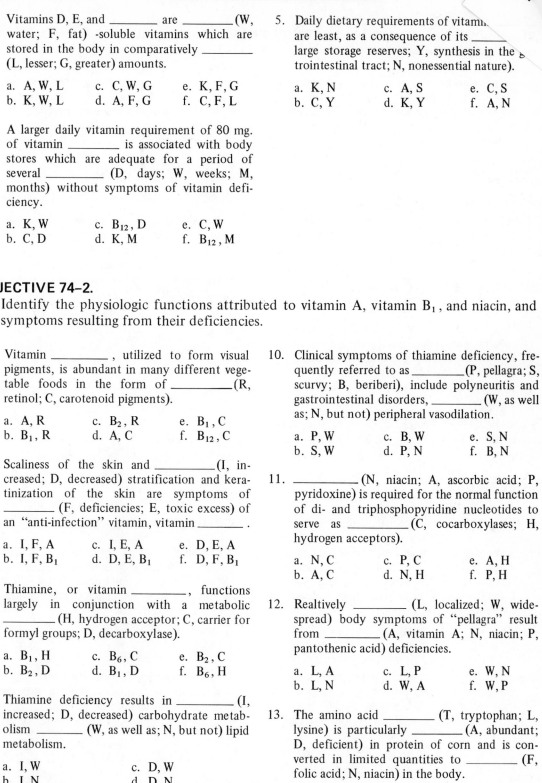

3. Vitamins D, E, and _____ are _____ (W, water; F, fat) -soluble vitamins which are stored in the body in comparatively _____ (L, lesser; G, greater) amounts.

a. A, W, L   c. C, W, G   e. K, F, G
b. K, W, L   d. A, F, G   f. C, F, L

4. A larger daily vitamin requirement of 80 mg. of vitamin _____ is associated with body stores which are adequate for a period of several _____ (D, days; W, weeks; M, months) without symptoms of vitamin deficiency.

a. K, W   c. B_{12}, D   e. C, W
b. C, D   d. K, M   f. B_{12}, M

5. Daily dietary requirements of vitamin ___ are least, as a consequence of its _____ large storage reserves; Y, synthesis in the gastrointestinal tract; N, nonessential nature).

a. K, N   c. A, S   e. C, S
b. C, Y   d. K, Y   f. A, N

OBJECTIVE 74–2.

Identify the physiologic functions attributed to vitamin A, vitamin B_1, and niacin, and the symptoms resulting from their deficiencies.

6. Vitamin _____ , utilized to form visual pigments, is abundant in many different vegetable foods in the form of _____ (R, retinol; C, carotenoid pigments).

a. A, R   c. B_2, R   e. B_1, C
b. B_1, R   d. A, C   f. B_{12}, C

7. Scaliness of the skin and _____ (I, increased; D, decreased) stratification and keratinization of the skin are symptoms of _____ (F, deficiencies; E, toxic excess) of an "anti-infection" vitamin, vitamin _____ .

a. I, F, A   c. I, E, A   e. D, E, A
b. I, F, B_1 d. D, E, B_1 f. D, F, B_1

8. Thiamine, or vitamin _____ , functions largely in conjunction with a metabolic _____ (H, hydrogen acceptor; C, carrier for formyl groups; D, decarboxylase).

a. B_1, H   c. B_6, C   e. B_2, C
b. B_2, D   d. B_1, D   f. B_6, H

9. Thiamine deficiency results in _____ (I, increased; D, decreased) carbohydrate metabolism _____ (W, as well as; N, but not) lipid metabolism.

a. I, W   c. D, W
b. I, N   d. D, N

10. Clinical symptoms of thiamine deficiency, frequently referred to as _____ (P, pellagra; S, scurvy; B, beriberi), include polyneuritis and gastrointestinal disorders, _____ (W, as well as; N, but not) peripheral vasodilation.

a. P, W   c. B, W   e. S, N
b. S, W   d. P, N   f. B, N

11. _____ (N, niacin; A, ascorbic acid; P, pyridoxine) is required for the normal function of di- and triphosphopyridine nucleotides to serve as _____ (C, cocarboxylases; H, hydrogen acceptors).

a. N, C   c. P, C   e. A, H
b. A, C   d. N, H   f. P, H

12. Realtively _____ (L, localized; W, widespread) body symptoms of "pellagra" result from _____ (A, vitamin A; N, niacin; P, pantothenic acid) deficiencies.

a. L, A   c. L, P   e. W, N
b. L, N   d. W, A   f. W, P

13. The amino acid _____ (T, tryptophan; L, lysine) is particularly _____ (A, abundant; D, deficient) in protein of corn and is converted in limited quantities to _____ (F, folic acid; N, niacin) in the body.

a. T, A, F   c. T, D, N   e. L, D, F
b. T, A, N   d. L, D, N   f. L, A, F

OBJECTIVE 74–3.

Identify the physiologic functions attributed to riboflavin, vitamin B_{12}, and folic acid, and the symptoms resulting from their deficiences.

14. Deficiencies of vitamin _____ , a constituent of FMN and FAD, result in symptoms which are similar in more respects to those occurring for _____ (N, niacin; F, folic acid; A, ascorbic acid) deficiencies.

 a. B_1, N   c. B_{12}, A   e. B_2, N
 b. B_2, F   d. B_1, A   f. B_{12}, F

15. Inflammation and cracking at the corners of the mouth, or _____ (C, cheilosis; B, beriberi), is a more common symptom of _____ (K, vitamin K; R, riboflavin; P, pantothenic acid) deficiency in man.

 a. C, K   c. C, P   e. B, R
 b. C, R   d. B, K   f. B, P

16. Vitamin _____ is a _____ (P, copper; B, cobalt) - containing vitamin which functions as a hydrogen-accepting coenzyme and as a co-enzyme for _____ (O, oxidizing; R, reducing) ribonucleotides to deoxyribonu-cleotides.

 a. B_6, P, O   c. B_6, B, O   e. B_{12}, B, O
 b. B_6, P, R   d. B_{12}, B, R   f. B_{12}, P, R

17. Vitamin _____ deficiency results in a form of _____ (H, hemolytic; M, maturation failure) anemia and impaired function of par-ticularly _____ (L, large myelinated; S, small unmyelinated) nerve fibers of the spinal cord.

 a. B_6, H, L   c. B_6, M, L   e. B_{12}, M, L
 b. B_6, H, S   d. B_{12}, M, S   f. B_{12}, H, S

18. Folic acid, or _____ (T, pteroylglutamic; P, pantothenic) acid, functions as an important _____ (G, carrier of hydroxymethyl and formyl groups; C, coenzyme in transamination reactions; O, oxidizing agent for tyrosine and phenylalanine).

 a. T, G   c. T, O   e. P, C
 b. T, C   d. P, G   f. P, O

OBJECTIVE 74–4.

Identify the physiologic functions attributed to pyridoxine, pantothenic acid, and ascorbic acid, and the symptoms resulting from their deficiencies.

19. A primary function of folic acid is to _____, while pyridoxine functions to _____ . (T, transaminate amino acids; G, promote growth; C, enhance calcium ion transport; B, promote oxidative phosphorylation)

 a. T, B   c. G, T   e. C, T
 b. B, C   d. G, B   f. B, G

20. _____ (T, thiamine; P, pantothenic acid; V, vitamin B_6) is incorporated into coenzyme A and performs a major role in carbohydrate _____ (W, as well as; N, but not) lipid metabolism.

 a. T, W   c. V, W   e. P, N
 b. P, W   d. T, N   f. V, N

21. Diminished rates of formation of interstitial matrix of subcutaneous tissues, _____ (W, as well as; N, but not) bone, and a failure of wounds to heal are characteristic of vitamin _____ deficiencies.

 a. W, C   c. W, E   e. N, D
 b. W, D   d. N, C   f. N, E

22. Platelet deficiencies and vitamin _____ deficiencies are similar in that they both _____ (I, increase; D, decrease) _____ (F, capillary fragility; E, erythropoiesis).

 a. B_{12}, I, E   c. B_{12}, D, F   e. C, D, F
 b. B_{12}, D, E   d. C, D, E   f. C, I, F

23. Vitamin C or _____ (C, citric; A, ascorbic; P, pantothenic) acid is considered a _____ (S, cause; R, cure) for _____ (V, scurvy; B, beriberi).

 a. C, S, V   c. P, R, B   e. A, R, V
 b. A, S, B   d. C, R, B   f. P, S, V

OBJECTIVE 74–5.

Identify the physiologic functions attributed to vitamins D, E, and K, and the symptoms resulting from their deficiencies.

24. Vitamin _____ functions in the regulation of calcium absorption from the gastrointestinal tract _____ (W, as well as; N, but not) calcium deposition in bone.

    a. C, W   c. E, W   e. D, N
    b. D, W   d. C, N   f. E, N

25. Ultraviolet irradiation of _____ (C, cholecalciferol; G, ergocalciferol; H, 7-dehydocholesterol) in the skin results in the formation of the natural vitamin _____ , _____ (C, cholecalciferol; G, ergocalciferol; H, 7-dehydrocholesterol).

    a. C, D, G   c. G, D, C   e. H, D, C
    b. C, E, H   d. G, E, H   f. H, E, G

26. Deficiencies of vitamin _____ , sometimes termed the "anti-sterility vitamin," are relatively _____ (R, rare; C, common) forms of vitamin deficiency.

    a. D, R   c. K, R   e. E, C
    b. E, R   d. D, C   f. K, C

27. Vitamin _____ is thought to play a major role in lipid metabolism, since its deficiency is associated with _____ (I, increased; R, reduced) cellular ratios of unsaturated to saturated fatty acids.

    a. D, I   c. K, I   e. E, R
    b. E, I   d. D, R   f. K, R

28. Dicumarol _____ (R, replaces; I, inhibits) major functions of vitamin _____ utilization in _____ (P, protein; L, lipid) synthesis.

    a. R, E, P   c. R, K, L   e. I, K, P
    b. R, E, L   d. I, K, L   f. I, E, P

29. Antibiotic drugs more frequently result in _____ (M, muscular dystrophy-like; H, hemorrhagic) symptoms as a consequence of vitamin _____ deficiencies.

    a. M, D   c. M, K   e. H, E
    b. M, E   d. H, D   f. H, K

OBJECTIVE 74–6.

Identify the relative daily requirements for major mineral constituents of the body. Identify the characteristics of magnesium, calcium, phosphorus, and iron utilization in the body.

30. Daily requirements for minerals include about 3.5 grams of _____ and 12 mg. of _____ .

    a. K, Ca   c. Ca, PO$_4$ e. Cl, K
    b. Cl, Fe   d. K, PO$_4$ f. Ca, Fe

31. Elevation of the normal extracellular concentration of magnesium of _____ mEq./liter results in _____ (I, increased; D, decreased) activity of the nervous system, while an abnormally low concentration is associated with peripheral _____ (VC, vasoconstriction; VD, vasodilation).

    a. 1.8 to 2.5, I, VC   d. 18 to 25, D, VD
    b. 1.8 to 2.5, D, VC   e. 18 to 25, I, VD
    c. 1.8 to 2.5, D, VD   f. 18 to 25, I, VC

32. _____ is a more abundant mineral of the body, primarily as a consequence of its high concentration in _____ (I, intracellular fluids; E, extracellular fluids; M, mineralized tissue).

    a. Na, M   c. Ca, M   e. Na, E
    b. Fe, I   d. K, I   f. Ca, I

33. _____ (I, increased; D, decreased) levels of extracellular calcium ion concentrations result in tetany and _____ (E, enhanced; R, reduced) cardiac contractility.

    a. I, E   c. D, E
    b. I, R   d. D, R

34. _____ (C, chloride; B, bicarbonate; P, phosphate) is the major anion of intracellular fluids and performs _____ (J, major; N, only minor) roles in metabolic functions of the cell.

    a. C, J   c. P, J   e. B, N
    b. B, J   d. C, N   f. P, N

35. The major fraction of body iron exists in the form of:

    a. Haptoglobin   d. Myoglobin
    b. Ferritin   e. Hemoglobin
    c. Hemosiderin   f. Cytochromes

OBJECTIVE 74–7.

Identify the important trace elements in the body, their functions, and the consequences of their deficiences.

36. Cobalt is more directly involved in the formation of the _____ (H, hemoglobin content; S, structure) of red blood cells, since cobalt excess is associated with _____ (I, increased; N, comparatively little change in; D, decreased) blood hemoglobin and _____ (I, increased; N, comparatively little change in; D, decreased) numbers of circulating erythrocytes.

    a. H, I, N   c. H, D, I   e. S, N, I
    b. H, N, D   d. S, I, I   f. S, D, D

37. _____ (Zn, zinc; Co, cobalt; Cu, copper) is an important inorganic constituent of carbonic anhydrase _____ (W, as well as; N, but not) lactic dehydrogenase.

    a. Zn, W   c. Cu, W   e. Co, N
    b. Co, W   d. Zn, N   f. Cu, N

38. _____ (Mn, manganese; Cu, copper) is an important inorganic constituent of cytochrome oxidase, whose deficiency results in a _____ (I, microcytic; A, macrocytic), _____ (R, hyperchromic; N, normochromic; O, hypochromic) type of anemia.

    a. Mn, I, O   c. Mn, A, O   e. Cu, I, R
    b. Mn, A, N   d. Cu, I, N   f. Cu, A, R

39. _____ (Zn, zinc; Co, cobalt; Mn, manganese) is required for the action of the enzyme arginase and _____ (A, ammonia; H, hemoglobin; U, urea) synthesis.

    a. Zn, H   c. Mn, A   e. Co, H
    b. Co, U   d. Zn, A   f. Mn, U

40. Thyroxine, containing _____ (F, fluorine; I, iodine; Mn, manganese), has an endocrine target action upon the _____ (T, thyroid gland; N, neurohypophysis; E, entire body).

    a. F, N   c. Mn, E   e. I, E
    b. I, T   d. F, T   f. Mn, N

41. _____ (I, iodine; F, fluorine; Br, bromine) is considered an enzyme _____ (A, activator; N, inactivator), whose presence in trace amounts is utilized to combat dental caries.

    a. I, A   c. Br, A   e. F, N
    b. F, A   d. I, N   f. Br, N

75

Introduction to Endocrinology; and the Pituitary Hormones

OBJECTIVE 75-1.

Identify the general characteristics of hormones.

1. Hormones of the posterior pituitary and adrenal _____ (C, cortex; M, medulla) are among a _____ (J, majority; N, minority) of endocrine tissues in that their secretion is predominantly under the control of direct _____ (U, neural; H, hormonal) regulation.

   a. C, J, U   c. C, N, H   e. M, N, U
   b. C, J, H   d. M, N, H   f. M, J, U

2. General _____ (W, as well as; N, but not) local hormones are usually secreted by specific endocrine glands, and include _____ (S, only steroid; P, only polypeptide; B, both steroid and polypeptide) substances.

   a. W, S   c. W, B   e. N, P
   b. W, P   d. N, S   f. N, B

3. Local hormones include _____ , whereas _____ are considered general hormones. (T, testosterone; A, acetylcholine; S, secretin; O, oxytocin; C, cholecystokinin)

   a. A, S, & C; T & O   d. T & A; S, O, & C
   b. A; T, S, O, & C   e. T, A, & S; O & C
   c. T, A, S, & C; O   f. T, S, & O; A & C

4. Secretion rates of general hormones, generally _____ (I, independent of; D, dependent upon) target cell functions, function largely to _____ (T, initiate; G, regulate) _____ (M, metabolic; C, contractile) functions of target tissues.

   a. I, T, M   c. I, G, M   e. D, T, C
   b. I, T, C   d. D, G, M   f. D, T, M

5. Radioimmunoassay techniques, _____ (L, lacking; P, possessing) the sensitivity required to assay the majority of hormone concentrations of the body, involves hormone _____ (C, competition for; S, saturation of) antibody binding sites by the hormone.

   a. L, C   c. P, C
   b. L, S   d. P, S

OBJECTIVE 75-2.

Identify the general hormone(s) produced by the various endocrine glands.

Directions: Place the name of the hormone(s) next to the endocrine gland that produces it.

6. _____ Anterior pituitary
7. _____ Posterior pituitary
8. _____ Adrenal cortex
9. _____ Thyroid
10. _____ Pancreas
11. _____ Ovaries
12. _____ Testes
13. _____ Parathyroid

a. Triiodothyronine
b. Insulin
c. Estrogen
d. TSH
e. Oxytocin
f. Aldosterone
g. PTH
h. ACTH
i. Cortisol
j. Testosterone

k. GH
l. ADH
m. FSH
n. Progesterone
o. LH
p. Thyroxine
q. Glucagon
r. Calcitonin
s. Prolactin
t. MSH

OBJECTIVE 75–3.

Identify the general mechanisms of target actions of hormones.

14. Cyclic AMP, _____ (W, as well as; N, but not) cyclic guanosine monophosphate, is a probable _____ (B, bloodborne; I, intracellular) _____ (P, "primary; S, "secondary) messenger" of hormone mediation.

   a. W, B, P   c. W, I, S   e. N, I, P
   b. W, B, S   d. N, I, S   f. N, B, P

15. Direct activation of a _____ (M, messenger RNA; C, cyclic AMP) mechanism is a known target action of catecholamines and glucagon, _____ (W, as well as; N, but not) luteninizing and follicle-stimulating hormones.

   a. M, W   c. C, W
   b. M, N   d. C, N

16. Prostaglandins, proposed as an intracellular "second messenger," are _____ (P, protein; L, lipid) compounds of _____ (R, largely the prostate; M, multiple types of) tissues with inhibitory _____ (W, as well as; N, but not) excitatory cellular functions.

   a. P, R, W   c. P, M, W   e. L, M, W
   b. P, R, N   d. L, M, N   f. L, R, N

17. _____ (P, polypeptide; S, steroid) hormones of the adrenal cortex, ovaries, and testes are thought to act upon _____ (M, cell membrane; C, cytoplasmic) receptor proteins to activate _____ (G, specific genes; A, adenylcyclase).

   a. P, M, G   c. P, C, A   e. S, C, G
   b. P, M, A   d. S, C, A   f. S, M, G

18. Proteins which increase _____ (K, potassium; Na, sodium) ion reabsorption by renal tubular cells are _____ (I, increased; D, decreased) in concentration by aldosterone after a latent period of about _____ .

   a. K, I, 2 to 3 min.   d. Na, D, 10 to 12 hrs.
   b. K, D, 2 to 3 min.   e. Na, I, 45 min.
   c. K, D, 45 min.   f. Na, I, 10 to 12 hrs.

OBJECTIVE 75–4.

Identify the major parts of the pituitary gland, their embryological origins, their endocrine secretions, and the general regulatory mechanisms of their secretion.

19. The pituitary gland, or _____ (H, hypophysis; HP, hypothalamus; M, mammillary body), weighs about _____ gram(s), and is anatomically _____ (S, separated from; C, connected to) the brain.

   a. H, 0.5, C   c. M, 0.5, S   e. HP, 50, C
   b. HP, 0.5, C   d. H, 50, S   f. M, 50, S

20. The pars intermedia, separating the posterior or _____ (A, adenohypophysis; N, neurohypophysis) from the anterior part of the pituitary, is _____ (W, relatively well developed; B, almost absent) in humans.

   a. A, W   c. N, W
   b. A, B   d. N, B

21. Rathke's pouch is an embryonic invagination of the _____ (H, hypothalamus; E, pharyngeal epithelium; I, infundibulum) which gives rise to the _____ (A, anterior; P, posterior) lobe of the pituitary.

   a. H, A   c. I, A   e. E, P
   b. E, A   d. H, P   f. I, P

22. The posterior pituitary is known to secrete _____ major hormones which include _____ (P, prolactin; O, oxytocin; C, cortisol).

   a. 2, P   c. 2, C   e. 6, O
   b. 2, O   d. 6, P   f. 6, C

23. _____ (A, adenohypophyseal; N, neurohypophyseal) hormones perform significant regulatory roles in carbohydrate and lipid, _____ (W, as well as; B, but not) protein, metabolism.

   a. A, W   c. N, W
   b. A, B   d. N, B

24. Adrenocorticotropin _____ (I, increases; D, decreases) liver gluconeogenesis and exerts a target action to _____ (I, increase; D, decrease) secretion rates of the adrenal _____ (M, medulla; C, cortex).

a. I, I, M   c. I, D, C   e. D, D, M
b. I, I, C   d. D, D, C   f. D, I, M

25. Gonadotropins _____ are secreted by the _____ (A, anterior lobe; P, posterior lobe) of the pituitary.

a. TSH & FSH, A   d. ADH & TH, P
b. LH & TSH, P   e. Oxytocin & FSH, A
c. LH & FSH, A   f. Oxytocin & ADH, P

26. Thyroid-stimulating hormone is secreted by the _____ (N, neurohypophysis; A, adenohypophysis) and _____ (D, directly; I, indirectly) _____ (E, enhances; R, reduces) metabolic rates of most body tissues.

a. N, D, E   c. N, I, R   e. A, I, E
b. N, D, R   d. A, I, R   f. A, D, E

27. Secretions of the posterior _____ (W, as well as; N, but not) the anterior lobe of the pituitary are largely controlled by signals transmitted directly from the _____ (T, target tissues; H, hypothalamus; I, pars intermedia).

a. W, T   c. W, I   e. N, H
b. W, H   d. N, T   f. N, I

28. Hypothalamic-hypophyseal portal vessels mediate releasing _____ (W, as well as; N, but not) inhibitory factors for _____ (A, only the anterior; P, only the posterior; B, both the anterior and the posterior) lobes of the pituitary.

a. W, A   c. W, B   e. N, P
b. W, P   d. N, A   f. N, B

29. The _____ (H, hypothalamus; P, pineal body) performs the major role in regulating pituitary secretion in response to _____ (S, select neural signals; D, diverse neural signals; E, endocrine secretions of target glands).

a. H, only S   d. P, only S
b. H, only S & E   e. P, only S & E
c. H, D & E   f. P, D & E

OBJECTIVE 75–5.

Using the following diagram, identify the various parts of the pituitary gland, including the hypothalamic-hypophyseal portal system.

Directions: Match the lettered headings with the diagram and numbered list of descriptive words and phrases.

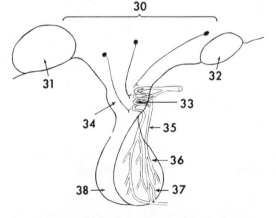

(From Guyton, A.C., Textbook of Medical Physiology, 5th ed. Philadelphia. W.B. Saunders Company, 1976.)

30. _____ Basal portion of the diencephalon.
31. _____ Contains the terminations of the fibers of the columns of the fornix.
32. _____ Area of partial decussation of the two optic nerves.
33. _____ Upper capillary network of the portal system.
34. _____ Enlarged vascular area on the infundibular stalk.
35. _____ Vessel or pathway connecting upper and lower capillary networks.
36. _____ Lower capillary network of the anterior lobe.
37. _____ Adenohypophysis, embryologically associated with Rathke's pouch.
38. _____ Neurohypophysis, outgrowth of the hypothalamus.

a. Sinuses
b. Anterior hypophysis
c. Primary capillary plexus

d. Median eminence
e. Hypothalamic-hypophyseal portal vessels
f. Hypothalamus

g. Optic chiasma
h. Mammillary bodies
i. Posterior pituitary

OBJECTIVE 75-6.

Identify the acidophils, basophils, and chromophobes of the adenohypophysis, and the mechanisms of action, function, and loci of control of hypothalamic releasing and inhibitory factors.

39. Basophils of the adenohypophysis are a functionally _____ (O, homogenous; E, heterogenous) group of cells which are characterized on the basis of their _____ (L, local hormone secretion; G, general hormone secretion; S, staining affinities).

    a. O, L   c. O, S   e. E, G
    b. O, G   d. E, L   f. E, S

40. _____ produce growth hormone and prolactin, _____ produce luteinizing, follicle-stimulating, and thyroid-stimulating hormones, and _____ are thought to produce adrenocorticotropin. (C, chromophobes; A, acidophils; B, basophils)

    a. C, B, A   c. A, C, B   e. B, A, C
    b. C, A, B   d. A, B, C   f. B, C, A

41. Releasing and inhibitory factors, formed by specialized neurons of the _____ , are secreted into the region of the _____ . (M, median eminence; A, anterior lobe of the pituitary; H, hypothalamus; P, posterior lobe of the pituitary)

    a. H, A   c. A, P   e. P, M
    b. H, M   d. A, M   f. P, A

42. The function of the releasing and inhibitory factors is to regulate the secretion of _____ (A, only the anterior; P, only the posterior; B, both the anterior and posterior) lobes of the pituitary by means of _____ (V, vascular portal; H, hypothalamic-hypophyseal neural) mechanisms of conveyance.

    a. A, V   c. B, V   e. P, H
    b. P, V   d. A, H   f. B, H

43. _____ (R, releasing; I, inhibitory) factors perform a more prominent regulatory role in the secretion of adenohypophyseal hormones, with the notable exception of _____ (FSH, follicle-stimulating hormone; GH, growth hormone; P, prolactin).

    a. R, FSH   c. R, P   e. I, GH
    b. R, GH   d. I, FSH   f. I, P

44. Pituitary transplants to ectopic loci decrease FSH secretion to less than 25% of normal. LTH, or _____ (O, oxytocin; L, luteinizing hormone; P, prolactin) secretion, however, increases to about three times normal as a more probable consequence of _____ (I, increased; D, decreased) target action of _____ (R, a releasing; H, an inhibitory) factor.

    a. O, I, R   c. P, I, R   e. L, D, H
    b. L, I, R   d. O, D, H   f. P, D, H

45. LRF and _____ releasing factors are thought to be formed by contiguous areas within the _____ (LVM, lateral ventro-medial nucleus; MPA, medial preoptic area) of the hypothalamus.

    a. GH, LVM   d. GH, MPA
    b. FSH, LVM   e. FSH, MPA
    c. Thyroxine, LVM   f. Thyroxine, MPA

OBJECTIVE 75-7.

Identify growth hormone, its effects upon growth of body tissues, its basic metabolic effects, and its specific effects upon protein metabolism. Identify "somatomedin" and its role in cartilage and bone growth.

46. Major anterior pituitary hormones, with the exception of _____ (T, TSH; P, prolactin; S, somatotropin), exert "target" actions upon _____ (R, relatively specific; W, widespread) body tissues.

    a. T, R   c. S, R   e. P,W
    b. P, R   d. T, W   f. S, W

47. Experimental excess of growth hormone administered during an early phase of animal growth results in _____ (P, proportionate; DP, disproportionate) body growth at _____ (I, increased; N, normal; D, decreased) rates of tissue growth.

    a. P, I   c. P, D   e. DP, N
    b. P, N   d. DP, I   f. DP, D

48. Somatotropin is a _____ (S, steroid; P, protein) hormone which promotes _____ (Z, only increased cell size; M, only increased mitotic rates; B, increased cell size and mitotic rates) of body tissues.

    a. S, Z   c. S, B   e. P, M
    b. S, M   d. P, Z   f. P, B

49. Somatomedin, formed in the _____ (E, epiphyses of long bones; L, liver), functions as an _____ (H, inhibitor; M, intermediate) for the effects of somatropin upon cartilage and bone growth.

    a. E, H   c. L, H
    b. E, M   d. L, M

50. Somatotropin administered to the adult generally results in significant increases in_____ of long bones and_____of membranous bones. (L, only the length; T, only the thickness; B, both the length and thickness)

    a. L, T   c. T, L   e. B, T
    b. L, B   d. T, B   f. B, L

51. Growth hormone _____ body protein synthesis, _____ body carbohydrate utilization, and _____ fat mobilization. (I, increases; D, decreases)

    a. I, D, D   c. I, I, D   e. D, D, D
    b. I, D, I   d. D, I, I   f. D, D, I

52. Growth hormone _____ amino acid transport mechanisms of cellular membranes, _____ rates of formation of RNA, and _____ net protein synthesis. (I, increases; N, does not significantly influence; D, decreases)

    a. I, I, I   c. I, D, D   e. N, N, I
    b. I, N, I   d. N, I, N   f. N, D, D

OBJECTIVE 75–8.

Identify the major effects of growth hormone upon lipid and carbohydrate metabolism and their consequences.

53. Growth hormone results in _____ (I, increased; D, decreased) plasma concentrations of free fatty acids and preferential utilization of cellular _____(G, glycogen; L, lipid; P, protein) for ATP production.

    a. I, G   c. I, P   e. D, L
    b. I, L   d. D, G   f. D, P

54. Ketosis occurs in association with _____ (D, deficient; E, excess) amounts of growth hormone, excessive _____ (M, mobilization; S, synthesis) of fat stores, and enhanced _____ (G, glycogenesis; L, glycogenolysis).

    a. D, M, G   c. D, S, G   e. E, M, G
    b. D, S, L   d. E, S, L   f. E, M, L

55. Chronically elevated levels of growth hormone _____ glucose uptake by cells, _____ glycogen stores, and _____ the utilization of glucose for energy purposes. (I, increases; D, decreases)

    a. D, I, D   c. D, D, D   e. I, D, D
    b. D, I, I   d. I, D, I   f. I, I, I

56. The administration of growth hormone results in an initial _____ (W, as well as; N, but not) persistent _____ (R, rise; F, fall) in levels of cellular uptake, _____ (W, as well as; N, but not) plasma concentrations of glucose.

    a. W, R, W   c. W, F, W   e. N, R, W
    b. W, F, N   d. N, F, N   f. N, R, N

57. Growth hormone possesses _____ (K, only ketogenic; D, only diabetogenic; B, ketogenic as well as diabetogenic) effects and _____ (S, does; N, does not) require insulin for its growth-promoting effects.

    a. K, S   c. B, S   e. D, N
    b. D, S   d. K, N   f. B, N

58. Maintained secretion of growth hormone _____ (W, as well as; N, but not) corticotropin tends to _____ (I, increase; D, decrease) the secretion of the beta cells of the islets of Langerhans via its effect to _____ (I, increase; D, decrease) blood glucose concentrations.

    a. W, I, I   c. W, D, I   e. N, D, I
    b. W, I, D   d. N, D, D   f. N, I, D

59. Pituitary diabetes, in contrast to diabetes mellitus, is associated with a _____ (G, greater; L, lesser) _____ (E, elevation; D, depression) of the rate of glucose utilization by cells, and a _____ (G, greater; L, lesser) sensitivity of blood glucose to insulin administration.

a. G, E, G   b. G, E, L   c. G, D, G   d. L, D, L   e. L, D, G   f. L, E, L

OBJECTIVE 75–9.
Identify the regulating mechanisms and abnormalities of growth hormone secretion.

60. Secretion rates of growth hormone _____ (D, all but disappear; L, are only slightly changed) following adolescence, and _____ (C, remain relatively constant; V, vary considerably) with day to day changes in nutritional status.

a. D, C   c. L, C
b. D, V   d. L, V

61. An acute decrease in the level of blood _____ (A, amino acid; G, glucose) concentration is more effective in _____ (I, increasing; D, decreasing) growth hormone secretion.

a. A, I   c. G, I
b. A, D   d. G, D

62. The _____ (H, high; L, low) levels of growth hormone that occur during starvation are more closely related to the degree of depletion of _____ (C, carbohydrate; P, protein) stores.

a. H, C   c. L, C
b. H, P   d. L, P

63. Altered growth hormone levels during starvation, functioning to _____ (R, reduce; P, preserve) lipid stores and to _____ (R, reduce; P, preserve) protein stores, return to essentially normal levels following restoration of _____ (T, only protein; E, either glycogen or protein) stores.

a. R, P, T   c. R, R, T   e. P, R, T
b. R, P, E   d. P, R, E   f. P, P, E

64. Congenital panhypopituitarism is a cause of _____ (A, acromegaly; G, giantism; D, dwarfism) and _____ (R, reduced; E, elevated) rates of secretion of somatotropin.

a. A, R   c. D, R   e. G, E
b. G, R   d. A, E   f. D, E

65. Mechanisms producing pituitary dwarfism generally _____ produce thyroid and adrenocortical deficient states, generally _____ produce mental retardation, and generally _____ permit the development of sexual maturity. (D, do; N, do not)

a. D, D, D   c. D, N, D   e. N, N, D
b. D, D, N   d. N, N, N   f. N, D, N

66. Acromegaly results as a consequence of _____ (E, preadolescent; O, postadolescent) _____ (A, acidophilic; C, chromophobe) tumors, and is associated with excessive growth of _____ (M, membranous bones and soft tissues; E, epiphyseal plates of long bones).

a. E, A, M   c. E, C, M   e. O, A, M
b. E, C, E   d. O, C, E   f. O, A, E

OBJECTIVE 75–10.
Identify melanocyte-stimulating hormone, its locus of secretion, and its target effects.

67. Melanocyte-stimulating hormone is secreted from the _____ (C, adrenal cortex; A, anterior pituitary; P, posterior pituitary) and results in a _____ (D, darker; L, lighter) skin color.

a. C, D   c. P, D   e. A, L
b. A, D   d. C, L   f. P, L

68. MSH has a greater effect upon individuals with a genetically _____ (L, light; D, dark) skin and has significant similarities both chemically and functionally with _____ (C, cortisol; S, somatotropin; A, ACTH).

a. L, C   c. L, A   e. D, S
b. L, S   d. D, C   f. D, A

69. _____ (C, Cushing's; A, Addison's) disease is frequently associated with _____ (I, increased; D, decreased) adrenocortical hormone secretion, _____ (I, increased; D, decreased) ACTH secretion, and increased pigmentation of the skin and mucous membranes.

   a. C, I, I   b. C, I, D   c. C, D, I   d. A, D, D   e. A, D, I   f. A, I, D

OBJECTIVE 75–11.

Identify the neurohypophysis, its relationship to the hypothalamus, and the mechanisms of secretion of neurohypophyseal hormones.

70. The neurohypophysis, or _____ (I, pars intermedia; P, posterior lobe) of the pituitary, is largely comprised of _____ (E, endocrine; N, nonendocrine) -secreting _____ (R, neuronal somae; G, glial-like cells) called pituicytes.

   a. I, E, R   c. I, N, R   e. P, E, R
   b. I, E, G   d. P, N, G   f. P, E, G

71. Neurohypophyseal hormones include _____ (A, antidiuretic hormone; P, prolactin; O, oxytocin), synthesized largely in cell bodies of the paraventricular nucleus, and a second hormone synthesized largely in the _____ (DM, dorsomedial; F, perifornical; S, supraoptic) nucleus.

   a. A, F   c. O, S   e. P, F
   b. P, DM   d. A, S   f. O, DM

72. Neurohypophyseal hormones, synthesized by cell bodies of the _____ (P, pineal body; H, hypothalamus; M, mammillary body) are transported to the neurohypophysis by means of _____ (S, a plasma; X, an axoplasmic) carrier protein called neurophysin.

   a. P, S   c. M, S   e. H, X
   b. H, S   d. P, X   f. M, X

73. Action potentials of the _____ (H, hypothalamic-hypophyseal; M, mammillary-thalamic) tract trigger the hormonal _____ (T, transport to; R, release from) the neurohypophysis.

   a. H, T   c. M, T
   b. H, R   d. M, R

OBJECTIVE 75–12.

Identify the regulatory mechanisms, influencing factors, target actions, and physiologic functions of ADH secretion.

74. Antidiuretic hormone _____ water excretion by the kidneys and _____ the permeability of the collecting duct epithelium to water. (I, increases; D, decreases)

   a. I, I   c. D, I
   b. I, D   d. D, D

75. ADH, with a half-life following its secretion of about 5 _____ (S, seconds; M, minutes; H, hours), acts to _____ (I, increase; B, block) the formation of cyclic AMP in target tissues.

   a. S, I   c. H, I   e. M, B
   b. M, I   d. S, B   f. H, B

76. Proposed osmoreceptors of the _____ (K, kidneys; M, medulla; H, hypothalamus) are thought to increase antidiuretic hormone secretion in response to _____ (D, dilute; C, concentrated) extracellular fluids.

   a. K, D   c. H, D   e. M, C
   b. M, D   d. K, C   f. H, C

77. Increased osmolarities of body fluids _____ discharge frequencies of neurons of the supraoptic nuclei, _____ ADH secretion rates, and _____ urine volume flow rates. (I, increase; D, decrease)

   a. I, I, I   c. I, D, I   e. D, D, I
   b. I, I, D   d. D, D, D   f. D, I, D

78. Major factors regulating sodium ion concentrations of body fluids, which _____ (D, do; N, do not) include ADH, control _____ (M, more; L, less) than 50% of the total osmotic pressure of extracellular fluids.

   a. D, M   c. N, M
   b. D, L   d. N, L

79. The secretion of vasopressin, _____ (T, an antagonist; S, a synonym) for ADH, is generally _____ (I, increased; D, decreased) by pain and anxiety, _____ (W, as well as; N, but not) alcohol.

   a, T, I, W   c. T, D, W   e. S, I, W
   b. T, D, N   d. S, D, N   f. S, I, N

80. A 25% _____ (I, increase; D, decrease) in blood volume increases ADH secretion by a _____ (G, slight; B, substantial) percentage as a consequence of alterations of vascular _____ (S, stretch receptors; O, osmolarity; C, blood gas concentrations).

a. I, G, O   c. I, B, C   e. D, B, C
b. I, G, S   d. D, B, S   f. D, G, O

81. Diabetes insipidus generally results from abnormal tissue mechanisms involved in the _____ (S, secretion; T, plasma transport; A, target action) of _____.

a. S, oxytocin   d. S, ADH
b. T, oxytocin   e. T, ADH
c. A, oxytocin   f. A, ADH

82. Insufflation of powdered vasopressin into the nostrils provides an effective and relatively _____ (E, efficient; I, inefficient) route of absorption for the treatment of _____ (DI, diabetes insipidus; IAS, idiopathic ADH syndrome).

a. E, DI   c. I, DI
b. E, IAS   d. I, IAS

83. An almost constant thirst, a specific gravity of urine between 1.002 and 1.006, and daily urine outputs of _____ liters frequently occur in _____ (DI, diabetes insipidus; IAS, idiopathic ADH syndrome).

a. 0.5 to 1.5, DI   d. 0.5 to 1.5, IAS
b. 4 to 6, DI   e. 4 to 6, IAS
c. 60 to 80, DI   f. 60 to 80, IAS

84. Larger doses of ADH cause direct _____ (E, excitation; I, inhibition) of smooth muscle of the intestine and uterus _____ (W, as well as; N, but not) blood vessels.

a. E, W   c. I, W
b. E, N   d. I, N

OBJECTIVE 75–13.

Identify the chemical nature, target effects, and functions of oxytocin.

85. Oxytocin _____ (W, as well as; N, but not) vasopressin is a(n) _____ (S, steroid; O, octapeptide; P, protein).

a. W, S   c. W, P   e. N, O
b. W, O   d. N, S   f. N, P

86. _____ (V, vasopressin; O, oxytocin) is a more potent _____ (S, stimulator; I, inhibitor) of smooth muscle of the pregnant uterus.

a. V, S   c. O, S
b. V, I   d. O, I

87. Oxytocin _____ (F, facilitates; I, inhibits) "milk letdown" by means of a reflex involving _____ (E, an endocrine; N, a neural) -mediated afferent limb and _____ (E, an endocrine; N, a neural) -mediated effector limb.

a. F, E, E   c. F, N, E   e. I, N, E
b. F, E, N   d. I, E, N   f. I, N, N

88. Oxytocin has a target action to _____ (F, facilitate; I, inhibit) _____ (M, myoepithelial cell contraction; D, development of the parenchyma and duct system; S, milk synthesis) of the mammary glands.

a. F, M   c. F, S   e. I, D
b. F, D   d. I, M   f. I, S

89. _____ (I, increased; D, decreased) secretion of _____ (V, vasopressin; X, oxytocin) during the female orgasm is a proposed mechanism of _____ (O, ovulation; S, sperm transport) in humans.

a. I, V, O   c. I, X, O   e. D, V, O
b. I, X, S   d. D, X, S   f. D, V, S

76

The Thyroid Hormones

OBJECTIVE 76–1.

Using the following diagram, locate the various gross anatomical and microscopic structures of the human thyroid.

Directions: Match the lettered headings with the diagrams and numbered list of descriptive words and phrases.

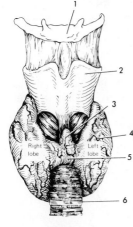

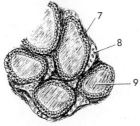

1. _____ Bony support for the larynx.
2. _____ Voice box.
3. _____ Lobe of the thyroid arising from the isthmus.
4. _____ A large ductless gland consisting of three lobes and lying close to the larynx.
5. _____ Bridge of tissue connecting the right and left lobes of the thyroid.
6. _____ Tube supported by incomplete cartilaginous rings.
7. _____ Cellular lining of the thyroid follicles.
8. _____ Route of endocrine transport.
9. _____ Intrafollicular substance composed mainly of thyroglobulin.

a. Thyroid gland
b. Isthmus
c. Hyoid bone
d. Blood vessel
e. Colloid
f. Trachea
g. Larynx
h. Pyramidal lobe
i. Follicular epithelium

(Top figure from Ganong, W.F.: Review of Medical Physiology, 8th Ed., Lange Medical Publications, 1977.)
(Bottom figure from Guyton, A.C., Textbook of Medical Physiology, 5th ed. Philadelphia, W.B. Saunders Company, 1976.)

OBJECTIVE 76–2.

Identify the hormones secreted by the thyroid gland, the thyroid gland utilization of iodide, and the fate of ingested iodides.

10. Thyroid gland secretion of larger quantities of _____ (MD, mono- and diiodotyrosine; T, tri- and tetraiodothyronine) is regulated by the _____ (F, facilitatory; I, inhibitory) influence of _____ (P, a posterior; A, an anterior) pituitary lobe hormone.

a. MD, F, P c. MD, I, P e. T, F, P
b. MD, I, A d. T, I, A f. T, F, A

11. _____ (P, parathormone; C, calcitonin) secretion by the thyroid gland is regulated by factors which control _____ (CM, calcium metabolism; BM, basal metabolic rates) and which act _____ (S, synergistically with; I, independent of) thyroxine secretion.

a. P, CM, S c. P, BM, S e. C, BM, S
b. P, CM, I d. C, BM, I f. C, CM, I

12. Thyroxine, when compared to triiodothyronine, is secreted in _____ amounts by the thyroid gland, and has a _____ potency for target actions. (G, greater; L, lesser).

a. G, G   c. L, G
b. G, L   d. L, L

13. Triiodothyronine, when compared to thyroxine, is present in blood in _____ (H, higher; L, lower) quantities and persists for _____ (G, longer; S, shorter) time intervals.

a. H, G   c. L, G
b. H, S   d. L, S

14. The thyroid gland, when compared to other tissues, has a relatively _____ (H, high; L, low) blood flow and is comprised largely of _____ (C, colloid follicles; P, parenchyma of compound acinar glands).

a. H, C   c. L, C
b. H, P   d. L, P

15. Thyroid hormones are largely stored in the thyroid gland in the form of _____ (TG, thyroglobulin; T, thyrotropin; TBM, thyroxine-binding prealbumin), and released into the general circulation by the action of _____ (CE, cuboidal epithelioid; RE, reticuloendothelial) cells.

a. TG, CE   c. TBM, CE   e. T, RE
b. T, CE   d. TG, RE   f. TBM, RE

16. Ingested iodides, when compared to chlorides, are absorbed from the gastrointestinal tract by _____ (D, distinctly different; S, similar) transport mechanisms, have a considerably _____ (H, higher; L, lower) renal plasma clearance, and normally are largely _____ (U, lost in the urine; S, stored in the thyroid gland).

a. D, H, U   c. D, H, S   e. S, H, U
b. D, L, S   d. S, L, S   f. S, L, U

17. Iodized table salt provides a means for obtaining the required 1 mg. per _____ (D, day; W, week; M, month) required to synthesize hormones of the _____ (T, thyroid; P, pituitary; A, adrenal) glands.

a. D, T   c. M, T   e. W, T
b. W, P   d. D, A   f. M, A

18. The active transport of iodide _____ (I, into; O, out of) the thyroid follicle, required for thyroid hormone _____ (R, release; S, synthesis), has a rather unique requirement for _____ (P, peroxidase; IA, iodoacetate).

a. I, R, P   c. I, S, P   e. O, R, P
b. I, S, IA   d. O, S, IA   f. O, R, IA

OBJECTIVE 76–3.

Identify thyroglobulin, its functional role, and the chemistry of thyroxine and triiodothyronine formation.

19. Thyroglobulin is a _____ (S, small; L, large) molecular weight _____ (LP, lipoprotein; GP, glycoprotein) which is synthesized within _____ (E, follicular epithelial cells; I, the interior of the thyroid follicle).

a. S, LP, E   c. S, GP, I   e. L, GP, E
b. S, LP, I   d. L, GP, I   f. L, LP, E

20. Thyroxine and triiodothyronine are synthesized _____ (S, separately from; W, within) thyroglobulin molecules from _____ (T, trypotophan; H, hydroxyproline; R, tyrosine) amino acid precursors.

a. S, T   c. S, R   e. W, H
b. S, H   d. W, T   f. W, R

21. The more direct utilization of _____ (D, iodides; N, elemental iodine) in the formation of thyroxine and triiodothyronine occurs largely within _____ (E, follicular epithelial cells; I, the interior of the thyroid follicle).

a. D, E   c. N, E
b. D, I   d. N, I

22. Thyroxine, or _____ (T, triiodothyronine; A, tetraiodothyronine), is formed by the direct combination of _____ (M, monoiodotyrosine; D, diiodotyrosine) units with the loss of _____ (I, iodide; L, alanine; M, monoiodotyrosine) units.

a. T, M, I   c. T, D, M   e. A, D, L
b. T, M, L   d. A, D, M   f. A, M, I

23. Thyroglobulin storage involves about _____ (E, equal; G, a nine to one; L, a one to nine) ratio(s) of triiodothyronine to thyroxine molecules, and about _____ thyroxine molecules per thyroglobulin molecule.

a. E, 2 to 3   b. G, 2 to 3   c. L, 2 to 3   d. E, 20 to 30   e. G, 20 to 30   f. L, 20 to 30

OBJECTIVE 76–4.

Identify the characteristics of thyroxine (T_4) and triiodothyronine (T_3) release from thyroglobulin, and their transport to tissues.

24. Thyroxine and triiodothyronine are released from thyroglobulin by the action of _____ (P, proteinases; I, iodases) within the _____ (E, follicular epithelial cells; I, interior of the follicle; B, blood).

a. P, E   c. P, B   e. I, I
b. P, I   d. I, E   f. I, B

25. A _____ (J, majority; N, minority) of the iodinated tyrosine units of thyroglobulin form a storage pool of iodine rather than active thyroid hormones through the releasing action of the enzyme _____ (I, iodinase; D, iodase).

a. J, I   c. N, I
b. J, D   d. N, D

26. A thyroid gland secretion ratio of thyroxine to triiodothyronine of _____ is subsequently _____ (E, enhanced; R, reduced) in blood by a slow rate of _____ (I, iodination of T_3; D, deiodination of T_4).

a. 1:3, E, I   d. 1:3, R, D
b. 3:1, E, I   e. 3:1, R, D
c. 9:1, E, I   f. 9:1, R, D

27. About two-thirds of the total thyroid hormone effects in target tissues are supplied by _____ , since the duration of action of _____ is about four times as long, and _____ is more potent in its intracellular target effects. (T_3, triiodothyronine; T_4, thyroxine)

a. T_3, T_3, T_4 d. T_4, T_4, T_3
b. T_3, T_4, T_3 e. T_4, T_3, T_4
c. T_3, T_4, T_4 f. T_4, T_3, T_3

28. Plasma proteins, particularly _____ (TG, thyroxine-binding globulin; TA, thyroxine-binding prealbumin; A, albumin) provide _____ (M, more; L, less) than 50% of the plasma transport of T_3 and T_4.

a. TG, M   c. A, M   e. TA, L
b. TA, M   d. TG, L   f. A, L

29. Thyroxine _____ (W, as well as; N, but not) triiodothyronine is stored in combination with proteins in significant amounts and released by _____ (G, the thyroid gland; P, plasma proteins; T, target cell intracellular proteins).

a. W, only G   d. N, only G
b. W, only G & P   e. N, only G & P
c. W, G, P, & T   f. N; G, P, & T

30. Triiodothyronine, when compared to thyroxine, has a _____ (H, higher; W, lower) concentration ratio of free to bound form in plasma, a _____ (L, longer; S, shorter) half-life in plasma, and a _____ (L, longer; S, shorter) latent period for tissue target actions.

a. H, L, S   c. H, S, L   e. W, L, L
b. H, S, S   dd. W, S, L   f. W, L, S

OBJECTIVE 76-5.

Identify the actions and underlying mechanisms of the effects of thyroid hormones upon metabolic rates and specific dietary substances.

31. The brain and retina, spleen, testes, and lungs are _____ (E, exceptions to; R, representative tissues for) the general case whereby thyroid hormones significantly _____ (I, increase; D, decrease) the metabolic rates of body tissues.

    a. E, I   c. R, I
    b. E, D   d. R, D

32. Thyroid hormone administration results in _____ average numbers of mitochondria per cell and _____ average size of mitochondria. (I, increased; N, no significant change in; D, decreased).

    a. I, I   c. D, I   e. N, D
    b. N, I   d. I, D   f. D, D

33. Thyroxine has a relatively _____ (S, specific; G, general) effect upon intracellular protein synthesis to _____ (I, increase; D, decrease) the rate of _____ (L, translation; C, transcription) processes.

    a. S, I, only L   d. G, D, only L
    b. S, D, only C   e. G, I, only C
    c. S, D, both L & C   f. G, I, both L & C

34. Increased levels of thyroid hormone favor _____ (R, reduced; A, accelerated) rates of bone growth, _____ (W, as well as; N, but not) soft tissue growth, and a _____ (D, delayed; P, premature) closure of epiphyseal plates.

    a. R, W, D   c. R, N, D   e. A, W, D
    b. R, N, P   d. A, N, P   f. A, W, P

35. Prolonged states of elevated body temperatures tend to _____ (I, increase; D, decrease) osteoclast activity, which is generally _____ (O, opposite; S, similar) to the effects of thyroxine excess.

    a. I, O   c. D, O
    b. I, S   d. D, S

36. Thyroxine decreases _____ (G, glycolysis; N, gluconeogenesis; I, insulin secretion; C, plasma cholesterol levels) and _____ (F, facilitates; H, inhibits) the conversion of cholesterol to bile salts.

    a. G, F   c. C, F   e. I, H
    b. I, F   d. C, H   f. N, H

37. Elevated levels of thyroid hormone secretion _____ (M, mobilize; E, enhance) fat stores, _____ (I, increase; D, decrease) the free fatty acid concentration of plasma, and _____ (I, increase; D, decrease) the phospholipid and triglyceride content of plasma.

    a. M, I, I   c. M, D, I   e. E, D, I
    b. M, I, D   d. E, D, D   f. E, I, D

OBJECTIVE 76-6.

Identify the major physiologic effects of thyroid hormone upon general body mechanisms.

38. The relationship between basal metabolic rate (BMR) and thyroid hormone concentration reflects a greater rate of change of BMR with _____ (L, lower; H, higher) than normal hormone concentration, and a BMR of about _____ % in the absence of thyroid hormones.

    a. L, -100   c. L, +60   e. H, -45
    b. L, -45   d. H, -100   f. H, +60

39. Greatly increased levels of thyroid hormone secretion _____ food intake and appetite, _____ body metabolic rates, and more frequently _____ body weight. (I, increase; N, do not significantly alter; D, decrease)

    a. I, I, N   c. I, D, I   e. D, I, I
    b. I, I, D   d. D, I, D   f. D, D, N

40. Hypothyroid states are generally associated with _____ body heights in the preadolescent individual and _____ body heights in the adult. Hyperthyroid states are generally associated with _____ body heights in the preadolescent and _____ body heights in the adult. (G, greater than average; A, average; L, less than average)

a. G, G, G, G   d. A, A, A, A
b. L, L, L, L   e. L, L, G, L
c. A, L, A, G   f. L, L, G, G

41. The effects of hyperthyroidism upon the cardiovascular system are generally _____ (S, similar; O, opposite) to the effects of hypothyroidism, and include _____ (I, increased; D, decreased) peripheral resistance to blood flow and _____ (I, increased; D, decreased) cardiac output.

a. S, I, D   c. S, D, D   e. O, D, D
b. S, I, I   d. O, D, I   f. O, I, I

42. Thyroxine generally produces a more substantial and consistent _____ (I, increase; D, decrease) in the _____ (HR, heart rate; V, blood volume; P, mean arterial pressure).

a. I, HR   c. I, P   e. D, V
b. I, V   d. D, HR   f. D, P

43. Hyperthyroid states generally tend to decrease _____ (P, pulse; S, systolic; D, diastolic) pressures and to _____ (I, increase; R, decrease) respiration.

a. P, I   c. D, I   e. S, R
b. S, I   d. P, R   f. D, R

44. Thyroid hormone tends to _____ (F, facilitate; H, inhibit) synaptic activity of neurons and to _____ (I, increase; N, not significantly alter; D, decrease) propagation velocities of action potentials in peripheral nerve axons.

a. F, I   c. F, D   e. H, N
b. F, N   d. H, I   f. H, D

45. Thyroid hormone _____ absorption rates of foodstuffs from the gastrointestinal tract, _____ appetite, and _____ gastrointestinal motility and secretion. (D, decreases; I, increases)

a. D, I, D   c. D, I, I   e. I, D, D
b. D, D, I   d. I, D, I   f. I, I, I

46. The hypothyroid state, as opposed to the hyperthyroid state, is characterized by:

a. Muscle tremors
b. Extreme nervousness
c. Extreme somnolence
d. Excessive protein catabolism
e. Frequent diarrhea
f. Physical exhaustion

OBJECTIVE 76–7.

Identify the regulatory mechanisms controlling thyroid hormone secretion, and the characteristics, mechanisms of secretion, and target actions of thyrotropin.

47. _____ (X, thyroxine; T, thyrotropin), a glycoprotein hormone with a molecular weight of about _____ , is secreted from the _____ (TG, thyroid gland; A, anterior pituitary; P, posterior pituitary).

a. X, 250, TG   d. T, 250, TG
b. X, 2500, P   e. T, 2500, P
c. X, 25,000, A   f. T, 25,000, A

48. TSH increases _____ (S, the endocrine secretion; N, the number of cells) of the thyroid gland as a probable consequence of its action to activate _____ (TRF, thyrotropin-releasing factor; AMP, cyclic AMP).

a. Only S, TRF   d. Only S, AMP
b. Only N, TRF   e. Only N, AMP
c. Both S & N, TRF   f. Both S & N, AMP

49. TSH _____ (I, increases; D, decreases) activity of the iodide pump, _____ (F, facilitates; H, inhibits) thyroglobulin proteolysis, and favors the conversion of thyroid glandular cells from _____ (B, columnar to cuboidal; L, cuboidal to columnar) form.

   a. I, F, B   c. I, H, B   e. D, F, B
   b. I, F, L   d. D, H, L   f. D, F, L

50. Electrical stimulation of the paraventricular area of the hypothalamus generally _____ (I, increases; D, decreases) activities of the thyroid gland as a consequence of the _____ (N, neural; A, adenohypophyseal) secretion of a tripeptide amide _____ (H, inhibitory; R, -releasing) factor.

   a. I, N, H   c. I, A, R   e. D, A, H
   b. I, N, R   d. D, A, R   f. D, N, H

51. TSH secretion is _____ by chronic exposure to cold environmental conditions, _____ by emotional states which enhance sympathetic activity, and _____ by experimental blockage of the portal system from the hypothalamus to the pituitary gland. (I, increased; N, not significantly affected; D, decreased)

   a. I, I, I   c. I, D, D   e. D, N, D
   b. I, N, D   d. D, I, I   f. D, D, N

52. Doubling the concentration of thyroid hormones in body fluids results in a _____ (L, slight; X, maximal) _____ (I, increase; S, suppression) of TSH secretion largely via _____ (D, direct; H, hypothalamic-mediated) effects upon the pituitary.

   a. L, I, D   c. L, S, D   e. X, I, H
   b. L, I, H   d. X, S, H   f. X, S, D

53. Diminished availability of thyroxine and triiodothyronine, associated with _____ (L, lowered; E, elevated) basal metabolic rates, tends to _____ (I, increase; D, decrease) thyrotropin secretion and subsequently to _____ (O, oppose; A, augment) thyroid hormone secretion.

   a. L, I, O   c. L, D, A   e. E, I, O
   b. L, I, A   d. E, D, O   f. E, I, A

54. Normally, doses of exogenous thyroid hormone that provide less than the amount secreted endogenously have no significant effect on metabolism due to inhibition of _____ released from the _____. (A, adrenals; P, pancreas; AH, adenohypophysis; N, neurohypophysis)

   a. TSH, AH   d. Insulin, A
   b. TSH, N   e. ACTH, AH
   c. Insulin, P   f. ACTH, N

OBJECTIVE 76–8.

Identify the effects and mechanisms of action of thiocyanate, propylthiouracil, and high concentrations of inorganic iodides upon thyroid secretion.

55. Propylthiouracil is more effective in blocking _____ (TG, thyroglobulin formation; TSH, thyrotropin secretion; C, coupling mechanisms of iodinated tyrosines), while thiocyanate inhibits _____ (I, the "iodide pump"; TG, thyroglobulin formation; TSH, thyrotropin secretion).

   a. TG, TSH   c. C, I   e. TSH, I
   b. TSH, TG   d. TG, I   f. C, TG

56. Administration of thiocyanate and propylthiouracil, _____ (W, as well as; N, but not) iodide excess, results in _____ (D, decreased; C, no significant change in; I, increased) circulating levels of TSH.

   a. W, D   c. W, I   e. N, C
   b. W, C   d. N, D   f. N, I

57. Iodides are frequently _____ (W, withheld; E, administered in excess) prior to surgical removal of the thyroid gland in order to _____ (I, increase; D, decrease) the size of the thyroid gland.

   a. W, I   c. E, I
   b. W, D   d. E, D

OBJECTIVE 76–9.

Identify the effects of thyroid gland secretion upon the secretion of other endocrine glands.

58. Increased thyroid hormone secretion has general effects to _____ tissue requirements for more hormones and to _____ the secretion rates of most endocrine glands of the body. (E, enhance; D, diminish)

    a. E, E   c. D, E
    b. E, D   d. D, D

59. Thyroid hormone acts to _____ ACTH secretion, to _____ glucocorticoid secretion by the adrenal cortex, and to _____ glucocorticoid conjugation and inactivation by the liver. (F, facilitate; H, inhibit)

    a. F, F, F   c. F, H, H   e. H, H, F
    b. F, F, H   d. H, F, F   f. H, H, H

60. Abnormal sexual functions are known to occur with _____ (E, hyperthyroid; O, hypothyroid) states for _____ (M, men; W, women).

    a. Only E, both M & W
    b. Only O, only W
    c. Both E & O, only W
    d. Only E, only M
    e. Only O, both M & W
    f. Both E & O, & both M & W

61. Hypothyroid states in the female are a more frequent cause of _____ (O, oligomenorrhea; M, menorrhagia) and _____ (I, increased; D, decreased) libido.

    a. O, I   c. M, I
    b. O, D   d. M, D

OBJECTIVE 76–10.

Identify the principal causes, symptoms, and consequences of hyperthyroidism.

62. Toxic goiter refers to conditions of _____ (R, reduced; E, enlarged) thyroid gland size, _____ (I, increased; D, decreased) rates of thyroid hormone secretion, and a more frequent _____ (I, increase; D, decrease) in TSH secretion.

    a. R, D, D.   c. R, I, I   e. E, I, D
    b. R, D, I   d. E, I, I   f. E, D, D

63. Long-acting thyroid stimulator (LATS) is a _____ (G, globulin of the IgG type; R, TSH-releasing factor; T, thyroxine inhibitor) which is present in high concentrations in _____ (X, thyrotoxicosis; A, thyroid adenomas; I, idiopathic nontoxic colloid goiter).

    a. G, X   c. T, I   e. R, I
    b. R, A   d. G, A   f. T, X

64. Graves' disease is associated with a greater intolerance to _____ (C, cold; H, heat), _____ (E, excessive amounts of; I, inability to) sleep, and mild to extreme amounts of weight _____ (G, gain; L, loss).

    a. C, E, G   c. C, I, G   e. H, I, G
    b. C, E, L   d. H, I, L   f. H, E, L

65. Exophthalmos, or ocular _____ (P, protrusion; I, intrusion), occurs to a major degree in about _____ out of every three _____ (E, hyperthyroid; O, hypothyroid) patients.

    a. P, 1, E   c. P, 3, E   e. I, 3, E
    b. P, 1, O   d. I, 3, O   f. I, 1, O

66. "Plasma-bound iodine," largely in a _____ (T, thyroxine and triiodothyronine form; C, soluble, chelated form of iodide; TG, thyroglobulin) is generally _____ (I, increased; N, not significantly altered; D, decreased) in hyperthyroid states.

    a. T, I   c. TG, N   e. C, N
    b. C, D   d. T, D   f. TG, I

67. Hyperthyroid individuals generally have a rate of radioactive iodine uptake by the thyroid gland of _____ than 5 to 10% of a small test dose administered intravenously, and a basal metabolic rate of _____ than +100. (M, more; L, less)

    a. M, M   c. L, M
    b. M, L   d. L, L

68. The administration of glucocorticoids and agents which _____ (I, increase; D, decrease) body temperature are generally _____ (N, indicated; C, contraindicated) in _____ (P, hyperthyroid; O, hypothyroid) patients in states of "thyroid crisis."

a. I, C, O   b. I, N, P   c. I, N, O   d. D, N, P   e. D, C, O   f. D, C, P

OBJECTIVE 76–11.
Identify the principal causes, symptoms, and consequences of hypothyroidism.

69. Goiters, associated with _____ (R, a reduction; E, an enlargement) of thyroid gland size, frequently occur with dietary _____ (X, excesses; D, deficiencies) of iodide.

a. R, X   c. E, X
b. R, D   d. E, D

70. Excessive, long-term dietary intake of a group of certain foods, including cabbages and turnips, tend to _____ (P, prevent; C, cause) _____ (E, endemic; I, idiopathic nontoxic) colloid goiter.

a. P, I   c. C & I
b. P, E   d. C & E

71. Hypothyroid states generally include symptoms of _____ (IS, insomnia; S, increased somnolence), _____ (C, constipation; R, diarrhea), _____ (I, increased; D, decreased) heart rates, and _____ (I, increased; D, decreased) rates of growth of body hair.

a. IS, C, I, D   c. IS, R, I, I   e. IS, C, I, I
b. S, R, D, D   d. S, R, D, I   f. S, C, D, D

72. Myxedema, _____ (W, as well as; N, but not) exophthalmos, is a symptom of severe _____ (E, hyperthyroidism; O, hypothyroidism) associated with _____ (I, increased; D, decreased) amounts of mucopolysaccharides in interstitial spaces.

a. W, E, I   c. W, O, D   e. N, O, I
b. W, E, D   d. N, O, D   f. N, E, I

73. Severe arteriosclerosis, occurring in association with _____ (I, increased; D, decreased) blood cholesterol levels, is particularly associated with _____ (E, hyperthyroid; O, hypothyroid) states occurring in _____ (C, conjunction with; A, the absence of) myxedema.

a. I, E, C   c. I, O, A   e. D, E, A
b. I, O, C   d. D, O, A   f. D, E, C

74. Cretinism is a _____ (E, hyperthyroid; O, hypothyroid) condition which results in _____ (G, greater; L, less) than normal body stature, and a disproportionately greater growth of _____ (S, soft; M, mineralized) tissues.

a. E, G, S   c. E, L, S   e. O, L, S
b. E, G, M   d. O, L, M   f. O, G, M

77

The Adrenocortical Hormones

OBJECTIVE 77–1.
Using the following diagram, identify the anatomical and histologic parts of the adrenal gland.

Directions: Match the lettered headings with the diagrams and the numbered list of descriptive words and phrases.

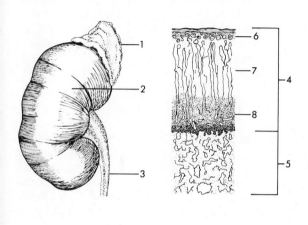

1. _____ Endocrine gland situated at the superior poles of the principal excretory organ of the body.
2. _____ Principal excretory organ of the body.
3. _____ A urine-containing fibromuscular tube.
4. _____ Corticosteroid-secreting portion of the endocrine gland.
5. _____ Highly vascularized, epinephrine- and norepinephrine-secreting portion of the endocrine gland.
6. _____ Aldosterone-secreting zone of the endocrine gland.
7. _____ Intermediate zone of corticosteroid-secreting portion of the endocrine gland.
8. _____ Inner zone of corticosteroid-secreting portion of endocrine gland.

a. Ureter
b. Zona fasciculata
c. Kidney
d. Adrenal medulla
e. Adrenal gland
f. Zona reticularis
g. Adrenal cortex
h. Zona glomerulosa

OBJECTIVE 77–2. *F 4 F*
Characterize the major types of endocrine secretions of the adrenal glands.

9. Endocrine secretions of the adrenal medulla, regulated by predominantly _____ (N, neural; P, pituitary endocrine; A, adrenal cortical endocrine) mechanisms, _____ (I, increase; D, decrease) blood glucose levels.

   a. N, I
   b. P, I
   c. A, I
   d. N, D
   e. P, D
   f. A, D

10. The adrenal medulla secretes _____ (SH, steroid hormones; EN, epinephrine and norepinephrine) and is innervated by _____ (S, sympathetic; P, parasympathetic) _____ (E, preganglionic; O, postganglionic) fibers.

    a. SH, S, E
    b. SH, P, O
    c. SH, P, E
    d. EN, P, O
    e. EN, S, E
    f. EN, S, O

11. _____ (A, androgenic; E, estrogenic; M, mineralocorticoid; G, glucocorticoid) hormones are highly significant types of adrenal _____ (D, medullary; C, cortical) secretions in the adult.

    a. A & G, D
    b. E & M, D
    c. M & G, D
    d. A & G, C
    e. E & M, C
    f. M & G, C

12. Mineralocorticoids have a more direct role upon _____ concentrations of extracellular fluids, while glucocorticoids _____ (I, increase; D, decrease) blood glucose levels and notably alter carbohydrate _____ (W, as well as; N, but not) protein and fat metabolism.

    a. Na & K, I, W
    b. Na & K, I, N
    c. Na & K, D, W
    d. Ca & phosphate, D, N
    e. Ca & phosphate, D, W
    f. Ca & phosphate, I, N

OBJECTIVE 77–3.

Identify the significance of adrenal mineralocorticoids, and the primary and secondary effects of aldosterone upon renal function.

13. Mineralocorticoids, principally _____ (C, corticosterone; D, deoxycorticosterone; L, aldosterone), _____ (A, are; N, are not) considered vital to life processes.

    a. C, A   c. L, A   e. D, N
    b. D, A   d. C, N   f. L, N

14. Aldosterone has an especially potent effect upon the _____ (P, proximal; S, distal) portion of the nephron to _____ (D, decrease; I, increase) the renal excretion of sodium _____ (W, as well as; N, but not) potassium.

    a. P, D, W   c. P, I, W   e. S, I, W
    b. P, D, N   d. S, I, N   f. S, D, N

15. Polydipsia is generally a compensatory mechanism for prevention of a severe state of _____ (ON, hyponatremia; EN, hypernatremia; EK, hyperkalemia) during conditions of mineralocorticoid _____ (D, deficiencies; X, excess).

    a. ON, D   c. EK, D   e. EN, X
    b. EN, D   d. ON, X   f. EK, X

16. Hyperpolarization of nerve and muscle membranes and _____ (I, increased; D, decreased) muscle strength occur as a consequence of alterations of extracellular _____ ion concentrations due to _____ (E, excessive; F, deficient) states of aldosterone secretion.

    a. I, Na, E   c. I, K, F   e. D, K, E
    b. I, Na, F   d. D, K, F   f. D, Na, E

17. Aldosterone promotes _____ anion reabsorption, _____ hydrogen ion reabsorption, and _____ potassium ion reabsorption by the renal tubules. (E, enhanced; D, diminished)

    a. E, E, E   c. E, D, D   e. D, E, D
    b. E, D, E   d. D, D, D   f. D, E, E

18. Potassium ion concentrations of extracellular fluids of about _____ times normal during conditions of aldosterone _____ (D, deficiency; E, excess) are generally considered threshold for cardiac toxic effects leading to death.

    a. 2, D   c. 10, D   e. 4, E
    b. 4, D   d. 2, E   f. 10, E

19. Aldosterone excess more frequently results in _____ (M, mild; S, severe) alterations of extracellular sodium concentration, an extracellular fluid _____ (C, acidosis; K, alkalosis), and _____ (I, increased; N, no significant change in; D, decreased) renal reabsorption of water.

    a. M, C, I   c. M, K, I   e. S, C, D
    b. M, K, N   d. S, K, D   f. S, C, N

OBJECTIVE 77–4.

Identify the effects of aldosterone upon fluid volumes, cardiovascular dynamics, sweat and salivary gland secretion, and intestinal absorption.

20. _____ (X, excessive; L, a lack of) aldosterone results in a _____ (M, mild; S, severe) reduction in extracellular fluid volume _____ (W, as well as; N, but not) plasma volume.

    a. X, M, W   c. X, S, N   e. L, S, W
    b. X, M, N   d. L, S, N   f. L, M, W

21. The phenomenon of "aldosterone escape," associated with _____ (X, excess; F, deficient) levels of aldosterone, has been attributed to _____ (H, higher; L, lower) than normal levels of water intake and _____ (H, higher; L, lower) than normal levels of urinary output.

    a. X, H, L   c. X, L, L   e. F, L, L
    b. X, H, H   d. F, L, H   f. F, H, H

22. More severe and detrimental consequences of states of aldosterone _____ (F, deficiency; X, excess) are associated with _____ (I, increased; D, decreased) levels of blood volume _____ (W, as well as; N, but not) cardiac output.

    a. F, I, W   c. F, D, W   e. X, I, W
    b. F, D, N   d. X, D, N   f. X, I, N

23. Moderate to severe hypertension is a more frequent consequence of _____ (E, early; L, long-term) effects of _____ (X, excessive; D, deficient) states of mineralocorticoid secretion.

    a. E, X   c. L, X
    b. E, D   d. L, D

24. Aldosterone increases _____ ion reabsorption and _____ ion secretion by salivary excretory ducts, _____ (S, similar; O, opposite) to its effects upon renal tubules.

    a. Na, K & Cl, S   d. Na, K & Cl, O
    b. Na & Cl, K, S   e. Na & Cl, K, O
    c. K & Cl, Na, S   f. K, Na & Cl, O

25. Diarrhea, accompanying _____ (I, increased; D, decreased) levels of sodium _____ (W, as well as; N, but not) anion absorption from the gastrointestinal tract, occurs as a consequence of aldosterone _____ (F, deficiencies; X, excess).

    a. I, W, F   c. I, N, F   e. D, W, F
    b. I, N, X   d. D, N, X   f. D, W, X

OBJECTIVE 77–5.

Identify the regulatory mechanisms of aldosterone secretion and the cellular mechanisms of action of aldosterone.

26. Aldosterone is a _____ (P, protein; S, steroid) which exerts target actions via _____ (M, cell membrane; I, intracellular) receptors to induce _____ (PS, protein synthesis; AMP, cyclic AMP formation).

    a. P, M, PS   c. P, I, PS   e. S, I, PS
    b. P, M, AMP   d. S, I, AMP   f. S, M, AMP

27. Aldosterone _____ (I, increases; R, decreases) renal reabsorption of Na ions after a latent period of about _____ , and exerts maximal effects after an interval of several _____ (H, hours; D, days).

    a. I, 5 to 10 min, H   d. R, 5 to 10 min, H
    b. I, 45 min, H   e. R, 45 min, D
    c. I, 5 to 6 hr, D   f. R, 5 to 6 hr, D

28. Factors essential for the regulation of aldosterone secretion include, in descending order of probable importance, _____ . (K, extracellular K ion concentrations; ACTH, adrenocorticotropic hormone; Na, quantities of body sodium; R, the renin-angiotensin system)

    a. K, Na, R, ACTH   d. R, K, Na, ACTH
    b. R, Na, K, ACTH   e. K, R, Na, ACTH
    c. ACTH, R, Na, K   f. Na, ACTH, K, R

29. Maintenance of extracellular potassium ion concentrations at a value of 1 mEq./liter _____ (A, above; W, below) normal levels results in a _____ (L, slight; B, substantial) and _____ (T, transient; P, persistent) increase in aldosterone secretion.

    a. A, L, T   c. A, L, P   e. W, L, T
    b. A, B, P   d. W, B, P   f. W, B, T

30. Potassium ion concentrations of extracellular fluids are thought to influence aldosterone secretion of cells of the zona _____ (R, reticularis; G, glomerulosa; F, fasciculata) via _____ (D, direct; I, indirect) actions.

    a. R, D   c. F, D   e. G, I
    b. G, D   d. R, I   f. F, I

31. Diminished states of extracellular fluid volume and arterial pressure _____ angiotensin formation and _____ aldosterone secretion. (A, augment; I, inhibit)

    a. A, A   c. I, A
    b. A, I   d. I, I

32. An absence of ACTH results in _____ (A, atrophic; H, hypertrophic) changes in the zona glomerulosa as a consequence of _____ (E, direct excitatory; I, direct inhibitory; P, permissive) effects upon aldosterone-secreting cells.

    a. A, E   c. A, P   e. H, I
    b. A, I   d. H, E   f. H, P

OBJECTIVE 77-6.

Identify the major glucocorticoids, their general significance, and the effects of cortisol upon carbohydrate metabolism.

33. Most of the glucocorticoid activity of adreno-cortical secretions results from the secretion of _____, also known as _____. (E, cortisone; L, cortisol; H, hydrocortisone; C, corticosterone)

    a. E, L   c. E, C   e. L, H
    b. E, H   d. L, E   f. L, C

34. Glucocorticoids are highly significant for their action to _____ (I, increase; D, decrease) _____ (T, glucose transport into cells; N, the rate of gluconeogenesis; E, the efficiency of ATP production from oxidative phosphorylation).

    a. I, T   c. I, E   e. D, N
    b. I, N   d. D, T   f. D, E

35. Cortisol results in mobilization of amino acids, particularly from _____ (H, hepatic; E, extrahepatic) tissues, and _____ (I, increased; D, decreased) stores of liver glycogen.

    a. H, I   c. E, I
    b. H, D   d. E, D

36. Cortisol tends to _____ (I, increase; D, decrease) blood glucose concentrations through its effects upon _____ (G, gluconeogenesis; U, rates of glucose utilization by cells).

    a. I, only G   d. D, only G
    b. I, only U   e. D, only U
    c. I, both G & U   f. D, both G & U

37. Glucose 6-phosphatase concentrations of liver, utilized to catalyze the _____ (U, uptake; R, release) of glucose by liver cells, are _____ (I, increased; D, decreased) in response to cortisol.

    a. U, I   c. R, I
    b. U, D   d. R, D

38. Adrenal diabetes is a condition of _____ (E, elevated; D, decreased) blood glucose, resulting from _____ (F, deficient; X, excess) levels of secretion of glucocorticoids.

    a. E, F   c. D, F
    b. E, X   d. D, X

39. Adrenal diabetes is considered to be _____ sensitive to insulin than pituitary diabetes and _____ sensitive to insulin than pancreatic diabetes. (M, more; L, less)

    a. M, M   c. L, M
    b. M, L   d. L, L

OBJECTIVE 77-7.

Identify the effects of cortisol upon protein and fat metabolism.

40. Cortisol _____ (I, increases; D, decreases) the rate of catabolism _____ (W, as well as; N, but not) the rate of synthesis of extrahepatic cellular proteins.

    a. I, W   c. D, W
    b. I, N   d. D, N

41. Cortisol _____ liver protein stores, _____ plasma proteins, and _____ the amino acid concentration of blood. (I, increases; D, decreases)

    a. I, I, I   c. I, D, D   e. D, D, I
    b. I, I, D   d. D, D, D   f. D, I, I

42. Cortisol _____ amino acid transport into muscle cells, _____ amino acid transport into liver cells, and _____ the rate of deamination of amino acids by the liver. (I, increases; D, decreases)

    a. I, I, I   c. I, D, D   e. D, D, I
    b. I, I, D   d. D, D, D   f. D, I, I

43. Cortisol is more potent in increasing fat _____ (S, synthesis; M, mobilization) by adipose tissues in the _____ (P, presence; A, absence) of growth hormone or ACTH.

    a. S, P   c. M, P
    b. S, A   d. M, A

44. Cortisol exerts _____ (M, more; L, less) rapid and _____ (M, more; L, less) powerful effects than similar shifts between carbohydrate and lipid metabolism elicited by _____ (I, increased; D, decreased) levels of insulin.

    a. M, M, I   c. M, L, I   e. L, L, I
    b. M, M, D   d. L, L, D   f. L, M, D

45. Ketogenic effects occur more readily in states of _____ cortisol secretion and _____ insulin secretion. (E, elevated; R, reduced)

    a. E, E   c. R, E
    b. E, R   d. R, R

OBJECTIVE 77-8.

Identify the actions and functions of cortisol during states of stress. Identify the effects of therapeutic doses of cortisol upon inflammatory responses and bodily functions.

46. ACTH secretion is effectively _____ (S, suppressed; E, enhanced) by states of _____ (P, physical; N, neurogenic) stress.

    a. S, only P   d. E, only P
    b. S, only N   e. E, only N
    c. S, both P & N   f. E, both P & N

47. Intense environmental conditions of _____ (H, heat; C, cold), _____ (W, as well as; N, but not) sympathomimetic agents, are effective in increasing ACTH secretion.

    a. Only H, W   d. Only H, N
    b. Only C, W   e. Only C, N
    c. Both H & C, W   f. Both H & C, N

48. _____ (I, increased; D, decreased) levels of glucocorticoid secretion, generally occurring during inflammatory states, are probably more significant in their _____ (M, general metabolic effects upon the body; D, direct effects upon the inflammatory process).

    a. I, M   c. D, M
    b. I, D   d. D, D

49. Administered cortisol exerts a _____ (L, labilizing; S, stabilizing) effect upon lysosomes and _____ (H, inhibits; F, facilitates) the formation of the _____ (C, vasoconstricting; D, vasodilating) substance bradykinin.

    a. L, H, C   c. L, F, C   e. S, H, C
    b. L, F, D   d. S, F, D   f. S, H, D

50. Cortisol in therapeutic doses reportedly _____ (D, diminishes; P, potentiates) the effects of norepinephrine and epinephrine upon peripheral vessels, _____ (D, diminishes; P, potentiates) the effects of histamine, and generally _____ (I, increases; R, reduces) capillary membrane permeabilities in areas of tissue inflammation.

    a. D, D, I   c. D, P, I   e. P, D, I
    b. D, P, R   d. P, P, R   f. P, D, R

51. Cortisol _____ (D, deficiencies; X, excesses) are associated with eosinopenia, _____ (LN, lymphocytopenia; LS, lymphocytosis), and _____ (A, anemia; P, polycythemia).

    a. D, LN, A   c. D, LS, A   e. X, LN, A
    b. D, LS, P   d. X, LS, P   f. X, LN, P

52. Therapeutic doses of cortisol _____ antibody production and _____ the intensity of the inflammatory state. (I, increase; D, decrease)

    a. I, I   c. D, I
    b. I, D   d. D, D

53. Cortisol therapy is effective in substantially altering the inflammatory state following a latent period of 1 to 2 _____ (M, minutes; H, hours; D, days) following its administration, and is a more effective means of _____ (A, preventing anaphylaxis; T, treating tuberculosis).

    a. M, A   c. D, A   e. H, T
    b. H, A   d. M, T   f. D, T

OBJECTIVE 77-9.
Identify the regulatory mechanisms of cortisol secretion.

54. The progressive sequence of events, in appropriate order, for a "neuroendocrine reflex" occurring in response to stress would involve _____ . (C, cortisol; H, hypophysis; A, adrenal cortex; G, gluconeogenesis; R, releasing factor; HT, hypothalamus; ACTH, corticotropin)

    a. C, H, A, R, HT, ACTH, & G
    b. R, ACTH, G, HT, A, C, & H
    c. H, R, C, A, ACTH, HT, & G
    d. HT, R, H, ACTH, A, C, & G
    e. HT, R, H, C, A, ACTH, & G
    f. None of the above

55. Glucocorticoid-secreting cells, _____ (L, like; U, unlike) aldosterone-secreting cells of the zona _____ (G, glomerulosa; F, fasciculata), are regulated largely by endocrine secretions of the _____ (A, adenohypophysis; N, neurohypophysis).

    a. L, G, A   c. L, F, N   e. U, F, A
    b. L, G, N   d. U, F, N   f. U, G, A

56. _____ (A, aldosterone; HC, hydrocortisone: C, cortisone) is a more important regulator of ACTH levels, while prepotent effects of pain _____ (E, enhance; S, suppress) ACTH secretion.

    a. A, E   c. C, E   e. HC, S
    b. HC, E   d. A, S   f. C, S

57. Cortisol _____ (H, inhibits; F, facilitates) ACTH secretion and _____ (I, increases; N, does not significantly alter; D, decreases) the formation of CRF.

    a. H, I   c. H, D   e. F, N
    b. H, N   d. F, I   f. F, D

58. Removal of one adrenal gland results in _____ (A, atrophic; H, hypertrophic) changes in the contralateral gland as a consequence of alterations of _____ secretion by the _____ (R, adrenal; P, pituitary) gland.

    a. A, ACTH, P   d. H, ACTH, P
    b. A, SH, P   e. H, SH, P
    c. A, ADH, R   f. H, ADH, R

59. A potential consequence of prolonged treatment with therapeutic doses of glucocorticoids involves prolonged atrophic changes of the adrenal _____ (M, medulla; C, cortex) as a consequence of _____ (I, increased; D, decreased) levels of a _____ (N, neurohypophyseal; A, adenohypophyseal) hormone.

    a. M, I, N   c. M, D, N   e. C, D, N
    b. M, I, A   d. C, D, A   f. C, I, A

60. Secretion rates of CRF _____ (W, as well as; N, but not) cortisol are reportedly highest in the _____ (M, early morning; A, late afternoon or evening) with a normal circadian rhythm.

    a. W, M   c. N, M
    b. W, A   d. N, A

OBJECTIVE 77-10.
Identify the loci and general chemical features of adrenocortical hormone synthesis, the chemical nature of CRF and ACTH, the mechanisms of target action of ACTH, and the mechanisms of transport and excretion of adrenocortical hormones.

61. Adrenal glucocorticoids_____ (W, as well as; N, but not) adrenal androgens are secreted by the zona _____ (G, glomerulosa; F, fasciculata and zona reticularis).

    a. W, G   c. N, G
    b. W, F   d. N, F

62. Adrenocortical hormones are _____ (S, steroid; PP, polypeptide; P, protein) substances synthesized largely in the adrenal cortex from _____ (A, acetyl CoA; AA, amino acids; C, preformed cholesterol).

    a. S, A   c. P, AA   e. PP, C
    b. PP, AA   d. S, C   f. P, A

63. Dexamethasone is a _____ (N, naturally occurring; S, synthetic) steroid which _____ (M, mimics; I, inhibits) the action of _____ (C, cortisol; A, aldosterone).

    a. N, M, C   c. N, I, C   e. S, M, C
    b. N, I, A   d. S, I, A   f. S, M, A

64. Spirolactones are utilized to _____ (E, enhance; B, block) the _____ (S, synthesis; T, target actions) of aldosterone.

    a. E, S   c. B, S
    b. E, T   d. B, T

65. Aldosterone is synthesized more directly from _____ , whereas progesterone is synthesized more directly from _____. (A, androstenedione; C, corticosterone; R, cortisol; P, pregnenolone)

    a. R, A   c. C, P   e. R, P
    b. A, C   d. C, R   f. P, A

66. ACTH is a _____ (S, steroid; PP, polypeptide) hormone which is thought to exert its primary target effects via _____ (M, cell membrane; I, intracellular cytoplasmic) receptors.

    a. S, M   c. PP, M
    b. S, I   d. PP, I

67. ACTH _____ cyclic AMP levels and _____ NADPH levels of target cells. (I, increases; D, decreases).

    a. I, I   c. D, I
    b. I, D   d. D, D

68. Plasma protein-bound cortisol, largely in a combined form with _____ (A, an albumin; G, a globulin) called transcortin, represents about _____ % of the total cortisol transported by blood.

    a. A, 10   c. A, 90   e. G, 50
    b. A, 50   d. G, 10   f. G, 90

69. Aldosterone, when compared to cortisol, has a _____ average concentration in blood and a _____ average secretion rate from the adrenal cortex. (H, higher; L, lower)

    a. H, H   c. L, H
    b. H, L   d. L, L

70. Adrenal steroids are inactivated in the _____ (K, kidneys; A, adrenal glands; L, liver) and largely excreted in _____ (F, feces; U, urine).

    a. K, F   c. L, F   e. A, U
    b. A, F   d. K, U   f. L, U

OBJECTIVE 77–11.

Identify Addison's disease, its causes, its consequences, and its means of treatment.

71. Addison's disease is generally characterized by rates of 17-hydroxysteroid excretion into urine which are _____ and exhibit _____ responsiveness to ACTH infusion. (H, higher than normal; N, about normal; L, lower than normal).

    a. H, H   c. L, H   e. N, L
    b. N, H   d. H, L   f. L, L

72. Addison's disease is associated with _____ levels of extracellular fluid volumes, _____ extracellular fluid potassium ion concentrations, and _____ extracellular fluid hydrogen ion concentrations. (E, elevated; R, reduced)

    a. E, E, E   c. E, R, E   e. R, E, E
    b. E, R, R   d. R, R, R   f. R, E, R

73. Addison's disease is more frequently associated with _____ levels of cortisol and _____ levels of ACTH. (H, higher than normal; N, about normal; L, lower than normal)

    a. H, H   c. H, L   e. L, N
    b. H, N   d. L, H   f. L, L

74. The administration of fludrocortisone, cortisol, and a considerably _____ (L, lowered; E, elevated) salt intake represents a more appropriate method of treatment of _____ (A, Addison's; C, Cushing's) disease.

    a. L, A   c. E, A
    b. L, C   d. E, C

75. Enhanced melanin deposition occurs more frequently in _____ (A, Addison's; C, Cushing's) disease in association with _____ (I, increased; D, decreased) levels of ACTH secretion.

    a. A, I   c. C, I
    b. A, D   d. C, D

76. The "Addisonian crisis" is a condition of _____ (X, excessive; I, inadequate) mobilization of tissue stores of proteins, _____ (W, as well as; N, but not) fats, and inadequate levels of _____ (G, glucocorticoid; P, pituitary) hormones during times of stress.

    a. X, W, G   c. X, N, G   e. I, W, G
    b. X, N, P   d. I, N, P   f. I, W, P

OBJECTIVE 77–12.

Identify Cushing's disease, primary aldosteronism, and the adrenogenital syndrome, and their causes, consequences, and treatment rationale.

77. Adrenocortical _____ (T, atrophy; H, hyperplasia) associated with Cushing's disease is _____ (G, generally, but not always; A, always) associated with _____ (D, decreased; I, increased) levels of ACTH secretion.

    a. T, G, D   c. T, A, I   e. H, G, I
    b. T, A, D   d. H, A, I   f. H, G, D

78. Cushing's disease is associated with disproportionately enhanced fat deposition in the _____ (L, lower; T, thoracic) region of the body, _____ (E, enhanced; D, diminished) collagen deposition in subcutaneous tissues, and _____ (E, enhanced; D, diminished) levels of muscle protein.

    a. L, E, E   c. L, D, E   e. T, D, E
    b. L, E, D   d. T, D, D   f. T, E, D

79. Cushing's disease is generally associated with _____ (I, increased; D, decreased) blood glucose levels and _____ (I, increased: D, decreased) stores of body tissue proteins, _____ (P, particularly for; E, with the exception of) the liver.

    a. I, I, P   c. I, D, P   e. D, I, P
    b. I, D, E   d. D, D, E   f. D, I, E

80. _____ (E, osteopetrosis; O, osteoporosis), accompanying Cushing's disease, occurs as a consequence of predominant abnormalities of _____ (M, mineral; P, protein) metabolism.

    a. E, M   c. O, M
    b. E, P   d. O, P

81. Puffiness of the face, a "buffalo torso," mild masculinizing effects, and _____ (I, increased; D, decreased) excretion of 17-hydroxysteroids into the urine are symptoms of _____ (O, hypoadrenalism; E, hyperadrenalism).

    a. I, O   c. D, O
    b. I, E   d. D, E

82. Primary aldosteronism, associated with _____ (E, hyperkalemia; O, hypokalemia) and _____ (I, increased; D, decreased) blood volumes, results from tumors of the _____ (P, pituitary; A, adrenal cortex).

    a. E, I, P   c. E, D, P   e. O, I, P
    b. E, D, A   d. O, D, A   f. O, I, A

83. The adrenogenital syndrome is associated with _____ (I, increased; N, no significant change in; D, decreased) levels of 17-ketosteroid excretions and general symptoms which are usually more difficult to detect in the _____ (E, prepubertal; O, postpubertal) _____ (M, male; F, female).

    a. I, E, M   c. D, E, F   e. N, O, M
    b. N, E, F   d. I, O, M   f. D, O, F

78

Insulin, Glucagon, and Diabetes Mellitus

OBJECTIVE 78–1.

Using the following diagram, identify the gross structure of the pancreas and associated organs, and the histologic composition of pancreatic tissue.

Directions: Match the lettered headings with the diagrams and the numbered list of descriptive words and phrases.

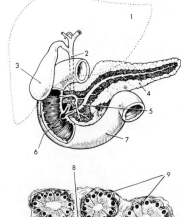

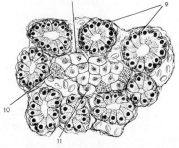

(Top figure from Bell, G.H., et al: Textbook of Physiology and Biochemistry. Churchill Liningstone, 1976.)
(Bottom figure from Guyton, A.C., Textbook of Medical Physiology, 5th ed. Philadelphia, W.B. Saunders Company, 1976.)

1. _____ A large dome-shaped organ, dark red in color, situated in the upper right side of the abdomen,
2. _____ Excretory duct of the liver and gall-bladder.
3. _____ Pear-shaped reservoir for bile.
4. _____ A large, racemose gland that secretes enzymes and hormones.
5. _____ Main excretory duct of the pancreas.
6. _____ A common orifice for the bile duct and pancreatic duct, located in the wall of the duodenum (papilla of Vater).
7. _____ The first or proximal portion of the small intestine.
8. _____ Cells in the pancreas that secrete insulin and glucagon directly into the blood.
9. _____ Cells that manufacture and secrete digestive enzymes.
10. _____ Pancreatic cells that secrete glucagon.
11. _____ Pancreatic cells that secrete insulin.

a. Gallbladder
b. Bile duct
c. Liver
d. Ampulla of the bile duct

e. Pancreas
f. Pancreatic acini
g. Beta cells

h. Pancreatic duct
i. Islets of Langerhans
j. Alpha cells
k. Duodenum

OBJECTIVE 78–2.

Recognize the endocrine and exocrine functions of the pancreas. Identify insulin and its actions upon glucose transport through cellular membranes.

12. The pancreas secretes the endocrine substances, _____ (G, glucagon; GC, glucocorticoids; I, insulin; E, epinephrine), _____ (W, as well as; N, but not) exocrine substances.

a. G & GC, W   c. I & G, W   e. GC & I, W
b. G & E, N   d. GC & E, N   f. E & I, N

13. Insulin is a _____ (S, steroid; LP, large molecular weight protein; SP, small molecular weight protein), secreted by the _____ (A, alpha; B, beta) cells of the _____ (L, islets of Langerhans; PA, pancreatic acini).

a. S, A, L   c. SP, A, PA   e. LP, B, L
b. LP, A, PA   d. S, B, PA   f. SP, B, L

14. The action of insulin upon glucose transport of cell membranes tends to increase _____ and decrease _____ (M. rates of glucose metabolism; B, blood glucose levels; G, tissue glycogen stores)

   a. M, B & G   c. G, B & M   e. B & G, M
   b. B, M & G   d. M & B, G   f. G & M, B

15. Insulin is thought to combine with _____ (C, cytoplasmic; M, cell membrane) receptors to _____ (I, increase; D, decrease) glucose transport into cells along _____ (F, a favorable; O, an opposing) concentration gradient.

   a. C, I, F   c. C, D, F   e. M, I, F
   b. C, D, O   d. M, D, O   f. M, I, O

16. The glucose transport process of most tissues of the body is largely that of _____ (P, a passive diffusion; F, a facilitated diffusion; A, an active transport) mechanism which is largely _____ (D, dependent upon; I, independent of) insulin.

   a. P, D   c. A, D   e. F, I
   b. F, D   d. P, I   f. A, I

17. Insulin enhances glucose transport by _____ (HR, neurons and red blood cells; IT, the intestinal mucosa and the renal tubular epithelium; MA, cardiac muscle, skeletal muscle, and adipose tissues) following a latent period closest to several _____ (M, minutes; H, hours; D, days).

   a. HR, M   c. MA, M   e. IT, M
   b. IT, H   d. HR, D   f. MA, H

OBJECTIVE 78–3.

Identify the extrahepatic effects of insulin upon glucose utilization, glycogen storage, and fat storage, and the hepatic effects of insulin upon glucose metabolism.

18. An absence of insulin results in rates of glucose transport for _____ (M, most, but not all; A, all) body cells of about _____ % of normal.

   a. M, 25   c. M, 175   e. A, 75
   b. M, 75   d. A, 0   f. A, 175

19. An excess of available glucose and insulin results in effects within skeletal muscle which include _____ (A, activation; I, inactivation) of glycogen synthetase, _____ (A, activation; I, inactivation) of phosphorylase, and _____ (F, facilitation; I, inhibition) of phosphofructokinase.

   a. A, I, F   c. A, A, F   e. I, A, F
   b. A, I, I   d. I, A, I   f. I, I, I

20. Liver cell membranes, when compared to most body tissues, are relatively _____ permeable to glucose diffusion in the presence of insulin and relatively _____ permeable to glucose diffusion in the absence of insulin. (M, more; L, less)

   a. M, M   c. L, M
   b. M, L   d. L, L

21. Insulin _____ (P, promotes; H, inhibits) glucose transport into liver cells primarily as a consequence of altered mechanisms of _____ (A, active transport; F, facilitated diffusion; G, glucose phosphorylation).

   a. P, A   c. P, G   e. H, F
   b. P, F   d. H, A   f. H, G

22. Insulin _____ (I, increases; D, decreases) the liver concentration of glucokinase following a latent period of several _____ (M, minutes; H, hours; Y, days).

   a. I, M   c. I, Y   e. D, H
   b. I, H   d. D, M   f. D, Y

23. A higher ratio of glycogen synthetase to phosphorylase activity is associated with predominant glucose conversion to liver _____ (G, glycogen; F, fat) during the _____ (P, presence; A, absence) of insulin secretion.

   a. G, P   c. F, P
   b. G, A   d. F, A

24. The liver plays a _____ (N, minor; J, major) body functional role in buffering a rise _____ (W, as well as; B, but not) a fall in blood glucose, and has saturated glycogen stores which represent about _____ % of the liver mass.

   a. N, W, 5   c. N, B, 25   e. J, W, 25
   b. N, B, 5   d. J, B, 25   f. J, W, 5

25. _____ (M, most, but not all; A, all) of the fat formed from glucose which is stored in adipose tissue is synthesized by _____ (AT, adipose tissue; L, the liver).

   a. M, AT   c. A, AT
   b. M, L   d. A, L

26. The _____ (P, presence; A, absence) of insulin promotes gluconeogenesis as well as _____ (I, increased; D, decreased) levels of cyclic AMP in liver cells.

   a. P, I   c. A, I
   b. P, D   d. A, D

27. The liver functions to buffer a _____ (R, rise; F, fall) in blood glucose during normal states of _____ (L, low; E, elevated) states of insulin secretion by means of glycogenolysis _____ (W, as well as; N, but not) gluconeogenesis.

   a. R, L, W   c. R, E, W   e. F, L, N
   b. R, E, N   d. F, E, W   f. F, L, W

28. Liver glycogen stores are reduced in the _____ (R, presence; A, absence) of insulin as a consequence of activation of _____ (P, phosphorylase; G, glucokinase; S, glycogen synthetase) and the inhibition of _____ (P, phosphorylase; G, glucokinase; S, glycogen synthetase).

   a. R, S, P   c. R, P, G   e. A, S, G
   b. R, S, G   d. A, P, S   f. A, G, P

OBJECTIVE 78–4.

Identify the effects of states of insulin lack or excess upon lipid metabolism.

29. Insulin has relatively _____ (W, weak; P, potent) effects upon adipose tissue and requires a relatively _____ (S, short; L, long) latent period for its actions.

   a. W, S   c. P, S
   b. W, L   d. P, L

30. The _____ (P, presence; L, lack) of insulin promotes increased alpha-glycerophosphate concentrations within fat cells and increased fat _____ (S, synthesis; B, breakdown).

   a. P, S   c. L, S
   b. P, B   d. L, B

31. Fatty acids are utilized by _____ (N, only a minority; S, most) body tissues for metabolic purposes, including _____ (C, cardiac muscle; B, the brain; L, the liver) but not _____ (C, cardiac muscle; B, the brain; L, the liver).

   a. N, B, C & L   d. S, B, C & L
   b. N, C & L, B   e. S, C & L, B
   c. N, B & L, C   f. S, B & L, C

32. A rise in the level of free fatty acids of blood is associated with insulin _____ (E, excess; L, lack) and increased utilization of _____ (F, free fatty acids; G, glucose) for body metabolic purposes.

   a. E, F   c. L, F
   b. E, G   d. L, G

33. _____ (I, increased; D, decreased) rates of catabolism of fatty acids by tissues approximately _____ (H, halves; B, doubles) the depressive action of insulin _____ (L, lack; X, excess) upon cellular uptake and utilization of glucose.

   a. I, H, L   c. I, B, L   e. D, H, L
   b. I, H, X   d. D, B, X   f. D, H, X

34. Ketone bodies are largely produced in _____ (L, the liver; A, adipose tissue) during states of predominantly _____ (G, glucose; F, fatty acid) catabolism resulting from the _____ (B, absence; P, presence) of insulin secretion.

   a. L, G, B   c. L, F, B   e. A, G, B
   b. L, F, P   d. A, F, P   f. A, G, P

35. Severe diabetes mellitus is characterized by states of _____ (I, increased; D, decreased) blood glucose levels, _____ (X, excess; L, a lack of) insulin secretion, and _____ (I, increased; D, decreased) blood levels of acetoacetic acid, beta-hydroxybutyric acid, and acetone.

   a. I, X, I   c. I, L, I   e. D, X, I
   b. I, L, D   d. D, L, D   f. D, X, D

36. The presence of excess ketone bodies leads to potential states of metabolic _____ (K, alkalosis; C, acidosis) and _____ (V, convulsions; M, coma).

a. K, V   b. K, M   c. C, V   d. C, M

OBJECTIVE 78–5.
Identify the effects of insulin upon protein metabolism and growth.

37. Insulin _____ (I, increases; D, decreases) the rate of amino acid transport into cells, and _____ (I, increases; D, decreases) the rate of protein synthesis by mechanisms which alter the rate of _____ (S, transcription of DNA; L, translation of messenger RNA).

     a. I, I, only L   d. D, D, only L
     b. I, I, both S & L   e. D, D, both S & L
     c. I, D, only S   f. D, I, only S

38. Untreated diabetics exhibit _____ than normal growth rates, _____ than normal body protein stores, and _____ than normal susceptibility to infectious processes. (H, higher; L, lower)

     a. H, H, L   c. H, L, L   e. L, L, H
     b. H, H, H   d. L, L, L   f. L, H, H

39. Insulin has a _____ (S, synergistic; C, competitive) action with somatotropin to _____ (F, facilitate; I, inhibit) growth rates of soft tissues _____ (W, as well as; N, but not) bone.

     a. S, F, W   c. S, I, W   e. C, I, W
     b. S, F, N   d. C, I, N   f. C, F, N

40. Insulin _____ (S, secretion; L, lack) promotes elevated plasma amino acid concentrations and _____ (E, enhanced; D, diminished) rates of gluconeogenesis.

     a. S, D   c. L, D
     b. S, E   d. L, E

41. Severe diabetes is associated with _____ (E, elevated; R, reduced) blood levels of glucose and free fatty acids, _____ (W, as well as; N, but not) amino acids, and _____ (H, higher; L, lower) than normal body stores of lipid _____ (W, as well as; N, but not) protein.

     a. E, W, H, W   c. E, N, L, N   e. R, N, H, W
     b. E, W, L, W   d. R, N, L, N   f. R, W, H, N

OBJECTIVE 78–6.
Identify and characterize the primary regulatory mechanism for insulin secretion and its functional significance. Identify other factors which enhance insulin secretion.

42. Insulin secretion by _____ (A, alpha; B, beta) cells of the _____ (D, adrenal glands; P, pancreas; H, adenohypophysis) is primarily regulated by variations in blood _____ (G, glucose; M, hormone) levels.

     a. A, D, G   c. A, H, G   e. B, P, G
     b. A, P, M   d. B, D, M   f. B, H, M

43. An increase in blood glucose concentrations above normal fasting levels of _____ mg./100 ml. results in _____ (I, increased; D, decreased) rates of insulin secretion.

     a. 20 to 30, I   d. 20 to 30, D
     b. 80 to 90, I   e. 80 to 90, D
     c. 220 to 320, I   f. 220 to 320, D

44. Considerably _____(E, elevated; L, lowered) levels of blood glucose result in _____ (G, slightly; B, substantially) elevated rates of insulin secretion following relatively _____ (S, short; P, prolonged) latent periods.

    a. E, G, S   c. E, B, S   e. L, G, S
    b. E, B, P   d. L, B, P   f. L, G, P

45. Elevated blood glucose levels augment endocrine secretion by_____ (A, alpha; B, beta) cells of the islets of Langerhans by probable mechanisms which include_____ (R, release of preformed endocrine stores; E, activation of enzyme systems for endocrine synthesis; H, cellular hypertrophy).

    a. A; only R   d. B; only R
    b. A; only R & E   e. B; only R & E
    c. A; R, E, & H   f. B; R, E, & H

46. Amino acids generally act to _____ (I, increase; D, decrease) insulin secretion, particularly when blood glucose levels are _____ (L, low; N, normal; H, high).

    a. I, L   c. I, H   e. D, N
    b. I, N   d. D, L   f. D, H

47. Alanine _____ (I, increases; D, decreases) insulin secretion and _____(I, increases; D, decreases) amino acid transport into body tissues through mechanisms which are _____ (O, opposite to; P, more potent than) those occurring for other amino acids.

    a. I, I, O   c. I, D, P   e. D, D, O
    b. I, I, P   d. D, D, P   f. D, I, O

48. Gastrin, secretin, and cholecystokinin are hormones which _____ (S, act synergistically with; O, oppose) the action of absorbed dietary glucose to _____ (I, increase; D, decrease) insulin secretion.

    a. S, I   c. O, I
    b. S, D   d. O, D

49. Glucagon _____ (W, as well as; N, but not) cortisol administration results in _____ (I, increased; D, decreased) rates of insulin secretion.

    a. W, I   c. N, I
    b. W, D   d. N, D

50. Elevated blood glucose levels result in _____ (I, increased; D, decreased) rates of insulin secretion and predominantly _____ (L, lipid; C, carbohydrate) catabolism.

    a. I, L   c. D, L
    b. I, C   d. D, C

51. Secretion of growth hormone _____ (W, as well as; N, but not) cortisol during states of _____(E, hyperglycemia; O, hypoglycemia), _____ (D, depresses; H, enhances) cellular utilization of glucose for energy purposes.

    a. W, E, D   c. W, O, H   e. N, E, H
    b. W, O, D   d. N, O, H   f. N, E, D

52. Epinephrine _____ (I, increases; D, decreases) the blood glucose concentration, _____ (I, increases; D, decreases) the free fatty acid concentration of blood, and more notably increases body tissue utilization of _____(G, glucose; F, free fatty acids) for energy purposes.

    a. I, I, G   c. I, D, G   e. D, D, G
    b. I, I, F   d. D, D, F   f. D, I, F

OBJECTIVE 78–7.

Identify glucagon, its target actions, factors which influence its secretion, and its means of regulation.

53. The most powerful hepatic glycogenolytic agent known is a _____(S, steroid; P, small protein) secreted by the _____ (AC, adrenal cortex; AM, adrenal medulla; EP, endocrine pancreas).

    a. S, AC   c. S, EP   e. P, AM
    b. S, AM   d. P, AC   f. P, EP

54. Glucagon, or _____ (O, hypoglycemic; E, hyperglycemic) factor, is secreted by _____ (A, alpha; B, beta) cells of the islets of Langerhans.

    a. O, A   c. E, A
    b. O, B   d. E, B

55. Glucagon has actions which are generally _____ (S, synergistic; O, opposite) to those occurring for insulin, and include increased _____ (L, glycogenolysis; N, glycogenesis) and _____ (D, diminished; E, enhanced) gluconeogenesis.

a. S, L, D   c. S, N, D   e. O, L, D
b. S, N, E   d. O, N, E   f. O, L, E

56. Glucagon has a more direct target action to _____ (T, activate; H, inhibit) _____ (C, adenyl cyclase; P, a protein kinase regulator), with subsequent actions which increase liver levels of _____ (A, phosphorylase a; B, phosphorylase b).

a. T, C, A   c. T, P, A   e. H, P, A
b. T, C, B   d. H, P, B   f. H, C, B

57. Glucagon has more significant effects to increase the rate of amino acid transport into _____ (H, hepatic; E, extrahepatic) cells, and to enhance adipose tissue _____ (N, lipogenesis; L, lipolysis) during states of _____ (S, saturation; D, depletion) of liver glycogen stores.

a. H, N, S   c. H, L, S   e. E, N, S
b. H, L, D   d. E, L, D   f. E, N, D

58. Exercise _____ (W, as well as; N, but not) starvation tends to _____ (R, reduce; E, elevate) blood glucose levels and subsequently to _____ (R, reduce; E, elevate) glucagon secretion rates.

a. W, R, R   c. W, E, E   e. N, E, R
b. W, R, E   d. N, E, E   f. N, R, R

59. Glucagon secretion rates are generally _____ (H, high; L, low) in severely diabetic patients as a consequence of _____ (E, elevated blood; D, diminished blood; I, altered intracellular) levels of glucose.

a. H, E   c. H, I   e. L, D
b. H, D   d. L, E   f. L, I

60. Amino acids tend to _____ the secretion of glucagon and _____ the secretion of insulin. (H, inhibit; E, enhance)

a. H, H   c. E, H
b. H, E   d. E, E

61. Glucagon, when compared to epinephrine, has a _____ potent action upon liver glycogenolysis, a _____ potent action upon muscle glycogenolysis, and a _____ potent action upon fat mobilization from adipose tissue. (M, more; L, less)

a. M, M, L   c. M, L, L   e. L, L, L
b. M, M, M   d. L, L, M   f. L, M, M

OBJECTIVE 78–8.

Identify the principal and supporting regulatory mechanisms for blood glucose levels, their significance, and the glucose buffer function of the liver.

62. Blood glucose concentrations are relatively _____ (P, poorly; W, well) regulated and typically average _____ mg. % in the fasting state.

a. P, 12 to 14   d. W, 12 to 14
b. P, 80 to 90   e. W, 80 to 90
c. P, 1200 to 1400   f. W, 1200 to 1400

63. Excessive elevations of plasma glucose result in intracellular _____ (H, hydration; D, dehydration), glucosuria, and renal _____ (U, diuresis; A, antidiuresis).

a. H, U   c. D, U
b. H, A   d. D, A

64. The brain, retina, and germinal epithelium are _____ (T, typical of; X, exceptions to) the general case, whereby most body tissues _____ (A, are; N, are not) obligatory users of glucose for metabolic energy purposes.

a. T, A
b. T, N
c. X, A
d. X, N

65. The more significant "glucose buffer" function of _____ (M, skeletal muscle; A, adipose tissue; L, the liver) occurs as a consequence of its role in preventing _____ (R, only a rise; F, only a fall; E, either a rise or a fall) in blood glucose levels.

a. M, E
b. A, R
c. L, E
d. M, R
e. A, F
f. L, R

66. Epinephrine secretion _____ (I, increases; D, decreases) blood glucose levels and is _____ (I, increased; N, not significantly affected; D, decreased) by hypoglycemic states.

a. I, I
b. I, N
c. I, D
d. D, I
e. D, N
f. D, D

67. Somatotropin _____ (W, as well as; N, but not) cortisol tends to _____ (I, increase; D, decrease) blood glucose concentrations and to _____ (E, enhance; M, diminish) the rate of glucose utilization by peripheral tissues.

a. W, I, E
b. W, I, M
c. W, D, M
d. N, D, M
e. N, D, E
f. N, I, E

68. Increased secretion of cortisol _____ (W, as well as; N, but not) somatotropin occurs in states of _____ (E, hyperglycemia; O, hypoglycemia).

a. W, E
b. W, O
c. N, E
d. N, O

69. The _____ (IG, insulin-glucagon; SC, somatotropin-cortisol) control system of blood glucose is a more powerful control system and has a relatively _____ (S, shorter; L, longer) latency for activation corresponding to several _____ (M, minutes; H, hours; D, days).

a. IG, S, M
b. IG, S, H
c. IG, L, H
d. SC, L, D
e. SC, L, H
f. SC, S, M

OBJECTIVE 78–9.

Identify diabetes mellitus, its potential causes, and major pathophysiologic consequences.

70. Surgical removal of a minimum of _____ % of the pancreas is utilized as a method of _____ (T, treatment; E, experimental induction) of diabetes _____ (M, mellitus; I, insipidus).

a. 25 to 40, T, M
b. 25 to 40, T, I
c. 25 to 40, E, I
d. 90 to 95, E, I
e. 90 to 95, E, M
f. 90 to 95, T, M

71. High carbohydrate dietary intakes are thought to _____ (O, oppose; S, act synergistically with) the actions of high levels of adenohypophyseal hormones to _____ (R, reduce; E, enhance) the incidence of diabetes.

a. O, R
b. O, E
c. S, R
d. S, E

72. Diabetes mellitus, generally caused by _____ (H, hereditary; N, nonhereditary) factors, occurs more frequently in a mild form in _____ (Y, younger; O, older) people who tend to be _____ (UW, underweight; OW, overweight).

a. H, Y, UW
b. H, Y, OW
c. H, O, OW
d. N, O, OW
e. N, O, UW
f. N, Y, UW

73. A _____ (R, recessive; D, dominant) genetic characteristic for diabetes mellitus is present in about _____ % of the general population.

a. R, 1
b. R, 4
c. R, 20
d. D, 1
e. D, 4
f. D, 20

74. Diabetes mellitus results in increased utilization of _____ (G, glucose; L, lipids; P, proteins) for energy purposes and _____ (I, increased; D, decreased) blood levels of glucose.

a. Only G, I
b. Only L, I
c. Both L & P, I
d. Only G, D
e. Only L, D
f. Both L & P, D

75. Glucose levels which exceed a blood "threshold" for the appearance of glucose in the urine of about _____ mg. % tend to promote intracellular _____ (H, hydration; D, dehydration) and extracellular _____ (H, hydration; D, dehydration).

a. 90, H, H
b. 180, D, D
c. 325, D, H
d. 90, D, D
e. 180, H, H
f. 325, H, D

76. Glucosuria, diuresis, a metabolic _____(C, acidosis; K, alkalosis), and _____ (O, hypoventilation; E, hyperventilation) accompanying a type of "Kussmaul respiration" are pathophysiologic symptoms of insulin _____ (L, lack; X, excess).

a. C, O, L   c. C, E, L   e. K, O, L
b. C, E, X   d. K, E, X   f. K, O, X

77. Ketone bodies have a relatively _____ (H, high; L, low) threshold for excretion by the kidneys, and a pK value averaging _____ .

a. H, 4   c. H, 8   e. L, 6
b. H, 6   d. L, 4   f. L, 8

78. The blood chemistry of a diabetic individual generally shows elevated concentrations of _____ and lower than normal concentrations of _____ . (K, keto acids; B, bicarbonate ions; H, hydrogen ions; C, cholesterol; L, chloride)

a. B & L; K, H, & C   d. K & C; B, H, & L
b. H & C; K, B, & L   e. K, H, C, & B; L
c. B & C; K, H, & L   f. K, H, & C; B & L

79. Symptoms of insulin lack include increased _____ and decreased _____ . (U, urine flows; W, water intake; F, food intake; BW, body weight)

a. U, W, & F; BW   d. U & W; F & BW
b. U & F; W & BW   e. W & F; U & BW
c. U; F, W, & BW   f. W & BW; U & F

OBJECTIVE 78–10.
Identify the diagnostic symptoms and means of detection of diabetes.

80. A _____ (H, higher; L, lower) than normal renal tubular maximum for glucose, a fasting blood glucose level of 80 to 90 mg %, and glycosuria are symptoms of _____ (D, diabetes mellitus; R, renal glycosuria).

a. H, only D   d. L, only D
b. H, only R   e. L, only R
c. H, both D & R   f. L, both D & R

81. The administration of 1 gram of glucose per kilogram of body weight to an individual with diabetes mellitus results in a glucose tolerance curve with a _____ (S, shorter; L, longer) rise time, a _____ (W, lower; G, greater) peak intensity, and a _____ (S, shorter; L, longer) decay time when compared to normal.

a. S, W, S   c. S, G, S   e. L, G, S
b. S, W, L   d. L, G, L   f. L, W, L

82. The administration of a test dose of insulin generally results in a greater _____ (F, fall; R, rise) in blood glucose levels in patients with _____ (E, excessive anterior pituitary hormone secretions; P, diabetes mellitus of pancreatic origin).

a. F, E   c. R, E
b. F, P   d. R, P

83. The presence of acetone in the breath of a diabetic individual during states of _____ (O, overadministration; L, a lack) of insulin reflects predominantly _____ (C, carbohydrate; P, protein; D, lipid) metabolism.

a. O, C   c. O, D   e. L, P
b. O, P   d. L, C   f. L, D

OBJECTIVE 78–11.
Identify the methods of treatment for diabetes and their effectiveness.

84. _____ (R, regular; P, protamine zinc; C, crystalline) insulin is a longer-acting form of insulin which continues to act for periods up to as long as _____ hours.

a. R, 6 to 8   c. C, 6 to 8   e. P, 36 to 48
b. P, 6 to 8   d. R, 36 to 48   f. C, 36 to 48

85. Subcutaneous injections of a _____ (L, lente; P, protamine zinc; C, crystalline) insulin are more effective in treating a _____ (R, rise; F, fall) in blood glucose occurring in the diabetic patient during his daily routine.

a. L, R   c. C, R   e. P, F
b. P, R   d. L, F   f. C, F

86. States of exercise_____(W, as well as; B, but not) states of fever and severe infection generally tend to _____ (I, increase; N, not significantly alter; D, decrease) the insulin requirements of a diabetic patient.

    a. W, I   c. W, D   e. B, N
    b. W, N   d. B, I   f. B, D

87. The treatment of diabetes is generally aimed at maintaining blood glucose levels at _____ values through the administration of insulin and _____ dietary levels of carbohydrate intake. (R, substantially lower than normal; N, normal; I, substantially higher than normal)

    a. R, R   c. I, I   e. N, R
    b. N, N   d. R, I   f. I, N

88. Increased susceptibility to infections, cataracts, chronic renal disease, and _____ (E, hypertension; O, hypotension) are symptoms resulting from the_____(N, nontreatment; T, insulin treatment) of diabetics.

    a. E, N   c. O, N
    b. E, T   d. O, T

89. Tolbutamide is _____ (A, an alloxan; S, a sulfonylurea) compound which may be administered orally to treat _____ (M, mild; V, severe) cases of diabetes effectively.

    a. A, only M   d. S, only M
    b. A, only V   e. S, only V
    c. A, both M & V   f. S, both M & V

OBJECTIVE 78–12.

Contrast the characteristics of diabetic coma with the characteristics of hyperinsulinism.

90. A fall in body pH to levels of _____ as a consequence of states of insulin _____ (L, lack; X, excess) is threshold for conditions of _____ (M, coma; V, convulsions).

    a. 7.2 to 7.3, L, M   d. 6.9 to 7.0, X, V
    b. 7.2 to 7.3, X, V   e. 6.9 to 7.0, L, M
    c. 7.2 to 7.3, X, M   f. 6.9 to 7.0, L, V

91. An acidic plasma is associated with _____ (G, greater; L, less) than normal states of insulin effectiveness and _____ (A, activation; H, inhibition) of an insulin antagonist within the _____ (B, albumin; AG, alpha globulin) fraction of plasma.

    a. G, A, B   c. G, H, B   e. L, A, B
    b. G, H, AG   d. L, H, AG   f. L, A, AG

92. The correction of diabetic coma generally requires the correction of states of body _____ (H, hydration; D, dehydration), the correction of states of acid-base imbalance with _____ (A, ammonium chloride; L, sodium lactate) administration, and the correction of _____ (E, elevated; F, deficient) body stores of potassium.

    a. H, A, E   c. H, L, E   e. D, L, E
    b. H, A, F   d. D, L, F   f. D, A, F

93. Instructions carried by diabetic individuals, which request that you administer a sweet drink and call a physician immediately if they are found in a dazed condition, are necessary for rapid correction of _____ (D, dehydration; O, hypoglycemia) as a consequence of insulin _____ (L, lack; X, excess).

    a. D, L   c. O, L
    b. D, X   d. O, X

94. "Insulin shock" is a condition associated with insulin _____ (L, lack; X, excess) which progresses to a loss of consciousness when the blood glucose concentration reaches threshold levels of about _____ mg %.

    a. L, 20 to 50   d. X, 20 to 50
    b. L, 60 to 70   e. X, 60 to 70
    c. L, 200 to 300   f. X, 200 to 300

95. Acetone breath and a rapid, deep type of breathing are generally _____ (P, present; A, absent) in diabetic _____ (W, as well as; N, but not) hypoglycemic comas.

    a. P, W   c. A, W
    b. P, N   d. A, N

79

Parathyroid Hormone, Calcitonin, Calcium and Phosphate Metabolism, Vitamin D, Bone and Teeth

OBJECTIVE 79-1.

Using the following diagrams, identify the gross anatomical and histologic structures of the parathyroid gland and adjacent structures.

Directions: Match the lettered headings with the diagrams and the numbered list of descriptive words and phrases.

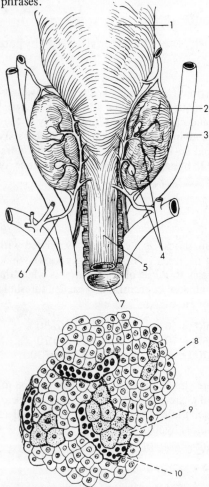

1. _____ The part of the alimentary canal between the oral cavity and the esophagus.
2. _____ Large ductless gland with two large lobes and one small lobe; secretes thyroxine.
3. _____ Chief artery that passes blood up to the head and neck.
4. _____ Small ovoid glands near or embedded in the thyroid gland, usually four in number; secrete parathormone.
5. _____ Commonly known as the food tube.
6. _____ Lower portion of the artery that supplies the thyroid with blood.
7. _____ Commonly known as the windpipe.
8. _____ Cells in the parathyroid that secrete most of the parathormone.
9. _____ Large granulose cells, function unknown.
10. _____ Erythrocytes.

a. Red blood cells
b. Parathyroid glands
c. Oxyphil cell
d. Pharynx
e. Chief cell
f. Thyroid gland
g. Trachea
h. Inferior thyroid artery
i. Common carotid artery
j. Esophagus

(Top figure from Ganong, W.F.: Review of Medical Physiology, 8th Ed., Lange Medical Publications, 1977 as redrawn from Goss, C.M., Ed., Gray's Anatomy of the Human Body, Lea & Febiger, 1973.)
(Bottom figure from Guyton, A.C., Textbook of Medical Physiology, 5th Ed. Philadelphia, W.B. Saunders Company, 1976.)

OBJECTIVE 79-2.

Contrast the dietary origins, intestinal absorption, and excretion characteristics for calcium and phosphate.

11. Milk and milk products represent a more significant dietary source of _____ (C, calcium; P, phosphate) for predominantly gastrointestinal absorption of _____ (I, ionized; IS, insoluble) forms of calcium and phosphate.

a. C, I   c. P, I
b. C, IS   d. P, IS

12. Phosphate absorption, occurring _____ (M, more; L, less) readily from the gastrointestinal tract than calcium absorption, is _____ (E, enhanced; N, not significantly altered; R, reduced) by excess dietary calcium levels.

a. M, E   c. M, R   e. L, N
b. M, N   d. L, E   f. L, R

13. The average adult has a net absorption of about _____ mg. of dietary calcium per day, representing about _____ % of the average dietary intake.

a. 10, 12   c. 1000, 50   e. 100, 78
b. 100, 12   d. 10, 50   f. 1000, 78

14. The renal plasma "threshold" for _____ (C, calcium; P, phosphate) is approximately 1 _____ (E, milliequivalent; M, millimol) per liter.

a. C, E   c. P, E
b. C, M   d. P, M

OBJECTIVE 79-3.

Identify the vitamin D compounds, the sequence of events leading to the formation of an active form of vitamin D_3, and its regulatory role in calcium absorption.

15. The natural vitamin D _____ (E, ergocalciferol; C, cholecalciferol) is largely formed from _____ (H, 7-dehydrocholesterol; X, 25-hydroxycholecalciferol) in the _____ (K, kidneys; L, liver; S, skin).

a. E, H, K   c. E, X, L   e. C, H, S
b. E, X, S   d. C, X, L   f. C, H, K

16. 25-hydroxycholecalciferol, synthesized in the _____ (L, liver; K, kidneys), is characterized by a relatively _____ (H, short; G, long) half-life in the body and a relatively _____ (V, variable; C, constant) concentration in plasma.

a. L, H, V   c. L, G, V   e. K, G, V
b. L, H, C   d. K, G, C   f. K, H, C

17. Parathyroid hormone _____ (I, inhibits; R, is required for) the _____ (L, liver; N, renal) formation of the active form of vitamin D_3 from a _____ (D, 1, 25-dihydroxycholecalciferol; H, 25-hydroxycholecalciferol) precursor.

a. I, L, D   c. I, N, D   e. R, N, D
b. I, L, H   d. R, N, H   f. R, L, H

18. _____ (H, 25-hydrocholecalciferol; D, 1, 25-dihydroxycholecalciferol) functions to _____ (E, enhance; R, reduce) calcium ion transport by the intestinal epithelial cells.

a. H, E   c. D, E
b. H, R   d. D, R

19. The active form of vitamin D_3 results in a _____ (B, brief; P, prolonged) _____ (D, decrease; I, increase) in the "calcium-binding protein" of intestinal epithelial cells and enhanced _____ (L, loss; A, absorption) of intestinal calcium.

a. B, D, L   c. B, D, A   e. P, D, A
b. B, I, L   d. P, I, A   f. P, I, L

20. Higher than normal levels of plasma calcium _____ parathyroid hormone rates of secretion and _____ plasma levels of 1, 25-dihydroxycholecalciferol. (I, increase; D, decrease)

a. I, I   c. D, I
b. I, D   d. D, D

OBJECTIVE 79–4.

Identify the forms of calcium and inorganic phosphate in extracellular fluids and their relative concentrations.

21. The calcium concentration of plasma normally ranges from _____ mg. %, or a value of about _____ mEq./liter.

   a. 2.1 to 2.3, 1.1   d. 2.1 to 2.3, 4.4
   b. 9.2 to 10.4, 5   e. 9.2 to 10.4, 20
   c. 80 to 90, 42   f. 80 to 90, 170

22. Plasma calcium consists of about _____ % of the calcium combined with plasma proteins, about _____ % combined with substances in a nonionized form which diffuses readily through the capillary membrane, and about _____ % in an ionized form.

   a. 70, 5, 25   c. 5, 25, 70   e. 50, 5, 45
   b. 25, 5, 70   d. 5, 50, 45   f. 50, 45, 5

23. Interstitial fluid _____ (W, as well as; N, but not) plasma has a normal calcium ion concentration of _____ mEq./liter.

   a. W, 0.5   c. W, 5.2   e. N, 2.3
   b. W, 2.3   d. N, 0.5   f. N, 5.2

24. An $H_2PO_4^-$ to HPO_4^{2-} ion concentration ratio of plasma closest to   is increased in states of _____ (C, acidosis; K, alkalosis).

   a. 1:10, C   c. 20:1, C   e. 1:10, K
   b. 10:1, C   d. 1:20, K   f. 20:1, K

25. Plasma concentrations of _____ ions of about _____ mEq./liter lie closer to the plasma concentrations of the _____ (I, calcium ion; T, total calcium) concentrations.

   a. HPO_4^{2-}, 0.26, I   d. $H_2PO_4^-$, 2.1, T
   b. HPO_4^{2-}, 2.1, T   e. $H_2PO_4^-$, 0.26, I
   c. HPO_4^{2-}, 2.1, I   f. $H_2PO_4^-$, 0.26, T

26. The average total quantity of inorganic phosphorus per 100 ml. of blood is about _____ mg., with a higher average occurring for _____ (C, children; A, adults).

   a. 4, C   c. 24, C   e. 8, A
   b. 8, C   d. 4, A   f. 24, A

OBJECTIVE 79–5.

Identify the major consequences of altered states of calcium and phosphate concentrations of the body fluids.

27. Prolonged states of hypophosphatemia, _____ (W, as well as; N, but not) hypocalcemia, result in _____ (I, increased; D, decreased) mineralization of bone.

   a. W, I   c. N, I
   b. W, D   d. N, D

28. _____ (I, increased; D, decreased) levels of particularly _____ (C, calcium; P, phosphate) ion concentrations of extracellular fluids are associated with extreme and immediate pathologic consequences.

   a. Only I, C   d. Only I, P
   b. Only D, C   e. Only D, P
   c. Both I & D, C f. Both I & D, P

29. Tetany occurs when extracellular calcium ion concentrations are altered to threshold levels about _____ % _____ (A, above; B, below) normal.

   a. 10, A   c. 80, A   e. 40, B
   b. 40, A   d. 10, B   f. 80, B

30. Overt states of tetany may be induced in patients with latent tetany due to the more direct effects of altered extracellular calcium ion concentrations upon _____ (P, membrane permeability; M, the contractile mechanism of skeletal muscle) and states of _____ (C, acidosis; K, alkalosis) induced by _____ (O, hypoventilation; E, hyperventilation).

   a. P, C, O   c. P, K, O   e. M, C, O
   b. P, K, E   d. M, K, E   f. M, C, E

31. Hypercalcemic states are more frequently associated with _____ (R, reduced; L, lengthened) QT intervals of the electrocardiogram, _____ (C, constipation; D, diarrhea), and _____ (E, enhanced; P, depressed) central nervous system activity.

   a. R, C, E   c. R, D, P   e. L, D, E
   b. R, C, P   d. L, D, P   f. L, C, E

32. Direct effects of altered levels of calcium to impair blood coagulation, reduce cardiac contractility, and cause dilation of the heart _____ (R, are; N, are not) considered primary causes of death in associated states of _____ (E, hypercalcemia; O, hypocalcemia).

a. R, E,   b. R, O   c. N, E   d. N, O

OBJECTIVE 79–6.
Identify the organic and inorganic components of bone, and their functional significance in states of tension and compression.

33. Organic components constitute _____ (M, more; L, less) than 50% of the weight of bone and are largely comprised of _____ (MP, mucoproteins; K, keratin; C, collagen).

a. M, MP   c. M, C   e. L, K
b. M, K   d. L, MP   f. L, C

34. A preferential orientation of collagen fibers _____ (L, parallel; N, perpendicular) to natural lines of force contribute more significantly to the _____ (T, tensional; C, compressional) strength of bone.

a. L, T   c. N, T
b. L, C   d. N, C

35. The mineral salts of bone are comprised largely of _____ (C, calcite; A, aragonite; H, hydroxyapatite) crystals of _____ (L, calcium; M, magnesium; P, phosphate; B, carbonate).

a. C, L & B   c. H, M & P   e. A, L & P
b. A, M & B   d. C, M & B   f. H, L & P

36. Inorganic crystals of bone are comprised of _____ (S, single large; N, numberous small) crystals with dimensions ranging from 10 to 400 _____ (A, angstroms; U, microns; MM, millimeters), with Ca/P ratios between _____ .

a. S, MM, 1.3 & 2.0   d. S, MM, 3.1 & 6.4
b. N, A, 1.3 & 2.0   e. N, A, 3.1 & 6.4
c. N, U, 1.3 & 2.0   f. N, U, 3.1 & 6.4

37. Magnesium, sodium, potassium, and carbonate ions, _____ (W, as well as; N, but not) strontium and lead ions, are thought to be readily _____ (D, adsorbed to; B, substituted for) the crystalline components of bone minerals.

a. W, D   c. N, D
b. W, B   d. N, B

38. The composite nature of bone is associated with an _____ (I, irregular arrangement; B, intimate and regular bonding) between collagen fibers and bone crystals and a higher _____ (T, tensional; C, compressional) strength attributed to properties of _____ (M, mineral salts; O, organic constituents).

a. I, T, M   c. I, C, O   e. B, C, M
b. I, T, O   d. B, C, O   f. B, T, M

OBJECTIVE 79–7.
Identify the factors involved in bone and ectopic calcification. Identify the "exchangeable" calcium fraction of bone and its significance.

39. Calcium salts in _____ (C, crystalline; A, amorphous) forms are thought to be initially formed _____ (D, during; F, following) the formation of corresponding osteoid tissue.

a. C, D   c. A, D
b. C, F   d. A, F

40. Purified collagen from bone _____ (W, as well as; N, but not) subcutaneous tissues _____ (F, facilitates; H, inhibits) hydroxyapatite crystallization, with pyrophosphate considered to be an _____ (A, activator; I, inhibitor) of hydroxyapatite nucleation.

a. W, F, A   c. W, H, I   e. N, H, A
b. W, F, I   d. N, H, I   f. N, F, A

41. Hydroxyapatite crystals, requiring a minimum of several _____ (D, days; W, weeks; M, months) for growth to 75% of their completed size, are formed from crystal nucleation sites occurring _____ (R, randomly; G, at regular repeating intervals) within collagen fibers of osteoid tissue.

a. D, R   c. M, R   e. W, G
b. W, R   d. D, G   f. M, G

42. The "exchangeable" calcium of bone, normally about _____ % of the total bone calcium, is highly significant in that it provides a buffer mechanism for _____ (L, lowered; E, elevated) levels of plasma calcium.

a. 0.4 to 1.0, only L
b. 0.4 to 1.0, only E
c. 0.4 to 1.0 both L & E
d. 4 to 10, only L
e. 4 to 10, only E
f. 4 to 10, both L & E

OBJECTIVE 79–8.

Identify the mechanisms, functional significance, and characteristics of bone remodeling.

43. The rate of bone remodeling, generally _____ (I, increasing; C, remaining constant; D, decreasing) as a function of increasing age, involves a major turnover of inorganic _____ (W, as well as; N, but not) the inorganic constituents of bone.

a. I, W   c. D, W   e. C, N
b. C, W   d. I, N   f. D, N

44. The exchangeable calcium fraction of bone is more closely correlated with loci of _____ (B, osteoblast; C, osteocyte; L, osteoclast) activities of bone _____ (F, formation; R, reabsorption).

a. B, F   c. L, F   e. C, R
b. C, F   d. B, R   f. L, R

45. Osteoclasts are thought to arise from mesenchymal stem cells _____ (W, as well as; N, but not) osteoblasts, and in turn give rise to _____ (B, osteoblasts; C, chondrocytes; F, fibroblasts).

a. W, B   c. W, F   e. N, C
b. W, C   d. N, B   f. N, F

46. Osteons, formed within reabsorption tunnels by the successive deposition of layers of new bone by _____ (C, chondrocytes; B, osteoblasts; L, osteoclasts), are _____ (I, initiated; T, completed) in the vicinity of haversian canals.

a. C, I   c. L, I   e. B, T
b. B, I   d. C, T   f. L, T

47. Loci of compressional stresses of bone are reportedly associated with _____ (N, negative; P, positive) piezoelectric potentials and local bone _____ (D, deposition; R, reabsorption).

a. N, D   c. P, D
b. N, R   d. P, R

48. A broken long bone of the leg which has healed at an irregular angle is subsequently subjected to compressional stress and net bone _____ (A, absorption; D, deposition) on the _____ (O, outer side; I, inside) of the angle.

a. A, O   c. D, O
b. A, I   d. D, I

49. "Callus" formation is associated with locally _____ (I, inhibited; A, activated) states of osteoblastic activity, while healing of fractured bones is generally _____ (I, inhibited; F, facilitated) by bone stresses in the presence of an immobilized fracture.

a. I, I   c. A, I
b. I, F   d. A, F

50. Blood alkaline phosphatase concentrations are generally _____ by enhanced rates of bone formation, _____ by diseases that cause bone destruction, and _____ by hypoparathyroid states. (I, increased; D, decreased)

a. I, D, D   c. I, I, D   e. D, I, D
b. I, D, I   d. D, I, I   f. D, D, I

OBJECTIVE 79-9.

Identify parathyroid hormone and its effects upon bone, the kidneys, the intestinal epithelium, and extracellular fluid concentrations of calcium and phosphate.

51. Parathyroid hormone is a _____ (P, protein; S, steroid) hormone which performs significant roles in calcium _____ (W, as well as; N, but not) phosphate metabolism.

    a. P, W   c. S, W
    b. P, N   d. S, N

52. Parathyroid hormone _____ (I, increases; D, decreases) the calcium ion concentration of extracellular fluids primarily as a consequence of its action upon _____ (K, the kidneys; E, the intestinal epithelium; B, bone).

    a. I, K   c. I, B   e. D, E
    b. I, E   d. D, K   f. D, B

53. Parathyroid hormone generally _____ (I, increases; D, decreases) the phosphate concentration of extracellular fluids primarily as a consequence of its action upon _____ (K, the kidneys; E, the intestinal epithelium; B, bone).

    a. I, K   c. I, B   e. D, E
    b. I, E   d. D, K   f. D, B

54. Removal of the parathyroid glands in humans is associated with _____ (I, increased; N, no significant change in; D, decreased) calcium ion concentrations of extracellular fluids as a consequence of a loss of secretion of _____ (P, parathormone; C, calcitonin).

    a. I, only C   d. I, only P
    b. N, only C   e. N, both P & C
    c. D, only P   f. D, both P & C

55. The administration of parathyroid hormone results in _____ (A, activation; I, inhibition) of osteoclast activity and _____ (E, an early; S, a secondary) phase of enhanced osteoblastic activity.

    a. A, E   c. I, E
    b. A, S   d. I, S

56. Parathyroid hormone _____ renal reabsorptive mechanisms for calcium ions and _____ renal reabsorptive mechanisms for phosphate ions. (E, enhances; N, does not significantly alter; I, inhibits)

    a. E, E   c. I, E   e. N, I
    b. N, E   d. E, I   f. I, I

57. Renal reabsorption of _____ and _____ is facilitated by the actions of parathyroid hormone. (S, sodium; K, potassium; H, hydrogen; A, amino acids; M, magnesium)

    a. K, H   c. S, H   e. S, A
    b. K, A   d. M, A   f. M, H

58. Parathyroid hormone _____ (F, facilitates; H, inhibits) 1,25-dihydroxycholecalciferol formation and _____ (I, increases; D, decreases) the intestinal absorption of calcium _____ (W, as well as; N, but not) phosphate.

    a. F, I, W   c. F, D, N   e. H, D, W
    b. F, I, N   d. H, D, N   f. H, I, W

59. Cyclic AMP is thought to be _____ (H, an inhibitor; S, a "second messenger") mechanism for the action of parathyroid hormone to _____ (E, enhance; R, reduce) osteoclast activities and to _____ (E, enhance; R, reduce) the calcium-binding protein content of intestinal and renal epithelia.

    a. H, E, E   c. H, R, E   e. S, E, E
    b. H, R, R   d. S, R, R   f. S, E, R

OBJECTIVE 79-10.

Identify the effects of vitamin D upon bone deposition and reabsorption. Identify the regulatory mechanisms for parathyroid secretion.

60. 1,25-dihydroxycholecalciferol, thought to perform major roles in bone reabsorption _____ (W, as well as; N, but not) bone deposition, tends to _____ (F, facilitate; H, inhibit) the action of parathyroid hormone upon osteoclast activity.

    a. W, F   c. N, F
    b. W, H   d. N, H

61. Increased levels of vitamin D in the diet tend to _____ the activity and size of the parathyroid glands and to _____ calcium ion concentrations of extracellular fluids. (I, increase; D, decrease)

    a. I, I   c. D, I
    b. I, D   d. D, D

62. As little as a 1% _____ (I, increase; D, decrease) in the plasma calcium concentration can increase parathyroid hormone secretion as much as 100% due to more sensitive _____ (A, acute; C, chronic) effects upon the parathyroid gland.

    a. I, A   c. D, A
    b. I, C   d. D, C

63. _____ (H, higher; L, lower) levels of parathyroid hormone secretion during lactation are associated with _____ (T, slight; B, substantial) _____ (I, increases; D, decreases) in extracellular fluid calcium concentrations.

    a. H, T, I   c. H, B, D   e. L, B, I
    b. H, T, D   d. L, B, D   f. L, T, I

64. Rickets is associated with parathyroid gland _____ (O, hypoplasia; E, hyperplasia) and a _____ (T, slight; B, substantial) _____ (I, increase; D, decrease) in extracellular fluid calcium levels.

    a. O, T, I   c. O, B, I   e. E, B, I
    b. O, T, D   d. E, B, D   f. E, T, D

OBJECTIVE 79-11.

Identify calcitonin, its target actions, its regulatory mechanisms, and its functional significance.

65. Calcitonin is a _____ (PP, polypeptide; S, steroid) secreted by the _____ (F, parafollicular; C, chief; O, oxyphil) cells of the human _____ (T, thyroid; P, parathyroid) gland.

    a. PP, F, T   c. PP, O, P   e. S, C, P
    b. PP, C, T   d. S, F, T   f. S, O, P

66. Calcitonin, when compared to parathyroid hormone, has a _____ (M, more; L, less) rapid action to _____ (A, augment; O, oppose) the effect of parathyroid hormone upon blood calcium ion concentrations.

    a. M, A   c. L, A
    b. M, O   d. L, O

67. Calcitonin has a more marked effect in _____ (A, adults; C, children) to _____ (I, increase; D, decrease) the calcium ion concentration of plasma.

    a. A, I   c. C, I
    b. A, D   d. C, D

68. Osteoclast absorption of bone, providing about _____ gram(s) of calcium to the extracellular fluid per day in the adult, is _____ (A, augmented; I, inhibited) by calcitonin.

    a. 0.8, A   c. 80, A   e. 8, I
    b. 8, A   d. 0.8, I   f. 80, I

69. Calcitonin secretion, augmented by _____ (O, hypocalcemic; E, hypercalcemic) states, has greater resultant effects upon plasma calcium ion concentrations when osteoclast activity is _____ (H, high; L, low).

    a. O, H   c. E, H
    b. O, L   d. E, L

70. Calcitonin, _____ (L, like; U, unlike) parathyroid hormone, acts mainly as a _____ (S, short-term; G, long-term) regulator of calcium ion concentrations of extracellular fluids.

    a. L, S   c. U, S
    b. L, G   d. U, G

71. Calcitonin results in a _____ (T, transient; P, prolonged) _____ (I, increase; D, decrease) in osteoblastic activity.

    a. T, I   c. P, I
    b. T, D   d. P, D

OBJECTIVE 79–12.

Identify the relative roles of the buffer system of exchangeable salts of bone, hormonal control, and intestinal and renal control in the regulation of the calcium ion concentration of the body fluids.

72. The total quantity of calcium in the extracellular fluids, closest to _____ gram(s), is in a state of reversible equilibrium with an "exchangeable" fraction of bone representing about _____ % of the total bone calcium.

    a. 1, 1   c. 100, 1   e. 10, 10
    b. 10, 1   d. 1, 10   f. 100, 10

73. A "buffer function" for _____ (R, only the release; U, only the uptake; B, the release as well as the uptake) of calcium by bone is served by a total blood flow to bone of about _____ % of the cardiac output.

    a. R, 5   c. B, 35   e. U, 35
    b. U, 15   d. R, 15   f. B, 5

74. Enhanced acidity of body fluids reduces the relative concentration of _____ (M, monohydrogen; D, dihydrogen) phosphate ions and thereby facilitates the _____ (A, absorption; P, deposition) of bone salts.

    a. M, A   c. D, A
    b. M, P   d. D, P

75. Reduced calcium ion concentrations of body fluids are opposed by a relatively _____ (W, weak; P, potent) and _____ (N, transient; R, persistent) mechanism, involving an increased endocrine secretion of the _____ (PT, parathyroid; T, thyroid) gland.

    a. W, N, PT   c. W, R, PT   e. P, R, PT
    b. W, N, T   d. P, R, T   f. P, N, T

76. Increased calcium ion concentrations of body fluids are opposed by a relatively _____ (W, weak; P, potent) and _____ (N, transient; R, persistent) mechanism in the adult, involving increased secretion of _____ (C, calcitonin; PT, parathyroid hormone).

    a. W, N, C   c. W, R, C   e. P, R, C
    b. W, N, PT   d. P, R, PT   f. P, N, PT

77. Intestinal absorption and renal reabsorption mechanisms, performing _____ (J, major; N, minor) roles in the long-term control of blood calcium ion concentrations, are more significantly altered by secretion levels of _____ (C, calcitonin; PT, parathyroid hormone).

    a. J, C   c. N, C
    b. J, PT   d. N, PT

OBJECTIVE 79–13.

Identify the consequences of hypoparathyroidism and hyperparathyroidism, and their means of treatment.

78. Hypoparathyroidism is characterized by _____ levels of osteoclast activity, _____ levels of osteoblast activity, and _____ levels of neural excitability. (I, increased; D, decreased)

   a. I, I, I   c. I, D, I   e. D, D, I
   b. I, I, D   d. D, D, D   f. D, I, D

79. Symptoms of tetany due to _____ (E, hyperparathyroidism; O, hypoparathyroidism) occur when the calcium concentration of plasma reaches threshold levels of about _____ mg.%.

   a. E, 1   c. E, 12   e. O, 6
   b. E, 6   d. O, 1   f. O, 12

80. Dihydrotachysterol acts to _____ (E, enhance; H, inhibit) the formation of 1,25-dihydroxycholecalciferol, _____ (I, increase; D, decrease) plasma calcium levels, and serves as a means of _____ (IT, induction; TT, treatment) for hypoparathyroidism.

   a. E, I, IT   c. E, D, TT   e. H, D, IT
   b. E, I, TT   d. H, D, TT   f. H, I, IT

81. Hyperparathyroidism occurs more frequently in _____ (M, men; W, women; C, children) as a consequence of _____ (P, pituitary; PT, parathyroid) gland tumors.

   a. M, P   c. C, P   e. W, PT
   b. W, P   d. M, PT   f. C, PT

82. Hyperparathyroid states are characterized by _____ (E, enhanced; R, reduced) rates of bone reabsorption, _____ (I, increased; D, decreased) plasma calcium levels, and a more frequent _____ (I, increase; D, decrease) in the plasma phosphate concentration.

   a. E, I, I   c. E, D, I   e. R, D, I
   b. E, I, D   d. R, D, D   f. R, I, D

83. Von Recklinghausen's disease, or _____ (C, cystic bone disease; P, osteopetrosis), is associated with _____ (L, elevated; R, reduced) levels of alkaline phosphatase in body fluids and _____ (O, hypoparathyroid; E, hyperparathyroid) states.

   a. C, L, O   c. C, R, E   e. P, R, O
   b. C, L, E   d. P, R, E   f. P, L, O

84. Renal calculi, with a lesser solubility in more _____ (C, acidic; K, alkaline) urine, and metastatic deposition of calcium phosphate crystals in body fluids are more frequent consequences of _____ (E, hyperparathyroidism; O, hypoparathyroidism).

   a. C, E   c. K, E
   b. C, O   d. K, O

85. Osteomalacia, _____ (W, as well as; N, but not) lactation, results in a _____ (P, primary; S, secondary) type of _____ (E, hyperparathyroidism; O, hypoparathyroidism).

   a. W, P, E   c. W, S, O   e. N, P, O
   b. W, S, E   d. N, S, O   f. N, P, E

OBJECTIVE 79–14.

Identify rickets, osteomalacia, and osteoporosis, and their causes, symptoms, and methods of treatment.

86. Rickets occurs more frequently in _____ (A, adults; C, children) in association with _____ (F, deficiencies; E, excesses) of _____ (CP, dietary calcium and phosphate; D, vitamin D).

   a. A, F, CP   c. A, E, CP   e. C, F, CP
   b. A, E, D   d. C, E, D   f. C, F, D

87. Rickets is generally associated with a comparatively greater _____ (E, elevation; D, depression) of plasma _____ (C, calcium; P, phosphate; H, parathormone) levels.

   a. E, C   c. E, H   e. D, P
   b. E, P   d. D, C   f. D, H

88. Rickets is generally associated with _____ (I, increased; D, decreased) parathyroid hormone levels of secretion, and a greater _____ (E, elevation; R, reduction) of plasma _____ (C, calcium; P, phosphate) levels.

a. I, E, C   c. I, E, P   e. D, E, C
b. I, R, P   d. D, R, P   f. D, R, C

89. Tetany is a more likely consequence of _____ (E, early; L, late) stages of rickets, and treatment regimens include vitamin D administration in conjunction with _____ (W, low; X, excess) levels of dietary calcium.

a. E, W   c. L, W
b. E, X   d. L, X

90. "Adult rickets," or _____ (M, osteomalacia; P, osteoporosis), may result from the effects of steatorrhea to _____ (D, decrease; I, increase) the intestinal absorption of calcium _____ (W, as well as; N, but not) vitamin D.

a. M, D, W   c. M, D, N   e. P, I, N
b. M, I, W   d. P, D, N   f. P, D, W

91. Congenital hypophosphatemia is a vitamin D _____ (T, treatable; R, resistant) type of rickets resulting from a primary defect of the _____ (I, intestinal epithelium; L, liver; RT, renal tubules).

a. T, I   c. T, RT   e. R, L
b. T, L   d. R, I   f. R, RT

92. "Renal rickets" is a form of _____ (M, osteomalacia; P, osteoporosis) thought to be caused by _____ (L, lower; H, higher) than normal levels of renal _____ (S, synthesis; I, inactivation) of 1,25-dihydroxycholecalciferol.

a. M, L, S   c. M, H, I   e. P, H, S
b. M, L, I   d. P, H, I   f. P, L, S

93. Osteoporosis, resulting from abnormal _____ (C, bone calcification; O, formation of organic matrix of bone), is generally associated with _____ (G, greater; L, lower) than normal levels of osteoblastic activity.

a. C, G   c. O, G
b. C, L   d. O, L

94. _____ (O, osteoporosis; M, osteomalacia) is a more common bone disorder, particularly among _____ (C, growing children; A, older adults).

a. O, C   c. M, C
b. O, A   d. M, A

95. Cushing's disease, acromegaly, malnutrition, physical inactivity, and vitamin _____ deficiencies are potential causes of _____ (O, osteoporosis; M, osteomalacia; P, osteopetrosis).

a. C, O   c. C, P   e. D, M
b. C, M   d. D, O   f. D, P

OBJECTIVE 79–15.
Using the following diagram, identify the major functional parts of a tooth.

Directions: Match the lettered headings with the diagram and the numbered list of descriptive words and phrases.

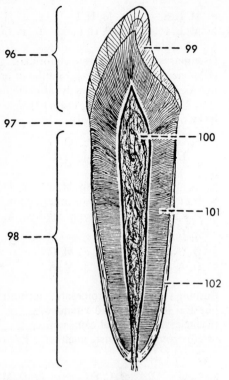

(From Guyton, A.C., Textbook of Medical Physiology, 5th
ed. Philadelphia, W.B., Saunders Company, 1976.)

96. _____ That portion of the tooth which functions within the oral cavity.
97. _____ That portion of the tooth which normally is attached to the gingival epithelium.
98. _____ That portion of the tooth which functions within the bony alveolus.
99. _____ The mineralized portion of the tooth which is formed by ameloblasts.
100. _____ The nonmineralized connective tissue portion of the tooth containing nerves and blood vessels.
101. _____ The mineralized portion of the tooth which is formed by odontoblasts.
102. _____ The mineralized attachment portion of the tooth for fibers of the periodontal membrane.

a. Dentine   c. Enamel   e. Cementum
b. Neck   d. Root   f. Pulp chamber
                                       g. Crown

OBJECTIVE 79–16.
Characterize the general characteristics of the mineralized constituents of the human dentition.

103. Maximal occlusal forces developed by _____ (A, anterior; P, posterior) teeth within the dental arcade range from _____ pounds.

a. A, 15 to 20   d. P, 15 to 20
b. A, 150 to 200   e. P, 150 to 200
c. A, 1500 to 2000   f. P, 1500 to 2000

104. The keratin-containing _____ (E, enamel; C, cementum; D, dentine) component is _____ (M, more; L, less) resistant to corrosive chemical agents and _____ (S, more elastic; H, harder) than other mineralized dental tissues.

a. E, M, S   c. D, M, H   e. C, L, S
b. C, L, H   d. E, M, H   f. D, M, S

105. _____ deciduous teeth provide for mastication in the young individual until replaced by a total of _____ permanent teeth.

   a. 10, 28 to 32   d. 10, 36 to 42
   b. 20, 28 to 32   e. 20, 36 to 42
   c. 30, 28 to 32   f. 30, 36 to 42

106. Calcium ions of body fluids are exchanged more readily with _____ and least readily with _____ . (D, dentine; E, enamel; C, cementum)

   a. D, E   c. E, D   e. C, E
   b. D, C   d. E, C   f. C, D

107. Tubular cellular processes of _____ (O, odontoblasts; A, ameloblasts) characterize the mineralized _____ (E, enamel; C, cementum; D, dentine) component of the dentition.

   a. O, E   c. O, D   e. A, C
   b. O, C   d. A, E   f. A, D

108. $Ca^{2+}_{10-x}(H_3O^+)_{2x} \cdot (PO_4)_6(OH^-)_2$ or _____ (H, hydroxyapatite; A, aragonite), comprises the predominantly crystalline mineral of _____(C, cementum; D, dentine; E, enamel).

   a. H, only E   d. A, only E
   b. H, only C & D   e. A, only C & D
   c. H; C, D, & E   f. A; C, D, & E

109. Dental caries, largely initiated upon _____ (E, enamel; D, dentinal) surfaces by _____ (B, bacterial; A, autoimmune) mechanisms, is _____ (L, slightly; S, substantially) reduced by a lifetime exposure to drinking water containing one part per million fluoride.

   a. E, B, L   c. E, A, S   e. D, A, L
   b. E, B, S   d. D, A, S   f. D, B, L

80

Reproductive Functions of the Male and the Male Sex Hormones

OBJECTIVE 80–1.
Identify the physiological anatomy of the male sexual organs.

Directions: Match the lettered headings with the diagram and the numbered list of descriptive words and phrases.

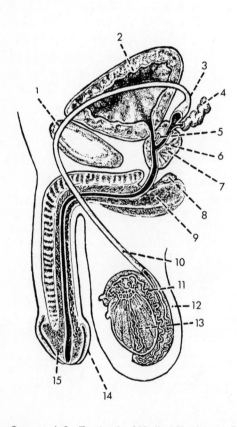

(From Guyton, A.C., Textbook of Medical Physiology, 5th ed. Philadelphia, W.B., Saunders Company, 1976.)

1. _____ The joint formed by union of the bodies of the pubic bone by a thick mass of fibrocartilage.

2. _____ A musculomembranous sac that serves as a reservoir for urine.

3. _____ A flask-like dilation of the vas deferens near the junction of the seminal vesicles.

4. _____ Paired sacculated pouches that secrete mucus and nutrients for ejaculated sperm.

5. _____ Small blind pouch arising in the prostatic substance.

6. _____ Tube or passageway into which the seminal vesicles empty their contents during ejaculation.

7. _____ A gland that surrounds the neck of the bladder and urethra, which adds its secretion to the seminal fluid.

8. _____ Mucus-secreting gland located near the origin of the urethra (Cowper's gland).

9. _____ Conical enlargement of the posterior portion of the corpus carvernosum urethrae.

10. _____ A vessel or canal for transporting sperm.

11. _____ Elongated coiled tube of the testes for sperm maturation.

12. _____ Pouch which contains the testes.

13. _____ Coiled tubules in the testis where sperm is formed.

14. _____ The fold of skin that covers the glans penis; foreskin.

15. _____ The conical vascular body which forms the extermity of the penis.

a. Bulbo-urethral gland
b. Vas deferens
c. Glans penis
d. Urinary bladder
e. Seminal vesicles

f. Scrotum
g. Symphysis pubis
h. Prepuce
i. Seminiferous tubules
j. Ejaculatory duct

k. Bulbus urethrae
l. Utriculus prostaticus
m. Prostate
n. Ampulla of the vas deferens
o. Epididymis

OBJECTIVE 80–2.

Characterize the events of spermatogenesis, the structure of human spermatozoa, and sperm maturation, storage, and motility.

16. Spermatogenesis is initiated at _____ (T, the time of testicular descent; P, an average age of 13) by means of _____ (I, increased; D, decreased) levels of secretion of _____ (AH, adenohypophyseal; NH, neurohypophyseal) hormones.

    a. T, I, AH   c. T, D, NH   e. P, I, NH
    b. T, D, AH   d. P, D, NH   f. P, I, AH

17. Germinal epithelial cells, or _____ (O, spermatogonia; D, spermatids) of the seminiferous tubules, give rise to _____ (P, primary; S, secondary) spermatocytes containing 23 chromosomes by the process of _____ (T, mitosis; E, meiosis).

    a. O, P, T   c. O, P, E   e. D, P, T
    b. O, S, E   d. D, S, E   f. D, S, T

18. Male offspring result from the presence of _____ chromosomes contributed by the _____ (S, sperm only; O, ovum only; SO, sperm as well as the ova).

    a. X, S   c. X, SO   e. Y, O
    b. X, O   d. Y, S   f. Y, SO

19. The _____ (N, neck; B, body) of human sperm contains the _____ (NM, nuclear material; M, mitochondria), whereas the acrosome forms the _____ (E, end piece of the tail; F, front of the head).

    a. N, NM, E   c. N, M, F   e. B, M, E
    b. N, NM, F   d. B, M, F   f. B, NM, E

20. Spermatids are converted into _____ (P, primary spermatocytes; Z, spermatozoa) in close association with sustentacular or _____ (S, Sertoli cells; L, interstitial cells of Leydig).

    a. P, S   c. Z, S
    b. P, L   d. Z, L

21. Mature sperm stored within the genital ducts, mostly in the _____ (V, vasa recta; E, epididymis; D, vas deferens), retain their fertility for a maximum period of about _____ days.

    a. V, 8   c. D, 42   e. E, 42
    b. E, 120   d. V, 120   f. D, 8

22. The fluid of the vas deferens is an _____ (C, acidic; K, alkaline) fluid which serves to _____ (E, enhance; H, inhibit) sperm motility.

    a. C, E   c. K, E
    b. C, H   d. K, H

23. The _____ (S, seminal vesicles; E, epididymis) is a _____ (C, secretory; NC, non-secreting) conduit where sperm develop the power of mobility _____ (W, as well as; N, but not) the capability of fertilization of the ovum.

    a. S, C, W   c. S, NC, W   e. E, C, W
    b. S, NC, N   d. E, NC, N   f. E, C, N

24. Normal sperm tend to travel in _____ (C, circuitous; L, linear) directions of travel at velocities of about 1 to 4 _____ (U, microns; MM, millimeters; CM, centimeters) per minute.

    a. C, U   c. C, CM   e. L, MM
    b. C, MM   d. L, U   f. L, CM

OBJECTIVE 80-3.

Identify the characteristics and functional significance of seminal vesicle secretions, prostate gland secretions, and semen.

25. The _____ (P, prostate glands; S, seminal vesicles) function as secretory, _____ (A, as well as; N, but not) sperm storage areas, and give semen a mucoid consistency.

    a. P, A   c. S, A
    b. P, N   d. S, N

26. _____ (S, seminal vesicle; P, prostate gland) secretions, containing fructose, ascorbic acid, inositol, ergothioneine, some amino acids, phosphorylcholine, _____ (W, as well as; N, but not) prostaglandins, are secreted into the ejaculatory duct just _____ (B, before; A, after) the vas deferens empties its sperm.

    a. S, W, B   c. S, N, A   e. P, N, B
    b. S, W, A   d. P, N, A   f. P, W, B

27. A thin milky _____ (C, acidic; K, alkaline) secretion of the _____ (S, seminal vesicles; P, prostate gland) is responsible for _____ (M, more; L, less) than 50% of the volume of semen.

    a. C, S, M   c. C, P, M   e. K, P, M
    b. C, S, L   d. K, P, L   f. K, S, L

28. Sperm mobility is _____ (F, facilitated; H, inhibited) by the _____ (K, alkaline; C, acidic) nature of vaginal secretions and _____ (F, facilitated; H, inhibited) by the _____ (K, alkaline; C, acidic) nature of the fluid of the vas deferens.

    a. F, K, F, K   c. F, C, H, K   e. H, C, F, K
    b. F, K, H, C   d. H, C, H, C   f. H, K, F, C

29. A clotting enzyme of _____ fluid acts upon fibrinogen of _____ fluid to form a weak coagulum which undergoes lysis by the action of a fibrinolysin formed from profibrinolysin of _____ fluid. (P, prostatic; S, seminal vesicle)

    a. P, P, P   c. P, S, S   e. S, P, S
    b. P, S, P   d. S, S, S   f. S, P, P

30. Human sperm have a relatively _____ (H, high; L, low) rate of mobility during the first few minutes following ejaculation as a more probable consequence of _____ (C, capacitation; V, semen viscosity; pH, semen pH).

    a. H, C   c. H, pH   e. L, V
    b. H, V   d. L, C   f. L, pH

31. A maximal life span of sperm ejaculated in semen of 1 to 3 _____ (H, hours; D, days; W, weeks) is _____ (L, significantly less than; E, about equal to; G, significantly greater than) their maximal life span within the male genital ducts.

    a. H, L   c. W, E   e. D, G
    b. D, L   d. H, E   f. W, G

OBJECTIVE 80-4.

Identify the significance of sperm morphology and numbers, acrosomal enzymes, and testicular temperature upon fertility in the male. Identify the commoner causes of infertility in men.

32. Orchiditis resulting from _____ (C, cryptorchidism; R, rheumatic fever; U, mumps) is a frequent cause of sterility in _____ (M, men; W, women).

    a. C, M   c. U, M   e. R, W
    b. R, M   d. C, W   f. U, W

33. The scrotum functions to _____ (I, increase; D, decrease) the temperature of the testes with respect to body temperature as testicular _____ (G, gametogenic; E, endocrine secretion) functions are impaired by lower thresholds of _____ (I, increased; D, decreased) testicular temperatures.

    a. D, G, I   c. D, E, I   e. I, E, I
    b. D, G, D   d. I, E, D   f. I, G, D

34. Cryptorchidism, or a condition of _____ (P, premature; F, failure of) testicular descent, generally _____ (D, does; N, does not) result in sterility.

a. P, D   c. F, D
b. P, N   d. F, N

35. Descent of the testes through the inguinal canals during _____ (L, late gestation; B, puberty) occurs as a consequence of _____ (T, testosterone; G, gonadotropin) secretion by the _____ (A, adrenal cortex; P, pituitary; S, testes).

a. L, T, S   c. L, G, A   e. B, T, S
b. L, T, A   d. B, G, P   f. B, T, A

36. A normal _____ -ml. volume of semen ejaculated during coitus contains an average total of _____ sperm.

a. 3 to 4, 4 x 10^3 d. 30 to 40, 4 x 10^3
b. 3 to 4, 4 x 10^6 e. 30 to 40, 4 x 10^6
c. 3 to 4, 4 x 10^8 f. 30 to 40, 4 x 10^8

37. Sterility resulting from a low sperm count is thought to be associated with inadequate _____ (L, lipolytic; F, fibrinolytic; A, acrosomal) enzymes required to penetrate the region of the uterine cervix _____, (W, as well as; N, but not) the external layers of the expelled ovum.

a. L, W   c. A, W   e. F, N
b. F, W   d. L, N   f. A, N

OBJECTIVE 80–5.

Identify the mechanisms of erection, emission, and ejaculation during the male sexual act.

38. Erection is initiated by increased _____ (P, parasympathetic; S, sympathetic) efferent activity to the vessels of the penis occurring in response to _____ (H, only psychic; A, only physical; E, either psychic or physical) stimuli.

a. P, H   c. P, E   e. S, A
b. P, A   d. S, H   f. S, E

39. The corpus covernosum of the penis is a highly _____ (V, vascular; C, contractile) tissue which functions to initiate _____ (E, emission; J, ejaculation; R, erection).

a. V, E   c. V, R   e. C, J
b. V, J   d. C, E   f. C, R

40. Contraction of the smooth muscle of the epididymus, vas deferens, and ampulla by enhanced _____ (P, parasympathetic; S, sympathetic) efferent activity from reflex centers within the _____ (M, medulla; C, spinal cord) is thought to initiate _____ (R, erection; J, ejaculation; E, emission).

a. P, M, J   c. P, C, R   e. S, C, E
b. P, M, E   d. S, C, J   f. S, M, R

41. Ejaculation results from _____ (M, smooth; K, skeletal) muscle reflex activity initiated by the filling of the _____ (V, vas deferens; U, internal urethra) with semen.

a. M, V   c. K, V
b. M, U   d. K, U

OBJECTIVE 80–6.

Identify the principal androgen in the male and the characteristics of its synthesis, metabolism, and chemical nature.

42. The principal androgen in the male, _____ (T, testosterone, A, androsterone) is secreted largely by the _____ (I, interstitial cells of Leydig; S, Sertoli cells; R, cells of the zona fasciculata).

a. T, I   c. T, R   e. A, S
b. T, S   d. A, I   f. A, R

43. Testosterone is a _____ (S, steroid; P, protein) hormone synthesized by cells of the _____ (A, adrenal gland only; T, testes only; B, adrenal glands as well as the testes).

a. S, A   c. S, B   e. P, T
b. S, T   d. P, A   f. P, B

44. The ovaries produce _____ estrogen and _____ testosterone, whereas the testes produce _____ estrogen and _____ testosterone. (M, major amounts of; S, only slight amounts of; N, no)

 a. M, S, N, M   d. S, N, N, M
 b. M, S, S, M   e. S, S, N, M
 c. M, N, N, M   f. S, N, S, S

45. Excess testosterone is inactivated largely by the _____, conjugated as glucuronides or sulfates by the _____, and excreted by the _____. (L, liver only; K, kidneys only; B, liver as well as the kidneys)

 a. L, B, K   c. L, K, L   e. K, B, K
 b. L, L, B   d. K, K, L   f. K, L, B

OBJECTIVE 80-7.

Characterize the secretion rates of testosterone as a function of age, and identify the prenatal functions of testosterone.

46. The presence or absence of _____ (T, testosterone; S, estrogen) in the fetus is the key determining factor in the development of _____ (M, only the male; F, only the female; E, either the male or the female) genital organs.

 a. T, M   c. T, E   e. S, F
 b. T, F   d. S, M   f. S, E

47. Testosterone secretion initiated during the _____ month of male fetal development occurs in response to _____ (P, placental; H, adenohypophyseal; C, adrenocortical) hormones.

 a. 2nd, P   c. 2nd, C   e. 7th, H
 b. 2nd, H   d. 7th, P   f. 7th, C

48. The highest rates of testosterone secretion in the male occurs between the ages of _____, whereas the lowest rates of testosterone secretion occur between the ages of _____.

 a. 11 & 13, 75 & 80   d. 11 & 13, 5 & 10
 b. 20 & 25, 75 & 80   e. 20 & 25, 5 & 10
 c. 35 & 40, 75 & 80   f. 35 & 40, 5 & 10

49. Testicular descent into the scrotum during _____ (F, the first two months of gestation; L, the last two months of gestation; P, puberty) occurs in response to endocrine secretion of the _____ (H, pituitary gland; T, testes).

 a. F, H   c. P, H   e. L, T
 b. L, H   d. F, T   f. P, T

OBJECTIVE 80-8.

Identify the postnatal functions and target actions of testosterone.

50. Testosterone is largely responsible for _____ (P, only primary; S, only secondary; B, both primary and secondary) sexual characteristics in the male and _____ (E, enlargement; C, a premature cessation of growth) of the larynx.

 a. P, E   c. B, E   e. S, C
 b. S, E   d. P, C   f. B, C

51. The castrate male adult is characterized by states of _____ (B, baldness only; F, reduced facial hair only; BF, reduced facial hair and baldness) as a consequence of altered levels of _____ (G, gonadotropins; T, testosterone).

 a. B, G   c. BF, G   e. F, T
 b. F, G   d. B, T   f. BF, T

52. Testosterone has target actions upon the skin to _____ skin thickness, _____ melanin deposition, and _____ sebaceous gland secretion. (I, increase; D, decrease)

 a. I, D, I   c. I, I, I   e. D, D, I
 b. I, I, D   d. D, D, D   f. D, I, D

53. Testosterone _____ (I, increases; D, decreases) the rate of growth of long bones, causes _____ (P, premature; Y, delayed) closure of epiphyseal plates, and _____ (I, increases; D, decreases) the rate of calcium salt deposition within the skeletal system.

 a. I, P, I   c. I, Y, I   e. D, Y, I
 b. I, P, D   d. D, Y, D   f. D, P, D

54. Testosterone functions to _____ basal metabolic rates, _____ red blood cell concentrations, and _____ blood and extracellular fluid volumes in the male. (I, increase; D, decrease)

   a. I, I, I   c. I, D, I   e. D, D, I
   b. I, I, D   d. D, D, D   f. D, I, D

55. Testosterone increases the RNA and protein, _____ (W, as well as; N, but not) the DNA, content of the prostate gland via its intracellular conversion to _____ (DE, dehydroepiandrosterone; DT, dihydrotestosterone).

   a. W, DE   c. N, DE
   b. W, DT   d. N, DT

OBJECTIVE 80–9.

Identify the gonadotropic hormones, and their target actions, functional roles, and regulatory mechanisms in the male. Identify the changes in gonadotropin secretion occurring with conditions of puberty and the male climacteric.

56. Interstitial cell-stimulating hormone, or_____, _____, (W, as well as; N, but not) follicle-stimulating hormone secreted by the _____ (P, posterior; A, anterior) pituitary, perform(s) major roles in the control of male sexual functions.

   a. LTH, W, P   c. LTH, N, P   e. LH, N, P
   b. LTH, W, A   d. LH, N, A   f. LH, W, A

57. Testosterone secretion by the _____ (L, interstitial cells of Leydig; S, Sertoli cells) occurs in approximately _____ (D, direct; I, inverse) proportion to the amount of available _____.

   a. L, D, FSH   c. L, I, LTH   e. S, I, LH
   b. L, D, LH   d. S, I, FSH   f. S, D, LTH

58. Chorionic gonadotropin has target actions upon the testes similar to those of _____ in initiating _____ (S, only spermatogenesis; T, only testosterone secretion; B, both spermatogenesis as well as testosterone secretion.

   a. LH, S   c. LH, B   e. FSH, T
   b. LH, T   d. FSH, S   f. FSH, B

59. The production of mature spermatozoa _____ (D, does; N, does not) require testosterone and is initiated by the target actions of _____.

   a. D, FSH   d. N, FSH
   b. D, LH   e. N, LH
   c. D, testosterone   f. N, LTH

60. The secretion of FSH, _____ (W, as well as; N, but not) LH, is regulated by gonadotropin-releasing factors secreted by the _____ (A, adenohypophysis; N, neurohypophysis; MA, median eminence and anterior hypothalamus).

   a. W, A   c. W, MA   e. N, N
   b. W, N   d. N, A   f. N, MA

61. The administration of testosterone into the male _____ (W, as well as; N, but not) the female results in _____ (I, increased; D, decreased) rates of secretion of particularly _____.

   a. W, I, LH   c. W, D, LH   e. N, I, LH
   b. W, D, FSH   d. N, D, FSH   f. N, I, FSH

62. Inhibin is a postulated _____ (T, testicular; P, pineal gland; H, hypothalamic) hormone which is thought to have a more pronounced _____ (F, facilitatory; I, inhibitory) action upon rates of _____ (S, testosterone secretion; M, spermatogenesis).

   a. T, F, S   c. H, F, M   e. P, I, M
   b. P, F, S   d. T, I, M   f. H, I, S

63. Puberty, culminating in full adult male sexual capability at an average age of _____, is thought to result from primary aging processes of the _____ (A, adenohypophysis; D, adrenal glands; H, hypothalamus; T, testes).

   a. 10, A   c. 17, H   e. 13, H
   b. 13, T   d. 10, T   f. 17, D

64. Increasing rates of gonadotropin secretion and _____ (I, increasing; D, decreasing) levels of testosterone secretion are characteristic of puberty _____ (W, as well as; N, but not) the male climacteric.

   a. I, W   c. D, W
   b. I, N   d. D, N

65. The male climacteric is generally associated with a rather _____ (A, abrupt; S, slow) decline in _____ (T, only testosterone; G, only gonadotropin; TG, testosterone as well as gonadotropin) rates of secretion.

   a. A, T   c. A, TG   e. S, G
   b. A, G   d. S, T   f. S, TG

OBJECTIVE 80–10.

Identify the principal causes and consequences of abnormalities of the prostate gland, hypogonadism, adiposogenital syndrome, and hypergonadism in the male.

66. Cryptorchidism, _____ (W, as well as; N, but not) eunuchism, is associated with _____ (I, infantile; R, regression of adult) _____ (F, female; M, male) sexual characteristics.

   a. W, I, F   c. W, R, M   e. N, R, F
   b. W, I, M   d. N, R, M   f. N, I, M

67. Prostate cancer results in approximately _____ % of all male deaths and is frequently inhibited by _____ (V, vasectomy; C, castration) _____ (W, as well as; N, but not) estrogen administration.

   a. 2 to 3, V, W   d. 10 to 12, C, N
   b. 2 to 3, C, N   e. 10 to 12, V, W
   c. 2 to 3, C, W   f. 10 to 12, V, N

68. Altered levels of _____ (G, gonadotropins; A, androgens) following castration in the male adult are generally associated with a loss of _____ (R, only erection; J, only ejaculation; B, both erection and ejaculation).

   a. G, R   c. G, B   e. A, J
   b. G, J   d. A, R   f. A, B

69. Fröhlich's syndrome is a condition of _____ (E, hypergonadism; O, hypogonadism) associated with _____ (H, higher than normal; N, normal; L, lower than normal) rates of secretion of gonadotropin-releasing factors.

   a. E, H   c. E, L   e. O, N
   b. E, N   d. O, H   f. O, L

70. _____ (O, obesity; I, inanition) is a _____ (C, a primary cause; A, an associated consequence) of disorders of the _____ (H, hypothalamus; P, adenohypophysis) resulting in the adiposogenital syndrome.

   a. O, C, H   c. O, A, P   e. I, C, P
   b. O, A, H   d. I, A, P   f. I, C, H

71. Interstitial cell tumors, generally more difficult to diagnose in the male _____ (C, child; A, adult), are associated with _____ (I, increased; D, decreased) levels of urinary excretion of testosterone end-products.

   a. C, I   c. A, I
   b. C, D   d. A, D

72. A relatively _____ (M, more; L, less) common condition of _____ (G, germinal; I, interstitial) cell tumors of the testes is a potential cause of teratomas.

   a. M, G   c. L, G
   b. M, I   d. L, I

81

Sexual Functions in the Female and the Female Hormones

OBJECTIVE 81-1.
Using the following diagram, review the physiologic anatomy of the female sexual organs.

Directions: Match the lettered headings with the diagram and numbered list of descriptive words and phrases.

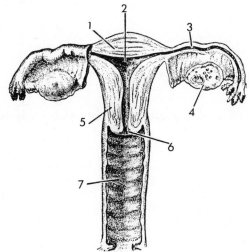

(From Guyton, A.C., Textbook of Medical Physiology, 5th ed. Philadelphia, W.B. Saunders Company, 1976.)

1. _____ A hollow, pear-shaped female organ; the womb.
2. _____ The hollow portion of the uterus that opens into the vagina and fallopian tubes.
3. _____ One of the two long slender tubes extending from the uterus to the region of the ovaries.
4. _____ One of the two female sexual glands in which the ova are formed; the female gonad.
5. _____ The mucous membrane lining the uterus.
6. _____ The neck or constricted portion of the uterus nearest the vagina.
7. _____ A sheath-like structure extending from the vulva to the cervix; receives the penis during coitus.

a. Endometrium
b. Fallopian tube
c. Vagina
d. Uterus
e. Uterine cavity
f. Cervix
g. Ovary

OBJECTIVE 81-2.
Characterize the development and fate of primordial follicles within the ovaries.

8. _____ (S, single; M, multiple) ova are expelled from ovarian follicles into the abdominal cavity at the _____ (D, middle; E, end) of each sexual cycle.

a. S, D
b. S, E
c. M, D
d. M, E

9. The largest number of primordial follicles are present at _____ , whereas the smallest number of primordial follicles are present at _____ . (B, birth; P, puberty; M, menopause)

a. B, P
b. B, M
c. P, B
d. P, M
e. M, P
f. M, B

10. During _____ (P, puberty; F, fetal life) primordial follicles differentiate from an _____ (O, outer covering; C, inner core) of germinal epithelia within the ovaries.

    a. P, O   c. F, O
    b. P, C   d. F, C

11. During the reproductive years in the female about _____ of the original _____ primordial follicles develop to expel their ova, while the remaining follicles _____ (I, remain dormant; A, become atretic).

    a. 450, 7,500, I   d. 4500, 7,500, A
    b. 450, 750,000, I   e. 4500, 7,500, I
    c. 450, 750,000, A   f. 4500, 750,000, A

OBJECTIVE 81–3.

Recognize the source of the various female hormones and releasing factors.

Directions: Match each of the following hormones and releasing factors with their source. (a, anterior pituitary; b, hypothalamus; c, ovaries)

12. FSH _____
13. Estrogen _____
14. LRF _____

15. LH _____
16. Progesterone _____
17. FRF _____

OBJECTIVE 81–4.

Identify the general characteristics of the female sexual cycle and its dependence upon gonadotropic hormones.

18. Cyclic ovarian changes occur in response to cyclic variations in the _____ (A, anterior; P, posterior) pituitary gonadotropins _____ (E, estrogen; R, progesterone; LH, luteinizing hormone; L, prolactin; FSH, follicle-stimulating hormone).

    a. A, E & R   d. P, E & R
    b. A, FSH & R   e. P, FSH & L
    c. A, FSH & LH   f. P, FSH & LH

19. Luteinizing hormone _____ follicle-stimulating hormone is a small glycoprotein with significant target effects upon the ovaries _____ the testes. (W, as well as; N, but not)

    a. W, W   c. N, W
    b. W, N   d. N, N

20. Interstitial cell-stimulating hormone or _____ stimulates ovarian cells via _____ (M, cell membrane; I, intracellular) receptors which _____ (E, enhance; R, reduce) cyclic AMP concentrations.

    a. FSH, M, E   c. FSH, I, E   e. LH, M, R
    b. FSH, I, R   d. LH, I, E   f. LH, M, E

21. Sexual cycles in the female are initiated between the ages _____ as a result of _____ (I, increased; D, decreased) secretion of _____ (P, pituitary; O, ovarian) hormones.

    a. 11 and 15, I, P   d. 16 and 20, D, O
    b. 11 and 15, I, O   e. 16 and 20, D, P
    c. 11 and 15, D, P   f. 16 and 20, I, O

22. The female sexual cycle, averaging _____ days, is associated with cyclic alterations of gonadotropins _____ (W, as well as; N, but not) ovarian hormones.

    a. 14, W   c. 40, W   e. 28, N
    b. 28, W   d. 14, N   f. 40, N

23. The female sexual cycle releases _____ (S, single; M, multiple) ova per cycle and prepares the _____ (F, fallopian tubes; E, endometrium; MS, mesentery) for possible implantation by a fertilized ovum.

    a. S, F   c. S, MS   e. M, E
    b. S, E   d. M, F   f. M, MS

OBJECTIVE 81–5.

Identify the sequence of events and corresponding functional roles of pituitary gonadotropins and ovarian hormones leading to the development of vesicular follicles, follicular atresia, and ovulation.

24. Primordial follicles, which _____ (C, continue to grow; R, remain inactive) during childhood, begin to mature during puberty in association with _____ (I, increasing; D, decreasing) levels of gonadotropins and _____(I, increasing; D, decreasing) levels of estrogens.

    a. C, I, I   c. C, D, D   e. R, D, I
    b. C, I, D   d. R, D, D   f. R, I, I

25. The concentration of FSH _____ (W, as well as; N, but not) LH _____ (I, increases; D, decreases) at the onset of each female sexual cycle.

    a. W, I   c. N, I
    b. W, D   d. N, D

26. Altered levels of gonadotropins at the beginning of each female sexual cycle cause accelerated growth of granulosa _____ (W, as well as; N, but not) theca cells of _____ (O, only one; T, about twenty; O, all) of the ovarian follicles.

    a. W, O   c. W, A   e. N, T
    b. W, T   d. N, O   f. N, A

27. Vesicular follicles contain an external capsule formed by _____ cells and an antrum formed by fluid secretion of _____ cells. (G, only granulosa; T, only theca; B, both theca and granulosa)

    a. G, T   c. B, G   e. T, B
    b. T, G   d. G, B   f. B, B

28. The formation of a vesicular follicle by FSH is _____ (A, augmented; O, opposed) by LH and _____ (A, augmented; O, opposed) by _____ (E, estrogen; P, progesterone) secretion into the antrum of the developing follicle.

    a. A, A, E   c. A, O, E   e. O, A, E
    b. A, O, P   d. O, O, P   f. O, A, P

29. The process of atresia of all of the developing follicles but one is thought to be associated with _____ (R, rising; F, falling) levels of FSH occurring _____ (P, prior to; D, during; W, following) the period of ovulation.

    a. R, P   c. R, W   e. F, D
    b. R, D   d. F, P   f. F, W

30. Ovulation is associated with expulsion of an ovum into the _____ (U, uterus; A, abdominal cavity) about _____ days following the onset of menstruation in an average _____ -day female cycle.

    a. U, 14, 38   c. U, 14, 14   e. A, 28, 28
    b. U, 28, 28   d. A, 14, 28   f. A, 14, 14

31. Rupture of the stigma is thought to be associated with the action of proteolytic enzymes produced by _____ (G, granulosa; T, theca) cells upon the _____ (G, granulosa; T, theca) cells of the _____ (C, corona radiata; F, capsule of the follicle).

    a. G, G, C   c. G, G, F   e. T, G, F
    b. G, T, C   d. T, T, F   f. T, T, C

32. A comparatively greater peak secretion of _____ occurs about _____ (D, 2 days; H, 18 hours) _____ (P, prior to; F, following) ovulation.

    a. FSH, D, P   c. FSH, H, P   e. LH, H, P
    b. FSH, D, F   d. LH, D, F   f. LH, H, F

33. _____ (E, estrogen; P, progesterone; L, LH; F, FSH) has a more specific effect to convert theca and granulosa cells into lutein cells which secrete greater amounts of _____ (P, progesterone; E, estrogen).

    a. E, P   c. F, P   e. L, E
    b. L, P   d. P, E   f. F, E

34. The time of ovulation is characterized by rising levels of _____ and falling levels of _____ . (P, progesterone; E, estrogen; FSH, follicle-stimulating hormone; LH, luteinizing hormone)

    a. P; E, FSH & LH   d. E & LH; P & FSH
    b. P, E, & LH; FSH   e. P, LH, & FSH; E
    c. E; P, FSH, & LH   f. E & P; LH & FSH

OBJECTIVE 81–6.

Characterize the events of the "luteal" phase of the ovarian cycle. Identify the corpus luteum, its endocrine dependence, its endocrine secretion, and its functional significance.

35. _____ is primarily responsible for the proliferation and enlargement, _____ (W, as well as; N, but not) the endocrine secretion, of the corpus _____ (A, albicans; L, luteum).

   a. FSH, W, A   c. FSH, N, L   e. LH, W, L
   b. FSH, N, A   d. LH, N, L   f. LH, W, A

36. The corpus luteum secretes large amounts of _____ but not _____.(LTH, luteotropic hormone; LH, luteinizing hormone; E, estrogen; P, progesterone; CG, chorionic gonadotropin)

   a. E, P, CG, & LTH; LH
   b. LH, E, & CG; P & LTH
   c. LTH, LH, & CG; E & P
   d. E, P, & CG; LTH & LH
   e. P & LTH; LH, E, & CG
   f. E & P; LTH, LH & CG

37. Peak secretions of progesterone during the _____ (F, first; S, second) half of ovarian cycles are associated with relatively _____ (H, high; L, low) levels of FSH and relatively _____ (H, high; L, low) levels of LH secretion.

   a. F, H, H   c. F, L, L   e. S, L, H
   b. F, H, L   d. S, L, L   f. S, H, H

38. The life span of the corpus luteum is _____ (S, shortened; L, lengthened) in the event of pregnancy by the action of _____ (P, a placental; O, an ovarian; H, a pituitary) hormone.

   a. S, P   c. S, H   e. L, O
   b. S, O   d. L, P   f. L, H

39. Increased levels of endocrine secretion of the corpus luteum function to _____ (A, augment; H, inhibit) the secretion of LH _____ (W, as well as; N, but not) FSH.

   a. A, W   c. H, W
   b. A, N   d. H, N

40. In the absence of pregnancy, the life span of the corpus luteum of about _____ days is completed by about day _____ of an average ovarian cycle.

   a. 6, 20   c. 24, 24   e. 6, 6
   b. 12, 26   d. 2, 16   f. 12, 40

41. Menstruation is initiated _____ (O, at the onset; M, during the middle) of the ovarian cycle in response to _____ (H, high; L, low) levels of secretion of _____ (P, pituitary; V, ovarian) hormones.

   a. O, H, P   c. O, L, V   e. M, L, P
   b. O, H, V   d. M, L, V   f. M, H, P

42. _____ (E, estrogen; P, progesterone) secretion during the second half of the ovarian cycle is thought to be particularly effective in _____ (F, facilitating; H, inhibiting) FSH _____ (W, as well as; N, but not) LH secretion.

   a. E, F, W   c. E, H, N   e. P, F, N
   b. E, H, W   d. P, H, N   f. P, F, W

OBJECTIVE 81–7.

Identify the major estrogens and "progestin" hormones and the characteristics of their chemical nature, mode of synthesis, and mode of degradation.

43. The major estrogen in the human female is _____ (E, estrone; B, beta-estradiol; I, estriol) as a consequence of its far greater _____ (C, concentration; P, potency).

   a. E, C   c. I, C   e. B, P
   b. B, C   d. E, P   f. I, P

44. Estrogens are _____ (S, steroid; R, protein) hormones secreted in major amounts by the _____ (O, ovaries; P, placenta; A, adrenals; H, hypophysis).

   a. S; O & P only   d. R; A & H only
   b. S; O, P, & A only   e. R; O & P only
   c. S; O & A only   f. R; O, P, A, & H

45. Ethynyl estradiol, _____ (W, as well as; N, but not) stilbestrol, is a _____ (T, naturally occurring; S, synthetic) substance with _____ (E, estrogenic; H, estrogenic inhibitory) activity.

a. W, T, E   c. W, S, H   e. N, T, H
b. W, S, E   d. N, S, H   f. N, T, E

46. _____ (E, estrone; B, beta-estradiol; I, estriol) is a relatively inactive estrogen formed by oxidation of _____ (E, estrone; B, beta-estradiol; I, estriol) largely in the _____ (K, kidneys; L, liver).

a. E, both B & I, K   d. E, only B, L
b. B, only E, K   e. B, both I & E, L
c. I, both E & B, K   f. I, both E & B, L

47. The most important progestin, _____ (D, pregnanediol; H, 17alpha-hydroxyprogesterone; P, progesterone), is secreted in significant quantities during _____ (F, only the first half of the; S, only the second half of the; E, the entire) ovarian cycle.

a. D, E   c. P, E   e. H, F
b. H, S   d. D, F   f. P, S

48. _____ (L, like; U, unlike) the estrogens, the _____ (V, liver; K, kidneys) is the major locus for metabolic degradation of circulating progestins.

a. L, V   c. U, V
b. L, K   d. U, K

OBJECTIVE 81–8.
Identify the functions of the estrogens and their target effects.

49. _____ (I, increasing; D, decreasing) levels of _____ (E, estrogens; P, progestins; G, gonadotropins) during puberty are more directly responsible for growth of the tissues of the female sexual organs.

a. I, E   c. I, G   e. D, P
b. I, P   d. D, E   f. D, G

50. Estrogens tend to change the vaginal epithelium into a _____ (C, cuboidal; S, stratified) type of epithelium and enhance the proliferation of glandular tissues of the uterine endometrium _____ (W, as well as; N, but not) the fallopian tubes.

a. C, W   c. S, W
b. C, N   d. S, N

51. Estrogen, _____ (L, like; U, unlike) testosterone, tends to _____ (D, decrease; I, increase) osteoblast activity and _____ (S, shorten; P, prolong) the period of epiphyseal growth of long bones.

a. L, D, S   c. L, I, S   e. U, D, S
b. L, I, P   d. U, I, P   f. U, D, P

52. Estrogen, when compared to testosterone, is _____ (M, more; L, less) effective in promoting a _____ (P, positive; N, negative) nitrogen balance and _____ (M, more; L, less) effective in _____ (I, increasing; D, decreasing) the metabolic rate.

a. M, P, L, I   c. M, N, M, D   e. L, P, L, I
b. M, N, L, D   d. L, P, M, I   f. L, N, L, D

53. Estrogens tend to _____ skin thickness, _____ skin vascularity, and _____ subcutaneous fat deposition. (I, increase; D, decrease)

a. I, I, I   c. D, D, I   e. I, D, D
b. D, I, I   d. D, D, D   f. I, I, D

54. Estrogens are primarily responsible for the initiation of _____ (G, growth of the breasts at puberty; L, lactation) and development of the _____ (D, ductile system; A, lobules and alveoli) of the breasts.

a. G, D   c. L, D
b. G, A   d. L, A

55. Development of hair in the female pubic region and axillae _____ (W, as well as; N, but not) a tendency for the development of acne is thought to be caused by increased secretion of _____ (O, ovarian; A, adrenal) _____ (D, androgens; E, estrogens) after puberty.

a. W, O, D   c. W, A, E   e. N, O, E
b. W, A, D   d. N, A, E   f. N, O, D

56. Estrogens are _____ (P, proteins; S, steroids) which exert their target actions by combination with _____ (M, cell membrane; I, intracellular) "receptor" proteins.

a. P, M   c. S, M
b. P, I   d. S, I

OBJECTIVE 81-9.

Identify the functions of progesterone and its target effects.

57. The most important effect of progesterone is to promote _____ (F, fat deposition; D, duct development; P, proliferation; S, secretory changes) in the _____ (M, mammary glands; E, endometrium).

    a. F, M   c. P, E
    b. D, M   d. S, E

58. Progesterone exerts a mild _____ (A, anabolic; C, catabolic) effect upon the body's proteins and has target actions upon the renal tubules which are _____ (S, similar; O, opposite) to those of aldosterone in _____ (P, promoting; R, reducing) renal sodium reabsorption.

    a. A, O, P   c. A, O, R   e. C, O, P
    b. A, S, R   d. C, S, R   f. C, S, P

59. Progesterone is more important in the development of _____ (C, ciliated epithelium; S, secretory changes) of the fallopian tubes and in the _____ (D, development of the duct system; A, development of the lobules and alveoli; F, deposition of fat) in the mammary glands.

    a. C, D   c. C, F   e. S, A
    b. C, A   d. S, D   f. S, F

60. Progesterone has a significant function to _____ (E, enhance; H, inhibit) _____ (C, uterine contractility; L, milk "letdown" during lactation).

    a. E, C   c. H, C
    b. E, L   d. H, L

OBJECTIVE 81-10.

Characterize the cyclic changes occurring in the uterine endometrium during the endometrial cycle, their functional significance, and their relationship to cyclic variations in estrogen and progesterone secretion.

61. The proliferative or _____ (P, progesterone; E, estrogen) phase of the endometrial cycle occurs in association with the _____ (F, first; M, middle; L, last) portion of the ovarian cycle.

    a. P, F   c. P, L   e. E, M
    b. P, M   d. E, F   f. E, L

62. The secretory phase of the endometrium, occurring during the _____ (F, first; S, second) half of the ovarian cycle, is largely due to the effects of _____ (P, progesterone; E, estrogen; O, oxytocin) upon the _____ (P, progesterone; E, estrogen) primed uterus.

    a. F, P, E   c. F, O, E   e. S, E, P
    b. F, E, P   d. S, P, E   f. S, O, P

63. The endometrium reaches a maximum thickness under the influence of high levels of _____ (E, only estrogen; P, only progesterone; PE, both estrogen and progesterone) secretion during the _____ (F, first; L, last) half of the endometrial cycle.

    a. E, F   c. PE, F   e. P, L
    b. P, F   d. E, L   f. PE, L

64. _____ (M, menstruation; O, ovulation) occurring at the end of the endometrial cycle is associated with a sharp _____ (I, increase; D, decrease) in circulating levels of progesterone _____ (W, as well as; N, but not) estrogen.

    a. M, I, W   c. M, D, N   e. O, I, N
    b. M, D, W   d. O, D, N   f. O, I, W

65. Menstrual fluid normally contains about _____ ml. of _____ (C, clotted; N, non-clotted) blood.

    a. 35, C   c. 300, C   e. 125, N
    b. 125, C   d. 35, N   f. 300, N

66. The uterus is generally _____ (M, more; L, less) resistant to infection during menstruation as a consequence of its associated _____ (K, leukorrhea; T, altered endometrial thickness; C, altered contractility).

    a. M, K   c. M, C   e. L, T
    b. M, T   d. L, K   f. L, C

OBJECTIVE 81–11.

Identify the interrelationships between ovarian and hypothalamic-pituitary hormones during the ovarian cycle.

67. High concentrations of _____ (E, estrogen; P, progesterone) have a more pronounced effect of inhibiting FSH _____ (W, as well as; N, but not) LH secretion.

a. E, W
b. E, N
c. P, W
d. P, N

68. The combined effect of estrogen and progesterone secretion during the latter half of the female sexual cycle _____ (I, inhibits; C, accelerates) the secretion of FSH and LH from the _____ (H, hypothalamus; A, anterior pituitary; P, posterior pituitary).

a. I, H
b. I, A
c. I, P
d. C, H
e. C, A
f. C, P

69. The preovulatory LH surge is thought to result from a _____ (P, positive; N, negative) feedback effect of _____ (G, progesterone; E, estrogen).

a. P, G
b. P, E
c. N, G
d. N, E

70. On the 8th through 10th days of an average ovarian cycle, FSH is _____ , LH is _____ , and estrogen is _____ . (I, increasing; D, decreasing)

a. I, I, D
b. I, I, I
c. I, D, D
d. D, D, I
e. D, D, D
f. D, I, I

71. On the 16th through 18th days of an average ovarian cycle, FSH is _____ , LH is _____ , estrogen is _____ , and progesterone is _____ . (I, increasing; D, decreasing)

a. I, I, D, I
b. I, I, D, D
c. I, D, I, D
d. D, D, I, D
e. D, D, I, I
f. D, I, D, I

72. A second peak of estrogen secretion, occurring at about day_____of an average ovarian cycle, would most probably be due to secretion of the _____ (F, ovarian follicles; L, corpus luteum; E, endometrium).

a. 14, F
b. 22, L
c. 28, E
d. 14, L
e. 22, E
f. 28, F

73. Anovulatory cycles are more generally associated with an absence of _____ (E, estrogen; FSH, follicle-stimulating hormone; P, progesterone) secretion during the _____ (S, same; N, next) cycle.

a. E, S
b. FSH, S
c. P, S
d. E, N
e. FSH, N
f. P, N

OBJECTIVE 81–12.

Characterize the events occurring in association with puberty and menopause.

74. Puberty is thought to result from a maturation process of the _____(O, ovaries; H, hypothalamus) involving _____ (I, increased; D, decreased) sensitivity of a _____(P, positive; N, negative) feedback mechanism.

a. O, I, N
b. O, I, P
c. O, D, N
d. H, D, P
e. H, D, N
f. H, I, P

75. An abrupt increase in and the highest rate of gonadotropin hormone secretion occurs in the _____ (M, male; F, female) as a result of _____ (P, puberty; M, menopause; C, the climacteric).

a. M, P
b. M, C
c. F, P
d. F, M

76. Menopause is caused by _____ (I, inadequate; E, excessive) levels of _____ (O, ovarian; G, gonadotropin) hormones.

a. I, O
b. I, G
c. E, O
d. E, G

77. Hot flashes, irritability, and psychic sensations of dyspnea, fatigue, and anxiety are symptoms of menopause, _____ (W, as well as; N, but not) puberty, caused by _____ (L, low; H, high) levels of _____ (E, estrogens; P, progestins).

a. W, L, E
b. N, L, E
c. N, L, P
d. N, H, P
e. W, H, P
f. W, H, E

OBJECTIVE 81-13.

Identify the relationship of the ovaries to adrenal gland secretion, the effects of testosterone administration in the female, and the consequences of hypogonadism, hypergonadism, and endometriosis in the female.

78. The adrenal glands normally secrete _____ (R, minor; J, major) quantities of estrogen _____ (W, as well as; N, but not) progesterone.

a. R, W   c. J, W
b. R, N   d. J, N

79. Estrogen administration results in _____ (A, atrophic; H, hypertrophic) changes in the adrenal cortex as a consequence of the effects of estrogen to _____ (I, increase; D, decrease) _____ (FSH, follicle-stimulating hormone; CH corticotropin) secretion.

a. A, I, FSH   c. A, D, CH   e. H, I, CH
b. A, D, FSH   d. H, D, CH   f. H, I, FSH

80. Testosterone administration tends to _____ estrogen secretion, _____ FSH secretion, and _____ LH secretion in the adult female. (E, enhance; H, inhibit)

a. E, E, E   c. E, E, H   e. H, E, E
b. E, H, H   d. H, H, H   f. H, H, E

81. Female eunuchism is associated with _____ (O, hypogonadism; R, hypergonadism), _____ (E, enhanced; D, reduced) development of secondary sexual characteristics, and _____ (L, less; G, greater) than normal length of long bones in the adult.

a. O, E, L   c. O, D, G   e. R, E, G
b. O, D, L   d. R, D, G   f. R, E, L

82. Hypothyroid states tend to produce states of _____ (O, hyposecretion; E, hypersecretion) of the ovaries, and _____ (I, increased; D, decreased) intervals between ovarian cycles.

a. O, I   c. E, I
b. O, D   d. E, D

83. Endometriosis refers to the proliferation _____ (W, as well as; N, but not) the secretion and desquamation of endometrial tissue of the _____ (U, uterus; O, ovaries; P, peritoneal cavity).

a. W, U   c. W, P   e. N, O
b. W, O   d. N, U   f. N, P

OBJECTIVE 81-14.

Identify the roles of local sexual and psychic stimuli in erection, lubrication, and orgasm during the female sexual act.

84. Erectile tissue of the _____ (H, hymen; C, clitoris, V, vagina), or female equivalent of the penis, is regulated largely by its _____ (P, parasympathetic; S, sympathetic) innervation.

a. H, P   c. V, P   e. C, S
b. C, P   d. H, S   f. V, S

85. _____ (P, parasympathetic; S, sympathetic) activity to the _____ (L, glands of Littre; B, Bartholin's glands) results in lubrication of the introitus by a _____ (M, mucus; R, serous) secretion.

a. P, L, M   c. P, B, M   e. S, B, M
b. P, L, R   d. S, B, R   f. S, L, R

86. _____ (A, antidiuretic hormone; R, relaxin; O, oxytocin) secretion by the _____ (N, anterior; P, posterior) pituitary during sexual intercourse has been postulated to _____ (E, enhance; H, inhibit) uterine contractility.

a. A, N, E   c. O, N, H   e. R, P, E
b. R, N, H   d. A, P, H   f. O, P, E

87. Autonomic reflexes resulting from sexual stimulation in women _____ (W, as well as; N, but not) in men are integrated in the _____ (T, thoracic; L, lumbar; S, sacral) portions of the spinal cord.

a. W, T   c. W, S   e. N, L
b. W, L   d. N, T   f. N, S

OBJECTIVE 81–15.

Identify the fertile period of the ovarian cycle, the rationale for the rhythm method, the "pill," and prostaglandins in contraception, and abnormal conditions causing female sterility.

88. _____ (S, ejaculated sperm; O, expelled ova) have a shorter period of viability within the female reproductive tract of about _____ day(s).

   a. S, 1   c. S, 7   e. O, 3
   b. S, 3   d. O, 1   f. O, 7

89. Ovulation would be expected to occur on about day _____ of a 21-day ovarian cycle, day _____ of a 28-day cycle, and day _____ of a 40-day cycle.

   a. 14, 14, 14   c. 10, 14, 20   e. 20, 27, 39
   b. 7, 14, 26   d. 20, 20, 20   f. 14, 20, 34

90. The "pill" utilizes _____ (N, natural; S, synthetic) forms of _____ (E, estrogens; G, gonadotropins) _____ (W, as well as; N, but not) progestins.

   a. N, E, W   c. N, G, N   e. S, E, N
   b. N, G, W   d. S, G, N   f. S, E, W

91. The "pill" acts as an antifertility device through its action of inhibiting _____ (L, luteinizing hormone; E, estradiol) secretion and thereby preventing _____ (U, proliferation of the endometrium; O, ovulation).

   a. L, U   c. E, U
   b. L, O   d. E, O

92. PGE_2 is a _____ (P, progestin; G, prostaglandin) which has been utilized to _____ (F, facilitate; R, prevent) implantation of the fertilized ovum.

   a. P, F   c. G, F
   b. P, R   d. G, R

93. An abrupt _____ (R, rise; F, fall) of body temperature occurs at the time of _____ (O, ovulation; M, menstruation) as a consequence of increased secretion of _____ (V, ovarian; G, gonadotropin) hormones.

   a. R, O, V   c. R, M, G   e. F, M, V
   b. R, O, G   d. F, M, G   f. F, O, V

94. Enhanced urinary excretions of pregnanediol during the _____ half of the ovarian cycle is associated with an _____ (A, anovulatory; O, ovulatory) cycle.

   a. 1st, A   c. 2nd, A
   b. 1st, O   d. 2nd, O

95. Human chorionic gonadotropin is _____ (I, an inhibitor; S, a stimulator) of ovulation whose target actions are similar to those of _____ .

   a. I, estrogen   d. S, estrogen
   b. I, LH   e. S, LH
   c. I, FSH   f. S, FSH

96. Infertile conditions occur in about one in every _____ marriages and are more frequently associated with sterility in the _____ (M, male; F, female).

   a. 6 to 10, M   d. 6 to 10, F
   b. 60 to 100, M   e. 60 to 100, F
   c. 600 to 1000, M   f. 600 to 1000, F

97. Salpingitis, or inflammation of the _____ (O, ovaries; C, cervix; F, fallopian tubes), _____ (W, as well as; N, but not) a failure of ovulation, is a common cause of female sterility.

   a. O, W   c. F, W   e. C, N
   b. C, W   d. O, N   f. F, N

82

Pregnancy and Lactation

OBJECTIVE 82–1.

Characterize the events of maturation, fertilization, transport, and implantation of the ovum.

1. The first polar body is expelled from the nucleus of the ovum shortly _____ (B, before; A, after) _____ (O, ovulation; F, fertilization), whereas the second polar body is expelled shortly _____ (B, before; A, after) _____ (O, ovulation; F, fertilization).

   a. B, O, B, F   c. A, O, A, F   e. A, O, B, F
   b. B, O, A, F   d. B, F, A, F   f. A, F, A, O

2. A _____ (J, majority; N, minority) of ejaculated sperm are transported to the ovarian end of the fallopian tubes and require a transport period of about _____ minutes.

   a. J, 5   c. J, 200   e. N, 45
   b. J, 45   d. N, 5   f. N, 200

3. _____ (S, single; M, multiple) sperm containing _____ chromosomes penetrate(s) the ovum during the process of fertilization.

   a. S, 23   c. S, 92   e. M, 46
   b. S, 46   d. M, 23   f. M, 92

4. A male offspring results from the contribution of the _____ chromosome by the _____ (S, sperm; O, ovum).

   a. X, S   c. Y, S
   b. X, O   d. Y, O

5. A slow fluid current flowing _____ (I, into; O, out of) the fimbriated end of the fallopian tubes results from activity of _____ (C, cilia; M, the uterine myometrium).

   a. I, C   c. O, C
   b. I, M   d. O, M

6. Fertilization of the ovum normally takes place in the _____ (A, abdominal cavity; F, fallopian tubes; U, uterus) _____ (B, before; T, after) dispersion of the cumulus oophorus has occurred.

   a. A, B   c. U, B   e. F, T
   b. F, B   d. A, T   f. U, T

7. Passage of the ovum through the fallopian tubes requires about _____ (H, 3 hours; A, 1 day; T, 3 days) during which the isthmus of the fallopian tubes remains _____ (D, dilated; C, contracted).

   a. H, D   c. T, D   e. A, C
   b. A, D   d. H, C   f. T, C

8. The fallopian tubes function to transport _____ (W, as well as; N, but not) to provide a nutrient supply for a fertilized ovum which _____ (D, does; DN, does not) undergo cell division in transit.

   a. W, D   c. N, D
   b. W, DN   d. N, DN

9. Implantation of the fertilized ovum in a _____ (Z, zygote; B, blastocyst stage) occurs on the _____ day following ovulation.

   a. Z, 1st or 2nd   d. B, 1st or 2nd
   b. Z, 3rd or 4th   e. B, 3rd or 4th
   c. Z, 7th or 8th   f. B, 7th or 8th

10. Implantation of the ovum results from the actions of the _____ (T, trophoblastic cells; C, cells of the cumulus oophorus) upon the _____ (E, endometrium; F, fallopian tube; V, vagina).

    a. T, E   c. T, V   e. C, F
    b. T, F   d. C, E   f. C, V

OBJECTIVE 82–2.

Identify the mechanisms of early intrauterine nutrition of the embryo and the developmental and physiologic anatomy of the placenta.

11. The early nutrition of the implanted fertilized ovum is provided by digestion of _____ (C, cytotrophoblast; D, decidual) cells of the _____ (B, blastocyst; E, endometrium) which are formed by the action of _____ (S, estrogen; P, progesterone).

    a. C, B, S   c. C, B, P   e. D, B, P
    b. C, E, S   d. D, E, P   f. D, E, S

12. The placenta begins to provide slight amounts of nutrition as early as the _____ day after fertilization, or about the _____ day following implantation.

    a. 16th, 8th   c. 96th, 88th   e. 56th, 41st
    b. 56th, 48th   d. 16th, 3rd   f. 96th, 81st

13. Placental villi develop from _____ (D, decidual; T, trophoblastic) cells and contain capillaries which transport _____ (F, fetal; M, maternal) blood.

    a. D, F   c. T, F
    b. D, M   d. T, M

14. Blood sinuses which surround the placental villi contain _____ (F, fetal; M, maternal) blood supplied by _____ (B, umbilical; T, uterine) _____ (A, arteries; V, veins).

    a. F, B, A   c. F, B, V   e. M, B, A
    b. F, T, A   d. M, T, A   f. M, T, V

15. The total villar surface area of the mature placenta approximates _____ (F, only a few; S, 70) square meters and represents a diffusing surface separating maternal from fetal blood by a minimal distance of _____ (T, 3.5 microns; I, 35 microns; M, 3.5 mm.).

    a. F, T   c. F, M   e. S, I
    b. F, I   d. S, T   f. S, M

16. Progressive enlargement of the fetus with time is associated with _____ surface area of the layers of the placental membrane, _____ thickness of the layers of the placental villi, and _____ placental permeability. (I, increased; C, a relatively constant; D, decreased)

    a. I, I, C   c. I, D, I   e. C, C, C
    b. I, C, I   d. C, I, D   f. C, D, I

OBJECTIVE 82–3.

Characterize the functional roles of the placenta in respiratory gas exchange, transport and storage of nutritional substances, and excretion of metabolic end-products.

17. The mean PO_2 pressure gradient for diffusion of oxygen through the placental membrane is about _____ mm. Hg and is associated with a PO_2 of umbilical venous blood of about _____ mm. Hg.

    a. 5, 30   c. 20, 73   e. 5, 50
    b. 10, 50   d. 20, 30   f. 10, 73

18. Fetal hemoglobin has _____ blood concentration and _____ percentage of oxyhemoglobin at comparable PO_2 values when compared to hemoglobin within the maternal circulation. (G, a greater; S, about the same; L, a lesser)

    a. G, G   c. L, G   e. S, L
    b. S, G   d. G, L   f. L, L

19. Carbon dioxide has a _____ solubility than oxygen in fetal fluids and a _____ diffusing capacity through the placenta. (G, greater; L, lesser)

    a. G, G   c. L, G
    b. G, L   d. L, L

20. Oxygen transport across the placental membrane is _____ (F, facilitated; I, inhibited) by a _____ (L, fall; R, rise) in pH of fetal blood entering the placenta and _____ (F, facilitated; I, inhibited) by pH changes of maternal blood entering the placenta.

    a. F, L, I   c. F, R, I   e. I, L, I
    b. F, R, F   d. I, R, F   f. I, L, F

21. Fetal blood when compared to maternal blood generally has lower concentrations of _____ and higher concentrations of _____ . (CP, calcium and phosphate; A, ascorbic acid; AA, amino acids; G, glucose)

   a. CP; A, AA, & G   d. A & AA; CP & G
   b. CP, AA, & G; A   e. G & AA; CP & A
   c. A & G; CP & AA   f. G; CP, A, & AA

22. _____ (U, urea; C, creatinine) has a greater concentration gradient across the placental membrane, largely as a consequence of differences in their _____ (S, rates of active transport; P, permeability coefficients; B, rates of fetal catabolism).

   a. U, S   c. U, B   e. C, P
   b. U, P   d. C, S   f. C, B

23. The _____ (P, placenta; L, fetal liver) provides for the earliest storage of protein, calcium, and iron, _____ (W, as well as; N, but not) glycogen, during fetal development.

   a. P, W   c. L, W
   b. P, N   d. L, N

OBJECTIVE 82–4.
Identify placental chorionic gonadotropin, its secretion, its target actions, and its functional significance.

24. When pregnancy occurs, a rise in _____ (P, pituitary; C, chorionic) gonadotropins to peak secretions about seven _____ (D, days; W, weeks; O, months) after ovulation functions to _____ (M, maintain; H, inhibit) secretion of the corpus luteum.

   a. P, D, M   c. P, O, H   e. C, W, M
   b. P, W, H   d. C, D, H   f. C, O, M

25. Chorionic gonadotropin is a _____ (S, steroid; G, glycoprotein) hormone synthesized by _____ (O, the ovaries; T, syncytial trophoblast cells; D, decidual cells).

   a. S, O   c. S, D   e. G, T
   b. S, T   d. G, O   f. G, D

26. The earliest detection of _____ (E, estrogens; P, progestins; C, chorionic gonadotropins) at about _____ days following ovulation forms a positive basis for clinical tests for pregnancy.

   a. E, 8   c. C, 8   e. P, 28
   b. P, 8   d. E, 28   f. C, 28

27. Chorionic gonadotropin exerts _____ (S, a stimulating; H, an inhibitory) effect upon _____ (L, interstitial cells of Leydig; R, Sertoli cells) to enhance their secretion of _____ (E, estrogen; T, testosterone).

   a. S, L, E   c. S, R, T   e. H, R, E
   b. S, L, T   d. H, R, T   f. H, L, E

OBJECTIVE 82–5.
Characterize the placental secretion and functional significance of estrogens and progesterone during fetal development. Identify human placental lactogen, its secretion, its target actions, and its functional significance.

28. Estrogen and progesterone _____ (W, as well as; N, but not) chorionic gonadotropin and human placental lactogen are thought to be secreted by fetal _____ (T, syncytial trophoblastic; D, decidual; A, adrenal cortical) cells.

   a. W, T   c. W, A   e. N, D
   b. W, D   d. N, T   f. N, A

29. Estrogen secreted during fetal development is largely _____ (T, estriol; B, beta-estradiol; N, estrone) synthesized from dehydroepiandrosterone formed by the _____ (P, placenta; A, adrenal glands).

   a. T, P   c. N, P   e. B, A
   b. B, P   d. T, A   f. N, A

30. Enlargement of the breasts and female external genitalia, _____ (W, as well as; N, but not) the uterus, and relaxation of the pelvic ligaments during pregnancy are attributed to _____ (P, progesterone; E, estrogen) secretion.

    a. W, P   c. N, P
    b. W, E   d. N, E

31. _____ (P, progesterone; E, estrogen) secretion functions to cause development of decidual cells, _____ (H, enhance; I, inhibit) contractility of the gravid uterus, and enhance _____ (R, proliferation; S, secretion) of the fallopian tubes and uterus.

    a. P, H, R   c. P, H, S   e. E, H, R
    b. P, I, S   d. E, I, S   f. E, I, R

32. Estrogen _____ (W, as well as; N, but not) progesterone secretion reaches a peak secretion during the _____ trimester of pregnancy.

    a. W, 1st   c. W, 3rd   e. N, 2nd
    b. W, 2nd   d. N, 1st   f. N, 3rd

33. Placental lactogen is a _____ (S, steroid; P, protein) hormone whose secretion is initiated about the _____ _____ (W, week; M, month) of pregnancy.

    a. S, 5th, W   c. S, 8th, M   e. P, 5th, M
    b. S, 5th, M   d. P, 5th, W   f. P, 8th, M

34. Human placental lactogen has effects upon the breasts which are similar to those of _____ (L, LH; P, prolactin; O, oxytocin) in _____ (I, inhibiting; F, facilitating) milk production.

    a. L, I   c. O, I   e. P, F
    b. P, I   d. L, F   f. O, F

OBJECTIVE 82–6.

Characterize the altered states of secretion of the nonsexual endocrine glands of the mother during pregnancy, and their functional significance.

35. Alterations of the size of the maternal anterior pituitary gland during pregnancy are attributed to _____ (E, elevated; L, lowered) rates of secretion of _____ (CTS, corticotropin, thyrotropin, and somatotropin; LF, luteinizing and follicle-stimulating hormone).

    a. E, only CTS   d. L, only CTS
    b. E, only LF   e. L, only LF
    c. E, both CTS & LF   f. L, both CTS & LF

36. Glucocorticoid _____ (W, as well as; N, but not) mineralocorticoid secretion by the maternal adrenal glands is _____ (R, reduced; E, enhanced) during pregnancy.

    a. W, R   c. N, R
    b. W, E   d. N, E

37. A higher drain of maternal calcium, occurring during _____ (G, fetal growth; L, lactation), is generally associated with _____ (E, an enlargement; R, a reduction) of the parathyroid gland size as a consequence of its altered rate of secretion of _____ (C, calcitonin; P, parathormone).

    a. G, E, C   c. G, R, C   e. L, E, P
    b. G, E, P   d. L, R, C   f. L, R, P

OBJECTIVE 82–7.

Identify the changes that occur in the circulatory system, metabolism, respiratory system, and urinary system of the mother during pregnancy. Characterize amniotic fluid and its formation.

38. A blood flow of about _____ ml. of blood through the maternal circulation of the placenta is associated with _____ (I, increased; N, no significant change in; D, decreased) peripheral resistance of the maternal circulatory system.

    a. 625, I   c. 625, D   e. 1800, N
    b. 625, N   d. 1800, I   f. 1800, D

39. The cardiac output of the maternal circulation _____ (H, reaches its highest level; N, falls to near normal levels) during the last eight weeks of pregnancy _____ (D, despite; C, as a consequence of) a _____ (L, low; H, high) uterine blood flow.

    a. H, D, L   c. H, C, H   e. N, D, H
    b. H, C, L   d. N, C, H   f. N, D, L

40. Shortly before term, the maternal blood volume is about _____ % _____ (H, higher; L, lower) than normal _____ (S, in spite; C, partly as a consequence) of the effects of altered rates of aldosterone and estrogen secretion upon the kidneys.

    a. 10, H, S   c. 50, H, C   e. 15, L, S
    b. 30, H, C   d. 5, L, S   f. 30, L, C

41. Full term weights of amniotic fluid, placenta, and fetal membranes of about _____ pounds, and the fetus of about _____ pounds, are included in an average weight gain of about _____ pounds for pregnant women.

    a. 7, 12, 60   c. 12, 7, 34   e. 4, 7, 12
    b. 12, 7, 60   d. 7, 12, 34   f. 4, 7, 24

42. Alterations of the maternal minute ventilation generally occurring during pregnancy are attributed largely to _____ (I, increased; D, decreased) _____ (T, tidal volumes; RR, respiratory rates) and occur in association with slightly _____ (E, elevated; R, reduced) arterial PCO_2 levels.

    a. I, T, E   c. I, RR, R   e. D, T, R
    b. I, RR, E   d. D, RR, R   f. D, T, E

43. Pregnancy is associated with _____ glomerular filtration rates, _____ intraureteral pressures, and _____ rates of aldosterone secretion. (I, increased; D, decreased)

    a. I, I, I   c. I, D, I   e. D, D, I
    b. I, I, D   d. D, D, D   f. D, I, D

44. The volume of amniotic fluid is normally between _____ and is completely replaced about once every three _____ (M, minutes; H, hours; D, days).

    a. 5 to 50 ml., M   d. 5 to 50 ml., H
    b. 0.5 to 1 liter, H   e. 0.5 to 1 liter, D
    c. 2 to 3 liters, H   f. 2 to 3 liters, D

OBJECTIVE 82–8.

Identify hyperemesis gravidarum and toxemia of pregnancy, their potential causes, and their potential consequences.

45. Hyperemesis gravidarum or "morning sickness" generally occurs during the _____ (E, early; L, late) months of pregnancy in association with the secretion of large amounts of _____ (LH, luteinizing hormone; HPL, human placental lactogen; CH, chorionic gonadotropin).

    a. E, LH   c. E, CH   e. L, HPL
    b. E, HPL   d. L, LH   f. L, CH

46. Eclampsia is a _____ (M, mild; S, severe) form of _____ (H, hyperemesis gravidarum; T, toxemia of pregnancy) occurring _____ (E, in the early stages of pregnancy; P, near parturition).

    a. M, H, E   c. M, H, P   e. S, H, P
    b. M, T, E   d. S, T, P   f. S, T, E

47. Toxemia of pregnancy occurs in about _____ % of all pregnant women and is generally associated with _____ (I, increased; D, decreased) renal blood flow and _____ (I, increased; D, decreased) glomerular filtration rates.

    a. 1, D, I   c. 32, I, D   e. 7, I, I
    b. 7, D, D   d. 1, I, I   f. 32, D, I

48. Toxemia of pregnancy is more probably associated with an underlying _____ (H, hormonal; A, autoimmune) etiology whose treatment generally includes _____ (I, increased; D, decreased) salt intake.

    a. H, I   c. A, I
    b. H, D   d. A, D

OBJECTIVE 82–9.

Identify parturition, its underlying mechanisms, and its associated characteristics.

49. Progesterone _____ (W, as well as; N, but not) estrogen functions to _____ (E, enhance; I, inhibit) uterine contractility during pregnancy.

    a. W, E   c. N, E
    b. W, I   d. N, I

50. The latter several months of pregnancy are thought to be characterized by _____ estrogen levels, _____ progesterone levels, and _____ ratios of estrogen to progesterone concentrations. (R, rising; F, falling)

    a. R, R, R   c. F, R, F   e. R, R, F
    b. R, F, R   d. F, F, F   f. F, F, R

51. Oxytocin secretion by the _____ (P, placenta; A, adenohypophysis; N, neurohypophysis) is thought to _____ (I, inhibit; E, enhance) uterine contractility.

    a. P, I   c. N, I   e. A, E
    b. A, I   d. P, E   f. N, E

52. Cervical irritation during labor is thought to _____ (F, facilitate; I, inhibit) uterine contractility and to _____ (F, facilitate; I, inhibit) oxytocin secretion, thereby participating in a _____ (P, positive; N, negative) feedback mechanism of uterine contractility.

    a. F, F, P   c. F, I, P   e. I, I, P
    b. F, F, N   d. I, I, N   f. I, F, N

53. _____ (I, intermittent; C, continuous tonic) Braxton-Hicks contractions of the _____ (P, perineal; U, uterine) musculature occur during _____ (M, most of the months; L, only the last month) of pregnancy.

    a. I, P, M   c. I, P, L   e. C, P, L
    b. I, U, M   d. C, U, L   f. C, U, M

54. Labor contractions with periodicities of 30 minutes have comparatively _____ (G, greater; S, lesser) amplitudes and occur _____ (E, earlier; L, later) during labor than labor contractions with periodicites of one to three minutes.

    a. G, E   c. S, E
    b. G, L   d. S, L

55. The first stage of labor, usually lasting about 8 to 24 _____ (M, minutes; H, hours) in the first pregnancy, involves a period of _____ (S, separation of the placenta from its implantation site; C, progressive cervical dilation).

    a. M, S   c. H, S
    b. M, C   d. H, C

56. Contractions of the abdominal _____ (W, as well as; N, but not) the uterine musculature generally force the _____ (H, head; B, buttocks) of the fetus forward as a wedge to open the structures of the birth canal.

    a. W, H   c. N, H
    b. W, B   d. N, B

57. A more severe type of pain mediated by _____ (S, somatic; H, hypogastric) nerves occurs during the _____ stage of labor.

    a. S, 1st   c. H, 1st
    b. S, 2nd   d. H, 2nd

58. Separation of the placenta from the uterus occurs within 10 to 45 _____ (M, minutes; H, hours) _____ (B, before; A, after) birth and results in an average blood loss of about _____ ml.

    a. M, B, 35   c. M, A, 35   e. H, A, 35
    b. M, B, 350   d. M, A, 350   f. H, A, 350

OBJECTIVE 82–10.

Identify the endocrine roles in development of the breasts, the initiation of lactation, and the ejection or "let-down" process in milk secretion. Identify the physiological effects of continued milk secretion following pregnancy upon the mother.

59. Development of the stoma, duct system, and deposition of fat within the breasts, _____ (W, as well as; N, but not) the growth of the lobules, alveoli, and secretory characteristics of the alveoli, are attributed to target actions of _____ (E, estrogens; P, progestins).

    a. W, E   c. N, E
    b. W, P   d. N, P

60. Adrenal corticosteroids and thyroid hormone, _____ (W, as well as; N, but not) insulin, are considered to have _____ (P, permissive; S, specific target) effects upon development of the breasts.

    a. W, P   c. N, P
    b. W, S   d. N, S

61. Lactation is generally _____ (E, enhanced; S, suppressed), during the latter months of pregnancy as a consequence of _____ (H, high; L, low) rates of prolactin secretion and _____ (H, high; L, low) rates of estrogen and progestin secretion.

    a. E, H, H   c. E, L, L   e. S, L, H
    b. E, L, H   d. S, H, H   f. S, L, L

62. Prolactin secretion by the _____ (A, anterior; P, posterior) pituitary is thought to be regulated by the _____ (L, placental; H, hypothalamic) secretion of a prolactin _____ (R, releasing; I, inhibitory) factor.

    a. A, L, R   c. A, H, I   e. P, H, I
    b. A, L, I   d. P, H, R   f. P, L, R

63. The ejection or milk "let-down" process of secretion by the breasts involves contractions of myoepithelial cells in response to _____ (P, prolactin; O, oxytocin) secretion by the _____ (A, anterior; S, posterior) lobe of the pituitary.

    a. P, A   c. O, A
    b. P, S   d. O, S

64. Milk is secreted in _____ (I, an intermittent; C, continuous) fashion into the alveoli of the breasts under the more probable direct action of _____ (O, oxytocin; P, prolactin; HPL, human placental lactogen).

    a. I, O   c. I, HPL   e. C, P
    b. I, P   d. C, O   f. C, HPL

65. Suckling on one breast, causing milk flow in that breast _____ (W, as well as; N, but not) the other breast, is _____ (F, facilitated; I, inhibited) by generalized sympathetic activity.

    a. W, F   c. N, F
    b. W, I   d. N, I

66. Women who do not nurse their offspring after parturition, when compared to those who do, have _____ (L, a longer; S, about the same; H, a shorter) interval of prolactin secretion and _____ (E, an earlier; T, a later) involution of the uterus.

    a. L, E   c. H, E   e. S, T
    b. S, E   d. L, T   f. H, T

67. Human milk when compared to cow's milk has a _____ concentration of lactose, a _____ concentration of protein, and a _____ mineral content. (H, higher; L, lower)

    a. H, H, H   c. H, L, L   e. L, L, H
    b. H, H, L   d. L, L, L   f. L, H, H

83

Special Features of Fetal and Neonatal Physiology

OBJECTIVE 83–1.

Characterize the fetal growth rate and the functional development of the fetal organ systems.

1. Fetal growth between the 6th and 36th week of gestation is characterized by a fetal length which increases approximately _____ the age of the fetus, and a fetal weight which increases approximately _____ the length of the fetus. (D, in direct proportion to; S, as the square of; C, as the cube of)

   a. D, S   c. C, S   e. S, C
   b. S, D   d. D, C   f. C, D

2. The fetal heart begins beating about the _____ week following fertilization, and approaches a heart rate of about _____ beats per minute just before birth.

   a. 4th, 65   c. 34th, 20   e. 20th, 65
   b. 20th, 20   d. 4th, 140   f. 34th, 140

3. The earliest _____ (N, nucleate; A, anucleate) red blood cells are formed in the yolk sac and mesothelial layers of the placenta at about the third _____ (D, day; W, week; M, month) of fetal development.

   a. N, D   c. N, M   e. A, W
   b. N, W   d. A, D   f. A, M

4. Fetal arterial blood when compared to the adult has an oxygen-hemoglobin dissociation curve shifted to the _____ (R, right; L, left) and a _____ (W, lower; H, higher) PO_2.

   a. R, W   c. L, W
   b. R, H   d. L, H

5. Peripheral reflexes first appear _____ (B, before; A, at; F, following) birth, _____ (F, following; B, before) completion of myelination of the major tracts of the central nervous system.

   a. B, F   c. F, F   e. A, B
   b. A, F   d. B, B   f. F, B

6. The latter three to four months of pregnancy are associated with net fluid _____ (A, absorption; S, secretion) by the alveolar epithelium and _____ (E, enhanced; D, diminished) pulmonary movements in the fetus.

   a. A, E   c. S, E
   b. A, D   d. S, D

7. Fetal kidneys _____ form urine and _____ regulate acid-base balance in utero. (D, do; N, do not)

   a. D, D   c. N, D
   b. D, N   d. N, N

8. The fetal _____ (K, kidneys; G, gastrointestinal tract) excrete(s) _____ (C, colostrum; M, meconium) into the amniotic fluid.

   a. K, C   c. G, C
   b. K, M   d. G, M

OBJECTIVE 83–2.

Recognize the fetal requirements and sources for calcium and phosphate, iron, and vitamins.

9. Ossification in the fetus, initially apparent from x-rays at about the _____ month of pregnancy, represents a _____ (M, more; L, less) substantial drain of maternal calcium and phosphate stores when compared to lactation.

   a. 2nd, M   c. 8th, M   e. 4th, L
   b. 4th, M   d. 2nd, L   f. 8th, L

10. Iron storage, largely in the fetal _____ (B, bone marrow; P, placenta; L, liver), is normally sufficient for iron requirements in the neonate for a minimum of several _____ (D, days; W, weeks; M, months) following birth.

    a. B, W   c. L, M   e. P, W
    b. P, D   d. B, M   f. L, W

11. Signs of vitamin deficiency appearing within the first several weeks of neonatal life are most frequently the result of insufficient vitamin:

    a. A   c. K   e. E
    b. B_{12} d. D   f. C

12. Vitamin _____ or the "antisterility vitamin" is thought to be required for the early development of the fetus, whereas vitamin _____ is required by the fetal _____ (S, spleen; L, liver) for the formation of factor VII and prothrombin.

    a. C, B_{12}, S   c. C, E, L   e. E, K, L
    b. C, K, S   d. E, B_{12}, L   f. E, C, S

OBJECTIVE 83–3.

Identify the respiratory, cardiovascular, and nutritive adjustments of the infant to extrauterine life.

13. The first few breaths of the newborn infant are generally _____ (W, weak; S, strong) breaths initiated _____ (I, almost immediately; F, about 4 to 5 minutes) following birth.

    a. W, I   c. S, I
    b. W, F   d. S, F

14. The initiation of respiration in the newborn infant is generally inhibited by _____ and facilitated by _____ . (A, obstetrical anesthesia; X, mild hypoxia; P, mild hypercapnia; D, prolonged fetal hypoxia during delivery; E, sudden exposure to the external world)

    a. A, X, & P; D & E   d. D & E; A, X, & P
    b. A, P, & E; X & D   e. X & D; A, P, & E
    c. A & D; X, P, & E   f. X, P, & E; A & D

15. Alveolar surfactant, functioning to _____ (I, increase; D, decrease) alveolar surface tension forces, _____ (S, is; N, is not) normally present at birth, and _____ (S, is; N, is not) present in excess in hyaline membrane disease.

    a. I, S, N   c. I, N, N   e. D, S, N
    b. I, N, S   d. D, N, S   f. D, S, S

16. The ductus arteriosus, connecting the pulmonary artery with the _____ (UV, umbilical vein; A, aorta; PV, pulmonary vein), contains primarily _____ (O, oxygenated; U, unoxygenated) blood in the fetus.

    a. UV, O   c. PV, O   e. A, U
    b. A, O   d. UV, U   f. PV, U

17. The ductus venosus contains _____ (O, well oxygenated; P, poorly oxygenated) blood of the placental circulation which largely _____ (B, bypasses; T, passes through) the liver.

    a. O, B   c. P, B
    b. O, T   d. P, T

18. Approximately _____ % of the blood volume flow of the fetal heart is transmitted to the placenta by the umbilical _____ (V, veins; A, arteries).

    a. 12, V   c. 100, V   e. 55, A
    b. 55, V   d. 12, A   f. 100, A

19. A loss of blood flow through the placenta at birth _____ the fetal systemic vascular resistance and _____ fetal aortic pressure. (I, increases; D, decreases)

    a. I, I
    b. I, D
    c. D, I
    d. D, D

20. The initiation of respiration in the newborn infant _____ pulmonary vascular resistance and _____ the pulmonary arterial pressure. (I, increases; D, decreases)

    a. I, I
    b. I, D
    c. D, I
    d. D, D

21. Closure of the small valve that overlies the foramen ovale, on the _____ (R, right; L, left) side of the _____ (A, atrial; V, ventricular) septum, is associated with _____ (D, a decreased; I, an increased) right atrial pressure.

    a. R, A, D
    b. R, V, I
    c. R, V, D
    d. L, V, I
    e. L, A, D
    f. L, A, I

22. The first few days of life of the newborn infant are generally characterized by _____ (I, increased; D, decreased) body weight, a comparatively _____ (H, high; L, low) rate of body fluid turnover, and predominantly _____ (C, carbohydrate; P, lipid and protein) utilization for energy purposes.

    a. I, H, C
    b. I, L, P
    c. I, L, C
    d. D, L, P
    e. D, H, C
    f. D, H, P

OBJECTIVE 83–4.

Identify the general functional characteristics and special problems of the cardiovascular, respiratory, and renal systems in the neonatal infant.

23. The newborn infant, with a respiratory rate of about _____ breaths per minute and a tidal volume of about_____ml., has a lower _____(M, minute respiratory volume; C, functional residual capacity) than the adult in relation to his body weight.

    a. 12, 16, M
    b. 22, 16, M
    c. 40, 16, C
    d. 12, 64, C
    e. 22, 64, C
    f. 40, 64, M

24. The newborn infant has average values of _____ ml. for blood volume, _____ ml./minute for cardiac output, and an arterial blood pressure of _____ .

    a. 75, 550, 70/50
    b. 300, 550, 70/50
    c. 1500, 550, 120/80
    d. 75, 1500, 120/80
    e. 300, 1500, 120/80
    f. 1500, 1500, 70/50

25. The newborn infant has a greater percentage _____ (I, increase; D, decrease) in blood _____ (E, erythrocyte; L, leukocyte) concentrations when compared to typical adult values.

    a. I, E
    b. I, L
    c. D, E
    d. D, L

26. _____ (I, increasing; D, decreasing) levels of bilirubin concentration of the newborn infant during the first few days of life are associated with _____ (B, bilirubin; L, a lack of bilirubin) transport by the placenta during fetal life and a _____ (W, low; H, high) rate of conjugation of bilirubin in the liver of the neonate.

    a. I, B, W
    b. I, B, H
    c. I, L, W
    d. D, L, H
    e. D, L, W
    f. D, B, H

27. Erythroblastosis fetalis is a condition associated with _____ (K, vitamin K deficiencies; Rh, Rh factor incompatability) which affects about one of every _____ newborn infants to some degree.

    a. K, 10 to 20
    b. K, 50 to 100
    c. K, 60,000 to 70,000
    d. Rh, 10 to 20
    e. Rh, 50 to 100
    f. Rh, 60,000 to 70,000

28. Postnatal bleeding tendencies, more frequently resulting from vitamin _____ deficiencies in the newborn, _____ (A, are; N, are not) prevented by administration of the vitamin to the mother prior to birth of the infant.

    a. D, A   c. K, A   e. C, N
    b. C, A   d. D, N   f. K, N

29. The newborn infant when compared with the adult has a _____ (G, greater; L, lesser) ability to concentrate urine and a tendency towards _____ (K, alkalosis; C, acidosis) of body fluids.

    a. G, K   c. L, K
    b. G, C   d. L, C

30. The newborn infant has an average body water content equal to _____ % of the body weight, which is generally _____ (H, higher; W, lower) than that of the adult.

    a. 42, H   c. 73, H   e. 58, W
    b. 58, H   d. 42, W   f. 73, W

31. Fetal and early neonatal life of the infant is associated with a ratio of intracellular to extracellular fluid volumes which is _____ (G, greater; L, less) than one, a condition which is _____ (S, similar; O, opposite) to that for the adult.

    a. G, S   c. L, S
    b. G, O   d. L, O

32. The newborn infant when compared to the adult has a higher total body _____ and a lower total body _____ in relation to body weight. (K, potassium, Na, sodium; Mg, magnesium; P, phosphate; Cl, chloride)

    a. Na & K; Mg, P, & Cl
    b. Mg & P; K, Na, & Cl
    c. K, Mg, & P; Na & Cl
    d. K, Na & Cl; Mg & P
    e. Na & Cl; K, Mg, & P
    f. Mg, P, & Cl; Na & K

OBJECTIVE 83–5.

Identify the general functional characteristics and special problems of metabolism, body temperature regulation, liver function, immunity, and endocrinology in the neonatal infant.

33. A _____ (H, high; W, low) level of liver function in the newborn infant is associated with _____ (H, high; W, low) rates of gluconeogenesis and _____ (E, elevated; N, normal; W, low) levels of plasma proteins.

    a. H, H, E   c. H, W, W   e. W, W, W
    b. H, H, N   d. W, W, E   f. W, H, W

34. The newborn infant generally has _____ (E, enhanced; I, impaired) fat absorption as compared to the older child and a deficient utilization of dietary _____ (M, monosaccharides; D, disaccharides; S, starches).

    a. E, M   c. E, S   e. I, D
    b. E, D   d. I, M   f. I, S

35. The newborn infant when compared to the adult has a _____ (H, higher; L, lower) ratio of body surface area to body mass, a _____ (H, higher; L, lower) metabolic rate in proportion to body mass, and _____ (W, well; P, poorly) regulated mechanisms of body temperature.

    a. H, H, W   c. H, L, P   e. L, L, W
    b. H, H, P   d. L, L, P   f. L, H, W

36. The vitamin _____ -deficient infant is generally _____ (L, less; M, more) susceptible to early development of rickets than the vitamin deficient adult.

    a. D, L   c. K, L   e. A, M
    b. A, L   d. D, M   f. K, M

37. Ascorbic acid, stored in _____ (I, insignificant; J, major) amounts by fetal tissues, is required for the synthesis of _____ (C, blood coagulation factors; IC, intercellular substances; D, 1,25-dihydroxycholecalciferol).

    a. I, C   c. I, D   e. J, IC
    b. I, IC   d. J, C   f. J, D

38. Allergic states such as eczema occur more frequently _____ (I, immediately; M, several months; Y, several decades) after birth in association with _____ (L, low; H, high) levels of immunoglobulins.

    a. I, L   c. Y, L   e. M, H
    b. M, L   d. I, H   f. Y, H

39. The newborn infant has a relatively _____ (W, well; P, poorly) developed endocrine system and is _____ (M, more; S, less) susceptible to hypoglycemic shock than the adult.

   a. W, M   c. P, M
   b. W, S   d. P, S

40. An infant born of a diabetic mother more frequently exhibits _____ (A, atrophy; H, hypertrophy) of the islets of Langerhans, a tendency for _____ (O, hypoglycemia; E, hyperglycemia), and a higher statistical probability of death resulting from _____ (G, altered glucose levels; R, respiratory distress syndrome).

   a. A, O, G   c. A, E, G   e. H, O, G
   b. A, E, R   d. H, E, R   f. H, O, R

OBJECTIVE 83–6.

Identify the special problems of the premature infant.

41. The premature infant, in comparison with the newborn infant carried to full term, requires a _____ dietary fat content and a _____ dietary calcium intake. (H, higher; L, lower)

   a. H, H   c. L, H
   b. H, L   d. L, L

42. A comparatively _____ (H, high; L, low) functional residual capacity of the premature infant in comparison with the normal neonate is frequently associated with a periodic, waxing and waning _____ (B, Biots; C, Cheyne-Stokes) type of breathing.

   a. H, B   c. L, B
   b. H, C   d. L, C

43. At normal room temperatures, the body temperature of the premature infant generally stabilizes at a temperature _____ (A, above; E, about equal; B, below) that of the normal adult which _____ (I, is; N, is not) conducive to his survival.

   a. A, I   c. B, I   e. E, N
   b. E, I   d. A, N   f. B, N

44. Edema in the premature infant is a more probable consequence of altered _____ (E, protein; K, potassium, G, blood glucose) levels as a consequence of immature _____ (R, kidney; L, liver; A, adrenal; P, pancreatic) development.

   a. E, L   c. G, P   e. K, A
   b. K, R   d. E, R   f. G, L

45. Vascular ingrowth into the vitreous humor of the eyes and subsequent retrolental fibroplasia occurs in response to a stimulus of _____ (I, increased; D, decreased) PO_2 levels associated with _____ (L, a lack of; T, the treatment of) the premature infant with oxygen therapy.

   a. I, L   c. D, L
   b. I, T   d. D, T

OBJECTIVE 83–7.

Identify the general growth and developmental characteristics of the child.

46. The nervous system _____ (I, is; N, is not) fully functional at birth and most reflexes are developed during _____ (P, prenatal; O, neonatal) life.

   a. I, P   c. N, P
   b. I, O   d. N, O

47. Girls usually attain a _____ (G, greater; S, shorter) final height than boys due to _____ (E, an earlier; L, a later) uniting of the epiphyses at an age of about _____ years.

   a. G, E, 11 to 13   d. S, L, 14 to 16
   b. G, L, 14 to 16   e. S, E, 11 to 13
   c. G, L, 11 to 13   f. S, E, 14 to 16

48. At an age of about _____ (A, 6 months; B, 12 months; Y, 2 years), when an infant can usually first walk alone, myelination of the major tracts of the central nervous system is _____ (I, incomplete; C, essentially complete), and the brain weight of the infant _____ (H, has; N, has not) attained adult proportions.

a. A, I, H   c. Y, C, H   e. B, C, N
b. B, I, H   d. A, C, N   f. Y, I, N

49. The infant is generally first able to roll over at about _____ months and stand alone at about _____ months.

a. 2, 6   c. 9, 14   e. 5, 14
b. 5, 11   d. 2, 11   f. 9, 24

ANSWERS TO QUESTIONS

The page references given here are to the fifth edition of the *Textbook of Medical Physiology* by Arthur C. Guyton (Philadelphia: W.B. Saunders, 1976).

CHAPTER 1

Number	Answer	Reference Page(s)	Number	Answer	Reference Page(s)
1.	f	2	13.	b	5
2.	c	2	14.	a	5
3.	f	2	15.	c	5
4.	c	2	16.	b	5
5.	b	3	17.	e	6
6.	b	3	18.	d	6
7.	c	3	19.	d	6
8.	d	3	20.	d	6
9.	c	3	21.	b	6
10.	e	3	22.	b	7
11.	e	3	23.	c	11
12.	a	3	24.	b	11

CHAPTER 2

Number	Answer	Reference Page(s)	Number	Answer	Reference Page(s)
1.	c		25.	f	17
2.	h		26.	d	18
3.	g		27.	e	18
4.	b		28.	b	18
5.	d		29.	f	18
6.	f		30.	c	19
7.	a		31.	c	20
8.	e		32.	c	20
9.	f	12	33.	f	20
10.	b	13	34.	e	21
11.	c	13	35.	c	21
12.	e	13	36.	a	21
13.	e	14	37.	c	21
14.	b	14	38.	a	22
15.	b	14	39.	f	23
16.	c	14	40.	c	23
17.	a	15	41.	a	24
18.	f	15	42.	f	23
19.	e	15	43.	c	23
20.	c	15, 16	44.	e	24, 25
21.	b	16	45.	c	25
22.	d	16	46.	d	25
23.	f	17	47.	c	25
24.	a	17			

CHAPTER 3

Number	Answer	Reference Page(s)	Number	Answer	Reference Page(s)
1.	c	28	14.	f	35
2.	b	28	15.	d	34
3.	e	29	16.	d	35
4.	e	29	17.	f	35
5.	f	30	18.	c	35
6.	b	31	19.	c	35
7.	c	31, 32	20.	a	37, 38
8.	e	31	21.	b	36
9.	f	32	22.	d	37
10.	f	32, 33	23.	f	37
11.	c	33	24.	d	38
12.	b	33	25.	a	38
13.	c	34	26.	b	38

CHAPTER 4

Number	Answer	Reference Page(s)	Number	Answer	Reference Page(s)
1.	e	40	22.	b	45
2.	d	40	23.	c	45
3.	f	40	24.	d	45
4.	d	40	25.	f	46
5.	b	40	26.	c	46
6.	a	41	27.	e	47
7.	b	41	28.	e	48, 49
8.	a	41	29.	c	49
9.	e	41	30.	b	49
10.	b	42	31.	c	49
11.	d	41, 42	32.	a	50
12.	c	42	33.	b	50, 51
13.	b	43	34.	a	51
14.	f	42	35.	b	52
15.	d	43	36.	d	52
16.	e	43	37.	c	52
17.	c	43, 44	38.	d	52
18.	e	44	39.	a	53
19.	b	44	40.	b	53
20.	a	44	41.	c	53
21.	e	44			

CHAPTER 5

Number	Answer	Reference Page(s)	Number	Answer	Reference Page(s)
1.	c	56	21.	b	58, 59
2.	d	56	22.	d	60
3.	b	56	23.	b	60
4.	e	56	24.	e	60
5.	b	56	25.	a	60, 61
6.	f	56	26.	c	61
7.	e	57	27.	e	61
8.	f	57	28.	c	57, 61
9.	f	57	29.	b	61
10.	c	56	30.	c	62
11.	a	57	31	a	62
12.	c	57	32.	c	63
13.	e	57	33.	c	63
14.	e	57	34.	a	63
15.	d	57	35.	f	63
16.	e	57, 60	36.	d	64
17.	f	57	37.	b	64, 65
18.	d	57, 58	38.	f	65
19.	c	58	39.	e	64
20.	d	59	40.	d	65

CHAPTER 6

Number	Answer	Reference Page(s)	Number	Answer	Reference Page(s)
1.	a	67	18.	d	71
2.	a	68	19.	d	72
3.	d	68	20.	c	72
4.	b	68, 69	21.	f	72
5.	f	67, 68	22.	d	73
6.	e	68	23.	a	73
7.	d	69	24.	e	73
8.	f	69	25.	c	73
9.	d	69	26.	d	74
10.	e	69	27.	d	74
11.	d	69, 70	28.	b	75
12.	e	70	29.	c	67, 68, 75
13.	d	70	30.	a	74
14.	c	70	31.	a	74
15.	d	70	32.	f	75
16.	e	70, 71	33.	a	75
17.	f	71			

CHAPTER 7

Number	Answer	Reference Page(s)	Number	Answer	Reference Page(s)
1.	b	77	24.	d	81
2.	d	77	25.	c	81, 82
3.	e	77	26.	b	82
4.	c	78	27.	e	82
5.	b	78	28.	d	82
6.	d	78	29.	d	83
7.	e	78	30.	f	83
8.	f	78	31.	a	83
9.	c	78	32.	e	83
10.	c	78	33.	b	83
11.	a	78, 79	34.	f	83, 84
12.	a	78	35.	d	84, 85
13.	c	80	36.	d	85
14.	f	80	37	c	83, 84
15.	c	80	38.	c	85
16.	b	80	39.	c	85
17.	c	80	40.	e	85
18.	a	80	41.	f	85
19.	b	81	42.	c	85, 86
20.	d	81	43.	c	86
21.	e	81	44.	d	86
22.	d	81	45.	b	86
23.	f	81	46.	d	86

CHAPTER 8

Number	Answer	Reference Page(s)	Number	Answer	Reference Page(s)
1.	f	86, 88	15.	e	90
2.	b	88	16.	b	90
3.	f	88	17.	d	90
4.	f	89	18.	b	90, 91
5.	f	89	19.	b	91
6.	b	89	20.	c	91
7.	e	89	21.	a	91
8.	b	89	22.	d	91
9.	c	89	23.	c	91
10.	b	89	24.	b	91
11.	a	89	25.	f	91, 92
12.	d	89	26.	f	91, 92
13.	c	89	27.	a	89, 91
14.	c	89, 90	28.	b	91

CHAPTER 8 (Continued)

Number	Answer	Reference Page(s)	Number	Answer	Reference Page(s)
29.	b	89, 91	40.	d	96
30.	e	92	41.	d	96
31.	b	92	42.	f	94
32.	c	93	43.	b	89, 95
33.	b	93	44.	b	96
34.	a	93	45.	e	96
35.	a	93	46.	b	96, 97
36.	c	93	47.	f	97
37.	b	94, 95	48.	b	98
38.	c	95	49.	d	97
39.	d	95			

CHAPTER 9

Number	Answer	Reference Page(s)	Number	Answer	Reference Page(s)
1.	c	99, 100	30.	c	105
2.	b	99	31.	e	105
3.	d	99	32.	c	105
4.	d	100	33.	d	106
5.	c	100	34.	b	106
6.	e	100	35.	b	106
7.	d	100	36.	e	106
8.	c	101, 102	37.	e	106
9.	b	61, 101	38.	c	106, 107
10.	b	101	39.	b	107
11.	d	102, 103	40.	e	107
12.	a	101	41.	a	107
13.	c	101	42.	d	107
14.	f	102	43.	a	104, 107
15.	d	102	44.	d	107
16.	c	102	45.	d	108
17.	d	102	46.	e	108
18.	e	103	47.	d	108
19.	c	103	48.	b	108
20.	c	103	49.	a	109
21.	d	101, 103	50.	f	109
22.	a	103, 105	51.	b	109
23.	d	103	52.	b	109
24.	b	105, 262	53.	d	109
25.	e	104, 105	54.	d	109
26.	a	104, 105	55.	b	110
27.	d	105	56.	b	110
28.	f	104, 105	57.	c	110
29.	d	103, 105	58.	e	109, 110

CHAPTER 10

Number	Answer	Reference Page(s)	Number	Answer	Reference Page(s)
1.	c	112	15.	c	113, 114
2.	a	112	16.	b	114
3.	f	112	17.	b	114
4.	d	112, 113	18.	e	40, 114
5.	c	112	19.	c	114
6.	b	112	20.	a	114
7.	f	113	21.	d	115
8.	e	113	22.	c	115
9.	f	113	23.	c	45, 115
10.	c	51, 113	24.	e	116
11.	b	113	25.	b	115
12.	f	113	26.	d	115, 116
13.	d	113. 114	27.	b	114, 115
14.	a	113	28.	b	115, 116

CHAPTER 10 (Continued)

Number	Answer	Reference Page(s)	Number	Answer	Reference Page(s)
29.	d	116, 117	54.	c	122
30.	c	116, 117	55.	c	122
31.	e	116, 117	56.	c	122
32.	d	116	57.	a	123
33.	d	117	58.	d	123
34.	f	118	59.	c	123
35.	e	118, 119	60.	e	123, 124
36.	b	118, 120	61.	d	123, 124
37.	b	119	62.	c	124
38.	a	118, 119	63.	a	125
39.	c	115, 116, 118, 119	64.	c	125
40.	d	116, 118	65.	a	125, 126
41.	a	116, 118	66.	a	126
42.	f	118	67.	e	126
43.	d	121	68.	d	126
44.	b	120	69.	b	126
45.	a	120	70.	e	126
46.	c	120, 121	71.	d	126
47.	c	120	72.	e	127
48.	e	120, 121	73.	d	127
49.	a	120, 121	74.	d	127
50.	a	121	75.	e	127, 128
51.	c	121	76.	d	128
52.	d	122	77.	c	128
53.	c	122			

CHAPTER 11

Number	Answer	Reference Page(s)	Number	Answer	Reference Page(s)
1.	d	131	36.	e	138
2.	f	130	37.	f	138
3.	g	132	38.	b	139
4.	e	131	39.	d	139
5.	a	130	40.	c	139
6.	c	131	41.	c	139
7.	b	130	42.	d	135, 139
8.	b	130	43.	a	140
9.	e	130	44.	a	126, 139, 140
10.	d	131	45.	a	132, 140
11.	c	130, 131	46.	d	140
12.	f	132	47.	a	140
13.	b	132	48.	a	140
14.	b	133	49.	d	140
15.	e	132, 133	50.	c	137, 140
16.	b	133	51.	f	141
17.	d	131, 133	52.	b	141
18.	a	133, 134	53.	b	141
19.	b	134	54.	d	141
20.	e	134	55.	e	141, 142
21.	b	134	56.	e	141
22.	f	135	57.	b	142
23.	b	135	58.	c	142
24.	d	135	59.	d	142
25.	a	136, 139, 145	60.	f	142
26.	d	136	61.	a	142, 143
27.	c	136	62.	b	143
28.	b	137	63.	f	140, 143
29.	b	137	64.	f	141, 143
30.	d	137	65.	b	143
31.	d	132, 138	66.	d	143
32.	e	137, 138	67.	a	144
33.	f	138	68.	d	144
34.	c	138	69.	c	144
35.	a	138	70.	f	144

CHAPTER 11 (Continued)

Number	Answer	Reference Page(s)	Number	Answer	Reference Page(s)
71.	d	139, 144	77.	c	145
72.	e	137, 145	78.	c	145, 146
73.	b	144	79.	c	146
74.	a	145	80.	a	146
75.	e	145	81.	b	146
76.	d	145	82.	d	146

CHAPTER 12

Number	Answer	Reference (Page(s))	Number	Answer	Reference Page(s)
1.	e	148	30.	c	151, 152
2.	g	148	31.	d	151
3.	b	148	32.	e	152
4.	i	148	33.	b	152
5.	c	148	34.	b	152
6.	f	148	35.	b	152, 153
7.	a	148	36.	f	153
8.	h	148	37.	d	153
9.	d	148	38.	d	153
10.	c	148	39.	a	153
11.	d	148, 149	40.	a	153
12.	c	149	41.	e	154
13.	d	149	42.	b	154
14.	b	149	43.	b	154
15.	f	149	44.	c	154
16.	d	23, 149	45.	d	154
17.	f	149	46.	e	155
18.	a	149	47.	b	154, 155
19.	b	149	48.	b	155
20.	d	149	49.	c	155, 156
21.	e	149	50.	d	156
22.	b	150	51.	e	156
23.	f	150	52.	b	156
24.	a	150	53.	f	156
25.	d	150	54.	f	157
26.	e	150	55.	d	157
27.	d	150	56.	d	157
28.	e	151	57.	d	157
29.	d	151	58.	b	157

CHAPTER 13

Number	Answer	Reference Page(s)	Number	Answer	Reference Page(s)
1.	f		20.	d	152, 161
2.	n		21.	c	161
3.	e		22.	a	161
4.	a		23.	d	161
5.	g		24.	d	161
6.	d		25.	c	161, 162
7.	j		26.	b	162
8.	i		27.	b	162
9.	h		28.	b	162
10.	l		29.	b	161
11.	b		30.	c	162, 163
12.	k		31.	a	163
13.	c		32.	T	163
14.	m		33.	a	163
15.	c	160	34.	c	167
16.	d	160	35.	T	164
17.	e	160	36.	F	164
18.	b	160	37.	F	164
19.	a	160	38.	d	165

CHAPTER 13 (Continued)

Number	Answer	Reference Page(s)	Number	Answer	Reference Page(s)
39.	a	165	58.	d	168
40.	c	165	59.	b	168
41.	f	165	60.	T	168
42.	d	165	61.	c	168
43.	c	164, 165	62.	e	169
44.	d	165	63.	T	169
45.	b	166	64.	f	169
46.	b	166, 167	65.	b	169
47.	c	166	66.	e	168
48.	d	166, 167	67.	c	169
49.	d	167	68.	a	171
50.	b	164, 167	69.	f	170
51.	f	164, 167	70.	b	171
52.	c	164	71.	d	170
53.	d	164	72.	c	172
54.	b	164	73.	f	172
55.	b	164	74.	d	173
56.	e	164	75.	a	173
57.	b	164			

CHAPTER 14

Number	Answer	Reference Page(s)	Number	Answer	Reference Page(s)
1.	b	177	22.	f	182
2.	e	177	23.	f	181
3.	c	177	24.	e	182
4.	a	177	25.	a	181, 182
5.	d	177	26.	c	182
6.	d	177	27.	f	183
7.	b	177	28.	a	183
8.	a	177	29.	c	183
9.	c	177	30.	a	183
10.	e	177	31.	f	183
11.	a	178	32.	b	171, 183
12.	c	177, 178	33.	b	183, 184
13.	b	178, 180	34.	f	183, 184
14.	e	179	35.	a	183
15.	f	179	36.	F	187
16.	a	180	37.	F	187
17.	f	180	38.	d	185
18.	d	163, 179	39.	b	162, 164, 187
19.	b	180	40.	d	187
20.	c	180	41.	c	188
21.	c	182	42.	c	162, 164, 188

CHAPTER 15

Number	Answer	Reference Page(s)	Number	Answer	Reference Page(s)
1.	a	191, 192	13.	b	193, 194
2.	b	190	14.	d	194
3.	f	191	15.	b	194
4.	b	191	16.	b	194
5.	f	191, 192	17.	e	194
6.	f	192	18.	a	194
7.	f	192	19.	e	195
8.	d	192	20.	c	195, 196
9.	a	192	21.	d	196
10.	e	192	22.	e	195
11.	c	190, 191	23.	d	196
12.	d	193, 194	24.	d	196

CHAPTER 16

Number	Answer	Reference Page(s)	Number	Answer	Reference Page(s)
1.	d	197	22.	f	203, 204
2.	b	198	23.	b	203
3.	d	198	24.	d	204
4.	b	198, 199	25.	b	203, 204
5.	c	198, 199	26.	c	204
6.	e	198, 199	27.	d	205
7.	f	198, 199	28.	b	204
8.	a	198, 199	29.	d	204
9.	d	198, 199	30.	e	207
10.	a	198, 199	31.	e	206
11.	b	199	32.	f	205, 207
12.	f	199	33.	e	207
13.	a	199	34.	b	207
14.	a	199	35.	e	208
15.	d	200	36.	d	208
16.	b	200, 201	37.	d	210
17.	f	200	38.	c	212
18.	b	201	39.	d	211
19.	e	202	40.	c	212
20.	c	200-203	41.	b	211
21.	a	203			

CHAPTER 17

Number	Answer	Reference Page(s)	Number	Answer	Reference Page(s)
1.	e	213	10.	c	216
2.	e	213	11.	a	216, 217
3.	d	213	12.	f	217
4.	d	214	13.	c	217
5.	d	214-216	14.	d	216
6.	c	215	15.	b	219
7.	b	214, 215	16.	f	218, 219
8.	e	214-216	17.	c	219
9.	f	184, 214	18.	c	218

CHAPTER 18

Number	Answer	Reference Page(s)	Number	Answer	Reference Page(s)
1.	b	222, 224	24.	a	230
2.	b	223	25.	a	230
3.	e	224	26.	b	231
4.	b	224	27.	d	231
5.	e	223	28.	d	231
6.	f	225	29.	a	230, 231
7.	c	224	30.	e	232
8.	a	225	31.	a	232
9.	e	225	32.	c	231, 232
10.	b	226, 227	33.	b	231, 232
11.	d	227, 228	34.	c	232
12.	d	228	35.	d	231, 232
13.	a	227, 228	36.	b	231, 232
14.	e	228	37.	b	232
15.	b	228	38.	c	232
16.	e	229, 230	39.	c	232, 233
17.	b	229, 230	40.	b	233
18.	b	230	41.	e	233, 234
19.	c	230	42.	b	234
20.	c	225, 230	43.	a	233, 234
21.	c	225, 230	44.	b	233
22.	a	230	45.	c	235
23.	a	230	46.	f	235

CHAPTER 19

Number	Answer	Reference Page(s)	Number	Answer	Reference Page(s)
1.	c	237	29.	b	241
2.	e	237	30.	c	242
3.	f	237	31.	a	241
4.	f	237	32.	b	242
5.	a	237	33.	d	242
6.	d	237	34.	c	242
7.	a	238	35.	f	242
8.	e	238	36.	c	243
9.	b	238	37.	e	244
10.	d	238	38.	c	244
11.	e	238, 242	39.	e	244
12.	c	238	40.	e	244
13.	d	238	41.	a	245
14.	c	238	42.	f	244
15.	d	238	43.	f	244, 245
16.	b	239	44.	a	245
17.	d	239	45.	d	245
18.	a	239	46.	b	245
19.	b	240	47.	a	245
20.	d	240	48.	b	245
21.	d	240	49.	d	246
22.	e	239	50.	e	247
23.	b	239, 240	51.	c	247
24.	c	239, 240	52.	b	247
25.	d	240, 241	53.	d	247
26.	f	240, 241	54.	a	247
27.	b	240, 241	55.	c	248
28.	c	240	56.	b	248

CHAPTER 20

Number	Answer	Reference Page(s)	Number	Answer	Reference Page(s)
1.	c	250	29.	a	258, 259
2.	b	251	30.	b	258, 259
3.	e	250	31.	c	259
4.	e	251	32.	d	259
5.	a	251	33.	c	259
6.	b	252	34.	d	259
7.	c	252	35.	b	260
8.	c	253	36.	b	260
9.	f	253	37.	c	260
10.	c	253	38.	a	260
11.	d	254	39.	f	260
12.	d	254	40.	b	261
13.	f	254	41.	f	260
14.	a	254	42.	c	261
15.	e	254	43.	f	261
16.	b	255	44.	a	261
17.	a	255	45.	a	261
18.	b	254	46.	c	261
19.	d	254, 255	47.	a	261
20.	e	255	48.	f	262
21.	d	256	49.	d	262
22.	b	256	50.	a	262, 263
23.	e	256	51.	d	262
24.	f	257	52.	c	262
25.	c	257	53.	e	262
26.	d	258	54.	f	263
27.	c	258	55.	b	263
28.	e	258	56.	b	263

CHAPTER 21

Number	Answer	Reference Page(s)	Number	Answer	Reference Page(s)
1.	f	265	31.	d	271
2.	a	265	32.	b	271, 272
3.	c	265	33.	b	272
4.	f	266	34.	e	272
5.	e	266	35.	a	272
6.	d	266	36.	c	272
7.	a	266, 267	37.	a	272
8.	d	266	38.	c	272
9.	f	266	39.	d	272, 273
10.	d	266	40.	c	273
11.	c	266	41.	d	272, 273
12.	d	266, 267	42.	c	269, 273
13.	b	267	43.	d	269, 273
14.	a	267	44.	d	273
15.	c	268	45.	d	273
16.	c	268	46.	b	273
17.	e	268, 269	47.	f	273
18.	d	268	48.	b	273
19.	a	269	49.	c	274
20.	c	269	50.	a	274
21.	c	269	51.	a	274
22.	a	269	52.	d	275
23.	c	269	53.	e	274
24.	b	183, 269	54.	a	275
25.	d	269	55.	f	275
26.	d	270	56.	f	276
27.	b	270	57.	e	275, 276
28.	d	271	58.	c	276, 277
29.	c	271	59.	d	277
30.	d	271	60.	c	277

CHAPTER 22

Number	Answer	Reference Page(s)	Number	Answer	Reference Page(s)
1.	a	279	15.	d	285
2.	d	279, 280	16.	d	287
3.	d	279	17.	b	288
4.	c	280	18.	b	288
5.	a	280	19.	d	288
6.	c	281	20.	b	289
7.	d	281	21.	c	289
8.	c	282	22.	b	289, 290
9.	d	283	23.	c	290
10.	d	283	24.	b	290, 291
11.	d	285	25.	b	291
12.	a	285	26.	c	291
13.	e	284	27.	a	291
14.	b	285			

CHAPTER 23

Number	Answer	Reference Page(s)	Number	Answer	Reference Page(s)
1.	d	295	10.	b	296
2.	d	295	11.	c	297
3.	b	295	12.	a	297
4.	a	295	13.	c	297
5.	d	295	14.	a	297
6.	b	296	15.	d	297
7.	b	296	16.	d	298
8.	c	296	17.	e	299, 300
9.	b	296	18.	a	299

CHAPTER 23 (Continued)

Number	Answer	Reference Page(s)	Number	Answer	Reference Page(s)
19.	e	299	36.	d	304
20.	e	300	37.	d	304
21.	a	300	38.	f	304
22.	a	300	39.	e	303, 304
23.	e	300	40.	a	304, 305
24.	d	300	41.	a	306
25.	b	301	42.	b	306
26.	d	301	43.	c	306
27.	e	301	44.	c	306
28.	d	301	45.	b	307
29.	c	302, 303	46.	d	307
30.	b	302	47.	c	308
31.	b	303	48.	e	308
32.	a	303	49.	f	309
33.	d	303	50.	b	309
34.	b	303	51.	c	308
35.	d	303, 304			

CHAPTER 24

Number	Answer	Reference Page(s)	Number	Answer	Reference Page(s)
1.	f	311	21.	c	314
2.	e	311	22.	b	314
3.	a	311	23.	c	315
4.	b	311	24.	d	244, 315
5.	d	312	25.	a	315, 316
6.	d	312	26.	e	315, 316
7.	e	312	27.	a	316
8.	d	313	28.	e	316
9.	c	312	29.	a	315
10.	b	313	30.	c	316, 317
11.	c	239, 313	31.	b	317
12.	b	313	32.	a	317
13.	b	313	33.	c	317
14.	b	313	34.	d	317
15.	b	313	35.	e	317
16.	c	313	36.	f	318
17.	a	313	37.	c	316, 318
18.	b	313	38.	d	319
19.	b	314	39.	d	319
20.	f	314	40.	a	318

CHAPTER 25

Number	Answer	Reference Page(s)	Number	Answer	Reference Page(s)
1.	d	320	17.	c	323
2.	e	320	18.	b	323
3.	b	320	19.	d	323
4.	c	321	20.	b	323
5.	a	321	21.	c	323
6.	c	321	22.	d	323, 324
7.	b	321	23.	b	324
8.	c	322	24.	c	324
9.	a	322	25.	c	324
10.	a	322	26.	b	325
11.	c	322	27.	b	326
12.	d	322	28.	c	325
13.	e	322	29.	d	326
14.	d	322	30.	d	327
15.	b	323	31.	b	327
16.	a	323	32.	e	328

CHAPTER 25 (Continued)

Number	Answer	Reference Page(s)	Number	Answer	Reference Page(s)
33.	b	329	36.	b	329
34.	c	329	37.	d	329, 330
35.	d	329			

CHAPTER 26

Number	Answer	Reference Page(s)	Number	Answer	Reference Page(s)
1.	d	332	23.	a	338
2.	b	332	24.	d	338
3.	f	332	25.	d	338
4.	a	333	26.	d	338
5.	c	333	27.	d	338
6.	b	333, 334	28.	c	339
7.	c	334	29.	c	339
8.	c	334	30.	d	340
9.	a	334	31.	b	340
10.	b	334	32.	e	340
11.	a	335	33.	e	340
12.	e	335	34.	a	340, 341
13.	b	335, 336	35.	c	341
14.	c	336	36.	e	341
15.	c	336	37.	c	341
16.	c	336	38.	a	341
17.	b	337	39.	d	341
18.	d	337	40.	c	342
19.	d	337	41.	b	342
20.	d	337	42.	c	343
21.	c	337	43.	f	343
22.	c	338			

CHAPTER 27

Number	Answer	Reference Page(s)	Number	Answer	Reference Page(s)
1.	c	345	20.	a	349
2.	b	345	21.	b	349
3.	a	345	22.	a	349
4.	e	346	23.	d	350
5	d	346	24.	c	350
6.	c	346	25.	f	351
7.	f	346	26.	b	350, 351
8.	b	346	27.	c	352
9.	f	347	28.	c	354
10.	b	347	29.	b	352
11.	d	347	30.	b	352
12.	b	347	31.	c	352, 353
13.	b	347	32.	d	353
14.	d	347, 348	33.	a	354
15.	c	348	34.	b	353
16.	c	348	35.	e	353
17.	d	348	36.	f	355
18.	b	348, 349	37.	b	355
19.	d	349	38.	d	355

CHAPTER 28

Number	Answer	Reference Page(s)	Number	Answer	Reference Page(s)
1.	c	357	23.	c	365
2.	b	357	24.	d	365
3.	c	358	25.	c	365
4.	a	358	26.	d	365
5.	d	358	27.	f	365
6.	a	358	28.	b	365
7.	c	358	29.	f	366
8.	f	359	30.	b	366
9.	a	359, 360	31.	a	367
10.	d	359	32.	d	366
11.	d	359	33.	b	367
12.	b	359	34.	d	367
13.	e	359	35.	b	367
14.	c	360, 361	36.	c	367
15.	b	362	37.	c	367, 368
16.	a	362, 363	38.	d	368
17.	b	360, 361	39.	d	368
18.	f	361	40.	d	368
19.	d	363, 937	41.	b	368
20.	e	364	42	f	369
21.	d	364	43.	c	369
22.	b	364			

CHAPTER 29

Number	Answer	Reference Page(s)	Number	Answer	Reference Page(s)
1.	e	130, 370	28.	a	377
2.	d	370, 371	29.	b	377
3.	d	371	30.	d	377
4.	b	371	31.	c	377
5.	a	371	32.	d	378
6.	b	371	33.	b	378
7.	c	371	34.	b	378
8.	b	372	35.	a	378
9.	e	372	36.	d	378, 379
10.	a	372	37.	d	379
11.	c	373	38.	d	379
12.	c	372	39.	a	379
13.	f	373	40.	a	379, 380
14.	a	373, 374	41.	d	380
15.	a	374	42.	c	380
16.	e	374, 375	43.	f	380
17.	a	374	44.	c	380
18.	c	374	45.	b	380
19.	b	375	46.	c	380, 381
20.	e	375	47.	a	381
21.	d	376	48.	b	381
22.	d	376	49.	d	381
23.	c	376	50.	d	381
24.	f	376	51.	f	381
25.	b	376	52.	a	381
26.	d	376	53.	b	381
27.	a	377	54.	b	382

CHAPTER 30

Number	Answer	Reference Page(s)	Number	Answer	Reference Page(s)
1.	b	386	18.	d	391
2.	b	386	19.	c	392
3.	e	387	20.	e	392
4.	d	386	21.	f	393
5.	f	386	22.	c	393
6.	d	387	23.	c	393
7.	a	388	24.	a	394
8.	c	388	25.	b	394
9.	c	389	26.	a	394
10.	b	388	27.	c	394
11.	b	389	28.	f	395
12.	f	389	29.	b	395
13.	a	389, 390	30.	a	395
14.	e	390	31.	c	395
15.	d	391	32.	a	395, 396
16.	b	391	33.	c	396
17.	b	390	34.	d	396

CHAPTER 31

Number	Answer	Reference Page(s)	Number	Answer	Reference Page(s)
1.	b	397	32.	e	405
2.	c	397	33.	e	405
3.	d	397	34.	e	405
4.	a	397	35.	c	406
5.	a	397	36.	d	406
6.	b	397, 398	37.	f	406
7.	b	398	38.	b	406
8.	e	398	39.	d	406
9.	b	398	40.	b	406, 407
10.	b	397, 398	41.	b	406
11.	b	398, 399	42.	f	407
12.	c	399	43.	a	408
13.	a	399	44.	a	407
14.	f	399	45.	c	408
15.	b	399	46.	b	407
16.	b	400	47.	c	407
17.	b	400	48.	f	409
18.	c	400	49.	b	409
19.	b	401	50.	b	409
20.	c	401	51.	f	409, 410
21.	a	402	52.	c	410
22.	c	402	53.	a	410
23.	b	402	54.	a	410
24.	c	402	55.	d	410
25.	b	402	56.	c	411
26.	e	403	57.	c	395, 410
27.	a	404	58.	e	412
28.	c	403	59.	d	412
29.	b	404	60.	c	412
30.	a	404	61.	d	412
31.	d	404			

CHAPTER 32

Number	Answer	Reference Page(s)	Number	Answer	Reference Page(s)
1.	a	414	22.	c	418
2.	b	414	23.	c	418
3.	c	415	24.	a	418
4.	d	415	25.	d	418
5.	d	415	26.	b	418
6.	c	415	27.	f	418
7.	b	416	28.	d	418, 419
8.	d	415, 416	29.	b	419
9.	c	416	30.	b	419
10.	b	416	31.	b	418, 420
11.	c	416	32.	f	420
12.	f	416	33.	d	420
13.	f	416	34.	b	421
14.	b	416	35.	e	420
15.	c	417	36.	d	421
16.	c	417	37.	b	421
17.	d	417	38.	d	422
18.	c	417	39.	b	422
19.	c	417	40.	f	422
20.	d	417	41.	c	422
21.	b	417			

CHAPTER 33

Number	Answer	Reference Page(s)	Number	Answer	Reference Page(s)
1.	c	424	31.	e	40, 430
2.	d	424	32.	b	431
3.	d	424	33.	d	431
4.	b	424	34.	c	431
5.	a	424	35.	d	432
6.	b	424	36.	c	432
7.	d	424	37.	a	432
8.	e	425	38.	f	433
9.	c	425	39.	e	433
10.	c	425	40.	c	433
11.	b	425	41.	d	433
12.	c	425	42.	e	433
13.	e	425	43.	d	434
14.	a	425, 426	44.	b	434
15.	b	425, 426	45.	e	434
16.	c	426	46.	d	434
17.	e	426	47.	a	433
18.	b	426	48.	d	434
19.	c	426, 427	49.	c	434, 435
20.	a	427	50.	d	435
21.	c	427	51.	a	434, 435
22.	b	427	52.	b	435
23.	b	428	53.	e	435
24.	b	429	54.	e	436
25.	b	429	55.	b	437
26.	e	429	56.	d	437
27.	a	430	57.	d	437
28.	f	430	58.	e	437
29.	c	430	59.	e	436
30.	e	430			

CHAPTER 34

Number	Answer	Reference Page(s)	Number	Answer	Reference Page(s)
1.	c	438	54.	d	448
2.	b	439	55.	b	448
3.	d	439	56.	d	448
4.	f	439	57.	e	449
5.	c	439	58.	e	449
6.	f	439	59.	a	449
7.	f	440	60.	e	449
8.	d	440	61.	d	450
9.	c	440	62.	a	450
10.	f	440	63.	e	450
11.	c	441	64.	c	450
12.	d	441	65.	b	450
13.	b	441	66.	d	450, 451
14.	d	441	67.	c	451
15.	c	442	68.	d	451
16.	b	442	69.	b	450, 451
17.	f	442	70.	a	451
18.	c	442	71.	c	451
19.	a	443	72.	a	451
20.	d	443	73.	a	451, 452
21.	d	443	74.	b	451
22.	a	443	75.	e	453
23.	f	443	76.	f	453
24.	c	443	77.	e	453
25.	d	440, 443	78.	b	453, 454
26.	b	440, 443	79.	d	452, 453
27.	b	444	80.	b	453
28.	f	444	81.	f	452
29.	b	443	82.	c	453
30.	d	443	83.	d	453
31.	e	443	84.	a	453
32.	d	444	85.	c	453
33.	f	444	86.	d	453
34.	f	444	87.	e	454
35.	d	444	88.	a	453, 454
36.	a	445	89.	d	453
37.	a	445	90.	b	454
38.	b	445	91.	c	453, 454
39.	e	445	92.	e	453
40.	a	445	93.	e	454
41.	c	445	94.	a	454
42.	f	446, 447	95.	d	454
43.	b	446	96.	a	454
44.	e	446	97.	a	454
45.	b	446	98.	b	454
46.	e	447	99.	e	454
47.	e	447	100.	a	454
48.	e	447	101.	b	454
49.	c	447	102.	d	454
50.	d	447	103.	a	454
51.	d	447	104.	d	455
52.	c	448	105.	a	455
53.	b	448	106.	b	455

CHAPTER 35

Number	Answer	Reference Page(s)	Number	Answer	Reference Page(s)
1.	d	456, 457	9.	c	458
2.	e	457	10.	f	458
3.	b	457	11.	b	458
4.	c	457	12.	b	458
5.	d	457	13.	e	458
6.	a	457	14.	d	458
7.	b	458	15.	e	458
8.	c	458	16.	e	458, 459

CHAPTER 35 (Continued)

Number	Answer	Reference Page(s)	Number	Answer	Reference Page(s)
17.	e	459	39.	f	462, 463
18.	d	459	40.	a	464
19.	d	460	41.	c	464
20.	f	460	42.	c	464
21.	f	461	43.	d	464
22.	f	461	44.	f	464
23.	f	461	45.	d	465
24.	b	461	46.	a	465
25.	a	461	47.	b	465, 466
26.	b	461	48.	c	466
27.	c	461	49.	c	466
28.	b	461	50.	a	467
29.	f	461, 462	51.	b	467
30.	d	461, 462	52.	c	468
31.	c	462	53.	a	468
32.	c	462	54.	d	468
33.	b	462, 463	55.	e	469
34.	f	462, 463	56.	c	469
35.	a	463	57.	b	470
36.	e	462	58.	a	470
37.	c	463	59.	a	470
38.	c	463			

CHAPTER 36

Number	Answer	Reference Page(s)	Number	Answer	Reference Page(s)
1.	e	472, 473	26.	d	477
2.	e	425, 472	27.	d	477, 478
3.	d	473	28.	b	478
4.	c	473	29.	e	478
5.	b	474	30.	c	478
6.	c	474	31.	d	478, 479
7.	c	474	32.	c	478
8.	c	474	33.	c	479
9.	f	474	34.	b	479
10.	a	474	35.	f	479
11.	e	474	36.	b	479
12.	a	474	37.	c	480
13.	b	475	38.	c	480
14.	c	475	39.	b	481
15.	c	475, 476	40.	f	481
16.	c	475	41.	c	481, 482
17.	b	475	42.	b	483
18.	b	476	43.	d	481, 482
19.	c	476	44.	d	482
20.	f	476	45.	d	483
21.	f	476	46.	d	483
22.	c	477	47.	b	483, 484
23.	c	477	48.	a	484
24.	d	477	49.	f	484
25.	a	476	50.	c	484

CHAPTER 37

Number	Answer	Reference Page(s)	Number	Answer	Reference Page(s)
1.	d	485	8.	c	486
2.	d	485	9.	a	486
3.	f	485	10.	e	486
4.	c	485	11.	c	486
5.	d	485	12.	b	487
6.	b	485	13.	f	487
7.	c	485	14.	a	487

CHAPTER 37 (Continued)

Number	Answer	Reference Page(s)	Number	Answer	Reference Page(s)
15.	d	487	47.	b	494
16.	b	485, 487	48.	b	494
17.	c	487	49.	b	494
18.	e	487	50.	f	494
19.	b	488	51.	e	494, 495
20.	f	489	52.	e	495
21.	a	489	53.	c	495
22.	d	488	54.	f	495
23.	d	488	55.	d	495
24.	d	488, 489	56.	a	495
25.	e	488	57.	c	496
26.	c	488	58.	c	496
27.	a	489	59.	f	496
28.	d	490	60.	d	496
29.	e	490	61.	f	496
30.	a	490	62.	d	497
31.	a	491	63.	a	497
32.	e	491	64.	f	497
33.	f	492	65.	d	497
34.	d	492	66.	a	497
35.	c	492	67.	c	498
36.	b	492	68.	c	498
37.	b	492	69.	c	498
38.	d	492	70.	c	498
39.	a	492	71.	d	498
40.	b	492, 493	72.	c	499
41.	a	493	73.	d	499
42.	c	493	74.	b	499
43.	b	493	75.	c	499
44.	b	493	76.	b	500
45.	d	494	77.	d	500
46.	c	494			

CHAPTER 38

Number	Answer	Reference Page(s)	Number	Answer	Reference Page(s)
1.	c	501	30.	f	507
2.	b	501	31.	c	508, 509
3.	c	501	32.	c	508
4.	e	502	33.	c	507
5.	b	502	34.	c	509
6.	b	502	35.	d	509
7.	b	502	36.	c	509
8.	d	502	37.	d	510
9.	d	502	38.	b	510
10.	c	503	39.	a	510
11.	e	503	40.	f	510
12.	f	503	41.	d	511
13.	e	503	42.	c	510
14.	f	503	43.	e	510
15.	b	504	44.	e	511
16.	c	504	45.	c	511
17.	a	504	46.	f	511
18.	f	505	47.	b	511
19.	c	505	48.	f	511
20.	b	505	49.	b	511, 512
21.	f	438, 506	50.	d	512
22.	f	506	51.	a	512
23.	a	506	52.	b	512
24.	e	506	53.	d	512
25.	e	506	54.	b	512
26.	f	506	55.	d	513
27.	d	507	56.	c	513
28.	c	507	57.	d	513
29.	d	507			

CHAPTER 39

Number	Answer	Reference Page(s)	Number	Answer	Reference Page(s)
1.	c	516	37.	d	521
2.	c	516	38.	e	521
3.	a	516	39.	a	521
4.	d	517	40.	b	522
5.	d	516	41.	d	522
6.	a	517	42.	e	522
7.	b	517	43.	d	522
8.	c	517	44.	a	522
9.	f	517	45.	e	523
10.	b	517	46.	a	523
11.	c	517	47.	c	523
12.	e	517	48.	c	523, 524
13.	d	517	49.	f	523, 524
14.	d	517, 518	50.	d	524
15.	c	518	51.	b	525
16.	a	518	52.	a	525
17.	c	518	53.	f	524, 525
18.	e	518	54.	d	525
19.	c	518	55.	e	525
20.	d	518	56.	d	525
21.	d	518	57.	d	524
22.	d	519	58.	b	523, 525
23.	b	519	59.	a	525
24.	a	519	60.	c	526
25.	d	519	61.	a	526
26.	b	519, 520	62.	b	526
27.	f	520	63.	f	525
28.	b	520	64.	d	527
29.	a	520	65.	d	526
30.	a	520	66.	d	526
31.	b	521	67.	d	527
32.	e	521	68.	f	527, 528
33.	b	521	69.	c	527
34.	d	521	70.	a	528
35.	c	521	71.	b	528
36.	e	521	72.	e	529

CHAPTER 40

Number	Answer	Reference Page(s)	Number	Answer	Reference Page(s)
1.	c	530	26.	d	535
2.	f	530	27.	d	535
3.	b	530	28.	c	535
4.	e	530, 531	29.	e	536
5.	d	531	30.	a	536
6.	b	531	31.	e	536
7.	c	531	32.	d	537
8.	b	531	33.	c	524, 536
9.	f	532	34.	d	537
10.	d	532	35.	b	537
11.	b	532, 533	36.	c	539
12.	e	533	37.	f	539
13.	e	532, 533	38.	f	539
14.	e	532	39.	c	540
15.	c	531, 532, 533	40.	b	540
16.	d	533	41.	d	540
17.	c	534	42.	c	540
18.	c	534	43.	d	540
19.	c	534	44.	f	532, 541
20.	d	534	45.	c	541
21.	b	534	46.	e	540
22.	a	534, 535	47.	e	540, 541
23.	e	534, 535	48.	a	541
24.	e	535	49.	a	541
25.	c	535	50.	d	541

CHAPTER 41

Number	Answer	Reference Page(s)	Number	Answer	Reference Page(s)
1.	d	543	45.	b	549
2.	a	543	46.	b	550
3.	b	543	47.	c	549, 550
4.	d	543	48.	a	550
5.	d	544	49.	b	550
6.	a	544	50.	c	550
7.	c	544	51.	d	550
8.	d	544	52.	a	550, 551
9.	c	544	53.	e	550
10.	d	544, 545	54.	d	551
11.	e	545	55.	d	551
12.	c	545	56.	d	551
13.	c	545	57.	b	552
14.	d	546	58.	b	552
15.	a	546	59.	b	552
16.	d	545, 546	60.	e	552
17.	c	546	61.	a	553
18.	c	546	62.	e	553
19.	b	546	63.	c	553
20.	d	546, 547	64.	e	553
21.	f	547	65.	e	553
22.	b	547	66.	b	553
23.	f	547	67.	a	553
24.	a	547	68.	c	554
25.	c	547	69.	d	553, 554
26.	d	547	70.	a	554
27.	a	547	71.	e	554
28.	a	547	72.	c	554
29.	c	547	73.	e	554
30.	e	547	74.	b	554
31.	f	547	75.	d	555
32.	b	548	76.	a	554
33.	b	548	77.	f	555
34.	c	548	78.	c	555
35.	b	548	79.	d	555
36.	b	308, 548	80.	a	555
37.	a	548	81.	c	555
38.	e	548	82.	b	555
39.	a	548	83.	b	556
40.	f	549	84.	a	535, 556
41.	c	549	85.	e	556
42.	d	549	86.	a	556
43.	d	549	87.	b	556
44.	b	550			

CHAPTER 42

Number	Answer	Reference Page(s)	Number	Answer	Reference Page(s)
1.	c	557, 558	18.	d	560, 561
2.	f	557, 558	19.	e	560
3.	d	558	20.	d	561
4.	a	525, 558	21.	b	561
5.	a	558	22.	d	561
6.	f	558	23.	b	561
7.	d	517, 558	24.	a	561
8.	c	559	25.	e	561
9.	e	559	26.	c	562
10.	d	559	27.	c	562
11.	c	559	28.	c	562
12.	a	559	29.	c	562
13.	a	559	30.	e	562
14.	e	559, 560	31.	b	562
15.	d	559	32.	d	563
16.	b	560	33.	d	563
17.	f	560	34.	a	563

CHAPTER 42 (Continued)

Number	Answer	Reference Page(s)	Number	Answer	Reference Page(s)
35.	a	564	48.	b	566, 567
36.	c	564	49.	b	567
37.	a	564	50.	b	566, 567
38.	c	564	51.	f	567
39.	c	564	52.	a	567
40.	a	564, 565	53.	c	568, 569
41.	b	565	54.	d	569
42.	f	566	55.	e	561, 569
43.	c	566	56.	f	569
44.	b	566	57.	f	569
45.	d	566	58.	a	569
46.	e	566	59.	d	571
47.	b	566, 567	60.	d	571

CHAPTER 43

Number	Answer	Reference Page(s)	Number	Answer	Reference Page(s)
1.	d	572	19.	e	579
2.	d	572	20.	f	579, 581
3.	a	572	21.	f	579
4.	f	572	22.	c	579, 581
5.	e	572	23.	c	579, 580
6.	f	574	24.	f	580
7.	e	574	25.	d	580
8.	a	574	26.	f	580
9.	e	574, 575	27.	b	580
10.	a	574	28.	a	581
11.	c	574	29.	d	581
12.	a	575	30.	b	581
13.	f	575	31.	b	581
14.	f	575	32.	e	582
15.	c	576	33.	a	582
16.	b	577	34.	c	582
17.	c	578	35.	b	582, 583
18.	d	578			

CHAPTER 44

Number	Answer	Reference Page(s)	Number	Answer	Reference Page(s)
1.	f	586	24.	e	591
2.	b	586	25.	e	591
3.	f	586	26.	c	592
4.	a	586, 587	27.	c	592
5.	c	586	28.	e	592
6.	b	586	29.	d	593
7.	e	587	30.	d	593
8.	b	588	31.	b	593
9.	d	587	32.	e	593
10.	f	588	33.	d	594
11.	c	588	34.	e	594
12.	b	588	35.	d	594
13.	d	588	36.	c	594
14.	a	588	37.	f	594
15.	b	589	38.	c	595
16.	f	589	39.	f	595
17.	b	589	40.	b	595
18.	b	589	41.	e	595, 596
19.	d	589, 590	42.	f	595, 596
20.	c	590	43.	d	596
21.	b	590	44.	e	596
22.	c	590, 591	45.	d	596
23.	e	591			

CHAPTER 45

Number	Answer	Reference Page(s)	Number	Answer	Reference Page(s)
1.	a	598	18.	e	602
2.	c	531, 598	19.	e	602, 603
3.	d	598	20.	d	602
4.	c	599	21.	e	603, 604
5.	f	599	22.	c	604
6.	a	599	23.	b	604
7.	d	600	24.	e	604
8.	a	600	25.	e	604
9.	b	600	26.	b	604
10.	a	600	27.	b	604
11.	d	600	28.	e	605
12.	e	600	29.	a	605
13.	c	601	30.	d	605
14.	f	601	31.	d	606
15.	e	601	32.	d	606
16.	d	601, 602	33.	d	606
17.	a	602			

CHAPTER 46

Number	Answer	Reference Page(s)	Number	Answer	Reference Page(s)
1.	a	608	39.	c	618
2.	d	608, 609	40.	d	618
3.	f	608	41.	d	618
4.	b	609	42.	c	618
5.	c	610	43.	b	618
6.	f	610	44.	c	619
7.	a	610	45.	a	619
8.	d	610	46.	c	619
9.	a	611	47.	e	619
10.	c	611	48.	f	619
11.	f	612	49.	f	619, 620
12.	d	612	50.	d	619, 620
13.	e	612	51.	c	619
14.	d	612	52.	e	620
15.	f	613	53.	b	620
16.	a	614	54.	e	620
17.	f	613, 614	55.	c	620
18.	b	614	56.	f	620
19.	b	614	57.	d	621
20.	d	614	58.	b	621
21.	c	614	59.	e	621
22.	e	614	60.	c	621
23.	d	149, 614	61.	b	621
24.	d	615	62.	c	622
25.	c	615	63.	c	622
26.	f	615	64.	a	622
27.	d	615	65.	f	622
28.	c	615	66.	f	623
29.	d	615	67.	c	623
30.	d	615	68.	a	623
31.	c	616	69.	d	623, 624
32.	d	616	70.	a	624
33.	a	616	71.	a	624
34.	c	616, 617	72.	e	624
35.	f	616, 617	73.	a	625
36.	e	617	74.	b	625
37.	a	616	75.	a	625
38.	d	617	76.	c	625

CHAPTER 47

Number	Answer	Reference Page(s)	Number	Answer	Reference Page(s)
1.	a	626	18.	d	633
2.	f	626	19.	b	634
3.	c	626, 627	20.	a	634
4.	d	627	21.	c	635
5.	d	627	22.	c	636
6.	b	627	23.	c	636
7.	c	628	24.	f	636
8.	e	628	25.	c	636
9.	b	628	26.	d	637
10.	b	629, 630	27.	d	637
11.	a	630	28.	b	637, 638
12.	c	630	29.	f	638
13.	d	630, 631	30.	c	638
14.	b	631	31.	a	638
15.	e	632	32.	c	638
16.	c	632, 633	33.	c	639
17.	c	633			

CHAPTER 48

Number	Answer	Reference Page(s)	Number	Answer	Reference Page(s)
1.	e	640	19.	c	644
2.	f	640, 641	20.	c	644
3.	a	641	21.	c	645
4.	e	641	22.	f	644, 645
5.	e	640	23.	c	645
6.	e	641	24.	d	645
7.	b	641	25.	c	643, 644, 646
8.	c	641	26.	c	627, 646
9.	c	641	27.	a	646
10.	c	642	28.	d	646
11.	e	643	29.	b	647
12.	c	642	30.	d	647
13.	e	642	31.	e	647
14.	b	643	32.	b	647
15.	e	643	33.	b	648
16.	c	643	34.	d	648
17.	b	644	35.	c	647, 648
18.	c	644	36.	b	647, 648

CHAPTER 49

Number	Answer	Reference Page(s)	Number	Answer	Reference Page(s)
1.	f	649	20.	f	652
2.	b	649	21.	a	652
3.	c	649	22.	a	653
4.	b	650	23.	b	653
5.	b	650	24.	d	653
6.	e	650	25.	a	653
7.	a	650	26.	e	653, 654
8.	c	650	27.	d	653
9.	f	650, 651	28.	f	655
10.	d	651	29.	b	654
11.	b	651	30.	d	654
12.	f	651	31.	b	654
13.	e	651	32.	d	655
14.	f	651	33.	b	655
15.	b	650	34.	c	655
16.	b	652	35.	a	655
17.	a	652	36.	d	656
18.	b	652	37.	b	656
19.	b	652	38.	b	656

CHAPTER 49 (Continued)

Number	Answer	Reference Page(s)	Number	Answer	Reference Page(s)
39.	d	656, 657	47.	f	659
40.	b	651	48.	c	660
41.	b	653, 658	49.	b	660
42.	f	658	50.	d	660
43.	e	659	51.	c	660
44.	b	659	52.	c	661
45.	e	652, 653, 659	53.	c	661
46.	e	659	54.	d	661

CHAPTER 50

Number	Answer	Reference Page(s)	Number	Answer	Reference Page(s)
1.	d	662	31.	d	669
2.	a	662	32.	c	669
3.	c	663	33.	e	669
4.	e	663	34.	f	671
5.	d	663	35.	f	670
6.	b	663	36.	c	670
7.	b	665	37.	a	670
8.	b	663	38.	f	670, 671
9.	d	664	39.	a	671
10.	e	664	40.	c	671
11.	c	664	41.	d	671
12.	b	664	42.	f	672
13.	f	664	43.	e	672
14.	e	664	44.	a	672
15.	d	664	45.	d	672
16.	b	665	46.	c	672
17.	c	665	47.	e	672
18.	d	658, 665	48.	d	673
19.	b	666	49.	b	673
20.	e	666	50.	c	673
21.	d	666, 667	51.	c	673
22.	b	666	52.	b	673, 674
23.	b	667	53.	d	674
24.	b	667	54.	d	674
25.	c	667	55.	f	675
26.	b	668	56.	f	675
27.	c	668	57.	d	676
28.	b	668	58.	a	676
29.	f	669	59.	a	676
30.	c	669	60.	c	676, 677

CHAPTER 51

Number	Answer	Reference Page(s)	Number	Answer	Reference Page(s)
1.	e	678	17.	f	681
2.	d	678	18.	f	681
3.	e	678, 679	19.	a	681
4.	d	679	20.	f	681
5.	b	679	21.	a	682
6.	c	679	22.	d	681, 682
7.	c	679, 680	23.	b	681
8.	b	679, 680	24.	b	682
9.	c	679	25.	a	683
10.	c	679, 680	26.	a	682
11.	a	680	27.	d	683
12.	c	680	28.	c	683
13.	e	680	29.	d	682, 683
14.	b	680	30.	d	684
15.	e	680, 681	31.	c	144, 683
16.	f	681	32.	d	684

CHAPTER 51 (Continued)

Number	Answer	Reference Page(s)	Number	Answer	Reference Page(s)
33.	d	685	50.	c	688, 689
34.	b	682, 685	51.	d	682, 689
35.	e	685	52.	f	688, 689
36.	e	686	53.	b	689
37.	c	686	54.	e	689
38.	c	686	55.	b	690
39.	d	648, 686	56.	e	690
40.	e	680, 686	57.	d	690
41.	d	686	58.	b	690
42.	d	686, 687	59.	d	691
43.	c	686, 687	60.	b	691
44.	a	687, 688	61.	d	691
45.	f	648, 687	62.	d	691, 692
46.	f	688	63.	f	692
47.	d	688	64.	f	692
48.	c	688	65.	a	692
49.	f	689			

CHAPTER 52

Number	Answer	Reference Page(s)	Number	Answer	Reference Page(s)
1.	c	694	33.	d	701
2.	c	694	34.	e	701
3.	d	695	35.	c	701
4.	b	694	36.	d	701
5.	b	695	37.	f	702
6.	c	695	38.	a	702
7.	c	695	39.	f	702
8.	b	695	40.	d	703
9.	c	696	41.	b	703
10.	a	696	42.	d	703
11.	e	696, 697	43.	f	703
12.	d	696, 698	44.	e	703
13.	c	696, 697	45.	d	703
14.	e	697	46.	c	704
15.	a	697	47.	c	704
16.	d	697	48.	d	704, 705
17.	f	697	49.	c	705
18.	b	698	50.	e	703
19.	f	698	51.	c	704
20.	a	698	52.	b	705
21.	e	698	53.	d	705, 706
22.	f	699, 700	54.	f	706
23.	b	699	55.	c	706
24.	d	699	56.	b	705
25.	b	699	57.	d	690, 707
26.	b	700	58.	b	706, 707
27.	c	700	59.	d	707
28.	e	700	60.	e	708
29.	c	700	61.	d	708
30.	a	701	62.	c	708
31.	a	701	63.	a	708
32.	c	701			

CHAPTER 53

Number	Answer	Reference Page(s)	Number	Answer	Reference Page(s)
1.	f	710	6.	f	679, 711
2.	a	710	7.	e	711
3.	b	710, 711	8.	b	712
4.	c	711	9.	c	711
5.	b	711	10.	b	712

CHAPTER 53 (Continued)

Number	Answer	Reference Page(s)	Number	Answer	Reference Page(s)
11.	e	712	45.	a	719
12.	c	713	46.	d	719
13.	a	713	47.	d	719
14.	b	713	48.	e	719
15.	c	713	49.	c	720
16.	a	713	50.	b	720
17.	b	142, 714	51.	b	720
18.	a	714	52.	f	720
19.	d	714, 715	53.	b	720
20.	b	715	54.	e	720
21.	d	715	55.	e	720
22.	b	715	56.	d	721
23.	b	715	57.	b	721
24.	d	715	58.	e	721
25.	b	715	59.	e	721
26.	c	716	60.	d	722
27.	d	716	61.	e	722
28.	d	716	62.	b	722
29.	d	716	63.	f	723
30.	c	716	64.	a	723
31.	d	716	65.	d	723
32.	c	716	66.	e	723, 724
33.	c	717	67.	d	724
34.	c	717	68.	d	724
35.	c	717	69.	b	724
36.	b	717	70.	c	724
37.	f	717	71.	f	724
38.	d	717, 718	72.	c	725
39.	c	718	73.	a	725
40.	d	718	74.	d	725
41.	a	718	75.	a	725
42.	b	718	76.	b	725, 726
43.	c	718	77.	d	726, 727
44.	e	718	78.	c	727

CHAPTER 54

Number	Answer	Reference Page(s)	Number	Answer	Reference Page(s)
1.	d	729	22.	d	735
2.	d	730	23.	f	735
3.	c	729	24.	b	735, 736
4.	c	731	25.	c	735
5.	c	731	26.	b	736
6.	c	731	27.	b	735, 736
7.	f	731	28.	f	736
8.	d	731, 732	29.	b	736, 737
9.	b	732	30.	d	737
10.	c	732	31.	e	737, 738
11.	d	733	32.	b	738
12.	e	733	33.	f	738
13.	c	733	34.	a	738
14.	b	734	35.	c	739
15.	d	734	36.	a	739
16.	e	734	37.	d	739
17.	a	734	38.	e	739, 740
18.	c	734	39.	f	741
19.	d	735	40.	b	741
20.	c	735	41.	e	741
21.	b	735	42.	b	741, 742

CHAPTER 55

Number	Answer	Reference Page(s)	Number	Answer	Reference Page(s)
1.	d	743	25.	a	749
2.	c	744	26.	d	750
3.	e	744	27.	d	750
4.	e	744	28.	a	751
5.	d	744	29.	d	752
6.	b	745	30.	d	751
7.	f	745	31.	c	751
8.	b	745	32.	b	752
9.	a	744, 746	33.	c	752
10.	d	746	34.	d	752
11.	f	746	35.	b	753
12.	d	746	36.	c	753
13.	d	746, 747	37.	a	753
14.	e	747	38.	b	753, 754
15.	d	747	39.	c	754
16.	d	747	40.	d	754
17.	b	747	41.	d	754
18.	e	747	42.	b	755
19.	d	748	43.	d	755
20.	b	748	44.	a	755
21.	b	748, 749	45.	e	756
22.	c	748	46.	c	756
23.	c	749	47.	b	757
24.	b	749			

CHAPTER 56

Number	Answer	Reference Page(s)	Number	Answer	Reference Page(s)
1.	c	758	21.	d	762
2.	e	758	22.	b	763
3.	c	758	23.	f	763
4.	c	758	24.	b	763
5.	e	758	25.	f	764
6.	f	758	26.	a	764
7.	b	760	27.	c	764
8.	e	760	28.	c	764
9.	c	760	29.	a	764
10.	b	761	30.	c	764
11.	e	761	31.	f	765
12.	d	761	32.	d	765
13.	a	761	33.	c	765
14.	d	761	34.	d	765
15.	c	761	35.	c	766
16.	f	761	36.	f	766
17.	e	761	37.	e	766
18.	c	762, 763	38.	d	766
19.	c	762	39.	e	766
20.	d	762			

CHAPTER 57

Number	Answer	Reference Page(s)	Number	Answer	Reference Page(s)
1.	d	768	11.	e	769
2.	b	768, 769	12.	c	770
3.	b	768	13.	b	769, 770
4.	b	768	14.	c	770
5.	b	768	15.	d	769
6.	c	768	16.	f	770
7.	e	768	17.	d	770
8.	a	768	18.	c	770
9.	e	769	19.	c	770
10.	e	769	20.	c	771

CHAPTER 57 (Continued)

Number	Answer	Reference Page(s)	Number	Answer	Reference Page(s)
21.	b	771	48.	c	775
22.	d	771	49.	b	773, 775
23.	e	771	50.	c	776
24.	f	771	51.	c	776
25.	b	771	52.	b	776
26.	a	771	53.	e	776
27.	a	771	54.	f	776, 777
28.	c	772	55.	e	776, 777
29.	c	772	56.	a	776, 777
30.	c	772	57.	b	769, 777
31.	a	772	58.	d	777
32.	c	772	59.	e	778
33.	d	772	60.	a	778
34.	b	772	61.	f	778
35.	d	773	62.	e	778
36.	d	772	63.	b	779
37.	d	772	64.	d	779
38.	c	774	65.	a	779
39.	c	774	66.	b	779
40.	c	774	67.	e	779
41.	a	774	68.	c	779
42.	a	774	69.	b	780
43.	c	772, 774	70.	b	780
44.	a	774, 775	71.	d	780
45.	b	775	72.	c	780
46.	b	775	73.	d	780
47.	b	775			

CHAPTER 58

Number	Answer	Reference Page(s)	Number	Answer	Reference Page(s)
1.	g	788	34.	e	788
2.	d	788	35.	a	788, 789
3.	c	789	36.	d	786, 789
4.	k	788	37.	e	786, 788
5.	e	789	38.	b	786, 788
6.	i	789	39.	d	789
7.	j	789	40.	e	789
8.	h	789	41.	b	789
9.	b	789	42.	e	789, 790
10.	a	789	43.	b	790
11.	f	810	44.	c	790, 791
12.	d	785	45.	c	790
13.	d	784	46.	c	791
14.	f	784	47.	b	791
15.	f	784	48.	f	791
16.	c	784	49.	d	791
17.	e	785	50.	d	791
18.	c	786, 787	51.	c	791
19.	d	786, 787	52.	e	791
20.	d	786	53.	d	791
21.	b	786	54.	b	791
22.	a	786, 787	55.	d	791, 792
23.	d	787	56.	d	822
24.	e	787	57.	a	791
25.	f	787	58.	b	791
26.	d	788	59.	c	792
27.	a	788	60.	e	790
28.	d	788	61.	d	793
29.	b	788	62.	e	793
30.	e	788	63.	b	793
31.	d	788	64.	c	794
32.	e	788, 789	65.	c	793, 794
33.	e	788	66.	d	794

CHAPTER 58 (Continued)

Number	Answer	Reference Page(s)	Number	Answer	Reference Page(s)
67.	d	794	70.	b	795
68.	c	794	71.	c	795
69.	d	794	72.	e	795

CHAPTER 59

Number	Answer	Reference Page(s)	Number	Answer	Reference Page(s)
1.	i	797	33.	e	803
2.	c	797	34.	b	803
3.	f	797	35.	b	803
4.	j	797	36.	f	803
5.	b	797	37.	b	805
6.	h	797	38.	c	797, 803
7.	e	797	39.	b	803
8.	a	797	40.	c	803
9.	g	797	41.	c	804
10.	d	797	42.	b	804
11.	e	640, 797	43.	d	804
12.	f	797	44.	f	804, 805
13.	e	797	45.	c	804, 805
14.	e	798	46.	b	805
15.	d	798	47.	a	805
16.	d	798, 799	48.	e	805
17.	b	799	49.	d	806, 807
18.	e	798	50.	a	805, 806
19.	e	799	51.	a	806
20.	a	799	52.	a	806
21.	b	799	53.	e	805, 806
22.	d	800	54.	f	806
23.	e	800	55.	e	806
24.	b	800	56.	f	806
25.	a	800	57.	c	806
26.	e	800	58.	c	803, 806
27.	b	800, 802	59.	b	808
28.	c	800, 802	60.	b	808
29.	b	801	61.	c	808
30.	d	802	62.	b	806, 808
31.	d	802	63.	c	808
32.	a	802			

CHAPTER 60

Number	Answer	Reference Page(s)	Number	Answer	Reference Page(s)
1.	f	810	19.	c	814
2.	a	810	20.	d	814
3.	c	810	21.	c	814
4.	f	810	22.	f	814
5.	a	810	23.	e	814
6.	c	810, 811	24.	e	814
7.	d	811	25.	d	814
8.	d	811	26.	e	815
9.	b	810	27.	e	815
10.	c	811	28.	c	815
11.	f	811, 812	29.	a	815, 816
12.	c	812	30.	a	816
13.	e	812	31.	c	816
14.	a	812	32.	e	817
15.	d	812	33.	a	817
16.	d	812	34.	f	817
17.	a	813, 814	35.	b	817
18.	b	814	36.	b	817, 818

CHAPTER 60 (Continued)

Number	Answer	Reference Page(s)	Number	Answer	Reference Page(s)
37.	c	818	57.	d	822
38.	c	818	58.	d	822
39.	c	818	59.	d	822
40.	d	818	60.	a	823
41.	a	818	61.	d	823
42.	a	818	62.	d	823
43.	b	818	63.	a	823
44.	b	818	64.	e	823
45.	c	818	65.	c	823
46.	e	819	66.	a	823, 824
47.	f	818	67.	a	823
48.	b	819	68.	b	823
49.	a	819	69.	a	824
50.	e	818	70.	a	824, 790
51.	e	820	71.	d	824
52.	b	820	72.	b	824
53.	c	820	73.	b	824
54.	e	821	74.	a	824
55.	a	821	75.	f	824
56.	d	822	76.	e	824

CHAPTER 61

Number	Answer	Reference Page(s)	Number	Answer	Reference Page(s)
1.	d	826	41.	b	829
2.	i	826	42.	a	829, 830
3.	h	826	43.	d	829, 830
4.	a	826	44.	a	830
5.	e	826	45.	c	830
6.	b	826	46.	d	830
7.	c	826	47.	d	830
8.	f	826	48.	b	831
9.	j	826	49.	c	831
10.	k	826	50.	a	831
11.	g	826	51.	b	831
12.	c	826	52.	a	831
13.	f	826, 827	53.	c	831
14.	d	827	54.	d	832
15.	a	827	55.	a	832
16.	d	827	56.	f	832
17.	c	827	57.	f	832
18.	e	827	58.	c	832
19.	c	827, 828	59.	f	832
20.	b	827	60.	e	832
21.	d	827	61.	c	832
22.	c	827	62.	a	832
23.	a	827	63.	e	832, 833
24.	c	827	64.	a	833
25.	d	827	65.	c	833
26.	c	828	66.	d	833
27.	f	828	67.	d	833
28.	h	828	68.	c	833
29.	j	828	69.	b	833
30.	g	828	70.	f	833
31.	i	830	71.	e	833
32.	a	830	72.	d	834
33.	e	830	73.	c	834
34.	b	828	74.	b	835
35.	d	830	75.	c	835
36.	e	828	76.	d	833, 835
37.	b	828	77.	d	836
38.	f	828	78.	d	836
39.	c	829	79.	e	836
40.	b	829	80.	b	836

CHAPTER 61 (Continued)

Number	Answer	Reference Page(s)	Number	Answer	Reference Page(s)
81.	e	836	85.	b	837
82.	c	837	86.	c	837
83.	c	837	87.	e	837
84.	e	837			

CHAPTER 62

Number	Answer	Reference Page(s)	Number	Answer	Reference Page(s)
1.	h	841	35.	b	841, 842
2.	f	841	36.	e	841, 842
3.	j	841	37.	a	841, 842
4.	c	841	38.	f	841, 842
5.	g	840	39.	e	842, 843
6.	a	840	40.	f	843
7.	k	840	41.	a	843
8.	b	840	42.	b	843
9.	e	840	43.	b	843
10.	d	840	44.	f	844
11.	i	840	45.	d	846
12.	b	839	46.	b	846
13.	b	839	47.	a	844
14.	e	839	48.	e	844
15.	d	839	49.	c	844
16.	d	839	50.	h	844
17.	c	839, 840	51.	k	844
18.	e	840	52.	l	844
19.	f	840	53.	g	844
20.	b	840	54.	i	846
21.	d	840	55.	j	846
22.	e	840	56.	d	844
23.	c	840	57.	c	844
24.	d	840	58.	b	844
25.	a	841	59.	b	844
26.	e	841	60.	d	844
27.	c	841	61.	e	845
28.	c	841	62.	d	845
29.	b	841	63.	a	845
30.	c	841, 842	64.	b	844, 845
31.	c	841, 842	65.	f	846
32.	b	842	66.	b	846
33.	f	841, 842	67.	d	846
34.	c	841, 842	68.	f	846

CHAPTER 63

Number	Answer	Reference Page(s)	Number	Answer	Reference Page(s)
1.	d	850	16.	i	850
2.	b	850	17.	n	850
3.	h	850	18.	e	851
4.	m	850	19.	k	851
5.	j	850	20.	h	851
6.	l	850	21.	a	851
7.	a	850	22.	d	851
8.	d	850	23.	g	851
9.	f	850	24.	i	851
10.	k	850	25.	l	851
11.	b	850	26.	b	851
12.	c	850	27.	f	851
13.	o	850	28.	j	851
14.	e	850	29.	c	851
15.	g	850	30.	d	851

CHAPTER 63 (Continued)

Number	Answer	Reference Page(s)	Number	Answer	Reference Page(s)
31.	a	851	87.	a	858
32.	b	851	88.	a	858
33.	d	851	89.	b	858
34.	c	851	90.	c	858
35.	c	851, 852	91.	b	859
36.	b	852	92.	d	858
37.	a	852	93.	e	858
38.	d	852	94.	d	859
39.	c	852	95.	e	859
40.	b	852	96.	a	859
41.	d	852	97.	d	859
42.	a	852	98.	e	860
43.	c	852	99.	f	860
44.	b	853	100.	d	860
45.	a	853	101.	d	860
46.	b	853	102.	c	860
47.	e	853	103.	c	860, 861
48.	a	853	104.	c	861
49.	d	853	105.	e	861
50.	d	853, 854	106.	f	861
51.	d	854	107.	d	861
52.	c	854	108.	d	861
53.	e	854	109.	d	861, 862
54.	a	854	110.	c	862
55.	f	854	111.	a	862
56.	c	855	112.	e	862
57.	d	855	113.	a	862
58.	c	855	114.	c	863
59.	e	855	115.	f	863
60.	e	855	116.	b	863
61.	f	855	117.	c	863
62.	e	855	118.	e	861, 863
63.	b	855	119.	f	863
64.	c	856	120.	c	863
65.	b	856	121.	a	863
66.	e	856	122.	b	863
67.	d	856	123.	a	863
68.	a	856	124.	b	863, 864
69.	b	856	125.	c	863, 864
70.	c	856, 859	126.	a	863, 864
71.	c	857	127.	d	863, 864
72.	b	857	128.	c	864
73.	c	857	129.	d	864
74.	d	857	130.	d	864
75.	d	857	131.	d	864
76.	d	857	132.	c	865
77.	d	857	133.	a	865
78.	f	858	134.	a	865, 866
79.	j	858	135.	a	865
80.	c	858	136.	c	865
81.	b	858	137.	f	852, 865
82.	i	858	138.	a	865
83.	e	858	139.	b	866
84.	h	858	140.	b	866
85.	d	858	141.	d	866
86.	g	858			

CHAPTER 64

Number	Answer	Reference Page(s)	Number	Answer	Reference Page(s)
1.	f	867	48.	b	874
2.	a	867	49.	e	874
3.	a	867	50.	d	874
4.	b	868	51.	b	874
5.	a	868	52.	a	874, 875
6.	a	868	53.	a	874
7.	a	868	54.	b	875
8.	d	868	55.	c	875
9.	f	868	56.	c	875
10.	a	868	57.	b	875, 876
11.	b	869	58.	a	876
12.	f	869	59.	c	876
13.	f	869	60.	c	876
14.	e	870	61.	d	876
15.	d	870	62.	a	876
16.	c	870	63.	e	876
17.	f	499, 870	64.	d	876
18.	c	870	65.	a	876
19.	d	870	66.	b	870, 877
20.	f	870	67.	a	877
21.	c	870	68.	c	877
22.	a	870, 871	69.	a	877
23.	c	870	70.	d	877
24.	f	871	71.	c	877
25.	b	871	72.	e	877
26.	c	871	73.	f	877
27.	a	871	74.	d	877
28.	c	870, 871	75.	a	863, 878, 879
29.	a	871	76.	e	878
30.	c	872	77.	f	878
31.	a	872	78.	d	878
32.	b	872	79.	d	878
33.	f	871	80.	d	878
34.	b	872	81.	d	878
35.	e	872	82.	a	878
36.	d	872	83.	c	879
37.	a	872	84.	e	879
38.	d	872	85.	f	879
39.	a	873	86.	a	879
40.	c	873	87.	b	879
41.	f	873	88.	f	879
42.	f	873	89.	a	879
43.	a	873	90.	b	879
44.	c	873	91.	f	880
45.	a	873	92.	c	880
46.	f	873, 874	93.	a	880
47.	b	874	94.	f	880

CHAPTER 65

Number	Answer	Reference Page(s)	Number	Answer	Reference Page(s)
1.	j	886	14.	b	881
2.	k	886	15.	e	881
3.	e	886	16.	b	881
4.	d	886	17.	e	881
5.	l	886	18.	d	882
6.	c	879, 886	19.	b	882
7.	g	886	20.	c	882
8.	f	886	21.	a	882
9.	h	886	22.	a	882
10.	b	886	23.	c	882
11.	a	886	24.	b	882
12.	i	886	25.	b	882
13.	b	881	26.	f	883

CHAPTER 65 (Continued)

Number	Answer	Reference Page(s)	Number	Answer	Reference Page(s)
27.	b	883	53.	a	888
28.	d	883	54.	d	889
29.	f	883	55.	f	889
30.	c	883	56.	d	889
31.	d	884	57.	b	889
32.	c	884	58.	d	889
33.	c	884	59.	b	889
34.	f	884	60.	f	889, 890
35.	e	885	61.	a	890
36.	b	885	62.	c	890
37.	f	885	63.	a	890
38.	b	425, 885	64.	b	890
39.	e	886	65.	d	890
40.	e	885	66.	b	890, 891
41.	b	886	67.	a	890
42.	a	887	68.	d	891
43.	a	887	69.	c	891
44.	d	887	70.	f	891
45.	d	887	71.	c	891
46.	e	888	72.	e	891
47.	f	888	73.	d	891
48.	c	888	74.	d	891
49.	e	888	75.	f	892
50.	d	888	76.	f	892
51.	b	888	77.	d	892
52.	f	888	78.	f	892

CHAPTER 66

Number	Answer	Reference Page(s)	Number	Answer	Reference Page(s)
1.	d	893	24.	b	897
2.	d	893	25.	e	897
3.	d	893	26.	d	897
4.	d	893	27.	c	897
5.	f	893, 894	28.	c	897
6.	b	893	29.	e	897
7.	f	894	30.	b	897, 898
8.	b	894	31.	d	891, 898
9.	c	894	32.	b	898
10.	a	894	33.	b	898
11.	f	894	34.	f	898, 899
12.	c	894	35.	a	899
13.	d	895	36.	b	899
14.	a	895	37.	d	899
15.	a	894, 895	38.	d	899
16.	d	895	39.	d	899
17.	a	895	40.	d	899
18.	b	895	41.	b	900
19.	b	895	42.	c	900
20.	a	895, 896	43.	d	901
21.	a	896	44.	d	901
22.	f	896	45.	c	901
23.	a	897			

CHAPTER 67

Number	Answer	Reference Page(s)	Number	Answer	Reference Page(s)
1.	a	904	27.	f	910
2.	d	904	28.	d	910
3.	e	904	29.	b	910
4.	c	905	30.	a	910, 911
5.	e	904	31.	b	910, 911
6.	d	905, 906	32.	c	912
7.	d	905	33.	d	912
8.	e	905	34.	e	912
9.	b	906	35.	a	912
10.	e	705, 906	36.	e	913
11.	b	906	37.	d	913
12.	a	906	38.	e	913
13.	b	906	39.	c	913
14.	d	906	40.	f	913, 914
15.	b	906	41.	a	914
16.	e	907	42.	a	914
17.	e	907	43.	b	914
18.	b	907	44.	f	914
19.	b	907	45.	c	914
20.	a	907	46.	c	914
21.	b	907, 908	47.	b	914, 915
22.	d	908	48.	b	914
23.	e	908, 909	49.	c	915
24.	b	909	50.	b	915
25.	c	909, 910	51.	e	915
26.	b	910			

CHAPTER 68

Number	Answer	Reference Page(s)	Number	Answer	Reference Page(s)
1.	c	916	33.	d	921
2.	d	916	34.	d	921
3.	b	917	35.	f	921
4.	f	917	36.	c	914, 921
5.	a	916	37.	d	922
6.	e	917	38.	b	922
7.	b	917	39.	d	921, 922
8.	b	917	40.	d	922
9.	d	917	41.	b	922
10.	e	917	42.	b	922, 923
11.	b	915, 917	43.	f	923
12.	d	917	44.	a	923
13.	c	917, 918	45.	b	923
14.	c	917, 918	46.	e	923
15.	b	917, 918	47.	b	923
16.	c	918	48.	e	923
17.	d	918	49.	a	923
18.	e	918	50.	a	924
19.	a	918	51.	b	924, 925
20.	c	918	52.	f	924
21.	c	919	53.	d	924
22.	b	918	54.	c	924, 925
23.	d	919	55.	b	925
24.	a	918	56.	d	925
25.	c	918	57.	a	925
26.	b	919	58.	c	925
27.	f	919	59.	f	925
28.	d	919	60.	c	926
29.	a	919	61.	e	927
30.	c	916, 920	62.	d	926
31.	c	920	63.	f	925, 926
32.	b	920, 921			

CHAPTER 69

Number	Answer	Reference Page(s)	Number	Answer	Reference Page(s)
1.	d	928	20.	a	932
2.	f	928	21.	c	932
3.	a	930	22.	f	932
4.	e	930	23.	b	929, 932
5.	c	930	24.	e	932
6.	d	929, 930	25.	a	933
7.	c	930	26.	b	933
8.	b	930	27.	b	933
9.	e	930	28.	a	933
10.	a	930	29.	b	934
11.	c	930	30.	e	934
12.	e	931	31.	c	934
13.	d	931	32.	c	934
14.	e	931	33.	c	934
15.	d	914, 931	34.	f	934
16.	b	101, 931	35.	c	934
17.	f	931	36.	a	934
18.	e	931, 932	37.	c	935
19.	b	932	38.	c	935

CHAPTER 70

Number	Answer	Reference Page(s)	Number	Answer	Reference Page(s)
1.	d	937	31.	a	938
2.	j	937	32.	e	938
3.	g	937	33.	b	939
4.	i	937	34.	c	939
5.	h	937	35.	f	939
6.	f	937	36.	c	939
7.	a	937	37.	b	939
8.	k	937	38.	d	939
9.	b	937	39.	c	939
10.	c	937	40.	e	940
11.	e	937	41.	d	940, 941
12.	e	932, 936	42.	e	941
13.	e	936	43.	c	941
14.	a	937	44.	d	941
15.	c	398, 937	45.	e	941
16.	d	937	46.	d	941
17.	a	937, 938	47.	a	939, 941
18.	c		48.	b	941, 942
19.	e		49.	d	942
20.	a		50.	f	942
21.	i		51.	a	942
22.	f		52.	f	942
23.	d		53.	b	943
24.	h		54.	b	943
25.	b		55.	d	943
26.	g		56.	f	943
27.	b	938	57.	e	939, 943
28.	b	938	58.	c	943
29.	e	938	59.	c	943
30.	a	938			

CHAPTER 71

Number	Answer	Reference Page(s)	Number	Answer	Reference Page(s)
1.	e	945	23.	f	950
2.	b	945	24.	e	950
3.	b	945	25.	a	950
4.	a	945	26.	b	951
5.	d	946	27.	e	951
6.	b	946	28.	b	951
7.	a	946	29.	b	951
8.	c	947	30.	f	951, 952
9.	d	947	31.	e	952
10.	c	947	32.	d	952
11.	e	947	33.	a	952
12.	e	947	34.	e	952
13.	f	947	35.	e	952
14.	f	948	36.	a	952
15.	d	913, 948	37.	b	953
16.	c	948	38.	f	953
17.	a	948	39.	b	953
18.	a	949	40.	c	953
19.	f	949	41.	b	953, 954
20.	c	949	42.	f	952, 953
21.	f	949	43.	e	953, 954
22.	c	950			

CHAPTER 72

Number	Answer	Reference Page(s)	Number	Answer	Reference Page(s)
1.	d	955	35.	d	961
2.	c	955	36.	e	961
3.	f	955	37.	f	961
4.	e	955	38.	c	961
5.	b	955	39.	d	962
6.	a	955, 956	40.	f	962
7.	d	956	41.	b	963
8.	b	955, 956	42.	b	962
9.	d	951, 956	43.	c	963
10.	a	956	44.	c	963
11.	c	952, 956	45.	d	963
12.	c	956	46.	d	963
13.	b	956	47.	a	963
14.	e	956	48.	c	964
15.	f	956	49.	a	964
16.	b	956, 984	50.	c	964
17.	a	957	51.	c	964, 965
18.	b	957	52.	e	965
19.	b	957	53.	a	965
20.	b	957	54.	c	965
21.	d	958	55.	b	965, 966
22.	f	958	56.	a	966
23.	b	958	57.	d	966
24.	d	958, 959	58.	d	966, 967
25.	e	959	59.	d	966, 967
26.	d	959	60.	a	966, 967
27.	f	959	61.	b	967
28.	a	959	62.	b	967
29.	b	959	63.	b	968
30.	c	959	64.	e	967
31.	e	959, 960	65.	b	968
32.	e	960	66.	d	968
33.	e	380, 961	67.	d	381, 549, 968
34.	d	961			

CHAPTER 73

Number	Answer	Reference Page(s)	Number	Answer	Reference Page(s)
1.	a	970	21.	a	973, 974
2.	a	970	22.	f	974, 975
3.	e	970	23.	d	974
4.	b	970	24.	d	974, 975
5.	b	970	25.	c	974, 975
6.	f	970	26.	a	975
7.	d	971	27.	b	975
8.	c	971	28.	d	976
9.	c	971, 972	29.	f	975
10.	b	972	30.	d	976
11.	d	972	31.	d	976
12.	d	972	32.	a	976
13.	d	972	33.	d	977
14.	a	972	34.	d	977
15.	c	972, 973	35.	b	977
16.	f	972	36.	b	977
17.	e	973	37.	c	977
18.	e	973	38.	f	977
19.	d	973	39.	a	977
20.	e	974			

CHAPTER 74

Number	Answer	Reference Page(s)	Number	Answer	Reference Page(s)
1.	b	979	22.	f	108, 983
2.	c	979	23.	e	983
3.	d	979	24.	b	983
4.	e	979	25.	e	984
5.	d	979	26.	b	984
6.	d	980	27.	e	984
7.	a	980	28.	e	984
8.	d	980	29.	f	984
9.	d	980	30.	b	985
10.	c	980, 981	31.	c	984, 985
11.	d	981	32.	c	985
12.	e	981	33.	d	985
13.	c	981	34.	c	985
14.	e	981	35.	e	985
15.	b	981	36.	e	985
16.	d	982	37.	a	985
17.	e	982	38.	d	985
18.	a	982	39.	f	985
19.	c	982	40.	e	985
20.	b	982	41.	e	985
21.	a	983			

CHAPTER 75

Number	Answer	Reference Page(s)	Number	Answer	Reference Page(s)
1.	e	988	13.	g	989
2.	f	988, 989	14.	c	991
3.	a	988	15.	c	991
4.	d	988, 989	16.	e	991
5.	c	989	17.	e	991
6.	d, h, k, m		18.	e	991
	o, s, t	989	19.	a	991, 992
7.	e, l	989	20.	d	992
8.	f, i	989	21.	b	992
9.	a, p, r	989	22.	b	992
10.	b, q	989	23.	a	992
11.	c, n	989	24.	b	992
12.	j	989	25.	c	992

CHAPTER 75 (Continued)

Number	Answer	Reference Page(s)	Number	Answer	Reference Page(s)
26.	e	992	58.	a	997
27.	b	992	59.	d	997
28.	a	993	60.	d	997
29.	c	993	61.	c	997
30.	f	993	62.	b	997
31.	h	993	63.	a	997, 998
32.	g	993	64.	c	998
33.	c	993	65.	d	998
34.	d	993	66.	e	999
35.	e	993	67.	b	999, 1000
36.	a	993	68.	f	1000
37.	b	993	69.	e	1000
38.	i	993	70.	d	1000
39.	f	993	71.	c	1000
40.	d	993	72.	e	1000
41.	b	994	73.	b	1000
42.	a	994	74.	c	1001
43.	c	994	75.	b	1001
44.	f	994	76.	f	1001
45.	e	994	77.	b	1001
46.	c	994	78.	a	1001
47.	a	995	79.	f	1002
48.	f	995	80.	d	1001
49.	d	995	81.	d	1002
50.	d	995	82.	c	1002
51.	b	995	83.	b	1002
52.	a	996	84.	a	1002
53.	b	996	85.	b	1003
54.	e	996	86.	c	1002
55.	a	996	87.	c	1003
56.	f	996	88.	a	1003
57.	c	997	89.	b	1003

CHAPTER 76

Number	Answer	Reference Page(s)	Number	Answer	Reference Page(s)
1.	c	1005	29.	c	1007, 1008
2.	g	1005	30.	b	1008
3.	h	1005	31.	a	1008
4.	a	1005	32.	a	1009
5.	b	1005	33.	f	1009
6.	f	526	34.	f	1009
7.	i	1005	35.	b	1009
8.	d	1005	36.	c	1009, 1010
9.	e	1005	37.	b	1010
10.	f	1005	38.	b	1010
11.	f	1005	39.	b	1010
12.	b	1005	40.	e	1010
13.	d	1005	41.	d	1010, 1011
14.	a	1005	42.	a	1011
15.	a	1005	43.	c	1011
16.	e	1006	44.	b	1011
17.	e	1005	45.	f	1011
18.	c	1006	46.	c	1011
19.	e	1006	47.	f	1011
20.	f	1006	48.	f	1012
21.	d	1006	49.	b	1012
22.	e	1006	50.	b	1012
23.	c	1007	51.	c	1012
24.	a	1007	52.	f	1012, 1013
25.	b	1007	53.	b	1012, 1013
26.	f	1007	54.	a	1012, 1013
27.	d	1007	55.	c	1013
28.	a	1007, 1008	56.	f	1013, 1014

CHAPTER 76 (Continued)

Number	Answer	Reference Page(s)	Number	Answer	Reference Page(s)
57.	d	1013, 1014	66.	a	1015
58.	a	1014	67.	b	1015
59.	a	1014	68.	d	1016
60.	f	1014	69.	d	1016
61.	d	1014	70.	c	1016
62.	e	1014	71.	f	1016
63.	a	1014	72.	e	1016, 1017
64.	d	1015	73.	b	1017
65.	a	1015	74.	e	1017

CHAPTER 77

Number	Answer	Reference Page(s)	Number	Answer	Reference Page(s)
1.	e	1019	43.	c	1026
2.	c	438	44.	d	1026
3.	a	438	45.	b	1026
4.	g	1030	46.	f	1026
5.	d	1030	47.	c	1026
6.	h	1030	48.	a	1027
7.	b	1030	49.	f	1027
8.	f	1030	50.	f	1027
9.	a	1019	51.	f	1027, 1028
10.	e	1019	52.	d	1027
11.	f	1019	53.	c	1027
12.	a	1019	54.	d	1027-1029
13.	c	1020	55.	f	1028
14.	f	1020, 1021	56.	b	1028, 1029
15.	e	1021	57.	c	1029
16.	e	1021	58.	d	1029
17.	c	1020, 1021	59.	d	1029
18.	a	1021	60.	a	1029
19.	c	1020, 1021	61.	b	1029
20.	e	1022	62.	d	1029
21.	b	1022	63.	e	1030
22.	c	1022	64.	d	1030
23.	c	1022	65.	c	1030
24.	b	1022	66.	c	1031
25.	e	1022	67.	a	1031
26.	e	1023	68.	f	1031
27.	b	1023	69.	d	1032
28.	e	1023	70.	f	1032
29.	b	1023	71.	f	1032, 1033
30.	b	1023	72.	e	1032
31.	a	1024	73.	d	1032
32.	c	1024	74.	c	1033
33.	e	1024	75.	a	1032
34.	b	1025	76.	e	1032, 1033
35.	c	1025	77.	e	1033
36.	c	1025	78.	d	1034
37.	c	1025	79.	b	1034
38.	b	1025	80.	d	1034
39.	b	1025	81.	b	1034
40.	b	1025	82.	f	1034
41.	a	1025, 1026	83.	d	1035
42.	f	1025, 1026			

CHAPTER 78

Number	Answer	Reference Page(s)	Number	Answer	Reference Page(s)
1.	c	938	49.	a	1043
2.	b	878	50.	b	1044
3.	a	938	51.	b	1044
4.	e	878	52.	b	1044
5.	h	878	53.	f	1044
6.	d	938	54.	c	1044
7.	k	938	55.	f	1044
8.	i	1036	56.	a	1044
9.	f	1036	57.	b	1044
10.	j	1036	58.	b	1044
11.	g	1036	59.	c	1045
12.	c	1036	60.	d	1045
13.	f	1036	61.	c	1045
14.	f	1036, 1037	62.	e	1045
15.	e	1037	63.	c	1046
16.	b	1038	64.	d	1046
17.	c	1038	65.	c	1046
18.	a	1038	66.	a	1046
19.	b	1039	67.	b	1046
20.	a	1039	68.	b	1046
21.	c	1039	69.	a	1046
22.	b	1039	70.	e	1047
23.	a	1039	71.	d	1047
24.	f	1039	72.	c	1047
25.	b	1040	73.	c	1047
26.	c	1040	74.	c	1047
27.	f	1040	75.	b	1047
28.	d	1040	76.	c	1047, 1048
29.	c	1040	77.	d	1048
30.	a	1040, 1041	78.	f	1048
31.	e	1040, 1041	79.	a	1048
32.	c	1041	80.	e	1048
33.	c	1041	81.	d	1048, 1049
34.	c	1041	82.	b	1049
35.	c	1042	83.	f	1049
36.	d	1042	84.	e	1049
37.	b	1042	85.	c	1049
38.	e	1042	86.	f	1049
39.	a	1042	87.	b	1049, 1050
40.	d	1042	88.	a	1050
41.	b	1040-1042	89.	d	1050
42.	e	1036, 1043	90.	e	1050
43.	b	1043	91.	f	1050
44.	c	1043	92.	d	1050
45.	f	1043	93.	d	1051
46.	c	1043	94.	d	1051
47.	b	1043	95.	b	1051
48.	a	1043			

CHAPTER 79

Number	Answer	Reference Page(s)	Number	Answer	Reference Page(s)
1.	d		13.	b	1052
2.	f		14.	d	1053
3.	i		15.	e	1053
4.	b		16.	b	1053, 1054
5.	j		17.	d	1053, 1054
6.	h		18.	c	1054
7.	g		19.	d	1054
8.	e	1061	20.	d	1054
9.	c	1061	21.	b	1055
10.	a	1061	22.	e	1055
11.	a	1052	23.	b	1055
12.	c	1052	24.	a	1055

CHAPTER 79 (Continued)

Number	Answer	Reference Page(s)	Number	Answer	Reference Page(s)
25.	c	1055	68.	d	1064
26.	a	1055	69.	c	1064
27.	b	1056	70.	c	1064
28.	c	1055	71.	a	1064
29.	e	1056	72.	a	1058, 1065
30.	b	1056	73.	f	1065
31.	b	1056	74.	a	1065
32.	d	1056	75.	e	1065
33.	f	1056	76.	a	1065
34.	a	1056, 1057	77.	b	1065
35.	f	1057	78.	e	1066
36.	b	1057	79.	e	1056, 1066
37.	a	1057	80.	b	1066
38.	e	1057	81.	e	1066
39.	d	1057, 1058	82.	b	1066
40.	b	1058	83.	b	1066
41.	d	1057, 1058	84.	c	1067
42.	c	1058	85.	b	1067
43.	c	1060	86.	f	1067
44.	f	1058, 1059	87.	e	1067
45.	d	1059	88.	b	1067
46.	e	1060	89.	c	1067, 1068
47.	a	1060	90.	a	1068
48.	d	1060	91.	f	1068
49.	d	1060	92.	a	1068
50.	c	1060	93.	d	1068
51.	a	1061	94.	b	1068
52.	c	1061	95.	a	1068
53.	d	1061	96.	g	1068
54.	c	1061	97.	b	1068
55.	b	1061, 1062	98.	d	1068
56.	d	1062	99.	c	1068
57.	f	1062	100.	f	1068
58.	a	1062	101.	a	1068
59.	e	1062	102.	e	1068
60.	a	1062	103.	e	1068
61.	c	1063	104.	d	1069
62.	d	1063	105.	b	1069
63.	b	1063	106.	e	1070
64.	f	1063	107.	c	1070
65.	a	1061	108.	c	1057, 1070
66.	b	1063	109.	b	1070
67.	d	1064			

CHAPTER 80

Number	Answer	Reference Page(s)	Number	Answer	Reference Page(s)
1.	g	1072	18.	d	1073
2.	d	1072	19.	d	1073
3.	n	1072	20.	c	1073
4.	e	1072	21.	c	1074
5.	l	1072	22.	b	1074
6.	j	1072	23.	e	1074
7.	m	1072	24.	e	1074
8.	a	1072	25.	d	1074
9.	k	1072	26.	b	1074
10.	b	1072	27.	d	1075
11.	o	1072	28.	d	1075
12.	f	1072	29.	b	1075
13.	i	1072	30.	e	1075
14.	h	1072	31.	b	1075
15.	c	1072	32.	c	1075
16.	f	1072, 1073	33.	a	1075
17.	b	1073	34.	c	1075, 1076

CHAPTER 80 (Continued)

Number	Answer	Reference Page(s)	Number	Answer	Reference Page(s)
35.	a	1076	54.	a	1080, 1081
36.	c	1076	55.	b	1081
37.	c	1076	56.	f	1081
38.	c	1077	57.	b	1081
39.	c	1077	58.	b	1081
40.	e	1077	59.	a	1081
41.	d	1077, 1078	60.	c	1082
42.	a	1078	61.	c	1082
43.	c	1078	62.	d	1082
44.	b	1078, 1079	63.	e	1083
45.	b	1079	64.	b	1079, 1082-83
46.	c	1079	65.	d	1083
47.	a	1079	66.	b	1083
48.	e	1079	67.	c	1083
49.	e	1079	68.	e	1083, 1084
50.	c	1080	69.	f	1084
51.	e	1080	70.	b	1084
52.	c	1080	71.	c	1084
53.	a	1080	72.	a	1084

CHAPTER 81

Number	Answer	Reference Page(s)	Number	Answer	Reference Page(s)
1.	d	1086	42.	b	1090
2.	e	1086	43.	e	1091
3.	b	1086	44.	a	1091
4.	g	1086	45.	b	1092
5.	a	1086	46.	f	1092
6.	f	1086	47.	f	1092
7.	c	1086	48.	a	1092
8.	a	1086	49.	a	1092
9.	b	1086	50.	c	1092, 1093
10.	c	1086	51.	c	1093
11.	c	1086	52.	e	1093
12.	a	1087	53.	a	1093
13.	c	1087	54.	a	1093
14.	b	1086	55.	b	1093
15.	a	1087	56.	d	1094
16.	c	1087	57.	d	1094
17.	b	1086	58.	f	1094
18.	c	1087	59.	e	1094
19.	a	1087	60.	c	1094
20.	f	1087	61.	d	1094, 1095
21.	a	1087	62.	d	1095
22.	b	1087	63.	f	1095
23.	b	1087	64.	b	1095
24.	f	1088	65.	d	1095
25.	a	1088	66.	a	1096
26.	b	1088	67.	a	1096
27.	e	1088	68.	b	1097
28.	a	1088	69.	b	1097
29.	d	1088	70.	d	1087, 1097
30.	d	1089	71.	e	1087, 1097
31.	d	1089	72.	b	1087, 1097
32.	e	1089	73.	c	1098
33.	b	1089	74.	e	1098
34.	a	1087, 1089	75.	d	1098
35.	e	1090	76.	a	1098
36.	f	1090	77.	b	1099
37.	d	1087, 1090	78.	a	1099
38.	d	1090	79.	e	1099
39.	c	1090	80.	d	1099
40.	b	1090	81.	c	1099
41.	c	1090	82.	a	1099

CHAPTER 81 (Continued)

Number	Answer	Reference Page(s)	Number	Answer	Reference Page(s)
83.	c	1100	91.	b	1101
84.	b	1100	92.	d	1102
85.	c	1100	93.	a	1102
86.	f	1101	94.	d	1102
87.	c	1100	95.	e	1102
88.	d	1101	96.	d	1102
89.	b	1101	97.	c	1102
90.	f	1101			

CHAPTER 82

Number	Answer	Reference Page(s)	Number	Answer	Reference Page(s)
1.	b	1104	35.	a	1111
2.	d	1104	36.	b	1111, 1112
3.	a	1104	37.	e	1112
4.	c	1105	38.	c	1112
5.	a	1105	39.	e	1112
6.	e	1105	40.	b	1112
7.	f	1105	41.	f	1113
8.	a	1105, 1106	42.	c	1113
9.	f	1106	43.	a	1114
10.	a	1106	44.	b	1114
11.	d	1106	45.	c	1114
12.	a	1106	46.	d	1115
13.	c	1107	47.	b	1114
14.	d	1107	48.	d	1114
15.	a	1107	49.	d	1115
16.	c	1107	50.	a	1115
17.	d	1108	51.	f	1115
18.	a	1108	52.	a	1115
19.	a	541, 1108, 1109	53.	b	1115, 1116
20.	b	1108	54.	c	1117
21.	f	1109	55.	d	1117
22.	e	1109	56.	a	1117
23.	a	1109	57.	b	1118
24.	e	1110	58.	d	1117
25.	e	1110	59.	c	1118
26.	c	1110	60.	a	1118
27.	b	1110	61.	e	1118, 1119
28.	a	1110	62.	c	1119
29.	d	1110	63.	d	1119
30.	b	1111	64.	e	1119
31.	b	1111	65.	b	1119
32.	c	1110, 1111	66.	f	1120
33.	d	1111	67.	c	1120
34.	e	1111			

CHAPTER 83

Number	Answer	Reference Page(s)	Number	Answer	Reference Page(s)
1.	d	1122	12.	e	984
2.	d	1122	13.	c	1124, 1125
3.	b	1123	14.	c	1124
4.	c	549, 1123	15.	e	1125
5.	d	1123	16.	e	1126
6.	d	1123	17.	a	1126
7.	b	1123	18.	e	1126
8.	d	1123	19.	a	1126
9.	e	1124	20.	d	1126, 1127
10.	c	1124	21.	e	1127
11.	c	1124	22.	f	1127

CHAPTER 83 (Continued)

Number	Answer	Reference Page(s)	Number	Answer	Reference Page(s)
23.	c	1128	37.	b	1130
24.	b	1128	38.	b	1130
25.	b	1128	39.	b	1130, 1131
26.	a	1128	40.	f	1130, 1131
27.	e	1128	41.	c	1131
28.	c	1128, 1129	42.	d	1131
29.	d	1129	43.	f	1131
30.	c	1129	44.	a	1131
31.	d	1129	45.	d	1131, 1132
32.	e	1129	46.	c	1132
33.	e	1129	47.	f	1132
34.	f	1129	48.	e	1132
35.	b	1129, 1130	49.	b	1132
36.	d	1130			